Pure Mathematics for Advanced Level

Complete Volume

Pure Mathematics for Advanced Level
is available either as a single volume or
in two parts:

Part A: Chapters 1–15
Part B: Chapters 16–28
Complete Volume: Chapters 1–28

Pure Mathematics for Advanced Level

Complete Volume

J. A. H. Shepperd

Department of Mathematics, University of Manchester
Institute of Science and Technology

C. J. Shepperd

Head of Mathematics, Hinchingbrooke School, Huntingdon

HODDER AND STOUGHTON

LONDON   SYDNEY   AUCKLAND   TORONTO

British Library Cataloguing in Publication Data

Shepperd, J. A. H.
  Pure mathematics for Advanced level.
  Complete volume
  1. Mathematics—1961–
  I. Title   II. Shepperd, C. J.
  510   QA39.2

  ISBN 0 340 33284 0

First printed 1983
Copyright © 1983 J. A. H. Shepperd and C. J. Shepperd

Typeset in Times New Roman (Monophoto) by Macmillan India Ltd., Bangalore

Printed in Great Britain for
Hodder and Stoughton Educational,
a division of Hodder and Stoughton Ltd.,
Mill Road, Dunton Green, Sevenoaks, Kent TN13 2YD,
by Clark Constable (1982) Ltd., Edinburgh

Contents

Preface   ix

Notation   xi

1 The algebra of functions   1

Sets, functions, linear functions.

2 Quadratic equations and quadratic functions   14

Algebraic manipulation, solution of quadratic equations (revision), the
graph of a quadratic function, a formula for the solution of a quadratic
equation, the discriminant and the location of the graph, symmetric
properties of the roots of a quadratic equation.

3 Trigonometry I   29

O-level revision, trigonometric ratios of angles of any magnitude, solution
of trigonometric equations, radian measure of angle, circular arc and
sector, effect of changes of scale and origin upon the functions $\cos x$ and
$\sin x$, the addition formulae, the triangle formulae.

4 Differentiation I   61

Linear relations and scale changes, speed, the gradient of a function,
estimating the gradient of a function using the tangent to its graph, use of
chords of the graph of a function to estimate its gradient, estimation of the
derivatives of $\cos x$ and $\sin x$, rules for finding the derivative of a function.

5 Vectors I   73

Displacements, vector and scalar quantities, origin and position vectors,
the ratio theorem, basis of a set of vectors in one plane, vector equation of a
line in three dimensions.

6 Integration I   92

A function and its derivative (revision), integration, rules for integration,
evaluation of the arbitrary constant of integration.

7 Polynomials   **97**

Definitions, algebra of polynomials, rational functions, division of a polynomial by a linear polynomial, factors of a polynomial, the sign of a polynomial function.

8 Composition of functions and applications to graphs   **107**

Composition of mappings, composition of real functions, translation of a graph, one-way stretches of a graph, reflections of a graph, composition of a function with a linear function, symmetries of a graph, inverse functions, conditions for an inverse function to exist.

9 Indices and logarithms   **125**

Powers of a number, rules of indices, extension to negative indices, extension to rational numbers, diversion, the power function 2^x and the logarithmic function $\log_2 x$, the logarithmic function $\log_a x$, rules for manipulating logarithms, change of base.

10 Linear laws   **138**

Linear relations (revision), mathematical models and linear laws, conversion of a law to a linear form, summary of substitutions.

11 Sketch graphs   **148**

Revision, the graph of a cubic function, useful ideas for graph sketching, the graph with equation $y = x^n$, $n \in \mathbb{N}$, the graph of $(x - h)^n$, scale change and the graph of $y = a(x - h)^n$, $n \in \mathbb{N}$, negative powers of x, rational powers of x.

12 Vectors II   **160**

Column vectors in the plane, linear combinations of vectors, Cartesian coordinates in three dimensions, the equation of a line in two dimensions, the vector equation of a line in three dimensions, scalar products, planes through the origin, the vector equation of a general plane, parametric equation of a plane, angles between lines and planes, distance of a point from a plane.

13 Inequalities   **207**

Ordering the real numbers, use of the modulus sign, solution of inequalities involving linear quotients, quadratic inequalities, inequalities using the modulus sign.

14 Numerical methods for solving equations   **220**

Graphical methods, decimal search, the midpoint method (interval bisection), linear interpolation the chord method, the use of tangents.

15 Complex numbers   232

Extensions of number systems, solution of polynomial equations, representation of complex numbers, polar coordinates, the triangle inequalities, complex conjugates, loci in the complex plane.

Answers to Part A   263

16 Lines and circles in the plane   285

The equation of a line (revision), the intersection of two lines, distance of a point from a line, the equation of a circle, intersection of a line and a circle.

17 Finite series I   299

Sequences, series, arithmetic progression, geometric progression, arithmetic and geometric means, how many choices?, permutations, combinations, the binomial series, the binomial theorem.

18 Differentiation II   320

Limits of functions, continuous functions, the tangent to the graph of a function, linear approximation to a function, the gradient of a function, the derived function, the derivative of x^n, derivative of a linear combination of two functions, derivative of the composition of functions, derivatives of products and quotients of functions.

19 Partial fractions   339

Algebraic identities, fractions with prime power denominators, partial fractions, no repeated factors in the denominator, repeated factors in the denominator, improper polynomial fractions, summary of the procedure.

20 Integration II   346

Revision of Chapter 6, the definite integral, the area under a curve, the definite integral as an area, limits of integration, numerical integration, the definite integral as a limit of a sum, mean values, centroids, volumes of revolution.

21 Trigonometry II   378

Further trigonometrical identities, the derivatives of cos x and sin x, derivatives of circular functions, inverse circular functions.

22   Differentiation techniques and applications   395

Derivative of an inverse function, implicit differentiation, parametric differentiation, higher derivatives, rates of change, approximations to small changes, stationary points, tangents and normals.

23   The exponential and logarithmic functions   417

Indices and logarithms (revision), the gradient of a^x, the natural logarithm function ln x, derivative and integral of the function a^x, alternative definitions for e^x and ln x, applications to integration.

24   Integration techniques   439

Indefinite integrals, integration by substitution, integration of rational functions, integration by parts.

25   Differential equations   456

Definitions, the differential equation of a family of curves, separable differential equations, problems involving differential equations.

26   Infinite series   469

Definitions, the geometric series, the binomial series.

27   Graphs and loci   479

Graphs of rational functions, the parabola, the ellipse, the hyperbola.

28   Finite series II   494

The method of differences, use of partial fractions, derivatives of the geometric series, mathematical induction.

Formulae   505

Answers to Part B   509

Index

Preface

Pure Mathematics for Advanced Level provides the pure mathematics half of an Advanced level G.C.E. course in mathematics, in which the other half may be applied mathematics (either mechanics or statistics, or a mixture of these) or further pure mathematics. The material has been selected and organised to provide the necessary pure mathematics required in the applied mathematics. It includes all the 'common core' in mathematics which, it is hoped, will soon be accepted by all the G.C.E. Boards. This core forms about 70 per cent of the material in the book. We have attempted to plot a path midway between the extremes of the 'traditional' and 'modern' versions of sixth-form mathematics, which have existed for nearly twenty years. The book, therefore, matches the new A-level mathematics syllabuses introduced by the University of London Board and the Joint Matriculation Board, which have removed the undesirable dichotomy between 'trad.' and 'mod.'. We have used the best of modern teaching ideas but have kept in mind the traditional skills. We believe that it is through the practice of mathematical skills that the desired understanding of fundamental concepts will be achieved.

The material is arranged for linear use by a student working alone, but a teacher will find it straightforward to vary the order of study. Many chapters begin with revision material and practice exercises. New work is illustrated by many worked examples and the exercises in the sections are carefully graded to lead the student towards an understanding in depth. It is intended that all these exercises should be completed. Miscellaneous exercises, at the end of the chapters, contain more practice material, which is not graded in difficulty and which may contain applications of work from any previous chapters. Many exercises are from the past papers of the Associated Examining Board (AEB), the University of Cambridge Local Examinations Syndicate (C), the Joint Matriculation Board (JMB), the University of London University Entrance and School Examinations Council (L), the Oxford and Cambridge Schools Examination Board School's Mathematics Project (SMP). We gratefully acknowledge these Boards' permission to reproduce their questions. We give our own brief answers to all the exercises.

The teacher, who is changing from a traditional mathematics course to one of the new courses, will find that all the new material required is contained in the first half of the book. As a result, the later chapters contain mostly traditional material, although the approach may give it a new look. Calculus is introduced initially in a practical manner and then is

dealt with more formally later. Other topics, such as vectors, numerical
solution of equations and complex numbers, are dealt with in depth in the
first half of the book. Some of these topics have been introduced in two
'bites' in order to aid the digestion.

We express our grateful thanks to many who have helped us in this
work, to a number of colleagues, and in particular to Miss Joyce Batty and
to the editors. Above all, we give tribute to the forbearance of Jenny and
Beryl without whose support this work would never have been completed.

C. J. Shepperd
J. A. H. Shepperd

Pure Mathematics for Advanced Level is
available either as a single volume or
in two parts.

Part A: Chapters 1–15
Part B: Chapters 16–28
Complete Volume: Chapters 1–28

Notation

$=$	is equal to
$\approx$	is approximately equal to
$\neq$	is not equal to
$<$	is less than
$\leqslant$	is less than or equal to
$>$	is greater than
$\geqslant$	is greater than or equal to
$\nless$	is not less than
$\ngtr$	is not greater than
$\sqrt{}$	the positive square root
$\Rightarrow$	implies that
$\Leftarrow$	is implied by
$\Leftrightarrow$	implies and is implied by
$\{a, b, c, \dots\}$	the set with elements $a, b, c, \dots$
$\in$	is an element of
$\notin$	is not an element of
$:$	such that
$\{x: a < x < b\}$	the set of x such that a is less than x and also x is less than b
$n(A)$	the number of elements in the set A
$\mathscr{E}$	the universal set
ϕ	the empty set, null set
S'	the complement of the set S
$\cup$	the union of
$\cap$	the intersection of
$\subseteq, \subset$	is a subset of, is a proper subset of, respectively
$\backslash$	the set difference, $A \backslash B = A \cap B'$
$\mathbb{N}$	the set of positive integers and zero $\{0, 1, 2, \dots\}$
$\mathbb{Z}$	the set of integers $\{0, \pm 1, \pm 2, \dots\}$
$\mathbb{Z}^+$	the set of positive integers
$\mathbb{Q}$	the set of rational numbers
$\mathbb{Q}^+$	the set of positive rational numbers
$\mathbb{R}$	the set of real numbers
$\mathbb{R}^+$	the set of positive real numbers
$\mathbb{C}$	the set of complex numbers
$f: A \to B$	the function f maps the set A into the set B
$f(x)$	the function value for x
$f: x \mapsto y$	f is the function under which x is mapped into y

f^{-1}	the inverse function of f

For functions f and g where domains and ranges are subsets of $\mathbb{R}$

$f+g$	is defined by $(f+g)(x) = f(x)+g(x)$
$f.g$	is defined by $(f.g)(x) = f(x)g(x)$
fg	is defined by $(fg)(x) = f(g(x))$
$\rightarrow$	approaches, tends to (in the context of a limit)
∞	infinity
$|\ \ |$	the unsigned part of a signed number, the modulus
$[x]$	the integer part of the number x
Σ	the sum of (precise limits may be given)
$n!$	n factorial
$\dbinom{n}{r}$	the binomial coefficient $n(n-1)(n-2)\ldots(n-r+1)/r!$
δx	a small increment of x
$f', f'', f^{(3)}, \ldots f^{(n)}$	the first, second, third, ... nth derivatives, respectively, of f
$\dfrac{dy}{dx}$	the derivative of y with respect to x
$\int y\,dx$	the indefinite integral of y with respect to x
$\displaystyle\int_a^b y\,dx$	the definite integral of y with respect to x between the limits $x=a$ and $x=b$
$\ln x$	the natural logarithm of x
e^x	the exponential function of x
$|\mathbf{r}|, r$	the magnitude of the vector $\mathbf{r}$
$\mathbf{i, j, k}$	unit vectors in the mutually perpendicular directions Ox, Oy, Oz
$\mathbf{a.b}$	the scalar product of the vectors $\mathbf{a}$ and $\mathbf{b}$
i	square root of -1
$|z|$	the modulus of the complex number z
$\arg z$	the argument of the complex number z
z^*	the conjugate of the complex number z
$\mathrm{Re}(z)$	the real part of the complex number z
$\mathrm{Im}(z)$	the imaginary part of the complex number z

1 The Algebra of Functions

1.1 Sets

In mathematics, we are concerned with collections of objects of various kinds, which form sets. In particular, we are concerned with sets of numbers, which are subsets of the set $\mathbb{R}$ of *real* numbers. We remind the reader of some definitions and state the set theory notation that we shall use.

Definitions and notation A *set* is a clearly defined collection of distinct objects, which are called the *members* or *elements* of the set. When x is a member of the set A, we write $x \in A$ meaning x belongs to A. If x is not a member of A, then we write $x \notin A$ meaning x does not belong to A. When we wish to list the members of a set, we write $A = \{a, b, c, d\}$ meaning that A is the set whose members are a, b, c and d; for example, $2 \in \{1, 2, 3\}$ and $3 \notin \{2, 4, 6\}$. These ideas may be shown in a *Venn diagram*, in which a set is indicated by a closed curve encircling all of its members. In Fig. 1.1, the elements a and d belong to both sets A and B, while b and c belong to A but not to B, and e and f belong to B and not to A.

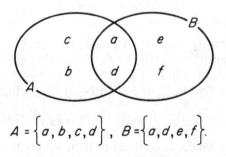

$$A = \left\{ a, b, c, d \right\}, \quad B = \left\{ a, d, e, f \right\}.$$

Fig. 1.1 Venn diagram of 2 sets.

Equality of sets

Two sets are *equal* when they consist of the same collection of elements, that is, when each member of one set belongs to the other set and conversely. Thus, $A = B$ means that $x \in A$ if and only if $x \in B$. When listing the members of a set, the order of the list is unimportant and so are any repetitions. For example,

$$\{1, 2, 3\} = \{2, 3, 1\} = \{3, 3, 1, 2, 1, 2\},$$

because each of these three sets contains just the three elements 1, 2 and 3, and hence they are equal sets.

Braces are also used when a set is described by means of some defining property. Thus

$$\{1, -1\} = \{x : x^2 = 1\},$$

where the colon means 'such that'. In this notation, the symbol x can be replaced by any other symbol, so

$$\{x : x^2 = 1\} = \{y : y^2 = 1\} = \{a : a^2 = 1\}.$$

Such a symbol is called a *dummy* variable.

Sets of numbers

We shall use the following notation for certain sets of numbers, which are all contained in the set $\mathbb{R}$ of real numbers.

$\mathbb{N}$ is the set of *natural* numbers, where $\mathbb{N} = \{0, 1, 2, 3, \ldots\}$;
$\mathbb{Z}$ is the set of *integers*, where

$$\mathbb{Z} = \{0, 1, -1, 2, -2, \ldots\} = \{\ldots, -2, -1, 0, 1, 2, \ldots\};$$

$\mathbb{Z}^+$ is the set of positive integers, where $\mathbb{Z}^+ = \{1, 2, 3, \ldots\}$;
$\mathbb{Q}$ is the set of *rational* numbers, where $\mathbb{Q} = \{p/q : q \neq 0 \text{ and } p, q \in \mathbb{Z}\}$;
$\mathbb{Q}^+$ is the set of positive rational numbers, where $\mathbb{Q}^+ = \{p/q : p, q \in \mathbb{Z}^+\}$;
$\mathbb{R}^+$ is the set of positive real numbers, where $\mathbb{R}^+ = \{x : x \in \mathbb{R} \text{ and } x > 0\}$.

Implications

When a statement Q can be deduced, by means of a mathematical argument, from another statement P, we say that Q is *implied* by P, or P implies Q, which is written $P \Rightarrow Q$. When both $P \Rightarrow Q$ and $Q \Rightarrow P$, then we say that P is *equivalent* to Q, or P if and only if Q, which is written $P \Leftrightarrow Q$. For example,

$$x \in \mathbb{N} \Rightarrow x \in \mathbb{Z}, \quad \text{and} \quad x^2 = 1 \Leftrightarrow x \in \{1, -1\}.$$

When Q cannot be deduced from P, then we say that P does not imply Q, which is written $P \nRightarrow Q$. Such a result would be proved by constructing a *counter-example*. For example, $x \in \mathbb{Z} \nRightarrow x \in \mathbb{N}$ because $-1 \in \mathbb{Z}$ but $-1 \notin \mathbb{N}$.

Subsets

We say that the set A is a *subset* of the set B when each element of A is also an element of B, and we use the notation $A \subseteq B$, meaning that

$$x \in A \Rightarrow x \in B.$$

From this definition, it will be seen that, for any set B, $B \subseteq B$, and we define a proper subset A of a set B to be a subset of B, not equal to B. We then use the notation $A \subset B$ meaning that

$$x \in A \Rightarrow x \in B \text{ and for some } b, b \notin A \text{ and } b \in B.$$

So $A \subset B$ means that $A \subseteq B$ and $A \neq B$.
A subset is indicated on a Venn diagram by marking part of that area which represents the whole set. An example is shown in Fig. 1.2(a).

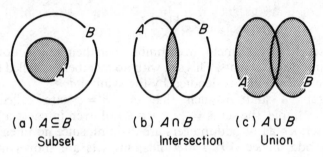

(a) $A \subseteq B$ (b) $A \cap B$ (c) $A \cup B$
    Subset   Intersection   Union

Fig. 1.2

Intersection

The *intersection* of two sets A and B is the set of all elements which belong to both the sets A and B, with the notation

$$A \cap B = \{x : x \in A \text{ and } x \in B\}.$$

Union

The *union* of two sets A and B is the set of all elements which belong to either the set A or to the set B or to both A and B, with the notation

$$A \cup B = \{x : x \in A \text{ or } x \in B\}.$$

Given two sets A and B, represented on a Venn diagram, their intersection, $A \cap B$, and their union, $A \cup B$, may be indicated by shading the appropriate area, as is shown in Fig. 1.2(b) and (c).

The universal set and the empty set

The set of all possible elements under consideration in a given piece of work is called the *universal* set and is denoted by $\mathscr{E}$. In many cases, we are concerned with the set of all real numbers and so $\mathscr{E} = \mathbb{R}$, but we might use $\mathscr{E} = \mathbb{Z}$ if, in some piece of work, we wished to restrict the argument to the integers.

On a Venn diagram, the universal set is usually represented by a large rectangle, inside which all the other sets occur. In Fig. 1.3(a), we show the universal set $\mathscr{E} = \mathbb{R}$, with subsets $\mathbb{Q}$ and $\mathbb{Z}$.

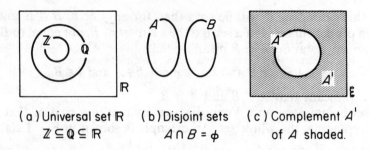

(a) Universal set $\mathbb{R}$ (b) Disjoint sets   (c) Complement A'
$\mathbb{Z} \subseteq \mathbb{Q} \subseteq \mathbb{R}$ $A \cap B = \phi$ of A shaded.

Fig. 1.3

If two sets have no elements in common, then their intersection will be a set containing no elements. This set, with no members, is called the *empty set*, or the *null set*, and is denoted by the symbol ϕ.

Two sets A and B such that $A \cap B = \phi$ are called *disjoint* sets. For example, the set $A = \{2n : n \in \mathbb{Z}\}$ of even integers and the set $B = \{2n + 1 : n \in \mathbb{Z}\}$ of odd integers are disjoint, since no integer is both even and odd, hence $A \cap B = \phi$. Disjoint sets are shown in a Venn diagram in Fig. 1.3(b).

The *complement* of a set A is the set of all elements in the universal set $\mathscr{E}$ which are not members of A. The complement of A is denoted by A'. Thus

$$A' = \{x : x \in \mathscr{E} \text{ and } x \notin A\}.$$

It follows that $\phi = A \cap A'$ and $\mathscr{E} = A \cup A'$. See Fig. 1.3(c).

Note: For every set A, $\phi \subseteq A$ and $A \subseteq \mathscr{E}$. Such a pair of subset relations can be written together as $\phi \subseteq A \subseteq \mathscr{E}$.

Set difference

It is useful to have a notation for the set of elements which belong to one set A and do not belong to another set B, and we shall call this the *difference* of the sets A and B with the notation $A \backslash B$. This means that

$$A \backslash B = \{x : x \in A \text{ and } x \notin B\} = A \cap B'.$$

For example, $\mathbb{R} \backslash \mathbb{Q}$ is the set of *irrational* real numbers.
With this notation, there is no need for the set B to be a subset of A; for example,

$$\{1, 2, 3, 4, 5\} \backslash \{4, 5, 6, 7, 8\} = \{1, 2, 3\}$$

and $\mathbb{Q} \backslash \mathbb{R} = \phi.$

EXAMPLE

Given two sets A and B, prove that

  (i) $A \cap B \subseteq A \subseteq A \cup B,$
 (ii) *if* $A \cup B \subseteq A \cap B$ *then* $A = B,$
(iii) $A \cap B = A \Leftrightarrow A \subseteq B.$

(i) Let $x \in A \cap B$, then $x \in A$ and $x \in B$ so $x \in A$. Hence $A \cap B \subseteq A$. Let $y \in A$, then $y \in A$ or $y \in B$ and so $y \in A \cup B$. Hence $A \subseteq A \cup B$.
(ii) If $A \cup B \subseteq A \cap B$, let $x \in A$, then $x \in A \cup B$ by (i), so $x \in A \cap B$ and thus $x \in B$. Hence $A \subseteq B$. Similarly, if $y \in B$ then $y \in A$. Therefore $A = B$.
(iii) First suppose that $A \cap B = A$. If $x \in A$, then $x \in A \cap B$ and so $x \in B$. Hence $A \subseteq B$. Conversely, suppose that $A \subseteq B$. Then, if $x \in A$, $x \in B$ so $x \in A \cap B$ and thus $A \subseteq A \cap B$. Also, by (i), $A \cap B \subseteq A$, and, therefore, $A = A \cap B$.

EXERCISE 1.1A

1 Given the sets $A = \{a, b, c, d\}$, $B = \{a, c, e, f\}$, $C = \{b, e, d\}$, list the elements of the sets: (i) $A \cup B$, (ii) $A \cap B$, (iii) $A \cap (B \cap C)$, (iv) $(A \cap B) \cap C$, (v) $A \cap (B \cup C)$, (vi) $(A \cap B) \cup C$, (vii) $A \backslash B$, (viii) $A \backslash C$, (ix) $(A \backslash B) \cup (A \backslash C)$, (x) $(A \backslash B) \cap (A \backslash C)$.
2 Using the sets given in question 1, describe all the subsets of $A \cup C$, which contain three elements, by listing their elements.
3 State whether the following statements are true, false, or invalid:
(i) $0 \in \mathbb{Z}$, (ii) $\{0, 1\} \subseteq \mathbb{Z}$, (iii) $0 \subseteq \mathbb{Z}$, (iv) $\phi \in \mathbb{Z}$, (v) $0 \in \phi$, (vi) $\phi \subseteq \mathbb{Z}$, (vii) $\mathbb{N} \subseteq \mathbb{Z}$, (viii) $\mathbb{Z} \subset \mathbb{N}$, (ix) $\mathbb{Z} \subset \mathbb{Q}$, (x) $1 \notin \mathbb{N}$.
4 List the members of the following subsets of the set $\mathbb{Z}$ of integers:
(i) $\{x : x^2 = 4\}$, (ii) $\{x : x^2 = -1\}$, (iii) $\{x : x^3 = -1\}$, (iv) $\mathbb{N} \cap \{x : x \leqslant 5\}$, (v) $\mathbb{Z} \cap \{x : 10 < x^2 < 50\}$.
5 List all the subsets of the set: (i) $\{1, 2\}$, (ii) $\{a, b, c, d\}$.
6 Given that the universal set is $\mathbb{R}$, $A = \{2n : n \in \mathbb{N}\}$, $B = \{x : x < 3\}$, and $C = \{x : x \in \mathbb{Z}, \ 2 \leqslant x\}$, describe the set: (i) $A \cap B$, (ii) $A \cap C$, (iii) B', (iv) $B \cap C'$, (v) $A \cap B \cap C$, (vi) $A \cap (B \backslash C)$, (vii) $(A \cap B) \backslash C$, (viii) $B \cup C \cup A$.
7 Prove that $A \cap (B \cup C) = (A \cap B) \cup (A \cap C)$, for any sets A, B and C.
Check your answers to these exercises (page 267). If you require more practice, work through the Exercise 1.1B below.

EXERCISE 1.1B

1 The set A is the set of all positive real numbers less than 1.5. Mark the following true, false or invalid:
(i) $1 \in A$, (ii) $1 \subseteq A$, (iii) $\frac{1}{2} \in A$, (iv) $0 \notin A$, (v) $\phi \in A$, (vi) $\{\frac{1}{2}, \frac{3}{8}, \frac{7}{8}\} \subset A$, (vii) $3/2 \in A$, (viii) $2 \in A$.
2 B is the set of plane quadrilaterals. State whether it is true or false that $x \in B$ when: (i) x is a square, (ii) x is a triangle, (iii) x is a trapezium, (iv) x is a hexagon, (v) x is a circle, (vi) x is a parallelogram, (vii) x is a rhombus, (viii) x is a pentagon, (ix) x is a rectangle.
3 Describe the following sets in the form $(x : P(x))$, where $P(x)$ is a defining statement:
(i) $\{1, 3, 5, 7\}$, (ii) $\{1, 8, 27, 64\}$, (iii) $\{-2, 0, 2\}$.
4 List the subset of $\mathbb{Z}$ of integers whose squares are less than 30.
5 State if each of the following statements is true or false for $x, y, z \in \mathbb{Z}$. If it is false, determine the largest subset of $\mathbb{Z}$ for which the statement is true:
(i) $x + y = y + x$, (ii) $(x + y) + z = x + (y + z)$, (iii) $x - y = y - x$, (iv) $(xy)/y = x$, (v) $y = 1/x \Rightarrow x = 1/y$, (vi) $(x + y)^2 = x^2 + y^2$,

(vii) $x^2 - y^2 = (x - y)(x + y)$, (viii) $x(yz) = (xy)z$, (ix) $x^2 x^3 = x^5$,
(x) $x^2 = y \Rightarrow y^2 = x$, (xi) $(x - 2)(x + 1) = 0$, (xii) $2(x - 1) = 2x - 2$,
(xiii) $x - 1 = 1 - x$, (xiv) $2x^2 - 5x + 2 = 0$, (xv) $2x \leqslant x + 1$.

6 Does every set contain at least two distinct subsets? If so, prove it. If not, list all the sets which contain less than two distinct subsets.

7 Given that $X = \{a, b, c, d\}$, find all possible sets Y and Z such that
   (i) $\{a\} \subseteq Y \subseteq X$, (ii) $\{a, b\} \subseteq Z \subseteq X$.

8 Given real numbers a and b, with $a < b$, define the interval (a, b) by $(a, b) = \{x : a < x < b\}$. State whether the following statements are true or false: (i) $(1, 2) \subseteq (0, 3)$, (ii) $(2, 5) \subseteq (3, 8)$, (iii) $(2, 5) \subset (1, 10)$,
   (iv) $(-1, 1) \subseteq (-1, 1)$, (v) $6 \in (5, 7)$, (vi) $(1, 2) \cap (2, 3) = \phi$,
   (vii) $(1, 3) \cap (2, 4) = (2, 3)$, (viii) $(1, 3) \backslash (2, 4) = (2, 3)$,
   (ix) $((2, 4) \cap (3, 5)) \cup (1, 3) = (2, 4) \cap ((3, 5) \cup (1, 3))$.
   Find three intervals, which are all different, and call them A, B, C, such that $(A \cap B) \cup C = A \cap (B \cup C)$.

9 Determine, for each of the statements P, Q, S, T, whether it implies, is implied by, or is equivalent to, the statement 'R: one member of the set A is also a member of the set B':
   (i) $P : A \subseteq B$, (ii) $Q : A = B$, (iii) $S : A \cap B \neq \phi$, (iv) $T : A \cup B \neq \phi$.

10 Given that $A \subseteq B, B \subseteq C, C \subseteq A$, prove that $A = B = C$. Hence, prove that if E is the set of even positive integers, S is the set of sums of two odd positive integers and Q is the set of positive integers whose squares are even, then $E = S = Q$.

1.2 Functions

A *function* is a mapping from one set A to another set B; that is, given any member a of the set A, the function determines one and only one member b of the set B. Functions may be defined in which the sets A and B are of any types, but we shall be concerned with sets of numbers.

Definitions and notation

f: $A \to B$ means that f is a function mapping the set A to the set B;
f: $a \mapsto f(a)$ means that, for $a \in A$, the function f assigns just one $f(a)$ in B;
the set A is called the *domain* of the function f;
the set B is called the *codomain* of the function f;
$f(a)$ is called the *image* of a under the function f, or
$f(a)$ is called the *value* of the function f at the point a of its domain A.

  The function is only defined when we know (i) the domain A of f, and (ii) the value f (a) of f for every point a in A. For example, if $A = B = \mathbb{Z}$ and the function f maps x to $2x$, then f: $\mathbb{Z} \to \mathbb{Z}$, f is defined by $f(x) = 2x$ for all $x \in \mathbb{Z}$. Thus $f(0) = 0$, $f(1) = 2$, $f(-3) = -6$, and so on.

  The function g: $\mathbb{N} \to \mathbb{N}$, defined by $g(x) = 2x$ for all $x \in \mathbb{N}$, takes the same values as does f for all points of $\mathbb{N}$. However, the functions f and g are not the same function since they have different domains.

Real functions

A *real function* is a function whose domain is a subset of the set $\mathbb{R}$ of real numbers and whose codomain is $\mathbb{R}$, so that the values of the function are also real numbers. When the domain of a real function is not stated, and the value of the function is given by some formula or formulae, then the domain of the function is assumed to be the *largest* subset of $\mathbb{R}$ for which the value of the function is defined. For example:

the function f with $f(x) = 1/x$ has domain $\mathbb{R}\backslash\{0\}$;

the function g with $g(x) = \sqrt{x}$ (the non-negative square root of x) has domain $\{x: x \in \mathbb{R}, 0 \leqslant x\}$, the set of non-negative real numbers.

The range of a function

Consider a real function $f: A \to \mathbb{R}$, with domain $A \subseteq \mathbb{R}$. For any subset C of A, the set of all the images of all the points in C is called the *image* of C under f and is denoted by

$$f(C) = \{f(x): x \in C\}.$$

The image under f of the domain A of f is called the *range* of f, and so the range of f is

$$f(A) = \{f(x): x \in A\},$$

a subset of $\mathbb{R}$, (Fig. 1.4).

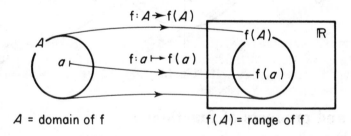

$$f: A \rightarrow f(A)$$

$$f: a \mapsto f(a)$$

A = domain of f $f(A)$ = range of f

Fig. 1.4

EXAMPLE 1
Find the domain and range of the function f given by
(i) $f(x) = 1/(1-x)$, (ii) $f(x) = \sqrt{(1-x^2)}$.

(i) $f(x)$ is defined for all real values of x except $x = 1$, so the domain of f is the set $\mathbb{R}\backslash\{1\}$. Putting $y = f(x) = 1/(1-x)$, we find $y - yx = 1$ and so $x = (y-1)/y$. All possible values of y can occur except 0, so the range of the function f is $\mathbb{R}\backslash\{0\}$.
(ii) $f(x)$ is defined for non-negative values of $1 - x^2$ so the domain of f is $\{x: -1 \leqslant x \leqslant 1\}$. Clearly, the range of f is $\{x: 0 \leqslant x \leqslant 1\}$.

EXAMPLE 2
Determine whether the given equation defines a function f such that $y = f(x)$ and, if so, state the domain and range of f: (i) $2y + 1 = x$, (ii) $x^2 - 2y^2 = 4x$.

(i) The equation is equivalent to $y = \frac{1}{2}(x-1)$ and so this defines the function f with $f(x) = \frac{1}{2}(x-1)$; the domain and range of f are both $\mathbb{R}$.

(ii) The equation may be rewritten $y^2 = \frac{1}{2}(x^2 - 4x)$ and this does not define $y = f(x)$ uniquely, since if $x = 8$, y can be either 4 or -4. Therefore, the equation **does not define a function**.

The Cartesian graph of a function

The function f with domain A is completely defined by the set of ordered pairs $\{(x, f(x)): x \in A\}$. For a real function f, these pairs may be plotted in the x–y plane by putting $y = f(x)$, and this gives the Cartesian graph of the function f. For example, we illustrate, in Fig. 1.5, the Cartesian graphs of the functions f, g, h, with a common domain $\mathbb{R}$, which are given by $f(x) = 4$, $g(x) = 3 - \frac{3}{2}x$, $h(x) = \frac{1}{2}x^2 - 4$.

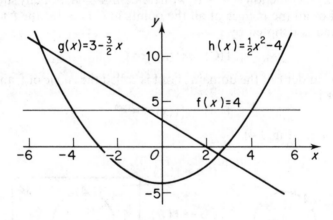

Fig. 1.5

Sums and products of functions

We often form new functions from given functions by using the usual processes of arithmetic on the values of the functions, which may be added, multiplied, and so on. As an example, consider the four functions:

f, with domain $\mathbb{R}$, given by $f(x) = 1 + x$;
g, with domain $\mathbb{R}$, given by $g(x) = x^3$;
h, with domain A, given by $h(x) = \sqrt{x}$, where $A = \{x: 0 \leqslant x\}$;
k, with domain B, given by $k(x) = 1/(x-1)$, where $B = \mathbb{R}\backslash\{1\}$.

For any $x \in \mathbb{R}$,

$$f(x) + g(x) = 1 + x + x^3$$

and this may be considered to be the value of a new function $(f + g)$, with the same domain $\mathbb{R}$, that is,

$$(f + g)(x) = 1 + x + x^3.$$

We may also form a new function as a sum of constant multiples of f and g, for example,

$$2f(x) - g(x) = 2 + 2x - x^3 = F(x),$$

where $F = 2f - g$ defines the new function F with domain $\mathbb{R}$.

If we form a similar combination of functions which have different domains, then the new function will have a domain which is the intersection of the domains of the original functions, since this is where it will be defined.

Thus,

$$3h(x) - 2k(x) = 3\sqrt{x} - \frac{2}{x-1} = G(x),$$

where $G = 3h - 2k$ with domain $A \cap B$.

We may also multiply functions by multiplying their values at common points of their domains, for example,

$$f(x) . k(x) = (1 + x)/(x - 1) = H(x),$$

and $H = f . k$ with domain $\mathbb{R} \cap B (= B)$. The quotient of two functions may also be formed provided that we also exclude from the new domain all points where the denominator vanishes, for example,

$$\frac{f(x)}{h(x)} = \frac{1 + x}{\sqrt{x}} = K(x),$$

where $K = f/h$, with domain $(\mathbb{R} \cap A)\backslash\{0\}$, that is, $\mathbb{R}^+$.

Arithmetic of functions

We use the above ideas to define arithmetical operations on real functions and, provided that we are careful to note the correct domains of the functions, all the usual rules of arithmetic will be satisfied.

Definitions Let f and g be real functions with domains A and B respectively, let $C = \{x : g(x) = 0\}$ be the set of zeros of the function g, and let a and b be real numbers. Then we define the function:

$h = f + g$	by $h(x) = f(x) + g(x)$	with domain $A \cap B$;
$h = f - g$	by $h(x) = f(x) - g(x)$	with domain $A \cap B$;
$h = af$	by $h(x) = af(x)$	with domain A;
$h = af + bg$	by $h(x) = af(x) + bg(x)$	with domain $A \cap B$;
$h = f . g$	by $h(x) = f(x) . g(x)$	with domain $A \cap B$;
$h = f/g$	by $h(x) = f(x)/g(x)$	with domain $A \cap B\backslash C$.

Note that the set $A \cap B\backslash C$ is unambiguous without the insertion of brackets because

$$A \cap (B\backslash C) = (A \cap B)\backslash C = A \cap B \cap C'.$$

EXAMPLE 3
Let f and g be the real functions defined by $f(x) = 2x$ *and* $g(x) = \sqrt{x}$. *Find the domain and the range of each of f and g.*

In each of the following cases, write down h(x) *and the domain and range of* h:
(i) $h = f + 3g$, (ii) $h = f.g$, (iii) $h = g.g$, (iv) $h = f/g$.

The domain of f is $\mathbb{R}$ and its range is also $\mathbb{R}$. The domain of g is $\mathbb{R}_0^+$, where $\mathbb{R}_0^+ = \{x : x \geqslant 0\}$, and its range is also $\mathbb{R}_0^+$.
(i) $h(x) = f(x) + 3g(x) = 2x + 3\sqrt{x}$. The domain of h is $\mathbb{R} \cap \mathbb{R}_0^+$, which is $\mathbb{R}_0^+$. For $x \in A$, $f(x) \geqslant 0$ and $g(x) \geqslant 0$, so $h(x) \geqslant 0$, and the range of h is $\mathbb{R}_0^+$.
(ii) $h(x) = f(x).g(x) = 2x\sqrt{x}$. Again the domain and the range of h is $\mathbb{R}_0^+$.
(iii) $h(x) = g(x).g(x) = \sqrt{x}\sqrt{x} = x$. Thus h is the identity function on the domain $\mathbb{R}_0^+ \cap \mathbb{R}_0^+$, which is $\mathbb{R}_0^+$. The range of h is also $\mathbb{R}_0^+$.
(iv) $h(x) = f(x)/g(x) = 2x/\sqrt{x} = 2\sqrt{x}$. The domain of h is $\mathbb{R} \cap (\mathbb{R}_0^+ \backslash \{0\})$, which is $\mathbb{R}^+$, and the range of h is also $\mathbb{R}^+$.

EXERCISE 1.2

1   The domain of the function f is $\{1, 2, 3, 4\}$. Find the range of f if:
    (i) $f(x) = 2x$, (ii) $f(x) = 5 - x$, (iii) $f(x) = x^2 - 4$, (iv) $f(x) = \sqrt{x}$.
2   State what is wrong with the following statements;
    (i) $f(x) = x^2 - 1$, f has domain $\mathbb{Z}$ and range $\mathbb{N}$;
    (ii) the function f has domain $\{1, 2, 3, 4\}$ and range $\{5, 6, 7, 8, 9\}$;
    (iii) the function f has domain $\mathbb{N}$ and $f(x)$ is the solution of the equation $f(x)^2 = x$;
    (iv) the function f has domain $\mathbb{Z}$ and $f(x) = \sqrt{(x^2 - 1)}$.
3   Determine whether the given equation defines a real function f such that $f(x) = y$ and, if it does, state the domain and range of f:
    (i) $y = 3x$, (ii) $3y + 4x = 5$, (iii) $2y = x^2$, (iv) $y^2 = 4x$, (v) $y^3 = 4x$, (vi) $x^2 + y^2 = 9$, (vii) $x^2 + y^2 = 0$, (viii) $\sqrt{y} = x$.
4   State the domain and the range of the real function f given by:
    (i) $f(x) = 5x$, (ii) $f(x) = x^2$, (iii) $f(x) = 2x - 1$, (iv) $f(x) = \sqrt{x}$, (v) $f(x) = 2x^2 - 1$, (vi) $f(x) = 1/(x - 2)$, (vii) $f(x) = \log_{10} x$, (viii) $f(x) = 2^x$, (ix) $f(x) = \sqrt{(x^2 - 4)}$, (x) $f(x) = \sqrt{(9 - x^2)}$.
5   Given the functions f and g, where $f(x) = x^2$ and $g(x) = \sqrt{(-x)}$, find h(x) and the domain and the range of the function h, if:
    (i) $h = f$, (ii) $h = g$, (iii) $h = 2f + g$, (iv) $h = 3f.g$, (v) $h = g.g$, (vi) $h = f/g$, (vii) $h = g/f$.
6   How many different functions can be defined:
    (i) with domain $\{a, b\}$ and range $\{x, y\}$,
    (ii) with domain $\{a, b, c\}$ and range $\{x, y\}$,
    (iii) with domain $\{a, b\}$ and range $\{x, y, z\}$?
7   Which of the Cartesian graphs in Fig. 1.6 represent a function?
8   The *identity function* f is defined by $f(x) = x$, with domain $\mathbb{R}$. The *modulus function* g is defined by $g(x) = |x| = x$ if $x \geqslant 0$, $|x| = -x$ if $x < 0$, with domain $\mathbb{R}$. The *integer-part function* h is defined by $h(x) = [x] =$ the largest integer which is less than or equal to x, with domain $\mathbb{R}$. Find the ranges of the following functions and sketch their Cartesian graphs:
    (i) g, (ii) h, (iii) $f + g$, (iv) $f - g$, (v) $f.g$, (vi) f/g, (vii) $f - h$.

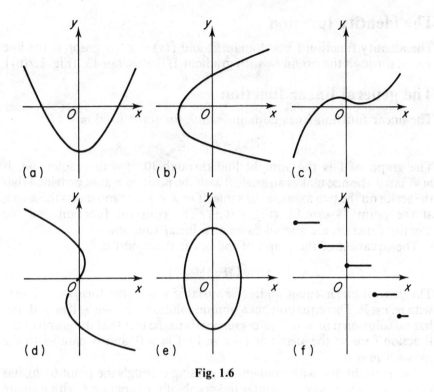

(a)   (b)   (c)

(d)   (e)   (f)

Fig. 1.6

1.3 Linear functions (revision)

We summarise some results concerning linear functions and their graphs.

The constant function

The constant function f has domain $\mathbb{R}$ and, for some fixed $c \in \mathbb{R}$, $f(x) = c$. The graph of the constant function is a line parallel to the axis Ox (Fig. 1.7(a)).

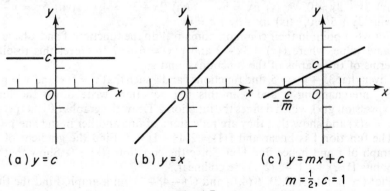

(a) $y = c$ (b) $y = x$ (c) $y = mx + c$
$m = \frac{1}{2}, c = 1$

Fig. 1.7

The identity function

The identity function f has domain $\mathbb{R}$ and $f(x) = x$. Its graph is the line $y = x$, through the origin O, with gradient 1, that is, $\tan 45°$ (Fig. 1.7(b)).

The general linear function

The linear function f has domain $\mathbb{R}$ and, for some fixed $m, c \in \mathbb{R}$,

$$f(x) = mx + c.$$

The graph of f is the straight line through $(0, c)$ with gradient m. If $m = \tan \theta$, the line makes an angle θ with the positive x-axis, provided that the scales on the two axes are the same. If $m \neq 0$, the graph meets the x-axis at the point $(-c/m, 0)$, (Fig. 1.7(c)). The constant function and the identity function are special cases of a linear function.

The equation of the graph of the linear function f is

$$y = f(x) = mx + c.$$

The general linear equation in one variable x is of the form $mx + c = 0$, with $m, c \in \mathbb{R}$. This equation has a unique solution $x = -c/m$ if $m \neq 0$, and has no solution if $m = 0$. This corresponds to the fact that the graph of the function f meets the x-axis at $(-c/m, 0)$ if $m \neq 0$, and is parallel to the x-axis if $m = 0$.

The straight line with gradient m, passing through the point (a, b), has equation $y - b = m(x - a)$ and is the graph of the function f, with domain $\mathbb{R}$, given by $f(x) = m(x - a) + b$. This is, of course, the same as the function f with $f(x) = mx + c$, where $c = b - ma$.

EXERCISE 1.3

1   Write down the form of $f(x)$ for the linear function f whose graph passes through the origin and has gradient: (i) 3, (ii) $-\frac{1}{2}$, (iii) m.
2   Find the form of $f(x)$ and hence write down the gradient of the graph of f, when the equation of this graph is: (i) $3y = x$, (ii) $3x + 5y = 0$, (iii) $\frac{1}{2}x = \frac{3}{4}y$, (iv) $3x - 2y = 0$,   (v) $ax = by$,   (vi) $2x + 3y = 4$,   (vii) $5x = y - 3$, (viii) $3y + 5 = 0$,   (ix) $ax + by + c = 0$.
3   For what point in their common domain $\mathbb{R}$ do the functions f and g have the same value, where $f(x) = 3x - 2$ and $g(x) = 6 - x$? Interpret this result in terms of the graphs of the functions f and g.
4   Given that $3x + 4y = 5$, find functions f and g such that $y = f(x)$ and $x = g(y)$. By interchanging x and y in this last equation, write down the linear expression $g(x)$, which defines the function g. Draw the graphs of $y = f(x)$ and $y = g(x)$ and show that they are reflections of one another in the line $y = x$.
5   The function f is linear and $f(1) = 2$, $f(-1) = 4$. Find the gradient of the graph of f and hence find $f(x)$. Solve the equation $f(x) = 0$. Show that the points $(1, 2)$, $(-1, 4)$, $(2, 1)$ are collinear.
6   Plot the points $A(2, 3)$, $B(4, 0)$ and $C(-4, -1)$, on a graph. Find the three linear functions whose graphs form the three sides of the triangle ABC.

Measure the angle BAC and calculate the product of the gradients of AB and AC. State what you expect to be the product of the gradients of any two perpendicular lines.

7   The graphs of the linear functions f and g are respectively the lines having equations $2x + 3y = 4$ and $y = 2x + 8$. Find the coordinates of the point P where these two lines intersect. Show that the graph of the function h, where h = 2f − g, is also a straight line through P.

8   Given linear functions f and g and real numbers a and b, with $a + b = 1$, prove that h, where h = af + bg, is also a linear function and that the graphs of f, g and h either meet at a point or are parallel.

2 Quadratic Equations and Quadratic Functions

2.1 Algebraic manipulation

Before proceeding further in the study of algebra, the reader is recommended to perform a self-test on some of the following exercises on algebraic manipulation. If it is found that practice is necessary, then all the exercises should be completed. First, we work through some examples.

EXAMPLE 1 Simplify (i) $(a+b)(a^2 - ab + b^2)$, (ii) $\dfrac{x}{y} + \dfrac{y}{x} - \dfrac{(x+y)^2}{xy}$.

(i) $(a+b)(a^2 - ab + b^2) = a^3 - a^2b + ab^2 + a^2b - ab^2 + b^3 = \boldsymbol{a^3 + b^3}$.

(ii) $\dfrac{x}{y} + \dfrac{y}{x} - \dfrac{(x+y)^2}{xy} = \dfrac{x^2}{xy} + \dfrac{y^2}{xy} - \dfrac{x^2 + 2xy + y^2}{xy} = \dfrac{x^2 + y^2 - x^2 - 2xy - y^2}{xy}$

$$= \dfrac{-2xy}{xy} = \boldsymbol{-2}.$$

EXAMPLE 2 Factorise (i) $a^2 - b^2$, (ii) $\dfrac{x}{y^2} + \dfrac{y}{x^2}$.

(i) $a^2 - b^2 = \boldsymbol{(a+b)(a-b)}$.

(ii) $\dfrac{x}{y^2} + \dfrac{y}{x^2} = \dfrac{x^3 + y^3}{x^2 y^2} = \dfrac{1}{x^2}\dfrac{1}{y^2}\boldsymbol{(x+y)(x^2 - xy + y^2)}$.   (cf. Example 1).

EXAMPLE 3 Simplify: (i) $\sqrt{(12)} \times \sqrt{2}$, (ii) $\sqrt{3}(2 - \sqrt{2}) + \sqrt{2}(3 - \sqrt{3})$, (iii) $\dfrac{\sqrt{3}}{1 + \sqrt{2}}$, (iv) $(\sqrt{2} + \sqrt{3})(2\sqrt{3} - \sqrt{2})$, (v) $\dfrac{\sqrt{3} + \sqrt{5}}{\sqrt{6} - \sqrt{5}}$.

(i) $\sqrt{(12)} \times \sqrt{2} = 2\sqrt{3} \times \sqrt{2} = \boldsymbol{2\sqrt{6}}$.

(ii) $\sqrt{3}(2 - \sqrt{2}) + \sqrt{2}(3 - \sqrt{3}) = 2\sqrt{3} - \sqrt{6} + 3\sqrt{2} - \sqrt{6}$

$$= \boldsymbol{2\sqrt{3} + 3\sqrt{2} - 2\sqrt{6}}.$$

(iii) $\dfrac{\sqrt{3}}{1 + \sqrt{2}} = \dfrac{\sqrt{3}(1 - \sqrt{2})}{(1 + \sqrt{2})(1 - \sqrt{2})} = \dfrac{\sqrt{3} - \sqrt{6}}{1 - 2} = \boldsymbol{\sqrt{6} - \sqrt{3}}$.

(iv) $(\sqrt{2} + \sqrt{3})(2\sqrt{3} - \sqrt{2}) = 2\sqrt{6} - 2 + 6 - \sqrt{6} = \boldsymbol{4 + \sqrt{6}}$.

(v) $\dfrac{\sqrt{3}+\sqrt{5}}{\sqrt{6}-\sqrt{5}} = \dfrac{(\sqrt{3}+\sqrt{5})(\sqrt{6}+\sqrt{5})}{(\sqrt{6}-\sqrt{5})(\sqrt{6}+\sqrt{5})} = \dfrac{\sqrt{3}\sqrt{2}\sqrt{3}+\sqrt{3}\sqrt{5}+\sqrt{5}\sqrt{6}+5}{6-5}$

$$= 3\sqrt{2}+\sqrt{(15)}+\sqrt{(30)}+5.$$

Note some useful factorisations: $p^2 - q^2 = (p-q)(p+q),$
$$p^3 - q^3 = (p-q)(p^2+pq+q^2),$$
$$p^3 + q^3 = (p+q)(p^2-pq+q^2).$$

EXERCISE 2.1

1   In the following expressions, remove the brackets and collect together like terms:

  (i) $x - (2x - y)$ (ii) $x^2 - x(x-2)$

  (iii) $3(a-2b)-2(a-3b)$ (iv) $2(x+y)-y(3-x)$

  (v) $x(x-2)+2(x+2)$ (vi) $5(x^2-4)+3(2x+7)$

  (vii) $(a+2b)(3a-6b)$ (viii) $(x-2)(x+3)-x(x+1)$

  (ix) $2(a+2b-3c)+3(c-2a+3b)$ (x) $(r+s-t)(r-s+t)$

  (xi) $(a+b)(a-b)-(a-b)^2$ (xii) $(x+y)^3$

  (xiii) $(2x-3)^3$ (xiv) $(a-b)(a^2+ab+b^2)$

  (xv) $(x-2y+3z)^2$ (xvi) $(x-1)^2(x+1)^2$

  (xvii) $(x^2+1)(x^4-x^2+1)$ (xviii) $(a-b)(b-c)(c+a).$

2   Simplify the following expressions:

  (i) $\dfrac{1}{x}+\dfrac{1}{y}-\dfrac{1}{xy}$ (ii) $\dfrac{a}{b}-\dfrac{b}{a}$ (iii) $\dfrac{2x-3}{4}-\dfrac{2(x-3)}{3}$

  (iv) $\dfrac{2}{x}-\dfrac{3}{x}+\dfrac{4}{x}$ (v) $\dfrac{1}{2x}-\dfrac{1}{3x}+\dfrac{1}{4x}$ (vi) $\dfrac{2y}{x}-\dfrac{y}{2x}$

  (vii) $\dfrac{2x}{y}-\dfrac{3y}{x}$ (viii) $\dfrac{3}{xy^2}-\dfrac{2}{x^2y}$ (ix) $\dfrac{x+1}{x-1}-\dfrac{x-1}{x+1}$

  (x) $x+\dfrac{x^2+xy}{x-y}$ (xi) $\dfrac{x}{x+y}+\dfrac{y}{x-y}+\dfrac{1}{x}$ (xii) $\dfrac{3xy}{x^2-y^2}-\dfrac{y}{x+y}.$

3   Factorise the following expressions:

  (i) $2a+2b$ (ii) x^2+3x (iii) x^3-x^2

  (iv) $ab+bc$ (v) a^2-b^2 (vi) x^2-4y^2

  (vii) $36t^2-4$ (viii) a^2b-bc^2 (ix) $2x^2+3x-2$

  (x) $2t^2+7t-30$ (xi) $(x+y)^2-z^2$ (xii) $2x^3+2x^2-x-1$

  (xiii) a^3-b^3 (xiv) $8x^3-27y^3$ (xv) $r^3-s^3t^3$

  (xvi) $2x^3+2y^3$ (xvii) a^4-b^4 (xviii) $y^6-1.$

4   Simplify the following expressions:

  (i) $\dfrac{1}{a+b}-\dfrac{1}{a-b}$ (ii) $\dfrac{2a-2b}{a^3-b^3}$

  (iii) $\dfrac{x^2-x-6}{9-x^2}$ (iv) $\dfrac{1}{x+y}-\dfrac{x}{x^2-y^2}$

(v) $\dfrac{p^2-q^2}{p^2+pq}\times\dfrac{pq}{p-q}$　　　　　(vi) $\dfrac{4}{x-2}+\dfrac{2x}{2-x}$

(vii) $\dfrac{2x+1}{12}-\dfrac{3x+2}{18}$　　　　(viii) $\dfrac{2}{x^2-2x+1}+\dfrac{3}{x^2-3x+2}$

(ix) $\dfrac{3}{2a+4}+\dfrac{a-1}{a^2-4}$　　　　(x) $\dfrac{2a}{a+b}+\dfrac{2b}{a-b}+\dfrac{a^2+b^2}{b^2-a^2}$

(xi) $\dfrac{1}{1-x}-\dfrac{1}{1+x}+\dfrac{2x}{1+x^2}$　　　(xii) $\dfrac{x+y}{y}-\dfrac{2x}{x+y}+\dfrac{x^2y-x^3}{y(x^2-y^2)}.$

5　Simplify the following:

(i) $2\sqrt{3}+3\sqrt{3}-4\sqrt{3}$　　　　(ii) $\sqrt{(12)}-\sqrt{3}+4$

(iii) $\sqrt{2}(3-2\sqrt{2})+2(3-\sqrt{2})$　　　(iv) $3\sqrt{2}+\sqrt{6}-\sqrt{(18)}$

(v) $2\sqrt{(18)}+\sqrt{(50)}-2\sqrt{2}$　　　(vi) $\sqrt{3}(\sqrt{2}+\sqrt{6})-\sqrt{6}(\sqrt{2}+\sqrt{3})$

(vii) $(\sqrt{2})^3$　　　　　　　　(viii) $3\sqrt{5}+\sqrt{5}(\sqrt{(10)}-\sqrt{2})$

(ix) $3\sqrt{2}\times2\sqrt{3}$　　　　　　(x) $\sqrt{(18)}\div\sqrt{2}$

(xi) $2\sqrt{2}\times3\sqrt{2}$　　　　　　(xii) $\sqrt{5}\div\sqrt{(20)}$

(xiii) $(1+\sqrt{2})^2$　　　　　　(xiv) $(\sqrt{2}-\sqrt{3})^2$

(xv) $(3\sqrt{2}+\sqrt{3})(2\sqrt{3}-\sqrt{2})$　　(xvi) $(\sqrt{2}+\sqrt{3})(\sqrt{2}-\sqrt{3})$

(xvii) $(\sqrt{a}+\sqrt{b})(\sqrt{a}-\sqrt{b})$　　(xviii) $(2+\sqrt{3})^3.$

6　Express in a form with a rational denominator:

(i) $\dfrac{1}{\sqrt{3}}$　　(ii) $\dfrac{2}{\sqrt{5}}$　　(iii) $\dfrac{\sqrt{3}}{\sqrt{2}}$　　(iv) $\dfrac{1}{3\sqrt{2}}$

(v) $\dfrac{1}{\sqrt{2}-1}$　　(vi) $\dfrac{1}{\sqrt{5}+\sqrt{2}}$　　(vii) $\dfrac{\sqrt{2}}{\sqrt{3}+1}$　　(viii) $\dfrac{2}{5+\sqrt{2}}$

(ix) $\dfrac{\sqrt{2}-1}{\sqrt{2}+1}$　　(x) $\dfrac{\sqrt{2}-1}{\sqrt{6}}$　　(xi) $\dfrac{(1+\sqrt{2})^2}{\sqrt{2}-1}$　　(xii) $\dfrac{\sqrt{5}-2\sqrt{3}}{\sqrt{6}-\sqrt{5}}.$

7　Given that $x=1+\sqrt{2}$ and $y=2-\sqrt{3}$, express in a form that has a rational denominator:

(i) x^2　　　　　　(ii) xy　　　　　　(iii) x/y

(iv) $x+\dfrac{1}{y}$　　　(v) $\dfrac{x-1}{y+1}$　　　(vi) $\dfrac{1}{x}+\dfrac{1}{y}$

(vii) $\dfrac{x}{y}-\dfrac{y}{x}$　　(viii) $\dfrac{xy}{x+y}$　　(ix) $\dfrac{x^2+y^2}{xy}.$

2.2 Solution of quadratic equations (revision)

For a real function f, a real number α such that $f(\alpha)=0$ is called a solution of the equation $f(x)=0$ or, alternatively, a *zero* of the function f. Thus the solution of the equation $f(x) = 0$ is the set $\{\alpha \in \mathbb{R}: f(\alpha) = 0\}$, called the *solution set* of the equation $f(x) = 0$, and also called the set of zeros of the function f.

In the case of a quadratic equation $ax^2 + bx + c = 0$, the solution may be obtained by *factorisation* or by *completing the square*.

EXAMPLE 1 *Solve the equation $x^2 + 5x + 6 = 0$.*

$$x^2 + 5x + 6 = 0 \Leftrightarrow (x+2)(x+3) = 0 \quad \text{(factorise by inspection)}$$
$$\Leftrightarrow x+2 = 0 \text{ or } x+3 = 0 \ (uv = 0 \Leftrightarrow u = 0 \text{ or } v = 0)$$
$$\Leftrightarrow x = -2 \text{ or } x = -3,$$

so the solution set of the equation is $\{-2, -3\}$.

EXAMPLE 2 *Solve the equation $2x^2 + 5x - 12 = 0$.*

Method I (by factorisation)

$$2x^2 + 5x - 12 = (2x-3)(x+4) \text{ and so}$$
$$2x^2 + 5x - 12 = 0 \Leftrightarrow 2x - 3 = 0 \quad \text{or} \quad x + 4 = 0$$
$$\Leftrightarrow x = \tfrac{3}{2} \text{ or } x = -4$$
$$\Leftrightarrow x \in \{\tfrac{3}{2}, -4\}.$$

Method II (by completing the square) We use the identity

$$(x+k)^2 = x^2 + 2kx + k^2$$

so that we may make the expression into a square by adding k^2. The first step is to obtain a coefficient of x^2 equal to 1, thus

$$2x^2 + 5x - 12 = 2(x^2 + \tfrac{5}{2}x) - 12 \quad (\text{so } k = \tfrac{5}{4})$$
$$= 2(x^2 + 2\tfrac{5}{4}x + (\tfrac{5}{4})^2) - 12 - 2(\tfrac{5}{4})^2$$
$$= 2(x+\tfrac{5}{4})^2 - \tfrac{121}{8}$$
$$= 2[(x+\tfrac{5}{4})^2 - \tfrac{121}{16}].$$

Hence
$$2x^2 + 5x - 12 = 0 \Leftrightarrow (x+\tfrac{5}{4})^2 = \tfrac{121}{16}$$
$$\Leftrightarrow x + \tfrac{5}{4} = \tfrac{11}{4} \quad \text{or} \quad x + \tfrac{5}{4} = -\tfrac{11}{4}$$
$$\Leftrightarrow x = \tfrac{6}{4} = \tfrac{3}{2} \quad \text{or} \quad x = -\tfrac{16}{4} = -4.$$
$$\Leftrightarrow x \in \{\tfrac{3}{2}, -4\}.$$

EXERCISE 2.2

1 Determine whether or not the given number is a solution of the given equation, by means of substitution in the equation:

(i) $\alpha = -3$, $x^2 - 2x - 15 = 0$

(ii) $\alpha = \sqrt{2}$, $x^2 + 2 = 0$

(iii) $\alpha = -\sqrt{5}$, $x^2 = 4x + 1$

(iv) $\alpha = 1 + \sqrt{2}$, $x^2 + 4x + 3 = 0$

(v) $\alpha = 2 + \sqrt{5}$, $x^2 = 4x + 1$

(vi) $\alpha = \sqrt{3} - 1$, $2x^2 + 4x = 4$.

2 Solve the following equations, by inspection:

  (i) $(x-1)(x+2) = 0$ (ii) $(y+4)(y+5) = 0$ (iii) $(2x-1)(x+3) = 0$
  (iv) $x(3x+2) = 0$ (v) $(t-4)^2 = 0$ (vi) $(3x-5)^2 = 0$
  (vii) $(2x-3)(3x+2) = 0$ (viii) $(x-a)(x-b) = 0$ (ix) $(x-1)^2 = 4$
  (x) $(2x-1)^2 = 25$ (xi) $(5-3x)^2 = 36$ (xii) $(ax-b)^2 = c^2$.

3 Solve the following equations, by factorisation:

  (i) $x^2 - 2x - 15 = 0$ (ii) $x^2 - x - 12 = 0$ (iii) $x^2 + 3x - 54 = 0$
  (iv) $x^2 + 13x + 42 = 0$ (v) $x^2 + 11x - 312 = 0$ (vi) $3x^2 - 4x + 1 = 0$
  (vii) $2x^2 - 5x - 12 = 0$ (viii) $5x^2 - 18x - 8 = 0$ (ix) $8x^2 + 24x + 10 = 0$
  (x) $28x^2 - 2x - 6 = 0$ (xi) $9x^2 - 27x + 20 = 0$ (xii) $x^2 - 33x + 272 = 0$
  (xiii) $12x^2 - 58x - 10 = 0$ (xiv) $6x^2 + 7x = 3$ (xv) $2x^2 = 3x + 14$
  (xvi) $x^2 + 4 = 4x$ (xvii) $(x-3)(x+2) = (x+2)(x-1)$
  (xviii) $(2x-1)(x-2) = 5$ (xix) $(2-x)(1-2x) = 5$ (xx) $x^2 - 1 = (x-1)^2$.

4 Complete the following:

  (i) $x^2 + 4x + \ldots = (x + \ldots)^2$ (ii) $x^2 - 10x + \ldots = (x - \ldots)^2$
  (iii) $s^2 - 3s + \ldots = (s - \ldots)^2$ (iv) $x^2 + 2ax + \ldots = (x + \ldots)^2$
  (v) $x^2 + x + \ldots = (x + \ldots)^2$ (vi) $x^2 + 4xy + \ldots = (x + \ldots)^2$.

5 What must be added to complete the square?

  (i) $x^2 + 12x$ (ii) $x^2 - 6x$ (iii) $t^2 + 2at$
  (iv) $y^2 - 3y$ (v) $z^2 + 4z$ (vi) $4x^2 + 4x$
  (vii) $2x^2 - 2x$ (viii) $3x^2 + 5x$ (ix) $ax^2 + bx$.

6 Solve the following equations, by completing the square:

  (i) $x^2 - 4x + 1 = 0$ (ii) $x^2 - 2x - 4 = 0$ (iii) $x^2 - 6x + 7 = 0$
  (iv) $x^2 - 3x + 1 = 0$ (v) $4x^2 - 12x + 7 = 0$ (vi) $9x^2 - 6x - 4 = 0$
  (vii) $x^2 - 2ax + b = 0$ (viii) $a^2x^2 - 2ax - 2 = 0$ (ix) $ax^2 + bx + c = 0$.

2.3 The graph of a quadratic function

The graph of the function f, given by $f(x) = x^2 - 3x$, may be drawn by plotting points (x, y) and joining them by a smooth curve, using a table of calculated values:

x:	-1	0	1	2	3	4
y:	4	0	-2	-2	0	4

This graph is shown in Fig. 2.1(a). The graph of any quadratic function is called a *parabola*. On completing the square for the function f,

$$f(x) = x^2 - 3x = x^2 - 2.\tfrac{3}{2}x + \tfrac{9}{4} - \tfrac{9}{4} = (x - \tfrac{3}{2})^2 - \tfrac{9}{4}.$$

Since, for all x, $(x - \tfrac{3}{2})^2 \geqslant 0$, $f(x) \geqslant -\tfrac{9}{4}$ and the least value of $f(x)$ is $-\tfrac{9}{4} = f(\tfrac{3}{2})$. Also

$$f(\tfrac{3}{2} + d) = d^2 - \tfrac{9}{4} = f(\tfrac{3}{2} - d),$$

for any d, and the function takes equal values at points equidistant on either side of the line $x = \tfrac{3}{2}$. Therefore the graph is *symmetrical* about the line $x = \tfrac{3}{2}$, reflection in this line being a symmetry of the graph.

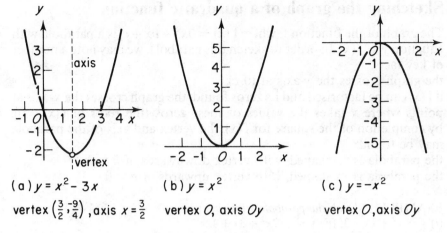

(a) $y = x^2 - 3x$

vertex $\left(\frac{3}{2}, \frac{-9}{4}\right)$, axis $x = \frac{3}{2}$

(b) $y = x^2$

vertex O, axis Oy

(c) $y = -x^2$

vertex O, axis Oy

Fig. 2.1 Parabolas.

The line of symmetry, $x = \frac{3}{2}$, is called the *axis* of the parabola. The lowest point $(\frac{3}{2}, -\frac{9}{4})$ is called the *vertex* of the parabola. For large values of x, both positive and negative, $f(x)$ becomes large and positive. We can express this fact, which tells us something about the shape of the graph at large distances from the origin, by saying that:
as x tends to positive infinity, $f(x)$ tends to positive infinity;
as x tends to negative infinity, $f(x)$ tends to positive infinity.
For these statements, we use the notation:

$$\text{as } x \to +\infty, \; f(x) \to +\infty; \quad \text{as } x \to -\infty, \; f(x) \to +\infty.$$

It must be understood that, in these statements, there is no suggestion that there is such a number as infinity; there is not. All that is implied is that $f(x)$ behaves in a certain way for large values of x.

The process of completing the square standardises a quadratic. From the standard form, the vertex and the axis of the parabola which is the graph of the function, may be found. Another approach is to start with a standard form. The parabola with equation $y = ax^2$ is symmetrical about the line $x = 0$, so the y-axis is the axis of this parabola, which touches the x-axis at the origin, O. If $a > 0$, O is the lowest point on the graph, so the origin is the vertex of this parabola, the parabola is $\cup$ shaped, with its vertex downwards. For $a < 0$, the vertex O is the highest point of the graph, which is $\cap$ shaped, with its vertex upwards (see Fig. 2.1(b), (c)).

If we translate the plane so that the origin moves to the point (p, q), the parabola $y = ax^2$ will be translated to become the curve with equation $y - q = a(x - p)^2$, since we have to replace x by $x - p$ and y by $y - q$. This new parabola has a vertex at (p, q) and has axis $x = p$, and its equation is

$$y = a(x - p)^2 + q = ax^2 - 2apx + ap^2 + q.$$

Sketching the graph of a quadratic function

The graph of the function f, where $f(x) = ax^2 + bx + c$, is a parabola with equation $y = f(x)$. In order to sketch this parabola, we may note a number of key facts:

the graph crosses the y-axis at $(0, c)$;

if $f(x)$ can be factorised and its zeros found, the graph crosses the x-axis at points where x takes the values of these zeros (the *roots* of $f(x) = 0$);

by completion of the square for $f(x)$, the vertex and axis of the parabola may be found;

the parabola is $\cup$ shaped, with vertex downwards, if $a > 0$;

the parabola is $\cap$ shaped, with vertex upwards, if $a < 0$.

EXAMPLE *Sketch the parabola with equation:*
(i) $y = x^2 + 2x - 3$; (ii) $y = -2x^2 + 3x + 4$.

(i) $x^2 + 2x - 3 = (x + 1)^2 - 4 = (x + 3)(x - 1)$, and so the parabola has vertex at the point $(-1, -4)$ and axis given by $x = -1$. It crosses the y-axis at $(0, -3)$ and the x-axis at $(-3, 0)$ and $(1, 0)$. The coefficient of x^2 is positive so y tends to positive infinity as x tends to positive infinity and to negative infinity. Thus the graph is $\cup$ shaped, with its vertex downwards. From this information, the graph is drawn in Fig. 2.2(a).

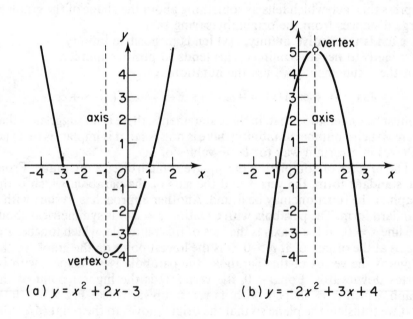

(a) $y = x^2 + 2x - 3$ (b) $y = -2x^2 + 3x + 4$

Fig. 2.2

(ii) Again we use the process of completing the square in order to change the equation into standard form and then to factorise it.

$$-2x^2 + 3x + 4 = -2\left(x^2 - \frac{3}{2}x\right) + 4 = -2\left\{x^2 - 2\left(\frac{3}{4}\right)x + \left(\frac{3}{4}\right)^2\right\} + 2\left(\frac{3}{4}\right)^2 + 4$$

$$= -2\left(x - \frac{3}{4}\right)^2 + \frac{41}{8} = -2\left\{\left(x - \frac{3}{4}\right)^2 - \frac{41}{16}\right\}$$

$$= -2\left(x - \frac{3}{4} - \frac{\sqrt{(41)}}{4}\right)\left(x - \frac{3}{4} + \frac{\sqrt{(41)}}{4}\right),$$

on factorising the difference of two squares. Thus the parabola has axis $x = \frac{3}{4}$, vertex $(\frac{3}{4}, \frac{41}{8})$ and meets the axes at $(0, 4)$, $(\{3 + \sqrt{(41)}\}/4, 0)$ and $(\{3 - \sqrt{(41)}\}/4, 0)$. Since the coefficient of x^2 is negative, the vertex is upwards, and the graph is shown in Fig. 2.2(b).

EXERCISE 2.3

1 Find the vertex and the axis of the parabola with the given equation and sketch the parabola: (i) $y = (x + 1)^2$,   (ii) $y = (x - 1)^2 - 4$,   (iii) $y = x^2 + x - 2$,   (iv) $y = -2x^2 + 3x + 14$,   (v) $y = 2x - 3x^2$,   (vi) $y = (2 - x)(1 + 2x)$.
2 Write down the equation of a parabola: (i) with vertex $(2, 3)$ and passing through $(0, 7)$,   (ii) with axis $x = -3$ and passing through $(-1, 0)$ and $(0, -5)$. In each case, state whether the parabola has its vertex upwards or downwards.

2.4 A formula for the solution of a quadratic equation

The process of completing the square, used in §2.2 and §2.3, can be applied to the general quadratic function f, where $f(x) = ax^2 + bx + c$, with $a \neq 0$.

$$f(x) = ax^2 + bx + c = a\left(x^2 + \frac{b}{a}x\right) + c$$

$$= a\left\{x^2 + 2\frac{b}{2a}x + \left(\frac{b}{2a}\right)^2\right\} - a\left(\frac{b}{2a}\right)^2 + c$$

$$= a\left(x + \frac{b}{2a}\right)^2 - \frac{b^2 - 4ac}{4a} = a\left(x + \frac{b}{2a}\right)^2 - \frac{\Delta}{4a},$$

where $\Delta = b^2 - 4ac$. The number Δ is called the *discriminant* of the quadratic function f. This leads to a formula for the solution of a quadratic equation.

$$ax^2 + bx + c = 0 \Leftrightarrow \left(x + \frac{b}{2a}\right)^2 = \frac{\Delta}{4a^2}$$

$$\Leftrightarrow x + \frac{b}{2a} = \pm\frac{\sqrt{\Delta}}{2a}$$

on taking square roots. Therefore, given the equation $ax^2 + bx + c = 0$,
if $\Delta < 0$, there are no solutions,
if $\Delta = 0$, there is just one solution, $x = -b/2a$,
if $\Delta > 0$, there are two solutions, $x = (-b + \sqrt{\Delta})/2a$ and
$x = (-b - \sqrt{\Delta})/2a$.

EXAMPLE   *Solve the equations* (i) $3x^2 + 7x + 3 = 0$; (ii) $3x^2 + 7x + 5 = 0$.

(i) Let $f(x) = 3x^2 + 7x + 3$,   then $\Delta = 7^2 - 4 \times 3 \times 3 = 13 > 0$,   so   there
are two solutions to the equation $f(x) = 0$, namely $x = (-7 + \sqrt{13})/6$ and
$x = (-7 - \sqrt{13})/6$.
(ii) Let $f(x) = 3x^2 + 7x + 5$, then $\Delta = 7^2 - 4 \times 3 \times 5 = -11 < 0$, so the equation
$f(x) = 0$ has no solutions; its solution set is the empty set ϕ.

EXERCISE 2.4

1   Use the formula method to solve the equation:
(i) $x^2 - 2x - 15 = 0$
(ii) $x^2 - x - 12 = 0$
(iii) $x^2 + 3x - 54 = 0$
(iv) $x^2 + 13x + 42 = 0$
(v) $x^2 + 11x - 312 = 0$
(vi) $3x^2 - 4x + 1 = 0$
(vii) $2x^2 - 5x - 12 = 0$
(viii) $5x^2 - 18x - 8 = 0$
(ix) $8x^2 + 24x + 10 = 0$
(x) $28x^2 - 2x - 6 = 0$
(xi) $9x^2 - 27x + 20 = 0$
(xii) $x^2 - 33x + 272 = 0$
(xiii) $12x^2 - 58x - 10 = 0$
(xiv) $6x^2 + 7x = 3$
(xv) $2x^2 = 3x + 14$
(xvi) $x^2 + 4 = 4x$
(xvii) $(x - 3)(x + 2) = (x + 2)(x - 1)$
(xviii) $(2x - 1)(x - 2) = 5$
(xix) $(2 - x)(1 - 2x) = 5$
(xx) $x^2 - 1 = (x - 1)^2$.

Check your answers by comparison with your answers to question **3** of
Exercise 2.2.

2.5 The discriminant and the location of the graph

The sign of the discriminant, $\Delta = b^2 - 4ac$, provides information about
the graph of the quadratic function f, where $f(x) = ax^2 + bx + c$.
If $\Delta < 0$, $f(x) = 0$ has no solutions, so the graph does not meet the x-axis;
if $\Delta > 0$, $f(x) = 0$ has two solutions, so the graph crosses the x-axis at
2 points;
if $\Delta = 0$, $f(x) = 0$ has one solution, so the graph touches the x-axis at
1 point.
For each of these three cases, the constant a, which is the coefficient of x^2,
may be positive or negative, corresponding to the graph having vertex
downwards or upwards. This gives a total of six possible cases, which are
illustrated in Fig. 2.3.

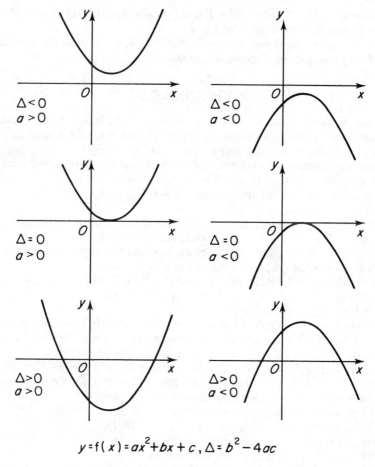

$$y = f(x) = ax^2 + bx + c, \Delta = b^2 - 4ac$$

Fig. 2.3

Range of the quadratic function

For the quadratic function f, given by $f(x) = ax^2 + bx + c$, the sign of a determines whether the vertex $(-b/2a, -\Delta/4a)$ is a minimum or a maximum point on the graph of f and so we see that

if $a > 0$, the range of f is $\{x : x \geqslant -\Delta/4a\}$,
if $a < 0$, the range of f is $\{x : x \leqslant -\Delta/4a\}$.

EXAMPLE *Find the range and the set of zeros of the function f given by:*
(i) $f(x) = x^2 - 4x + 3$,　(ii) $f(x) = -x^2 + 2x - 2$,　(iii) $f(x) = x^2 - 4x + 5$,
(iv) $f(x) = -x^2 + 2x - 2$.

(i) $f(x) = x^2 - 4x + 3 = (x-2)^2 - 1 = (x-3)(x-1), \Delta = 4, a = 1$, so the range of f is $\{x : x \geqslant -1\}$ and the set of zeros of f is $\{1, 3\}$.
(ii) $f(x) = x^2 - 4x + 4 = (x-2)^2, \Delta = 0, a = 1$, so the range of f is $\{x : x \geqslant 0\}$ and the set of zeros of f is $\{2\}$.

(iii) $f(x) = x^2 - 4x + 5 = (x - 2)^2 + 1, \Delta = -4, a = 1$, so the range of f is $\{x : x \geqslant 1\}$ and the set of zeros of f is ϕ.
(iv) $f(x) = -x^2 + 2x - 2 = -(x - 1)^2 - 1, \Delta = -4, a = -1$, so the range of f is $\{x : x \leqslant -1\}$ and the set of zeros of f is ϕ.

EXERCISE 2.5

1   Write down the value of the discriminant Δ of the function f and state the number of zeros of f, when: (i) $f(x) = x^2 + 4x + 4$,   (ii) $f(x) = 2x^2 + 1$, (iii) $f(x) = 9x^2 - 6x + 1$,   (iv) $f(x) = -2x^2 + 5x - 3$.

2   State the number of solutions, in the set of real numbers, of the equation:
(i) $x^2 + 3x - 2 = 0$,   (ii) $x^2 - 9 = 0$,   (iii) $x^2 + 4x + 8 = 0$,
(iv) $2x^2 = 4x - 2$,   (v) $a^2 x^2 - 3ax + 2 = 0, 0 \neq a \in \mathbb{R}$,
(vi) $x^2 - 4ax + 4a^2 = 0, a \in \mathbb{R}$.

3   By completing the square, or otherwise, find the range of the real function f given by $f(x) = x^2 - 6x + 12$. Sketch the graph of f.

4   For the quadratic function f as given, find the range of f and state at how many points the graph of f meets the x-axis:
(i) $f(x) = x^2 - 4x + 12$ (ii) $f(x) = 3x^2 - 7x + 2$ (iii) $f(x) = 2 - 3x + 4x^2$
(iv) $f(x) = 1 - 4x - 2x^2$ (v) $f(x) = (x - 2)(3 - x)$
(vi) $f(x) = (x - 2)(3 - x) - 1$.

5   For the curve given by the following equation, state in how many points the curve meets the y-axis and in how many points it meets the x-axis, and find the set of values of y for which there are points on the curve:
(i) $y = \frac{1}{2}x^2 + 5x + 6$ (ii) $y = 4x^2 - 30x + 25$ (iii) $y = 5 - 8x - 3x^2$
(iv) $y = 10x - x^2 - 25$ (v) $y = (x - 2)(x - 4)$ (vi) $y = (x - 3)(5 - x)$.

6   Find $f(x)$ for the quadratic function f, whose graph:
(i) passes through the points $(0, 2), (4, 0), (5, 0)$,
(ii) passes through the points $(0, -7), (-2, 0), (3, 0)$,
(iii) has vertex $(2, -1)$ and passes through $(0, 3)$,
(iv) has axis $x = 1$ and passes through $(0, 2), (3, -7)$.

7   The expression $ax^2 + bx + c$ takes the value 2 when $x = -2$ and when $x = 4$, and its minimum value is -7. Find the values of x for which the value of the expression is 9. *(L)*

8   Express $3x^2 - 12x + 14$ in the form $a(x - p)^2 + q$. Show how the graph of $y = 3x^2 - 12x + 14$ may be obtained from the graph of $y = x^2$ by appropriate translations and stretches.

9   Prove that, for all real values of x, the expression $-x^2 + 4x - 5$ is negative. *(L)*

10 Given that $f(x) \equiv 3x^2 + 2x - 8$, find $f(0)$, the values of x for which $f(x) = 0$, and the set of values of x for which $f(x) > 0$. *(L)*

2.6 Symmetric properties of the roots of a quadratic equation

Let α and β be the solutions of the quadratic equation $ax^2 + bx + c = 0$; α and β are also called the *roots* of the equation. Then

$$\alpha + \beta = -b/a \quad \text{and} \quad \alpha\beta = c/a.$$

This follows from §2.4, since

$$\alpha + \beta = (-b + \sqrt{\Delta} - b - \sqrt{\Delta})/2a = -b/a,$$
$$\alpha\beta = (b^2 - \Delta)/4a^2 = (b^2 - b^2 + 4ac)/4a^2 = c/a.$$

Alternatively, the same result may be obtained by factorisation,

$$ax^2 + bx + c = a(x - \alpha)(x - \beta) = a(x^2 - (\alpha + \beta)x + \alpha\beta).$$

The results are still true when $\alpha = \beta$ and $\Delta = 0$.

Summary: If the quadratic equation $ax^2 + bx + c = 0$ has one or two real roots, (i.e. $\Delta \geqslant 0$) then the *sum* of the roots is $-b/a$ and the *product* of the roots is c/a. Conversely, we may write down a quadratic equation whose roots have sum s and product p, namely $x^2 - sx + p = 0$.

Using these results, it is possible to find a quadratic equation whose roots are symmetric functions of the roots of a given quadratic equation, without actually solving this equation. This is illustrated by some examples.

EXAMPLE *Show that the quadratic equation $2x^2 + 5x + 1 = 0$ has two roots. If these roots are α and β, find quadratic equations with roots:*

(i) α^2 and β^2, (ii) $\dfrac{1}{\alpha}$ and $\dfrac{1}{\beta}$, (iii) α^3 and β^3.

The discriminant $\Delta = 25 - 8 = 17 > 0$, so the equation has two distinct real roots, α and β and $\alpha + \beta = -\frac{5}{2}$, $\alpha\beta = \frac{1}{2}$. We use these results to calculate the sum and the product of the roots of the new equation.

(i) $\alpha^2 + \beta^2 = \alpha^2 + 2\alpha\beta + \beta^2 - 2\alpha\beta = (\alpha + \beta)^2 - 2\alpha\beta = 25/4 - 1 = 21/4$, $\alpha^2\beta^2 = (\alpha\beta)^2 = \frac{1}{4}$, so the required equation is $x^2 - 21x/4 + \frac{1}{4} = 0$ or $4x^2 - 21x + 1 = 0$.

(ii) $\dfrac{1}{\alpha} + \dfrac{1}{\beta} = \dfrac{\alpha + \beta}{\alpha\beta} = -5$ and $\dfrac{1}{\alpha}\dfrac{1}{\beta} = \dfrac{1}{\alpha\beta} = 2$, so the required equation is

$x^2 + 5x + 2 = 0$. Alternatively, replace x by $\dfrac{1}{x}$ in the original equation.

(iii) $\alpha^3 + \beta^3 \doteq (\alpha + \beta)(\alpha^2 - \alpha\beta + \beta^2) = (\alpha + \beta)(\alpha^2 + 2\alpha\beta + \beta^2 - 3\alpha\beta)$

$= (\alpha + \beta)\{(\alpha + \beta)^2 - 3\alpha\beta\} = -\frac{5}{2}(\frac{25}{4} - \frac{3}{2}) = -\frac{95}{8}$,

and $\alpha^3\beta^3 = (\alpha\beta)^3 = \frac{1}{8}$. Thus, the equation is $8x^2 + 95x + 1 = 0$.

EXERCISE 2.6

1 Without finding the roots, show that the following equations have two real roots and write down their sum and their product:

(i) $x^2 + 3x - 5 = 0$, (ii) $3x^2 - 2x - 2 = 0$, (iii) $4x^2 + 10x + 3 = 0$,

(iv) $x^2 + 4dx - d^2 = 0$, $d \in \mathbb{R}$, (v) $x + \dfrac{1}{x} = 4$, (vi) $2r^2 - 3r - 4 = 0$.

2 For equations (i), (iii) and (v) in question **1**, if the roots are α and β, write down
the values of (a) $3(\alpha+\beta)$, (b) $(\alpha+\beta)^2$, (c) $(\alpha-\beta)^2$, (d) $\alpha^2\beta^2$, (e) $\dfrac{1}{\alpha}+\dfrac{1}{\beta}$,
(f) $2\alpha^2+2\beta^2+5\alpha\beta$, (g) $(\alpha+\beta)^3$, (h) $3\alpha\beta(\alpha+\beta)$, (i) $\alpha^3+\beta^3$.

3 Using the above results, in each of the 3 cases, write down a quadratic
equation whose roots are: $\alpha^2,\beta^2;\dfrac{1}{\alpha},\dfrac{1}{\beta};2\alpha+\beta,\alpha+2\beta;\alpha-\beta,\beta-\alpha;\alpha+\dfrac{1}{\beta},\beta+\dfrac{1}{\alpha};$
α^3,β^3.

4 The function
$$f(x)=x^2+px+1,$$
where p is a constant, is zero when $x=\alpha$ and $x=\beta$; and the function
$$g(x)=x^2-9x+q,$$
where q is a constant, is zero when $x=\alpha+2\beta$ and $x=\beta+2\alpha$. Find p and q,
and show that $f(3)=g(3)$. (*JMB*)

5 The equation $ax^2+bx+c=0$ has roots α, β. Express $(\alpha+1)(\beta+1)$ in terms
of a, b and c. (*L*)

MISCELLANEOUS EXERCISE 2

1 Given that the roots of the equation $ax^2+bx+c=0$ are β and $n\beta$, show that
$(n+1)^2ac=nb^2$. (*L*)

2 Given that α and β are the roots of the equation $x^2-px+q=0$ obtain the
quadratic equation whose roots are $1/\alpha^2$ and $1/\beta^2$. (*JMB*)

3 The equations $ax^2+bx+c=0$ and $bx^2+ax+c=0$, where $a\neq b$, $c\neq0$,
have a common root. Prove that $a+b+c=0$. (*L*)

4 Given that b and c are non-zero constants and that the equations
$x^2+bx+c=0$ and $7x^2+2bx-3c=0$ have a common root, prove that
$b^2=4c$. (*L*)

5 If a, b, c are constants and $a<0$, derive the condition under which
ax^2+bx+c will be negative for all real values of x. Given that
$$f(x)=px^2-2x+3p+2,$$
(a) find the two values of p for which the equation $f(x)=0$ has equal roots,
(b) find the set of values of p for which $f(x)$ is negative for all real values of x.
Sketch the graph of $y=f(x)$ for each of the cases $p=-2$, $p=1$. (*L*)

6 If the two roots of the quadratic equation
$$ax^2+2bx+c=0$$
differ by 4, show that $8a=-c\pm\sqrt{(c^2+16b^2)}$. Find the two values of a for
which the equation $ax^2+2x+3=0$ has roots differing by 4. (*L*)

7 The real roots of the equation $x^2+6x+c=0$ differ by $2n$, where n is real and
non-zero. Show that $n^2=9-c$. Given that the roots also have opposite
signs, find the set of possible values of n. (*JMB*)

8 The roots of the equation $9x^2+6x+1=4kx$, where k is a real constant, are
denoted by α and β.
(a) Show that the equation whose roots are $1/\alpha$ and $1/\beta$ is
$$x^2+6x+9=4kx.$$

(b) Find the set of values of k for which α and β are real.

(c) Find also the set of values of k for which α and β are real and positive. $\hfill (L)$

9 If a, b, c are constants and $a > 0$, derive the condition that $ax^2 + bx + c$ should be positive for all real values of x. Prove that the equation

$$x - 1 = k(x-2)(x+1)$$

has real roots for all non-zero values of k. $\hfill (L)$

10 Given that α and β are the roots of the equation

$$x^2 - px + q = 0,$$

prove that $\alpha + \beta = p$ and $\alpha\beta = q$.

Prove also that

(a) $\alpha^{2n} + \beta^{2n} = (\alpha^n + \beta^n)^2 - 2q^n$,

(b) $\alpha^4 + \beta^4 = p^4 - 4p^2 q + 2q^2$.

Hence, or otherwise, form the quadratic equation whose roots are the fourth powers of those of the equation $x^2 - 3x + 1 = 0$. $\hfill (L)$

11 Given that α, β are the roots of the equation

$$2x^2 - 7x - 17 = 0,$$

show that an equation with roots $(\alpha - 4)$, $(\beta - 4)$ is

$$2x^2 + 9x - 13 = 0.$$

Find the coordinates of the point of intersection of the curves whose equations are

$$y = 2x^2 - 7x - 17$$

and

$$y = 2x^2 + 9x - 13.$$

Calculate the coordinates of the vertex of each curve and sketch both curves on the same diagram, labelling each curve clearly. $\hfill (L)$

12 Find the ranges of values of k for which the equation

$$(x-5)(x+1) = k(x-7)$$

has real roots.

Deduce the range of the function

$$g: x \mapsto \frac{(x-5)(x+1)}{(x-7)},$$

(for all real $x \neq 7$). $\hfill (JMB)$

13 Given that the roots of the equation $x^2 + px + q = 0$ are α and β, express $(\alpha - 2\beta)(\beta - 2\alpha)$ in terms of p and q. Hence, or otherwise, show that the condition for one root of the equation to be double the other is $2p^2 = 9q$. $\hfill (JMB)$

14 Given that the equation $x^2 + px + q = 0$ has roots α and β, express p and q in terms of α and β. Find the equation, with coefficients expressed in terms of p and q, whose roots are $3\alpha - \beta$ and $3\beta - \alpha$. $\hfill (JMB)$

15 Given that one root of the equation $x^2 + px + q = 0$ is twice the other root,

show that $2p^2 = 9q$. The roots of the equation

$$x^2 + (2k+4)x + (k^2+3k+2) = 0$$

are non-zero and one root is twice the other root. Calculate the value of k.
(*JMB*)

16   If α and β are the roots of the equation $x^2 + bx + c = 0$, express $(\alpha - \beta^2)(\alpha^2 - \beta)$ in terms of b and c. Hence, or otherwise, show that, if one root of the equation is the square of the other, then $b^3 + c^2 + c = 3bc$.
(*JMB*)

17   The roots of the equation

$$x^2 + px + q = 0$$

are α, β. Show that the equation whose roots are α^2, β^2 is

$$x^2 + (2q - p^2)x + q^2 = 0.$$

Without finding the roots of either equation, determine all possible pairs of values of p, q for which the two equations have the same roots.   (*JMB*)

3 Trigonometry I

3.1 O-level revision

For acute angles, the trigonometric ratios are defined in terms of the ratios of the sides of a right-angled triangle. Let the triangle ABC have a right-angle at B and let angle $BAC = \theta°$, see Fig. 3.1(a). Then $\cos \theta° = AB/AC$, $\sin \theta° = BC/AC$, $\tan \theta° = BC/AB$, where we use the usual abbreviations cos, sin, tan, for cosine, sine, tangent.

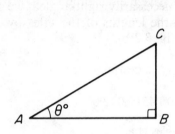
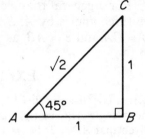

(a) Right-angled triangle   (b) Isosceles right-angled triangle

Fig. 3.1

Pythagoras' theorem, $AB^2 + BC^2 = AC^2$, then gives the result

$$\cos^2 \theta° + \sin^2 \theta° = 1.$$

For some particular angles, the trigonometric ratios can be found exactly, in terms of rational numbers or surds, otherwise we have to approximate to them using tables or calculators. For example, consider the right-angled isosceles triangle ABC, with $AB = 1 = BC$, of Fig. 3.1(b). The angle $BAC = 45°$ and $AC = \sqrt{2}$. Thus

$$\cos 45° = 1/\sqrt{2} = \sin 45°, \quad \tan 45° = 1.$$

Next, consider the equilateral triangle ACD with an altitude CB and sides AC, AD, CD each of length 2, as shown in Fig. 3.2(a). Then $AB = 1$, $BC = \sqrt{3}$ and angle BAC is $60°$, angle ACB is $30°$, and so

$$\cos 60° = \frac{1}{2}, \quad \sin 60° = \frac{\sqrt{3}}{2}, \quad \tan 60° = \sqrt{3},$$

$$\cos 30° = \frac{\sqrt{3}}{2}, \quad \sin 30° = \frac{1}{2}, \quad \tan 30° = \frac{1}{\sqrt{3}}.$$

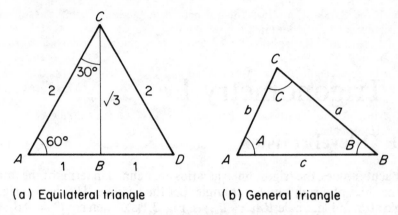

(a) Equilateral triangle   (b) General triangle

Fig. 3.2

Notation   For a general triangle ABC, not necessarily right-angled, we denote the three angles by A, B and C and the lengths of the sides by $a = BC$, $b = CA$ and $c = AB$, as shown in Fig. 3.2(b).

EXERCISE 3.1

1   Verify that $\cos^2 \theta + \sin^2 \theta = 1$, for $\theta = 0°, 30°, 45°, 60°, 90°$.
2   Let the triangle ABC be right-angled at B.
   (a) Given that $AB = 3$, $BC = 4$, find the angles A and C;
   (b) Given that $AC = 6$, $A = 40°$, find AB, BC and angle C;
   (c) Given that $AB = 5$, $A = 20°$, find BC and AC.
3   A ship sails on a triangular route, 100 km on a bearing 010° E. of N., then 200 km on a bearing 020° E. of N., and finally returning to its starting point. Find the bearing on which the ship must sail for the return leg, and the total distance travelled.
4   (a) Find the height of a tree if the elevation of the top of the tree from a point on horizontal ground, at a distance 10 m from the tree, is 60°.
   (b) Find the height of the tree if the ground slopes at 10° down from the base of the tree to the point of observation. Assume that this point is still at a distance 10 m from the base of the tree, measured along the ground, and that the observed elevation of the top of the tree is still 60°.
5   Find the radius of a circle which has a chord of length 0.1 m subtending an angle 120° at the centre.
6   Given that the radius of the Earth is R and that the angle of depression (below the horizontal) of the horizon, as seen from the top of a cliff at a height h above sea-level, is $\theta°$, prove that $\cos \theta° = R/(R + h)$.
7   A pendulum bob P is suspended from a point O by a string of length 0·8 m, and the point O is 1 m above the floor. If the pendulum swings through a total angle 2θ find, in terms of θ, the greatest height of P above the floor.
8   Given a triangle ABC, not necessarily right-angled, where the lengths of BC and AC are a and b respectively and the acute angle ACB is C, prove that the area of the triangle is $\frac{1}{2}ab \sin C$. Prove that the same formula gives the area when the angle C is obtuse.

9 With the standard notation for a general triangle ABC, use the result of question **8** to show that $a/\sin A = b/\sin B = c/\sin C$.

10 Find the ratio of the lengths of the sides of a rectangle whose diagonals cross at an angle θ.

11 The base of a right circular cone has radius r and the height of the cone is h. Find, in terms of r and h,
(i) the slant height k of the cone;
(ii) $\sin \phi$, where ϕ is the angle between a line on the curved surface of the cone and its base;
(iii) $\sin \theta$, where θ is the angle at the top of the cone between the axis and a line on the curved surface;
(iv) the area of the curved surface of the cone.

3.2 Trigonometric ratios of angles of any magnitude

On a Cartesian graph $O(x, y)$, draw a circle of unit radius, centre O, meeting the axis Ox at the point $A(1, 0)$, as shown in Fig. 3.3. An angle $\theta°$, of any magnitude, can be represented on this graph by a rotation of Ox through $\theta°$ measured positively towards Oy. If, after such a rotation $\theta°$, the new position of Ox meets the circle at $P(x, y)$, then the angle AOP is $\theta°$, and the point P corresponds to the angle $\theta°$. A positive angle $\theta°$ is represented by an anticlockwise rotation of OA about O through $\theta°$.

A negative angle $-\theta°$ is represented by a clockwise rotation of OA about O through the angle $+\theta°$. Thus in Fig. 3.3, the points $P_1(-\frac{1}{2}, \sqrt{3}/2)$, $P_2(\frac{1}{2}, -\sqrt{3}/2)$, $P_3(-1, 0)$, $P_4(-\sqrt{3}/2, -\frac{1}{2})$ correspond to angles of $120°$, $-60°$, $180°$, $570°$, respectively.

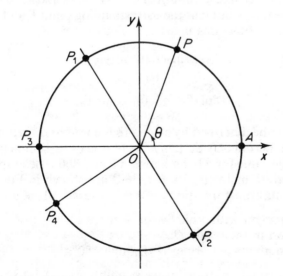

Fig. 3.3

EXERCISE 3.2A

1   Draw a unit circle with centre O and axes $O(x, y)$ and, on the circle mark the
points $P_1, P_2, \ldots, P_6$, which correspond to angles $40°$, $230°$, $-20°$, $-200°$,
$400°$, $1000°$. Without calculating the coordinates of these six points, state the
signs of each of their x and y coordinates.

When $P(x, y)$ is the point on the unit circle corresponding to the angle $\theta°$,
so that angle AOP is $\theta°$ measured positively in an anticlockwise sense, then
we define the circular functions:

$$\cos \theta° = x \qquad\qquad \sin \theta° = y \qquad\qquad \tan \theta° = y/x, \text{ if } x \neq 0$$
$$\sec \theta° = 1/x, (x \neq 0) \quad \operatorname{cosec} \theta° = 1/y, (y \neq 0) \quad \cot \theta° = x/y, (y \neq 0).$$

Thus the cosine and sine of $\theta°$ are respectively the projections of OP on to
the axes Ox and Oy, with the sign taken into account. Referring to Fig. 3.3,
for example, we see that

$$\cos 120° = -\tfrac{1}{2} \quad \cos(-60°) = \tfrac{1}{2} \qquad \cos 180° = -1 \quad \cos 570° = -\sqrt{3}/2,$$
$$\sin 120° = \sqrt{3}/2 \quad \sin(-60°) = -\sqrt{3}/2 \quad \sin 180° = 0. \qquad \sin 570° = -\tfrac{1}{2}.$$

If $P(x, y)$ corresponds to the positive rotation $\theta°$ and Q corresponds to
the corresponding negative rotation $-\theta°$, then clearly Q is the reflection
of P in the axis Ox, so that Q is the point $(x, -y)$. Hence

$$\cos(-\theta°) = \quad x \quad = \cos \theta°,$$
$$\sin(-\theta°) = -y \quad = -\sin \theta°,$$
$$\tan(-\theta°) = -y/x = -\tan \theta°.$$

The addition of any integer multiple of $360°$ to $\theta°$ (positive or negative)
will not change the position of the corresponding point P and so leaves all
the trigonometric functions unaltered. Thus for $n \in \mathbb{Z}$,

$$\cos(\theta + n360)° = \cos \theta°,$$
$$\sin(\theta + n360)° = \sin \theta°,$$
$$\tan(\theta + n360)° = \tan \theta°.$$

This result may be described by saying that $\cos \theta°$ and $\sin \theta°$ are *periodic*
functions with period 360. So the value of one of these functions may be
found from the value for θ lying between 0 and 360. The graphs of $\cos \theta°$
and $\sin \theta°$, with domain $\{\theta : 0 \leqslant \theta \leqslant 360\}$ are shown in Fig. 3.4.
   The following transformations of the Cartesian plane are equivalent:

   (i) an enlargement with scale factor -1;
   (ii) a reflection in the origin O;
   (iii) a rotation about the origin O through $180°$.

Such a transformation transforms a point $P(x, y)$ into the point Q
$(-x, -y)$, and if P corresponds to the angle $\theta°$ then Q corresponds to the

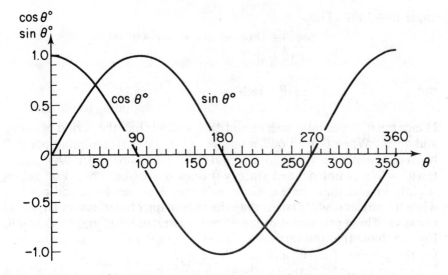

Fig. 3.4 Graphs of cos $\theta°$ and sin $\theta°$, $0 \leqslant \theta \leqslant 360$.

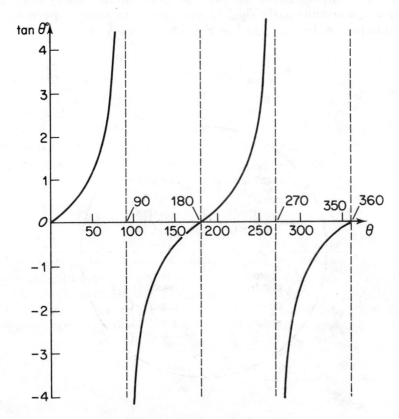

Fig. 3.5 Graph of tan $\theta°$, $0 \leqslant \theta \leqslant 360$.

angle $(\theta + 180)°$. Thus

$$\cos (\theta + 180)° = -x = -\cos \theta°,$$
$$\sin (\theta + 180)° = -y = -\sin \theta°,$$

but
$$\tan (\theta + 180)° = \frac{-y}{-x} = \frac{y}{x} = \tan \theta°.$$

Hence $\tan \theta°$ is periodic with period 180, which is half the period of $\cos \theta°$ and $\sin \theta°$. When θ is an odd multiple of 90, the corresponding point P has coordinates $x = 0 = \cos \theta°$, and $y = \pm 1 = \sin \theta°$. In such cases, $\tan \theta° = y/x$ is not defined since $y/0$ does not make sense. For angles slightly smaller than these critical angles $\tan \theta°$ is very large and positive while for angles slightly larger than the critical ones $\tan \theta°$ is very large and negative. These properties of $\tan \theta°$ are illustrated in its graph, shown in Fig. 3.5. Note that the vertical scale for the graph of $\tan \theta°$ is double that for the graphs of $\cos \theta°$ and $\sin \theta°$ in Fig. 3.4.

For any angle $\theta°$, corresponding to a point P on the unit circle, there is an angle $\phi°$ lying between $0°$ and $360°$, with $\phi = \theta + 360n$, for some $n \in \mathbb{Z}$, and P also corresponds to the angle $\phi°$. We can then classify angles into four sets, according to the signs of their cosines and sines. The results are summarised in Table 3.2 and in Fig. 3.6.

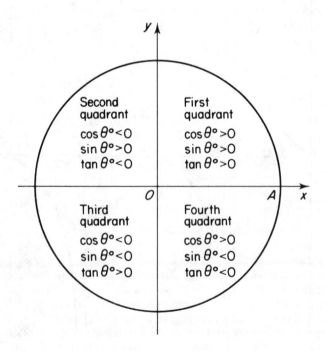

Fig. 3.6 Signs of the circular functions.

Table 3.2 The signs of the circular functions

First quadrant	Second quadrant	Third quadrant	Fourth quadrant
$0 < \phi < 90$	$90 < \phi < 180$	$180 < \phi < 270$	$270 < \phi < 360$
$x = \cos \theta° > 0$	$x = \cos \theta° < 0$	$x = \cos \theta° < 0$	$x = \cos \theta° > 0$
$y = \sin \theta° > 0$	$y = \sin \theta° > 0$	$y = \sin \theta° < 0$	$y = \sin \theta° < 0$
$\dfrac{y}{x} = \tan \theta° > 0$	$\dfrac{y}{x} = \tan \theta° < 0$	$\dfrac{y}{x} = \tan \theta° > 0$	$\dfrac{y}{x} = \tan \theta° < 0$

EXAMPLE *Find in terms of* cos $\theta°$ *and* sin $\theta°$: (a) cos $(180 + \theta)°$,
(b) sin $(180 - \theta)°$, (c) cos $(90 + \theta)°$, (d) sin $(90 - \theta)°$.

Let $P(x, y)$ be the point on the unit circle which corresponds to the angle $\theta°$.
(a) The point corresponding to $(180 + \theta)°$ is $(-x, -y)$ so $\cos(180 + \theta)° = -x$
$= -\cos \theta°$.
(b) The point corresponding to $-\theta°$ is $(x, -y)$ so the point corresponding to
$(180 - \theta)°$ is $(-x, y)$ and $\sin(180 - \theta)° = y = \sin \theta°$.
(c) The point corresponding to $(90 + \theta)°$ lies in the next quadrant to P and has
coordinates $(-y, x)$ so $\cos(90 + \theta)° = -y = -\sin \theta°$.
(d) Since $90 - \theta = 180 - (90 + \theta)$, the point corresponding to $(90 - \theta)°$ has
coordinates (y, x), on using (b) and (c), so $\sin(90 - \theta)° = x = \cos \theta°$.
Note that the method of proof used here is independent of the quadrant in which
the angle $\theta°$ lies. The same result will always be obtained by assuming that $\theta°$ lies
in the first quadrant. Thus in (b) if $\theta°$ lies in the first quadrant, x and y are positive,
$-\theta°$ lies in the fourth quadrant and so $(180 - \theta)°$ lies in the second quadrant, with
corresponding point $(-x, y)$.

EXERCISE 3.2B

1 Complete the following table of quadrants in which the angles lie:

Angle	$\theta°$	$-\theta°$	$(180 + \theta)°$	$(180 - \theta)°$	$(90 + \theta)°$	$(90 - \theta)°$
Quadrant	1	4				1
	2	3		1		
	3		1			
	4				1	

2 Find, in terms of $\cos \theta°$ and $\sin \theta°$, the following:
    (i) $\cos(-\theta)°$ (ii) $\sin(-\theta)°$ (iii) $\tan(-\theta)°$
   (iv) $\cos(180 + \theta)°$ (v) $\sin(180 + \theta)°$ (vi) $\tan(180 + \theta)°$
  (vii) $\cos(180 - \theta)°$ (viii) $\sin(180 - \theta)°$ (ix) $\tan(180 - \theta)°$
    (x) $\cos(90 + \theta)°$ (xi) $\sin(90 + \theta)°$ (xii) $\tan(90 + \theta)°$
 (xiii) $\cos(90 - \theta)°$ (xiv) $\sin(90 - \theta)°$ (xv) $\tan(90 - \theta)°$.

3 State whether the equation is true or false in general:
    (i) $\cos(-\theta)° = \cos(180 + \theta)°$ (ii) $\cos(\theta - 270)° = -\sin \theta°$,
  (iii) $\tan(\theta - 180)° = -\tan \theta°$ (iv) $\sin(270 + \theta)° = \sin \theta°$,
   (v) $\sin(\theta - 90)° = \cos \theta°$ (vi) $\sin(\theta + 45)° = \cos(45 - \theta)°$.

4   (a) Given that $\sin \theta° = 4/5$, find the possible values of the other two trigonometric ratios for $\theta°$;
    (b) Given that $\tan \theta° = -5/12$, find the possible values of $\cos \theta°$ and $\sin \theta°$.

Other trigonometrical ratios

You will have found, in question **2** of Exercise **3.2B** that $\tan (90 + \theta)° = -1/\tan \theta°$ and $\tan (90 - \theta)° = 1/\tan \theta°$. The function $1/\tan$ and the reciprocals of the sine and cosine functions are of sufficient importance to be given their own names:
the reciprocal of cosine is secant and is written sec,
the reciprocal of sine is cosecant and is written cosec,
the reciprocal of tangent is cotangent and is written cot.
Thus we have three more circular functions:

$$\sec \theta° = \frac{1}{\cos \theta°}, \text{ with domain } \mathbb{R} \backslash \{(2n+1)90 : n \in \mathbb{Z}\},$$

$$\operatorname{cosec} \theta° = \frac{1}{\sin \theta°}, \text{ with domain } \mathbb{R} \backslash \{180n : n \in \mathbb{Z}\},$$

$$\cot \theta° = \frac{1}{\tan \theta°}, \text{ with domain } \mathbb{R} \backslash \{180n : n \in \mathbb{Z}\}.$$

Note that, since $\tan \theta° = \dfrac{\sin \theta°}{\cos \theta°}$, $\cot \theta° = \dfrac{\cos \theta°}{\sin \theta°}$.

Consequences of Pythagoras' theorem

$$\cos^2 \theta° + \sin^2 \theta° = 1,$$
$$1 + \tan^2 \theta° = \sec^2 \theta°, \text{ when } \cos \theta° \neq 0,$$
$$\cot^2 \theta° + 1 = \operatorname{cosec}^2 \theta°, \text{ when } \sin \theta° \neq 0.$$

Proof Since $\cos \theta° = \pm x$ and $\sin \theta° = \pm y$, $\cos^2 \theta° = x^2$ and $\sin^2 \theta° = y^2$. Now, for every point $P(x, y)$ on the unit circle, $x^2 + y^2 = 1$, so, for all values of θ, $\cos^2 \theta° + \sin^2 \theta° = 1$. The other results are obtained by dividing by $\cos^2 \theta°$ and $\sin^2 \theta°$ respectively, for values of θ in their domains.

EXERCISE 3.2C

1   For each of the functions (a) $\sec \theta°$, (b) $\operatorname{cosec} \theta°$, (c) $\cot \theta°$, state the domain, the range and the period of the function.
2   Using the graphs of $\cos \theta°$, $\sin \theta°$ and $\tan \theta°$, of Figs 3.4 and 3.5, sketch the corresponding graphs of $\sec \theta°$, $\operatorname{cosec} \theta°$ and $\cot \theta°$.
3   Prove that (i) $\sin^3 \theta + \sin \theta \cos^2 \theta = \sin \theta$, (ii) $\dfrac{\cos^2 \theta}{1 - \sin \theta} = 1 + \sin \theta$.

4 Simplify: (i) $\sec\theta - \sec\theta\sin^2\theta$, (ii) $\cos\theta\,\mathrm{cosec}\,\theta\tan\theta$,
(iii) $\cos^2\theta(1 + \tan^2\theta)$, (iv) $(\sin\theta + \cos\theta)^2 + (\sin\theta - \cos\theta)^2$,
(v) $\tan\theta + \dfrac{\cos\theta}{1 + \sin\theta}$, (vi) $\dfrac{\sin\theta - \cos\theta + 1}{\sin\theta + \cos\theta - 1}$.

3.3 Solution of trigonometric equations

Because of the periodic nature of the circular functions, there is generally an infinite number of solutions to a trigonometric equation, unless the domain is restricted and solutions are limited to some fixed set of values.

EXAMPLE 1 *Solve the equation* $\sin 2\theta° = 0$:
(*a*) *for* $0 \leqslant \theta < 360$, (*b*) *for general values of* θ.

(a) Since $\sin\phi° = 0$ for $\phi \in \{0, 180, 360, 540\}$, on putting $\theta = \frac{1}{2}\phi$ we see that the solution set of $\sin 2\theta° = 0$ is $\{\mathbf{0, 90, 180, 270}\}$.
(b) Using the same argument as in (a) the solution of $\sin 2\theta° = 0$ for general values of θ is the set $\{\mathbf{90n:n \in \mathbb{Z}}\}$.

EXAMPLE 2 *Solve the equation; for* $0 \leqslant \theta \leqslant 360$:
(*i*) $\cos\theta° = \cos\alpha°$; (*ii*) $\sin\theta° = \sin\alpha°$; (*iii*) $\tan\theta° = \tan\alpha°$, *where* $0 < \alpha < 90$.

We use the symmetries of the graphs of the circular functions, as indicated in Fig. 3.7, to solve these equations.

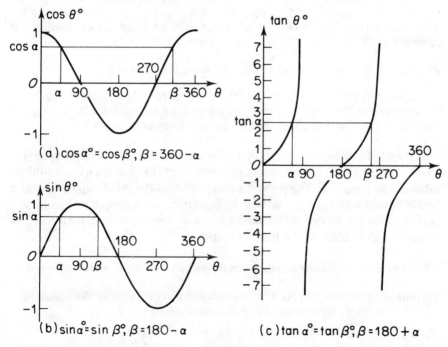

Fig. 3.7

(i) In (a) we see that $\cos \alpha° = \cos (360 - \alpha)°$ and so the solution set of the equation $\cos \theta° = \cos \alpha°$ is $\{\alpha, \mathbf{360 - \alpha}\}$.
(ii) In (b) we see that $\sin \alpha° = \sin (180 - \alpha)°$ giving the solution set $\{\alpha, \mathbf{180 - \alpha}\}$.
(iii) Similarly from (c) $\tan \alpha° = \tan (180 + \alpha)°$ and the solution of the equation is
$\theta = \alpha \text{ or } \theta = 180 + \alpha$.

As in other equations, the method of factorising may be used to solve trigonometric equations, using the fact that a product is zero when one of the factors is zero.

EXAMPLE 3   *Solve* $\sin \theta° + 2 \cos \theta° \sin \theta° = 0, 0 \leqslant \theta \leqslant 360$.

Since $\sin \theta° + 2 \cos \theta° \sin \theta° = \sin \theta° (1 + 2 \cos \theta°)$, the equation is satisfied when either $\sin \theta° = 0$ or when $\cos \theta° = -\frac{1}{2}$. In the required range of values of θ, the graphs of Fig. 3.7 indicate that these equations require that $\theta = 0$ or 180 or that $\theta = 120$ or 240. Each of these values satisfies the equation, so the solution set is $\{\mathbf{0, 120, 180, 240}\}$.

When more than one circular function appears in the equation, one strategy for the solution is to eliminate one of the functions, and this may be done by using the consequences of Pythagoras' theorem.

EXAMPLE 4   *Solve the equation* $3 \tan^2 \theta° - 2 \sec^2 \theta° = 1, -180 < \theta < 180$.

Obtain an equation in $\tan \theta°$ by using the relation $\sec^2 \theta° = 1 + \tan^2 \theta°$. $3 \tan^2 \theta° - 2(1 + \tan^2 \theta°) = 1$, so $\tan^2 \theta° = 3$, an $\tan \theta° = \pm \sqrt{3}$. This gives the solution set $\{\mathbf{60, 120, -60, -120}\}$, since each of these values satisfies the equation.

EXAMPLE 5   *Solve the equation* $\tan \theta° = \sec \theta°, 0 \leqslant \theta \leqslant 360$.

Multiplying by $\cos \theta°$, we find that $\sin \theta° = 1$, which is only satisfied by $\theta = 90$, in the given range. However, this does not satisfy the equation since 90 is not in the domain of $\tan \theta°$ nor of $\sec \theta°$, and so the **solution set is empty**.

It should be noted that, in the solutions above, the procedure was to find a set of values in which θ must lie and then to check that each of the values satisfies the equation. The procedure used to find the set of values may give rise to values which do not satisfy the equation, as happened in Example 5. It is necessary to be on one's guard against obtaining such false roots of an equation, as is seen in the next example.

EXAMPLE 6   *Solve the equation* $\cos \theta° - \sin \theta° = 1, 0 \leqslant \theta < 360$.

To obtain an equation in $\cos \theta°$, by eliminating $\sin \theta°$, rewrite the equation as $\cos \theta° - 1 = \sin \theta°$, square and use Pythagoras' theorem.

Then $\cos^2 \theta° - 2 \cos \theta° + 1 = \sin^2 \theta° = 1 - \cos^2 \theta°$,

so $2 \cos^2 \theta° - 2 \cos \theta° = 0$ or $\cos \theta° (\cos \theta° - 1) = 0$.

Thus either $\cos\theta° = 0$ and $\theta = 90$ or 270,

or $\cos\theta° = 1$ and $\theta = 0$.

On testing these values of θ in the original equation,

$$\cos 0° - \sin 0° \quad = 1 - 0 = 1,$$
$$\cos 90° - \sin 90° = 0 - 1 = -1,$$
$$\cos 270° - \sin 270° = 0 - (-1) = 1,$$

so the solution set is $\{\mathbf{0, 270}\}$, and the 'false solution', $\theta = 90$, is discarded. It arose from the process of squaring.

EXAMPLE 7   *Find the general solution of the equation* $\cos\theta° = \sin\alpha°$; *for a given number* α.

We have to satisfy the equations

$$\cos\theta° = \sin\alpha° = \cos(90 - \alpha)°$$

and, on using the results of Example 2, either $\theta = 90 - \alpha + 360n$ or $\theta = -(90 - \alpha) + 360n$, where n is an integer. All these values satisfy the equation and so the solution set is $\{\mathbf{360n + 90 - \alpha,\ 360n - 90 + \alpha : n \in \mathbb{Z}}\}$.

EXERCISE 3.3

1   Solve the following equations for $0 \leqslant \theta < 360$:
    (i) $2\sin\theta° + 1 = 0$ (ii) $\sin\theta° \cos 2\theta° = 0$
   (iii) $(2\cos\theta° - 1)^2 = 0$ (iv) $(1 + 3\sin\theta°)(4 - \cos\theta°) = 0$.

2   Find the general solutions of the equations:
    (i) $\sin 2\theta° \cos\theta° = 0$ (ii) $\sin 2\theta° = \sin 3\theta°$
   (iii) $\sin 2\theta° = \cos\theta°$ (iv) $\sin\theta° + \cos\theta° = 0$.

3   By factorisation, solve the equations for $-180 < \theta \leqslant 180$:
    (i) $\cos^2\theta° + \cos\theta° - 2 = 0$ (ii) $\cos\theta° \sin\theta° - \cos\theta° - \sin\theta° + 1 = 0$
   (iii) $\cos\theta° \sin\theta° - \cos\theta° + \sin\theta° = 1$ (iv) $2\cos\theta° \sin\theta° - \sin\theta° + 2\cos\theta° = 1$.

4   Use Pythagoras' theorem to solve for $0 \leqslant \theta \leqslant 180$:
    (i) $2\sec^2\theta° + 3\tan\theta° = 1$ (ii) $\cos^2\theta° = 3\sin^2\theta°$
   (iii) $\sin\theta° + \cos\theta° = 1$ (iv) $\tan\theta° + \cot\theta° = 2\sec\theta°$.

5   Find all the values of x between 0 and 360 inclusive such that
    (a) $\cos x° = -\sqrt{3}/2$,   (b) $\cos x° = -\sin x°$.   (L)

6   Obtain the general solution of the equation $\sin\theta° = \cos\alpha°$ for θ in terms of α.   (L)

7   Given that $7\sin^2\theta - 5\sin\theta + \cos^2\theta = 0$, find the possible values of $\sin\theta$.   (L)

3.4 Radian measure of angle

Although the measurement of angles in degrees is quite common, the division of a full circle into 360° is, in fact, quite arbitrary. It is connected with the division of a day into 24×60 minutes, so that the Earth rotates through one degree of longitude every four minutes.

There is a more natural way of measuring angles, which is more appropriate in the case of circular functions with real domains. Consider the unit circle, on which the point P corresponds to the angle x, equal to the angle AOP (see Fig. 3.8). Then the arc length AP on the circle is proportional to x. We make the factor of proportionality unity and define the measurement of angle in *radians*.

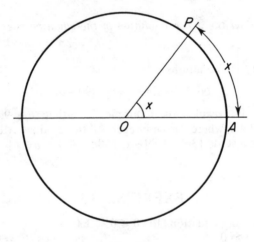

Fig. 3.8

Definition   The angle AOP is x *radians* if the arc length AP on the unit circle, centre O, is x units.

Since the circumference of the unit circle is of length 2π, we find the corresponding angular measure in degrees and radians:

Degrees	Radians
360	2π
180	π
90	$\frac{1}{2}\pi$
θ	$\dfrac{2\pi\theta}{360} = \dfrac{\pi\theta}{180} = \dfrac{\pi}{180}\theta.$

Note: The radian measure x of an angle is a real number, with no dimensions. The notation x^c is sometimes used for x radians, but we shall only use x.

When referred to radian argument, the circular functions are real functions. The properties of these functions follow from the properties of the functions with degree argument, obtained in §3.2. We summarise some of these properties, beginning with a table of their values for some key values of the argument.

Table 3.4A   *Values of circular functions*

x	$\cos x$	$\sin x$	$\tan x$	$\sec x$	$\operatorname{cosec} x$	$\cot x$
0	1	0	0	1	—	—
$\dfrac{\pi}{6}$	$\dfrac{\sqrt{3}}{2}$	$\dfrac{1}{2}$	$\dfrac{1}{\sqrt{3}}$	$\dfrac{2}{\sqrt{3}}$	2	$\sqrt{3}$
$\dfrac{\pi}{4}$	$\dfrac{1}{\sqrt{2}}$	$\dfrac{1}{\sqrt{2}}$	1	$\sqrt{2}$	$\sqrt{2}$	1
$\dfrac{\pi}{3}$	$\dfrac{1}{2}$	$\dfrac{\sqrt{3}}{2}$	$\sqrt{3}$	2	$\dfrac{2}{\sqrt{3}}$	$\dfrac{1}{\sqrt{3}}$
$\dfrac{\pi}{2}$	0	1	—	—	1	0

EXERCISE 3.4A

1   Continue the above table for $x = \dfrac{2\pi}{3}, \dfrac{3\pi}{4}, \dfrac{5\pi}{6}, \pi.$

The functions $\cos x$ and $\sin x$ have domain $\mathbb{R}$ and are periodic with period 2π; $\tan x$ has domain $\mathbb{R}\backslash\{(n+\tfrac{1}{2})\pi : n \in \mathbb{Z}\}$ and has period π. The graphs of $\cos x$, $\sin x$ and $\tan x$ are shown in Fig. 3.9, and the results of question **2** of Exercise 3.2B may be rewritten

$$\cos(-x) = \cos x, \quad \sin(-x) = -\sin x, \quad \tan(-x) = -\tan x,$$
$$\cos(\pi + x) = -\cos x, \quad \sin(\pi + x) = -\sin x, \quad \tan(\pi + x) = \tan x,$$
$$\cos(\pi - x) = -\cos x, \quad \sin(\pi - x) = \sin x, \quad \tan(\pi - x) = -\tan x,$$
$$\cos(\tfrac{1}{2}\pi + x) = -\sin x, \quad \sin(\tfrac{1}{2}\pi + x) = \cos x, \quad \tan(\tfrac{1}{2}\pi + x) = -\cot x,$$
$$\cos(\tfrac{1}{2}\pi - x) = \sin x, \quad \sin(\tfrac{1}{2}\pi - x) = \cos x, \quad \tan(\tfrac{1}{2}\pi - x) = \cot x.$$

Repeating Table 3.2, for radian measure, shows, in Table 3.4B, how the real numbers, other than integer multiples of $\tfrac{1}{2}\pi$, can be divided into four sets, corresponding to the four quadrants:
in the first quadrant all the circular functions are positive;
in the second quadrant $\sin x$ is positive, $\cos x$ and $\tan x$ are negative;
in the third quadrant $\tan x$ is positive, $\sin x$ and $\cos x$ are negative;
in the fourth quadrant $\cos x$ is positive, $\sin x$ and $\tan x$ are negative.

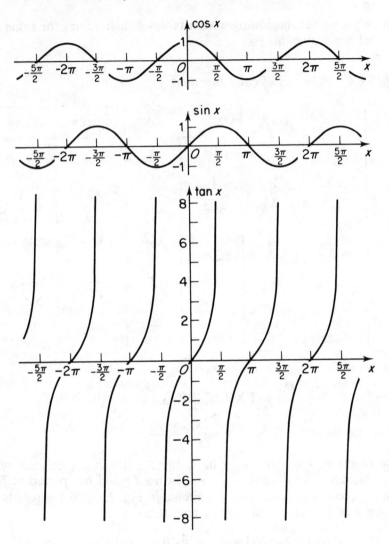

Fig. 3.9 Graphs of cos x, sin x and tan x.

Table 3.4B Signs of the trigonometric functions   For $n \in \mathbb{Z}$:

First quadrant	Second quadrant	Third quadrant	Fourth quadrant
$2n\pi < x < (2n+\frac{1}{2})\pi$	$(2n+\frac{1}{2})\pi < x < (2n+1)\pi$	$(2n+1)\pi < x < (2n+\frac{3}{2})\pi$	$(2n+\frac{3}{2})\pi < x < (2n+2)\pi$
$\cos x > 0$	$\cos x < 0$	$\cos x < 0$	$\cos x > 0$
$\sin x > 0$	$\sin x > 0$	$\sin x < 0$	$\sin x < 0$
$\tan x > 0$	$\tan x < 0$	$\tan x > 0$	$\tan x < 0$

EXAMPLE *Solve the equation* $\cos x + \cos 5x = 0$
(i) for $-\pi \leqslant x \leqslant \pi$, *(ii) for general values of x.*

Since we require $\cos 5x = -\cos x = \cos(\pi - x) = \cos(\pi + x)$, the solutions of the equation satisfy

either $\qquad\qquad 5x = 2n\pi + \pi - x \quad$ or $\quad 5x = 2n\pi + \pi + x$

that is, either $\qquad 6x = (2n+1)\pi \qquad$ or $\quad 4x = (2n+1)\pi,$

so the solution set is $\left\{ \dfrac{2n+1}{6}\pi, \dfrac{2n+1}{4}\pi : n \in \mathbf{Z} \right\}$, in case (ii). In case (i) we choose those values of n which give a solution in the range $-\pi \leqslant x \leqslant \pi$. That is for $-3 \leqslant n \leqslant 2$ giving a solution set $\left\{ \dfrac{-5\pi}{6}, \dfrac{-\pi}{2}, \dfrac{-\pi}{6}, \dfrac{\pi}{6}, \dfrac{\pi}{2}, \dfrac{5\pi}{6}, \dfrac{-3\pi}{4}, \dfrac{-\pi}{4}, \dfrac{\pi}{4}, \dfrac{3\pi}{4} \right\}.$

Apart from working in radians, the next exercise can be done using the same methods as were used in §3.3.

<h3 style="text-align:center">EXERCISE 3.4B</h3>

1 Find all the solutions in the interval $0 \leqslant x \leqslant \pi$ of the equation $\sin 2x - \sin x = 0.$ *(JMB)*
2 Find all solutions in the interval $0 \leqslant x < 2\pi$ of the equation $\sin x = \cos 2x.$ *(JMB)*
3 Find, in radians, the general solution of the equation $\sin 2x = \cos x.$ *(L)*
4 Find, in radians, the general solution of the equation $\sin 2\theta + \sin \theta = 0.$ *(L)*
5 Leaving all answers as multiples of π, find the values of x for which $0 < x < \pi$ and $\cos x + \cos 3x = 0.$ *(L)*
6 Find all the values of x in the range; $0 < x < \pi$ for which $\sin 5x + \sin 3x = 0.$ *(L)*
7 In $\triangle ABC$, $BC = 12$ cm, $AB = 4$ cm and angle C is acute with $\sin C = \frac{1}{6}$. Find, in radians, the two possible values of the angle A, leaving your answer in terms of π. *(L)*
8 Giving your answers in terms of π, solve for $0 \leqslant x \leqslant 2\pi$ the equations
  (a) $\tan 2x = -1,$
  (b) $\cos 2x = \cos x.$ *(L)*
9 Eliminate θ from the pair of simultaneous equations

$$x \cos \theta + y \sin \theta = 1$$
$$y \cos \theta + x \sin \theta = 2.$$ *(JMB)*

3.5 Circular arc and sector

Consider a circle of radius r, having an arc BQ which subtends an angle θ at the centre O of the circle. Let OB and OQ meet the corresponding unit circle with centre O at the points A and P (see Fig. 3.10).

The circle of radius r is obtained from the unit circle by an all-round enlargement with scale factor r, so that all linear dimensions are multiplied by r. The arc length AP is θ and so the arc length BQ is $s = r\theta$. Notice that this is the fraction $\theta/2\pi$ of the circumference $2\pi r$ of the circle.

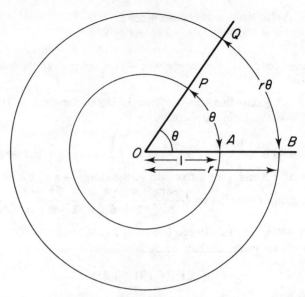

Fig. 3.10

To find the area S of the sector OBQ, we note that this is the same fraction $\theta/2\pi$ of the area of the circle, and so

$$S = (\theta/2\pi).\pi r^2 = \tfrac{1}{2}r^2\theta.$$

EXAMPLE 1 *Find the area and the length of the curved boundary of a sector of a circle: (i) of radius 3 cm and angle 2/3; (ii) of radius 10 m and angle 20°.*

The area $S = \tfrac{1}{2}r^2\theta$ and the arc length, s, is $r\theta$, where θ is measured in radians.

(i) $r = 3$ cm, $\theta = \tfrac{2}{3}$, so $S = \tfrac{1}{2}. 9. \tfrac{2}{3} = \mathbf{3\,cm^2}$, and $s = 3. \tfrac{2}{3} = \mathbf{2\,cm}$.
(ii) $r = 10$ m, $\theta = 20.\pi/180 = \pi/9$, so $S = \tfrac{1}{2}.100.\pi/9 = \mathbf{17\cdot5\,m^2}$, and $s = 10\pi/9$ = **3·49 m**.

EXAMPLE 2 *Find the area and the length of the boundary of a segment of a circle of radius r, bounded by a chord AB and the minor arc AB, where AB subtends an angle 2θ at the centre.*

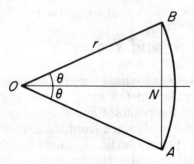

Fig. 3.11

Refer to Fig. 3.11, and let N be the midpoint of AB. The required area S is the area of the sector OAB less the area of the triangle, so

$$S = \tfrac{1}{2}r^2 \cdot 2\theta - ON \cdot AN = r^2\theta - r^2\cos\theta\sin\theta = r^2(\theta - \cos\theta\sin\theta).$$

The length of the boundary $b = r2\theta + 2r\sin\theta = 2r(\theta + \sin\theta)$.

EXAMPLE 3 *Prove that, for $0 < \theta < \tfrac{1}{2}\pi$, $\sin\theta < \theta < \tan\theta$. Deduce that as $\theta \to 0$, $\sin\theta/\theta \to 1$, which is a notation for*
'*as θ approaches zero, $\sin\theta/\theta$ approaches 1*'.

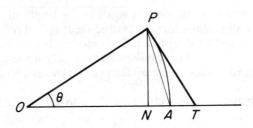

Fig. 3.12

In Fig. 3.12, let AP be an arc of a unit circle subtending an angle θ at the centre O. Let PN be the perpendicular from P on to OA and let PT be the tangent to the circle at P, meeting OA produced at T. Then the figure shows the situation for an acute angle θ and the area of the triangle OPA is less than the area of the sector OAP which is itself less than the area of the triangle OPT.

$$\tfrac{1}{2}OA \cdot PN = \tfrac{1}{2}\sin\theta < \tfrac{1}{2}\theta < \tfrac{1}{2}OT \cdot PN = \tfrac{1}{2}\sec\theta\sin\theta$$

and so, on dividing by $\tfrac{1}{2}$,

$$\sin\theta < \theta < \tan\theta.$$

Since θ and $\theta\sec\theta$ are both positive, we may divide the two inequalities by these numbers, when we obtain the inequalities

$$\cos\theta < \sin\theta/\theta < 1.$$

Now, as θ approaches zero, $\cos\theta$ approaches 1 and $\sin\theta/\theta$ is sandwiched between $\cos\theta$ and 1, so it also must approach 1, as was to be proved.

Note Referring to Fig. 3.11, the ratio of the lengths of the chord AB and the arc AB is $\dfrac{2r\sin\theta}{2r\theta} = \dfrac{\sin\theta}{\theta}$, and thus, by the result of the above example, as the point B approaches A along the circle of radius r, the ratio of the lengths of the chord AB and the arc AB approaches 1. These results will be used in Chapter 21.

EXERCISE 3.5

1 An arc of a circle of radius a subtends an angle θ at the centre. Find the length of the arc and the area of the corresponding sector when:

(i) $a = 5\,\text{m}$, $\theta = 2\frac{1}{2}$;   (ii) $a = 8\,\text{cm}$, $\theta = \frac{3}{4}\pi$;   (iii) $a = 2\cdot5\,\text{m}$, $\theta = 120°$;
(iv) $a = 1\cdot32\,\text{m}$, $\theta = 12°$.

2   Determine the angle, in degrees, subtended at the centre of a circle by an arc, given that the area of the corresponding sector is equal to the square of the length of the arc.

3   Two circles, of radii 3 m and 4 m respectively, lie in a plane with their centres a distance 5 m apart. Find the area common to these circles and the length of the boundary of this area.

4   Find the ratio of the areas of the two segments into which a circle is divided by a chord:
    (i) when the radius of the circle is equal to the length of the chord;
    (ii) when the radius of the circle is 2 m and the length of the chord is $2\sqrt{3}\,\text{m}$;
    (iii) when the radius of the circle is 5 m and the chord is of length 8 m.

5   Find the ratio of the areas of two segments into which a circle is divided by a chord, if the chord divides the circumference into two arcs whose lengths are in the ratio $a:b$.

6   Find the area between three circles, each of radius a, which all touch each other externally.

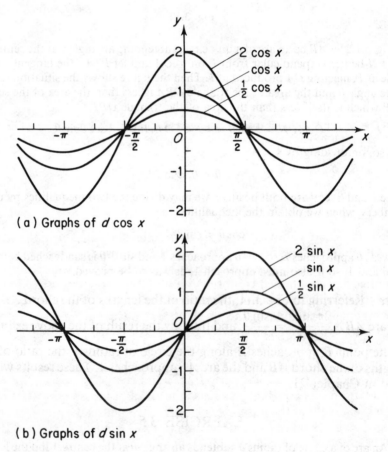

(a) Graphs of $d\cos x$

(b) Graphs of $d\sin x$

Fig. 3.13

3.6 Effect of changes of scale and origin upon the functions cos *x* and sin *x*

The graph of the function cos *x* is a curve with equation $y = \cos x$. A change in the scale for *y* is obtained by multiplying cos *x* by some constant, say *d*, giving the curve with equation $y = d \cos x$. The new function $d \cos x$ has the same period 2π as the function cos *x*, since $d \cos(x + 2\pi) = d \cos x$. The new curve has its peaks at the same values of *x* as the original curve but the height of the peaks (the amplitude of the cosine oscillation) is multiplied by the factor *d*. Fig. 3.13(a) shows the effect of this scale change for the values $d = 2$ and $d = \frac{1}{2}$, and Fig. 3.13(b) shows the corresponding graphs for $d \sin x$.

EXERCISE 3.6A

1 Write down the values of (a) $2 \cos x$, (b) $\frac{1}{2} \sin x$, (c) $3 \cos x + 4 \sin x$, for:
(i) $x = 0$, (ii) $x = \frac{\pi}{6}$, (iii) $x = \frac{\pi}{4}$, (iv) $x = \frac{\pi}{3}$, (v) $x = \frac{\pi}{2}$, (vi) $x = \frac{2\pi}{3}$,
(vii) $x = \frac{3\pi}{4}$, (viii) $x = \frac{5\pi}{6}$, (ix) $x = \pi$.
2 Sketch the graphs of (a) $2 \cos x$, (b) $\frac{1}{2} \sin x$, (c) $3 \cos x + 4 \sin x$, with domain $-2\pi < x < 2\pi$, using your results from question 1. Comment on the shape and the amplitude of the graph (c).
3 Tabulate the values of cos 2*x* and sin 2*x*, for the values of *x* given in question 1, (i) to (ix), and sketch the graphs of cos 2*x* and sin 2*x* with domain $-2\pi < x < 2\pi$. Use the graphs to solve the equation $\cos 2x = \sin 2x$.
4 In each of the following cases, state which of (a) or (b) is true:
(i) the graph of cos 2*x* has (a) twice, (b) half as many peaks as the graph of cos *x*;
(ii) the period of cos 2*x* is (a) twice, (b) half the period of cos *x*;
(iii) the graph of sin *dx* has (a) *d*, (b) 1/*d* times as many peaks as the graph of sin *x*;
(iv) the period of sin *dx* is (a) $2\pi d$, (b) $2\pi/d$.

Questions 3 and 4 above give a clue to the effect of a change in the scale for *x*. The graph of the function cos 2*x* is the same as the graph of cos *x* with the *x*-scale halved. This means that for graphs on the same scale, the graph of cos 2*x* is obtained from the graph of cos *x* by a one-way stretch parallel to O*x* with scale factor $\frac{1}{2}$. The period of cos 2*x* is half of the period of cos *x*, namely π. Similarly, the graph of cos *ax* arises from the graph of cos *x* by a one-way stretch parallel to O*x* with scale factor 1/*a*, so the period of cos *ax* is $2\pi/a$. The same results apply to sin *ax*.

These results are illustrated in Fig. 3.14, by showing the graphs of cos *x*, cos 3*x* and cos $\frac{1}{2}x$. Cos 3*x* has period $2\pi/3$ and cos $\frac{1}{2}x$ has period 4π.

The addition of a constant *c* to *x* has the effect of moving the origin to the point (*c*, 0) so that the graph of f(*x* + *c*) is the same as the graph of f(*x*) but translated a distance $(-c)$ in the direction O*x*. One illustration of this is that the graph of cos *x* is the graph of sin *x* translated a distance $\frac{1}{2}\pi$ in the negative *x*-direction because $\cos x = \sin(x + \frac{1}{2}\pi)$. The graph of $\cos(x + 1)$

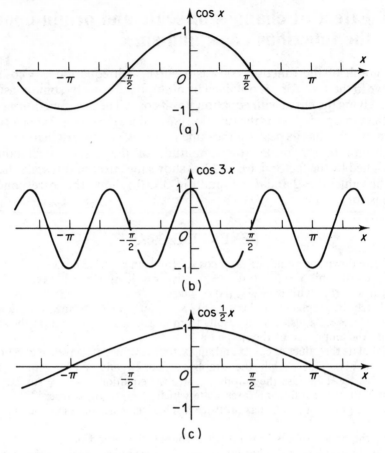

Fig. 3.14 Graphs of (a) cos x, (b) cos $3x$, (c) cos$\frac{1}{2}$$x$.

is the graph of cos x translated a unit distance parallel to the negative x-axis, as shown in Fig. 3.15(b).

Finally, combining scale change and origin shift, the graph of cos $(ax + b)$, that is, cos $a(x + b/a)$, is the graph of cos x stretched by a scale factor $1/a$ parallel to Ox and then translated $(-b/a)$ in the direction Ox. This is shown for the graph of cos $(3x + 2)$ in Fig. 3.15(c).

EXAMPLE   *Find a cosine function which has maximum and minimum values* $+2$ *and* -2 *respectively, period* $\pi/3$ *and the graph of which passes through the origin. What sine function has the same graph?*

The general form of a cosine function is $f(x) = d \cos a(x + c)$ and this function has maximum and minimum values $\pm d$. So we choose $d = 2$. The period of $2 \cos a(x + c)$ is $2\pi/a$ since

$$\cos a\left(x + c + \frac{2\pi}{a}\right) = \cos\left[a(x + c) + 2\pi\right] = \cos a(x + c),$$

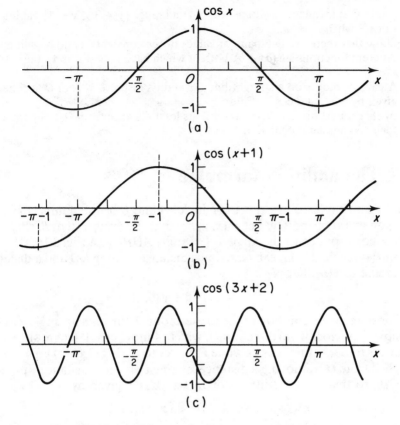

Fig. 3.15 Graphs of (a) cos *x*, (b) cos (*x* + 1), (c) cos (3*x* + 2)

so we require $2\pi/a = \pi/3$ and $a = 6$. Finally, when $x = 0$,

$$2\cos 6(x + c) = 2\cos 6c = 0$$

when $6c$ is an odd multiple of $\frac{1}{2}\pi$. There are two essentially different cases, either $6c = \frac{1}{2}\pi$, or $6c = \frac{3}{2}\pi$, that is, $c = \pi/12$ or $c = \pi/4$, giving two functions which satisfy the given conditions, namely $\mathbf{2\cos(6x + \frac{1}{2}\pi)}$ **and** $\mathbf{2\cos(6x + \frac{3}{2}\pi)}$. The corresponding sine functions are $\mathbf{2\sin 6x}$ **and** $\mathbf{2\sin(6x + \pi)}$.

EXERCISE 3.6B

1 For the following functions, write down: (a) the maximum value, (b) the minimum value, (c) the period, (d) the coordinates of the three points, nearest to the origin, at which the graph of the function meets the coordinate axes Ox and Oy:

(i) $\cos 2x$　　　　(ii) $\sin(x - 1)$　　　　(iii) $\cos(2 + x)$

(iv) $2\sin\dfrac{x}{3}$　　　　(v) $6\cos\frac{1}{2}(5x - 4)$　　　　(vi) $5\cos\dfrac{\pi}{4}(x + 2)$.

2   Find a cosine function, with period $\pi/4$ and range $\{y: -3 \leqslant y \leqslant 3\}$, taking the value 3 when $x = \frac{1}{2}\pi$.

3   Show that there are two pairs of values of the constants a and b, giving two different functions $\sin(ax + b)$, both of which have period π and vanish when $x = 1$.

4   A particle moves on a straight line Ox and its distance x from O at time t is given by $x = 3 + 2\cos 3t$. Find:
(i) its greatest distance from O,   (ii) its least distance from O,   (iii) the time for one complete oscillation.

3.7 The addition formulae

On the unit circle, centre at the origin, let A and B be the points $(1, 0)$ and $(0, 1)$ on axes Ox and Oy. Let $P(x, y)$ be given by $(x, y) = (\cos \beta, \sin \beta)$, so that P corresponds to the angle β ($=$ angle AOP), (see Fig. 3.16(a)). The point P is reached from O by moving a distance x along OA and a distance y parallel to OB, that is

$$(x, y) = x(1, 0) + y(0, 1).$$

Consider now a rotation of the plane about O, through an angle α in the positive (anticlockwise) direction, in which A moves to A' ($\cos \alpha, \sin \alpha$) and B moves to B' ($-\sin \alpha, \cos \alpha$), while P moves to P' (Fig. 3.16(b)). Then P' is reached from O by moving a distance x along OA' and a distance y parallel to OB', so that the coordinates of P' are (x', y'), given by

$$\begin{aligned}
(x', y') &= x(\cos \alpha, \sin \alpha) + y(-\sin \alpha, \cos \alpha) \\
&= (x \cos \alpha - y \sin \alpha, \, x \sin \alpha + y \cos \alpha) \\
&= (\cos \beta \cos \alpha - \sin \beta \sin \alpha, \, \cos \beta \sin \alpha + \sin \beta \cos \alpha).
\end{aligned}$$

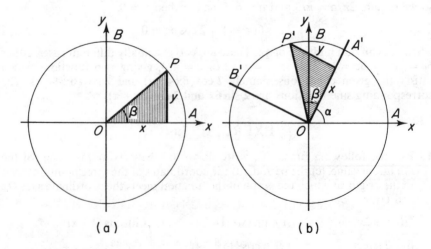

(a)   (b)

Fig. 3.16

But angle $AOP' = (\alpha + \beta)$ and so P' represents the angle $(\alpha + \beta)$ and has coordinates $(\cos\{\alpha + \beta\}, \sin\{\alpha + \beta\})$. Comparing these two forms for the coordinates of P', we find that

$$\cos(\alpha + \beta) = \cos\alpha\cos\beta - \sin\alpha\sin\beta,$$
$$\sin(\alpha + \beta) = \sin\alpha\cos\beta + \cos\alpha\sin\beta.$$

These formulae are the addition formulae for cosine and sine.

For negative angles, since $\cos(-\beta) = \cos\beta$ and $\sin(-\beta) = -\sin\beta$:

$$\cos(\alpha - \beta) = \cos\alpha\cos\beta + \sin\alpha\sin\beta,$$
$$\sin(\alpha - \beta) = \sin\alpha\cos\beta - \cos\alpha\sin\beta.$$

If we now put $\alpha = \beta = \theta$ into the addition formulae we obtain the double angle formulae

$$\cos 2\theta = \cos^2\theta - \sin^2\theta,$$
$$\sin 2\theta = 2\sin\theta\cos\theta.$$

Also, since $\cos^2\theta + \sin^2\theta = 1$, we may write $\sin^2\theta = 1 - \cos^2\theta$ and $\cos^2\theta = 1 - \sin^2\theta$ to obtain

$$\cos 2\theta = 2\cos^2\theta - 1$$
$$= 1 - 2\sin^2\theta.$$

These formulae are now summarised, using capital italic letters for the angles, instead of greek. They should be learned by heart so that they can be instantly recalled when needed. Access to tables of formulae is not a satisfactory alternative to knowing these formulae thoroughly.

The addition and double angle formulae

$$\cos(A + B) = \cos A\cos B - \sin A\sin B$$
$$\sin(A + B) = \sin A\cos B + \cos A\sin B$$
$$\cos(A - B) = \cos A\cos B + \sin A\sin B$$
$$\sin(A - B) = \sin A\cos B - \cos A\sin B$$
$$\cos 2A = \cos^2 A - \sin^2 A$$
$$= 2\cos^2 A - 1$$
$$= 1 - 2\sin^2 A$$
$$\sin 2A = 2\sin A\cos A$$

EXAMPLE 1   *Express* $\cos 3A$ *in terms of* $\cos A$.

$$\cos 3A = \cos(2A + A) = \cos 2A\cos A - \sin 2A\sin A$$
$$= \cos A\,(2\cos^2 A - 1) - 2\sin^2 A\cos A$$
$$= 2\cos^3 A - \cos A - 2(1 - \cos^2 A)\cos A$$
$$= \mathbf{4\cos^3 A - 3\cos A}.$$

EXAMPLE 2   *Simplify* $\dfrac{\sin 2A\,(2\cos 2A+1)}{2\sin 3A}$

$$\begin{aligned}\sin 3A &= \sin(2A+A) = \sin 2A\cos A + \cos 2A\sin A\\ &= 2\sin A\cos^2 A + (2\cos^2 A - 1)\sin A\\ &= \sin A(4\cos^2 A - 1).\end{aligned}$$

So

$$\frac{\sin 2A\,(2\cos 2A+1)}{2\sin 3A} = \frac{2\sin A\cos A\,(2(2\cos^2 A - 1)+1)}{2\sin A\,(4\cos^2 A - 1)}$$

$$= \frac{4\cos^2 A - 1}{4\cos^2 A - 1}\cos A = \mathbf{cos\,A}.$$

EXERCISE 3.7

1 Verify that a rotation α takes the point P $(\cos\beta,\sin\beta)$ into the point P' $(\cos\alpha\cos\beta - \sin\alpha\sin\beta,\ \cos\alpha\sin\beta + \sin\alpha\cos\beta)$ in the cases:
   (i) $\alpha = \pi/3,\ \beta = \pi/3$; (ii) $\alpha = \pi/3,\ \beta = \pi/6$; (iii) $\alpha = \pi/3,\ \beta = \pi/2$;
   (iv) $\alpha = 2\pi/3,\ \beta = 5\pi/6$; (v) $\alpha = 5\pi/6,\ \beta = 5\pi/3$;
   (vi) $\alpha = -\pi/3,\ \beta = 4\pi/3$.

2 Check that the sum and difference formulae for cosine and sine give the correct results in the cases when:
   (i) $A = 2\pi,\ B = 0$; (ii) $A = \pi,\ B = 0$; (iii) $A = \tfrac{1}{2}\pi,\ B = 0$;
   (iv) $A = 0,\ B = \pi$; (v) $A = 0,\ B = \tfrac{1}{2}\pi$.

3 Use the addition formulae for sine and cosine to prove that:
   (i) $\tan(A+B) = \dfrac{\tan A + \tan B}{1 - \tan A\tan B}$; (ii) $\tan(A-B) = \dfrac{\tan A - \tan B}{1 + \tan A\tan B}$;
   (iii) $\tan 2A = \dfrac{2\tan A}{1 - \tan^2 A}$.

4 Simplify the following:
   (i) $\sin(A+B) + \sin(A-B)$; (ii) $\cos(A+B) + \cos(A-B)$;
   (iii) $\cos A\cos 2A - \sin A\sin 2A$; (iv) $\cos A\cos 2A + \sin A\sin 2A$;
   (v) $\cos A\cos(A+B) + \sin A\sin(A+B)$; (vi) $\dfrac{\cos(A-B) - \cos(A+B)}{\sin(A+B) + \sin(A-B)}$;
   (vii) $\dfrac{\cos(A+B) + \cos(A-B)}{\sin(A+B) - \sin(A-B)}$;
   (viii) $\sin A + \sin B\cos(A+B) + \cos B\sin(B-A)$;
   (ix) $\tfrac{1}{2}\tan 4A\,(1 - \tan^2 2A)$; (x) $\dfrac{\cos 2A}{\cos A + \sin A}$; (xi) $\dfrac{\sin 2A}{1 + \cos 2A}$;
   (xii) $\dfrac{\sin 2A + \cos 2A + 1}{\sin 2A - \cos 2A + 1}$; (xiii) $\dfrac{\sin 3A + \sin A}{2\sin 2A}$; (xiv) $\dfrac{\cos 3A + \cos A}{2\cos 2A}$.

5 Find, without using tables or calculators, the values of $\cos 2A$, $\sin 2A$ and $\tan 2A$ in each of the following cases:
   (i) $\sin A = 4/5$, and A is acute; (ii) $\sin A = 4/5$, and A is obtuse;

(iii) $\cos A = 5/13$, and $0 < A < \pi$; (iv) $\cos A = 5/13$, $-\pi < A < 0$;
(v) $\tan A = 9/40$, and A is acute.

6 Prove that
$$\frac{\sin\theta + \sin 2\theta}{1 + \cos\theta + \cos 2\theta} = \tan\theta,$$

where $\cos\theta \neq -\frac{1}{2}$ and $\cos\theta \neq 0$.   (*L*)

7 Prove that $\sin 3\theta = 3\sin\theta - 4\sin^3\theta$.
    Given that $\sin 3\theta = \sin^2\theta$, find three possible values for $\sin\theta$. Hence find all the solutions of the equation $\sin 3\theta = \sin^2\theta$ for $90° \leqslant \theta \leqslant 270°$.
    Using the same axes, sketch the graphs of the functions $\sin 3\theta$ and $\sin^2\theta$ for $90° \leqslant \theta \leqslant 270°$. Find the subset of values of θ for which $\sin 3\theta > \sin^2\theta$.
                                                        (*L*)

8 Prove that $\tan 3A = \dfrac{3\tan A - \tan^3 A}{1 - 3\tan^2 A}$.

9 Find the solutions in the range $0° \leqslant \theta \leqslant 180°$ of each of the following equations
    (i) $\sin 2\theta = \cos\theta$,
    (ii) $2\cos 2\theta = 1 + 4\cos\theta$.   (*JMB*)

10 When $\cos\theta \neq 0$, simplify $(\cos 2\theta + \tan\theta\sin 2\theta)^{-1}$.   (*L*)

11 Without using tables evaluate $\{\sin(7\pi/12) + \cos(7\pi/12)\}^2$.   (*L*)

12 Simplify $\dfrac{\sin 4\theta(1 - \cos 2\theta)}{\cos 2\theta(1 - \cos 4\theta)}$, where $\cos 2\theta(1 - \cos 4\theta) \neq 0$.   (*L*)

13 Express $\tan(45° + \theta)$ in terms of $\tan\theta$. Hence, or otherwise,
    (i) express $\tan 75°$ in the form $a + b\sqrt{3}$ where a and b are integers;
    (ii) express $\tan(45° + \theta) + \cot(45° + \theta)$ in terms of $\cos 2\theta$.   (*JMB*)

14 Given that X is the acute angle such that $\sin X = 4/5$ and Y is the obtuse angle such that $\sin Y = 12/13$, find the exact value of $\tan(X + Y)$.   (*L*)

3.8 The triangle formulae

Reminder   The three angles of the triangle ABC are denoted by A, B and C, and the sides BC, CA, AB are denoted by a, b, c respectively. Let BN be the altitude from B and consider the two cases in Fig. 3.17, (a) when the angle A is acute and (b) when A is obtuse. Of course, A is a right-angle if and only if $A = N$.

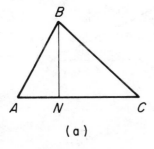
(a)

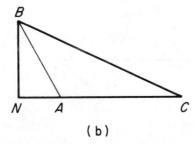
(b)

Fig. 3.17

Let the area of the triangle ABC be Δ, then $\Delta = \frac{1}{2} AC \cdot BN$ and in (a)

$$BN = AB \sin A = c \sin A,$$

while in (b)

$$BN = AB \sin BAN = c \sin (\pi - A).$$

But $\sin (\pi - A) = \sin A$ and so in each case

$$\Delta = \tfrac{1}{2} bc \sin A.$$

Similar results are obtained by permuting a, b, c and A, B, C, that is, by replacing a by b, b by c, c by a, A by B, B by C and C by A. We then obtain the sine formulae for a triangle ABC, the area of which is Δ:

$$\Delta = \tfrac{1}{2} ab \sin C = \tfrac{1}{2} bc \sin A = \tfrac{1}{2} ca \sin B,$$

and, on dividing these equations by $\frac{1}{2} abc$,

$$\frac{\sin A}{a} = \frac{\sin B}{b} = \frac{\sin C}{c}.$$

These ratios are also connected to the radius of the circumcircle of the triangle. Let this circumcircle have centre O and let BO meet the circle again at Q (see Fig. 3.18). Then angle BCQ is a right angle, being the angle in a semicircle, and angle BQC equals angle A in (a) and angle $(\pi - A)$ in (b), since these angles are subtended on the circle by BC. Applying sine formula to triangle BCQ, we find that $\sin A/a = 1/2R$ where R is the radius of the circumcircle, and so

$$\frac{a}{\sin A} = \frac{b}{\sin B} = \frac{c}{\sin C} = 2R.$$

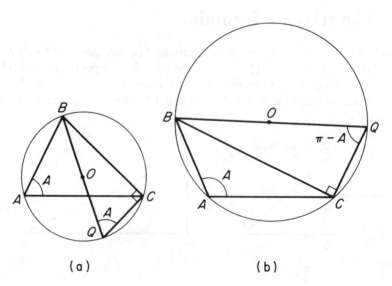

(a)   (b)

Fig. 3.18

Returning now to Fig. 3.17, $AN = AB \cos BAN$, so in (a) $AN = AB \cos A$ and in (b) $AN = AB \cos(\pi - A) = -AB \cos A$. We now use Pythagoras' theorem on the right-angled triangles, in the two cases.

(a)
$$a^2 = BC^2 = BN^2 + NC^2 = AB^2 - AN^2 + (AC - AN)^2$$
$$= AB^2 - AN^2 + AC^2 - 2AC \cdot AN + AN^2$$
$$= AC^2 + AB^2 - 2AC \cdot AB \cos A = b^2 + c^2 - 2bc \cos A.$$

(b)
$$a^2 = BC^2 = BN^2 + NC^2 = AB^2 - AN^2 + (AC + AN)^2$$
$$= AB^2 - AN^2 + AC^2 + 2AC \cdot AN + AN^2$$
$$= AC^2 + AB^2 - 2AC \cdot AB \cos A = b^2 + c^2 - 2bc \cos A.$$

Both cases give rise to the cosine formula for the triangle,

$$a^2 = b^2 + c^2 - 2bc \cos A.$$

Note that when the angle BAC is a right-angle, $A = \pi/2$, $\cos A = 0$, and the cosine formula becomes Pythagoras' theorem, $a^2 = b^2 + c^2$.

Solution of triangles

The sine and cosine formulae are useful in calculating unknown angles and sides of a triangle. The list shows the appropriate formula, according to the given information, and the methods are then illustrated by examples.

Information given	Formula to use
2 angles and 1 side	sine formula
2 sides and non-included angle	sine formula
2 sides and included angle	cosine formula
3 sides	cosine formula

EXAMPLE 1 *In the triangle ABC, the angles $A = 42°$, $B = 73°$ and the side AB is of length 0·5 m. Solve the triangle and find the radius R of its circumcircle.*

Since $A + B + C = 180°$, $C = 180° - 42° - 73° = 65°$ and, on using the sine formula, since $AB = c = 0.5$ m,

$$\frac{a}{\sin 42°} = \frac{b}{\sin 73°} = \frac{0.5}{\sin 65°} = 2R \quad \text{(see Fig. 3.19(a))}.$$

Hence
$$a = \frac{0.5 \sin 42°}{\sin 65°} = 0.37 \text{ m}, \quad b = \frac{0.5 \sin 73°}{\sin 65°} = 0.53 \text{ m},$$

and $R = \dfrac{1}{2} \dfrac{0.5}{\sin 65°} = 0.28$ m.

When a triangle is to be determined given two sides and a non-included angle, there is some difficulty, since it is possible that two, one or no triangles fit the conditions. Suppose that, for triangle ABC, we are given the angle A and the sides $a = BC$ and $c = AB$ (Fig. 3.19(b)). Then C must

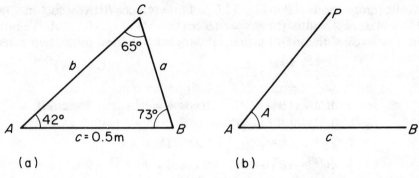

Fig. 3.19

lie on the line *AP*, such that angle *BAP* is the required angle *A*. Also *C* must lie on the circle, centre *B* and radius $a(= BC)$. Now this circle may meet *AP* in one or two points, or it may touch *AP* at one point, or it may not meet *AP* at all. The four possibilities are shown in Fig. 3.20, (a), (b), (c) and (d).

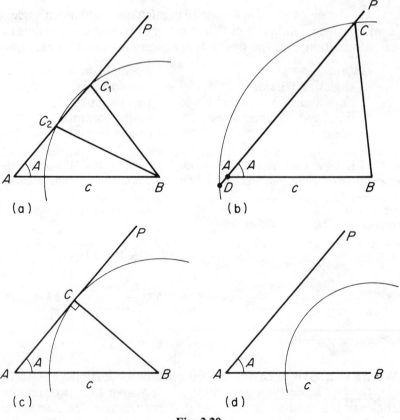

Fig. 3.20

(a) The circle meets AP in C_1 and C_2 and there are two possible triangles ABC_1 and ABC_2, satisfying the conditions.
(b) The circle meets AP at C and PA produced at D, so there is only one possible triangle ABC.
(c) The circle touches AP at C, giving a right-angled triangle ABC.
(d) The circle does not meet AP at all and no triangle exists.

The sine formula allows us to calculate the angle C, using $\sin C = (c/a) \sin A$, and the various possibilities are shown in Example 2.

EXAMPLE 2 *Solve the triangle ABC, given that $A = 40°$, $c = 10$ and:*
(a) $a = 9$, (b) $a = 11$, (c) $a = 6.428$, (d) $a = 6$.

(a) $\sin C = (10/9) \sin 40° = 0.7142$, so $C = 45.6°$ **or 134.4°**. Then $B = 94.4°$ **or 5.6°** and $b = a (\sin B / \sin A)$ so $b = 14.0$ **or 1.4**. This is a case where there are two possible triangles (see Fig. 3.20(a)).
(b) $\sin C = (10/11) \sin 40° = 0.5843$, so $C = 35.8°$ or $144.2°$. In the latter case, $A + C = 184.2°$, which is impossible since this is greater than $180°$. There is one triangle, as in Fig. 3.20(b), with $C = 35.8°$, $B = 104.2°$ and $b = 16.6$.
(c) $\sin C = (10/6.428) \sin 40° = 1$, so $C = 90°$ and we have the case (c) of Fig. 3.20. Then $B = 50°$ and $b = 7.660$.
(d) $\sin C = (10/6) \sin 40° = 1.071$, which is greater than 1 so **no solution** is possible, as in Fig. 3.20(d).

EXAMPLE 3 *Solve the triangle ABC where $b = 9$, $c = 8$ and $A = 50°$.*

Using the cosine formula,

$$a^2 = 9^2 + 8^2 - 2.9.8 \cos 50° = 81 + 64 - 92.56 = 52.44,$$

so $a = 7.24$. The other angles may now be calculated using the sine formula.

$$\sin B = (b/a) \sin A = 0.9521$$

and so $B = 72.2°$ or $107.8°$.

$$\sin C = (c/a) \sin A = 0.8463$$

and so $C = 57.8°$ or $122.2°$. Since $A + B + C = 180°$, the only possible solution is $B = 72.2°$, $C = 57.8°$.

EXAMPLE 4 *Solve the triangle ABC when $a = 8$, $b = 9$, $c = 10$, and find its area.*

By the cosine formula,

$$\cos A = \frac{b^2 + c^2 - a^2}{2bc} = \frac{81 + 100 - 64}{180} = 0.65$$

so $A = 49.5°$. The triangle is clearly acute angled, so by the sine formula,

$$\sin B = (9/8) \sin 49.5° = 0.855$$

and $B = 58.8°$. Then

$$C = 180° - 58.8° - 49.5° \text{ and } C = 72.7°.$$

The area of the triangle is

$$\Delta = \tfrac{1}{2}bc \sin A = \tfrac{1}{2}.9.10.\sin 49.5° = 34.2.$$

EXERCISE 3.8

1 Solve the triangle ABC given that:
   (i) $A = 75°, B = 55°, c = 9$; (ii) $B = 68°, C = 82°, c = 15$;
   (iii) $A = 121°, C = 13°, a = 3$; (iv) $A = 32°, C = 87°, b = 22$.
2 Solve the triangle ABC given that:
   (i) $A = 15°, a = 2, b = 3$; (ii) $B = 82°, b = 5, c = 4$;
   (iii) $B = 38°, a = 12, b = 9$; (iv) $C = 150°, a = 6, c = 10$;
   (v) $A = 93°, a = 7, c = 2$; (vi) $A = 41°, b = 13, a = 18$.
3 Solve the triangle ABC given that:
   (i) $A = 34°, b = 5, c = 7$; (ii) $C = 58°, a = 12, b = 8$;
   (iii) $B = 111°, a = 9, c = 3$; (iv) $a = 4, b = 5, c = 6$;
   (v) $a = 13, b = 5, c = 12$; (vi) $a = 3, b = 7, c = 5$.
4 The area of an acute-angled triangle ABC is $12\,\text{m}^2$. Given that $AB = 5\,\text{m}$, $BC = 8\,\text{m}$, find the length of CA.   (L)
5 With the usual notation in a triangle ABC, $a = 7$, $b = 5$ and $c = 4$. Without using tables, find $\cos A$ as a fraction in its lowest terms, and prove that
$$\sin B = \frac{2\sqrt{6}}{7}.$$ (L)
6 Calculate in degrees and minutes the smallest angle of a triangle whose sides are of length 6, 7 and 8 cm.   (L)
7 In $\triangle ABC$, $BC = 10\,\text{cm}$ and $\angle BAC = 60°$.
   (a) Calculate the radius of the circle passing through A, B and C.
   (b) Given that AC is a diameter of this circle, calculate the area of $\triangle ABC$.
   (Answers may be left in surd form.)   (L)
8 (i) In $\triangle ABC$, M is the mid-point of BC.
   (a) Show that $\cos \angle AMB + \cos \angle AMC = 0$.
   (b) By applying the cosine rule in $\triangle ABM$ and in $\triangle ACM$, or otherwise, prove that
$$4AM^2 = 2CA^2 + 2AB^2 - BC^2.$$

   (ii) Giving your answers to one decimal place, solve, for $0 \leqslant \theta \leqslant 360$, the equations
   (a) $\sin \theta° = \tfrac{3}{5}$,
   (b) $\sin \theta° = -\tfrac{3}{4}$.   (L)
9 Three forts A, B and C are situated in a flat desert region. A is 8 km due west of B, and C is 3 km due east of B. An oasis O is situated to the north of the line ABC and is 7 km from both A and C. Calculate
   (a) the distance between O and B,
   (b) the bearing of O from B.
   A mine M is situated to the south of ABC. The bearing of M from A is 135° and the bearing of M from C is 210°.
   (c) Calculate the distance between M and B.   (L)

MISCELLANEOUS EXERCISE 3

1 (i) Find the angles between $0°$ and $360°$ which satisfy the equations
(a) $\cot 2x = 2 + \cot x$,
(b) $\cos 3x - 3\cos x = \cos 2x + 1$.
(ii) Find the smallest angle of the triangle in which $a = 12\cdot45$, $b = 8\cdot76$ and
$c = 13\cdot89$. (L)

2 Prove that in *any* triangle ABC, with the usual notation,

$$a^2 = b^2 + c^2 - 2bc\cos A.$$

A convex quadrilateral $ABCD$ has $AB = 5\,\text{cm}$, $BC = 8\,\text{cm}$, $CD = 3\,\text{cm}$, $DA = 3\,\text{cm}$ and $BD = 7\,\text{cm}$. Find the angles DAB, BCD and show that the quadrilateral is cyclic.
Show also that $AC = 39/7\,\text{cm}$. (L)

3 The points A, B, C lie on a circle with centre O.
Given that $AB = 11\,\text{m}$, $BC = 13\,\text{m}$, $CA = 20\,\text{m}$, find the angles AOB, BOC, COA to the nearest tenth of a degree and the radius of the circle to the nearest tenth of a metre. (L)

4 In a triangle ABC, $AB = x$, $BC = (x + 2)$ and $AC = (x - 2)$, where $x > 4$. Prove that
$$\cos A = \frac{x - 8}{2(x - 2)}.$$

Find the set of integral values of x for which A is obtuse. (L)

5 In the triangle ABC, given that $A - B = 90°$ and $a + b = 2c$,
(a) prove that $\cos 2A - \cos 2B + 2\sin C = 0$,

(b) show that $\dfrac{a+b}{c} = \dfrac{\sin A + \sin B}{\sin C}$ and hence prove that $\sin\dfrac{C}{2} = \dfrac{1}{2\sqrt{2}}$. (L)

6 In the triangle ABC, AB is of unit length and $BC = CA = p$. The point P lies on AB at a distance x from A and is such that $\angle ACP = \theta$ and $\angle BCP = 2\theta$. By using the sine rule, or otherwise, show that

$$\cos\theta = \frac{1-x}{2x}.$$

State the range of possible values of θ as p varies and deduce that $\frac{1}{3} < x < \frac{1}{2}$. Express $\cos 3\theta$ in terms of p. Hence, or otherwise, find the value of x correct to two decimal places when $p = 1/\sqrt{2}$. (JMB)

7 The triangle OMN is horizontal with angles NOM and ONM equal to θ and ϕ respectively. MA and NB are equal lines drawn vertically upwards and angles AOM and BON are α and β respectively. Prove that
$$\cot\phi = \tan\alpha\cot\beta\,\text{cosec}\,\theta - \cot\theta.$$

8 Use standard formulae to prove that, for any angles A, B,

$$\cos(A+B) - \sin(A-B) = (\cos A - \sin A)(\cos B + \sin B).$$

If $\cos(A+B)$ is equal to $\sin(A-B)$, what can you deduce about the values, in degrees, of A and/or B? (SMP)

9 Solve, for $0 \leqslant \theta \leqslant 2\pi$, the equation $\sin\left(\theta + \dfrac{\pi}{4}\right) = 2\cos\left(\theta - \dfrac{\pi}{4}\right)$. (L)

10 Find all the angles θ between $0°$ and $360°$ such that $2\cos 2\theta = 1$.   *(SMP)*

11 Given that $\sin x = 0\cdot6$ and that $\cos x = -0\cdot8$, evaluate $\cos(x + 270°)$ and $\cos(x + 540°)$.   *(L)*

12 (i) Solve, for $0 \leqslant \theta \leqslant 2\pi$, the equation $6\sin^2\theta + \cos\theta - 5 = 0$.
(ii) Given that $5\cos\theta° - 12\sin\theta° \equiv R\cos(\theta° + \alpha°)$, where $R > 0$ and $0 < \alpha < 90$, calculate
(a) the value of R,
(b) the value of α to 1 decimal place.
Hence solve, for $0 \leqslant \theta \leqslant 360$, the equation $5\cos\theta° - 12\sin\theta° = 6\frac{1}{2}$.   *(L)*

13 (a) Starting with the formulae for $\sin(A - B)$ and $\cos(A - B)$ prove that
$$\tan(A - B) = \frac{\tan A - \tan B}{1 + \tan A \tan B}.$$

(b) State the values of $\tan\dfrac{\pi}{4}$ and $\tan\dfrac{\pi}{6}$ and hence prove that $\tan\dfrac{\pi}{12} = \dfrac{\sqrt{3} - 1}{\sqrt{3} + 1}$.

(c) Given that $\sin\left(\theta - \dfrac{\pi}{12}\right) = 2\cos\left(\theta + \dfrac{\pi}{12}\right)$, prove that $\tan\theta = \dfrac{3\sqrt{3} + 1}{3\sqrt{3} - 1}$.
   (L)

14 The depth of water at the entrance to a harbour t hours after high tide is D metres, where $D = p + q\cos(rt°)$ for suitable constants p, q, r. At high tide the depth is $7\,\text{m}$; at low tide 6 hours later the depth is $3\,\text{m}$.
(a) Show that $r = 30$ and find the values of p and q.
(b) Sketch the graph of D against t for $0 \leqslant t \leqslant 12$.
(c) Find how soon after low tide a ship which requires a depth of at least $4\,\text{m}$ of water will be able to enter harbour.   *(L)*

15 Students have been known to make the mistake of writing $\sin A + \sin B$ in place of $\sin(A + B)$.
Prove that these two expressions are equal only if
(a) $\sin\frac{1}{2}A = 0$ or (b) $\sin\frac{1}{2}B = 0$ or (c) $\sin\frac{1}{2}(A + B) = 0$.
What can you deduce about the values of A and B in these three cases?
   (SMP)

4 Differentiation I

4.1 Linear relations and scale changes

Linear functions (see Chapter 1)

A linear function f is given by $f(x) = mx + c$, where m and c are constants. The graph of f is a straight line. The equation of a line in two-dimensional Cartesian coordinates is

$$ax + by + d = 0,$$

with a, b, d constant, and, if $a \neq 0 \neq b$, this equation gives a linear relation between x and y.

Gradient

The *gradient* of a linear function is defined to be the gradient, or slope, of its graph. Let f be a linear function with $f(x) = mx + c$; then the graph of f is the line with equation $y = mx + c$ and gradient m. On this graph consider the two points:

P with coordinates (x, y) where $y = mx + c$

and $\qquad P_1$ with coordinates (x_1, y_1) where $y_1 = mx_1 + c$, (see Fig. 4.1).

From P to P_1 the change in the variable x is $(x_1 - x)$ and the change in y is given by $(y_1 - y) = m(x_1 - x)$, namely m times the change in x. Thus the function f implies a change of scale between x and y. We say that:

$\qquad m$ is the *rate of change* of y with respect to x.

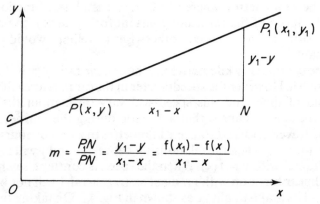

Fig. 4.1 Graph of f, where $f(x) = mx + c$, $m > 0$.

Note: This graph is drawn with the gradient m of f positive so that y increases as x increases. When m is negative, y decreases as x increases.

EXERCISE 4.1

1   Redraw the graph of Fig. 4.1 in the cases:
    (i) $c < 0 < m$;   (ii) $m < 0 < c$;   (iii) $c < 0$ and $m < 0$.
2   The points A, B, C have coordinates $(-1, 2), (3, 5), (2, -3)$, respectively. Draw the three lines AB, BC, CA on a graph and find their gradients.
3   Plot the points A $(1, 1)$ and $B(4, 2)$ on a graph and find the gradient of the line AB. Draw lines through A and B with gradients 1.5 and -2 respectively and find the coordinates of the point C where these two lines intersect.
4   Given that the x-coordinates of P and Q are 1 and 2 respectively, find the gradient of the chord PQ of the curve with equation:
    (i) $y = 3x - 2$;   (ii) $y = 5x^2 + 3x - 4$;   (iii) $y = 4 - 3x^2$;   (iv) $y = x^2 - x^3$.
5   Sketch the graph of the function f where $f(x) = 2x^2 - 1$ for values of x between 1 and 3. Let P be the point $(2, 7)$ on the graph and Q the point $(x, f(x))$ on the graph. Complete the table:

x	1·0	1·3	1·6	1·8	1·9	2·0	2·1	2·2	2·4	2·7	3·0
$f(x)$	1·0					7·0					17·0
Gradient PQ	6·0					—					10·0

From this table, deduce the value that you think the gradient of PQ approaches as the value of x tends to 2·0. Use a ruler to draw a tangent to the graph of f at P, as carefully as you can, judging the position by eye. Measure the gradient of this tangent and compare your result with the limit of the gradient of PQ which you previously deduced.

4.2 Speed

If a car is travelling at a steady speed of 60 km/h, then it will travel 60 kilometres in 1 hour, or 1 kilometre in 1 minute. The distance travelled in x hours will be y kilometres, where $y = 60x$, so y is a linear function of x. In practice, the speed of the car is not constant for a journey; the car has to start and stop at zero speed, otherwise travelling would be most uncomfortable!

When the car travels y kilometres in x hours, the ratio y/x is the average speed in km p h. However the speedometer in the car measures the speed at each instant of time, and this speed will vary throughout the journey. Whatever the speed changes that are made during the journey, suppose that the car travels a total distance y kilometres in a time x hours after the start of the journey. Then the relationship between x and y will be given by some function f, with $y = f(x)$. When the speed is constant, say m km p h, then f is a linear function with gradient m. In general, f will not be a linear function and its graph might be as shown in Fig. 4.2. On taking two points, P (x, y) and P_1 (x_1, y_1), on the graph, the car is seen to travel the distance

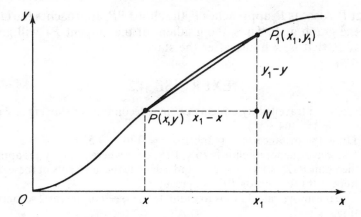

Fig. 4.2 Distance-time graph for a car journey.

$(y_1 - y)$ km in the time $(x_1 - x)$ hours, so that the average speed in this time interval is $(y_1 - y)/(x_1 - x)$ km p h. Now

$$\frac{y_1 - y}{x_1 - x} = \frac{f(x_1) - f(x)}{x_1 - x}$$

and this is the slope (or gradient) of the line PP_1.

The speed of the car at P, that is, x hours after the start of the journey, measured by the speedometer, may be interpreted from the graph in the following way. Keep x (and so P) fixed and consider a succession of points P_n (x_n, y_n) on the graph, $n = 1, 2, 3, \ldots$, approaching P (Fig. 4.3).

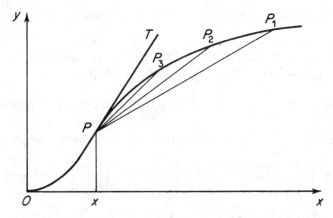

Fig. 4.3 Graph of $y = f(x)$ with chords PP_n and tangent PT.

In the time interval $(x_n - x)$ hours, the car moves $(y_n - y)$ kilometres, so its average speed in km p h is $(y_n - y)/(x_n - x)$, the gradient of the chord PP_n. As P_n approaches P along the graph, the time interval $x_n - x$ becomes very small and the speed in this time interval approaches the instantaneous speed at P. Also, as P_n approaches P, the chord PP_n approaches the tangent PT to the graph at P. Thus, the gradient of the tangent PT will give the speed at P, that is, x hours after the start.

EXERCISE 4.2

1   The distance travelled by a racing car in x seconds after starting is y metres where $y = \frac{1}{4}x^2 \,(64 - x^2)$.
(a) Draw the distance-time graph for $x = 0$ to $x = 5$.
(b) Draw a sequence of chords $PP_1, PP_2, \ldots$, where $P_n\,(x_n, y_n)$ is approaching the point $P\,(2, 60)$ with $x_n > 2$, and calculate the gradients of these chords.
(c) Repeat (b) for chords PP_n, where $x_n < 2$.
(d) Use the above calculations to estimate the speed of the car 2 seconds after starting.

4.3 The gradient of a function

The *gradient* of the function f at the point $P(x, f(x))$ is defined to be the slope (or gradient) of the tangent at P to the graph of f and is denoted by $f'(x)$.

If $y = f(x)$ then $f'(x)$ is denoted by $\dfrac{dy}{dx}$ and is called the *derivative* of y with respect to x or, alternatively, the *differential coefficient* of y with respect to x.

In the case of a linear function f (see §4.1), $f'(x) = \dfrac{dy}{dx}$ is the rate of change of y with respect to x. For a general function, $\dfrac{dy}{dx}$ is the instantaneous rate of change of y with respect to x, but the word 'instantaneous' is usually omitted. The speed of the car in §4.2 at a time x hours after the start of the journey is $f'(x)$ km p h.

Note: the symbol $\dfrac{dy}{dx}$ is read as 'dee y by dee x' and is *not a fraction*. In fact, the three letters $\dfrac{d}{dx}$ form a differential operator which changes y into its derivative. Thus, when $y = 2x^2 + x$ we write $\dfrac{dy}{dx} = \dfrac{d}{dx}\,(2x^2 + x)$.

4.4 Estimating the gradient of a function using the tangent to its graph

If the graph of the function f is carefully drawn, the value of $f'(x)$ can be estimated by drawing a tangent to the graph at $(x, f(x))$ and measuring its slope.

EXAMPLE *Draw a graph of the function* f, *with domain* $-2 \leqslant x \leqslant 4$ *and* $f(x) = x^3$. *Draw a tangent to the graph at each point for which* $x \in \{-1, 0, 1, 2, 3\}$ *and estimate the values of* $f'(x)$, $x \in \{-1, 0, 1, 2, 3\}$. *Compare these estimates with the corresponding values of* $g(x)$, *where* $g(x) = 3x^2$, *by drawing the graph of* g *and on it marking the points* $(x, f'(x))$, $x \in \{-1, 0, 1, 2, 3\}$.

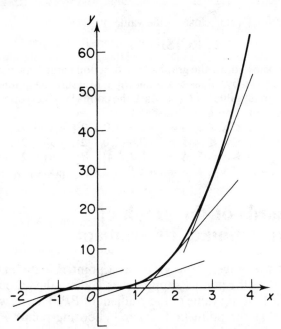

(a) Graph of x^3 with tangents used to measure gradient

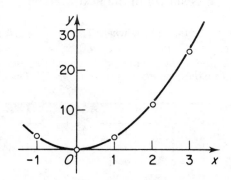

(b) Graph of $3x^2$, with measured values of the gradient of x^3

Fig. 4.4

The results are displayed in the form of a table overleaf and Fig. 4.4.

x	-2	-1	0	1	2	3	4
$f(x) = x^3$	-8	-1	0	1	8	27	64
Gradient of tangent $= f'(x)$ (from graph)	—	3.2	0	3·1	11·3	25	—
$g(x) = 3x^2$	—	3	0	3	12	27	—

Note that the estimate for $f'(x)$ is close to the value of $g(x)$.

EXERCISE 4.4

1   In each of the cases below, draw the graph of the function f and draw tangents at each of the points given by values of x in the set S. Measure the gradients of these tangents to estimate values of $f'(x)$. Mark the points $(x, f'(x))$ on a graph of $g(x)$ for the given function g.
  (i) $f(x) = x^2$, $-3 \leqslant x \leqslant 3$, $S = \{-2, -1, 0, 1, 2\}$, $g(x) = 2x$;
  (ii) $f(x) = x^2 - 4$, $-3 \leqslant x \leqslant 3$, $S = \{-2, -1, 0, 1, 2\}$, $g(x) = 2x$;
  (iii) $f(x) = x^2 - 4x$, $-1 \leqslant x \leqslant 5$, $S = \{0, 1, 2, 3, 4\}$, $g(x) = 2x - 4$;
  (iv) $f(x) = x^3 - 3x^2$, $-2 \leqslant x \leqslant 4$, $S = \{-1, 0, 1, 2, 3\}$, $g(x) = 3x^2 - 6x$.

4.5 Use of chords of the graph of a function to estimate its gradient

We saw in §4.3 that the tangent to a graph at a point P is the limiting position of chords PP_n as the point P_n approaches P along the graph, and the gradient at P is the limiting value of the gradient of PP_n. Hence, we can find $f'(x)$ by taking P as the point $(x, f(x))$ and choosing points P_1, P_2, P_3, ... near to P on the graph and plotting a graph of the slopes of the chords PP_1, PP_2, PP_3, ... as is shown in the next example.

EXAMPLE   *On the graph of the function f, where $f(x) = x^3$, mark the points $P(x, f(x))$, $P_n(x_n, f(x_n))$, where $x = 2$, $x_1 = -1$, $x_2 = 0$, $x_3 = 1$, $x_4 = 3$, $x_5 = 4$. For each of the five values of n, find y_n the gradient of the chord PP_n and on another graph plot the points. (x_n, y_n). Join these points by a smooth curve and hence estimate $f'(x)$; the gradient of f at P.*

Point	P	P_1	P_2	P_3	P_4	P_5
x	2	-1	0	1	3	4
$f(x) = x^3$	8	-1	0	1	27	64
$f(x) - f(2)$	0	-9	-8	-7	19	56
$x - 2$	0	-3	-2	-1	1	2
Gradient PP_n $\dfrac{f(x) - f(2)}{x - 2}$	—	3	4	7	19	28

The graph of the gradient of PP_n passes through $(2, 12)$, so $f'(2) = 12$, as can be seen in Fig. 4.5.

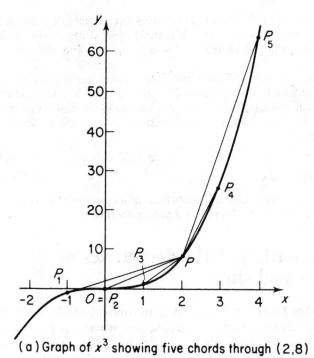

(a) Graph of x^3 showing five chords through (2,8)

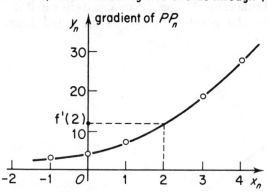

(b) Graph of the gradient of PP_n plotted against x_n

Fig. 4.5

EXERCISE 4.5

1   Repeat the above example taking P to be the given point and P_1, P_2, P_3, P_4, P_5, to be the other 5 points from the set of points $\{(-1, -1), (0, 0), (1, 1), (2, 8), (3, 27)\}$ for the following:
   (i) P is $(-1, -1)$;   (ii) P is $(0, 0)$;   (iii) P is $(1, 1)$;   (iv) P is $(3, 27)$.
2   Repeat the above example using the functions given in Exercise 4.4, (i), (ii), (iii) and (iv), and using the x-coordinate chosen as each member in turn of the

appropriate set S. This exercise involves five choices of P for each of four functions, that is, twenty cases. If a group of students are doing the exercise, the labour could be shared out with advantage and the results collected together.

3　The results of Exercise 4.4 and question **2** above should give estimates of $f'(x)$ which are close to the values of $g(x)$ in each case. The method of using the gradients of chords is more accurate than the method of measuring slopes of the tangents. The values of the functions used in these exercises are now tabulated:

$f(x)$:	x^2	x^3	$x^2 - 4$	$x^2 - 4x$	$x^3 - 3x^2$
$g(x)$:	$2x$	$3x^2$	$2x$	$2x - 4$	$3x^2 - 6x$

Consider the above table and given that for all the entries $g(x) = f'(x)$, try and discover a rule for finding $f'(x)$ when given $f(x)$.

4.6　Estimation of the derivatives of cos x and sin x

The method of §4.4 is used to estimate the derivative of $\cos x$ by measuring the gradients to the graph of the function at various points. Remember that x is always measured in radians when dealing with trigonometrical functions as real functions. Values of $\cos x$ and $\sin x$ and the gradients of some tangents to the graph of $\cos x$, as measured from Fig. 4.6 are tabulated.

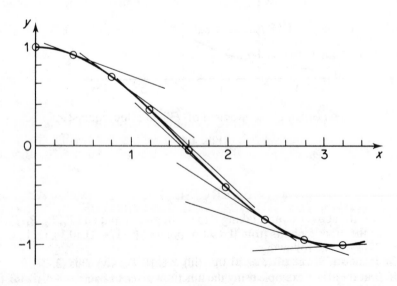

Fig. 4.6 Tangents to the graphs of cos x.

x	0	0·2	0·4	0·6	0·8	1·0	1·2	1·4	1·6
cos x	1·00	0·98	0·92	0·83	0·70	0·54	0·36	0·17	−0·03
sin x	0	0·20	0·39	0·56	0·72	0·84	0·93	0·9	1·00
Gradient of tangent to graph of cos x	0		−0·37		−0·73		−0·89		−1·00

x	1·8	2·0	2·2	2·4	2·6	2·8	3·0	3·2	3·4
cos x	−0·23	−0·42	−0·59	−0·74	−0·86	−0·94	−0·99	−1·00	−0·97
sin x	0·97	0·91	0·81	0·68	0·52	0·33	0·14	−0·06	−0·26
Gradient of tangent to graph of cos x		−0·92		−0·62		−0·31		−0·07	

EXERCISE 4.6

1 Using the above table, draw the graph of sin x and draw tangents to this graph at intervals of 0·4 radians. Measure the slopes of these tangents to estimate $\dfrac{d}{dx}$ sin x and compare the values found with the corresponding values of cos x.

The above results should suggest that $\dfrac{d}{dx}\cos x = -\sin x$ and $\dfrac{d}{dx}\sin x = \cos x$, and this is the case. In elementary kinematics, the derivatives of cos nx and sin nx, where n is constant, are also required. Since $\dfrac{dy}{dx}$ is the rate of change of y with respect to x and is therefore a (local) change of scale between y and x (cf. §4.1), the replacement of x by nx gives a further change of scale and so we may expect that $\dfrac{d}{dx}f(nx) = nf'(x)$, which is, in fact, the case.

The addition of a constant to x changes the position of the graph of a function with respect to the origin but does not change the shape of the graph, so it leaves the derivative of a function unaltered, that is, if $f'(x) = g(x)$ then $f'(x+c) = g(x+c)$.

4.7 Rules for finding the derivatives of a function

The previous sections suggest that the following rules are plausible. They are stated without proof in order that they can be used in examples. Proof of these rules will be given in Chapter 18.

Rule 1 The derivative of a linear combination of functions is the corresponding linear combination of their derivatives. If a and b are constants and f and g are functions then

$$\text{if } h = af + bg \quad \text{then} \quad h' = af' + bg',$$

$$\text{if } y = au + bv \quad \text{then} \quad \frac{dy}{dx} = a\frac{du}{dx} + b\frac{dv}{dx}.$$

Rule 2 For any constant n,

$$\text{if } f(x) = x^n \text{ then } f'(x) = nx^{n-1},$$

$$\text{if } y = x^n \quad \text{then} \quad \frac{dy}{dx} = nx^{n-1},$$

$$\frac{d}{dx}\cos(nx + c) = -n\sin(nx + c),$$

$$\frac{d}{dx}\sin(nx + c) = n\cos(nx + c).$$

These rules may be used to find the derivatives of many functions.

EXAMPLE 1   *Find the derivative f' of the function f where $f(x) = 3x^2 - 2x + 1$.*

Using rule 2, $\dfrac{d}{dx}x^2 = 2x, \dfrac{d}{dx}x = 1, \dfrac{d}{dx}1 = \dfrac{d}{dx}x^0 = 0$, so by rule 1,
$f'(x) = 3(2x) - 2 = \mathbf{6x - 2}$.

EXAMPLE 2   *Differentiate the function $(x - 1/x)^2$.*

Expanding the square, $(x - 1/x)^2 = x^2 - 2 + x^{-2}$, and by rule 2 $\dfrac{d}{dx}x^{-2} = -2x^{-3}$.
Hence,

$$\frac{d}{dx}\left(x - \frac{1}{x}\right)^2 = 2x - 0 - 2x^{-3} = \mathbf{2x - \frac{2}{x^3}}.$$

EXAMPLE 3   *Find the equation of the tangent to the curve given by $y = x + \sin x$ at the point $(\pi, 0)$.*

Using the rules,

$$\frac{dy}{dx} = 1 + \cos x,$$

so the gradient of the tangent when $x = \pi$ is given by

$$m = 1 + \cos\pi = 0.$$

Thus the equation of the tangent is $y = \pi$.

EXAMPLE 4 *The displacement of a particle moving in a straight line is y metres after it has been moving t seconds, where $y = f(t) = t^3 - 2t^2 + t$. Find the time elapsed and the position of the particle at each instant when the particle is at rest. During what interval of time is the particle moving backwards towards its initial position?*

In our previous formula, x is now replaced by t, so the speed v of the particle at time t is given by

$$v = \frac{dy}{dt} = 3t^2 - 4t + 1,$$

on using the two rules. Hence

$$v = (3t - 1)(t - 1)$$

and $v = 0$, that is, the particle is stationary, at times $t = \frac{1}{3}$ and $t = 1$, that is, after the particle has been moving for **1/3 second** and **1 second**. The particle moves backwards when $v < 0$ and this occurs **between 1/3 second and 1 second** after the motion starts.

EXERCISE 4.7

1 Differentiate with respect to x:
  (i) x^5, (ii) x^{-4}, (iii) $2x^7$, (iv) $3/x$, (v) $2/x^2$, (vi) $x^2 + 2x$, (vii) $x^2 + 3$,
  (viii) $1 + 3x + 2x^2$, (ix) $3x^3 + 2x$, (x) $x^4 - 4x^3 + 2x^2$, (xi) $2 - 3x - 4x^2$,
  (xii) $x^5 - 5x^4$, (xiii) $2 \sin x - 3 \cos x$, (xiv) $\cos x - \sin x$, (xv) $\sin 3x$.
2 Expand and then differentiate with respect to x:
  (i) $(2x - 3)^2$, (ii) $(x + 1)(x - 2)$, (iii) $x^2(2x + 3)$, (iv) $(x^2 + 1)/x$,
  (v) $(x^2 + 1)(x^2 - 1)$, (vi) $\left(x + \dfrac{1}{x}\right)^2$.
3 Find $f'(x)$ when:
  (i) $f(x) = x^3 - 2x + 1$, (ii) $f(x) = 2 \cos 3x + x$, (iii) $f(x) = (x + 1)^2/x$,
  (iv) $f(x) = 2x^3 - 5x$, (v) $f(x) = 4x - 2 + x^{-2}$, (vi) $f(x) = 2x - \cos 2x$,
  (vii) $f(x) = 3x^2 - 2x^4$, (viii) $f(x) = (2x + 1)^3$, (ix) $f(x) = 2x + \dfrac{2}{x}$,
  (x) $f(x) = \left(x - \dfrac{1}{x}\right)^3$.

4 (i) Given that $v = 2 + t - 3t^2$, find $\dfrac{dv}{dt}$;

   (ii) Given that $V = \frac{4}{3}\pi r^3$, find $\dfrac{dV}{dr}$;

   (iii) Given that $s = ut + \frac{1}{2}at^2$, find $\dfrac{ds}{dt}$, u, a are constants;

   (iv) Given that $PV = 25$, find $\dfrac{dP}{dV}$.

5 Find the gradient of the tangent to the curve $y = 2x^3 - 3x^2$ at the point where
   (i) $x = 2$, (ii) $x = 1$, and, hence, write down the equation of the tangent at the
   given point.

6 Find the equation of the tangent to the curve $y = f(x)$ at the point where
   $x = b$, given that:
   (i) $f(x) = x^2 - 2x - 1$, $b = 2$;   (ii) $f(x) = (x - 1)^3$, $b = -1$;
   (iii) $f(x) = \left(x + \dfrac{1}{x}\right)^2$, $b = 1$.

7 The displacement of a particle, moving in a straight line, is x metres after it has
   been moving for t seconds, where

$$x = f(t) = 2t^2 - 10t.$$

   Find the speed of the particle when $t = 0, 1, 2, 3$ and 4. Find at what time the
   particle is stationary, at what time it returns to its starting point and at what
   speed it is then travelling.

8 The height of the tide x hours after midnight is h metres, where

$$h = 8 + 5 \cos \frac{\pi}{12}(x + 3).$$

   Find, in terms of x, the rate of change of height of the tide at time x. Find also
   the maximum height of the tide.

9 A particle P moves in a horizontal straight line and passes through a fixed
   point O. After t seconds the distance s metres of P from O is given by

$$s = 12t - t^3, \text{ where } t \geqslant 0.$$

   Calculate
   (a) the value of t when P is instantaneously at rest,
   (b) the speed of P at the instant when it again passes through O.   (L)

5 Vectors I

In this chapter, we consider the elementary algebra of vectors, based on vector addition and multiplication by a scalar, with the vectors represented by displacements. In Chapter 12, Vectors II, we shall consider units of length and direction, Cartesian frames of reference and column vectors.

5.1 Displacements

Suppose that we wish to describe a movement, or *displacement*, from a point P to a point Q. It is necessary to specify two things:

(i) a distance PQ explained by reference to a known unit of length,
(ii) a direction explained by reference to a known set of directions.

For example, if the known directions are compass bearings, then we could describe a displacement by (a) 3 km North West or (b) 200 miles on a bearing 060°.

It is useful to have a notation which combines the two pieces of information, distance and direction, and we use the arrow notation:
$\overrightarrow{PQ}$ meaning a displacement through the distance PQ in the direction of PQ.
$\overrightarrow{PQ}$ is a vector quantity and the distance PQ is called the *magnitude* of the vector $\overrightarrow{PQ}$ (see Fig. 5.1 (a)).

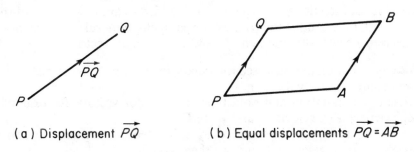

(a) Displacement $\overrightarrow{PQ}$ (b) Equal displacements $\overrightarrow{PQ} = \overrightarrow{AB}$

Fig. 5.1

Two displacements are *equal* when they have the same magnitude and the same direction. Thus in Fig. 5.1 (b) $\overrightarrow{PQ} = \overrightarrow{AB}$ when $PQ = AB$ and PQ is parallel to AB, that is, when $PQBA$ is a parallelogram.

If a displacement $\overrightarrow{PQ}$ is followed by another $\overrightarrow{QR}$, then the total displacement is $\overrightarrow{PR}$ and this is defined as the sum of $\overrightarrow{PQ}$ and $\overrightarrow{QR}$, thus

$$\overrightarrow{PQ} + \overrightarrow{QR} = \overrightarrow{PR}.$$

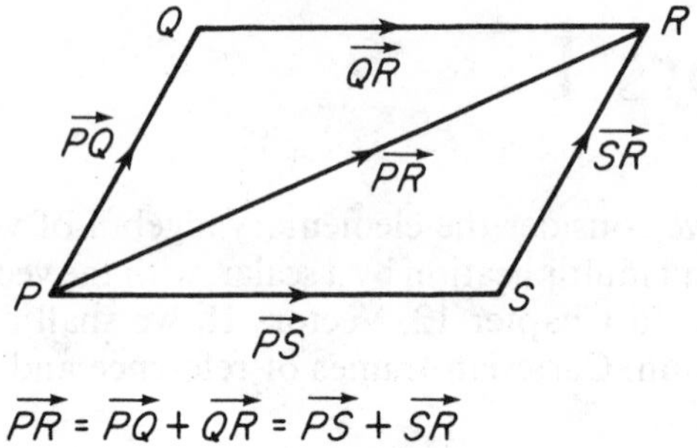

$$\overrightarrow{PR} = \overrightarrow{PQ} + \overrightarrow{QR} = \overrightarrow{PS} + \overrightarrow{SR}$$

Fig. 5.2

Let $PQRS$ be a parallelogram, then $\overrightarrow{PQ} = \overrightarrow{SR}$ and $\overrightarrow{PS} = \overrightarrow{QR}$, hence in Fig. 5.2.

$$\overrightarrow{PR} = \overrightarrow{PQ} + \overrightarrow{QR} = \overrightarrow{PS} + \overrightarrow{SR}$$

The *sum* of two displacements is equal to the third side of a triangle whose two sides represent the given displacements (*the triangle rule*).

The *sum* is also represented by the diagonal of a parallelogram whose edges represent the given displacements (*the parallelogram rule*).

5.2 Vector and scalar quantities

A scalar quantity or *scalar* is a quantity which is completely specified by its magnitude, in terms of a given unit. For example, mass, distance, speed, area, volume and energy are all scalars.

A vector quantity or *vector* is a quantity which is completely specified by its magnitude in terms of a given unit and its direction relative to some agreed set of directions, and which can be added to another vector, of the same kind, by means of the triangle, or parallelogram, rule. For example, displacement, velocity, momentum and force are all vectors.

Notation   Italic letters are used for scalars, for example mass m, distance s, speed v, area S and volume V.

The arrow notation and bold type are used for vectors, for example, displacement $\overrightarrow{PQ}$, velocity $\mathbf{v}$ and force $\mathbf{F}$.

A vector, $\mathbf{a}$, may be represented by a directed line segment $\overrightarrow{PQ}$, with respect to some given unit, and we then write $\mathbf{a} = \overrightarrow{PQ}$.

Equality of vectors

Let $\mathbf{a} = \overrightarrow{PQ}$ and $\mathbf{b} = \overrightarrow{RS}$ be two vectors, then $\mathbf{a} = \mathbf{b}$ means that $\overrightarrow{PQ}$ and $\overrightarrow{RS}$ have the same magnitude and direction so that $PQSR$ is a parallelogram, as seen in Fig. 5.3(a).

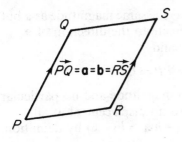

(a) Equal vectors

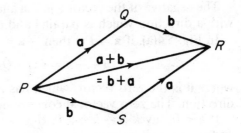

(b) Vector addition

Fig. 5.3

Addition of vectors

In Fig. 5.3(b), let $\mathbf{a} = \overrightarrow{PQ}$ and $\mathbf{b} = \overrightarrow{QR}$, then we add $\mathbf{a}$ and $\mathbf{b}$ by the triangle rule

$$\mathbf{a} + \mathbf{b} = \overrightarrow{PQ} + \overrightarrow{QR} = \overrightarrow{PR}.$$

Let $PQRS$ be a parallelogram, then $\mathbf{a} = \overrightarrow{PQ} = \overrightarrow{SR}$ and $\mathbf{b} = \overrightarrow{QR} = \overrightarrow{PS}$ so that

$$\mathbf{a} + \mathbf{b} = \overrightarrow{PQ} + \overrightarrow{QR} = \overrightarrow{PR} = \overrightarrow{PS} + \overrightarrow{SR} = \mathbf{b} + \mathbf{a}$$

and so we see that vector addition is commutative, $\mathbf{a} + \mathbf{b} = \mathbf{b} + \mathbf{a}$.

Multiplication of a vector by a scalar

Suppose that $\overrightarrow{PQ} = \mathbf{a} = \overrightarrow{QR}$ (Fig. 5.4(a)); then Q is the midpoint of PR. Thus

$$\overrightarrow{PR} = \overrightarrow{PQ} + \overrightarrow{QR} = \mathbf{a} + \mathbf{a} = 2\mathbf{a},$$

and so $2\mathbf{a}$ is the vector with magnitude twice the magnitude of $\mathbf{a}$ and with the same direction as $\mathbf{a}$. We extend this idea and define, for a vector $\mathbf{a}$ and a scalar $k, k > 0$:

$k\mathbf{a}$ is the vector with the same direction as $\mathbf{a}$ and of magnitude k times the magnitude of $\mathbf{a}$.

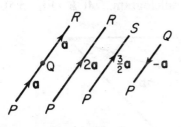

(a) Product by scalar

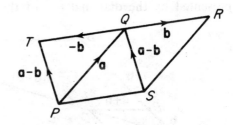

(b) Vector difference

Fig. 5.4

The negative of the vector **a** is − **a** and has the same magnitude as **a** but with a direction which is parallel and opposite to the direction of **a**.

In Fig. 5.4(a), if $\mathbf{a} = \overrightarrow{PQ}$ then $-\mathbf{a} = \overrightarrow{QP}$ and

$$\mathbf{a} + -\mathbf{a} = \overrightarrow{PQ} + \overrightarrow{QP} = \overrightarrow{PP} = \mathbf{0},$$

where **0** is the zero vector, which has zero magnitude and no particular direction. The zero vector **0** corresponds to no displacement at all.

If $k < 0$, say $k = -h, h > 0$, then $k\mathbf{a} = (-h)\mathbf{a} = h(-\mathbf{a})$ by definition, and then it will be seen that

$$h(-\mathbf{a}) = -(h\mathbf{a}).$$

Subtraction of vectors

We define $\mathbf{a} - \mathbf{b} = \mathbf{a} + -\mathbf{b}$ and, therefore, in Fig. 5.4(b), if $\overrightarrow{PQ} = \mathbf{a}$ and $\overrightarrow{QR} = \mathbf{b}$, let $\overrightarrow{QT} = -\mathbf{b} = \overrightarrow{RQ}$ so that Q is the midpoint of TR. Then

$$\mathbf{a} - \mathbf{b} = \overrightarrow{PQ} - \overrightarrow{QR} = \overrightarrow{PQ} + \overrightarrow{QT} = \overrightarrow{PT} = \overrightarrow{SQ}.$$

Thus the sum and difference of two vectors are represented by the two diagonals of the parallelogram whose sides represent the two vectors.

Distributive and associative laws

Since multiplication by a positive scalar only changes the magnitude of a vector and multiplication by a negative scalar changes the magnitude and reverses the direction, the associative law

$$h(k\mathbf{a}) = (hk)\mathbf{a}$$

holds for any vector **a** and scalars h, k.

Similarly, one distributive law follows at once from the definition

$$h\mathbf{a} + k\mathbf{a} = (h + k)\mathbf{a}.$$

The other distributive law that, for any scalar h and vectors **a**, **b**,

$$h\mathbf{a} + h\mathbf{b} = h(\mathbf{a} + \mathbf{b})$$

is proved by using corresponding displacements. Let **a** and **b** be represented by displacements $\overrightarrow{OA}$ and $\overrightarrow{OB}$ so that $\mathbf{a} + \mathbf{b}$ is represented by the diagonal $\overrightarrow{OC}$ of the parallelogram $OACB$ (Fig. 5.5).

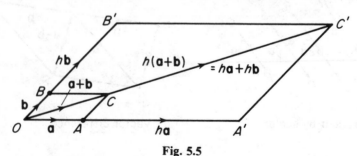

Fig. 5.5

Multiplication by h corresponds to an enlargement by a scale factor h, $h\mathbf{a}$ and $h\mathbf{b}$ being represented by $\overrightarrow{OA'}$ and $\overrightarrow{OB'}$ and then $h\mathbf{a} + h\mathbf{b}$ is represented by

$$\overrightarrow{OA'} + \overrightarrow{OB'} = \overrightarrow{OC'} = h(\mathbf{a} + \mathbf{b})$$

since $OA'C'B'$ is a parallelogram obtained from $OACB$ by an all-round enlargement with scale factor h.

The proof of the associative law of addition is given as question 1 in Exercise 5.2. It states that for any three vectors $\mathbf{a}$, $\mathbf{b}$ and $\mathbf{c}$,

$$\mathbf{a} + (\mathbf{b} + \mathbf{c}) = (\mathbf{a} + \mathbf{b}) + \mathbf{c}$$

so we may omit the brackets and write the sum $\mathbf{a} + \mathbf{b} + \mathbf{c}$.

EXERCISE 5.2

1   Given three vectors $\mathbf{a}$, $\mathbf{b}$ and $\mathbf{c}$, represent them on a diagram by $\overrightarrow{PQ} = \mathbf{a}$, $\overrightarrow{QR} = \mathbf{b}$, $\overrightarrow{RS} = \mathbf{c}$, and hence show that

$$\mathbf{a} + (\mathbf{b} + \mathbf{c}) = (\mathbf{a} + \mathbf{b}) + \mathbf{c}.$$

2   Two sets of equally spaced parallel lines are shown in Fig. 5.6 with a number of their intersections labelled by letters. Represent each of the following vectors as a displacement from O given that $\mathbf{a} = \overrightarrow{OA}$ and $\mathbf{b} = \overrightarrow{OB}$; for example, $\mathbf{a} + \mathbf{b} = \overrightarrow{OC}$:

(i) $2\mathbf{a}$,   (ii) $3\mathbf{b}$,   (iii) $2\mathbf{a} + \mathbf{b}$,   (iv) $3\mathbf{a} + 2\mathbf{b}$,   (v) $-\mathbf{b}$,   (vi) $-\mathbf{a} + 3\mathbf{b}$, (vii) $2\mathbf{a} + -\mathbf{b}$,   (viii) $4\mathbf{a} + \mathbf{b}$,   (ix) $\mathbf{b} - \mathbf{a}$.

3   Using the same diagram, Fig. 5.6, as in question 2 above, express in terms of

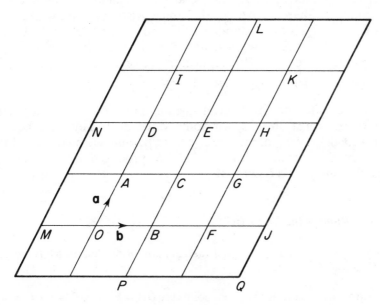

Fig. 5.6

a and **b** the vector represented by the following displacements:
(i) $\overrightarrow{AI}$, (ii) $\overrightarrow{CL}$, (iii) $\overrightarrow{AG}$, (iv) $\overrightarrow{CK}$, (v) $\overrightarrow{PJ}$, (vi) $\overrightarrow{PA}$, (vii) $\overrightarrow{AQ}$,
(viii) $\overrightarrow{GO}$, (ix) $\overrightarrow{JM}$, (x) $\overrightarrow{KP}$.

4 Again using the same diagram, Fig. 5.6, as in question **2**, write down three displacements equal to the following vectors, for example, $\mathbf{a}+2\mathbf{b}=\overrightarrow{OG}=\overrightarrow{MC}=\overrightarrow{DK}$:
(i) $\mathbf{a}+\mathbf{b}$, (ii) $\mathbf{b}+-\mathbf{a}$, (iii) $-2\mathbf{a}+-\mathbf{b}$, (iv) $-\mathbf{a}-2\mathbf{b}$.

5 Let $\overrightarrow{OS}=\mathbf{s}$, $\overrightarrow{OT}=\mathbf{t}$ and let T be the mid-point of SU. Find, in terms of **s** and **t**:
(i) $\overrightarrow{ST}$, (ii) $\overrightarrow{TS}$, (iii) $\overrightarrow{OU}$, (iv) $\overrightarrow{US}$, (v) $\overrightarrow{TU}$.

5.3 Origin and position vectors

Again, refer to Fig. 5.6. In Exercise 5.2, several questions involved the point O which was used as a reference point for displacements. Thus $\overrightarrow{OA}=\mathbf{a}$ gives the position of the point A as a displacement **a** from the reference point O. A point O, used in this way as a reference point, is called the *origin* and **a** is the *position vector* for the point A. In other words, once an origin O is fixed, the position of every other point R in space can be defined by a unique position vector $\mathbf{r}=\overrightarrow{OR}$.

Thus, in Fig. 5.6, the position vectors of F, G, H and K are given by

$$\overrightarrow{OF}=2\mathbf{b},\quad \overrightarrow{OG}=2\mathbf{b}+\mathbf{a},\quad \overrightarrow{OH}=2\mathbf{b}+2\mathbf{a},\quad \overrightarrow{OK}=2\mathbf{b}+3\mathbf{a}.$$

Given a scalar parameter λ, ($\lambda\in\mathbb{R}$), consider the set of points

$$S=\{R:\overrightarrow{OR}=\mathbf{r}=2\mathbf{b}+\lambda\mathbf{a}\}.$$

Clearly, the points F, G, H and K are all elements of the set S, the corresponding values of λ being 0, 1, 2 and 3. Also, the point with position vector $\mathbf{r}=2\mathbf{b}+\frac{1}{2}\mathbf{a}$ is the midpoint of FG, and $\mathbf{r}=2\mathbf{b}+2\frac{3}{4}\mathbf{a}$ is the position vector of the point R between H and K such that $\overrightarrow{HR}=\frac{3}{4}\overrightarrow{HK}$.

In general, if the point R has position vector $\mathbf{r}=2\mathbf{b}+\lambda\mathbf{a}$, then

$$\overrightarrow{FR}=\overrightarrow{OR}-\overrightarrow{OF}=\mathbf{r}-2\mathbf{b}=\lambda\mathbf{a}=\lambda\overrightarrow{FG},$$

and R lies on the line FG. Conversely, any point on FG has a position vector of the form $2\mathbf{b}+\lambda\mathbf{a}$, for some value of λ. Thus, the equation $\mathbf{r}=2\mathbf{b}+\lambda\mathbf{a}$ defines an infinite set of points, parameterised by λ, and these points form the *line* through F, G, H and K, produced in both directions, that is, the line FG. We say that

$$\mathbf{r}=2\mathbf{b}+\lambda\mathbf{a}$$

is *the equation of the line FG*.

EXAMPLE 1 *Using Fig. 5.6, find the equation of the line through the points P, B, C, E and L.*

The position vectors of the five given points are $\overrightarrow{OP}=\mathbf{b}-\mathbf{a}$, $\overrightarrow{OB}=\mathbf{b}$, $\overrightarrow{OC}=\mathbf{b}+\mathbf{a}$, $\overrightarrow{OE}=\mathbf{b}+2\mathbf{a}$, $\overrightarrow{OL}=\mathbf{b}+4\mathbf{a}$. Hence the equation of the line PL is $\mathbf{r}=\mathbf{b}+\lambda\mathbf{a},\lambda\in\mathbf{R}$.

EXAMPLE 2   *Again with reference to Fig. 5.6, find the locus of points with position vector* **r**, *satisfying the equation*

$$\mathbf{r} = \mathbf{a} + \lambda(\mathbf{a} + \mathbf{b}), \lambda \in \mathbb{R}.$$

Consider the values of **r** for some values of λ:
if $\lambda = 0$, $\mathbf{r} = \mathbf{a}$ so A lies on the locus;
if $\lambda = 1$, $\mathbf{r} = \mathbf{a} + (\mathbf{a} + \mathbf{b}) = 2\mathbf{a} + \mathbf{b}$ so E lies on the locus;
if $\lambda = 2$, $\mathbf{r} = \mathbf{a} + 2(\mathbf{a} + \mathbf{b}) = 3\mathbf{a} + 2\mathbf{b}$ so K lies on the locus,
if $\lambda = -1$, $\mathbf{r} = \mathbf{a} - (\mathbf{a} + \mathbf{b}) = -\mathbf{b}$ so M lies on the locus.
We know that the equation is the equation of a line which must be the line through
M, A, E and **K**.

Note: The equation $\mathbf{r} = \mathbf{a} + \lambda(\mathbf{a} + \mathbf{b})$, of Example 2, has two main parts in it. Firstly, the part independent of λ, obtained by putting $\lambda = 0$, gives a value $\mathbf{r} = \mathbf{a} = \overrightarrow{OA}$, so that A lies on the line. The second part, which contains the factor λ, is $\lambda(\mathbf{a} + \mathbf{b})$, which shows that the line is in the direction of the vector $(\mathbf{a} + \mathbf{b}) = \overrightarrow{OC} = \overrightarrow{AE}$. Similarly, in Example 1, the term independent of λ gives $\mathbf{r} = \mathbf{b}$, showing that B lies on the line. The other term $\lambda\mathbf{a}$ shows that the line is parallel to $\mathbf{a} = \overrightarrow{OA} = \overrightarrow{BC}$.

EXAMPLE 3   *Using Fig. 5.6, write down one point on the line with equation* $\mathbf{r} = 2\mathbf{a} + \mathbf{b} + \lambda\,(\mathbf{b} - \mathbf{a})$ *and find a vector displacement which is parallel to the line.*

The line passes through the point E with position vector $\mathbf{r} = 2\mathbf{a} + \mathbf{b} = \overrightarrow{OE}$, corresponding to $\lambda = 0$. The term $\lambda(\mathbf{b} - \mathbf{a})$ in the equation shows that the line is parallel to $\mathbf{b} - \mathbf{a} = \overrightarrow{OP}$.

EXERCISE 5.3

1 Use the diagram of Fig. 5.7 to find the equation of the line which passes through the following set of points:
   (i) $\{E, C, F, G\}$; (ii) $\{G, H, I\}$; (iii) $\{G, J, L, K\}$; (iv) $\{M, H, L\}$;
   (v) $\{J, F, H\}$; (vi) $\{E, N\}$.
2 Again using Fig. 5.7, find the locus of the points R, with position vector **r** satisfying the equation:
   (i) $\mathbf{r} = 2\mathbf{c} + \mathbf{d} + \lambda\mathbf{c}, \lambda \in \mathbb{R}$; (ii) $\mathbf{r} = \mathbf{c} + 2\mathbf{d} + \lambda(\mathbf{c} + \mathbf{d}), \lambda \in \mathbb{R}$;
   (iii) $\mathbf{r} = \lambda(3\mathbf{c} + 2\mathbf{d}), \lambda \in \mathbb{R}$; (iv) $\mathbf{r} = \mathbf{c} + \mu(2\mathbf{d} - \mathbf{c}), \mu \in \mathbb{R}$;
   (v) $\mathbf{r} = \mathbf{d} + k(\mathbf{c} - \mathbf{d}), k \in \mathbb{R}$.
3 Let A and B be two points, with position vectors **a** and **b** respectively, with respect to an origin O. Find the equation of the line:
   (i) through A parallel to $\overrightarrow{OB}$, (ii) through O parallel to $\overrightarrow{BA}$, (iii) through the midpoint of OB parallel to $\overrightarrow{OA}$, (iv) through A and B.
4 The points P and Q have position vectors **p** and **q**, relative to an origin O. Describe the locus of the points R with position vectors **r**, given by $\mathbf{r} = \lambda\mathbf{p} + \mu\mathbf{q}$, when:
   (i) $\lambda = \mu$; (ii) $\lambda = 1$, $\mu \in \mathbb{R}$; (iii) $\lambda = -\mu$; (iv) $\lambda + \mu = 1$; (v) there is no restriction on the values of λ and μ, which take all real values.

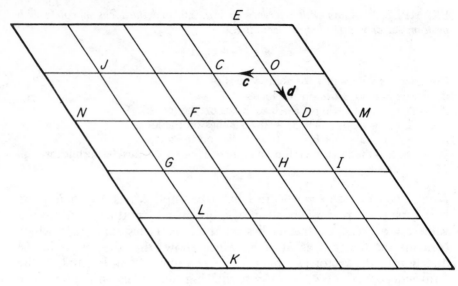

Fig. 5.7

5 (i) Three points P, Q and R have position vectors $\mathbf{p}$, $\mathbf{q}$ and $k(2\mathbf{p}+\mathbf{q})$ respectively, relative to a fixed origin. Find the numerical value of k if
   (a) $\overrightarrow{QR}$ is parallel to $\mathbf{p}$,
   (b) $\overrightarrow{PR}$ is parallel to $\mathbf{q}$,
   (c) P, Q and R are collinear.
   (ii) A hexagon $OABCDE$ has its pairs of opposite sides parallel and equal. The position vectors of A, B and C relative to O are $\mathbf{a}$, $\mathbf{a}+\mathbf{b}$, $\mathbf{a}+\mathbf{b}+\mathbf{c}$ respectively. Find the position vectors of D and E relative to O.   (L)
6 The points P and Q have position vectors $\mathbf{p}$ and $\mathbf{q}$ respectively relative to an origin O, which does not lie on PQ. Three points R, S, T have respective position vectors $\mathbf{r}=\frac{1}{4}\mathbf{p}+\frac{3}{4}\mathbf{q}$, $\mathbf{s}=2\mathbf{p}-\mathbf{q}$, $\mathbf{t}=\mathbf{p}+3\mathbf{q}$. Show in one diagram the positions of O, P, Q, R, S and T.   (JMB)
7 The points A, B, C have position vectors $\mathbf{a}$, $\mathbf{b}$, $\mathbf{c}$ respectively with respect to an origin O. Write down, in terms of $\mathbf{a}$ and $\mathbf{b}$, the vector $\overrightarrow{BA}$ and find the position vector of the fourth vertex D of the parallelogram $ABCD$.   (L)

5.4 The ratio theorem

Let the two points A and B have position vectors $\mathbf{a}$ and $\mathbf{b}$ with respect to an origin O and let M be the midpoint of AB with position vector $\mathbf{m}$ (Fig. 5.8(a)). Then $\overrightarrow{AB}=\overrightarrow{AO}+\overrightarrow{OB}=\mathbf{b}-\mathbf{a}$, and so $\overrightarrow{AM}=\frac{1}{2}(\mathbf{b}-\mathbf{a})$. Now

$$\mathbf{m}=\overrightarrow{OM}=\overrightarrow{OA}+\overrightarrow{AM}=\mathbf{a}+\tfrac{1}{2}(\mathbf{b}-\mathbf{a})=\mathbf{a}+\tfrac{1}{2}\mathbf{b}-\tfrac{1}{2}\mathbf{a}=\tfrac{1}{2}(\mathbf{a}+\mathbf{b}).$$

The same result is demonstrated in Fig. 5.8(b). If A_1 and B_1 are the respective midpoints of OA and OB, then

$$\overrightarrow{OA_1}=\tfrac{1}{2}\mathbf{a},\ \overrightarrow{A_1M}=\overrightarrow{OB_1}=\tfrac{1}{2}\mathbf{b},$$

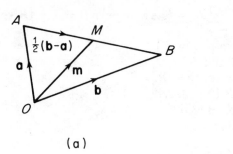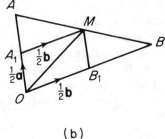

(a)   (b)

Fig. 5.8

so $$\overrightarrow{OM} = \overrightarrow{OA_1} + \overrightarrow{A_1M} = \tfrac{1}{2}(\mathbf{a} + \mathbf{b}).$$

This result can be generalised by using any real number instead of the fraction $\tfrac{1}{2}$. Let R be a point on AB with $\overrightarrow{AR} = \mu\overrightarrow{AB}$.

Then $\overrightarrow{RB} = \overrightarrow{AB} - \overrightarrow{AR} = \overrightarrow{AB} - \mu\overrightarrow{AB} = (1 - \mu)\overrightarrow{AB}.$

Put $1 - \mu = \lambda$, then

$$\overrightarrow{AR} = \mu\overrightarrow{AB}, \quad \overrightarrow{RB} = \lambda\overrightarrow{AB}, \quad \lambda + \mu = 1.$$

If $\lambda = 0$, then $R \equiv B$. If $\lambda \neq 0$, let $k = \dfrac{\mu}{\lambda}$, then, since

$$\overrightarrow{AR} = \mu\overrightarrow{AB} = \frac{\mu}{\lambda}\overrightarrow{RB} = k\overrightarrow{RB},$$

taking its sign into account, k represents the ratio $AR:RB$ and fixes the position of R on the line AB. When $k > 0$, $\overrightarrow{AR}$ and $\overrightarrow{RB}$ are in the same direction and, when $k < 0$, $\overrightarrow{AR}$ and $\overrightarrow{RB}$ are in opposite directions. As an example, in Fig. 5.9, three values of k are chosen:

(a) if $k = 3$, $\mu = 3\lambda$ so $\mu = \tfrac{3}{4}$, $\lambda = \tfrac{1}{4}$, $\overrightarrow{AR} = 3\overrightarrow{RB} = \tfrac{3}{4}\overrightarrow{AB}$, so R lies between A and B;

(b) if $k = -3$, $\mu = -3\lambda$ so $\mu = \tfrac{3}{2}$, $\lambda = -\tfrac{1}{2}$, $\overrightarrow{AR} = -3\overrightarrow{RB} = \tfrac{3}{2}\overrightarrow{AB}$, so R lies on AB produced;

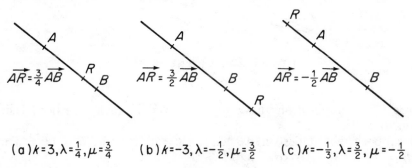

(a) $k = 3$, $\lambda = \tfrac{1}{4}$, $\mu = \tfrac{3}{4}$ (b) $k = -3$, $\lambda = -\tfrac{1}{2}$, $\mu = \tfrac{3}{2}$ (c) $k = -\tfrac{1}{3}$, $\lambda = \tfrac{3}{2}$, $\mu = -\tfrac{1}{2}$

Fig. 5.9

(c) if $k = -\frac{1}{3}, \mu = -\frac{1}{3}\lambda$ so $\mu = -\frac{1}{2}, \lambda = \frac{3}{2}, \overrightarrow{AR} = -\frac{1}{3}\overrightarrow{RB} = -\frac{1}{2}\overrightarrow{AB}$, so R lies on BA produced.

The general situation is shown in Fig. 5.10.
If $k > 0$, then R lies between A and B;
if $k = 0$, then $R = A$, and if $k = \infty$, then $R = B$;
if $-1 < k < 0$, then R lies on BA produced;
if $k < -1$, then R lies on AB produced.

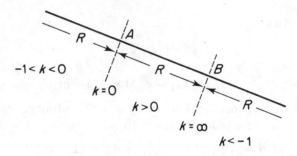

k is the ratio $AR : RB$, given by $\overrightarrow{AR} = k\overrightarrow{RB}$

Fig. 5.10

These results provide a vector theorem which fixes the position of a point R on the line AB in terms of the ratio k of AR to RB.

The Ratio Theorem
Let $\mathbf{a} = \overrightarrow{OA}$ and $\mathbf{b} = \overrightarrow{OB}$ be the position vectors of points A and B with respect to an origin O, and let $\mathbf{r} = \overrightarrow{OR}$ be the position vector of any point R in space. Then

R lies on AB with $\overrightarrow{AR} = \mu\overrightarrow{AB}, \overrightarrow{BR} = \lambda\overrightarrow{BA} \Leftrightarrow \mathbf{r} = \lambda\mathbf{a} + \mu\mathbf{b}$ and $\lambda + \mu = 1$.

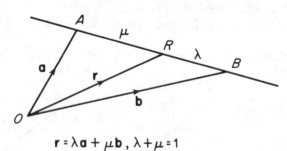

$\mathbf{r} = \lambda\mathbf{a} + \mu\mathbf{b}, \ \lambda + \mu = 1$

Fig. 5.11

Proof The implications have to be proved in each direction. Using Fig. 5.11:
(i) suppose that R lies on AB with $\overrightarrow{AR} = \mu\overrightarrow{AB}$ and $\overrightarrow{BR} = \lambda\overrightarrow{BA}$. Then $\overrightarrow{RB} = -\lambda\overrightarrow{BA} = \lambda\overrightarrow{AB}$, so

$$\overrightarrow{AB} = \overrightarrow{AR} + \overrightarrow{RB} = \mu\overrightarrow{AB} + \lambda\overrightarrow{AB} = (\lambda + \mu)\overrightarrow{AB},$$

and thus $\lambda + \mu = 1$. Also

$$\mathbf{r} = \overrightarrow{OR} = \overrightarrow{OA} + \overrightarrow{AR} = \mathbf{a} + \mu\overrightarrow{AB} = \mathbf{a} + \mu(\mathbf{b} - \mathbf{a})$$
$$= (1 - \mu)\mathbf{a} + \mu\mathbf{b} = \lambda\mathbf{a} + \mu\mathbf{b};$$

(ii) conversely, suppose that $\mathbf{r} = \lambda\mathbf{a} + \mu\mathbf{b}$ and $\lambda + \mu = 1$. Then $\lambda = 1 - \mu$. Then

$$\overrightarrow{AR} = \mathbf{r} - \mathbf{a} = \lambda\mathbf{a} + \mu\mathbf{b} - \mathbf{a} = (1 - \mu)\mathbf{a} + \mu\mathbf{b} - \mathbf{a} = \mu(\mathbf{b} - \mathbf{a}) = \mu\overrightarrow{AB},$$

so R lies on AB. Also

$$\overrightarrow{BR} = \mathbf{r} - \mathbf{b} = \lambda\mathbf{a} + \mu\mathbf{b} - \mathbf{b} = \lambda\mathbf{a} + (\mu - 1)\mathbf{b} = \lambda(\mathbf{a} - \mathbf{b}) = \lambda\overrightarrow{BA}.$$

Note that the equation relating $\mathbf{r}$, $\mathbf{a}$ and $\mathbf{b}$ depends only upon the position of R on the line AB. It is independent of the origin O.

EXAMPLE 1   *Find the position of the point R on the line AB, given that the position vectors of A and B are* $\mathbf{a}$ *and* $\mathbf{b}$ *and that* $AR:RB = 3:2$.

The given ratio means that $2\overrightarrow{AR} = 3\overrightarrow{RB}$ so that $\overrightarrow{AR} = \frac{3}{5}\overrightarrow{AB}$ and $\overrightarrow{BR} = \frac{2}{5}\overrightarrow{BA}$. By the ratio theorem, $\mathbf{r} = \frac{2}{5}\mathbf{a} + \frac{3}{5}\mathbf{b}$. See Fig. 5.12(a).

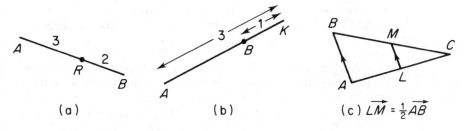

(a)   (b)   (c) $\overrightarrow{LM} = \frac{1}{2}\overrightarrow{AB}$

Fig. 5.12

EXAMPLE 2   *Describe the position of the point K with respect to the points A and B given that the position vectors* $\mathbf{k}$, $\mathbf{a}$ *and* $\mathbf{b}$ *of the three points satisfy the equation* $2\mathbf{k} = 3\mathbf{b} - \mathbf{a}$.

We note that $\mathbf{k} = \frac{3}{2}\mathbf{b} - \frac{1}{2}\mathbf{a}$, and $\frac{3}{2} + (-\frac{1}{2}) = 1$, so by the ratio theorem, with $\lambda = \frac{3}{2}$ and $\mu = -\frac{1}{2}$, K is the point on the line AB produced with $\overrightarrow{AK} = \frac{3}{2}\overrightarrow{AB}$ and $\overrightarrow{BK} = \frac{1}{2}\overrightarrow{AB}$. See Fig. 5.12(b).

EXAMPLE 3   *Given that L and M are the midpoints of the sides AC and BC of the triangle ABC show that LM is parallel to AB and LM is half the length of AB.*

By the ratio theorem, $\overrightarrow{AM} = \frac{1}{2}\overrightarrow{AB} + \frac{1}{2}\overrightarrow{AC} = \frac{1}{2}\overrightarrow{AB} + \overrightarrow{AL}$.

So, $\overrightarrow{LM} = \overrightarrow{AM} - \overrightarrow{AL} = \frac{1}{2}\overrightarrow{AB} + \overrightarrow{AL} - \overrightarrow{AL} = \frac{1}{2}\overrightarrow{AB}$.

Therefore, LM is parallel to AB and is half its length. See Fig. 5.12(c).

EXERCISE 5.4

1   In the parallelogram $ABCD$, E is the midpoint of CD and F is the point of trisection of DB. Let $\overrightarrow{AB} = \mathbf{b}$ and $\overrightarrow{AD} = \mathbf{d}$. Express $\overrightarrow{AC}, \overrightarrow{AE}, \overrightarrow{AF}$ in terms of $\mathbf{b}$ and $\mathbf{d}$ and deduce that F is the point of trisection of EA.

2   In the quadrilateral $ABCD$, E, F, G, H are the midpoints of AB, BC, CD, DA respectively. The position vectors of A, B, C, D are $\mathbf{a}, \mathbf{b}, \mathbf{c}, \mathbf{d}$ respectively. Express the position vectors of E, F, G, H in terms of $\mathbf{a}, \mathbf{b}, \mathbf{c}, \mathbf{d}$ and prove that:
   (i) $EFGH$ is a parallelogram;
   (ii) the diagonals of $EFGH$ meet at the point with position vector $\frac{1}{4}(\mathbf{a} + \mathbf{b} + \mathbf{c} + \mathbf{d})$.

3   Given three non-parallel vectors in space, $\mathbf{a} = \overrightarrow{OA}$, $\mathbf{b} = \overrightarrow{OB}$ and $\mathbf{c} = \overrightarrow{OC}$:
   (i) if C lies in the plane OAB, use a parallelogram with diagonal OC and edges parallel to $\mathbf{a}$ and $\mathbf{b}$ to show that, for some real numbers h, k, $\mathbf{c} = h\mathbf{a} + k\mathbf{b}$;
   (ii) conversely, prove that if $\mathbf{c} = h\mathbf{a} + k\mathbf{b}$ then C lies in the plane OAB.

4   The six points P, Q, R, S, T, U are equally spaced on a circle, centre O, so that they form the vertices of a regular hexagon. Given that $\overrightarrow{OP} = \mathbf{p}$, $\overrightarrow{OQ} = \mathbf{q}$, express in terms of $\mathbf{p}$ and $\mathbf{q}$ the vectors $\overrightarrow{OR}, \overrightarrow{OS}, \overrightarrow{OT}, \overrightarrow{OU}, \overrightarrow{RU}, \overrightarrow{RT}$.

5   $PQRS$ is a parallelogram and $\overrightarrow{PR} = \mathbf{a}$, $\overrightarrow{QS} = \mathbf{b}$. Express the edges $\overrightarrow{PQ}, \overrightarrow{QR}, \overrightarrow{RS}, \overrightarrow{SP}$ of the parallelogram in terms of $\mathbf{a}$ and $\mathbf{b}$.

6   In the triangle ABC, the points P, Q and R are respectively the midpoints of the sides BC, CA and AB. Find the position vectors of P, Q and R in terms of the position vectors $\mathbf{a}, \mathbf{b}$ and $\mathbf{c}$ of A, B and C and prove that the three medians AP, BQ and CR intersect at a point which is two-thirds down the length of each median from its corresponding vertex.

7   The centroid of n points $A_1, A_2, \ldots, A_n$ is defined to be the point with position vector $\frac{1}{n}(\mathbf{a}_1 + \mathbf{a}_2 + \ldots + \mathbf{a}_n)$, where $\mathbf{a}_i$ is the position vector of A_i
   (i) Prove that the centroid of two points is the point midway between them.
   (ii) Prove that the centroid of three non-collinear points A, B and C is the point G where the three medians of triangle ABC intersect (see question 6).
   (iii) Generalise the above to the case of four points A, B, C and D which do not lie in one plane.

8   In the quadrilateral $OABC$, D is the mid-point of BC and G is the point on AD such that $AG:GD = 2:1$. Given that $\overrightarrow{OA} = \mathbf{a}$, $\overrightarrow{OB} = \mathbf{b}$ and $\overrightarrow{OC} = \mathbf{c}$, express $\overrightarrow{OD}$ and $\overrightarrow{OG}$ in terms of $\mathbf{a}, \mathbf{b}$ and $\mathbf{c}$. *(L)*

9   The points A, B, C have position vectors $\mathbf{a}, \mathbf{b}, \mathbf{c}$ respectively referred to an origin O. State the position vector of P, the mid-point of AB.
   Find the position vector of the point which divides the line segment PC internally in the ratio $3:1$. *(L)*

10   The position vectors of the vertices of a triangle ABC referred to a given origin are $\mathbf{a}, \mathbf{b}, \mathbf{c}$. P is a point on AB such that $AP/PB = \frac{1}{2}$, Q is a point on AC such that $AQ/QC = 2$, and R is a point on PQ such that $PR/RQ = 2$. Prove that the position vector of R is $\frac{4}{9}\mathbf{a} + \frac{1}{9}\mathbf{b} + \frac{4}{9}\mathbf{c}$.
   Prove that R lies on the median BM of the triangle ABC, and state the value of BR/RM. *(SMP)*

11   Points X and Y are taken on the sides QR and RS, respectively, of a parallelogram $PQRS$, so that $\overrightarrow{QX} = 4\overrightarrow{XR}$ and $\overrightarrow{RY} = 4\overrightarrow{YS}$. The line XY cuts the line PR at Z. Prove that $\overrightarrow{PZ} = \frac{21}{25}\overrightarrow{PR}$. *(JMB)*

12   (i) In the quadrilateral $ABCD$, X and Y are the mid-points of the diagonals

AC and BD respectively. Show that

(a) $\overrightarrow{BA} + \overrightarrow{BC} = 2\overrightarrow{BX}$,

(b) $\overrightarrow{BA} + \overrightarrow{BC} + \overrightarrow{DA} + \overrightarrow{DC} = 4\overrightarrow{YX}$.

(ii) The point P lies on the circle through the vertices of a rectangle $QRST$. The point X on the diagonal QS is such that $\overrightarrow{QX} = 2\overrightarrow{XS}$. Express $\overrightarrow{PX}$, $\overrightarrow{QX}$ and $(\overrightarrow{RX} + \overrightarrow{TX})$ in terms of $\overrightarrow{PQ}$ and $\overrightarrow{PS}$.

(L)

5.5 Basis of the set of vectors in one plane

A plane Π through the origin O may be defined by any two non-parallel vectors **a** and **b** which lie in Π. This was the situation in Exercise 5.2, in those questions which used Fig. 5.6, and again in Exercise 5.3 in those questions which used Fig. 5.7.

In Fig. 5.13, let $\overrightarrow{OA} = \mathbf{a}$ and $\overrightarrow{OB} = \mathbf{b}$, then Π is the plane of the triangle OAB. Also, since **a** and **b** are not parallel,

$$r\mathbf{a} = s\mathbf{b} \Rightarrow r = 0 = s.$$

We now prove that

[a point P lies in Π] $\Leftrightarrow$ [$\mathbf{r} = \overrightarrow{OP} = \lambda\mathbf{a} + \mu\mathbf{b}$, for some scalars λ, μ].

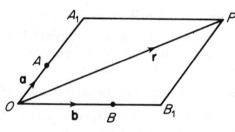

Fig. 5.13

Proof

(i) Suppose that P is a point in space with $\overrightarrow{OP} = \mathbf{r} = \lambda\mathbf{a} + \mu\mathbf{b}$, for some scalars λ and μ. Then define points A_1 and B_1, on OA and OB respectively, produced if necessary, such that $\overrightarrow{OA_1} = \lambda\overrightarrow{OA}$ and $\overrightarrow{OB_1} = \mu\overrightarrow{OB}$. Then

$$\overrightarrow{OP} = \lambda\mathbf{a} + \mu\mathbf{b} = \lambda\overrightarrow{OA} + \mu\overrightarrow{OB} = \overrightarrow{OA_1} + \overrightarrow{OB_1},$$

so A_1PB_1O is a parallelogram and hence P lies in the plane OA_1B_1, which is the plane OAB, that is, Π.

(ii) Conversely, let P be any point in the plane Π. Construct two lines through P, one parallel to **b** meeting OA at A_1, and one parallel to **a** meeting OB at B_1, where OA and OB are produced if necessary. Then OA_1PB_1 is a parallelogram and so $\overrightarrow{OP} = \overrightarrow{OA_1} + \overrightarrow{OB_1}$. Because $\overrightarrow{OA_1}$ and $\overrightarrow{OB_1}$ are respectively parallel to $\overrightarrow{OA}$ and $\overrightarrow{OB}$, $\overrightarrow{OA_1} = \lambda\overrightarrow{OA}$ and $\overrightarrow{OB_1} = \mu\overrightarrow{OB}$ for some scalars λ and μ. Hence

$$\mathbf{r} = \overrightarrow{OP} = \overrightarrow{OA_1} + \overrightarrow{OB_1} = \lambda\overrightarrow{OA} + \mu\overrightarrow{OB} = \lambda\mathbf{a} + \mu\mathbf{b}.$$

We can take this idea one step further and prove that, for every point P in Π, $\mathbf{r}$ is expressed *uniquely* as a sum of multiples of $\mathbf{a}$ and $\mathbf{b}$.

Suppose that $\overrightarrow{OP} = \mathbf{r} = \lambda\mathbf{a} + \mu\mathbf{b} = \lambda_1\mathbf{a} + \mu_1\mathbf{b}$. Then

$$\lambda\mathbf{a} - \lambda_1\mathbf{a} = \mu_1\mathbf{b} - \mu\mathbf{b}$$

so that

$$(\lambda - \lambda_1)\mathbf{a} = (\mu_1 - \mu)\mathbf{b}.$$

But $\mathbf{a}$ and $\mathbf{b}$ are not parallel, so only their zero multiples are equal, hence

$$(\lambda - \lambda_1) = 0 = (\mu_1 - \mu)$$

and so

$$\lambda = \lambda_1, \mu = \mu_1,$$

and the expression for $\mathbf{r}$ as a sum of multiples of $\mathbf{a}$ and $\mathbf{b}$ is unique.

We combine these two results in the following theorem.

Theorem 5.5 Let $\mathbf{a}$ and $\mathbf{b}$ be non-parallel vectors in the plane Π through O and let $\overrightarrow{OP} = \mathbf{r}$. Then

P lies in $\Pi \Leftrightarrow$ there exist unique λ, μ with $\mathbf{r} = \lambda\mathbf{a} + \mu\mathbf{b}$.

Note: In the proof of the theorem, we proved a result which is useful in its own right:

For non-parallel vectors $\mathbf{a}$ and $\mathbf{b}$,

$$[\lambda\mathbf{a} + \mu\mathbf{b} = \lambda_1\mathbf{a} + \mu_1\mathbf{b}] \Leftrightarrow [\lambda = \lambda_1 \text{ and } \mu = \mu_1].$$

The situation of theorem **5.5**, when every point P in a plane Π has a position vector $\mathbf{r}$ which is the sum of unique multiples of $\mathbf{a}$ and $\mathbf{b}$, is described by saying that the pair of vectors

$\{\mathbf{a}, \mathbf{b}\}$ is a *basis* for the set of all position vectors of points in Π.

Since we began with any two non-parallel vectors $\mathbf{a}$ and $\mathbf{b}$ in the plane Π, it will follow that any two non-parallel vectors in a plane form a basis for the set of position vectors of all points in the plane.

In three dimensions, we can choose three vectors, not in one plane, to form a similar basis for all the position vectors in three dimensional space. This we shall consider in Chapter 12.

EXAMPLE 1   *Relative to an origin O, the points A, B and C have position vectors $\mathbf{a}$, $\mathbf{b}$ and $\mathbf{c}$, respectively. Given that C lies in the plane OAB with $\mathbf{c} = \lambda\mathbf{a} + \mu\mathbf{b}$, and that BC is not parallel to OA, find, in terms of $\mathbf{a}$, λ, μ, the position vector $\mathbf{d}$ of D, the point of intersection of OA and BC (Fig. 5.14).*

$\overrightarrow{BC} = \overrightarrow{OC} - \overrightarrow{OB} = \mathbf{c} - \mathbf{b} = \lambda\mathbf{a} + \mu\mathbf{b} - \mathbf{b} = \lambda\mathbf{a} + (\mu - 1)\mathbf{b} \neq k\mathbf{a}$, since $\overrightarrow{BC}$ is not parallel to $\overrightarrow{OA}$, and so $\mu \neq 1$. Now, for some constant ν,

$$\overrightarrow{BD} = \nu\overrightarrow{BC} = \nu\lambda\mathbf{a} + \nu(\mu - 1)\mathbf{b}$$

and so $\overrightarrow{OD} = \mathbf{d} = \overrightarrow{OB} + \overrightarrow{BD} = \mathbf{b} + \nu\lambda\mathbf{a} + \nu(\mu - 1)\mathbf{b} = \nu\lambda\mathbf{a} + (1 + \nu\mu - \nu)\mathbf{b} = k\mathbf{a}$.

Hence $1 + \nu\mu - \nu = 0$ and $k = \nu\lambda$. Solving $1 - \nu + \mu\nu = 0$ for ν, gives $\nu = \dfrac{1}{1 - \mu}$

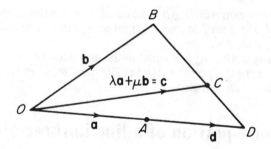

Fig. 5.14

and so

$$\mathbf{d} = v\lambda\mathbf{a} = \frac{\lambda}{1-\mu}\mathbf{a}.$$

EXAMPLE 2   *With reference to an origin O in a plane and a basis* $\{\mathbf{a}, \mathbf{b}\}$, *the equations of two lines are* $\mathbf{r} = \mathbf{a} + \mathbf{b} + \lambda(\mathbf{a} - \mathbf{b})$ *and* $\mathbf{r} = 2\mathbf{a} - 3\mathbf{b} + \lambda(\mathbf{a} + 2\mathbf{b})$. *Find the position vector of the point of intersection of the two lines.*

In the two equations of the lines, the same symbol λ is used for the parameter. Since the point P of position vector $\mathbf{r}$, where the lines intersect, will have different parameters in the equations of the two lines, we use two symbols λ and μ. Then we have to solve the equation

$$\mathbf{r} = \mathbf{a} + \mathbf{b} + \lambda(\mathbf{a} - \mathbf{b}) = 2\mathbf{a} - 3\mathbf{b} + \mu(\mathbf{a} + 2\mathbf{b}).$$

We separate the terms containing $\mathbf{a}$ and $\mathbf{b}$, to give

$$\mathbf{a} + \lambda\mathbf{a} - 2\mathbf{a} - \mu\mathbf{a} = -\mathbf{b} + \lambda\mathbf{b} - 3\mathbf{b} + 2\mu\mathbf{b},$$

that is, $(1 + \lambda - 2 - \mu)\mathbf{a} = (-1 + \lambda - 3 + 2\mu)\mathbf{b}$ and, since $\mathbf{a}$ and $\mathbf{b}$ are not parallel, ($\{\mathbf{a}, \mathbf{b}\}$ being a basis for the plane), we must have

$$1 + \lambda - 2 - \mu = 0 \quad \text{and} \quad -1 + \lambda - 3 + 2\mu = 0.$$

Eliminating λ from these two equations, $\lambda = 1 + \mu$ and $\lambda = 4 - 2\mu$, so $1 + \mu = 4 - 2\mu$, $3\mu = 3$, $\mu = 1$ and $\lambda = 1 + 1 = 2$. Putting these values of the parameters into the equations of the two lines, we find the position vector $\mathbf{r}$ of their common point to be $\mathbf{r} = 3\mathbf{a} - \mathbf{b}$.

EXERCISE 5.5

1   In the parallelogram $ABCD$, E is the midpoint of CD, and AE meets BD at F. Let $\overrightarrow{AB} = \mathbf{b}$ and $\overrightarrow{AD} = \mathbf{d}$. Express $\overrightarrow{AC}$, $\overrightarrow{AE}$, $\overrightarrow{AF}$ in terms of $\mathbf{b}$ and $\mathbf{d}$ and deduce that F is the point of trisection of DB. (Note: This question is a converse of question 1 of Exercise 5.4).

2   The position vectors, relative to O, of the vertices A and B of the rectangle $OACB$ are $\mathbf{a}$ and $\mathbf{b}$, respectively. The point P divides OC internally in the ratio $3:1$ ($= OP:PC$). The lines BP and OA meet at Q and the lines AP and OB meet at R. In terms of $\mathbf{a}$ and $\mathbf{b}$, find the position vectors of P, Q and R.

3   In the parallelogram $ABCD$, $\overrightarrow{AB} = \mathbf{b}$ and $\overrightarrow{AC} = \mathbf{c}$. L is the midpoint of AD and $\overrightarrow{AM} = 2\overrightarrow{AB}$. The lines LM and BC meet at N. Find, in terms of $\mathbf{b}$ and $\mathbf{c}$, the vectors $\overrightarrow{AL}$, $\overrightarrow{AM}$, $\overrightarrow{AN}$.

4   Given a triangle ABC, with R equal to the mid point of AB, and Q the point such that C is the midpoint of AQ. Let RQ meet BC at P. Express the vector $\overrightarrow{AP}$ in terms of $\overrightarrow{AB}$ and $\overrightarrow{AC}$.

5.6 Vector equation of a line in three dimensions

Most examples considered so far have been two dimensional. However, the theory applies to a three-dimensional situation. We shall consider briefly the equations of lines in three dimensions. They will be dealt with in more detail in §12.5. The vector equation of a line, which was used in §5.3, applies equally well to a line in space.

Theorem 5.6   Given a point A, with position vector $\mathbf{a} = \overrightarrow{OA}$, and any vector $\mathbf{u}$, the equation of the line L through A in the direction of $\mathbf{u}$ is $\mathbf{r} = \mathbf{a} + t\mathbf{u}$, where t is a scalar parameter.

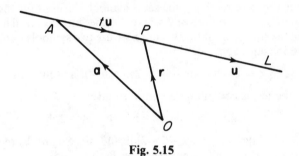

Fig. 5.15

Proof
Consider Fig. 5.15. Let P be any point in space, with position vector $\mathbf{r} = \overrightarrow{OP}$. Then

$$P \text{ lies on } L \Leftrightarrow \overrightarrow{AP} \text{ is parallel to } \mathbf{u}$$
$$\Leftrightarrow \overrightarrow{AP} = \mathbf{r} - \mathbf{a} = t\mathbf{u}, \text{ for some scalar } t,$$
$$\Leftrightarrow \mathbf{r} = \mathbf{a} + t\mathbf{u}.$$

So the equation of L is $\mathbf{r} = \mathbf{a} + t\mathbf{u}$.

Intersecting lines

Consider two coplanar lines which are not parallel. One line passes through A in the direction $\mathbf{u}$, and the other passes through B in the direction $\mathbf{v}$. If $\overrightarrow{OA} = \mathbf{a}$ and $\overrightarrow{OB} = \mathbf{b}$, the two lines have equations $\mathbf{r} = \mathbf{a} + t\mathbf{u}$ and $\mathbf{r} = \mathbf{b} + s\mathbf{v}$. See Fig. 5.16. Note that we use different scalar parameters t and s for the two lines. Let the two lines intersect at Q, where $\overrightarrow{OQ} = \mathbf{q}$, then

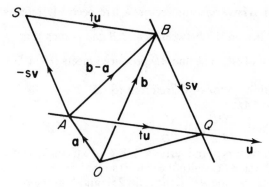

Fig. 5.16

$\mathbf{q}$ satisfies both equations, $\mathbf{q} = \mathbf{a} + t\mathbf{u} = \mathbf{b} + s\mathbf{v}$, for some numbers t and s. Therefore, $\overrightarrow{AB} = \mathbf{b} - \mathbf{a} = t\mathbf{u} - s\mathbf{v}$ and, as we saw in theorem 5.5, t and s can be found by construction of a parallelogram with diagonal AB and edges parallel to $\mathbf{u}$ and $\mathbf{v}$.

Note that if the lines are parallel, $\mathbf{u} = k\mathbf{v}$, then it is no longer possible to find t and s such that $\overrightarrow{AB} = t\mathbf{u} - s\mathbf{v}$. This corresponds to the fact that parallel lines do not meet.

Skew lines

In three dimensions, two lines which are not parallel may not intersect, and then they are said to be *skew* (Fig. 5.17).

This is indicated in the algebra. There will be no values of t and s such that $\mathbf{a} + t\mathbf{u} = \mathbf{b} + s\mathbf{v}$, because $(\mathbf{b} - \mathbf{a})$, $\mathbf{u}$ and $\mathbf{v}$ do not lie in one plane.

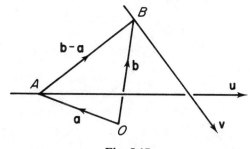

Fig. 5.17

MISCELLANEOUS EXERCISE 5

1 *AOB* is a triangle with $\overrightarrow{OA} = \mathbf{a}$, $\overrightarrow{OB} = \mathbf{b}$, and P, Q are the mid-points of OB, OA, respectively. Find the position vector of
(a) a general point on AP, (b) a general point on BQ, each in terms of $\mathbf{a}$, $\mathbf{b}$ and a scalar parameter. Find also the position vector of the point in which AP and BQ intersect.   *(L)*

2 The points A, B have position vectors $\mathbf{a}$, $\mathbf{b}$ respectively, referred to an origin O. The point C lies on AB between A and B and is such that $\dfrac{AC}{CB} = \dfrac{2}{1}$, and D is the mid-point of OC. The line AD produced meets OB at E. Find, in terms of $\mathbf{a}$ and $\mathbf{b}$,

   (i) the position vector of C (referred to O),
   (ii) the vector $\overrightarrow{AD}$.

   Find the values of $\dfrac{OE}{EB}$ and $\dfrac{AE}{ED}$. $\hspace{2cm}$ (C)

3 The position vectors of the vertices A, B, C of a triangle relative to an origin O are $\mathbf{a}$, $\mathbf{b}$, $\mathbf{c}$. The side BC is produced to D so that $BC = CD$. We define X as the point dividing the side AB in the ratio $2:1$, and Y as the point dividing AC in the ratio $4:1$. Express the position vectors $\mathbf{d}$, $\mathbf{x}$, $\mathbf{y}$ of D, X, Y relative to O in terms of $\mathbf{a}$, $\mathbf{b}$, $\mathbf{c}$, and use these to prove that the three points D, X, Y are in a straight line. $\hspace{2cm}$ (SMP)

4 $ABCDEF$ is a regular hexagon. If $\overrightarrow{AB} = \mathbf{a}$ and $\overrightarrow{BC} = \mathbf{b}$, express the following vectors on terms of $\mathbf{a}$ and $\mathbf{b}$:
   (a) $\overrightarrow{CD}$, $\hspace{0.4cm}$ (b) $\overrightarrow{CE}$, $\hspace{0.4cm}$ (c) $\overrightarrow{EB}$. $\hspace{2cm}$ (L)

5 The points A, B, C have position vectors $\mathbf{a}$, $\mathbf{b}$, $\mathbf{c}$ respectively referred to an origin O. State the position vector of P, the mid-point of AB.

   Find the position vector of the point which divides the line segment PC internally in the ratio $3:1$. $\hspace{2cm}$ (L)

6 The points A, B and C have position vectors $\mathbf{a}$, $\mathbf{b}$ and $\mathbf{c}$ with respect to an origin O. The point R in the plane ABC has position vector $\mathbf{r}$ where $\mathbf{r} = \frac{1}{2}(\frac{2}{3}\mathbf{a} + \frac{1}{3}\mathbf{b}) + \frac{1}{2}\mathbf{c}$. Use the ratio theorem to obtain a geometrical description of the position of R with reference to the points A, B and C. Illustrate your answer with a diagram.

   By writing $\mathbf{r}$ in the form $\lambda \mathbf{a} + \mu(m\mathbf{b} + n\mathbf{c})$ where $\lambda + \mu = 1$, obtain an alternative description of the position of R. Given that the point S has position vector $\frac{2}{3}\mathbf{a} + \frac{1}{12}\mathbf{b} + \frac{1}{4}\mathbf{c}$, show that S bisects the line segment AR. $\hspace{1cm}$ (JMB)

7

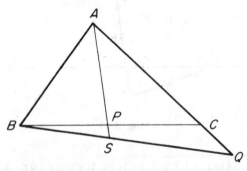

   In the diagram P is the mid-point of the side BC of the triangle ABC and Q lies on AC produced, so that $AC:CQ = 3:1$. Find vector equations for the lines AP and BQ in terms of $\mathbf{a}$, $\mathbf{b}$ and $\mathbf{c}$, which are, respectively, the position vectors of the points A, B and C relative to an origin O. Hence obtain, in terms of $\mathbf{a}$, $\mathbf{b}$ and $\mathbf{c}$, the position vector of the point S at which the lines AP and BQ intersect. $\hspace{2cm}$ (JMB)

8   The tetrahedron $OABC$ has edges $\overrightarrow{OA} = \mathbf{a}$, $\overrightarrow{OB} = \mathbf{b}$, $\overrightarrow{OC} = \mathbf{c}$. The points L, M, N are midpoints of OA, OB, OC, respectively, and P, Q, R are the midpoints of BC, CA, AB, respectively. Prove that PL and QM lie in one plane and find the position of their point of intersection, in terms of $\mathbf{a}$, $\mathbf{b}$ and $\mathbf{c}$. Deduce that the lines joining the midpoints of opposite edges of a tetrahedron all meet at their common midpoint.

9   The position vectors of P, Q with reference to an origin O are $\mathbf{p}$, $\mathbf{q}$, and M is the point on PQ such that

$$\beta \overrightarrow{PM} = \alpha \overrightarrow{MQ}.$$

Prove that the position vector of M is $\mathbf{m}$, where

$$\mathbf{m} = \frac{\beta \mathbf{p} + \alpha \mathbf{q}}{\alpha + \beta}.$$

The vectors $\mathbf{p}$, $\mathbf{q}$ are given by $u\mathbf{a}$, $v\mathbf{b}$ respectively, where u, v are positive scalars and $\mathbf{a}$, $\mathbf{b}$ are unit vectors. Prove that the position vector of any point on the internal bisector of $P\hat{O}Q$ has the form $r(\mathbf{a} + \mathbf{b})$, and deduce that, if M is the point where this bisector meets PQ, then

$$\frac{\alpha}{\beta} = \frac{u}{v}. \qquad\qquad (C)$$

6 Integration I

6.1 A function and its derivative (revision)

EXERCISE 6.1

1 Complete the following table of functions and derivatives:

$f(x)$	$f'(x)$
x^4	$4x^3$
$x^3 + 2x$	
	$3x^2$
	$2x + 1$

2 Find the derivative function $f'(x)$ for the function $f(x)$:

    (i) x^2 (ii) $\frac{1}{3}x^3$ (iii) $x + 2$
    (iv) $x^3 + 3$ (v) $x^3 - 1$ (vi) 2
    (vii) $x^4 + c$ (viii) $x^4 - c$ (ix) $x^{3/2}$
    (x) $x^2 + 2x$ (xi) $2x^{1/2} - 2x^{-1/2}$ (xii) $x^{1/4} + x^{-3/4}$.

3 Find the function $f(x)$ with the following derivative:

    (i) x^2 (ii) x^3 (iii) $x^{1/2}$
    (iv) $x - x^{1/2}$ (v) $x^2 + 7x + 7$ (vi) $x^2 + 7x$
    (vii) 7 (viii) $\frac{1}{4}x^3 + c$ (ix) $x^{3/4} + x^{-1/4}$.

You will see that, when you are given $f(x)$, there is a unique expression for $f'(x)$. However, when $f'(x)$ is given, then $f(x)$ is not known uniquely because any constant may be added to $f(x)$ without affecting $f'(x)$. For example, if $f'(x) = 3x^2$, then $f(x)$ can be any function of the form $f(x) = x^3 + c$, where c is some real constant.

6.2 Integration

The reverse process of differentiation, that is to find $f(x)$ given $f'(x)$, is called *integration*. Given a function $g(x)$, then a function $f(x)$, such that $g(x) = f'(x)$, is called an *integral*, or *antiderivative*, of $g(x)$. When the integral includes the addition of an arbitrary constant c, then it is written

$\int g(x)dx$ = the (*indefinite*) integral of $g(x)$ with respect to x.

(The word 'indefinite' is often omitted.)

Thus $$\int 3x^2 dx = x^3 + c.$$

In reading this expression, the symbols have the following meaning:

$\int$ means   'the integral of',

dx means   'with respect to x'.

When the constant c is given a particular value, then we obtain a function whose derivative is $g(x)$, thus

$x^3 + c$ is the indefinite integral of $3x^2$

and   if $f(x) = x^3 + 2$ then $f'(x) = 3x^2$

so $x^3 + 2$ is an integral of $3x^2$.

6.3 Rules for integration

The rules for differentiation, given in §4.7, lead to corresponding rules for integration, that is, the process of finding the integral of a function.

Rule 1 The integral of a linear combination of functions is the corresponding linear combination of their integrals, (with the linear combination of the constants of integration combined into one arbitrary constant). If a and b are constants and f and g are functions,

$$\int [af(x) + bg(x)]dx = a \int f(x)dx + b \int g(x)dx.$$

Rule 2 For any constants n and a,

$$\int x^n dx = \frac{1}{n+1} x^{n+1} + c, \text{ provided } n \neq -1,$$

$$\int \cos(nx + a)dx = \frac{1}{n}\sin(nx + a) + c,$$

$$\int \sin(nx + a)dx = \frac{-1}{n}\cos(nx + a) + c.$$

EXAMPLE 1 *Find the integral with respect to x of:*
(i) $2 + 3x$;   (ii) $x^4 - x^2$;   (iii) $3\cos(2x + 5)$.

Using the rules:
(i) $\int (2 + 3x)dx = 2\int 1dx + 3\int xdx = 2x + \frac{3}{2}x^2 + c$;
(ii) $\int (x^4 - x^2)dx = \int x^4 dx - \int x^2 dx = \frac{1}{5}x^5 - \frac{1}{3}x^3 + c$;
(iii) $\int 3\cos(2x + 5)dx = 3\int \cos(2x + 5)dx = \frac{3}{2}\sin(2x + 5) + c.$

Sometimes it is necessary to change the form of the *integrand*, that is, the function being integrated. The independent variable may also be something other than x.

EXAMPLE 2   *Integrate* (i) $(1+2x)^3$, (ii) $\cos^2 \theta$.

(i) $\int (1+2x)^3 dx = \int (1+6x+12x^2+8x^3)dx = x+3x^2+4x^3+2x^4+c.$

(ii) $\int \cos^2 \theta\, d\theta = \int \frac{1}{2}(1+\cos 2\theta)d\theta = \frac{1}{2}\theta + \frac{1}{4}\sin 2\theta + c.$

EXERCISE 6.3

1 Check the results of the above examples, by differentiation.
2 Write down
   (i) $\int 6x^2 dx$ (ii) $\int (4x-3)dx$ (iii) $\int x^4 dx$
   (iv) $\int \cos 2x\, dx$ (v) $\int (x^3-2x)dx$ (vi) $\int x^6 dx$.
3 Find the integral of:
   (i) $\frac{3}{2}x$ (ii) $\frac{-2}{x^2}$ (iii) $5/x^2 + 1$
   (iv) $(x^2-1)/x^2$ (v) $(1+x)^2$ (vi) $(2+x)(x-3)$.
4 Integrate, with respect to x, the following:
   (i) $3x^2+2x+1$ (ii) $(3x-4)(2x-1)$ (iii) $(\cos x+\sin x)^2$
   (iv) $(x^2+1/x^2)^2$ (v) ax^2+bx+c (vi) $2x^{-\frac{1}{2}}+3x^{\frac{1}{2}}$.
5 Find
   (i) $\int (x^2+2x)dx$ (ii) $\int (t^2+2t)dt$ (iii) $\int \sin 2\theta\, d\theta$
   (iv) $\int (2u+u^{\frac{1}{2}})du$ (v) $\int (s+4)^3 ds$ (vi) $\int (u+u^{-1})^2 du$.

6.4 Evaluation of the arbitrary constant of integration

The process of integrating $f'(x)$ to obtain $f(x)$ involves an arbitrary constant c. Then c may be chosen so as to satisfy a condition that f takes a particular value for a given value of x. The graphs of the functions obtained by giving some set of values to the arbitrary constant of integration c are all translations of one another parallel to the y-axis (see Fig. 6.1). By choosing a suitable value of c we are choosing one particular graph, and we make this graph pass through some point (a, b).

EXAMPLE   *The gradient of the function* f *is given by* $f'(x) = 2x+1$. *Integrate* f'(x) *and, by choosing the arbitrary constant* c *from the set* $\{-1, 1, 3\}$, *draw the graphs of three functions which all have the same gradient as* f *at any given value of* x. *Draw also the graph of* f, *given that* $f(1) = 2$.

Since $f'(x) = 2x+1$, $f(x) = x^2+x+c$, for some constant c. The graphs of the functions corresponding to $c = -1$, $c = 1$, and $c = 3$ are shown in Fig. 6.1. For the required function f, $f(1) = 1+1+c = 2$, so $c = 0$, $f(x) = x^2+x$ and the graph of this function is also shown in Fig. 6.1.

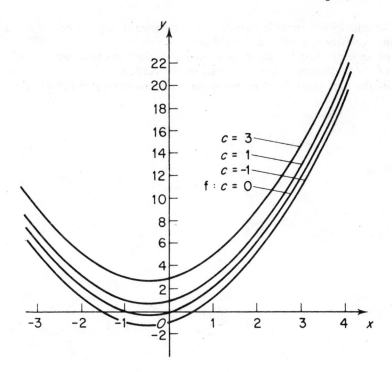

Fig. 6.1 Graphs of the functions $f(x) = x^2 + x + c$.

EXERCISE 6.4

1   Write down $\int 4x\,dx$ and for $c \in \{0, 1, 2\}$ draw the graph of the indefinite integral, where c is the constant of integration. For what value of c does the graph pass through $(3, 4)$?

2   The equation of a curve is $y = f(x)$. Find $f(x)$ given that:

   (i) $\dfrac{dy}{dx} = 4x$ and the curve passes through $(0, 4)$;

   (ii) $\dfrac{dy}{dx} = 3x^2 - 6x + 3$ and the curve passes through $(1, 0)$;

   (iii) $\dfrac{dy}{dx} = 3(x + 2)^2$ and the curve passes through $(0, 8)$;

   (iv) $\dfrac{dy}{dx} = 2\cos 2x - 2\sin 2x$ and the curve passes through $(0, 1)$.

3   At time t, the velocity of a particle moving in a straight line is $v = 3t^2 - 2t$. Given that the displacement s of the particle takes the value 2 when $t = 1$, find the value of s when $t = 2$.

4   At time t, the velocity of a particle moving in a straight line is $v = t^3 + 1$.
   (i) Through what distance does the particle move between $t = 3$ and $t = 4$?
   (ii) What is the average velocity between $t = 1$ and $t = 3$?

5   The velocity of a particle thrown vertically upwards is given by $v = u - gt$, at time t, where u, g are constants. Find:
   (i) an expression for the displacement s of the particle at time t,
   (ii) the maximum height h reached by the particle,
   (iii) the time elapsed before the particle returns to its point of projection.

7 Polynomials

7.1 Definitions

A *polynomial* is an algebraic expression consisting of a linear combinaton of positive integer powers of some indeterminate x. The *degree* of the polynomial is the highest exponent of x occurring in the expression.

For example, the polynomials $2x - 3$, $x^3 + 4x$, $ax^2 + bx + c$, have degrees 1, 3, 2, respectively. The general polynomial of degree n is

$$p(x) = a_n x^n + a_{n-1} x^{n-1} + \ldots + a_1 x + a_0, a_n \neq 0.$$

Then $a_n, a_{n-1}, \ldots, a_1, a_0$ are the *coefficients* of $p(x)$,
$\quad a_n$ is the *leading coefficient* of $p(x)$,
$\quad a_0$ is the *constant term* of $p(x)$,
$\quad a_i x^i$ is the *term of degree i* of $p(x)$.
If p is a function whose value $p(x)$ is a polynomial then p is a *polynomial function*. The polynomial functions of low degree are:

constant function : $\quad n = 0, \quad p(x) = a_0$;
linear function $\quad$: $\quad n = 1, \quad p(x) = a_1 x + a_0$;
quadratic function : $\quad n = 2, \quad p(x) = a_2 x^2 + a_1 x + a_0$.

Two polynomials are *equal* if they have the same degree and the same term of each degree. Equal polynomial functions are, of course, equal as functions.

EXERCISE 7.1

1 For the following polynomials state (a) the degree, (b) the constant term, (c) the leading coefficient, (d) $p(0)$, (e) $p(2)$:

$\quad$ (i) $p(x) = 2x + 3$ $\qquad$ (ii) $p(x) = x^2 - 4x$ $\qquad$ (iii) $p(x) = x^3 + x - 1$
$\quad$ (iv) $p(x) = 4$ $\qquad\qquad$ (v) $p(x) = x - 2$ $\qquad\quad$ (vi) $p(x) = x^2 - 4$
$\quad$ (vii) $p(x) = 5 - 2x$ $\quad$ (viii) $p(x) = 4x - 3x^2 - 7$ $\quad$ (ix) $p(x) = x^3 + 2x^4 - x^5$.

2 In question **1**, your answers to (b) and (d) should be the same in each case. Explain why this is so.

7.2 Algebra of polynomials

Polynomials may be added, subtracted or multiplied by a constant, and a linear combination of polynomials is also a polynomial. A linear combination of polynomials is obtained by multiplying each polynomial by a real number and adding the results. For example

$$(x^2 + 2x + 3) + (2x^2 - 4x - 2) = 3x^2 - 2x + 1.$$

The calculation of linear combinations may be laid out so that terms of the same degree are placed under each other.

EXAMPLE 1   *Given* $p(x) = x^3 - 2x^2 + 3x - 4$ *and* $q(x) = 2x^3 + x^2 + 2$, *find* $r(x) = 3p(x) - 2q(x)$.

$$
\begin{aligned}
3p(x) = \quad & 3x^3 - 6x^2 + 9x - 12 \\
-2q(x) = \ & -4x^3 - 2x^2 \qquad\quad - 4 \\
\hline
r(x) = \quad & -x^3 - 8x^2 + 9x - 16
\end{aligned}
$$

The product of two polynomials is also a polynomial and is calculated by multiplying one polynomial by each term of the other polynomial in turn, and then adding. Again the work is assisted by using the long multiplication layout.

EXAMPLE 2   *Find* $r(x) = p(x).q(x)$ *where* $p(x) = x^2 + 3x - 4$ *and* $q(x) = 2x - 1$.

$$
\begin{aligned}
p(x) = \quad & x^2 + 3x - 4 \\
q(x) = \quad & \underline{\qquad\quad 2x - 1} \\
(-1)p(x) = \quad & -x^2 - 3x + 4 \\
2xp(x) = \quad & \underline{2x^3 + 6x^2 - 8x} \\
p(x)q(x) = \quad & 2x^3 + 5x^2 - 11x + 4
\end{aligned}
$$

In Example 3 note that a gap is left in a line where a term of a given degree is absent, that is, has zero coefficient.

EXAMPLE 3   *Find* $p(x).q(x)$, *where* $p(x) = 2x^3 - x^2 + 4$, $q(x) = x^2 - 2x + 2$.

$$
\begin{aligned}
p(x) = \quad & 2x^3 - x^2 \qquad\quad + 4 \\
q(x) = \quad & \underline{\qquad\qquad x^2 - 2x + 2} \\
2p(x) = \quad & 4x^3 - 2x^2 \qquad\quad + 8 \\
-2xp(x) = \quad & -4x^4 + 2x^3 \qquad\quad - 8x \\
x^2 p(x) = \quad & \underline{2x^5 - x^4 \qquad\qquad + 4x^2} \\
p(x).q(x) = \quad & 2x^5 - 5x^4 + 6x^3 + 2x^2 - 8x + 8
\end{aligned}
$$

Another method, which we might call the 'loops method', allows the multiplication to be done mentally, with a minimum of writing. Each step consists of picking out all the terms in the product of a certain degree by multiplying appropriate terms in each bracket. These terms are indicated by connecting loops and each time a term is used it is marked with a tick. When finished, each term should have as many ticks as there are terms in the other bracket. The steps for the product of Example 3 are as follows:

$$(2x^3 - x^2 + 4)(x^2 - 2x + 2) \qquad\qquad 2x^3 x^2 = \quad 2x^5$$

$$(2x^3 - x^2 + 4)(x^2 - 2x + 2) \qquad\qquad 2x^3(-2x) - x^2 x^2 = -5x^4$$

$$(2x^3 - x^2 + 4)(x^2 - 2x + 2) \qquad\qquad 2x^3 2 + x^2 2x = \quad 6x^3$$

$$(2x^3 - x^2 + 4)(x^2 - 2x + 2) \qquad\qquad -2x^2 + 4x^2 = \quad 2x^2$$

$$(2x^3 - x^2 + 4)(x^2 - 2x + 2) \qquad\qquad 4(-2x) = -8x$$

$$(2x^3 - x^2 + 4)(x^2 - 2x + 2) \qquad\qquad 4.2 = \quad 8.$$

Now sum the terms on the right-hand side for the answer.

Note 1 The product of two polynomials of degrees m and n has degree $m + n$, since its leading coefficient is the product of their leading coefficients.

Note 2 The degree of a linear combination of two polynomials will be less than or equal to the larger of the two degrees. For example:

$$2(x^2 - 4) - (2x^2 - 6x) = 6x - 8,$$
$$2(x^2 - 4) - 3(x - 8) = 2x^2 - 3x + 16.$$

EXERCISE 7.2

1 Express as a polynomial the linear combination:
   (i) $(2x + 4) - (x + 3)$, (ii) $2(x + 5) + 3(x^2 - 2)$,
   (iii) $4(x^2 - 2x + 1) - 2(2x^2 + 3)$, (iv) $4(x - 2) - 3(x^2 - 2x + 5)$,
   (v) $5(1 - 3x + x^3) - 2(3x^2 - 2x + 4)$, (vi) $9(3x^2 + 4x + 1) - 7(3 - x - 2x^2)$.

2 Using long multiplication, express the product as a polynomial:
   (i) $(x - 2)(x^2 + x + 3)$, (ii) $(x^2 - 2x - 5)(x^3 + x)$, (iii) $(2x^2 + x + 1)^2$,
   (iv) $(x + 2)^2(x - 3)$, (v) $(x + 4)^3$, (vi) $(x^2 + 2)(x^3 - 3x)$.

3 Multiply the following polynomials, using the loop method:
   (i) $(x^2 + 2)(3x - 1)$, (ii) $(2x + 1)(3x^2 - 4x + 1)$, (iii) $(4x^3 - 2x^2)(3x^2 - 1)$,
   (iv) $(5x + 2)(5x^2 - x - 3)$, (v) $(4x^2 - 2x + 1)(1 + 3x - 2x^2)$,
   (vi) $(1 - 3x - 6x^2)(x + 2x^2 - 5x^3)$, (vii) $(3x - 1)(x^4 - 2x^3 + 3x^2 - 4x + 5)$,
   (viii) $(x^3 - 3x^2 + 2x - 1)(4 - 2x + 3x^2)$.

4 Given that $p(x) = 2x^2 - 3x + 4$, $q(x) = x^2 + 1$, find the polynomial:
   (i) $p(x) + q(x)$, (ii) $2p(x) + 3q(x)$, (iii) $3p(x) - 4q(x)$, (iv) $p(x)q(x)$,
   (v) $p(x)p(x)$, (vi) $\{q(x)\}^2$, (vii) $\{p(x) + q(x)\}\{p(x) - q(x)\}$.

5 Find the product of $1 + 2x + 3x^2 + 4x^3$ and $1 - 2x + x^2$.

6 (i) Find the coefficient of x^3 in the product $(3x^2 - 2x - 1)(2x^2 + 3x - 4)$.
   (ii) Find the coefficient of x^4 in the product $(3 + 2x - 4x^2)(2 - x^2 + 3x^3 - x^4)$.

7 Expand in ascending powers of x, as far as x^3:
   (i) $(1 - x + 2x^2 - x^3 + x^4)^2$, (ii) $(1 + x)^2(2 - x)(1 + 3x - x^2)$,
   (iii) $(2 + x)(1 - x^2)(3x - 4)^2$, (iv) $(3x^2 - 2x + 4)(4x - 2)(x^2 - x)$.

7.3 Rational functions

When polynomials are divided the result may not always be a polynomial. For example, $(x+2)/(x+1)$ is not a polynomial. Such fractions are called *rational functions* and may be manipulated in the usual way.

EXAMPLE   *Simplify* (i) $\dfrac{x+2}{x+1} - \dfrac{x+1}{x+2}$,   (ii) $\dfrac{x^2-1}{x+1}$.

(i) Use a common denominator:

$$\frac{x+2}{x+1} - \frac{x+1}{x+2} = \frac{(x+2)^2 - (x+1)^2}{(x+1)(x+2)} = \frac{x^2+4x+4-x^2-2x-1}{(x+1)(x+2)}$$

$$= \frac{2x+3}{(x+1)(x+2)}.$$

(ii) $\dfrac{x^2-1}{x+1} = \dfrac{(x-1)(x+1)}{x+1} = x-1$. In this case, the quotient of the two polynomials is also a polynomial. However, as a function, we must remember that the domain must exclude the point $x = -1$.

EXERCISE 7.3

1   Express as a rational function:

(i) $\dfrac{1}{x} - \dfrac{1}{x^2}$

(ii) $\dfrac{x}{1+x} + \dfrac{1}{1-x}$

(iii) $2x + \dfrac{1}{x+1}$

(iv) $\dfrac{2x}{3-x} + \dfrac{4}{x-3}$

(v) $x^2 + x - 1 + \dfrac{2}{x+3}$

(vi) $x - 1 + \dfrac{2x}{x^2+1}$

(vii) $\left(1 - \dfrac{7}{x+3}\right)\left(2 + \dfrac{16}{x-5}\right)$

(viii) $\dfrac{x^3+8}{x^3-8} - \dfrac{x^2}{x^2+2x+4}$

7.4 Division of a polynomial by a linear polynomial

A polynomial $p(x)$ of degree n can be divided by a linear polynomial $(x+a)$ to give a quotient $q(x)$ of degree $(n-1)$ and a constant remainder r. This is achieved by long division, in a sequence of steps. One step is described and the whole process is shown in an example.

Division step

When we are left with a polynomial $t(x)$ of degree m, with $m \geqslant 1$, we divide the leading term of $t(x)$ by x, giving $a_m x^{m-1}$ which then forms part of the quotient q(x). Then subtract $a_m x^{m-1}(x+a)$ from t(x) to give a new polynomial of degree less than m. This polynomial becomes t(x) for the next step, unless it is constant, when it becomes the remainder r. Initially t(x) = p(x). The work may be laid out as a long division.

EXAMPLE 1 *Divide $x^3 + 3x^2 - 3$ by $x+1$.*

$$
\begin{array}{r}
x^2 + 2x - 2 \longleftarrow \text{quotient} \\
x+1\overline{)x^3 + 3x^2 + 0x - 3}
\end{array}
$$

Step 1
$$\underline{x^3 + x^2}$$
$$2x^2 + 0x$$

Step 2
$$\underline{2x^2 + 2x}$$
$$-2x - 3$$

Step 3
$$\underline{-2x - 2}$$
$$-1 \longleftarrow \text{remainder}$$

Thus $x^3 + 3x^2 - 3 = (x+1)(x^2+2x-2) - 1$. **The quotient is $x^2 + 2x - 2$ and the remainder is -1.**

The 'loop method' can also be used, the terms of the dividend being ticked off as they are accounted for, as in Example 2 below.

EXAMPLE 2 *Divide $x^3 - 4x + 5$ by $x - 2$.*

$$x^3 + 0x^2 - 4x + 5 \quad :(x-2)x^2 = x^3 - 2x^2$$

$$x^3 + 0x^2 - 4x + 5 \quad :(x-2)(x^2+2x) = x^3 - 4x$$

$$x^3 + 0x^2 - 4x + 5 = (x-2)(x^2+2x) + 5,$$

so **the quotient is $x^2 + 2x$ and the remainder is 5.**

EXERCISE 7.4

1 Find the quotient q(x) and the remainder r on dividing:
   (i) $x^2 + 4x - 5$ by $x + 2$ (ii) $x^3 + 1$ by $x + 1$
   (iii) $x^2 + 3x + 2$ by $x + 1$ (iv) $3x^2 - 11x + 8$ by $x - 3$
   (v) $x^3 - 2x^2 + 3x - 6$ by $x + 2$ (vi) $2x^3 + 4x$ by $x + 1$
   (vii) $x^3 + x^2 - 9x + 9$ by $x + 3$ (viii) $x^4 - 3x^2$ by $x - 2$
   (ix) $x^3 - x^2 - 5x + 6$ by $x - 2$ (x) $2x^3 - x + 8$ by $2x - 1$
   (xi) $3x^4 + 2x^3 - x^2 + 4x - 5$ by $x + 3$ (xii) $x^4 - 3x^2 + 2x$ by $2x - 3$.

2 Check your results in question **1** by multiplying the quotient q(x) by the linear divisor and adding the remainder.

7.5 Factors of a polynomial

If a polynomial $p(x)$ can be written $p(x) = p_1(x) \cdot p_2(x)$ where $p_1(x)$ and $p_2(x)$ are polynomials of degree at least one, then p_1 and p_2 are factors of the polynomial p.

If we divide $p(x)$ by a linear polynomial $(x - \alpha)$ and obtain a quotient $q(x)$ and remainder r, then

$$p(x) = (x - \alpha) q(x) + r$$

so clearly $(x - \alpha)$ is a factor of $p(x)$ if $r = 0$.

Regarding the polynomials as functions we see that

$$p(\alpha) = (\alpha - \alpha) q(\alpha) + r = r.$$

These results give us two useful theorems.

Remainder theorem

The remainder on dividing the polynomial $p(x)$ by $(x - \alpha)$ is $p(\alpha)$.

Factor theorem

$[(x - \alpha)$ is a factor of the polynomial $p(x)] \Leftrightarrow p(\alpha) = 0$.

When $p(\alpha) = 0$, α is called a *zero* of p, or a *root* of $p(x) = 0$.
Note the signs: $(x - \alpha)$ is a factor when $p(\alpha) \ = 0$,
$(x + \alpha)$ is a factor when $p(-\alpha) = 0$.

As a slight modification, suppose

$$p(x) = (ax + b) q(x) + r.$$

then $p(-b/a) = (-b + b)\ q(-b/a) + r = r,$

so the remainder on dividing $p(x)$ by $(ax + b)$ is $p(-b/a)$ and the remainder on dividing $p(x)$ by $(ax - b)$ is $p(b/a)$.

EXAMPLE 1   *Find the remainder on dividing $x^3 + 2x^2 - 3x - 4$ by:* (i) $x - 1$, (ii) $x + 2$.

(i) The remainder is $p(1) = 1 + 2 - 3 - 4 = -4$.
(ii) The remainder is $p(-2) = -8 + 8 + 6 - 4 = 2$.

EXAMPLE 2   *Solve the equation $x^3 - 2x^2 - 5x + 6 = 0$.*

Since we cannot solve a cubic equation in general, we look for a factor $(x - \alpha)$ of $p(x) = x^3 - 2x^2 - 5x + 6$. Since the factors of 6 are $\pm 1, \pm 2, \pm 3, \pm 6$ we try some of these values for α.

$$p(-1) = -1 - 2 + 5 + 6 = 8,$$
$$p(1) = 1 - 2 - 5 + 6 = 0,$$

so $(x - 1)$ is a factor of $p(x)$.

On dividing p(x) by ($x-1$) we find

$$p(x) = x^3 - 2x^2 - 5x + 6 = (x-1)(x^2 - x - 6)$$
$$= (x-1)(x+2)(x-3)$$

on factorising the quadratic factor.

Hence p(1) = 0, p(-2) = 0, p(3) = 0 and the solution set of p(x) = 0 is $\{1, -2, 3\}$.

EXERCISE 7.5

1 Use the remainder theorem to find the remainder on dividing:
  (i) $x^3 - 2x^2 + x - 1$ by $x-1$ (ii) $x + x^2 + 1$ by $x+2$,
  (iii) $3x^2 + 2x - 1$ by $x-2$ (iv) $x^3 + x^2 + x + 1$ by $x+1$
  (v) $4x^3 + x - 2$ by $2x-1$ (vi) $x^4 + x^2 + 1$ by $x+2$.
  Check your answers by long division.

2 State whether the linear polynomial is a factor of the other polynomial:
  (i) $x-1$, $2x^3 + x^2 - 2x - 1$ (ii) $x-1$, $x^4 - 2x + 1$
  (iii) $x+2$, $x^3 - 3x^2 + 5x - 10$ (iv) $x+4$, $x^3 + 2x^2 + 5x - 12$
  (v) $x+1$, $2x^3 + 3x^2 - 1$ (vi) $2x-1$, $4x^3 + 3x - 2$
  (vii) $2x-3$, $x^4 - 2x - 3$ (viii) $3x+1$, $6x^3 - 13x^2 + 13x + 6$.

3 Factorise the polynomial completely (note that some quadratic factors can not themselves be factorised into two linear factors):
  (i) $2x^3 + x^2 - 2x - 1$ (ii) $2x^3 + 3x^2 - 1$ (iii) $x^3 + 3x^2 + 3x + 1$
  (iv) $x^4 - 1$ (v) $x^3 - 1$ (vi) $x^3 - x^2 - x + 1$
  (vii) $x^3 + x^2 - x + 2$ (viii) $x^3 - x^2 + x - 1$ (ix) $4x^3 + 3x - 2$
  (x) $6x^3 - 13x^2 + 13x + 6$ (xi) $3x^3 - 7x^2 + 4$ (xii) $x^3 - 4x^2 + x + 6$
  (xiii) $x^3 - 21x + 20$ (xiv) $4x^3 - 8x^2 + x + 3$ (xv) $6x^3 + 17x^2 - 4x - 3$.

4 Find the value of a, given that:
  (i) $x^3 - 2x^2 + 3x - a$ is divisible by $x-2$,
  (ii) $x^3 + x^2 + ax + 8$ is divisible by $x-1$,
  (iii) $x^3 + ax^2 - 2x + 1$ is divisible by $x+2$,
  (iv) $x^3 - 3x^2 + ax + 5$ has remainder 8 when divided by $x-2$,
  (v) $3x^3 + 2x^2 - ax + 3$ has remainder 1 when divided by $2x-1$.

5 Find the values of a and b and factorise the polynomial p(x) completely, given that:
  (i) p(x) = $x^3 + ax^2 - x + b$ and p(x) is divisible by $x-1$ and by $x+3$,
  (ii) p(x) = $x^3 + ax^2 + 2x + b$ and p(x) is divisible by $(x+1)(x-2)$,
  (iii) p(x) = $4x^3 + ax + b$ and p(x) is divisible by $(x+2)(2x-1)$.

6 Show that, for any constant a and any natural number n, $x - a$ is a factor of $x^n - a^n$. Find conditions (if any) on n that are required in order that:
  (i) $x + a$ is a factor of $x^n + a^n$,
  (ii) $x + a$ is a factor of $x^n - a^n$,
  (iii) $x - a$ is a factor of $x^n + a^n$.

7 Use the results of question 6 to state a linear factor of p(x) or to show that p(x) has no linear factors:
  (i) p(x) = $x^3 - 8$ (ii) p(x) = $x^4 + 16$
  (iii) p(x) = $x^3 + 27$ (iv) p(x) = $8x^3 - 125$.

8 The polynomial $x^3 + ax + b$, where a, b are constants, leaves remainders of 11 and 41 when divided by $(x-3)$ and $(x-4)$ respectively. Find the remainder when this polynomial is divided by $(x-5)$.   (L)

9   Given that $(x+3)$ is a factor of $f(x)$ where
$$f(x) \equiv 2x^3 - ax + 12,$$
find the constant a. Express $f(x)$ as a product of linear factors.   *(L)*

10   Let $f(x) = 2x^4 + ax^2 + bx - 60$. The remainder when $f(x)$ is divided by $(x-1)$ is -94. One factor of $f(x)$ is $(x-3)$. Determine the constants a and b.   *(L)*

11   $f(x) \equiv x^3 - x^2 + ax + b$, where a and b are constants. When $f(x)$ is divided by $(x-1)$ the remainder is 1, and when $f(x)$ is divided by x the remainder is 0. Find the remainder when $f(x)$ is divided by $(x-2)$.   *(L)*

12   Use the factor theorem to find a linear factor of $P(x)$, where $P(x) = x^3 + 6x^2 - 7x - 60$. Hence express $P(x)$ as a product of three linear factors.

13   Show that if $(x+t)$ is a common factor of $x^3 + px^2 + q$ and $ax^3 + bx + c$, then it is also a factor of $apx^2 - bx + aq - c$. Show that $x^3 + \sqrt{7}x^2 - 14\sqrt{7}$ and $2x^3 - 13x - \sqrt{7}$ have a common factor and hence find all the roots of the equation $2x^3 - 13x - \sqrt{7} = 0$.   *(L)*

7.6 The sign of a polynomial function

When a polynomial function can be factorised into a product of linear factors, the sign of its value at any point will be given by the signs of its factors. This may be used to give a rough sketch of the graph of the function.

EXAMPLE   *Find the roots of the polynomial function* $p(x) = x^3 - 2x^2 - 5x + 6$ *and sketch its graph.*

Using the factor theorem we evaluate $p(x)$ for values of x dividing 6.

$p(1) = 1 - 2 - 5 + 6 = 0$ $p(-1) = -1 - 2 + 5 + 6 = 8$
$p(2) = 8 - 8 - 10 + 6 = -4$ $p(-2) = -8 - 8 + 10 + 6 = 0$
$p(3) = 27 - 18 - 15 + 6 = 0$ $p(-3) = -27 - 18 + 15 + 6 = -24$,

and so $p(x) = c(x-1)(x+2)(x-3)$, where c is a constant. By comparing coefficients of x^3 we see that $cx^3 = x^3$ and $c = 1$.

So $p(x) = (x-1)(x+2)(x-3)$ and the graph of p crosses the x-axis where $x = -2, 1, 3$.

We obtain the signs of x from the signs of its factors:

	$(x+2)$	$(x-1)$	$(x-3)$	$p(x)$
for $x < -2$	$-$ ve	$-$ ve	$-$ ve	$-$ ve
for $-2 < x < 1$	$+$ ve	$-$ ve	$-$ ve	$+$ ve
for $1 < x < 3$	$+$ ve	$+$ ve	$-$ ve	$-$ ve
for $3 < x$	$+$ ve	$+$ ve	$+$ ve	$+$ ve.

Thus, the graph of p is of the form shown in Fig. 7.1.

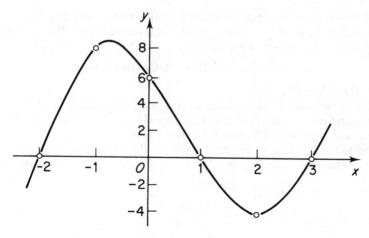

Fig. 7.1

EXERCISE 7.6

1 Using the signs of the factors of p(x), sketch its graph:
   (i) $p(x) = (x+1)(x-2)$ (ii) $p(x) = (x+2)(x-3)(x+4)$
  (iii) $p(x) = (2-x)(3-x)$ (iv) $p(x) = (2x+1)(2-x)$
   (v) $p(x) = (4x+1)(x-2)(x+1)$ (vi) $p(x) = (x+3)(2x-1)(x+1)(x-4)$.
2 Find the points where the graph of p(x) crosses the axes $O(x, y)$ and hence sketch the graph:
   (i) $p(x) = x^3 - 3x^2 + 2x$ (ii) $p(x) = x^4 - 2x^2 + 1$
  (iii) $p(x) = x - x^3$ (iv) $p(x) = 2x^3 - 3x^2 - 3x + 2$.
3 The cubic polynomial (that is, of degree 3) p(x) has a graph which passes through the four points $(-2, 0)$, $(1, 0)$, $(4, 0)$ and $(0, 16)$. Find p(x).
4 State a possible polynomial which has roots: (a) $\{-2, 4\}$; (b) $\{-4, -1, 3\}$; (c) $\{-3, 0, 2, 5\}$.

MISCELLANEOUS EXERCISE 7

1 Given that $f(x) = x^4 + 3x^3 - 5x^2 - 9x - 2$, find by trial two integer solutions of the equation $f(x) = 0$.
    Hence factorise $f(x)$ and solve the equation completely.   (*SMP*)
2 The cubic polynomial $x^3 + ax^2 + bx + c$ when divided by x, $(x-1)$, $(x-2)$ has remainders of 1, 2, 5 respectively. Find the values of a, b, and c.   (*L*)
3 Prove that the remainder on dividing a polynomial $f(x)$ by $(x-a)$ is $f(a)$.
    The remainder when a polynomial $f(x)$ is divided by $(x-2)$ is 3 and the remainder when it is divided by $(x+1)$ is 6. If the remainder when $f(x)$ is divided by $(x-2)(x+1)$ is $px+q$, find the numerical values of p and q.   (*JMB*)

4 The polynomial $P(x)$ leaves a remainder of 2 when divided by $(x-1)$ and a remainder of 3 when divided by $(x-2)$. The remainder when $P(x)$ is divided by $(x-1)(x-2)$ is $ax+b$. By writing

$$P(x) \equiv (x-1)(x-2)Q(x)+ax+b,$$

find the values of a and b.

Given also that $P(x)$ is a cubic polynomial with coefficient of x^3 equal to unity, and that -1 is a root of the equation $P(x)=0$, obtain $P(x)$. Show that the equation $P(x)=0$ has no other real roots. (*JMB*)

5 Find the quotient and the remainder (of degree less than or equal to 1) when the polynomial $x(x+1)(2x-3)$ is divided by the polynomial x^2+1. (*SMP*)

6 Given that the equations

$$x^3-2x+4=0$$

and

$$x^2+x+k=0$$

have a common root, show that

$$k^3+4k^2+14k+20=0.$$ (*JMB*)

7 Show that, when $f(x)$, a polynomial in x, is divided by $(x-a)(x-b)$, where $a \neq b$, the remainder is

$$\left(\frac{f(a)-f(b)}{a-b}\right)x+\left(\frac{af(b)-bf(a)}{a-b}\right).$$

Verify that $(2x-1)$ is a factor of

$$4x^3-3x+1$$

and find the other linear factors.

Determine the remainder when $4x^3-3x+1$ is divided by x^2+x-6. (*L*)

8 Composition of Functions and Applications to Graphs

8.1 Composition of mappings

Suppose that A, B and C are sets and that two mappings of functions are defined:

$$f : A \to B, \quad \text{with domain } A \text{ and codomain } B;$$
$$g : B \to C, \quad \text{with domain } B \text{ and codomain } C.$$

The corresponding mappings of any element x in A and any element y in B are

$$f : x \mapsto f(x), \quad g : y \mapsto g(y),$$

and these are composed to form the *composite function* gf of f and g, by

$$gf : A \to C, \text{ for all } x \in A, \quad gf : x \mapsto gf(x) = g(f(x))$$

so that $\qquad gf(x) = g(y), \text{ where } y = f(x)$.

It may be helpful to consider the function composition in terms of a Venn diagram showing the mappings (Fig. 8.1).

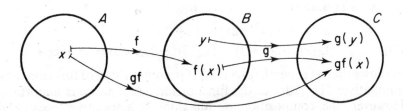

Fig. 8.1 Function composition: $gf(x) = g(f(x))$.

Note　The composite function gf is sometimes referred to as a *function of a function*, because $gf(x) = g(f(x))$ may be read as g of f of x, and sometimes the notation used for gf is $g \circ f$, the small circle between g and f being read as 'of'. However, it is not good practice to refer to gf as a function of a function because this phrase suggests a function whose domain is a set of functions, which is not at all the case here. We shall always use the term 'composite function'. It is quite a different way of combining functions from the product of functions (Chapter 1).

It is important to remember that the function gf means that you perform the mapping f first and follow it by g, the reverse of the order in which the functions are written.

EXERCISE 8.1

1   Given the functions $f: A \rightarrow B$ and $g: B \rightarrow C$, state the meaning of $gf(x)$, where:
(i) A is the set of children in a town, B is the set of all classes, C is the set of all schools in the town and x is John Smith, if f and g have the meanings 'the class containing' and 'the school of' respectively;
(ii) A, B and C are sets of people, x is you, f means 'the father of' and g means 'the mother of';
(iii) A is a set of football teams, B is a set of strips, C is a set of colours, f means 'the strip of', g means 'the colour of', and x is Manchester United.
2   In question 1 (ii), write down the meaning of $\{ff(x), fg(x), gf(x), gg(x)\}$.
3   Using question 1 (i) and (iii) as examples, show that gf can have a meaning when fg has no meaning.
    Give a condition to be satisfied by the sets A, B and C in order that fg has a meaning.
4   The domains of f and g are the set of real numbers $\mathbb{R}$. Given that $f(x) = 2x - 1$ and $g(x) = x^2$, write down:
(i) $fg(0)$, (ii) $gf(0)$, (iii) $ff(4)$, (iv) $fg(4)$, (v) $fgf(4)$, (vi) $fg(x)$, (vii) $gf(x)$, (viii) $fg(t)$.

8.2 Composition of real functions

When f and g are real functions, their domains and their codomains are all subsets of $\mathbb{R}$. For example, let f and g have domain $\mathbb{R}$, and for $x \in \mathbb{R}$ let $f(x) = x^2 - 2x$ and $g(x) = x + 3$. Then

$$gf(x) = g(x^2 - 2x) = x^2 - 2x + 3,$$
whilst $$fg(x) = f(x + 3) = (x + 3)^2 - 2(x + 3) = x^2 + 4x + 3.$$

This shows that, in general, $fg \neq gf$, that is, composition of functions is not commutative. The order of the functions in a composite is important.

However, the composition of functions is associative, because the composition of mappings generally is associative. Let f, g, h be real functions, then

$$(fg)h = f(gh) = fgh,$$

the second equality being one of notation, since there is no ambiguity. In other words, the brackets are unimportant in the composition of functions and so we omit them. To prove the result of associativity, for all x in the domain of fgh,

$$(fg)h(x) = fg(h(x)) = f(g(h(x))) = f(gh(x)) = f(gh)(x).$$

We recall from Chapter 1 that the domain of a real function may be a subset and not the whole of $\mathbb{R}$. In fact, when f is defined by means of a formula for $f(x)$, we assume that the domain of f is as large as possible and that it includes all the points where $f(x)$ is defined. When considering the domain of a composite function, we again assume that it is as large as possible, as a convention. Thus, if the domain of f is A and its range is B, $f(A) = B$; let the domain of g be C, then the domain of gf will be the subset of A which is mapped on to $B \cap C$. In words, the domain of gf is that subset of the domain of f which is mapped into the domain of g.

Notation   Since we do not need to use brackets when writing down the composite of three or more functions, there is only one meaning to fff or to ffff, and so we write $ff = f^2$, $fff = f^3$, $ffff = f^4$, and the composite of n copies of the function f is written f^n.

EXAMPLE 1   *Find the domain and the form of the function fgh given that*
$$f(x) = \frac{1}{x-1}, \; g(x) = 2x - 1, \; h(x) = x^2.$$

Following the convention that the domain of a real function is as large a subset of the real numbers as possible, the domains of g and h are each $\mathbb{R}$ and the domain of f is $\mathbb{R}\backslash\{1\}$. Then

$$fg(x) = f(2x-1) = \frac{1}{(2x-1)-1} = \frac{1}{2(x-1)} \quad \text{for } x \in \mathbb{R}\backslash\{1\}$$

$$(fg)h(x) = fg(x^2) = \frac{1}{2(x^2-1)} \quad \text{for } x \in \mathbb{R}\backslash\{1, -1\}.$$

Note also $f(gh)(x) = f(2x^2 - 1) = \frac{1}{2(x^2-1)} = (fg)h(x)$, as expected.

Hence, $fgh(x) = \frac{1}{2(x^2-1)}$ and fgh has domain $\mathbb{R}\backslash\{1, -1\}$.

When finding the value of a composite function, it is sometimes helpful to introduce some intermediate variables. For instance, if we put

$$h(x) = y \text{ and } g(y) = z \text{ then } fgh(x) = fg(y) = f(z).$$

In Example 1, $h(x) = y = x^2$, $g(y) = z = 2y - 1$,

$$fgh(x) = f(z) = \frac{1}{2y-1} = \frac{1}{2(x^2-1)}.$$

EXAMPLE 2   *The real functions f, g and h are given by the equations*
$$f(x) = 1/x, \quad g(x) = \sqrt{x}, \quad h(x) = x^2 - 2x - 3.$$

Find the domain and the range of the real function:
(*i*) fg, (*ii*) gf, (*iii*) gh, (*iv*) gfh.

(i) $fg(x) = f(\sqrt{x}) = 1/\sqrt{x}$, g has domain $\mathbb{R}^+ \cup \{0\}$, f has domain $\mathbb{R}\backslash\{0\}$ and so fg has domain $\mathbb{R}^+$. Clearly, the range of fg is also $\mathbb{R}^+$.

(ii) $gf(x) = g(1/x) = \sqrt{(1/x)} = 1/\sqrt{x} = fg(x)$, so $gf = fg$, with domain and range each equal to $\mathbb{R}^+$, as in (i).

(iii) $gh(x) = g(x^2 - 2x - 3) = \sqrt{\{(x-3)(x+1)\}}$, which will only be defined when $(x-3)(x+1) \geq 0$, that is, when either $x \geq 3$ or $x \leq -1$. The domain of h is $\mathbb{R}$ so the domain of gh is $\{x \in \mathbb{R}: x \geq 3\} \cup \{x \in \mathbb{R}: x \leq -1\}$. Since, for all x in this domain, $gh(x) \geq 0$, $gh(3) = 0$ and $gh(x)$ can be made as large as we like by taking x large enough, the range of gh is $\mathbb{R}^+ \cup \{0\}$.

(iv) $gfh(x) = \sqrt{\dfrac{1}{(x-3)(x+1)}}$, so gfh has domain

$\{x \in \mathbb{R}: x > 3\} \cup \{x \in \mathbb{R}: x < -1\}$, since the points 3, -1 do not lie in the domain of fh. Also, since $(x-3)(x+1)$ can be made as small as we like as long as x is near enough to 3, and can be made as large as we like for large enough x, the range of gfh is $\mathbb{R}^+$.

EXERCISE 8.2

1   Find the domain and range of (a) fg, (b) gf, when the real functions f and g are given by:
   (i) $f(x) = 1/x$, $g(x) = \sqrt{x}$; (ii) $f(x) = x^2$, $g(x) = \sqrt{x}$;
   (iii) $f(x) = x - 3$, $g(x) = 1/(x^2 - 1)$;   (iv) $f(x) = \sqrt{(x^2 - 4)}$, $g(x) = 2/(x+1)$.

8.3 Translation of a graph

Consider the graph of the function f, which is the curve with the equation $y = f(x)$. If we wish to translate the graph by a distance b parallel to the y-axis, then this means that we add b to each value of y since the point (x, y) is to become $(x, y + b)$. Thus the equation of the translated graph will be $y = f(x) + b$. Let g be the real function given by $g(x) = x + b$, then $f(x) + b = gf(x)$ so the new graph is the graph of the composite function gf. Note that the equation of gf could be written $y - b = f(x)$ so we can regard the translation as given by either adding b to $f(x)$ or subtracting b from y in the equation. This translation with $b = 3$ and $f(x) = x^2$ is shown in Fig. 8.2(a).

Now consider the translation of the graph of f a distance a parallel to the x-axis, so that the point (x, y) becomes $(x + a, y)$. Since (x, y) satisfies the equation $y = f(x)$, the point $(x + a, y) = (x_1, y)$ must satisfy the equation $y = f(x_1 - a)$. This means that, to find the height of the translated graph at x, we move back a distance a to the point $(x - a)$ and find the height of the original graph at $(x - a)$, which is $f(x - a)$. Therefore, the equation of the translated graph is $y = f(x - a) = fh(x)$, where $h(x) = x - a$. This translation of the parabola $y = x^2$, with $a = 2$ is shown in Fig. 8.2(b).

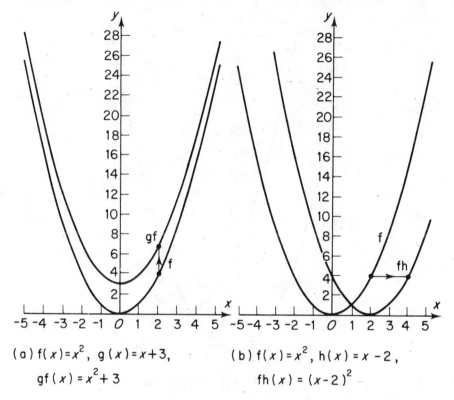

(a) $f(x)=x^2$, $g(x)=x+3$,
$gf(x)=x^2+3$

(b) $f(x)=x^2$, $h(x)=x-2$,
$fh(x)=(x-2)^2$

Fig. 8.2 Translations of graphs

The above two translations may be combined, and we may translate the graph of f by a distance a parallel to Ox and a distance b parallel to Oy. This means that the origin moves to (a, b) and every point (x, y) moves to $(x+a, y+b) = (x_1, y_1)$. The new coordinates (x_1, y_1), where $x_1 = x+a$ and $y_1 = y+b$, will satisfy an equation obtained from the equation $y = f(x)$ by replacing x, y by $x_1 - a$, $y_1 - b$ respectively and so $y_1 - b = f(x_1 - a)$. Therefore, the equation of the translated graph is

$$y - b = f(x - a) \quad \text{or} \quad y = f(x - a) + b = gfh(x).$$

In the case when $f(x) = x^2$, $gfh(x) = (x - 2)^2 + 3$ and the parabola with this equation is shown in Fig. 8.3.

Summary   Let $h(x) = x - a$ and $g(x) = x + b$, then a translation of the graph of the function f:

distance a along Ox is the graph of fh;
distance b along Oy is the graph of gf;
a along Ox and b along Oy is the graph of gfh.

Note that the different signs of the constants in h and g occur because in

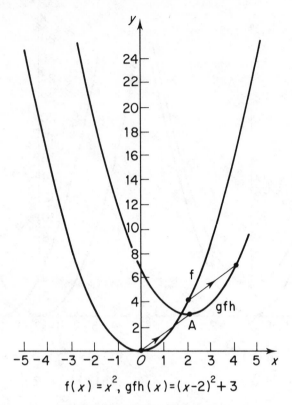

$$f(x) = x^2, \, gfh(x) = (x-2)^2 + 3$$

Fig. 8.3 Translation of a graph

order to translate the point (x, y) on the graph of f to the point $(x + a, y + b)$ it is necessary to replace x and y in the equation of the graph by $x - a$ and $y - b$ respectively. In Fig. 8.3, the vertex O of the graph of f becomes the vertex A of the graph of gfh, since $f(0) = 0$, and

$$gfh\,(a) = gf\,(a - a) = gf\,(0) = g\,(0) = b.$$

It is, thus, necessary to subtract a from x and add b to $f(x)$.

8.4 One-way stretches of a graph

In order to stretch the graph of $f(x)$ in the Oy direction, with a scale factor 2, the point on the graph with coordinates $(x, f(x))$ has to become the point $(x, 2f(x))$. Hence, if g is the function given by $g(x) = 2x$, then $gf(x) = 2f(x)$ and the graph of gf will be the graph of f stretched with scale factor 2 in the Oy direction. In the case when $f(x) = x^2$, $gf(x) = 2x^2$, and this stretch is shown in Fig. 8.4(a).

On the other hand, in order to stretch the graph of f with scale factor 2 in the Ox direction, the point of the graph with coordinates $(x, f(x))$

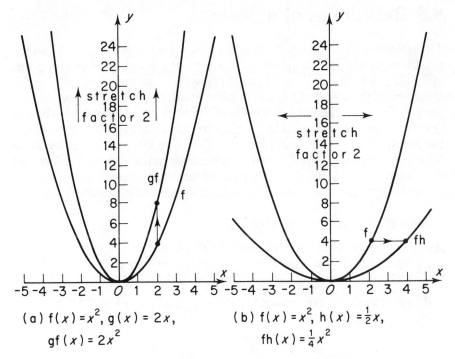

Fig. 8.4 One-way stretches of a graph.

becomes $(2x, f(x))$ and the point $(\frac{1}{2}x, f(\frac{1}{2}x))$ becomes $(x, f(\frac{1}{2}x))$. This is achieved by the composite function fh, where $h(x) = \frac{1}{2}x$ so $fh(x) = f(\frac{1}{2}x)$, and this stretch is shown in Fig. 8.4(b) for the parabola $y = x^2$.

The combination of the above two one-way stretches means that the graph of the function gfh, with $g(x) = 2x$, $h(x) = \frac{1}{2}x$, is obtained from the graph of f by stretch with scale factors 2 parallel to Ox and Oy, that is an all-round stretch, or magnification, of scale factor 2. Of course, stretches with different scale factors can be combined, as shown in the summary.

Summary Let $h(x) = \dfrac{1}{a}x$ and $g(x) = bx$, then

the graph of fh is the graph of f stretched along Ox with scale factor a,
the graph of gf is the graph of f stretched along Oy with scale factor b,
the graph of gfh is the graph of f stretched both along Ox with scale factor
   a and along Oy with scale factor b.

Again, note that x is divided by a in h and multiplied by b in g, because this corresponds to dividing y by b in the equation $y = f(x)$.

Also, note that, instead of obtaining a new graph by one of these one-way stretches, the same graph could be used with a corresponding change of scale on the appropriate axis.

8.5 Reflections of a graph

The reflection of the point (x, y) in the axis Ox is the point $(x, -y)$, so the reflection in Ox of the point $(x, f(x))$ on the graph of f is the point $(x, -f(x))$. Let $g(x) = -x$. Then $-f(x) = gf(x)$, so the graph of gf is the reflection of the graph of f in the axis Ox. Similarly, $fg(x) = f(-x)$ and the point $(x, f(-x))$ is the reflection in Oy of the point $(x, f(x))$, and the graph of fg is the graph of f reflected in Oy. Combining these two transformations, the graph of gfg is the graph of f first reflected in Ox and then reflected in Oy (or the combination of these two reflections in the opposite order), and this can also be expressed as a rotation of the graph about O through $180°$ or as a reflection in the origin O, or as an enlargement with scale factor -1.

Summary   Let $g(x) = -x$, then

the graph of gf is the reflection of the graph of f in Ox,
the graph of fg is the reflection of the graph of f in Oy,
the graph of gfg is the graph of f rotated $180°$ about 0.

These reflections are shown for the graph of 2^x in Fig. 8.5.

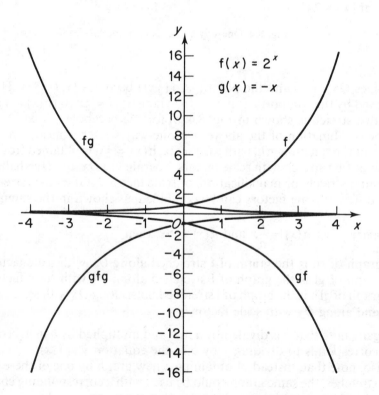

Fig. 8.5 Reflections of the graph of 2^x

8.6 Composition of a function with a linear function

The result of the composition of a function f with a linear function g can be found from the previous results as a combination of translations, stretches and reflections.

We illustrate this by an example where f and g are given by $f(x) = 2^{(\frac{1}{2}x+1)}$ and $g(x) = -2x+3$ (Fig. 8.6). Then $g = g_1 g_2 g_3$, where

$$g_1(x) = x+3, \quad g_2(x) = -x, \quad g_3(x) = 2x.$$

Thus, the graph of $gf = g_1 g_2 g_3 f$ is obtained from the graph of f by a succession of three transformations:

(i) $g_3 f$: a stretch with scale factor 2 in the direction Oy,
(ii) $g_2 g_3 f$: a reflection in the axis Ox,
(iii) gf: a translation of 3 units parallel to Oy.

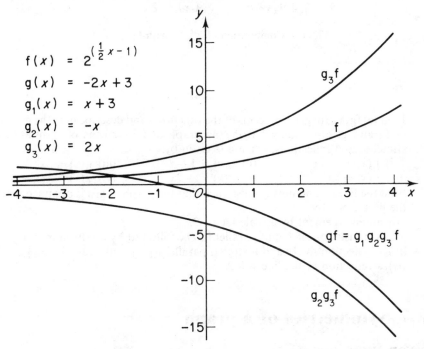

$$f(x) = 2^{(\frac{1}{2}x-1)}$$
$$g(x) = -2x+3$$
$$g_1(x) = x+3$$
$$g_2(x) = -x$$
$$g_3(x) = 2x$$

Fig. 8.6 Composition of $2^{(\frac{1}{2}x+1)}$ with $(-2x+3)$.

Similarly, the graph of the function fg can be obtained from the graph of f through three stages:

(i) fg_1: a translation by (-3) units in the Ox direction,
(ii) $fg_1 g_2$: a reflection in the axis Oy,
(iii) fg: a stretch with scale factor $\frac{1}{2}$ in the direction of Ox.

This is demonstrated in Fig. 8.7.

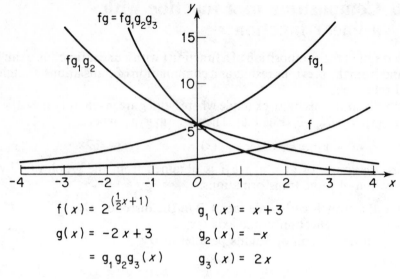

$$f(x) = 2^{\left(\frac{1}{2}x+1\right)}$$

$$g(x) = -2x+3$$

$$= g_1 g_2 g_3 (x)$$

$$g_1(x) = x+3$$

$$g_2(x) = -x$$

$$g_3(x) = 2x$$

Fig. 8.7 Composition of $2^{\left(\frac{1}{2}x+1\right)}$ with $(-2x+3)$.

EXERCISE 8.6

1 Find (a) fg(x), (b) gf(x) in terms of the function f and describe how the graphs of fg and gf are obtained from the graph of f by means of translations, stretches and reflections, when g is given by:
   (i) $g(x) = x-1$ (ii) $g(x) = 1-x$ (iii) $g(x) = 3x$
   (iv) $g(x) = \frac{1}{2}x$ (v) $g(x) = -2x$ (vi) $g(x) = \frac{1}{2}(1-x)$.

2 Find the linear functions g and h such that the graph of gfh is obtained from the graph of f by:
   (i) an enlargement by a scale factor 3,
   (ii) a translation of 1 unit parallel to Ox followed by a reflection in Oy,
   (iii) a reflection in Oy and a stretch parallel to Oy with scale factor $\frac{1}{2}$,
   (iv) a reflection in the line $y = 2$.

8.7 Symmetries of a graph

Even functions

The graph of an even power of x, say x^{2n}, $n \in \mathbb{N}$, is symmetrical about the axis Oy, since $(-x)^{2n} = x^{2n}$. Any function $f: x \mapsto f(x)$ whose graph is symmetrical about Oy is called an *even* function.

f is an even function $\Leftrightarrow f(-x) = f(x)$, for all x in the domain of f.

Examples of even functions are x^2, $3x^4$, $x^4 + 2x^2 - 1$, $3\cos 2x$. Let $g(x) = -x$, then $f(-x) = fg(x)$ and so f is even if $fg = f$.

Odd functions

The graph of an odd power of x, say x^{2n+1}, $n \in \mathbb{N}$, is symmetrical about the origin O, that is, it is unaltered by a rotation through $180°$ about O, which displacement can also be described as an enlargement with scale factor -1. Any function f with the same symmetry satisfies the equation $f(-x) = -f(x)$, and is called an *odd* function.

   f is an odd function $\Leftrightarrow f(-x) = -f(x)$, for all x in the domain of f.

Examples of odd functions are $x, 2x^3, x^5 - 4x^3, 2 \sin 3x, \tan 2x$. Using the same function g as before, with $g(x) = -x$, then f is an odd function if $fg = gf$. Fig. 8.8 shows graphs of an even function and an odd function.

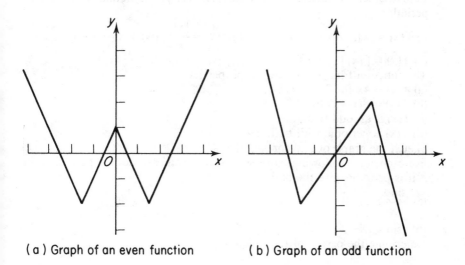

(a) Graph of an even function   (b) Graph of an odd function

Fig. 8.8

Periodic functions

Suppose that, for some fixed number k, $f(x+k) = f(x)$ for all x in the domain of f.

Then if $g(x) = x + k$, $fg = f$ and the graph of f is unaltered by a translation of amount k parallel to the x-axis. Then $fg^n = fgg^{n-1} = fg^{n-1} = \ldots = fg = f$, and the graph of f is unaltered by a translation of amount nk, parallel to Ox, with $n \in \mathbb{Z}$. Let k be the smallest positive number such that

$$f(x+k) = f(x), \text{ for all } x \text{ in the domain of f,}$$

then we say f is *periodic*, with period k.

For example, $\cos x$ and $\sin x$ are periodic, with period 2π,
   $\tan x$ and $\cos 2x$ are periodic, with period π,
   $\sin 3x$ is periodic with period $\tfrac{2}{3}\pi$.

EXERCISE 8.7

1 State which of the properties, odd, even, periodic or none of these, are held by the function:
   (i) $x^2 - 2$, (ii) $(x - 2)^2$, (iii) $\sqrt[3]{x}$, (iv) $\cos x$, (v) $\tan(x - \pi)$,
   (vi) $\sin x \cos x$, (vii) $x \sin x$, (viii) $x^3 \cos x$, (ix) $\sin(x^2)$.

2 Define the following real functions:
   (a) modulus of x: $|x| = \sqrt{(x^2)} =$ the positive square root of x^2, so $|x| = x$ for $x \geqslant 0$ and $|x| = -x$ for $x < 0$;
   (b) integer part of x: $[x] =$ the largest integer which is not greater than x. For example, $|2| = 2$, $|-3| = 3$, $[2\frac{1}{2}] = 2, [-2\frac{1}{2}] = -3 = [-3]$. Write down: (i) $[4\frac{1}{4}]$, (ii) $|-4\frac{7}{8}|$, (iii) $[4\frac{5}{8}]$, (iv) $[-4\frac{5}{8}]$.

3 Using the definitions of the functions in question **2**, sketch the function f and state whether the function f is (a) odd, (b) even, (c) periodic and, if so, state the period:

   (i) $f(x) = |x|$, (ii) $f(x) = [x]$, (iii) $f(x) = \dfrac{|x|}{x}$, (iv) $f(x) = x - [x]$,

   (v) $f(x) = [|x|]$, (vi) $f(x) = |[x]|$, (vii) $f(x) = |x| - [|x|]$.

4 The function f has the following properties:
   (a) $f(x) = 4x$ for $0 \leqslant x < 2/3$,
   (b) $f(x) = 8(1 - x)$ for $2/3 \leqslant x \leqslant 1$,
   (c) $f(x)$ is an odd function,
   (d) $f(x)$ is periodic with period 2.
   Sketch the graph of $f(x)$ for $-3 \leqslant x \leqslant 3$. $\hspace{1cm}$ *(L)*

5 For each of the following expressions state whether or not it is periodic, and, if it is periodic, give the period.

   (a) $\dfrac{\sin x}{x}$, $x \neq 0$,

   (b) $|\sin x|$. $\hspace{1cm}$ *(L)*

6 Determine the period of the function

   $$f: x \to \sin^2\left(\frac{x}{3}\right), \qquad x \in \mathbb{R}. \hspace{1cm} (JMB)$$

7 The function f is periodic with period π and

   $$f(x) = \sin x \quad \text{for } 0 \leqslant x \leqslant \pi/2,$$
   $$f(x) = 4(\pi^2 - x^2)/(3\pi^2) \quad \text{for } \pi/2 < x < \pi.$$

   Sketch the graph of $f(x)$ in the range $-\pi \leqslant x \leqslant 2\pi$. $\hspace{1cm}$ *(L)*

8 A function $g(x)$ of period 2π is defined by

   $$g(x) = x^2 \quad \text{for } 0 \leqslant x \leqslant \pi/2,$$
   $$g(x) = \pi^2/4 \quad \text{for } \pi/2 < x \leqslant \pi.$$

   Given also that $g(x) = g(-x)$ for all x, sketch the graph of $g(x)$ for $-2\pi \leqslant x \leqslant +2\pi$. $\hspace{1cm}$ *(L)*

9 The function f is periodic with period 2 and
   $$f(x) = 4/x \hspace{1.5cm} \text{for } 1 < x \leqslant 2,$$
   $$f(x) = 2(x - 1) \hspace{1cm} \text{for } 2 < x \leqslant 3.$$
   Sketch the graph of f when $-3 \leqslant x \leqslant 3$. $\hspace{1cm}$ *(L)*

10 The functions f and g are defined by
$$f: x \mapsto \sin 2x, \qquad x \in \mathbb{R},$$
$$g: x \mapsto \cot x, \qquad x \in \mathbb{R}, \quad x \neq k\pi \, (k \in \mathbb{Z})$$
State the periods of f and g.
Find the period of the function f.g.
On separate axes, sketch the graphs of f, g and f.g for the interval $\{x: -\pi < x < \pi, x \neq 0\}$.
Find the range of the function f.g. *(JMB)*

11 The function f is defined by
  (i) f has domain $\mathbb{R}$,
  (ii) f is even,
  (iii) f is periodic with period 2,
  (iv) for $0 \leqslant x \leqslant 1$, $f(x) = x^3$.
Sketch the graph of $f(x)$ for $-4 \leqslant x \leqslant 4$.

12 The function f is periodic with period 3 and
$$f(x) = \sqrt{(9 - 4x^2)} \quad \text{for } 0 < x \leqslant 1 \cdot 5,$$
$$f(x) = 2x - 3 \qquad \text{for } 1 \cdot 5 < x \leqslant 3.$$
Sketch the graph of $f(x)$ in the range $-3 \leqslant x \leqslant 6$. *(L)*

8.8 Inverse functions

Identity function

The function e, given by $e(x) = x$ for all $x \in \mathbb{R}$, has a graph with equation $y = x$, which is the line through O with unit gradient; e is an odd function and, for every real function f,

$$fe(x) = f(e(x)) = f(x) = e(f(x)) = ef(x)$$

for all x in the domain of f. Therefore, $fe = f = ef$ and so e is the identity of the operation of composition of real functions. The function e is called the *identity function*. In cases when we wish to restrict the domain to some subset X of $\mathbb{R}$, we shall still use the symbol e for the identity function with domain X.

Inverse function

Let the real function f have domain X and range Y,

$$f: X \rightarrow Y = f(X), \, X \subseteq \mathbb{R}, \, Y \subseteq \mathbb{R}.$$

We say that f has an inverse function f^{-1} if

$$f^{-1}: Y \rightarrow X, \text{ such that for all } x \in X, \, f^{-1}f(x) = x,$$
which means that if $y = f(x)$ then $f^{-1}(y) = x$, so that $f^{-1}f$ is the identity function with domain X. If such an f^{-1} exists, then for every $y \in Y$, there exists $x = f^{-1}(y) \in X$ such that

$$y = f(x) = f(f^{-1}(y)) = ff^{-1}(y) = e(y),$$

and so ff^{-1} is the identity function with domain Y.

Inverse image

The inverse of a function f may not exist. For example, if f has domain $\mathbb{R}$ and $f(x) = x^2$, then if $y = f(x)$, $y = f(-x)$, and we would not be able to give a unique meaning to $f^{-1}(y)$, since both x and $-x$ would be possible candidates. However, a meaning can still be given to f^{-1} by defining a function whose domain is the set of all subsets of Y. We define the *inverse image* of a subset A of Y as $f^{-1}(A)$ where

$$f^{-1}(A) = \{x \in X : f(x) \in A\},$$

then f^{-1} is a function with domain the set of all subsets of Y and range the set of all subsets of X.

There will not be any ambiguity in using the same symbol f^{-1} for the inverse image of a set, as well as for the inverse function of f when this happens to exist. If $y \in Y$, then the inverse image of y is $f^{-1}(\{y\})$ whereas, if f has an inverse function, then the inverse of y is $f^{-1}(y)$.

EXAMPLE 1   *The real function f is given by* $f(x) = 2x - 1$. *Show that f has an inverse function* f^{-1}, *given by* $f^{-1}(x) = \frac{1}{2}(x+1)$. *Write down:* (i) $f(3)$; (ii) $f(\{0, 1, 2\})$; (iii) $f(\mathbb{Z})$; (iv) $f^{-1}(-1)$; (v) $f^{-1}(\{0, 1, 2\})$; (vi) $f^{-1}(\mathbb{Z})$.

Let $2x - 1 = y$, then $2x = y + 1$ and $x = \frac{1}{2}(y + 1)$. Therefore, f has domain and range equal to $\mathbb{R}$. The inverse function f^{-1} is obtained by interchanging x and y in the last equation. Then, for the inverse function $y = \frac{1}{2}(x+1)$ and so f has an inverse function f^{-1}, with $f^{-1}(x) = \frac{1}{2}(x+1)$.
(i) $f(3) = \mathbf{5}$;   (ii) $f(\{0,1,2\}) = \{-\mathbf{1}, \mathbf{1}, \mathbf{3}\}$;   (iii) $f(\mathbb{Z})$ is **the set of odd integers**;
(iv) $f^{-1}(-1) = \mathbf{0}$;   (v) $f^{-1}(\{0,1,2\}) = \{\frac{1}{2}, \mathbf{1}, \frac{3}{2}\}$;   (vi) $f^{-1}(\mathbb{Z}) = \{\frac{1}{2}n : n \in \mathbb{Z}\}$.

EXAMPLE 2   *Find the domain and range of the function* f *where* $f(x) = 1/(x^2 - 1)$. *Show that an inverse function* f^{-1} *does not exist. Write down:*
(i) $f(3)$;   (ii) $f(-1)$;   (iii) $f(0)$;   (iv) $f(-\frac{1}{2})$;   (v) $f^{-1}(\{-1\})$;   (vi) $f^{-1}(\{0\})$;
(vii) $f^{-1}(\{\frac{1}{3}, \frac{1}{8}, \frac{1}{15}\})$.

Since $x^2 - 1$ vanishes for $x \in \{1, -1\}$, the domain of f is $\mathbb{R}\backslash\{\mathbf{1}, -\mathbf{1}\}$. For $x > 1$ and for $x < -1$, $f(x)$ is positive, $f(x)$ tends to zero as x tends to infinity and tends to infinity as x tends to 1 or -1. For $-1 < x < 0$ and for $0 < x < 1$, $f(x)$ is negative and is less than or equal to -1 and it tends to negative infinity as x tends to 1 or to -1. Thus the range of f is $\{x: x > 0\} \cup \{x: x \leqslant -1\}$. Since $f(2) = f(-2) = \frac{1}{3}$, an inverse function cannot exist, since $f^{-1}(\frac{1}{3})$ cannot be defined.
(i) $f(3) = \frac{1}{8}$;   (ii) $f(-1)$ **does not exist**:   (iii) $f(0) = -\mathbf{1}$;   (iv) $f(-\frac{1}{2}) = -\frac{3}{4}$;
(v) $f^{-1}(\{-1\}) = \{\mathbf{0}\}$;   (vi) $f^{-1}(\{0\})$ **does not exist** (or is ϕ);
(vii) $f^{-1}(\{\frac{1}{3}, \frac{1}{8}, \frac{1}{15}\}) = \{\mathbf{2}, -\mathbf{2}, \mathbf{3}, -\mathbf{3}, \mathbf{4}, -\mathbf{4}\}$.

EXERCISE 8.8

1   The following functions all have the domain $\mathbb{N}$ and are defined by the given value $f(n)$ for $n \in \mathbb{N}$:

(i) $1-n$, (ii) n^2, (iii) $1/(n+1)$, (iv) n^3.

In each case find the range of f and show that f has an inverse function f^{-1}.

2   Given that the functions, defined in (i)–(iv) of question 1, are real functions, find, in each case, the domain and the range of the function and determine whether it has an inverse.

3   The linear function f is given by $f(x) = ax + b$, with domain $\mathbb{R}$. State the range of f, whether f has an inverse function f^{-1}, and the form of $f^{-1}(x)$ when the inverse exists. Distinguish between $a = 0$ and $a \neq 0$.

4   Prove that the following real functions have no inverse functions. In each case give a restricted domain such that the function then has an inverse and state the form of the inverse function:

(i) $2x^2 - 1$, (ii) $(x^2 - 1)^{-2}$, (iii) $\sin x$, (iv) $\cos 4x$, (v) $|x|$, (vi) $[x] - x$.

8.9 Conditions for an inverse function to exist

Suppose that a real function f has domain X and range Y and an inverse function f^{-1} with domain Y and range X. If $a, b \in X$ and $f(a) = f(b)$, then

$$a = f^{-1}f(a) = f^{-1}(f(a)) = f^{-1}(f(b)) = f^{-1}f(b) = b.$$

This implies that for every y in Y there is one and only one x in X such that $y = f(x)$. In other words, the inverse image of every one-element subset of Y is a one-element subset of X. We call such a function f *one-one*.

Definition   The function $f: X \rightarrow Y = f(X)$ is one-one if, for every $y \in Y$, there is a unique $x \in X$ such that $y = f(x)$.

We have proved that if f^{-1} exists then f is one-one. The converse is also true, namely that if f is one-one then f^{-1} exists.

Proof   Suppose that the function $f: X \rightarrow Y = f(X)$ is a one-one function. Then, given any element y of Y, there is a unique element x of X such that $f(x) = y$. So let the equation $x = g(y)$ define a function g with domain Y. Then the range of g is X and $gf(x) = g(y) = x$ for every x in X. Hence g is the inverse function f^{-1} of f.

We have now proved the following theorem.

Theorem   Let $f: X \rightarrow Y$ be a real function with range $Y = f(X)$. Then f is one-one $\Leftrightarrow$ f has an inverse function f^{-1}.

Another way of stating this theorem is:

the condition that the real function f is one-one is both necessary and sufficient for f to possess an inverse function f^{-1}.

Suppose that $f: X \rightarrow Y$ is a real one-one function with inverse function $f^{-1}: Y \rightarrow X$. Then the graph of f has equation $y = f(x)$. But $y = f(x)$ if $x = f^{-1}(y)$, so the graph of f^{-1}, which has equation $y = f^{-1}(x)$ is obtained from the graph of f by a reflection in the line $y = x$ since this reflection transforms (x, y) to (y, x). The equation of f^{-1} is obtained from the equation of f by interchanging x and y. This is done in §9.6 for the particular function 2^x, and the graphs of 2^x and its inverse function are shown in Fig. 9.2.

EXAMPLE   *Determine whether or not the function f has an inverse function* f^{-1}, *and illustrate your argument with Cartesian graphs, in the cases*
(i) $f(x) = x^2$ *and f has domain* $\mathbb{R}$,
(ii) $f(x) = x^2$ *and f has domain* $X = \{x \in \mathbb{R}: x < 0\}$.

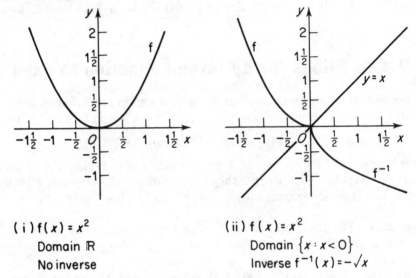

(i) $f(x) = x^2$

Domain $\mathbb{R}$

No inverse

(ii) $f(x) = x^2$

Domain $\{x : x < 0\}$

Inverse $f^{-1}(x) = -\sqrt{x}$

Fig. 8.9

The graphs of the two functions defined in (i) and (ii) are shown in Fig. 8.9 (i) and (ii). In (i), the range of f is $\mathbb{R}^+ \cup \{0\}$ and $f(1) = 1 = f(-1)$. So f is not one-one and has no inverse function. In fact, apart from 0, each value in the range of f arises from two points in its domain.

In case (ii), with the restricted domain f becomes a one-one function and its inverse function f^{-1}, with domain $\mathbb{R}^+$, is given by $f^{-1}(x) = -\sqrt{x}$.

EXERCISE 8.9

1   The function f is defined by $f(x)$ and a given domain, in each of the following cases. Find an expression for $f^{-1}(x)$, where f^{-1} is the inverse function of f. State the domain of f^{-1} and show that $f^{-1}f(x) = x$, for all x in the domain of f.
   (i) $f(x) = 2x$, domain $\mathbb{R}$ (ii) $f(x) = x - 3$, domain $\mathbb{Z}$
   (iii) $f(x) = 3x + 2$, domain $\mathbb{R}$ (iv) $f(x) = (x + 2)^2$, domain $\mathbb{R}^+$
   (v) $f(x) = x^3$, domain $\mathbb{R}$ (vi) $f(x) = \sqrt{x}$, domain $\mathbb{R}^+$
   (vii) $f(x) = x|x|$, domain $\mathbb{R}$ (viii) $f(x) = [x] - x$, domain $\{x : 2 \leqslant x < 3\}$
   (ix) $f(x) = \sqrt{(1 - x^2)}$, domain $\{x : 0 \leqslant x \leqslant 1\}$.

2 The function f, with domain ℝ, is given by $f(x) = x^2 - 4x + 5$. Find the range of f and sketch the graph of f. Find the real number d such that the following two conditions are both satisfied:
(i) the function g which is f with domain restricted to $\{x : d \leqslant x\}$ is a one-one function,
(ii) the function h which is f with domain restricted to $\{x : x \leqslant d\}$ is a one-one function.
State the domains of g^{-1} and h^{-1} and find $g^{-1}(x)$ and $h^{-1}(x)$.

3 Given that the real functions f, g and h, with domain ℝ, satisfy the condition that $fg = hf = e$ (the identity function), prove that $g = h$.

4 A condition P is a necessary condition for a function f to be one-one means that

'f is one-one' $\Rightarrow$ 'P is true'.

P is a sufficient condition for a function f to be one-one means that

'P is true' $\Rightarrow$ 'f is one-one'

For the following four conditions, state which is necessary, which is sufficient, which is both, and which is neither, for the function f to be one-one:
P(1) for all a, b in the domain of f, $f(a) = f(a) \Rightarrow a = b$;
P(2) for all a, b in the domain of f, $f(a) \neq f(b) \Rightarrow a \neq b$;
P(3) for all a, b in the domain of f, $a = b \Rightarrow f(a) = f(b)$;
P(4) for all a, b in the domain of f, $a \neq b \Rightarrow f(a) \neq f(b)$.

5 Given two functions, f with domain A and range B, and g with domain B and range A, such that gf is the identity function with domain A, prove that $g = f^{-1}$ and that fg is the identity function with domain B.

6 The function f, with domain D and range R, is $1-1$. Prove that its inverse function f^{-1}, with domain R and range D, is also $1-1$.

7 The function f, with domain D, and the function g, with domain C, satisfy the condition that gf is the identity function with domain D. Prove that f is one-one. Prove that g is not necessarily the inverse function of f by considering the case when $f(x) = \sqrt{x}$, $g = x^2$, $D = ℝ^+$, $C = ℝ$.

MISCELLANEOUS EXERCISE 8

1 Given that $f(x) = \dfrac{x+2}{x-1}$, where the domain of f is all real values of x except 1, sketch the graph of $f(x)$. Find $f^{-1}(x)$, where f^{-1} is the inverse of the function f, and state the domain of f^{-1}.

2 The functions f and g are defined by

$$f : x \mapsto \cos x \quad (0 \leqslant x \leqslant \pi),$$
$$g : x \mapsto \tan x \quad (0 \leqslant x < \tfrac{1}{2}\pi).$$

Sketch separate graphs of each of f, g and f^{-1}. State, giving your reasons, whether $f^{-1}g$ and $g^{-1}f$ exist. Solve the equation $f(x) = g(x)$, giving two decimal places in your answer. *(C)*

3 Find the ranges of the following functions:
(i) $f : x \mapsto x^2 - x$, $(x \in ℝ)$,
(ii) $g : x \mapsto \dfrac{1}{x^2 - x}$, $(x \in ℝ, x \neq 0, x \neq 1)$.

Sketch, on separate axes, the graphs of f and g.   (*C*)

4   The functions f and g are defined by

$$f: x \mapsto 1 - x^2, \ x \in \mathbb{R};$$
$$g: x \mapsto 2/(3 - 4x), \ x \in \mathbb{R} \text{ but } x \neq \tfrac{3}{4}.$$

(a) Find a formula for gf(*x*). What is the largest possible domain for the function gf?

(b) Find a formula for $g^{-1}(x)$ for $x \neq 0$.   (*SMP*)

5   For each of the following functions determine whether it has the same sign for all values of *x*, in which case state the sign. Give reasons for your results.

$$f(x) = 3x^2 - 7x + 5,$$
$$g(x) = 5 + 4x - x^2,$$
$$h(x) = 2x^2 - x - 10.$$

For each function which does not have the same sign for all values of *x* determine the range or ranges of values of *x* for which it is positive.   (*C*)

6   The function f from $\mathbb{R}$ to $\mathbb{R}$ is such that $f(x) = px^2 + qx + r$, where *p*, *q* and *r* are real constants.

(i) Find the range of f given that $p = q = r = -1$.

(ii) Find the possible values of *q* given that $p = 1, r = 3$ and the range of f is $\{y : y \geqslant -1\}$.   (*C*)

7   The functions f, g and h are defined by

$$f: x \mapsto 1/\sqrt{(1 + x^2)}, \ x \in \mathbb{R},$$
$$g: x \mapsto \sin x, \ x \in \mathbb{R},$$
$$h: x \mapsto \sqrt{(1 - x^2)}, \ x \in \mathbb{R}, |x| \leqslant 1.$$

(a) For each function, state whether it is even, odd or neither. State also whether or not it is periodic, giving the period where appropriate. (All your statements should be justified.)

(b) Find the value of hg($\pi/3$).

8   A function f has domain the set of real numbers. For $0 \leqslant x \leqslant 1$ it is given by the equation

$$f(x) = 1 - x.$$

Given also that f is an even function with period 2, draw its graph over the interval $-3 \leqslant x \leqslant 3$. Write down equations for the function for (i) $-1 \leqslant x \leqslant 0$, (ii) $2 \leqslant x \leqslant 3$.   (*SMP*)

9   The functions f and g have for domain the real numbers. What is meant by saying that (i) f is an odd function, (ii) g is an even function?

Assuming that neither function is zero for all *x*, which of the following statements is true of the function

$$h: x \mapsto f(x) + g(x)?$$

(a) h is an odd function.

(b) h is an even function.

(c) h is neither odd nor even.

(d) You cannot tell between (a), (b) or (c).

Justify your answer.   (*SMP*)

9 Indices and Logarithms

9.1 Powers of a number

The number formed by multiplying 2 by itself n times is denoted by 2^n, so that $2^1 = 2$, $2^2 = 4$, $2^3 = 8$, and so on. This means that we can define a function f, with domain equal to the set $\mathbb{N}$ of natural numbers, by $f(x) = 2^x$, for $x \in \mathbb{N}$. We tabulate the values of $f(x)$ for a number of values of x:

x	1	2	3	4	5	6
$f(x)$	2	4	8	16	32	64

The graph of f is a set of points with coordinates $(x, 2^x)$ and if we attempt to draw a smooth curve through these points, we obtain a curve which can be interpreted as the graph of a real function g, with $g(x) = 2^x$ (Fig. 9.1).

In this chapter, we shall extend the definition of a^x, for a positive number a, from the domain where x is a natural number to a domain where x is a rational number. Since any real number can be approximated

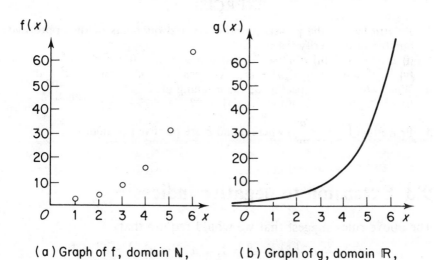

(a) Graph of f, domain $\mathbb{N}$,
$f(x) = 2^x$

(b) Graph of g, domain $\mathbb{R}$,
$g(x) = 2^x$

Fig. 9.1

by a rational number (by using a terminating decimal as an approximation) the definition of a^x can then be extended to the domain $\mathbb{R}$, by continuity.

9.2 Rules of indices

Consider the simple rules of indices, when the index (or power) is a positive integer. These are all proved by counting the number of occurrences of a. For all positive real numbers a and b, and for positive integers n and m,

$$a^n = a \times a \times a \times \ldots \times a \ n \text{ times,}$$
$$a^n \times a^m = a^{n+m},$$
$$(a^n)^m = a^{nm},$$
$$(a^n)/(a^m) = a^{n-m}, \text{ providing that } m < n, \text{ (by cancellation of } m \ a\text{'s).}$$
$$(ab)^n = a^n b^n.$$

Note $ab^3 = a \times b \times b \times b$,   whereas $(ab)^3 = (ab) \times (ab) \times (ab) = aaabbb$ $= a^3 b^3$, using the associativity and commutativity of multiplication.

Consider how we may fit a value for a^0 into the above. Putting $m = n$, we need the equation $1 = (a^n)/(a^n) = a^{n-n} = a^0$, and so we *define*

$$a^0 = 1.$$

We then find that all the above rules hold for all natural numbers n and m.

EXERCISE 9.2

1 By writing out the powers as products and quotients of the appropriate number of a's, verify that:
  (i) $a^2 a^3 = a^5$,   (ii) $a^4 a^5 = a^9$,   (iii) $a^2 a^3 a^4 = a^9$,   (iv) $a^4 a^4 = a^8$,
  (v) $(a^2)^4 = a^8$,   (vi) $a^8/a^3 = a^5$,   (vii) $a^2 a^3/a^5 = a^0$,   (viii) $a^2/a^2 = a^0$.

2 What would you expect to be the meaning of:
  (i) a^{3-4},   (ii) a^{5-8},   (iii) a^{-1},   (iv) a^{-4}?

3 Prove that $\left(\dfrac{a}{b}\right)^3 = \dfrac{a^3}{b^3}$, where a and b are positive real numbers.

9.3 Extension to negative indices

The above rules suggest that we would require that

$$a^{-1}a = a^{-1+1} = a^0 = 1, \ a^3 a^{-3} = a^{3-3} = a^0 = 1,$$

so that

$$a^{-1} = \frac{1}{a}, \ a^{-3} = \frac{1}{a^3},$$

so we extend the domain of a^x to the whole of the integers by defining

$$a^{-n} = \frac{1}{a^n}, \text{ for positive integers } n.$$

EXERCISE 9.3A

1 Using the rules of indices for positive integers, prove that:
(i) $a^2a^{-4} = a^{-2}$, (ii) $a^{-5}a^4 = a^{-1}$, (iii) $a^6a^{-4} = a^2$, (iv) $(a^3)^{-2} = a^{-6}$,
(v) $(a^{-2})^4 = a^{-8}$, (vi) $(a^{-3})^{-2} = a^6$, (vii) $a^2/a^{-3} = a^5$,
(viii) $a^{-5}/a^3 = a^{-8}$, (ix) $a^{-4}/a^{-7} = a^3$.

2 Using the rules of indices for positive integers, prove that
(i) $a^na^m = a^{n+m}$, (ii) $a^n/a^m = a^{n-m}$, (iii) $(a^n)^m = a^{nm}$, where n and m are any
integers, by considering separately the cases: (a) $0 \leqslant n \leqslant m$, (b) $n < 0 \leqslant m$,
(c) $m < 0 \leqslant n$, (d) $n \leqslant m \leqslant 0$, (e) $m < n \leqslant 0$.

3 Tabulate the values of 2^x for all integer values of x, with $-6 \leqslant x \leqslant 6$, and plot
the graph of the function f with domain equal to
$\{-6, -5, -4, -3, -2, -1, 0, 1, 2, 3, 4, 5, 6\}$, where $f(x) = 2^x$.
    By drawing a smooth curve through the points of this graph, draw the curve
which could be the graph of 2^x for real values of x between -6 and 6.
Estimate, from the graph, the values of $2^{\frac{1}{2}}, 2^{-\frac{1}{2}}, 2^{\frac{3}{4}}, 2^{-\frac{7}{8}}$. Can you guess anything
about the sign of 2^x for all real values of x?

4 Simplify the following, for example, $(a^3a^5)^2 = a^{16}$, $(a^2b^{-3})^{-2} = a^{-4}b^6$:
(i) $10^3 \times 10^4$, (ii) $10^5 \times 10^{-3}$, (iii) $x^{-2}x^3$, (iv) $x^2x^3x^4$, (v) $10^4 \div 10^7$,
(vi) $2^5 \div 2^3$, (vii) $a^6 \div a^2$, (viii) $x^2 \div x^{-4}$, (ix) $(2^2 \times 3^{-1})^3$, (x) $(2^3)^4$,
(xi) $a^0 \times a^7$, (xii) $(2^{-4})^3$, (xiii) $(2^{-4})^{-3}$, (xiv) $(2^{-4})^0$,
(xv) $(2^2 \times 3^{-1})^{-1}$, (xvi) $(9^2 \times 3^{-5})^{-2}$, (xvii) $(10^{-2})^3/(10^{-3})^2$,
(xviii) $(a^2b^3)^2$, (xix) $a^3(b^{-2}a)^3$, (xx) $(a^{-2} \times b^{-3})^2$, (xxi) $(a^{-3}b^{-2})^{-4}$.

Scientific notation

By using a multiplier, which consists of the base 10 raised to a positive
or negative index, we can express every positive real number in the
form $a \times 10^n$, where $1 \leqslant a < 10$ and $n \in \mathbb{Z}$. This is then called the
standard form of the number. For example, $24\,600 = 2\cdot46 \times 10^4$ and
$0\cdot002\,46 = 2\cdot46 \times 10^{-3}$. This is the *scientific notation* of number in
standard form and it enables us to express very large or very small numbers
conveniently. It is used in some pocket calculators to handle and display
numbers between 10^{99} and 10^{-99}. The standard form may be used to
indicate the intended number of significant figures in a number, which is
important since it shows the accuracy. The number between 1 and 10 in the
standard form is assumed to be as accurate as the number of digits it
displays. Thus $2\cdot46 \times 10^4$ indicates $24\,600$, correct to three significant
figures, whereas $2\cdot4600 \times 10^4$ indicates $24\,600$, correct to 5 significant figures.
    The index of 10 in the standard form of a number is called its
characteristic index, and it is equal to the number of digits between the
position of the decimal point in the standard form and in the decimal form
of the number, positive for numbers greater than one and negative for
numbers less than one.

EXERCISE 9.3B

1 Write in scientific notation:
   (i) 27, (ii) 270, (iii) 0·270, (iv) 0·027, (v) 3456, (vi) 34000, (vii) 34·56,
   (viii) 3456000, (ix) 0·00003456.
2 Write in decimal notation:
   (i) $2·7 \times 10^3$, (ii) $2·789 \times 10^4$, (iii) $1·234 \times 10^{-3}$, (iv) $1·2 \times 10^{21}$,
   (v) $2·5 \times 10^{-10}$, (vi) $3·45 \times 10^6$.

9.4 Extension to rational numbers

The rational numbers $\mathbb{Q}$ consist of all fractions m/n, where $m \in \mathbb{Z}$ and
$n \in \mathbb{Z}^+$. The rational number $m/1$, with denominator 1, is identified with
the integer m. Also $\frac{1}{2} = \frac{2}{4} = \frac{3}{6}$; and so on, and, therefore, in order to have a
single fraction to represent the rational number m/n, we assume that m and
n have no common factor.

For rational indices of the real number a, we shall always assume that
$a > 0$. We first consider $a^{\frac{1}{2}}$. The rules of indices require that

$$a^{\frac{1}{2}}a^{\frac{1}{2}} = a^{\frac{1}{2}+\frac{1}{2}} = a^1 = a, \text{ and so } a^{\frac{1}{2}} = \sqrt{a}.$$

Since a is a positive number, it has two square roots, one positive and one
negative. We avoid any ambiguity by the following definition.

Definition $a^{\frac{1}{2}} = \sqrt{a} =$ the positive square root of a.

In a similar manner, for any natural number n, the product of n terms,
each equal to $a^{1/n}$ must be

$$a^{1/n} \times a^{1/n} \times a^{1/n} \times \ldots \times a^{1/n} = a^{(1/n + 1/n + 1/n + \cdots + 1/n)} = a^{n/n} = a^1 = a,$$

and so

Definition $a^{1/n} = \sqrt[n]{a}$, the positive nth root of a.

Since the nth root of a product of numbers is equal to the product of
their nth roots, for $m \in \mathbb{Z}$ and $n \in \mathbb{Z}^+$,

$$(a^m)^{1/n} = (a \times a \times a \times \ldots \times a)^{1/n} = a^{1/n} \times a^{1/n} \times a^{1/n} \times \ldots \times a^{1/n} = (a^{1/n})^m.$$

We, therefore, satisfy the rules of indices by the

Definition $a^{m/n} = (a^m)^{1/n} = (a^{1/n})^m$

and then $\sqrt[n]{(a^m)} = (\sqrt[n]{a})^m = a^{m/n}.$

It is then easy, but tedious, to show that, for such rational indices, all the
rules of indices hold. That is, for all positive numbers a and b and all

rational numbers p and q:

$$a^p \times a^q = a^{p+q},$$

$$a^{-p} = \frac{1}{a^p},$$

$$(a^p)/(a^q) = a^{p-q},$$

$$(a^p)^q = a^{pq},$$

$$a^p \times b^p = (ab)^p,$$

$$\frac{a^p}{b^p} = \left(\frac{a}{b}\right)^p.$$

We have now extended the domain of the function a^x to $\mathbb{Q}$. The graph of 2^x is the graph you drew in Exercise 9.3A, question 3. From this graph, the values of such numbers as $2^{\frac{1}{2}}$, $2^{-\frac{1}{2}}$, $2^{\frac{3}{8}}$ were estimated.

EXAMPLE *Simplify:* (i) $(\frac{4}{9})^{-\frac{1}{2}}$, (ii) $(3\frac{1}{2})^3(27\frac{1}{3})^2(81)^{-1}$.

(i) $(\frac{4}{9})^{-\frac{1}{2}} = 1/(\frac{4}{9})^{\frac{1}{2}} = 1/\sqrt{(\frac{4}{9})} = 1/(\frac{2}{3}) = \frac{3}{2}$,

(ii) $(3\frac{1}{2})^3(27\frac{1}{3})^2(81)^{-1} = (\sqrt{3})^3(3)^2/3^4 = 3\sqrt{3}/9 = 1/\sqrt{3}$.

EXERCISE 9.4

1 Write down the value of:
(i) 3^3, (ii) $27^{\frac{1}{3}}$, (iii) $64^{\frac{1}{6}}$, (iv) $(\frac{4}{9})^{\frac{1}{2}}$, (v) $(2.25)^{\frac{1}{2}}$, (vi) 7^0, (vii) 3^{-1},
(viii) 6^{-2}, (ix) $(\frac{1}{8})^{\frac{1}{3}}$, (x) $(\frac{1}{4})^{-\frac{1}{2}}$, (xi) $(\frac{81}{16})^{\frac{1}{4}}$, (xii) $(\frac{16}{81})^{-\frac{1}{4}}$.

2 Simplify the following:
(i) $8^{\frac{2}{3}}$, (ii) $4^{-\frac{3}{2}}$, (iii) $25^{\frac{3}{2}}$, (iv) $5^{\frac{2}{3}}5^{\frac{2}{3}}$, (v) $4^{\frac{1}{4}}4^{\frac{1}{6}}$, (vi) $6^{\frac{1}{2}}24^{\frac{1}{2}}$, (vii) $12^{\frac{1}{3}}6^{\frac{2}{3}}$,

(viii) $\dfrac{12^{\frac{1}{3}}3^{\frac{2}{3}}}{2^{\frac{2}{3}}}$, (ix) $12^{\frac{1}{6}}6^{\frac{1}{3}}81^{-\frac{1}{6}}$, (x) $3^n \div 27^{\frac{2}{3}n}$, (xi) $(\frac{1}{32})^{\frac{4}{5}}(64)^{\frac{5}{6}}$.

3 By squaring, verify the following special cases of rules of indices:

(i) $\sqrt{x}\sqrt{y} = \sqrt{(xy)}$, (ii) $\dfrac{\sqrt{x}}{\sqrt{y}} = \sqrt{\left(\dfrac{x}{y}\right)}$.

4 Express in the form a^p the following (remember that $x^{-\frac{1}{n}} = \dfrac{1}{x^{1/n}} = \dfrac{1}{\sqrt[n]{x}}$, and that this is quite different from $-x^{1/n}$, which is $-(\sqrt[n]{x})$):
(i) $3\frac{1}{2} \times 3\frac{2}{3}$, (ii) $4^{\frac{1}{4}} \times 4^{\frac{1}{2}}$, (iii) $(5\frac{1}{2})^3$, (iv) $(5^5)^{\frac{1}{2}}$, (v) $a^{\frac{3}{4}}a^{-\frac{1}{4}}$, (vi) $x^{-\frac{2}{3}}x$, ?
(vii) $a^{\frac{4}{5}}a^{\frac{2}{5}}$, (viii) $x^{1.2}x^{0.8}$, (ix) $a^{-1.5} \times a^{3.5}$, (x) $a^{-\frac{1}{4}}a^{\frac{2}{4}}$, (xi) $2^{-5} \div 2^{-7}$,
(xii) $\sqrt{(a^4)}$, (xiii) $x^{-\frac{2}{3}}x^{\frac{1}{4}}/x^{\frac{1}{6}}$, (xiv) $3^{-n}8^n/27^{n/3}$, (xv) $2^44^{-3}8^2$,
(xvi) $(4\frac{1}{2})^3(32\frac{1}{5})^2(256)^{-1}$.

5 Complete the following table of values of the function $f: x \mapsto 16^x$.

x	-2	$-1\frac{1}{2}$	-1	$-\frac{1}{2}$	$-\frac{1}{4}$	0	$\frac{1}{4}$	$\frac{1}{2}$	$\frac{3}{4}$	1	$1\frac{1}{2}$	2
16^x			$1/16$			1				16		

Sketch a graph of this function and state whether you think that the following statements are true or false:
 (i) f is a monotonic increasing function, that is, if $x < y$ then $f(x) < f(y)$;
 (ii) $f(x)$ is always positive and tends to zero as x tends to $-\infty$.

9.5 Diversion

Even if a is a positive rational number, fractional powers of a, such as $a^{\frac{1}{2}}$, $a^{\frac{7}{8}}$, are usually not rational numbers, and are then called *surds*.

EXAMPLE   *Prove that $2^{\frac{1}{2}} = \sqrt{2}$ is not rational.*

The proof is by the method of contradiction. That is, we assume that $2^{\frac{1}{2}}$ is a rational number and show that this leads to a contradiction, and so our assumption was false and $2^{\frac{1}{2}}$ is not rational.

Suppose that m and n are natural numbers with no common factors and that $2^{\frac{1}{2}} = m/n$. Then $n2^{\frac{1}{2}} = m$, so that $2n^2 = m^2$. Now the square of an odd integer is odd and, since m^2 is even, m is also even. Hence $m = 2p$ for some natural number p. But then $m^2 = 2n^2 = 4p^2$, which implies $n^2 = 2p^2$. As with m, n must be even because n^2 is even, but then m and n have a common factor 2 which contradicts our assumption. This assumption must, therefore, be false and hence $2^{\frac{1}{2}}$ is not rational.

Any real number can be thought of as an infinite decimal, where integers and finite decimals are regarded as infinite decimals ending in recurring zero. When an integer m is divided by an integer n, there can only be at most n different remainders so any rational number m/n will be represented by a recurring decimal with at most n digits recurring. Conversely, it can be shown (see §26.2) that a recurring decimal represents a rational number, so that the non-recurring decimals are real numbers which are not rational, and are, therefore, called irrational. Irrational numbers include surds such as $2^{\frac{1}{2}}$ and other numbers such as π.

In practice, we approximate to a real number by a finite decimal. A number may be given to four decimal places in a table or to six decimal places by a calculator. For example, we might use $3\cdot1415926$ for π, accurate to 8 significant figures, or $1\cdot414$ for $\sqrt{2}$, accurate to three decimal places. This means that we are, in fact, using rational approximations, namely $\dfrac{31415926}{10000000}$ or $\dfrac{1414}{1000}$. In practice, we use rational approximations to real numbers and, by using more and more decimal places, we can achieve greater and greater accuracy of approximation. In this way we can define a^x, for any real number x, as a limit of approximations.

9.6 The power function 2^x and the logarithmic function $\log_2 x$

The graph of 2^x is shown in Fig. 9.2 and, from this graph, it is seen that 2^x is monotonic increasing, that is,

$$\text{if } x_1 < x_2 \quad \text{then} \quad 2^{x_1} < 2^{x_2}.$$

The domain of 2^x is the set $\mathbb{R}$ of all real numbers and its range is $\mathbb{R}^+$, the set of all positive real numbers.

$$\text{As } x \to +\infty, \quad 2^x \to +\infty, \quad \text{as} \quad x \to -\infty, \quad 2^x \to 0.$$

Since 2^x takes as its value each positive real number just once, we see that

2^x **is a one-one function of $\mathbb{R}$ on to $\mathbb{R}^+$.**

As a result, we can introduce an inverse function to 2^x, with domain $\mathbb{R}^+$

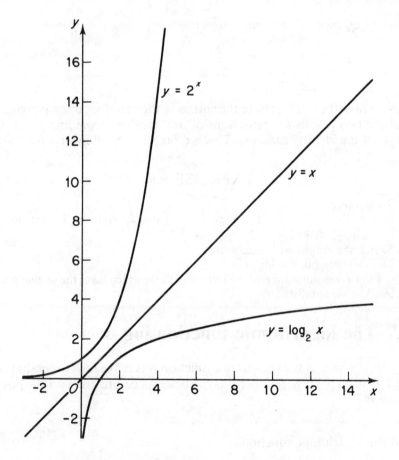

Fig. 9.2 Graphs of the power, identity and logarithmic functions.

and range $\mathbb{R}$. This function is called

$$\log_2 x = \text{the } logarithm \text{ of } x \text{ to the } base \text{ 2.}$$

Logarithm is another name for index or power, since $\log_2 x$ is the index of 2 which gives that power of 2 equal to x.

Definition $y = \log_2 x \Leftrightarrow 2^y = x$

and then we have the two functions, which are functional inverses of each other.

Power function   Logarithmic function

$2^x: \mathbb{R} \to \mathbb{R}^+$ $\log_2 x: \mathbb{R}^+ \to \mathbb{R}$

$x \mapsto 2^x$ $x \mapsto \log_2 x$

The graphs of these two functions have been drawn in Fig. 9.2, using the following tables of values:

x	-2	-1	0	1	2	3	4
2^x	$\frac{1}{4}$	$\frac{1}{2}$	1	2	4	8	16
x	$\frac{1}{4}$	$\frac{1}{2}$	1	2	4	8	16
$\log_2 x$	-2	-1	0	1	2	3	4

Note how these tables have the entries in the two rows interchanged, so that the two graphs are reflections of one another in the line $y = x$, the graph of the identity function. This is a property of all inverse functions.

EXERCISE 9.6

1   Write down:
   (i) $\log_2 4$,   (ii) $\log_2 8$,   (iii) $\log_2 32$,   (iv) $\log_2 2$,   (v) $\log_2 1$,   (vi) $\log_2 \frac{1}{2}$,
   (vii) $\log_2 \frac{1}{4}$,   (viii) $\log_2 \sqrt{2}$,   (ix) $\log_2 \sqrt[3]{2}$.
2   Sketch the graphs with equations
   (i) $y = (\frac{1}{2})^x$,   (ii) $y = \log_2 x$.
   Find a transformation of the Cartesian plane which takes the second graph into the first graph.

9.7 The logarithmic function $\log_a x$

In §9.6, the base of the logarithmic function was 2. We now generalise and use any fixed positive number a as the base. As before, we have the power function

$$a^x: \mathbb{R} \to \mathbb{R}^+, \quad x \mapsto a^x$$

and the logarithmic function

$$\log_a x: \mathbb{R}^+ \to \mathbb{R}, \quad x \mapsto \log_a x.$$

Definition $\quad y = \log_a x \Leftrightarrow a^y = x, \quad$ for all $x, a \in \mathbb{R}^+, y \in \mathbb{R}.$

Thus, $\log_a x$, the logarithm of x to the base a, is the power to which a must be raised in order to give x. It follows immediately from the definition that

$$a^{\log_a x} = x = \log_a(a^x),$$

which is the usual property of inverse functions.

The logarithm is the functional inverse of the power function and this is why on some calculators the $\log_a x$ key is named (INV)(a^x).

EXERCISE 9.7

1 Simplify:
   (i) $\log_5 5$, (ii) $\log_{10} 100$, (iii) $\log_5 125$, (iv) $\log_7 343$, (v) $\log_3 81$,
   (vi) $\log_{13}(13)^2$, (vii) $\log_2 1024$, (viii) $\log_4 2$, (ix) $\log_4 8$, (x) $\log_4 32$,
   (xi) $\log_8 32$, (xii) $\log_{10} 0 \cdot 0001$, (xiii) $\log_{16} 4$, (xiv) $\log_a \sqrt{a}$, (xv) $\log_b b^3$,
   (xvi) $\log_c c^{\frac{1}{4}}$, (xvii) $\log_{27} 9$, (xviii) $\log_4 1$, (xix) $.\log_{0 \cdot 1} 10$.
2 Express in index notation the equation:
   (i) $\log_t s^2 = 2$, (ii) $\log_3(\frac{1}{9}) = -2$, (iii) $\log_x y = z$.
3 Express in logarithmic notation, (for example, $3^4 = 81 \Rightarrow \log_3 81 = 4$):
   (i) $10^6 = 1\,000\,000$, (ii) $125 = 5^3$, (iii) $1024 = 2^{10}$, (iv) $\frac{1}{2} = 16^{-\frac{1}{4}}$,
   (v) $9^{\frac{3}{2}} = 27$, (vi) $a = b^c$.
4 In compound interest, the amount £A which is in the account at the beginning of a year is increased by an interest of $x\%$ at the end of the year, the interest being added to the account. Prove that an initial principal of £P will increase to an amount £A after n years, where $A = P\left(1 + \dfrac{x}{100}\right)^n$. Express n in terms of A, P and x, using logarithms to the base 10.

9.8 Rules for manipulating logarithms

The rules for indices give rise to corresponding rules for logarithms. Suppose that

$$a^x = b \quad \text{so that} \quad x = \log_a b$$
$$a^y = c \quad \text{so that} \quad y = \log_a c.$$

Then $\qquad\qquad a^{x+y} = a^x a^y = bc$

so that $\qquad\qquad x + y = \log_a(bc)$

and thus $\quad \log_a b + \log_a c = \log_a(bc).$

Also $\qquad\qquad a^{x-y} = a^x a^{-y} = a^x/a^y = b/c$

so $\qquad\qquad x - y = \log_a(b/c)$

and $\qquad \log_a b - \log_a c = \log_a(b/c).$

Consider $\qquad b^d = (a^x)^d = a^{xd} = a^{dx}$

and so $\qquad \log_a(b^d) = dx = d\log_a b.$

Also, from the definitions of logarithms, $\log_a a = 1$ and

$$\log_a(a^b) = b = a^{\log_a b}.$$

These rules are summarised:

$$\log_a b + \log_a c = \log_a(bc)$$
$$\log_a b - \log_a c = \log_a(b/c)$$
$$\log_a(b^d) = d\log_a b$$
$$\log_a a = 1$$
$$\log_a(a^b) = b = a^{\log_a b}.$$

EXAMPLE 1   *Solve for x the equation $4 \times 3^x = 7$, leaving your answer in logarithmic form.*

Taking logarithms of both sides to the base 3,

$$\log_3(4 \times 3^x) = \log_3 4 + \log_3(3^x) = \log_3 7.$$

Hence, $x = \log_3(3^x) = \log_3 7 - \log_3 4 = \mathbf{\log_3(7/4)}.$

EXERCISE 9.8

1   Express, in terms of $\log_a x$, $\log_a y$ and $\log_a z$:
(i) $\log_a\left(\dfrac{xy}{z}\right)$,   (ii) $\log_a(xyz)$,   (iii) $\log_a(x^2 y^3 z^4)$,   (iv) $\log_a\sqrt{(xz)}$,
(v) $\log_a(1/\sqrt[3]{y})$.

2   Express as a single logarithm to the base a:
(i) $\log_a 18 - \log_a 9$,   (ii) $\log_a x + \log_a y - \log_a z$,   (iii) $3\log_a y - \log_a x$,
(iv) $1 + \log_a 3$,   (v) $5 - \log_a 5$.

3   Simplify:
(i) $\frac{1}{2}\log_3 81$,   (ii) $\log_4(\frac{1}{8})$,   (iii) $\log_{25}(\frac{1}{5})$,   (iv) $3\log_9 3 - \log_4\frac{1}{4}$,
(v) $\log_8 16 + \log_{27} 9$,   (vi) $\log_{100} 10 - \log_{1000} 100$,   (vii) $\log_5 45 - \log_5 9$,
(viii) $\log_a(b^2) + \log_a(b^3) - \log_a(b^4)$,   (ix) $\log_a(b/c) + \log_a(c/b)$.

4   Find a value of x satisfying the equation:
(i) $\log_x 3 = \log_4 4$,   (ii) $\log_3 x = 3$,   (iii) $\log_x 16 = 2$,   (iv) $\log_x 16 = 8$,
(v) $\log_2 x = 3\log_2 5 - 2\log_2 3$,   (vi) $\log_{10} x = \log_{10} 4 - 2\log_{10} 5 + 3\log_{10} 6$.

5   Solve, in terms of logarithms, the equation:
(i) $3^x = 7$,   (ii) $10^x = 4$,   (iii) $(0\cdot1)^x = 7$,   (iv) $5^{1-x} = 3$,   (v) $5 \times 2^x = 120$.

6   Solve the equations:
(i) $3^{x-1} = 3^{2x+1}$,   (ii) $3^{2x+1} = 9^{2x}$,   (iii) $2^{2x+6} = 4^{5-x}$.

9.9 Change of base

To see the effect of changing the base of a logarithm, suppose that $b = a^x$ and $c = b^y = (a^x)^y = a^{xy}$. Then $x = \log_a b$, $y = \log_b c$ and $xy = \log_a c$.

Hence, we have proved that

$$\log_a b \times \log_b c = \log_a c.$$

This can be used in order to change logarithms from one base a to another base b, since

$$\log_b c = \log_a c \times \frac{1}{\log_a b},$$

so $\dfrac{1}{\log_a b}$ is a common scale factor for changing logarithms to the base a into logarithms to the base b.

In particular,

$$\log_a b = \frac{1}{\log_b a}$$

since $\log_a a = 1$.

EXERCISE 9.9

1 Simplify:
   (i) $\log_a b \times \log_b c \times \log_c a$, (ii) $\log_a b / \log_a c + \log_c (b^2)$.
2 Solve the equation $\log_9 2 = \log_3 x$.
3 Given that $\log_{10} 2 = 0.3010$, $\log_{10} 5 = 0.6990$, solve the equation $5 = 2^x$.

Other notations

The above rules of logarithms together with tables of logarithms to the base 10, can be used to perform the arithmetical operations of multiplication and division in terms of addition and subtraction. Such tables were commonly used before the advent of electronic calculators and the logarithms to the base 10 are called common logarithms. As a notation, the base 10 is often omitted in common logarithms, and the letter 'o' can also be omitted, thus

$$\log_{10} x = \log x = \lg x.$$

In calculus (Chapter 23), logarithms to another base are used. The logarithms are called natural logarithms and the base is denoted by e. The number e ≈ 2.7182818 and can be defined as that base such that the function $\log_e x$ has unit gradient when $x = 1$. The notation for $\log_e x$ is

$$\log_e x = \ln x.$$

The function ln x has domain $\mathbb{R}^+$ and is called the *logarithmic* function; ln x has range $\mathbb{R}$ and is a one-one function. Its inverse function is called the *exponential* function and is denoted by e^x. The exponential function has domain $\mathbb{R}$ and range $\mathbb{R}^+$. Values of the two functions ln x and e^x are available in tables and on scientific calculators. Their properties will be discussed in detail in Chapter 23. The only property of these functions, which we use at present, is that

$$y = \ln x \Leftrightarrow x = e^y, \text{ or } x = \ln y \Leftrightarrow y = e^x,$$

and that e^x satisfies the usual rules of indices.

EXAMPLE 1   *Simplify* $\dfrac{e^{2x} - \ln e}{e^x - e^{\ln 1}}$.

Since $e = e^1$, $\ln e = 1$. Also $\ln 1 = 0$ so that $e^{\ln 1} = e^0 = 1$.

Hence $\dfrac{e^{2x} - \ln e}{e^x + e^{\ln 1}} = \dfrac{e^{2x} - 1}{e^x + 1} = \dfrac{(e^x)^2 - 1}{e^x + 1} = \dfrac{(e^x + 1)(e^x - 1)}{e^x + 1} = \mathbf{e^x - 1}$.

EXAMPLE 2   *Solve the equation* $5e^x - 2e^{-x} + 3 = 0$.

Let $e^x = y$ so that $e^{-x} = \dfrac{1}{y}$ and the equation becomes $5y - \dfrac{2}{y} + 3 = 0$, or $5y^2 + 3y - 2 = 0$. On factorising, this becomes $(5y - 2)(y + 1) = 0$ so $y = -1$ or $y = \frac{2}{5}$. However, $y = e^x > 0$ for all x, so there is only one solution, $e^x = y = \frac{2}{5}$, that is, $x = \ln \frac{2}{5}$.

MISCELLANEOUS EXERCISE 9

1   Use the substitution $y = 5^x$ to solve $25^x = 5^{x+1} - 6$.
2   Given that $\log 2 = 0.301$, $\log 3 = 0.477$, $\log 5 = 0.699$, find: (i) $\log_2 3$; (ii) $\log_3 2$; (iii) $\log_2 5$; (iv) $\log_2 6$; (v) $\log (2.4)$.
3   Find the value of $\log_4 (2\sqrt{2})$.   (L)
4   Solve the equation $\log_2 x + 4 \log_x 2 = 5$.   (L)
5   Solve the equation

$$e^{\ln x} + \ln e^x = 8.$$ (L)

6   Given that $\lg y = 3 - \frac{3}{4} \lg x$, express y in terms of x in a form not involving logarithms.

$$[\lg \equiv \log_{10}.]$$ (L)

7   Find, without using tables or a calculator, the value of x given that

$$\frac{4^{3+x}}{8^{10x}} = \frac{2^{10-2x}}{64^{3x}}.$$ (L)

8   Solve, for real x, the equation $\log_2 (x + 4) = 2 - \log_2 x$.   (L)

9   Simplify $\dfrac{e - (e^{(x + \frac{1}{2})} \div e^{-2x})^2}{e^{3x} + \ln e}$. (L)

10   If $\log_a 45 + 4 \log_a 2 - \frac{1}{2} \log_a 81 - \log_a 10 = \frac{3}{2}$, find the base a.   (L)
11   Find the real solution of the equation $e^{3x} - 2e^x - 3e^{-x} = 0$.   (L)
12   Solve the equation $2^{2x} - 5(2^{x+1}) + 16 = 0$.   (L)
13   You are given that positive variables x, y are related by the equation

$$2 \ln y = c - \ln x,$$

where c is constant, and that $y = 3$ when $x = 4$. Find, in as simple a form as possible (and not involving logarithms), an expression for y in terms of x. *(SMP)*

14 Given that
$$\ln y = 1 - 3\ln x,$$
write an equation, expressing y in terms of x, which does not involve logarithms. *(SMP)*

15 (i) Solve the equation $10^{x-3} = 2^{10+x}$ giving your answer correct to two significant figures. *(L)*

16 Given that $2e^x - 2e^{-x} = 3$, find the value of x. *(L)*

17 (a) Solve the equation $2^{2/x} = 32$.
(b) Solve the equation $\log_x 2 \times \log_x 3 = 5$,
giving your answers correct to two decimal points. *(L)*

18 Establish the formula $\log_y x = \dfrac{1}{\log_x y}$.

Solve the simultaneous equations
$$\log_x y + 2\log_y x = 3,$$
$$\log_9 y + \log_9 x = 3. \quad (JMB)$$

19 If x and y are positive numbers, prove that $(\log_x y)(\log_y x) = 1$.

Find the possible values of $\log_x y$ such that $\log_x y = 1 + 2\log_y x$.

Hence, or otherwise, solve the simultaneous equations
$$\log_x y = 1 + 2\log_y x,$$
$$x + y = 12. \quad (JMB)$$

20 If
$$2\log_y x + 2\log_x y = 5,$$
show that $\log_y x$ is either $\frac{1}{2}$ or 2. Hence find all pairs of values of x and y which satisfy simultaneously the equation above and the equation
$$xy = 27. \quad (JMB)$$

21 (a) Given that $\log 2 = 0.301\,030\,0$ and $\log 3 = 0.477\,121\,3$, find, without using tables or calculators,

(i) $\log 16$, (ii) $\log\left(\dfrac{1}{9}\right)$, (iii) $\log\sqrt[3]{12}$.

(b) Without using tables or calculators, find the value of

(i) $\dfrac{\log 125}{\log 5}$, (ii) $(\log_7 7) \times (\log_7 49)$.

(c) Given that $\log(x^3 y^2) = 9$ and $\log\left(\dfrac{x}{y}\right) = 2$, find $\log x$ and $\log y$ without using tables or calculators.

(d) Solve the equation $3^x . 3^{2x+3} = 10$, giving the answer correct to two decimal places. *(JMB)*

10 Linear laws

10.1 Linear relations (revision)

It often happens in scientific experiments that two variables are proportional to each other. For example, if the pressure P and the temperature T of a fixed volume of a gas are measured (in appropriate units) and if (P_1, T_1) and (P_2, T_2) are two pairs of measurements, then it is found that P and T are proportional, that is, $P_1/T_1 = P_2/T_2$. This result is modelled mathematically by saying that P is a *function* of T, given by $P = aT$, for some constant a, where $a = P_1/T_1 = P_2/T_2$. More generally, two variables may be directly related by a linear relation. If the length L of an elastic string is measured when the string is stretched by a force F, it is found that L is related to F by an equation $L = aF + b$, where a and b are constants.

In both these examples, one variable, say y, is a *linear function* of another variable, x, given by the equation

$$y = ax + b, \quad \text{where } a \text{ and } b \text{ are constants.}$$

The special case of proportional variables occurs when $b = 0$ and $y = ax$. The relation holds for the particular units in which the quantities are measured, and this determines the units of a and b. However, the question of units will not be considered here. The values of x, y, a and b will be regarded as real numbers. At the end of any calculations, the units can be considered and put into the answers, if this is needed.

When $y = ax + b$, the graph of y as a function of x is a straight line, with gradient a, passing through the point $(0, b)$. If a number of pairs of values of x and y are obtained by measurements, the corresponding points (x, y) may be plotted on a Cartesian graph and these points should lie on a straight line. The points may not lie exactly on a line, due to experimental error or to rounding off of numbers. However, a ruler may be used to draw the best possible line through (or near) the points. Then, by using this line, the values of the constants a and b can be estimated. Finally the relation (or *law* as it is called), $y = ax + b$, can be used to estimate a value of y for a given value of x.

EXAMPLE   *The quantities x and y are related by the linear law $y = ax + b$. Use the given data to estimate the values of a and b by drawing a straight line graph, approximately passing through the set of points $\{(x, y)\}$:*

x	2·0	4·0	6·0	8·0	10·0
y	2·8	4·3	5·7	7·2	8·8

Plot the five points on a graph and use a ruler to draw the best straight line through these points (see Fig. 10.1).

The line meets the y-axis at $(0, 1\cdot3)$, so $b = 1\cdot3$. Also the line passes through $(10, 8\cdot7)$ and hence $8\cdot7 = 10a + 1\cdot3$, so that $a = 0\cdot74$.

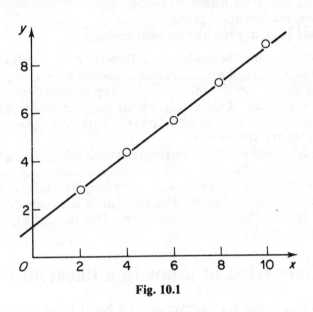

Fig. 10.1

EXERCISE 10.1

1 Plot the following sets of points, draw the best straight line through or near them, then fit the given data to the linear law $y = mx + c$:

(i)

x	0·8	2·0	3·2	4·2	5·6	6·8
y	3·6	4·0	4·2	4·6	5·0	5·4

(ii)

x	3	13	22	31	37
y	0·22	0·18	0·14	0·11	0·08

(iii)

x	3·0	3·1	3·2	3·3	3·4	3·5
y	81	82	83·5	85	86	88

10.2 Mathematical models and linear laws

When mathematics is used to solve a problem, the problem must first be translated into a mathematical form. This is called making a *mathematical model* of the problem, and will usually involve a number of variables and various relations between them. Certain assumptions will be made which will often involve approximations. Mathematical arguments and processes are then used to solve the mathematical problem of the model, and finally the results are translated back into a solution of the original problem. Thus, the steps are:

I state the original problem,
II translate into a mathematical model,
III solve the mathematical problem,
IV translate back to solve the original problem.

In science generally, the model uses information previously obtained in order to predict the results of future measurements or experiments. If these predictions agree with the results (within experimental error), then the model is satisfactory. If later experiments produce results which differ from the predictions, then the model must be improved and modified until its predictions are satisfactory.

The relations used in a scientific theory may be referred to as *laws* of the particular science and such laws often contain constants which have to be determined from experimental data. This may be difficult unless the relations concerned are linear. Therefore, it is necessary to convert a functional relationship into a linear form. This may often be done by means of a change of variables.

10.3　Conversion of a law to a linear form

Once a given law has been converted to a linear form, a graph may be plotted using the new variables, which are linearly related, and this graph should be a straight line. The equation of the line may then be obtained as in §10.1 and, from this, the values of the constants in the original law are determined.

EXAMPLE 1　*The e.m.f. of a cell is E volts and its internal resistance is r ohms. When the cell has a load of R ohms, its external voltage is V volts. These quantities are related by the equation*

$$ER = V(R+r).$$

For various loads, the voltage V is measured, with the following results:

R :	120	180	270	410	2200
V :	1·591	1·604	1·613	1·619	1·629

Convert the relationship between R and V to a linear form, by a suitable change of variables, and hence determine E and r.

Divide the equation of the law by ERV, so that it becomes

$$\frac{1}{V} = \frac{r}{ER} + \frac{1}{E} \quad \text{or} \quad y = \frac{r}{1000E}x + \frac{1}{E}$$

if we put $x = \dfrac{1000}{R}$ and $y = \dfrac{1}{V}$.

Using the given data, the corresponding values of x and y are

$$\frac{1000}{R} = x: \qquad 8\cdot33 \qquad 5\cdot55 \qquad 3\cdot70 \qquad 2\cdot44 \qquad 0\cdot455$$

$$\frac{1}{V} = y: \qquad 0\cdot6285 \qquad 0\cdot6234 \qquad 0\cdot6200 \qquad 0\cdot6177 \qquad 0\cdot6139.$$

These points (x, y) are plotted in Fig. 10.2 and a line is drawn through them. Reading from the graph, we see that, when $x = 0$, $y = 0\cdot6132$ (with some doubt about the last digit) so $1/E = 0\cdot6132$ and $E = 1/0\cdot6132 = \mathbf{1\cdot631}$.

In order to find the slope m of the graph, where $y = mx + 0\cdot6132$, we note that, when $x = 5$, $y = 0\cdot6224$, again with some doubt about the last digit. From these figures,

$$m = \frac{r}{1000E} = \frac{0\cdot6224 - 0\cdot6132}{5} = \frac{0\cdot0092}{5}$$

and so $r = 200 \times 0\cdot0092E$ giving $r = \mathbf{3\cdot001}$.

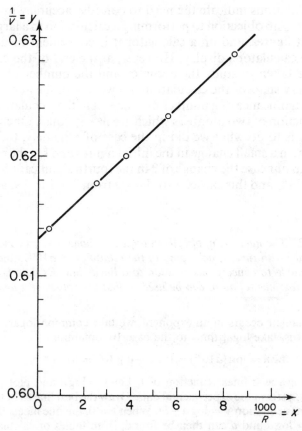

Fig. 10.2 Graph for Example 1 $y = \dfrac{1}{V}$ and $x = \dfrac{1000}{R}$.

Accuracy   The above figures have doubtful accuracy, and this needs investigating. In calculating E, we noted that the value of y for $x = 0$ had some doubt in the last digit. In fact, the most accurate value that we can read from the graph for $y(0)$ is 0·613 which gives $E = 1·63$, indicating accuracy to two decimal places. Then, in calculating m, we read two numbers from the graph, and each had at most three figure accuracy, so we take them as 0·622 and 0·613. This gives

$$m = \frac{0·622 - 0·613}{5} = \frac{0·009}{5} = 0·0018,$$

with considerable doubt in the last digit. Finally,

$$r = 1000 \times 0·0018 \times 1·63 = 2·9,$$

accurate to at most one digit. We, therefore, find that the e.m.f. of the cell is E volts, that is, 1·63 volts and its internal resistance is r ohms, that is, 3 ohms.

These calculations indicate the need to consider accuracy in numerical results. There is no objection to performing a calculation to a large number of significant figures, and on a calculator it is convenient to use the full range of the calculator's display. However, at the end of the calculation, care must be taken to study the accuracy and the number of significant figures at every stage of the calculation. As we saw in the above example, most of the significance of a number may be lost in the calculation because of the subtraction of two numbers which are nearly equal. One way to test the accuracy is to do what we did in the case of r, namely, to repeat the calculation with a small change in the initial figures and see the effect upon the answer. In our case the change of 2 in the fourth significant figure was a change of 0·03 % and this caused r to change from 3·001 to 2·9, a change of 3 %.

EXAMPLE 2   *The number, n, of cells in a given volume of a growing medium is assumed to vary with time, t, according to the relation $n = a\,10^{bt}$. Find a suitable change of variable to convert this relation to a linear law. Explain how a graph, drawn from experimental data, can be used to find the values of a and b.*

Since the variable t occurs in an exponent, we take common logarithms of the equation, that is, take logarithms to the base 10, obtaining

$$\log n = \log(a\,10^{bt}) = \log a + \log 10^{bt} = \log a + bt,$$

which gives $\log n$ as a linear function of t. Let $y = \log n$ and plot points with coordinates (t, y) from the experimental data. Draw the best line to fit these points and this line has equation $y = \log a + bt$. When $t = 0$, the line meets the y-axis at $(0, y_o)$ so $y_o = \log a$ and a can then be found, from tables or calculator, as the antilog of y_o, that is, $a = 10^{y_o}$. The gradient of the line is the other required constant b, so, if the line also meets the t-axis at $(t_o, 0)$, then $b = -y_o/t_o$.

10.4 Summary of substitutions

Some of the substitutions, that is, changes of variable, which are useful to convert a given relation into linear form, are listed under the types of relations concerned.

(i) **Products** Suppose that $xy = a$, then two types of substitution are possible:

(a) let $X = \dfrac{1}{x}$, $Y = y$, so that $Y = aX$;

(b) let $X = \log x$, $Y = \log y$, so that $Y + X = \log a$, assuming that x, y and a are all positive, using logarithms to any convenient base.

(ii) **Sums of functions of x and of y** Some examples are:
(a) if $ax^2 + by^2 = c$, let $X = x^2$, $Y = y^2$, then $aX + bY = c$;

(b) if $\dfrac{a}{x} + \dfrac{b}{y} = c$, let $X = \dfrac{1}{x}$, $Y = \dfrac{1}{y}$, then $aX + bY = c$;

(c) if $y = \sqrt{\{a + b \sin x\}}$, let $X = \sin x$, $Y = y^2$, then $Y = a + bX$.

(iii) **Products involving powers or exponents**
(a) if $yx^q = p$, let $Y = \log y$, $X = \log x$, then $Y + qX = \log p$;
(b) if $y^2 10^{kx} = a$, let $Y = \log y$, $X = x$, then $2Y + kX = \log a$;
(c) if $s = a 10^{-bt}$, let $Y = \log s$, $X = t$, then $Y = \log a - bX$;
(d) if $2Lf = \sqrt{(T/M)}$, where L and M are constants, let $Y = f^2$, $X = T$, then $X = 4L^2 MY$.

EXERCISE 10.4

1 Given that $ax + by = 1$, use the following approximate values of x and y to estimate the values of a and b graphically.

x	0·10	0·20	0·30	0·40	0·50
y	0·42	0·36	0·29	0·24	0·18

2 The variables x and y are related by the equation $xy = c + dx$, and five measurements gave the following results:

x	1·2	2·1	3·3	4·6	6·1
y	8·8	5·3	3·5	2·7	2·1

Using a new variable $X = 1/x$, plot an X–y graph to find c and d.

3 A mathematical model assumes that x and y are related by the equation $\dfrac{a}{x} + \dfrac{b}{y} = 1$, for some constants a and b. By converting to new variables $X = 1/x$ and $Y = 1/y$ and drawing a straight line graph, estimate the values of a

and *b* from the data:

x	12·5	14·5	18·0	22·0	32·5
y	14·9	16·5	18·5	20·5	24·5

4 Assuming that $s = ut + \frac{1}{2}ft^2$ and by changing to a new variable $y = s/t$, find the values of *u* and *f*, given the data:

s	4·8	15·6	37·1	59	99
t	1	2	3	4	5

5 The concentrations *C* of an acid in water and *D* of the acid in another solvent are related by the equation $D = kC^n$, for a constant *k* and an integer *n*. By finding a corresponding linear law and drawing a straight line graph, find the values of *n* and *k* from the data, where both *C* and *D* are numbers of grams per litre:

C	245	332	447	681	900
D	13·2	23·5	44·7	82·2	148

6 The concentration *C* (in 10^{-3} mols per litre) at time *t* (in seconds) of a gas during its thermal decomposition was found to be:

C	22·9	12·29	6·89	3·68	2·16
t	0	20	40	60	80

Given that *C* and *t* are related by the equation $C = ae^{-bt}$, find a corresponding linear law, draw a graph from the data, and estimate the values of the constants *a* and *b*. (Use natural logarithms, base e.)

7 Using the ideal gas law, $PV = RT$, estimate the value of *R* from the following set of values of *P* and *V*, measured for *T* = 273:

V	28·4	23·7	17·9	16·2
P	80	100	120	140

8 Assume that *r* is proportional to s^n, where *n* is a positive integer, use the given values to draw a straight line graph and find *n*. Deduce the function f, where $r = f(s)$.

s	1·2	1·3	1·4	1·5
r	4·1	5·3	6·6	8·1

9 Given that $y = ab^x$, use the data to find estimates of *a* and *b*.

x	2·0	2·4	2·8	3·2	3·6
y	69	88	101	131	138

MISCELLANEOUS EXERCISE 10

1 A relation of the form $y = ae^x + b$ is known to exist between two variables x and y. By plotting y against e^x, use the following table of experimental values of x and y to estimate the constants a and b to 1 significant figure.

x	1	2	3	4
y	24	56·7	145·6	387·2

(L)

2

x	1·5	2·5	3	4	5
y	−6·0	11·0	6·2	4·1	3·3

The table shows values of a variable y corresponding to certain values of another variable x. By drawing a suitable linear graph relating the reciprocals of x and y, verify that the values of x and y satisfy approximately a relationship of the form

$$\frac{1}{x} + \frac{1}{y} = \frac{1}{a},$$

where a is a constant. Use your graph to estimate the value of a to one decimal place. Estimate also the value of x when $y = 5\cdot1$. (L)

3 The values of x and y given in the table below satisfy approximately the relationship $e^y = Kx^a$. By drawing a suitable linear graph determine the values of the constants a and K.

x	1	2	3	4	5	6	7
y	2	8·9	13·0	15·9	18·1	19·9	21·5

(L)

4 Given that the values of x and y in the table below are experimental values of variables that satisfy $y = ax + b$, estimate graphically the constants a and b.

x	2	4	7	8	10
y	1·8	1·2	0·5	0·3	−0·2

(L)

5 A relation of the form $y = ax^n$ is known to exist between x and y. By plotting $\ln y$ against $\ln x$, or otherwise, use the following table of approximate values to estimate the constants a and n to 2 significant figures.

x	0·3	0·5	1	1·5	2
y	1·4	6·3	50	169	401

(L)

6

x	0	1	2	3	4	5
y	2·28	3·45	5·17	7·76	11·65	17·46

The table shows approximate values of a variable y corresponding to certain values of another variable x. By drawing a suitable linear graph, verify that the values of x and y satisfy approximately a relationship of the form $y = ab^x$.

Use your graph to estimate values of a and b, giving answers to one decimal place. (L)

7   In an experiment to find the focal length f of a lens, the following object distances u and image distances v were recorded:

u	15	20	25	30	35
v	59	31	23	20	18

By plotting the reciprocal of u against the reciprocal of v estimate a value for f from the formula

$$\frac{1}{u}+\frac{1}{v}=\frac{1}{f}.$$ (L)

8   Rewrite the following equations in suitable form to display a linear relationship in each case between two of the variables, x, $\ln x$, y, and $\ln y$.

(a) $\dfrac{2}{x}+\dfrac{3}{y}=\dfrac{4}{xy}$,

(b) $5x = 6^y$. (L)

9   Two variables are related by the law $y = a + b \ln x$. Using the measured values of x and y given in the table draw a suitable graph and use it to estimate, to the nearest whole number, the values of a and b.

x	2	3	4	5	6	7
y	57	98	127	149	167	183

(L)

10   An experiment to estimate constants a and b in a relationship of the form $y = ab^x$ produced the following results:

x	5	15	25	30	35	40
y	2·4	6·3	16·3	26·2	42·2	67·9

By plotting $\lg y$ against x, using suitable scales, estimate a and b correct to 2 significant figures. $[\lg y \equiv \log_{10} y.]$ (L)

11

x	0·5	1·5	2·5	3·5	4·0	4·5
y	3	10	31	92	162	280

The table shows approximate values of a variable y corresponding to certain known values of another variable x. It is believed that the values of x and y satisfy a relationship of the form $y = ab^x$. By taking logarithms and by drawing a suitable linear graph, estimate values of a and b to one decimal place.

Estimate also the value of y when $x = 4·2$. (L)

12

x	1	2	3	4	5
y	3	21	59	110	178

The table shows approximate values of a variable y corresponding to certain values of another variable x. The equation $y = ax + bx^2$, where a and b are constant integers, is an approximation to the relation between x and y. By plotting a graph of y/x against x, estimate the values of a and b.   (L)

13   The table below gives approximate values for y corresponding to the given values for x. Assuming that x and y satisfy an equation of the form $y = ab^x$, estimate, by drawing a straight line graph relating x and $\log_{10} y$, values for a and b, giving two places of decimals in your answers.

x	0·2	0·4	0·7	0·9	1·3
y	0·4885	0·5967	0·8055	0·9838	1·4677

Estimate the value of y when $x = 0.82$.   (L)

14   The table below gives values of the variables u and v, which are related by an equation of the form $v = au + b/u$. Find values of the constants a and b by drawing a straight-line graph relating uv and u^2.

u	1	2	3	4	5
v	12·5	7·0	5·5	5·0	4·9

(L)

11 Sketch Graphs

One method of drawing a graph of a function is to calculate the value of the function at a large number of points of its domain and to plot the corresponding points. The graph is then obtained by joining these points by a curve. This can be very laborious, and the object of this chapter is to find methods of *graph sketching* involving less calculation, and yet to retain all the important features of the graph. We begin by recalling some results.

11.1 Revision

Linear functions (Chapter 1)

The graph of a linear function is a straight line and is fixed when two points on it are fixed. The straight line graph with equation:

(i) $y = mx + c$ passes through $(0, c)$ and has gradient m, for example, Fig. 11.1(a);

(ii) $\dfrac{y}{b} + \dfrac{x}{a} = 1$ passes through the points $(a, 0)$ and $(0, b)$, for example, Fig. 11.1(b).

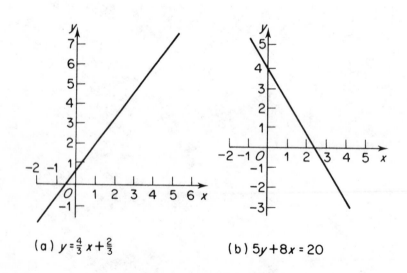

(a) $y = \frac{4}{3}x + \frac{2}{3}$

(b) $5y + 8x = 20$

Fig. 11.1 Graphs of linear functions.

Quadratic functions (Chapter 2)

The graph of a quadratic function is a parabola with its axis parallel to the
y-axis, and the sign of the coefficient of x^2 determines the orientation:
coefficient of x^2 positive, curve is $\cup$ shaped with minimum at the vertex;
coefficient of x^2 negative, curve is $\cap$ shaped with maximum at the vertex.
   The equation may be given in factorised form or may be rearranged by
completing the square. If the equation of the graph is:
(i) $y = a(x - \alpha)(x - \beta)$, $\alpha < \beta$, then the graph crosses the axis at $(\alpha, 0)$ and
$(\beta, 0)$ and between these points the graph is below the x-axis if $a > 0$
and above the x-axis if $a < 0$, the vertex is at $((\alpha + \beta)/2, -\frac{1}{4}a(\alpha - \beta)^2)$,
Fig. 11.2(a);
(ii) $y = a(x - d)^2 + c$, then the vertex is at (d, c), Fig. 11.2(b). In Fig. 11.2,
we have shown one parabola (a) with $a > 0$ and one (b) with $a < 0$.

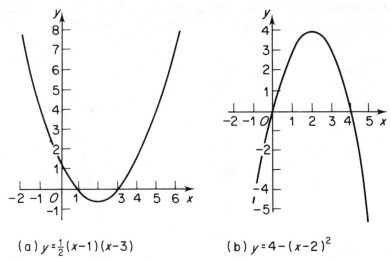

(a) $y = \frac{1}{2}(x-1)(x-3)$ (b) $y = 4 - (x-2)^2$

Fig. 11.2 Graphs of quadratic functions.

EXERCISE 11.1

1 Sketch the graph of the function f, given that:
   (i) $f(x) = 2x - 1$, (ii) $f(x) = 2(x - 1)(x - 2)$, (iii) $f(x) = -3(x - 1)^2 + 4$.
2 Sketch the graph with equation:
   (i) $3y + 2x = 6$, (ii) $2y = 5$, (iii) $y = (3 - x)(x + 2)$,
   (iv) $y - 2x^2 - 3x + 1 = 0$.

11.2 The graph of a cubic function

Suppose that we wish to sketch the graph with equation

$$y = x^3 - x^2 - 6x + 1.$$

We begin by calculating the values of y for a number of values of x, and show the calculation in a table, thus:

x	-4	-3	-2	-1	0	1	2	3	4	5
x^3	-64	-27	-8	-1	0	1	8	27	64	125
$-x^2$	-16	-9	-4	-1	0	-1	-4	-9	-16	-25
$-6x$	24	18	12	6	0	-6	-12	-18	-24	-30
y	-55	-1	1	5	1	-5	-7	1	25	70.

Note   The formula $y = \{(x-1)x - 6\}x + 1$ can be useful to calculate y on a calculator or computer. It involves fewer multiplications and does not need the storage of intermediate results, compared with the method of calculating the separate terms and adding. This saves computing time. Expand the expression to see that it gives the correct form for y. Check the results in the table by using this formula with a calculator.

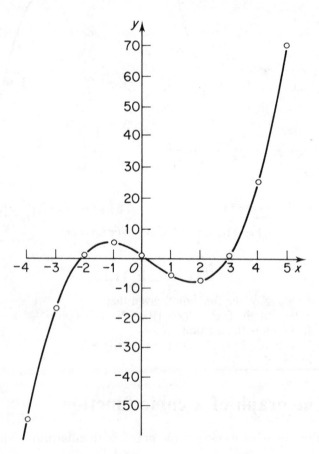

Fig. 11.3 Graph: $y = x^3 - x^2 - 6x + 1$

The points (x, y), found from the table, are plotted in Fig. 11.3 and the graph is drawn as a curve through these points. What are the general features of the graph and could they have been discovered without having to do the above calculations?

Large values of $|x|$

As x (or $-x$) becomes large, the term in x^3 becomes the most important term because the higher powers of x become larger much faster than the lower powers do. We say that x^3 is the *dominant term* and, for large values of x, and of $-x$, the shape of the graph will be similar to the graph of x^3. In a polynomial of degree n, the dominant term will be the term $a_n x^n$. Therefore, it is useful to know the shape of the graph of x^n, for different values of n, and we consider these graphs in §11.4.

Intersections with the axes

Obviously, by putting $x = 0$, we see that the graph of $y = f(x)$ crosses the y-axis at $(0, f(0)) = (0, 1)$ and this point on a graph can always be found. It is seen in Fig. 11.3 that the graph crosses the x-axis near to the points where $x \in \{-2, 0, 3\}$. If the expression for $f(x)$ can be factorised, then the factorised form can indicate where the graph meets the x-axis. For example, the graph of the function g where

$$g(x) = x^3 - x^2 - 6x = x(x+2)(x-3)$$

crosses the x-axis at $(-2, 0)$, $(0, 0)$, $(3, 0)$, giving three useful points of the graph.

Small values of x

When x is very small (either positive or negative), the higher powers of x will be very small and the dominant terms will be the linear terms. In the general polynomial these are the terms $a_1 x + a_0$. In our example, this means that, for small values of x, the graph will approximate to the graph of the linear function $-3x + 1$, which is the tangent to the polynomial graph when $x = 0$. Thus the graph crosses the y-axis at the point $(0, 1)$, as we noted before, but now we also know that, at this point, the gradient of the graph is -3.

Turning points

The graph in Fig. 11.3 clearly has a peak and a trough, and these are called *turning points* of the graph, since at these points the graph turns over, either from rising to falling, or from falling to rising, as x increases. Near the peak, the values of the function are less than the values at the peak and so this peak is called a *(local) maximum* of the graph. Similarly, near the trough, the function takes larger values than the value at the turning point and so this trough is called a *(local) minimum* of the graph. Of course, at other parts of the graph, y may take values which are greater than the maximum and other values which are less than the minimum. That is why the word 'local' has been used, but it has been written in parentheses because it is usually omitted and we just refer to maxima and minima.

The gradient f' of the function f can be used to find the turning points on the graph $y = f(x)$. At such a point, the tangent is parallel to Ox and so the gradient is zero. At a maximum, the gradient changes from positive to negative as x increases and at a minimum it changes from negative to positive. This is easily seen by considering the shape of the graph on each side of the turning point. The gradient may vanish at a point without changing sign on passing through that point. In this case, the point is a *point of inflexion* (Chapter 22). The maximum and minimum points and the points of inflexion are collectively called *stationary points* since, at all such points, the tangent to the graph is parallel to Ox and the rate of change of the value of the function is zero.

In the case of our cubic example where $f(x) = x^3 - x^2 - 6x + 1$ the gradient f' is given by $f'(x) = 3x^2 - 2x - 6$. The stationary points are given by $f'(x) = 0$, and we may solve the quadratic equation by means of the formula, which gives the turning points when $x = -1 \cdot 09$ and when $x = 1 \cdot 78$. These values agree with the position of the turning points seen on the graph in Fig. 11.3.

11.3   Useful ideas for graph sketching

We summarise the ideas that arose in §11.2, when we considered the cubic function. Not all these ideas will be useful in every case, but many of them will be found useful when graph sketching the graph $y = f(x)$.

1   Use the dominant terms in $f(x)$ to find the shape of the graph for large values of x and of $-x$.

2   When $f(x)$ is a polynomial, the linear terms in $f(x)$ give the tangent to the graph at $(0, f(0))$ and so this gives the intersection with the y-axis and the gradient at this point.

3   Any zeros of f will give the points of intersection with the x-axis, and these will easily be spotted if $f(x)$ factorises.

4   Any zeros of f' will give stationary points on the graph, and consideration of the sign of $f'(x)$ on each side of such stationary points will distinguish between a maximum, a minimum or an inflexion.

Note: In the example of §11.2, the cubic graph has a maximum and a minimum but this may not always be so for a cubic graph. It is possible for $f'(x)$ to have only one zero (equal roots of the quadratic equation) when the maximum and the minimum coincide at a point of inflexion (where $f'(x) = 0$ but $f'(x)$ has the same sign each side of the point). It is also possible that the quadratic equation $f'(x) = 0$ has no solution. In that case, $f'(x)$ is either always positive or always negative and the graph of f has no turning points.

EXERCISE 11.3

1 By calculating a number of points on the graph and plotting them, draw the graph with the equation:
(i) $y = x$, (ii) $y = 5 - 3x$, (iii) $y = -x^2$, (iv) $y = x^2 + x + 1$,
(v) $y = -x^3$, (vi) $y = 3 - x^3$, (vii) $y = 2x^3$.

2 Use the hints to graph sketching to sketch the graph of the function f, where f(x) is:
(i) $x^3 + x$, (ii) $x^3 + 3x^2 + 3x + 1$, (iii) $x^3 - 4x$, (v) $(x-1)(x-2)(x-3)$,
(v) $2x^3 - 9x^2 + 12x - 5$.

11.4 The graph with equation $y = x^n$, $n \in \mathbb{N}$

The graph of f, where $f(x) = x^n$, $n \in \mathbb{N}$, takes one of two forms depending on whether n is odd or even.

n odd In this case, $f(-x) = (-x)^n = -x^n = -f(x)$, so f is an odd function and its graph is unaltered by a rotation of a half turn about the origin O. For every point (x, y) on the graph, there is another point $(-x, -y)$. The graph passes through $(-1, -1)$, $(0, 0)$ and $(1, 1)$. For $n = 1$, the graph is the line $y = x$. For $n > 1$, the gradient $f'(x)$ is nx^{n-1}, which is always positive except at the origin, since $n - 1$ is even. Thus, the curve touches the x-axis at the origin, where there is a point of inflexion.

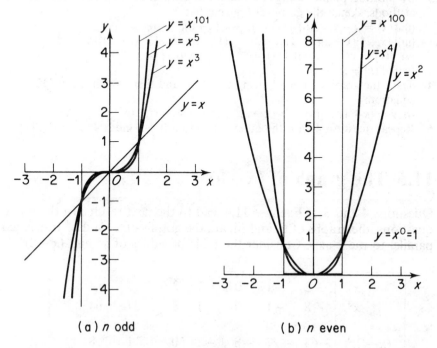

(a) n odd   (b) n even

Fig. 11.4 Graphs of $y = x^n$

For increasing values of n, the graph becomes steeper for $x > 1$ and $x < -1$ and it approaches the x-axis for $-1 < x < 1$. In Fig. 11.4(a), these features are shown for the cases $n = 1, 3, 5, 101$.

n even $f(x) = x^n = (-x)^n = f(-x)$, so f is an even function with a graph which is unaltered by a reflection in the y-axis. The graph passes through $(-1, 1)$ and $(1, 1)$ and, except for $n = 0$, through $(0, 0)$, where it touches the x-axis at a minimum point. The effect of increasing n is to make the graph steeper for $|x| > 1$ and nearer to the x-axis for $|x| < 1$. In Fig. 11.4(b) the graphs are shown for $n = 0, 2, 4, 100$.

EXERCISE 11.4

1   State the degree of $f'(x)$ if $f(x)$ is a polynomial of degree:
    (i) 3, (ii) 4, (iii) n.
    In each case, state the largest possible number of maxima and of minima that the graph of f can have.
2   Indicate in a rough sketch a graph of a polynomial of degree n, with r maxima and s minima:
    (i) $n = 2, r = 1$; (ii) $n = 2, r = 0$; (iii) $n = 3, r = 0 = s$,
    (iv) $n = 3, r = 1 = s$, (a) coefficient of x^3 positive, (b) coefficient of x^3 negative;
    (v) $n = 4, r = 1, s = 0$; (vi) $n = 4, r = 1, s = 2$; (vii) $n = 4, r = 0, s = 1$;
    (viii) $n = 4, r = 2, s = 1$.
3   By plotting points, using the factors of $f(x)$, and using any other of our methods, sketch the graph of f, given that
    (i) $f(x) = x - x^3$, (ii) $f(x) = (x-1)(x+2)(x-3)$,
    (iii) $(x+1)(x+2)(x-1)(x-2) = f(x)$, (iv) $f(x) = (x-2)^2(x+3)^3$,
    (v) $f(x) = x^3 - x^5$, (vi) $f(x) = x^4 - x^2$, (vii) $f(x) = x^4 + x^2 + 1$,
    (viii) $f(x) = x^4 + x^3 + x^2 + x + 1$.
4   Trace the graph of $y = x^3$, using Fig. 11.4, and then sketch the graphs with equation:
    (i) $y = (x+1)^3$, (ii) $y = (x-2)^3$, (iii) $y = -x^3$, (iv) $y = \frac{1}{2}x^3$.
5   Repeat question **4** with the index 3 replaced by an index 4.

11.5 The graph of $(x-h)^n$

Questions 4 and 5 of Exercise 11.4 lead to the next results. In these two questions, the graphs of (i) and (ii) are the graphs of x^3 and x^4 translated parallel to the x-axis. Consider the table of values of x^3 and $(x-2)^3$:

x	-2	-1	0	1	2	3	4
x^3	-8	-1	0	1	8	27	64
$(x-2)^3$	-64	-27	-8	-1	0	1	8

The effect of subtracting 2 from x, to give $f(x-2)$ instead of $f(x)$ is to translate (shift) the graph along the x-axis by 2 units, so that the centre of the graph is at $(2, 0)$ instead of $(0, 0)$.

Generally, if the graph of $f(x)$ is translated a distance h along the x-axis, that is, a distance h parallel to Ox, then the new curve will be the graph of $f(x-h)$. This was explained in terms of the composition of functions in §8.3.

Summary The graph of $(x-h)^n$ is the same as the graph of x^n translated a distance h parallel to Ox, with the sign of h taken into account.

For odd n, the graph of x^n is symmetrical about the origin O $(0,0)$, and the corresponding point of symmetry of the graph of $(x-h)^n$ is $(h, 0)$.

For even n, the graph of x^n is symmetrical about the axis $x = 0$, and the corresponding axis of symmetry of the graph of $(x-h)^n$ is $x = h$.

Note: There is a sign change in the correspondence. The graph of $(x-2)^5$ is symmetrical about $(2, 0)$ but the graph of $(x+2)^5$ is symmetrical about $(-2, 0)$.

11.6 Scale change and the graph of $y = a(x-h)^n$, $n \in \mathbb{N}$

The graph of $y = \frac{1}{2}x^3$ can be obtained from the graph of $y = x^3$ by halving the distance of all the points on the graph from the x-axis, which can be described by applying to the graph a one-way stretch with factor $\frac{1}{2}$ parallel

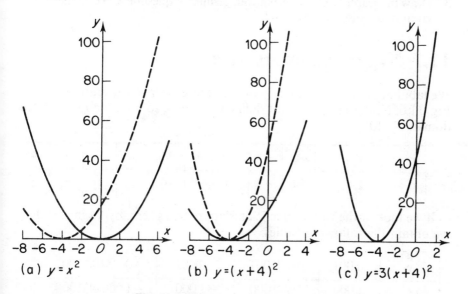

(a) $y = x^2$ (b) $y = (x+4)^2$ (c) $y = 3(x+4)^2$

Fig. 11.5 Change of origin and scale.

to the axis Oy. Similarly the graph of $3x^4$ is obtained from the graph of x^4 by a one-way stretch with factor 3 parallel to Oy. We may superimpose the transformations of §11.5 and scale change to obtain a sketch of the graph of $a(x-h)^n$, as we saw in §8.6.

EXAMPLE   *Sketch the graph with equation $y = 3(x+4)^2$.*

We obtain the required graph in three steps, as shown in Fig. 11.5. Start, in (a), with the graph of x^2 and shift the whole graph a distance of 4 units to the left, in the *negative x-axis direction*, to give the graph with equation $y = (x+4)^2$. This means that the vertex of the graph is shifted from $(0, 0)$ to $(-4, 0)$ because $x = -4$ makes $(x+4) = 0$. Finally, the scale is changed in the y-axis direction by a one way stretch of factor 3 to give the graph $y = 3(x+4)^2$, shown in Fig. 11.5(c).

Summary   The graph of $a(x-h)^n$ is obtained from the graph of x^n by
(i) a translation through a (signed) distance h parallel to Ox, to give the graph of $(x-h)^n$ followed by
(ii) a one way stretch parallel to Oy with factor a.

EXERCISE 11.6

1   Sketch the graph with equation:
   (i) $y = 2(x-1)^2$,   (ii) $y = -3x^3$,   (iii) $y = 3(x+2)^3$,   (iv) $y = -2(x+3)^4$,
   (v) $y = 4(3-x)^3$.
2   Explain the transformation which takes the graph $y = x^n$ into the graph:
   (i) $y = 2x^n$,   (ii) $y = x^n - 2$,   (iii) $y = x^n + 3$,   (iv) $y = 3 - x^n$,
   (v) $y = (x-3)^n$,   (vi) $y = (x+2)^n$,   (vii) $y = 2x^n + 5$,   (viii) $y = 4 - 3x^n$,
   (ix) $y = 4(x-3)^n$,   (x) $y = 4(x+1)^n$,   (xi) $y = 4 + (x+1)^n$.
3   Draw the graph of x^3 and, using your results for question **2**, sketch the graphs given in question **2** with $n = 3$.

11.7 Negative powers of x

We tabulate some values of x^{-1} and use these to draw the graph $y = x^{-1}$, Fig. 11.6. Note that x^{-1} is not defined at $x = 0$, so the function x^{-1} has domain $\mathbb{R} \backslash \{0\}$.

x	-4	-2	-1	$-\frac{1}{2}$	-0.1	0	0.1	$\frac{1}{2}$	1	2	4
x^{-1}	$-\frac{1}{4}$	$-\frac{1}{2}$	-1	-2	-10	$-$	10	2	1	$\frac{1}{2}$	$\frac{1}{4}$

If we look at the values of x^{-1} as x approaches zero, by taking small x, we obtain the following values:

x	0.001	0.000001	-0.001	-0.000000001
x^{-1}	$1\,000$	$1\,000\,000$	$-1\,000$	$-1\,000\,000\,000.$

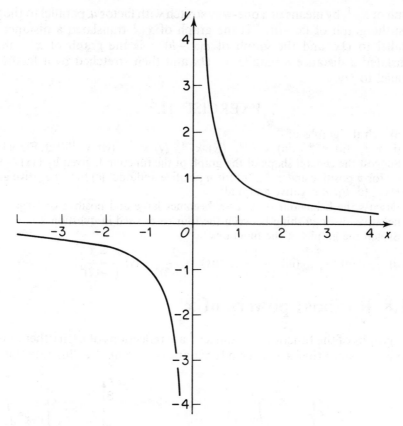

Fig. 11.6 Graph of $y = x^{-1}$

If we choose a number N, however large, we can find a small value h of x so that h^{-1} is larger than N. Similarly, if we choose M to be large and negative, we can find a negative number $-h$ such that $(-h)^{-1} < M$. This manner in which the graph behaves near to $x = 0$ is described by saying

as x tends to zero from above x^{-1} tends to infinity
as x tends to zero from below x^{-1} tends to negative infinity.

This is a description of the behaviour of x^{-1} and must not be interpreted as meaning that a number infinity (or negative infinity) exists. It does not! In symbols, we write:

$$x^{-1} \to \infty \text{ as } x \to 0+, \quad \text{and} \quad x^{-1} \to -\infty \text{ as } x \to 0-.$$

Note that the graph is not continuous, in the sense that it can not be drawn without lifting pen from paper, but it consists of two branches. The lines $x = 0$ and $y = 0$, which the graph approaches 'at infinity' are called *asymptotes* of the graph.

Using the methods of §11.6, the graph of ax^{-1} can be drawn from the graph of x^{-1}, by means of a one-way stretch with factor a, parallel to the y-axis; the graph of $(x-h)^{-1}$ is the graph of x^{-1} translated a distance h parallel to Ox; and the graph of $a(x-h)^{-1}$ is the graph of x^{-1} first translated a distance h parallel to Ox and then stretched by a factor a parallel to Oy.

EXERCISE 11.7

1   Sketch the graphs of
    (i) x^{-2}, (ii) x^{-4}, (iii) x^{-100}, (iv) x^{-3}, (v) x^{-5}, (vi) x^{-101}; cf. Fig. 11.4.
2   Suggest the general shape of the graph of the function f, given by $f(x) = x^n$:
    (a) for n positive and even, (b) for n positive and odd, (c) for n negative and even, (d) for n negative and odd.
3   Discuss the behaviour of x^n as x becomes large and positive or large and negative, distinguishing between the four cases used in question 2.
4   Sketch the graphs of the functions:

   (i) $\dfrac{2}{x^2}$, (ii) $\dfrac{-3}{x^3}$, (iii) $\dfrac{1}{(x-1)^3}$, (iv) $\dfrac{2}{(x+3)^3}$, (v) $\dfrac{-3}{(x-2)^4}$.

11.8 Rational powers of x

The graphs of the functions x^3 and $x^{1/3}$ are reflections of each other in the line $y = x$, since they are inverse functions (see §8.8). We illustrate this in

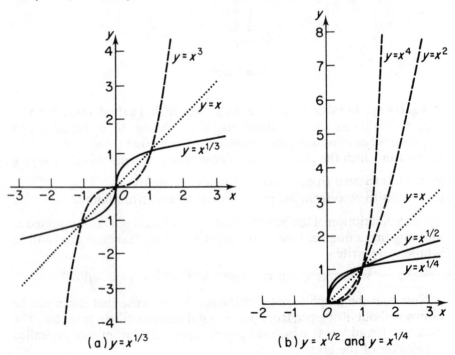

(a) $y = x^{1/3}$ (b) $y = x^{1/2}$ and $y = x^{1/4}$

Fig. 11.7

Fig. 11.7(a). A similar situation holds for x^n and $x^{1/n}$ for any odd integer n. For even n, the function x^n is not one-one so we restrict its domain to $\mathbb{R}^+ \cup \{0\}$ and then it has an inverse function $x^{1/n}$ with the same domain, where

$$x^{1/n} = \sqrt[n]{x} = \text{the positive } n\text{th root of } x.$$

The graphs for $n = 2$ and $n = 4$ are shown in Fig. 11.7(b).

If the domain of the functions is limited to the positive real numbers then the graphs of the functions x^n all pass through $(1, 1)$ and are spaced out according to the size of n, as can be seen from Figs. 11.4 and 11.7. The graphs for $n > 0$ all have increasing gradient, and those with $n < 0$ have decreasing gradient, with increasing x. A graph may be drawn for any rational value of n, for instance, the graph for $n = 3/2$ lies between the graphs for $n = 1$ and $n = 2$.

MISCELLANEOUS EXERCISE 11

1 Sketch the graph of the function f, with domain $\mathbb{R}^+$, where $f(x)$ is given by:
   (i) $\sqrt[4]{x}$, (ii) $x^{2/3}$, (iii) $\frac{1}{4}x^{5/2}$, (iv) $\sqrt{(x+1)}$.
2 Draw, on the same graph the curves with equation $y = x^n$ for:
   (i) $n = 0$, (ii) $n = \frac{1}{4}$, (iii) $n = \frac{1}{2}$, (iv) $n = 1$, (vi) $n = 1\frac{1}{2}$, (vii) $n = 2$.
3 Repeat question 2 for the corresponding negative values of n.
4 Sketch the graph:
   (i) $y = 1 - x^3$, (ii) $y = -2\sqrt{x}$, (iii) $y = -x^4/6$, (iv) $y = 3x^2 - 1$,
   (v) $y = -\frac{1}{4}\sqrt[3]{x}$, (vi) $y = 2/(1+x)$, (vii) $y = 1/(2+3x)$.
5 The function f is periodic with period 2 and
   $$f(x) = 4/x \qquad \text{for } 1 < x \leqslant 2,$$
   $$f(x) = 2(x-1) \quad \text{for } 2 < x \leqslant 3.$$
   Sketch the graph of f when $-3 \leqslant x \leqslant 3$. $\hfill (L)$

12 Vectors II

12.1 Column vectors in the plane

In Chapter 5 we considered vectors in terms of displacements. We now extend the definitions, referring to Cartesian axes $O(x, y)$ in the plane. A point P, with coordinates (x, y), has position vector $\mathbf{r}$, which we write as a column vector,

$$\mathbf{r} = \overrightarrow{OP} = \begin{pmatrix} x \\ y \end{pmatrix}.$$

The length OP of the vector $\mathbf{r}$ is called the *magnitude* or *modulus* of $\mathbf{r}$. We denote this by means of the modulus sign, and if OP is of length r,

$$OP = r = |\mathbf{r}| = |\overrightarrow{OP}| = \left| \begin{pmatrix} x \\ y \end{pmatrix} \right|.$$

By Pythagoras' theorem,

$$OP^2 = x^2 + y^2 \quad \text{so} \quad r = |\mathbf{r}| = \sqrt{(x^2 + y^2)} = \left| \begin{pmatrix} x \\ y \end{pmatrix} \right|.$$

For example, $\left| \begin{pmatrix} 1 \\ 1 \end{pmatrix} \right| = \sqrt{2}$ and $\left| \begin{pmatrix} 3 \\ -4 \end{pmatrix} \right| = 5$.

Note For every vector $\mathbf{a}$, $|\mathbf{a}| \geqslant 0$ and $|\mathbf{a}| = 0$ only when $\mathbf{a} = \mathbf{0} = \begin{pmatrix} 0 \\ 0 \end{pmatrix}$.

Unit vector

A vector of unit modulus is called a *unit vector*. Thus

$$\mathbf{u} \text{ is a vector} \Leftrightarrow |\mathbf{u}| = 1.$$

For example, $\begin{pmatrix} \frac{1}{\sqrt{2}} \\ \frac{1}{\sqrt{2}} \end{pmatrix}$ and $\begin{pmatrix} \frac{3}{5} \\ -\frac{4}{5} \end{pmatrix}$ are unit vectors.

The unit vector $\mathbf{u}$ in the direction of the vector $\mathbf{a}$ is given by $\mathbf{u} = \dfrac{\mathbf{a}}{|\mathbf{a}|}$. Clearly $\mathbf{u}$ is parallel to $\mathbf{a}$ and

$$|\mathbf{u}| = \left| \frac{\mathbf{a}}{|\mathbf{a}|} \right| = \frac{1}{|\mathbf{a}|}|\mathbf{a}| = 1.$$

We use a special notation for the Cartesian unit vectors, parallel to the axes Ox and Oy, namely

$$\begin{pmatrix} 1 \\ 0 \end{pmatrix} = \mathbf{i}, \quad \begin{pmatrix} 0 \\ 1 \end{pmatrix} = \mathbf{j}.$$

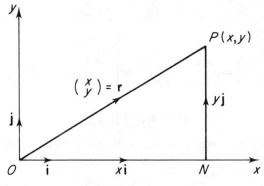

Fig. 12.1

In Fig. 12.1, let P be the point (x, y), so that $\overrightarrow{OP} = \mathbf{r} = \begin{pmatrix} x \\ y \end{pmatrix}$. Let N be the foot of the perpendicular from P on to Ox, then $ON = x$, $PN = y$ and, so, $\overrightarrow{ON} = x\mathbf{i}$ and $\overrightarrow{NP} = y\mathbf{j}$. Therefore,

$$\mathbf{r} = \overrightarrow{OP} = \begin{pmatrix} x \\ y \end{pmatrix} = x\mathbf{i} + y\mathbf{j} = x\begin{pmatrix} 1 \\ 0 \end{pmatrix} + y\begin{pmatrix} 0 \\ 1 \end{pmatrix}.$$

The properties of vector addition and of multiplication by a scalar, and the fact that the unit vectors $\mathbf{i}$ and $\mathbf{j}$ are not parallel, immediately lead to the following rules for manipulating column vectors.

Theorem 12.1 Let the point $P(x, y)$ have position vector $\mathbf{r}$, with $\mathbf{r} = \begin{pmatrix} x \\ y \end{pmatrix}$. Then

$$\begin{pmatrix} x_1 \\ y_1 \end{pmatrix} = \begin{pmatrix} x_2 \\ y_2 \end{pmatrix} \Leftrightarrow x_1 = x_2 \quad \text{and} \quad y_1 = y_2,$$

$$\begin{pmatrix} x_1 \\ y_1 \end{pmatrix} + \begin{pmatrix} x_2 \\ y_2 \end{pmatrix} = \begin{pmatrix} x_1 + x_2 \\ y_1 + y_2 \end{pmatrix},$$

$$\lambda \begin{pmatrix} x \\ y \end{pmatrix} = \begin{pmatrix} \lambda x \\ \lambda y \end{pmatrix}.$$

EXAMPLE   *Given that* $\mathbf{a} = \begin{pmatrix} 2 \\ 3 \end{pmatrix}$ *and* $\mathbf{b} = \begin{pmatrix} -1 \\ 5 \end{pmatrix}$, *write down:*
(i) $\mathbf{a} + \mathbf{b}$, (ii) $\mathbf{a} - \mathbf{b}$, (iii) $2\mathbf{a}$, (iv) $3\mathbf{a} - 2\mathbf{b}$.

(i) $\begin{pmatrix} 2 \\ 3 \end{pmatrix} + \begin{pmatrix} -1 \\ 5 \end{pmatrix} = \begin{pmatrix} 2-1 \\ 3+5 \end{pmatrix} = \begin{pmatrix} 1 \\ 8 \end{pmatrix}$,

(ii) $\begin{pmatrix} 2 \\ 3 \end{pmatrix} - \begin{pmatrix} -1 \\ 5 \end{pmatrix} = \begin{pmatrix} 2-(-1) \\ 3-5 \end{pmatrix} = \begin{pmatrix} 3 \\ -2 \end{pmatrix}$,

(iii) $2\begin{pmatrix} 2 \\ 3 \end{pmatrix} = \begin{pmatrix} 4 \\ 6 \end{pmatrix}$,

(iv) $3\begin{pmatrix} 2 \\ 3 \end{pmatrix} - 2\begin{pmatrix} -1 \\ 5 \end{pmatrix} = \begin{pmatrix} 3\times 2 + 2\times 1 \\ 3\times 3 - 2\times 5 \end{pmatrix} = \begin{pmatrix} 8 \\ -1 \end{pmatrix}$.

EXERCISE 12.1

1   By expressing each vector in terms of the Cartesian unit vectors, prove theorem 12.1.

2   Write down:

(i) $\begin{pmatrix} 1 \\ -2 \end{pmatrix} + \begin{pmatrix} 2 \\ -1 \end{pmatrix}$,   (ii) $2\begin{pmatrix} -1 \\ 2 \end{pmatrix} - \begin{pmatrix} 2 \\ 1 \end{pmatrix}$,   (iii) $4\begin{pmatrix} 3 \\ -2 \end{pmatrix} - 3\begin{pmatrix} 3 \\ 1 \end{pmatrix}$,

(iv) $a\begin{pmatrix} 1 \\ 2 \end{pmatrix} - b\begin{pmatrix} 2 \\ 1 \end{pmatrix}$,   (v) $3\begin{pmatrix} x \\ y \end{pmatrix} - 4\begin{pmatrix} -x \\ y \end{pmatrix}$,   (vi) $a\begin{pmatrix} x \\ y \end{pmatrix} + b\begin{pmatrix} u \\ v \end{pmatrix}$.

3   Verify that:

(i) $\begin{pmatrix} 2 \\ 1 \end{pmatrix} + \begin{pmatrix} 3 \\ 4 \end{pmatrix} = \begin{pmatrix} 3 \\ 4 \end{pmatrix} + \begin{pmatrix} 2 \\ 1 \end{pmatrix}$,   (ii) $2\left\{ 3\begin{pmatrix} 1 \\ 2 \end{pmatrix} \right\} = 6\begin{pmatrix} 1 \\ 2 \end{pmatrix}$,

(iii) $\begin{pmatrix} 1 \\ 2 \end{pmatrix} + \left\{ \begin{pmatrix} 4 \\ 3 \end{pmatrix} + \begin{pmatrix} 2 \\ 5 \end{pmatrix} \right\} = \left\{ \begin{pmatrix} 1 \\ 2 \end{pmatrix} + \begin{pmatrix} 4 \\ 3 \end{pmatrix} \right\} + \begin{pmatrix} 2 \\ 5 \end{pmatrix}$.

4   Verify that column vectors satisfy the rules of §5.2, by using $\mathbf{a} = \begin{pmatrix} x \\ y \end{pmatrix}$,

$\mathbf{b} = \begin{pmatrix} u \\ v \end{pmatrix}$, $\mathbf{c} = \begin{pmatrix} w \\ z \end{pmatrix}$ in the following:

(i) commutativity of addition $\mathbf{a} + \mathbf{b} = \mathbf{b} + \mathbf{a}$;

(ii) zero multiplication $0\mathbf{a} = \mathbf{0} = \begin{pmatrix} 0 \\ 0 \end{pmatrix}$;

(iii) associativity of addition $(\mathbf{a} + \mathbf{b}) + \mathbf{c} = \mathbf{a} + (\mathbf{b} + \mathbf{c})$;

(iv) associativity of products $(hk)\mathbf{a} = h(k\mathbf{a})$;

(v) distributivity $h\mathbf{a} + k\mathbf{a} = (h+k)\mathbf{a}$;

(vi) distributivity $h(\mathbf{a} + \mathbf{b}) = h\mathbf{a} + h\mathbf{b}$.

12.2 Linear combinations of vectors

The splitting of $\mathbf{r}$ into the sum of $x\mathbf{i}$ and $y\mathbf{j}$ is called *resolving* $\mathbf{r}$ into *vector components* $x\mathbf{i}$ and $y\mathbf{j}$, and x and y are called the *Cartesian components* of $\mathbf{r}$.

Any vector can be resolved into Cartesian component vectors. These components are represented by the sides of a rectangle, parallel to the axes $O(x, y)$, with the given vector represented by the diagonal of the rectangle.

Using the language of §5.5, the pair of Cartesian unit vectors $\{\mathbf{i}, \mathbf{j}\}$ forms a basis for the set of all vectors in the plane.

Definition   A *linear combination* of a set of vectors is a sum of scalar multiples of the vectors.

Theorem 5.5 may be restated: every vector in the plane is uniquely expressed as a linear combination of a pair of given basis vectors. For example, using the basis $\{\mathbf{i}, \mathbf{j}\}$, the vector $\begin{pmatrix} 5 \\ 12 \end{pmatrix}$ is a unique linear combination $5\mathbf{i} + 12\mathbf{j}$ of the basis vectors. The modulus of this vector is given in terms of the multiples of the basis vectors, namely $\sqrt{(5^2 + 12^2)}$, by Pythagoras' theorem. This is because the basis vectors $\mathbf{i}$ and $\mathbf{j}$ are perpendicular unit vectors. The same is not always true. Consider the basis $\{\mathbf{a}, \mathbf{b}\}$, where $\mathbf{a} = \begin{pmatrix} 1 \\ 0 \end{pmatrix}$, $\mathbf{b} = \begin{pmatrix} 1 \\ 2 \end{pmatrix}$, which are not perpendicular unit vectors. Then

$$\begin{pmatrix} 5 \\ 12 \end{pmatrix} = -\mathbf{a} + 6\mathbf{b} \quad \text{and} \quad 1^2 + 6^2 \neq 5^2 + 12^2.$$

EXAMPLE 1   *Express the vector* $\begin{pmatrix} 3 \\ 4 \end{pmatrix}$: *(i) as a linear combination of the Cartesian unit vectors* $\mathbf{i}$ *and* $\mathbf{j}$, *(ii) as a linear combination of the two vectors* $\begin{pmatrix} 2 \\ 1 \end{pmatrix}$ *and* $\begin{pmatrix} 1 \\ -2 \end{pmatrix}$.

(i) $\mathbf{a} = \begin{pmatrix} 3 \\ 4 \end{pmatrix} = 3\begin{pmatrix} 1 \\ 0 \end{pmatrix} + 4\begin{pmatrix} 0 \\ 1 \end{pmatrix} = 3\mathbf{i} + 4\mathbf{j}.$

(ii) Suppose that $\mathbf{a} = \begin{pmatrix} 3 \\ 4 \end{pmatrix} = s\mathbf{b} + t\mathbf{c} = s\begin{pmatrix} 2 \\ 1 \end{pmatrix} + t\begin{pmatrix} 1 \\ -2 \end{pmatrix}$, then we have to solve two simultaneous equations in s and t,

$$3 = 2s + t \quad \text{and} \quad 4 = s - 2t.$$

Eliminate t by adding twice the first equation to the second,

$$6 + 4 = 4s + 2t + s - 2t = 5s, \quad \text{or} \quad 10 = 5s,$$

so $s = 2$ and then

$$t = 3 - 2s = 3 - 4 = -1.$$

Thus $\mathbf{a} = 2\mathbf{b} - \mathbf{c}$, Fig. 12.2(a).

EXAMPLE 2   *Find the unit vector which bisects the angle between the vectors* $\begin{pmatrix} 4 \\ 3 \end{pmatrix}$ *and* $\begin{pmatrix} -3 \\ 4 \end{pmatrix}$.

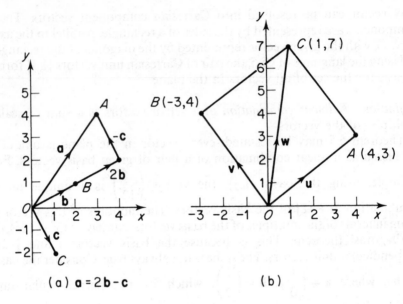

(a) **a** = 2**b** − **c**   (b)

Fig. 12.2

The unit vectors **u** and **v** in the given directions are given by $\mathbf{u} = \begin{pmatrix} \frac{4}{5} \\ \frac{3}{5} \end{pmatrix}$ and

$\mathbf{v} = \begin{pmatrix} -\frac{3}{5} \\ \frac{4}{5} \end{pmatrix}$. The parallelogram with sides **u** and **v** is a rhombus and its diagonal

u + **v** bisects the angle between the sides. Hence, the required vector **w** is in the direction of (**u** + **v**). So

$$\mathbf{w} = k(\mathbf{u}+\mathbf{v}) = \frac{k}{5}\begin{pmatrix} 1 \\ 7 \end{pmatrix}.$$

This vector has modulus $\frac{k}{5}\sqrt{(1+49)}$, so $k = \frac{5}{\sqrt{50}}$ and $\mathbf{w} = \frac{1}{5\sqrt{2}}\begin{pmatrix} 1 \\ 7 \end{pmatrix}$. This

method of solution is a general one, which would apply to other problems of this type. However, in this particular example, the initial vectors are perpendicular and equal in length, and so another method of solution is possible. In

Figure 12.2(b), let A and B be the points with position vectors given by $\mathbf{a} = \begin{pmatrix} 4 \\ 3 \end{pmatrix}$

and $\mathbf{b} = \begin{pmatrix} -3 \\ 4 \end{pmatrix}$ respectively. Then it will be seen that the point C, with position

vector given by $\mathbf{c} = \mathbf{a}+\mathbf{b} = \begin{pmatrix} 1 \\ 7 \end{pmatrix}$, is the fourth vertex of the square $OACB$. Hence

OC bisects the angle between OA and OB and the required unit vector

$$\mathbf{w} = \frac{\overrightarrow{OC}}{|OC|} = \frac{1}{5\sqrt{2}}\begin{pmatrix} 1 \\ 7 \end{pmatrix}.$$

EXERCISE 12.2

1 Find the modulus of the vector:

(i) $\begin{pmatrix} -3 \\ -4 \end{pmatrix}$, (ii) $\begin{pmatrix} 5 \\ 12 \end{pmatrix}$, (iii) $\begin{pmatrix} 1 \\ 1 \end{pmatrix}$, (iv) $\begin{pmatrix} -8 \\ 6 \end{pmatrix}$, (v) $3\mathbf{i} - 4\mathbf{j}$, (vi) $\mathbf{i} + 2\mathbf{j}$,
(vii) $3\mathbf{j}$.

2 For each of the vectors in question **1**, find a unit vector in that direction.

3 Given the four points $A(1, 3)$, $B(4, 5)$, $C(5, 3)$, $D(2, 1)$, find the column vectors representing $\overrightarrow{AB}$ and $\overrightarrow{CD}$ and prove that $ABCD$ is a parallelogram.

4 Express as a linear combination of the vectors $\mathbf{a}$ and $\mathbf{b}$, where $\mathbf{a} = \begin{pmatrix} 2 \\ 1 \end{pmatrix}$ and $\mathbf{b} = \begin{pmatrix} -3 \\ 4 \end{pmatrix}$, the vectors:

(i) $\begin{pmatrix} -1 \\ 5 \end{pmatrix}$, (ii) $\begin{pmatrix} 6 \\ 3 \end{pmatrix}$, (iii) $\begin{pmatrix} 0 \\ 1 \end{pmatrix}$, (iv) $\begin{pmatrix} 1 \\ 0 \end{pmatrix}$, (v) $\begin{pmatrix} x \\ y \end{pmatrix}$, (vi) $\mathbf{i} + 2\mathbf{j}$,
(vii) $11\mathbf{j}$.

5 Express as a linear combination of $\mathbf{a}$ and $\mathbf{b}$, where $\mathbf{a} = \mathbf{i} - 2\mathbf{j}$ and $\mathbf{b} = 2\mathbf{i} + \mathbf{j}$, the vectors:
(i) $\mathbf{i} + 4\mathbf{j}$, (ii) $\mathbf{i}$, (iii) $\mathbf{j}$, (iv) $\mathbf{i} + \mathbf{j}$, (v) $\mathbf{i} - \mathbf{j}$, (vi) $x\mathbf{i} + y\mathbf{j}$,
(vii) $\begin{pmatrix} 2 \\ 5 \end{pmatrix}$, (viii) $\begin{pmatrix} s \\ t \end{pmatrix}$.

6 Write down the ratio theorem, §5.4, in terms of Cartesian vectors, using $\mathbf{r} = \begin{pmatrix} x \\ y \end{pmatrix}$, $\mathbf{a} = \begin{pmatrix} p \\ q \end{pmatrix}$, $\mathbf{b} = \begin{pmatrix} s \\ t \end{pmatrix}$. Rewrite the theorem in terms of the Cartesian coordinates (x, y), (p, q), (s, t) of the points concerned.

7 The coordinates of the points A and B are $(3, 4)$ and $(12, 5)$ respectively. If O is the origin, find the coordinates of the point on the line-segment AB which is equidistant from the lines OA and OB. *(L)*

8 Points A, B, C have coordinates $(1, 3)$, $(3, -1)$, $(5, 7)$ respectively. The triangle ABC has medians AD, BE, CF. Write down, as column vectors, the position vectors of A, B, C, D, E, F.

Given that $3\overrightarrow{OG} = \overrightarrow{OA} + \overrightarrow{OB} + \overrightarrow{OC}$, find the position vector of G. Prove that DA, EB, FC intersect at G, and, by using the ratio theorem, prove that G is the point of trisection of each of these line segments.

9 Find unit vectors in the directions of $\overrightarrow{OA}$ and $\overrightarrow{OB}$, where A and B are points $(4, 3)$ and $(3, -4)$. Use the sum and difference of the unit vectors to find points on the unit circle, centre O, equidistant from the diameters through A and B.

10 The points A and B have position vectors $3\mathbf{i} + 2\mathbf{j}$ and $-\mathbf{i} + 4\mathbf{j}$ respectively, referred to O as origin. The point D has position vector $11\mathbf{i} - 2\mathbf{j}$ and is such that
$$\overrightarrow{OD} = m\overrightarrow{OA} + n\overrightarrow{OB}.$$
Calculate the values of m and n. *(L)*

11 Given that $\overrightarrow{OA} = (-2\mathbf{i} - 3\mathbf{j})$ and $\overrightarrow{OB} = (5\mathbf{i} + t\mathbf{j})$, where t is a constant, determine the value of t when the points O, A and B are collinear. *(L)*

12 The vectors $\overrightarrow{OP}$ and $\overrightarrow{OQ}$ are given by
$$\overrightarrow{OP} = (\lambda\mathbf{i} + \mu\mathbf{j}), \quad \overrightarrow{OQ} = (-\mu\mathbf{i} + \lambda\mathbf{j}),$$

where λ and μ are non-zero constants. Show that the angle between $\overrightarrow{OP}$ and $\overrightarrow{OQ}$ is $\pi/2$ and find the length of PQ in terms of λ and μ.

Given that R is the mid-point of PQ, express $\overrightarrow{OR}$ in terms of λ, μ, **i** and **j**. (L)

12.3 Cartesian coordinates in three dimensions

In two dimensions, we use two Cartesian coordinates (x, y) to fix the position of a point relative to axes $O(x, y)$. In three-dimensional space, we need a third coordinate, giving us coordinates (x, y, z) to fix the position of a point, relative to axes $O(x, y, z)$. The third axis Oz is perpendicular to Ox and to Oy, so it is perpendicular to the plane $O(x, y)$. There are two directions perpendicular to this plane and we choose Oz such that, on looking along Oz, a clockwise rotation of one right-angle will bring the axis Ox into the position of Oy. This is shown in Fig. 12.3, in (a) where the axis Oz is 'going into the paper', and in (b) where the axis Oz is 'coming out of the paper'. Note that, in each case, Ox is brought into the position of Oy by a rotation through 90° clockwise when looking along Oz. In (b), the axes $O(x, y)$ are drawn in the usual position for plane coordinates, and it appears that an anti-clockwise rotation is needed to bring Ox to Oy. However, this anticlockwise rotation is about an axis going *into* the paper and this is clockwise about Oz which is coming *out* of the paper.

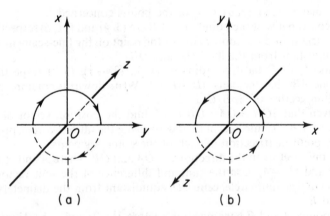

Rotation Ox to Oy is clockwise about Oz

Fig. 12.3 Three-dimensional axes $O(x, y, z)$.

Fig. 12.4, shows how the position of a point P is fixed by its coordinates (x, y, z), so that we can identify P with (x, y, z).

Choose unit vectors **i**, **j**, **k**, along Ox, Oy, Oz respectively. Let OP be the diagonal of a rectangular box, which has vertices A, B, C on Ox, Oy, Oz respectively and L, M, N in the planes $O(y, z)$, $O(z, x)$, $O(x, y)$ respectively.

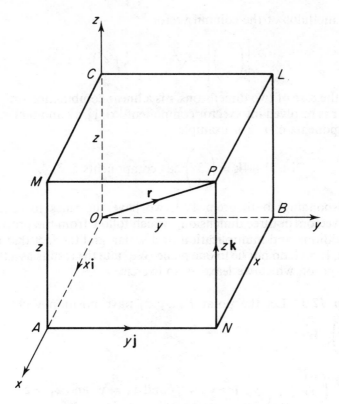

P has coordinates x, y, z, all positive
The arrows show the positive directions of Ox, Oy, Oz.

Fig. 12.4 Three-dimensional coordinates

Then
$$\overrightarrow{OP} = \overrightarrow{OA} + \overrightarrow{AN} + \overrightarrow{NP} = x\mathbf{i} + y\mathbf{j} + z\mathbf{k}.$$

We introduce three-dimensional column vectors, extending the ideas of §12.1. For the point $P(x, y, z)$ the position vector of P is $\overrightarrow{OP}$ where

$$\overrightarrow{OP} = \mathbf{r} = \begin{pmatrix} x \\ y \\ z \end{pmatrix} = x\begin{pmatrix} 1 \\ 0 \\ 0 \end{pmatrix} + y\begin{pmatrix} 0 \\ 1 \\ 0 \end{pmatrix} + z\begin{pmatrix} 0 \\ 0 \\ 1 \end{pmatrix},$$

so that $\mathbf{i} = \begin{pmatrix} 1 \\ 0 \\ 0 \end{pmatrix}$, $\mathbf{j} = \begin{pmatrix} 0 \\ 1 \\ 0 \end{pmatrix}$, $\mathbf{k} = \begin{pmatrix} 0 \\ 0 \\ 1 \end{pmatrix}$, and $\mathbf{r} = x\mathbf{i} + y\mathbf{j} + z\mathbf{k}$.

By two applications of Pythagoras' theorem,
$$OP^2 = ON^2 + NP^2 = OA^2 + AN^2 + NP^2,$$
so
$$|\mathbf{r}|^2 = r^2 = OP^2 = x^2 + y^2 + z^2$$

and the modulus of the column vector

$$\left| \begin{pmatrix} x \\ y \\ z \end{pmatrix} \right| = \sqrt{(x^2 + y^2 + z^2)}.$$

As in the case of two dimensions, $\mathbf{r}$ is a linear combination of $\mathbf{i}, \mathbf{j}, \mathbf{k}$. We say that $\mathbf{r}$ is resolved into vector components $x\mathbf{i}, y\mathbf{j}, z\mathbf{k}$ and that x, y, z are the components of $\mathbf{r}$. For example,

$$2\mathbf{i} + 3\mathbf{j} + 4\mathbf{k} = \begin{pmatrix} 2 \\ 3 \\ 4 \end{pmatrix} \text{ has components 2, 3, 4.}$$

Corresponding to theorem 12.1, we have the rules for combining column vectors in three dimensions, which follow from the properties of vector addition and multiplication by a scalar, and the fact that the unit vectors $\mathbf{i}, \mathbf{j}$ and $\mathbf{k}$ do not lie in one plane. We state the results as a theorem, without proof, which we leave as an exercise.

Theorem 12.3   Let the point $P(x, y, z)$ have position vector $\mathbf{r}$, with $\mathbf{r} = \begin{pmatrix} x \\ y \\ z \end{pmatrix}$. Then

$$\begin{pmatrix} x_1 \\ y_1 \\ z_1 \end{pmatrix} = \begin{pmatrix} x_2 \\ y_2 \\ z_2 \end{pmatrix} \Leftrightarrow x_1 = x_2 \text{ and } y_1 = y_2 \text{ and } z_1 = z_2,$$

$$\begin{pmatrix} x_1 \\ y_1 \\ z_1 \end{pmatrix} + \begin{pmatrix} x_2 \\ y_2 \\ z_2 \end{pmatrix} = \begin{pmatrix} x_1 + x_2 \\ y_1 + y_2 \\ z_1 + z_2 \end{pmatrix},$$

$$\lambda \begin{pmatrix} x \\ y \\ z \end{pmatrix} = \begin{pmatrix} \lambda x \\ \lambda y \\ \lambda z \end{pmatrix}.$$

Other bases in three dimensions

More generally, a vector can be resolved into components in any three non-coplanar directions. *Non-coplanar* means not parallel to one plane and so (in particular) non-parallel. This corresponds to the resolution of a vector into components in two directions in the plane (§5.5). Let $\overrightarrow{OP} = \mathbf{r}$, and let $\mathbf{u}, \mathbf{v}$ and $\mathbf{w}$ be any three non-coplanar unit vectors. Construct three planes through O, each containing two of the three vectors $\mathbf{u}, \mathbf{v}, \mathbf{w}$, and three more parallel planes through P, Fig. 12.5.

Then these six planes form the faces of a parallelepiped, with edges parallel to $\mathbf{u}, \mathbf{v}, \mathbf{w}$ and diagonal $\overrightarrow{OP}$. Let the three edges $\overrightarrow{OA} = a\mathbf{u}, \overrightarrow{AF} = b\mathbf{v}, \overrightarrow{FP} = c\mathbf{w}$, then

$$\mathbf{r} = \overrightarrow{OP} = a\mathbf{u} + b\mathbf{v} + c\mathbf{w}.$$

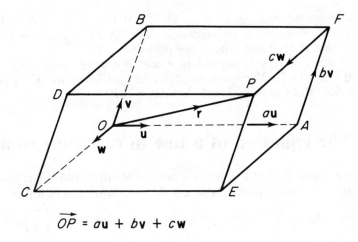

$$\overrightarrow{OP} = a\mathbf{u} + b\mathbf{v} + c\mathbf{w}$$

Fig. 12.5

In this way, $\overrightarrow{OP}$ is resolved into three unique components $a\mathbf{u}$, $b\mathbf{v}$, $c\mathbf{w}$, in the directions of $\mathbf{u}$, $\mathbf{v}$, $\mathbf{w}$, and $\overrightarrow{OP}$ is called the resultant of $a\mathbf{u}$, $b\mathbf{v}$, $c\mathbf{w}$. This means that $\mathbf{u}$, $\mathbf{v}$, $\mathbf{w}$ is a basis for vectors in space. Algebraically, a, b and c are found by solving three simultaneous equations, which is beyond the scope of our present study.

EXERCISE 12.3

1 State which of the alternatives (a) or (b) is true in each case:
   (i) Oz is vertically (a) up, (b) down, when Ox is North and Oy is East;
   (ii) Ox is vertically (a) up, (b) down, when Oy is West and Oz is South;
   (iii) If O is at the centre of the Earth and Oz is its axis pointing to the North Pole, then the rotation of the Earth is (a) positive, (b) negative, about Oz.

2 Find the magnitudes of **a** and **b**, where $\mathbf{a} = \begin{pmatrix} 1 \\ 2 \\ -2 \end{pmatrix}$ and $\mathbf{b} = \begin{pmatrix} 3 \\ -4 \\ 5 \end{pmatrix}$, and of

   $\mathbf{a} + \mathbf{b}$, $2\mathbf{a} - 3\mathbf{b}$. Write down the unit vectors in the directions of each of these four vectors.

3 Given the two points $A(1, -2, 3)$ and $B(-1, 3, 4)$, find the column vector $\overrightarrow{AB}$ and, hence, find the distance between A and B.

4 Given that $\mathbf{a} = \begin{pmatrix} 3 \\ 2 \\ 1 \end{pmatrix}$, $\mathbf{b} = \begin{pmatrix} -2 \\ 1 \\ -1 \end{pmatrix}$, $\mathbf{c} = \begin{pmatrix} 4 \\ 2 \\ -1 \end{pmatrix}$, write down the vectors:

   (i) $\mathbf{a} + \mathbf{b} - \mathbf{c}$, (ii) $2\mathbf{a} + 3\mathbf{b}$, (iii) $-\mathbf{b} + 3\mathbf{c}$, (iv) $3\mathbf{a} + 2\mathbf{b} + \mathbf{c}$.

5 For the following sets of coordinates for the four points A, B, C, D, respectively, find the column vectors $\overrightarrow{AB}$ and $\overrightarrow{CD}$ and determine whether or not the four points lie at the vertices of a parallelogram:
   (i) $(1, 0, 2)$, $(2, 2, 5)$, $(-1, -1, -1)$, $(0, 1, 2)$;
   (ii) $(2, -3, 5)$, $(0, -1, 2)$, $(1, 0, 1)$, $(-1, 1, -2)$;
   (iii) $(1, -2, 3)$, $(3, -1, -1)$, $(4, 0, 1)$, $(2, -1, 5)$.

6 The coordinates of P, Q, R, S are $(1, 2, -1)$, $(2, 4, 2)$, $(-1, 2, 1)$, $(0, 3, 2)$, respectively. Write down the column vectors $\overrightarrow{PQ}$, $\overrightarrow{PR}$, $\overrightarrow{PS}$, $2\overrightarrow{PQ}+3\overrightarrow{PR}$, and deduce that P, Q, R and S lie in one plane.

7 The position vectors of the points A and B relative to an origin O are $5\mathbf{i}+4\mathbf{j}+\mathbf{k}$, $-\mathbf{i}+\mathbf{j}-2\mathbf{k}$ respectively. Find the position vector of the point P which lies on AB produced such that $AP = 2BP$.   *(L)*

12.4 The equation of a line in two dimensions

In §5.3, we found that, if $\mathbf{a}$ is the position vector of A and if $\mathbf{b}$ is any non-zero vector, then the equation $\mathbf{r} = \mathbf{a} + t\mathbf{b}$ is the equation of a line, with a parameter t. In the Cartesian plane, let $\mathbf{a} = \begin{pmatrix} 2 \\ 1 \end{pmatrix}$ and $\mathbf{b} = \begin{pmatrix} 4 \\ 3 \end{pmatrix}$, then we have the equation

$$\mathbf{r} = \begin{pmatrix} x \\ y \end{pmatrix} = \begin{pmatrix} 2 \\ 1 \end{pmatrix} + t\begin{pmatrix} 4 \\ 3 \end{pmatrix} = \begin{pmatrix} 2+t4 \\ 1+t3 \end{pmatrix}$$

giving a pair of scalar equations for the line, with one parameter t,

$$x = 2 + 4t,$$
$$y = 1 + 3t.$$

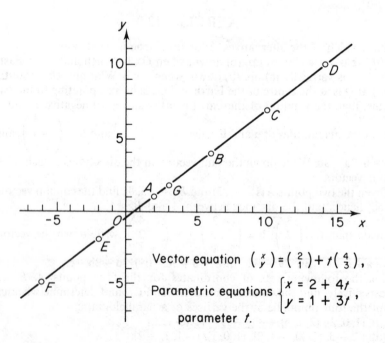

Fig. 12.6 Equations of a line.

For example, in Fig. 12.6, if we give t the values 0, 1, 2, 3, . . . , we obtain points A, (2, 1), B, (6, 4), C, (10, 7), D, (14, 10), ... Similarly, taking t equal to $-1, -2, \ldots$ gives points E, $(-2, -2)$, F, $(-6, -5)$, . . . , and all these points lie on the line through A with gradient $\frac{4}{3}$. Positive values of t give points on the line 'above' A and negative values give points 'below' A. Any real value of t can be used, for instance $t = \frac{1}{4}$ gives the point G, (3, $1\frac{3}{4}$). Each point on the line corresponds to one, and only one, value of t.

If we eliminate t from the parametric equations of the line, by equating the values of t found from the two equations,

$$t = \frac{x-2}{4} = \frac{y-1}{3}$$

we obtain

$$y - 1 = \tfrac{3}{4}(x - 2),$$

the standard equation of the line through (2, 1) with gradient $\frac{3}{4}$.

On the other hand, the equation of the line through (a, b) with gradient m is $y - b = m(x - a)$ (see §1.3), and this may be written

$$\frac{y-b}{m} = \frac{x-a}{1} = t$$

so the line has parametric equations, and a vector equation,

$$\begin{array}{c} x = a + t \\ y = b + mt \end{array}, \qquad \mathbf{r} = \begin{pmatrix} x \\ y \end{pmatrix} = \begin{pmatrix} a \\ b \end{pmatrix} + t \begin{pmatrix} 1 \\ m \end{pmatrix}.$$

EXERCISE 12.4

1   Find parametric equations and a Cartesian equation for the line:

(i) $\mathbf{r} = \begin{pmatrix} 2 \\ 3 \end{pmatrix} + t \begin{pmatrix} 5 \\ 1 \end{pmatrix}$,   (ii) $\mathbf{r} = \begin{pmatrix} 1 \\ -3 \end{pmatrix} + t \begin{pmatrix} -2 \\ 5 \end{pmatrix}$,   (iii) $\mathbf{r} = \mathbf{i} + t(\mathbf{j} - 2\mathbf{i})$.

2   Find the vector equation and the Cartesian equation of the line through:
(i) A, $(4, -3)$ and B, (3, 5); (ii) P and Q, with position vectors $\overrightarrow{OP}$ and $\overrightarrow{OQ}$, where $\overrightarrow{OP} = 2\mathbf{i} + 3\mathbf{j}$ and $\overrightarrow{OQ} = -\mathbf{i} + \mathbf{j}$.

3   Find a vector equation of the line L which passes through $(\frac{5}{8}, \frac{3}{4})$ and has gradient $\frac{1}{2}$.

4   The line L has equation $\mathbf{r} = \begin{pmatrix} 3 \\ 1 \end{pmatrix} + t \begin{pmatrix} -4 \\ 3 \end{pmatrix}$. Find a unit vector in the direction of L and hence find the coordinates of the points on L whose distance from $\begin{pmatrix} 3 \\ 1 \end{pmatrix}$ is 15 units.

5   Find the vector equation of a line through O parallel to $y + 3x = 10$.

6   The points $A(2, 3)$, $B(-1, 2)$, $C(0, 5)$ are vertices of a rhombus $ABCD$. Find:
(i) the coordinates of D, (ii) the vector equations of AC and of BD, (iii) the position vector of the point where AC meets BD.

12.5 The vector equation of a line in three dimensions

Consider the vector equation

$$\mathbf{r} = \begin{pmatrix} x \\ y \\ z \end{pmatrix} = \begin{pmatrix} 4 \\ 2 \\ 1 \end{pmatrix} + t \begin{pmatrix} 5 \\ 10 \\ 9 \end{pmatrix}. \tag{12.5}$$

When $t = 0, \mathbf{r} = \mathbf{a} = \begin{pmatrix} 4 \\ 2 \\ 1 \end{pmatrix}$, corresponding to the point A, $(4, 2, 1)$. When t takes values 1, 2, . . . the x, y, z components of $\mathbf{r}$ increase by 5, 10, 9 respectively, for each increase of 1 in t, giving a succession of position vectors of the points $(9, 12, 10)$, $(14, 22, 19)$, ... Similarly the values -1, -2, . . . of t give the points $(-1, -8, -8)$, $(-6, -18, -17)$, . . . These points all lie on the line through $A\,(4, 2, 1)$, parallel to the vector $\begin{pmatrix} 5 \\ 10 \\ 9 \end{pmatrix}$. If B is the point $(9, 12, 10)$, then the line is the line AB, and equation 12.5 is the vector equation of AB. The vector $\begin{pmatrix} 5 \\ 10 \\ 9 \end{pmatrix}$, $= \mathbf{v}$, defines the direction of

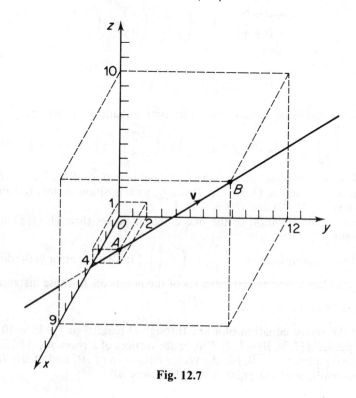

Fig. 12.7

the line *AB* and it is called a *direction vector* of the line *AB* (see Fig. 12.7). Any other vector parallel to **v** is also a direction vector of the line.

The scalar form of the vector equation of the line *AB* gives us the parametric equations of the line; in terms of the parameter *t*.

$$x = 4 + 5t$$
$$y = 2 + 10t$$
$$z = 1 + 9t$$

Elimination of *t* from these three equations gives the two Cartesian equations of the line in standard form:

$$\frac{x-4}{5} = \frac{y-2}{10} = \frac{z-1}{9}.$$

Note that from this equation, the coordinates of *A*, (4, 2, 1) can be read off from the numerators and the direction vector $\begin{pmatrix} 5 \\ 10 \\ 9 \end{pmatrix}$ can be read off from the denominators.

Note also that, in three dimensions, a line is given by one vector equation, or by three parametric equations, or by two Cartesian equations. The Cartesian equations of a line can not always be written in the above standard form. For example, the parametric equations

$$x = 3 + 4t, \quad y = 2, \quad z = 1 + 5t$$

become two Cartesian equations

$$\frac{x-3}{4} = \frac{z-1}{5}, y = 2.$$

In the general case, suppose that a line *L* passes through a point *A*, with position vector **a**. Then a general point *P*, with position vector **r**, lies on *L* if, and only if, $\overrightarrow{AP}$ is parallel to a direction vector of *L*. Let **v** be a direction

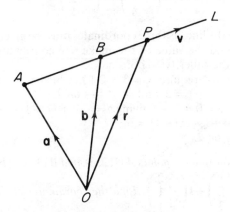

Fig. 12.8

174 Pure Mathematics for Advanced Level Part A

vector of L (see Fig. 12.8), then

$$P \text{ lies on the line } L \Leftrightarrow \overrightarrow{AP} = t\mathbf{v}$$
$$\Leftrightarrow \overrightarrow{OP} = \mathbf{r} = \overrightarrow{OA} + t\mathbf{v} = \mathbf{a} + t\mathbf{v}$$

and the vector equation of L is $\mathbf{r} = \mathbf{a} + t\mathbf{v}$.

Let B be another point on the line L, with $\overrightarrow{OB} = \mathbf{b}$. Then B will be given by some value, say s, of the parameter t, and then $\mathbf{b} = \mathbf{a} + s\mathbf{v}$. Two points on L with parameters of the same sign will lie on the same side of A, and two points with parameters of opposite signs will lie on opposite sides of A.

Warning   It is very important to distinguish between:
 (i) the direction vector $\mathbf{v}$ of the line AB; and
(ii) the vector $\mathbf{r}$, which is the position vector $\mathbf{r} = \overrightarrow{OP}$ of a general point P on the line AB;
and also between
(iii) the vector $\overrightarrow{AB}$, parallel to $\mathbf{v}$, (since $\overrightarrow{AB} = \mathbf{b} - \mathbf{a} = s\mathbf{v}$); and
(iv) the vector equation of AB, namely $\mathbf{r} = \mathbf{a} + t\mathbf{v}$, which has direction vector $\mathbf{v}$.

In the example above, for the line L, the vector $\begin{pmatrix} 5 \\ 10 \\ 9 \end{pmatrix}$ is a direction vector, but so is $\begin{pmatrix} -10 \\ -20 \\ -18 \end{pmatrix}$ and so is $\begin{pmatrix} 2\frac{1}{2} \\ 5 \\ 4\frac{1}{2} \end{pmatrix}$.

Another point which is important to remember is that a vector equation is an equation, and both sides of the '$=$' sign must be present, and both sides must be vectors, even if one side of the equation is the zero vector $\mathbf{0}$.

EXAMPLE 1   *Determine whether or not the points $A(3, 1, 3)$ and $B(0, -5, 2)$ lie on the line L with equation* $\mathbf{r} = \begin{pmatrix} 2 \\ -1 \\ 4 \end{pmatrix} + t\begin{pmatrix} 1 \\ 2 \\ -1 \end{pmatrix}$.

For a point to lie on the line, its three coordinates must be given by the same value of t. We therefore find the value of t to make one coordinate correct and then check the value of the other two coordinates.

For the point A, we require $x = 3 = 2 + t$, which is given by $t = 1$. Then $y = -1 + 2t = 1$, $z = 4 - t = 3$, and so A lies on L.

For B, we require $x = 0 = 2 + t$, giving $t = -2$, $y = -1 + (-2)2 = -5$, which is correct, and $z = 4 + (-2)(-1) = 6 \neq 2$.

So B **does not lie on** L.

EXAMPLE 2   *Show that the points $A(1, 2, 3)$, and $B(3, 3, 2)$ both lie on the line L with equation* $\mathbf{r} = \begin{pmatrix} 1 \\ 2 \\ 3 \end{pmatrix} + t\begin{pmatrix} 2 \\ 1 \\ -1 \end{pmatrix}$. *Find the coordinates of the point C where L meets the coordinate plane $y = 0$. Write down the ratio of the lengths AB and AC.*

Clearly A is the point on L with parameter $t = 0$, and B is the point with parameter $t = 1$. When $y = 0$, $t = -2$, so $x = -3$, $z = 5$, and the coordinates of C are $(-3, 0, 5)$. The distances along L from A are proportional to the value of the parameter t. Hence B and C lie on opposite sides of A and $AB:AC = 1:2$. Note that had we been asked to find the ratio of the vectors $\overrightarrow{AB}$ and $\overrightarrow{AC}$, with direction counting, it would be $-1:2$.

EXAMPLE 3 *Write down a direction vector for the line AB, where $\overrightarrow{OA} = -2\mathbf{i} + 3\mathbf{j} + \mathbf{k}$ and $\overrightarrow{OB} = 5\mathbf{k}$. Find a vector equation for AB. Given that $\overrightarrow{OC} = 2\mathbf{i} - 3\mathbf{j} + 9\mathbf{k}$ and $\overrightarrow{OD} = -4\mathbf{i} + 6\mathbf{j} - 3\mathbf{k}$, use a similar method to find a vector equation of the line CD. Show that these two equations represent the same line and deduce that A, B, C, D are collinear.*

$\overrightarrow{AB} = \overrightarrow{OB} - \overrightarrow{OA} = 5\mathbf{k} - (-2\mathbf{i} + 3\mathbf{j} + \mathbf{k}) = 2\mathbf{i} - 3\mathbf{j} + 4\mathbf{k}$, which is a direction vector for the line AB. Therefore, the equation of AB is

$$\mathbf{r} = 5\mathbf{k} + t(2\mathbf{i} - 3\mathbf{j} + 4\mathbf{k}).$$

$\overrightarrow{CD} = \overrightarrow{OD} - \overrightarrow{OC} = -4\mathbf{i} + 6\mathbf{j} - 3\mathbf{k} - (2\mathbf{i} - 3\mathbf{j} + 9\mathbf{k}) = -6\mathbf{i} + 9\mathbf{j} - 12\mathbf{k}$, which is a direction vector for the line CD. Therefore the equation of CD is

$$\mathbf{r} = 2\mathbf{i} - 3\mathbf{j} + 9\mathbf{k} + s(-6\mathbf{i} + 9\mathbf{j} - 12\mathbf{k}).$$

Since $-6\mathbf{i} + 9\mathbf{j} - 12\mathbf{k} = -3(2\mathbf{i} - 3\mathbf{j} + 4\mathbf{k})$, the lines AB and CD are parallel. Also, for $s = \frac{1}{3}$,

$$\mathbf{r} = 2\mathbf{i} - 3\mathbf{j} + 9\mathbf{k} + \tfrac{1}{3}(-6\mathbf{i} + 9\mathbf{j} - 12\mathbf{k}) = 5\mathbf{k},$$

so B lies on CD. Thus the two lines are the same line and A, B, C, D are collinear.

EXERCISE 12.5

1 Find parametric and Cartesian equations of the line having a vector equation:

(i) $\mathbf{r} = \begin{pmatrix} 2 \\ 5 \\ 4 \end{pmatrix} + t\begin{pmatrix} -3 \\ 1 \\ -2 \end{pmatrix}$, (ii) $\mathbf{r} = \begin{pmatrix} 0 \\ 2 \\ -1 \end{pmatrix} + t\begin{pmatrix} 2 \\ 1 \\ 4 \end{pmatrix}$,

(iii) $\mathbf{r} = \begin{pmatrix} 4 \\ 1 \\ 2 \end{pmatrix} + t\begin{pmatrix} -3 \\ 12 \\ -5 \end{pmatrix}$, (iv) $\mathbf{r} = \begin{pmatrix} 2 \\ -2 \\ 3 \end{pmatrix} + s\begin{pmatrix} 1 \\ 1 \\ 1 \end{pmatrix}$.

2 Find a vector equation for the line:

(i) through $(3, -1, 4)$ in the direction $\begin{pmatrix} -1 \\ 3 \\ 2 \end{pmatrix}$;

(ii) through $(1, 1, -1)$ in the direction $\begin{pmatrix} 4 \\ 3 \\ 5 \end{pmatrix}$;

(iii) through the points $(2, -3, 4)$ and $(1, -3, 2)$;
(iv) through the points $(2, -4, 1)$ and $(1, -5, 7)$;
(v) through the points $(-7, 3, -1)$ and $(2, -3, 5)$.

3 Find a vector equation of the line with parametric equations:
(i) $x = 2 + t$, $y = -1 + 3t$, $z = 3 - 4t$;   (ii) $x = t$, $y = 2 - 3t$, $z = 4t$;
(iii) $x = 4 - 2t$, $y = t$, $z = 0$;   (iv) $x = 3$, $y = 2 - t$, $z = t$.

4 Use the parameter indicated to find a vector equation of the line:
(i) $y = 2x + 3$, $z = -4x - 2$, (parameter x);
(ii) $x = 3y - 7$, $z = 2y - 1$, (parameter y);
(iii) $x = 3z - 4$, $y = z$, (parameter z);
(iv) $x = 2y + 6$, $z = 2$, (parameter y);
(v) $x + y = 3$, $z - x = 2$, (parameter x);
(vi) $2y = 3x + 4$, $5z = 4 - x$, (parameter x).

5 Find a direction vector of the line with equations:

(i) $\dfrac{y-3}{2} = \dfrac{x-1}{4} = \dfrac{z-6}{3}$,   (ii) $\dfrac{x}{2} = \dfrac{2y-4}{3} = z$,

(iii) $\dfrac{3x+1}{7} = \dfrac{4-y}{5} = \dfrac{2z}{3}$,   (iv) $x = 2y - 6$, $z = 4 - 3y$,

(v) $x + 2y = 9$, $3 - z = x$,   (vi) $4x - 8 = y$, $x = 3z - 8$.

6 Find Cartesian equations of the line:

(i) with vector equation $\mathbf{r} = \begin{pmatrix} 2 \\ 4 \\ 3 \end{pmatrix} + t \begin{pmatrix} -3 \\ -14 \\ 9 \end{pmatrix}$;

(ii) through the origin O and the point $A(2, -3, 6)$;
(iii) with parametric equations $x = 3 - t$, $y = 5 + 2t$, $z = 7$;
(iv) through the points $A(5, 7, -3)$ and $B(2, 4, 5)$;
(v) with equation $\mathbf{r} = 3\mathbf{i} - \mathbf{k} + t(2\mathbf{j} - 4\mathbf{k})$.

7 Determine whether the point $A(-6, +5, -2)$ lies on the line L with

equation $\mathbf{r} = \begin{pmatrix} 2 \\ 3 \\ -5 \end{pmatrix} + t \begin{pmatrix} -4 \\ 1 \\ 2 \end{pmatrix}$.

8 Show that $A(3, -1, 4)$, $B(9, 11, -11)$ and $C(-5, -17, 24)$ all lie on the line $\mathbf{r} = 3\mathbf{i} - \mathbf{j} + 4\mathbf{k} + t(2\mathbf{i} + 4\mathbf{j} - 5\mathbf{k})$. State the corresponding values of t and find the ratio of the lengths of AB and AC.

9 The points A and B are given by $t = 0$ and $t = 1$, respectively, on the line L

with vector equation $\mathbf{r} = \begin{pmatrix} 2 \\ 7 \\ 3 \end{pmatrix} + t \begin{pmatrix} 1 \\ 3 \\ 6 \end{pmatrix}$. Determine the coordinates of the

point C on L such that A lies between B and C and $CA : AB = 2 : 1$.

10 Determine whether the pair of vector equations represent the same line:

(i) $\mathbf{r} = \begin{pmatrix} -3 \\ 2 \\ 5 \end{pmatrix} + t \begin{pmatrix} 1 \\ 3 \\ -4 \end{pmatrix}$ and $\mathbf{r} = \begin{pmatrix} 0 \\ 11 \\ 7 \end{pmatrix} + s \begin{pmatrix} -1 \\ -3 \\ 4 \end{pmatrix}$,

(ii) $\mathbf{r} = \begin{pmatrix} 2 \\ -1 \\ 4 \end{pmatrix} + t \begin{pmatrix} \frac{1}{2} \\ \frac{1}{3} \\ \frac{1}{4} \end{pmatrix}$ and $\mathbf{r} = \begin{pmatrix} 8 \\ 3 \\ 6 \end{pmatrix} + s \begin{pmatrix} 6 \\ 4 \\ 3 \end{pmatrix}$.

11 State a vector equation of the line passing through the points A and B whose position vectors are $\mathbf{i} - \mathbf{j} + 3\mathbf{k}$ and $\mathbf{i} + 2\mathbf{j} + 2\mathbf{k}$ respectively.
Determine the position vector of the point C which divides the line-segment AB internally such that $AC = 2CB$.   (L)

12 Two loci are given by Cartesian equations (i) $x + 2y + 3z = 18$, and (ii) $x = 2y = 3z$. Show that one locus is a line and that the other is not. State what you think the other locus might be. Find points which lie on both loci and interpret this result.

12.6 Scalar products

So far, we have dealt with linear combinations of vectors, using the operations of addition and multiplication by a scalar. If we wish to define some sort of product of two vectors, we could consider the product of their moduli. However, this would not take into account the directions of the vectors, and it is not very useful. What turns out to be useful is the product of the moduli of the two vectors multiplied by the cosine of the angle between them. This product takes into account their moduli and their directions. It is called the *scalar product* of the vectors and the notation is to place a dot between the two vectors, so it is also called 'the dot product'.

Definition Let the vectors **a** and **b** have moduli (lengths) a and b and let θ be the angle between **a** and **b**, then the *scalar product* of **a** and **b** is the real number **a** . **b**, where

$$\mathbf{a} . \mathbf{b} = |\mathbf{a}| |\mathbf{b}| \cos \theta = ab \cos \theta.$$

If two vectors happen to be represented by line segments, which do not originate from the same point, then we define the angle between them as the angle between two position vectors (originating from O) which are equal to the given vectors. Thus, in Fig. 12.9, the vectors **a** and **b** are represented by $\overrightarrow{PQ}$ and $\overrightarrow{RS}$ and also by $\overrightarrow{OA}$ and $\overrightarrow{OB}$ and the angle between $\overrightarrow{PQ}$ and $\overrightarrow{RS}$, that is, the angle between **a** and **b** is $\theta = \angle AOB$. Note that, if θ is acute then **a** . **b** is positive, if θ is obtuse then **a** . **b** is negative.

We now show that many of the properties of products are satisfied by scalar products. The one property that cannot apply to scalar products is associativity since, because **a** . **b** is a scalar, (**a** . **b**) . **c** has no meaning,

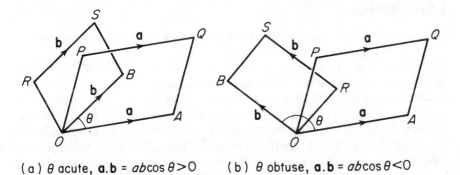

(a) θ acute, **a.b** = $ab \cos \theta > 0$ (b) θ obtuse, **a.b** = $ab \cos \theta < 0$

Fig. 12.9 Scalar product.

whereas $(\mathbf{a} \cdot \mathbf{b})\mathbf{c}$ is the multiple of $\mathbf{c}$ by $\mathbf{a} \cdot \mathbf{b}$ so the brackets can not be moved. The following shows that scalar products commute and distribute with addition. Let $|\mathbf{a}| = a$, $|\mathbf{b}| = b$, $|\mathbf{c}| = c$.

Commutativity

$$\mathbf{a} \cdot \mathbf{b} = ab\cos\theta = ba\cos\theta = \mathbf{b} \cdot \mathbf{a}.$$

Multiplication by scalars

Let k, l be scalars. Then if $k \geqslant 0, l \geqslant 0$,

$$(k\mathbf{a}) \cdot (l\mathbf{b}) = |k\mathbf{a}||l\mathbf{b}|\cos\theta = kl|\mathbf{a}||\mathbf{b}|\cos\theta = (kl)\mathbf{a} \cdot \mathbf{b}.$$

If either, or both, of k and l is negative it still follows that

$$(k\mathbf{a}) \cdot (l\mathbf{b}) = (kl)\mathbf{a} \cdot \mathbf{b}$$

on using two facts:

(i) if θ is the angle between $\mathbf{a}$ and $\mathbf{b}$, then $(\pi - \theta)$ is the angle between $\mathbf{a}$ and $(-\mathbf{b})$ and also between $(-\mathbf{a})$ and $\mathbf{b}$, and θ is the angle between $(-\mathbf{a})$ and $(-\mathbf{b})$ (see Fig. 12.10);
(ii) $\cos(\pi - \theta) = -\cos\theta$.

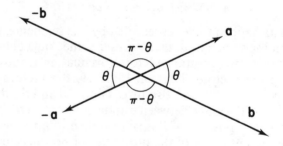

Fig. 12.10

Distributivity

$$\mathbf{a} \cdot (\mathbf{b} + \mathbf{c}) = \mathbf{a} \cdot \mathbf{b} + \mathbf{a} \cdot \mathbf{c}$$

Proof In Fig. 12.11, let $\mathbf{a} = \overrightarrow{OA}$, $\mathbf{b} = \overrightarrow{OB}$ and $\mathbf{c} = \overrightarrow{BC}$, so that $(\mathbf{b} + \mathbf{c}) = \overrightarrow{OC}$. Let BL and CM be the perpendiculars from B and C on to OA, produced, if necessary, in either direction. Let $\overrightarrow{BD}$ be a line through B parallel to $\mathbf{a}$ and let CN be the perpendicular from C on to BD. Note that, because BND and $OLMA$ are both parallel to $\mathbf{a}$, these seven points all lie in one plane, but the point C may not lie in this plane. If it does, then, because CN and CM are both perpendicular to $\mathbf{a}$, C lies on NM and $BLMN$ is a rectangle. If C does not lie in the plane OAB, then, as shown in Fig. 12.11, since CN and CM are each perpendicular to $\mathbf{a}$, the plane CMN is

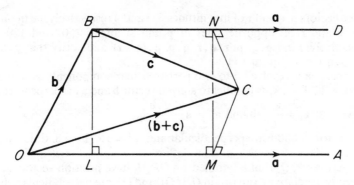

Fig. 12.11

perpendicular to **a** and again *BLMN* is a rectangle. Therefore, in each case, $\vec{BN} = \vec{LM}$. Now, with the signs taken into account,

$$\mathbf{a} . \mathbf{b} + \mathbf{a} . \mathbf{c} = a\,OB \cos \angle BOL + a\,BC \cos \angle CBN$$

$$= a\,OL + a\,BN = a\,OL + a\,LM = a\,OM$$

$$= a\,OC \cos \angle AOC = \mathbf{a} . \vec{OC}$$

$$= \mathbf{a} . (\mathbf{b} + \mathbf{c}).$$

Parallel vectors

If **a** and **b** are parallel, they can be in the same or in opposite directions to each other, corresponding to the angle between them being 0 or π.

a and **b** in the same direction, $\theta = 0$, $\mathbf{a} . \mathbf{b} = ab$.

a and **b** in opposite directions, $\theta = \pi$, $\mathbf{a} . \mathbf{b} = -ab$.

In particular, $\mathbf{a} . \mathbf{a} = aa = a^2$ and hence $|\mathbf{a}| = \sqrt{(\mathbf{a} . \mathbf{a})}$.

Perpendicular vectors

If **a** and **b** are perpendicular, $\theta = \frac{1}{2}\pi$, $\cos \theta = 0$, and so $\mathbf{a} . \mathbf{b} = 0$.

Conversely, if $\mathbf{a} . \mathbf{b} = 0$, then either $\mathbf{a} = \mathbf{0}$, or $\mathbf{b} = \mathbf{0}$, or **a** and **b** are perpendicular.

EXERCISE 12.6A

1   Given vectors **a** and **b**, with moduli a and b, and with an angle θ between them, the projection of **a** along the direction of **b** is defined as $a \cos \theta$. Show that the scalar product $\mathbf{a} . \mathbf{b}$ is equal to b times the projection of **a** in the direction of **b** and is also equal to a times the projection of **b** in the direction of **a**.
2   The unit vector **n** is in the direction North. Vectors **a** and **b** have lengths 5 km and 2 km and are in the directions of bearings 60° and 330° respectively. Draw a diagram showing **n**, **a**, **b**, $(\mathbf{a} + \mathbf{b})$. Find $\mathbf{n} . \mathbf{n}$, $\mathbf{a} . \mathbf{a}$, $\mathbf{b} . \mathbf{b}$, $\mathbf{a} . \mathbf{n}$, $\mathbf{b} . \mathbf{n}$, $\mathbf{a} . \mathbf{b}$, $(\mathbf{a} + \mathbf{b}) . \mathbf{n}$.

3   Three vectors $\mathbf{p}$, $\mathbf{q}$ and $\mathbf{r}$ of magnitudes 5, 3 and 4 respectively, lie in one plane, with $\mathbf{q}$ and $\mathbf{r}$ on opposite sides of $\mathbf{p}$ and at angles $150°$ and $120°$ with $\mathbf{p}$ respectively. Find $\mathbf{p} \cdot \mathbf{q}$, $\mathbf{p} \cdot \mathbf{r}$, $\mathbf{q} \cdot \mathbf{r}$, $\mathbf{q} \cdot \mathbf{p}$, $\mathbf{p} \cdot (\mathbf{q} + \mathbf{r})$, and verify that $\mathbf{p} \cdot \mathbf{q} = \mathbf{q} \cdot \mathbf{p}$ and $\mathbf{p} \cdot \mathbf{q} + \mathbf{p} \cdot \mathbf{r} = \mathbf{p} \cdot (\mathbf{q} + \mathbf{r})$.

4   The vector $\mathbf{a}$ is resolved into two perpendicular components $\mathbf{a}_1$ and $\mathbf{a}_2$, such that $\mathbf{a} = \mathbf{a}_1 + \mathbf{a}_2$, $\mathbf{a}_1$ is parallel to a given vector $\mathbf{b}$ and $\mathbf{a}_2$ is perpendicular to $\mathbf{b}$.
Prove that $\mathbf{a}_1 = \dfrac{\mathbf{a} \cdot \mathbf{b}}{b^2} \mathbf{b}$ and $\mathbf{a}_2 = \mathbf{a} - \dfrac{\mathbf{a} \cdot \mathbf{b}}{b^2} \mathbf{b}$.

5   The vectors $\mathbf{a}$ and $\mathbf{b}$ are perpendicular and $|\mathbf{a}| = 15$, $|\mathbf{b}| = 8$. Calculate $|\mathbf{a} + \mathbf{b}|$ and $|\mathbf{a} - \mathbf{b}|$.   (L)

6   The vertices P, Q, R of a tetrahedron $OPQR$ have position vectors $\mathbf{a}$, $\mathbf{b}$ and $\mathbf{c}$ respectively relative to an origin O. If OP and QR are perpendicular, show that $\mathbf{a} \cdot (\mathbf{b} - \mathbf{c}) = 0$. If also OQ and RP are perpendicular, prove that OR and PQ are perpendicular.   (L)

7   The vectors $\mathbf{a}$ and $\mathbf{b}$ are such that $|\mathbf{a}| = 3$ and $|\mathbf{b}| = 5$. Calculate the magnitude of $\mathbf{a} + \mathbf{b}$ given that the angle between $\mathbf{a}$ and $\mathbf{b}$ is $\pi/3$.   (L)

Unit vectors

For any unit vector $\mathbf{u}$, $\mathbf{u} \cdot \mathbf{u} = 1 \times 1 \times \cos 0 = 1$.

For the Cartesian basis unit vectors, $\mathbf{i}$, $\mathbf{j}$, $\mathbf{k}$,

$$\mathbf{i} \cdot \mathbf{i} = \mathbf{j} \cdot \mathbf{j} = \mathbf{k} \cdot \mathbf{k} = 1,\ \mathbf{i} \cdot \mathbf{j} = \mathbf{j} \cdot \mathbf{k} = \mathbf{k} \cdot \mathbf{i} = \mathbf{j} \cdot \mathbf{i} = \mathbf{k} \cdot \mathbf{j} = \mathbf{i} \cdot \mathbf{k} = 0.$$

Cartesian components

Using the above and the distributive rule, if $\mathbf{a} = a_1 \mathbf{i} + a_2 \mathbf{j} + a_3 \mathbf{k}$,

$$\mathbf{a} \cdot \mathbf{i} = a_1 \mathbf{i} \cdot \mathbf{i} + a_2 \mathbf{j} \cdot \mathbf{i} + a_3 \mathbf{k} \cdot \mathbf{i} = a_1,\ \mathbf{a} \cdot \mathbf{j} = a_2,\ \mathbf{a} \cdot \mathbf{k} = a_3.$$

In two dimensions,

$$\mathbf{a} = a_1 \mathbf{i} + a_2 \mathbf{j},\ \mathbf{a} \cdot \mathbf{i} = a_1,\ \mathbf{a} \cdot \mathbf{j} = a_2.$$

The scalar product of $\mathbf{a}$ with a Cartesian unit vector is the component of $\mathbf{a}$ in that direction.

Component form

Again using the distributive rule and the results for unit vectors,

if $\mathbf{a} = a_1 \mathbf{i} + a_2 \mathbf{j} + a_3 \mathbf{k}$, $\mathbf{b} = b_1 \mathbf{i} + b_2 \mathbf{j} + b_3 \mathbf{k}$,

then $\mathbf{a} \cdot \mathbf{b} = (a_1 \mathbf{i} + a_2 \mathbf{j} + a_3 \mathbf{k}) \cdot (b_1 \mathbf{i} + b_2 \mathbf{j} + b_3 \mathbf{k}) = a_1 b_1 + a_2 b_2 + a_3 b_3$.

Vector form

$$\mathbf{a} \cdot \mathbf{b} = \begin{pmatrix} a_1 \\ a_2 \\ a_3 \end{pmatrix} \cdot \begin{pmatrix} b_1 \\ b_2 \\ b_3 \end{pmatrix} = a_1 b_1 + a_2 b_2 + a_3 b_3.$$

Thus, if θ is the angle between the vectors **a** and **b**, then $\mathbf{a} . \mathbf{b} = ab \cos \theta$ and

$$\cos \theta = \frac{a_1 b_1 + a_2 b_2 + a_3 b_3}{\sqrt{(a_1^2 + a_2^2 + a_3^2)} \sqrt{(b_1^2 + b_2^2 + b_3^2)}},$$

a very important result.

The scalar product can be used to establish geometrical properties such as distances from lines and angles between lines. The acute angle between two intersecting lines is either the angle, or the supplement of the angle, between their direction vectors (see Fig. 12.12). Also, the angle between two non-intersecting lines is defined in the same way, so that it is the angle between one line and a line which meets that first line and is parallel to the second line. This is also shown in Fig. 12.12. When two lines do not intersect, and are not parallel, they are called *skew*.

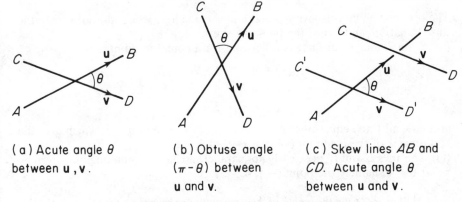

(a) Acute angle θ between **u**, **v**.

(b) Obtuse angle $(\pi - \theta)$ between **u** and **v**.

(c) Skew lines *AB* and *CD*. Acute angle θ between **u** and **v**.

Fig. 12.12 Angle θ between lines *AB*, *CD*

EXAMPLE 1 *Find the angle between the lines with equations*

$$\mathbf{r} = \begin{pmatrix} 3 \\ 1 \\ 2 \end{pmatrix} + t \begin{pmatrix} 3 \\ 5 \\ 4 \end{pmatrix}, \quad \text{and} \quad \mathbf{r} = \begin{pmatrix} 3 \\ 5 \\ 1 \end{pmatrix} + s \begin{pmatrix} -6 \\ 3 \\ 2 \end{pmatrix}.$$

The lines have direction vectors $\mathbf{v}_1 = \begin{pmatrix} 3 \\ 5 \\ 4 \end{pmatrix}$ and $\mathbf{v}_2 = \begin{pmatrix} -6 \\ 3 \\ 2 \end{pmatrix}$, and

$$\mathbf{v}_1 . \mathbf{v}_2 = -18 + 15 + 8 = 5 = |\mathbf{v}_1||\mathbf{v}_2| \cos \theta.$$

Also, $|\mathbf{v}_1| = \sqrt{(16 + 25 + 9)} = \sqrt{(50)}$, $|\mathbf{v}_2| = \sqrt{(36 + 9 + 4)} = \sqrt{(49)} = 7$, and so $\cos \theta = \dfrac{5}{7\sqrt{(50)}} = \dfrac{1}{7\sqrt{2}} = 0.101$, giving $\theta \approx 1.47$.

EXAMPLE 2 *Three lines L, L', L'', have vector equations*

$$\mathbf{r} = \begin{pmatrix} 3 \\ 1 \\ 2 \end{pmatrix} + t \begin{pmatrix} 5 \\ 3 \\ 4 \end{pmatrix}, \mathbf{r} = \begin{pmatrix} -1 \\ 4 \\ -3 \end{pmatrix} + t \begin{pmatrix} -2 \\ -3 \\ 1 \end{pmatrix}, \mathbf{r} = \begin{pmatrix} -1 \\ 4 \\ -3 \end{pmatrix} + t \begin{pmatrix} -2 \\ -3 \\ -1 \end{pmatrix},$$

respectively. Find the common point of the lines, or show that the lines do not meet, in the case of the two lines:
(i) *L and L'*, (ii) *L and L"*, (iii) *L' and L".*

The same parameter t is used in the equations of the three lines. For a common point on two lines, the parameter will generally take different values for the two lines. Therefore, to avoid confusion, it is necessary to use different symbols for the parameters on the three lines. We shall therefore use parameters t for L, s for L', q for $L"$.
(i) Since the direction vectors are not multiples of each other, the lines L and L' are not parallel. For a common point, we need values of s and t satisfying the three equations

$$3 + 5t = -1 - 2s$$
$$1 + 3t = 4 - 3s$$
$$2 + 4t = -3 + s.$$

The first two equations give $t = -2$ and $s = 3$. These values do not satisfy the third equation, and hence **the lines are skew**.
(ii) Again L and $L"$ are not parallel, and for a common point

$$3 + 5t = -1 - 2q$$
$$1 + 3t = 4 - 3q$$
$$2 + 4t = -3 - q$$

and now all three equations are satisfied by $t = -2$ and $q = 3$, which give the coordinates of the common point as **$(-7, -5, -6)$**.
(iii) The lines L' and $L"$ are clearly not parallel and have a common point given by $s = 0$, $q = 0$, namely the point **$(-1, 4, -3)$**.

EXAMPLE 3   *Find the length of the perpendicular from the origin O on to the line with equation* $\mathbf{r} = \begin{pmatrix} 4 \\ -1 \\ 0 \end{pmatrix} + t \begin{pmatrix} 1 \\ 0 \\ 1 \end{pmatrix}.$

Let P be the foot of the perpendicular OP, given by the parameter t. Then $\overrightarrow{OP} = \begin{pmatrix} 4+t \\ -1 \\ t \end{pmatrix}$ and this vector must be perpendicular to the direction vector $\mathbf{v} = \begin{pmatrix} 1 \\ 0 \\ 1 \end{pmatrix}$, of the line. So $0 = \overrightarrow{OP}.\mathbf{v} = 4 + t + 0 + t$, whence $t = -2$ and the coordinates of P are $(2, -1, -2)$. The required perpendicular distance is $|\overrightarrow{OP}| = \sqrt{(4+1+4)} = \mathbf{3}.$

Resolution of vectors

Let $\mathbf{u}$ be a unit vector along a line L. Let $\mathbf{a} = \overrightarrow{QP}$ be a vector which makes an angle θ with $\mathbf{u}$. Let L' be the line parallel to L through Q. Let PN, QM, PN' be the perpendiculars from P, Q, P on to L, L, L' respectively (see Fig. 12.13). The diagram shows two cases (a) when θ is acute and (b) when θ is obtuse.

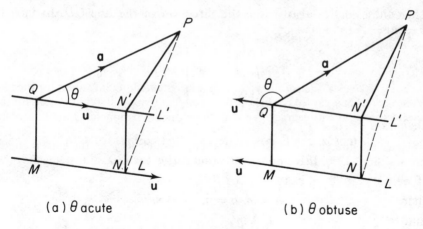

(a) θ acute   (b) θ obtuse

Fig. 12.13

The projection of $\mathbf{a} = \overrightarrow{QP}$ on to the line L is $\overrightarrow{MN} = \overrightarrow{QN'}$ and this is called the resolved part of $\mathbf{a}$ in the direction $\mathbf{u}$. The length of MN is

$$MN = QN' = QP \cos\theta = \mathbf{a} \cdot \mathbf{u}$$

in case (a) and is

$$MN = QN' = -QP \cos\theta = -\mathbf{a} \cdot \mathbf{u}$$

in case (b). In case (a) $\overrightarrow{MN} = MN\mathbf{u} = (\mathbf{a} \cdot \mathbf{u})\mathbf{u}$. In case (b) $\overrightarrow{MN} = -MN\mathbf{u} = -(-\mathbf{a} \cdot \mathbf{u})\mathbf{u}$. So, in either case, the resolved part of $\mathbf{a}$ in the direction $\mathbf{u}$ is $(\mathbf{a} \cdot \mathbf{u})\mathbf{u}$.

EXAMPLE 4 *Find the resolved part of the vector* $2\mathbf{i} + 3\mathbf{j} - 4\mathbf{k} = \mathbf{a}$ *in the direction of the vector* $\mathbf{i} - \mathbf{j} + \mathbf{k}$.

The unit vector $\mathbf{u}$ in the required direction is $\mathbf{u} = \dfrac{1}{\sqrt{3}}(\mathbf{i} - \mathbf{j} + \mathbf{k})$ so the resolved part of $\mathbf{a}$ in this direction is

$$(\mathbf{a} \cdot \mathbf{u})\mathbf{u} = \left(2 \times \frac{1}{\sqrt{3}} + 3 \times \frac{-1}{\sqrt{3}} - 4 \times \frac{1}{\sqrt{3}}\right)\mathbf{u} = \frac{2 - 3 - 4}{\sqrt{3}}\mathbf{u} = \frac{-5}{\sqrt{3}}\mathbf{u}$$

$$= -\frac{5}{3}\mathbf{i} + \frac{5}{3}\mathbf{j} - \frac{5}{3}\mathbf{k}.$$

Direction cosines

Let the point D have coordinates (a, b, c) and position vector $\mathbf{d}$, where $\mathbf{d} = \overrightarrow{OD} = \begin{pmatrix} a \\ b \\ c \end{pmatrix}$; then $\mathbf{d}$ is a direction vector for the line $\overrightarrow{OD}$. Let this line $\overrightarrow{OD}$ make angles α, β, γ, with the Cartesian axes Ox, Oy, Oz, respectively.

Then these angles also define the direction of the line $\overrightarrow{OD}$. In fact, if
$d = |\mathbf{d}| = \sqrt{(a^2 + b^2 + c^2)}$,

$$d \cos \alpha = \mathbf{d} \cdot \mathbf{i} = a,$$
$$d \cos \beta = \mathbf{d} \cdot \mathbf{j} = b,$$
$$d \cos \gamma = \mathbf{d} \cdot \mathbf{k} = c.$$

Definition

$(\cos \alpha, \cos \beta, \cos \gamma)$ are *direction cosines* for $\overrightarrow{OD}$,

(a, b, c) are *direction ratios* for $\overrightarrow{OD}$.

If we put $\qquad \cos \alpha = l, \cos \beta = m, \cos \gamma = n,$

then $\qquad\qquad a = ld, b = md, c = nd,$

and $\qquad\qquad l^2 + m^2 + n^2 = 1.$

EXAMPLE 5   *Find the direction cosines for the vector:* (i) $\begin{pmatrix} 2 \\ -1 \\ 2 \end{pmatrix}$, (ii) $3\mathbf{j} - 4\mathbf{k}$.

(i) The direction ratios for the given vector are $(2, -1, 2)$. The length of the vector is $\sqrt{(4+1+4)} = 3$. Hence the required direction cosines are $(\frac{2}{3}, -\frac{1}{3}, \frac{2}{3})$.
(ii) The direction ratios are $(0, 3, -4)$ and the vector has length 5. The direction cosines of the vector are $(0, \frac{3}{5}, \frac{4}{5})$.

EXERCISE 12.6B

1 Find the angle between:
   (i) the vectors $\mathbf{i} + \mathbf{j} - \mathbf{k}$ and $\mathbf{i} - 2\mathbf{j} - 2\mathbf{k}$;
   (ii) $\mathbf{i} + 2\mathbf{j} - 3\mathbf{k}$ and $\mathbf{k} - 2\mathbf{i} - 3\mathbf{j}$;
   (iii) the lines OA and OB, where A and B have coordinates $(2, -1, 2)$ and $(2, -2, 2)$;
   (iv) the lines PQ and RS, where the points P, Q, R, S have coordinates $(4, -3, 2)$, $(3, 5, 6)$, $(1, -1, -1)$, $(-5, 2, 4)$;

   (v) the lines with equations: $\mathbf{r} = \mathbf{a} + t\mathbf{b}$ and $\mathbf{r} = 2\mathbf{b} - 3t\mathbf{a}$,

where $\mathbf{a} = \begin{pmatrix} 1 \\ 3 \\ -2 \end{pmatrix}$ and $\mathbf{b} = \begin{pmatrix} 2 \\ 4 \\ 5 \end{pmatrix}$.

2 The co-ordinates of the points A, B and C are $(2, 3)$, $(4, 7)$, and $(5, 4)$ respectively. Calculate the acute angle between the lines AB and AC.   (L)
3 Expand the scalar product $(\mathbf{a} - \mathbf{b}) \cdot (\mathbf{a} - \mathbf{b})$. In the triangle ABC, $\overrightarrow{CA} = \mathbf{a}$, $\overrightarrow{CB} = \mathbf{b}$, use this scalar product to deduce the cosine formula for the triangle.
4 The vectors $\mathbf{a}$ and $\mathbf{b}$ are inclined to one another at an angle of $60°$ and $|\mathbf{a}| = 5$, $|\mathbf{b}| = 8$. Calculate $|\mathbf{a} - \mathbf{b}|$.   (L)
5 Find a unit vector $\hat{\alpha}$ parallel to the vector $\alpha = 2\mathbf{i} - \mathbf{j} - 3\mathbf{k}$. Determine the length of the resolved part of the vector $\beta = 2\mathbf{i} + 3\mathbf{j} - \mathbf{k}$ in the direction of α.   (L)
6 Determine whether the two lines, with the given equations, intersect. If they

do intersect, state the coordinates of the common point:

(i) $\mathbf{r} = \begin{pmatrix} 0 \\ 2 \\ -1 \end{pmatrix} + t \begin{pmatrix} 4 \\ -2 \\ 3 \end{pmatrix}$ and $\mathbf{r} = \begin{pmatrix} 4 \\ 1 \\ -2 \end{pmatrix} + s \begin{pmatrix} -1 \\ 3 \\ 4 \end{pmatrix}$,

(ii) $\mathbf{r} = \begin{pmatrix} 2 \\ -1 \\ 0 \end{pmatrix} + t \begin{pmatrix} -2 \\ 3 \\ 4 \end{pmatrix}$ and $\mathbf{r} = \begin{pmatrix} 4 \\ 1 \\ -2 \end{pmatrix} + s \begin{pmatrix} -1 \\ 4 \\ 3 \end{pmatrix}$,

(iii) $\mathbf{r} = \begin{pmatrix} 4 \\ 3 \\ -2 \end{pmatrix} + t \begin{pmatrix} 5 \\ 1 \\ 2 \end{pmatrix}$ and $\mathbf{r} = \begin{pmatrix} 2 \\ -5 \\ 2 \end{pmatrix} + t \begin{pmatrix} 3 \\ 2 \\ -1 \end{pmatrix}$.

7 Given that A and B are the points whose position vectors referred to the origin O are $(\mathbf{i} + 2\mathbf{j} + \mathbf{k})$ and $(3\mathbf{i} + 4\mathbf{j} + 2\mathbf{k})$ respectively, determine $|\overrightarrow{AB}|$ and direction cosines for the line AB. (L)

8 Find the length of the perpendicular from O to the line with the given equation. State the coordinates of the foot of the perpendicular:

(i) $\mathbf{r} = \begin{pmatrix} 6 \\ -2 \\ 1 \end{pmatrix} + t \begin{pmatrix} -4 \\ 3 \\ 5 \end{pmatrix}$; (ii) $\mathbf{r} = \begin{pmatrix} 4 \\ -2 \\ 3 \end{pmatrix} + t \begin{pmatrix} 2 \\ 1 \\ 1 \end{pmatrix}$.

9 Verify that the point A, $(1, 3, 4)$, is the foot of the perpendicular from O to the line $L: \mathbf{r} = \begin{pmatrix} 1 \\ 3 \\ 4 \end{pmatrix} + t \begin{pmatrix} 2 \\ 6 \\ -5 \end{pmatrix}$. Given that B is the reflection of O in L, write down the coordinates of B. Given that C and D are the points of L with parameters $t = 1$ and $t = -1$ respectively, calculate the area of the rhombus $OCBD$, giving the answer in surd form.

10 Using the fact that a diameter of a sphere subtends a right angle at a point on the surface, show that the equation of the sphere with diameter AB, where A and B are the points (a_1, a_2, a_3) and (b_1, b_2, b_3), is

$$(x - a_1)(x - b_1) + (y - a_2)(y - b_2) + (z - a_3)(z - b_3) = 0.$$

11 Find the coordinates of the foot of the perpendicular from $(1, 10, -2)$ to the line $\mathbf{r} = \begin{pmatrix} 4 \\ 1 \\ -3 \end{pmatrix} + t \begin{pmatrix} -3 \\ 2 \\ 5 \end{pmatrix}$.

12 Let $\mathbf{a} = \mathbf{i} - 2\mathbf{j} + \mathbf{k}$, $\mathbf{b} = 2\mathbf{i} + \mathbf{j} - \mathbf{k}$. Given that $\mathbf{c} = \lambda\mathbf{a} + \mu\mathbf{b}$ and that $\mathbf{c}$ is perpendicular to $\mathbf{a}$, find the ratio of λ to μ.

Let A, B be the points with position vectors $\mathbf{a}$, $\mathbf{b}$ respectively with respect to an origin O. Write down, in terms of $\mathbf{a}$ and $\mathbf{b}$, a vector equation of the line l through A, in the plane of O, A and B, which is perpendicular to OA.

Find the position vector of P, the point of intersection of l and OB. (JMB)

13 Given that the two lines

$$\frac{x-1}{p} = \frac{y+6}{5} = \frac{z-3}{-1} \quad \text{and} \quad \frac{x-2}{-2} = \frac{y-3}{4} = \frac{z-5}{3}$$

intersect, find the value of p and also the coordinates of their point of intersection. (JMB)

14 Show that the lines $L: \mathbf{r} = \begin{pmatrix} 3 \\ 2 \\ 1 \end{pmatrix} + t \begin{pmatrix} 5 \\ 4 \\ 3 \end{pmatrix}$ and $M: \mathbf{r} = \begin{pmatrix} 16 \\ -10 \\ 2 \end{pmatrix} + s \begin{pmatrix} 3 \\ 2 \\ -1 \end{pmatrix}$

are skew. Verify that both lines L and M are perpendicular to $\mathbf{n}$, where

$\mathbf{n} = \begin{pmatrix} 5 \\ -7 \\ 1 \end{pmatrix}$.

 The point A on L has coordinates $(3, 2, 1)$. Write down a vector equation of the line N, through A with direction vector $\mathbf{n}$.

 Show that the lines M and N intersect and find the coordinates at their point of intersection, B. Find, in surd form, the length AB.

15 The points P and Q are given by $t = p$ and $s = q$ on the lines

$$\mathbf{r} = \begin{pmatrix} 3 \\ 1 \\ 2 \end{pmatrix} + t \begin{pmatrix} 2 \\ -2 \\ 0 \end{pmatrix} \text{ and } \mathbf{r} = \begin{pmatrix} 1 \\ 0 \\ 3 \end{pmatrix} + s \begin{pmatrix} 3 \\ 4 \\ 7 \end{pmatrix},$$

respectively. If PQ is perpendicular to both lines, show that $\overrightarrow{PQ}$ may be

written as $k \begin{pmatrix} 1 \\ 1 \\ -1 \end{pmatrix}$, where k is a scalar. Express $\overrightarrow{PQ}$ also in terms of p and q

and deduce the value of k and the shortest distance between the lines.

(JMB)

16 The lines L_1 and L_2 have vector equations given by $\mathbf{r}_1 = 3\mathbf{i} + \mathbf{j} + t(2\mathbf{j} + \mathbf{k})$ and $\mathbf{r}_2 = 4\mathbf{k} + s(\mathbf{i} + \mathbf{j} - \mathbf{k})$ where t and s are parameters. Show that L_1 and L_2 intersect, and find in surd form the cosine of the angle between these lines.

(JMB)

12.7 Planes through the origin

Given a direction vector $\mathbf{n}$, there is just one plane Π through the origin O perpendicular to $\mathbf{n}$. The plane Π contains the set of all lines through O perpendicular to $\mathbf{n}$. Every line in Π, whether through O or not, is also perpendicular to $\mathbf{n}$, so we say that the direction of $\mathbf{n}$ is perpendicular, or normal, to the plane Π. This direction $\mathbf{n}$ is called

the direction *normal* to Π,

$\mathbf{n}$ is a *normal vector* to the plane Π.

Clearly, any non-zero multiple of $\mathbf{n}$ is also a normal vector to Π.

 For example, with Cartesian axes $O(x, y, z)$, the axis Oz is normal to the

x, y-plane, the plane on which $z = 0$. Thus any vector $\begin{pmatrix} 0 \\ 0 \\ k \end{pmatrix}$, $k \neq 0$, is a

normal vector to the plane $z = 0$.

 In the general case, if Π is a plane through O with normal $\mathbf{n}$, then a point P lies in Π if and only if its position vector $\overrightarrow{OP}$ is perpendicular to $\mathbf{n}$. Thus, if $\overrightarrow{OP} = \mathbf{r}$

$$P \text{ lies in } \Pi \Leftrightarrow \mathbf{r} \cdot \mathbf{n} = 0,$$

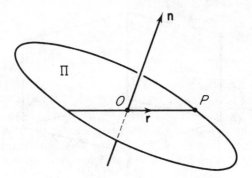

Fig. 12.14

so the equation of Π is $\mathbf{r} \cdot \mathbf{n} = 0$, (Fig. 12.14).

If $\mathbf{r} = \begin{pmatrix} x \\ y \\ z \end{pmatrix}$ and $\mathbf{n} = \begin{pmatrix} a \\ b \\ c \end{pmatrix}$, we obtain the Cartesian equation of Π,

$$ax + by + cz = 0.$$

EXAMPLE 1   *Find an equation of the plane through O with normal $\mathbf{n} = \begin{pmatrix} 3 \\ -1 \\ 2 \end{pmatrix}$.*

The point P (x, y, z) lies in the plane if $\mathbf{r}\,(= \overrightarrow{OP})$ is perpendicular to $\mathbf{n}$. So the equation of the plane is $\mathbf{r} \cdot \mathbf{n} = 3x - y + 2z = 0$.

EXAMPLE 2   *Find the equation of the plane Π, through the origin O and perpendicular to the two planes Π_1, $3x + 2y - z = 0$ and Π_2, $x - 2y + z = 0$.*

Let the normal to Π be $\mathbf{n}$, where $\mathbf{n} = \begin{pmatrix} a \\ b \\ c \end{pmatrix}$. Then $\mathbf{n}$ lies in each plane Π_1 and Π_2

and so $\mathbf{n}$ is perpendicular to the normals to each of these planes. Therefore,

$$\begin{pmatrix} a \\ b \\ c \end{pmatrix} \cdot \begin{pmatrix} 3 \\ 2 \\ -1 \end{pmatrix} = 0 \text{ and } \begin{pmatrix} a \\ b \\ c \end{pmatrix} \cdot \begin{pmatrix} 1 \\ -2 \\ 1 \end{pmatrix} = 0,$$

(see Fig. 12.15). Hence, $3a + 2b - c = 0$ and $a - 2b + c = 0$. On addition, $4a = 0$,

so $c = 2b$. As a possible vector $\mathbf{n}$, we take $b = 1, c = 2, a = 0$, and then $\mathbf{n} = \begin{pmatrix} 0 \\ 1 \\ 2 \end{pmatrix}$.

The equation of Π is

$$\mathbf{r} \cdot \mathbf{n} = y + 2z = 0.$$

Note that the two equations we solved for a, b, c showed that the point A, (a, b, c) lies in both planes Π_1 and Π_2, so that these two planes intersect in the line OA, as shown in Fig. 12.15.

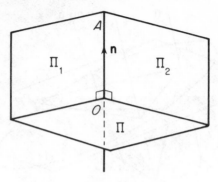

Fig. 12.15

EXERCISE 12.7

1 Find the equation of the plane through the origin O, perpendicular to:

(i) $\begin{pmatrix} 0 \\ 1 \\ 0 \end{pmatrix}$, (ii) $\begin{pmatrix} 2 \\ 3 \\ 4 \end{pmatrix}$, (iii) $\begin{pmatrix} 4 \\ -2 \\ 6 \end{pmatrix}$, (iv) $\mathbf{i} - \mathbf{j}$, (v) $5\mathbf{i} + 3\mathbf{j} - 2\mathbf{k}$.

2 Find a normal vector to the plane with equation:
   (i) $x + 2y = 0$, (ii) $x + y + z = 0$, (iii) $3x - 4y + 2z = 0$.

3 Find the equation of the plane through the origin O such that the two points, $A\,(2, 0, 4)$ and $B\,(0, 4, -2)$, are reflections of each other in the plane.

4 Find normal vectors for the two planes through O with equations $2x + y - 3z = 0$ and $3y - 2x + z = 0$. Find the equations of the lines through O perpendicular to these planes. Find also the angle between these lines.

12.8 The vector equation of a general plane

For a given direction vector $\mathbf{n}$, there is a whole family of parallel planes for which $\mathbf{n}$ is a normal vector. One member of the family will pass through O (cf. §12.7), and the others are all parallel to this plane. Let the plane Π be the one member of the family which passes through a given point A. The position vector of A is $\overrightarrow{OA}(= \mathbf{a})$. Then Π is the plane through A, perpendicular to $\mathbf{n}$, as is shown in Fig. 12.16.

Let ON be the perpendicular from O on to Π. Two cases are shown in Fig. 12.16(a) and (b), differing in the side of Π on which O lies. In (a), $\overrightarrow{ON}$ is in the same direction as $\mathbf{n}$ and in (b) $\overrightarrow{ON}$ is in the opposite direction to $\mathbf{n}$.

Vector equation of Π

Let P be any point in Π with position vector $\mathbf{r}(= \overrightarrow{OP})$. Then $\overrightarrow{AP}$ is perpendicular to $\mathbf{n}$, so

$$0 = \overrightarrow{AP} \cdot \mathbf{n} = (\overrightarrow{OP} - \overrightarrow{OA}) \cdot \mathbf{n} = (\mathbf{r} - \mathbf{a}) \cdot \mathbf{n}.$$

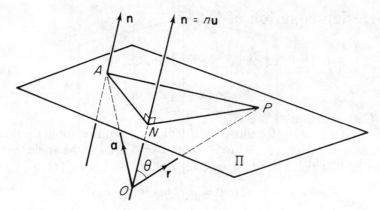

(a) $\overrightarrow{ON}.\mathbf{u} = p > 0$

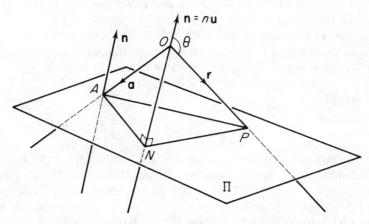

(b) $\overrightarrow{ON}.\mathbf{u} = p < 0$

Fig. 12.16

Conversely, if P is any point whose position vector $\mathbf{r}$ satisfies the equation $(\mathbf{r} - \mathbf{a}).\mathbf{n} = 0$, then

$$0 = (\mathbf{r} - \mathbf{a}).\mathbf{n} = \overrightarrow{AP}.\mathbf{n}$$

and so $\overrightarrow{AP}$ is perpendicular to $\mathbf{n}$ and P lies in Π. Hence the vector equation of Π is

$$(\mathbf{r} - \mathbf{a}).\mathbf{n} = 0.$$

Using the distributative rule for scalar products $(\mathbf{r} - \mathbf{a}).\mathbf{n} = \mathbf{r}.\mathbf{n} - \mathbf{a}.\mathbf{n}$, and, if we put $\mathbf{a}.\mathbf{n} = d$ then the equation of the plane Π is

$$\mathbf{r}.\mathbf{n} = \mathbf{a}.\mathbf{n} = d.$$

Note that in the particular case when $\mathbf{a} = \mathbf{0} = \begin{pmatrix} 0 \\ 0 \\ 0 \end{pmatrix}$, $d = 0$ and the equation of a plane through the origin is $\mathbf{r}.\mathbf{n} = 0$, as in §12.7.

Cartesian equation of Π

If $\mathbf{r} = \begin{pmatrix} x \\ y \\ z \end{pmatrix}$ and $\mathbf{n} = \begin{pmatrix} a \\ b \\ c \end{pmatrix}$, the equation becomes

$$ax + by + cz = d,$$

the Cartesian equation of Π.

To find a meaning for the scalar d, let $\mathbf{u}$ be the unit vector in the direction $\mathbf{n}$. Then $\mathbf{n} = n\mathbf{u}$, where $n = |\mathbf{n}| = \sqrt{(a^2 + b^2 + c^2)}$. Let the angle between $\mathbf{u}$ and $\mathbf{r}$ be θ, which is acute in case (a) and obtuse in case (b) in Fig. 12.16. Then

$$d = \mathbf{r} \cdot \mathbf{n} = \mathbf{r} \cdot (n\mathbf{u}) = n(\mathbf{r} \cdot \mathbf{u}) = nOP\cos\theta = np,$$

where $p = OP\cos\theta$, so that $\overrightarrow{ON} = p\mathbf{u}$.

Now $|p|$ is the length of the perpendicular ON from O on to Π, that is, the perpendicular distance from the origin O to Π.

The sign of p indicates on which side of Π the origin O lies.

Case (a): $p > 0$ and O lies on the opposite side of Π from $\mathbf{n}$.

Case (b): $p < 0$ and O lies on the same side of Π as $\mathbf{n}$.

If $p = 0$, then $N = O$ and P lies in the plane Π.

In conclusion,

$$p = \frac{d}{n} = \frac{d}{\sqrt{(a^2 + b^2 + c^2)}} = \mathbf{r} \cdot \frac{\mathbf{n}}{|\mathbf{n}|} = \mathbf{r} \cdot \mathbf{u},$$

so, in terms of the unit normal $\mathbf{u}$ and the distance p from O to the plane Π, measured positively in the direction $\mathbf{u}$, the equation of the plane is

$$\mathbf{r} \cdot \mathbf{u} = p.$$

General linear equation in three dimensions

Suppose that, instead of being given the plane Π, we wish to find the locus of a point $P(x, y, z)$ given by the equation

$$ax + by + cz = d. \tag{12.8}$$

Let $\mathbf{r} = \overrightarrow{OP} = \begin{pmatrix} x \\ y \\ z \end{pmatrix}$ and $\mathbf{n} = \begin{pmatrix} a \\ b \\ c \end{pmatrix} = n\mathbf{u}$, where $n = |\mathbf{n}| = \sqrt{(a^2 + b^2 + c^2)}$,

so that $\mathbf{u}\left(= \dfrac{1}{|\mathbf{n}|}\mathbf{n} \right)$ is the unit vector in the direction $\mathbf{n}$. Then equation 12.8 can be written

$$d = \mathbf{r} \cdot \mathbf{n} = \mathbf{r} \cdot (n\mathbf{u}) = n(\mathbf{r} \cdot \mathbf{u})$$

or

$$\mathbf{r} \cdot \mathbf{u} = p, \text{ where } p = \frac{d}{n}.$$

Let N be the point with position vector $p\mathbf{u}$, as is shown in Fig. 12.17, in case (a) with $p > 0$ and in case (b) with $p < 0$.

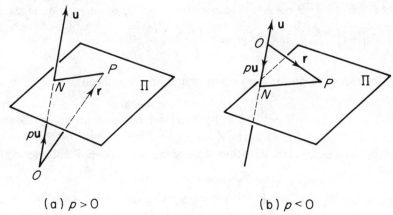

(a) $p > 0$ (b) $p < 0$

Fig. 12.17

Then   **r** satisfies equation $12.8 \Leftrightarrow \mathbf{r} \cdot \mathbf{u} = p = p\mathbf{u} \cdot \mathbf{u}$

$$\Leftrightarrow (\mathbf{r} - p\mathbf{u}) \cdot \mathbf{u} = 0$$

$$\Leftrightarrow (\overrightarrow{OP} - \overrightarrow{ON}) \cdot \mathbf{u} = \overrightarrow{NP} \cdot \mathbf{u} = 0$$

$$\Leftrightarrow \overrightarrow{NP} \text{ is perpendicular to } \mathbf{u}$$

$$\Leftrightarrow P \text{ lies in } \Pi,$$

where Π is the plane through N, perpendicular to **n**. Hence Π is the locus given by equation 12.8.

Summary $\mathbf{r} \cdot \mathbf{n} = ax + by + cz = d$ is the equation of the plane Π, perpendicular to **u**, the unit vector in the direction $\begin{pmatrix} a \\ b \\ c \end{pmatrix} (= \mathbf{n})$. The perpendicular $\overrightarrow{ON}$ from the origin O to Π is given by $\overrightarrow{ON} = p\mathbf{u}$, and Π is also given by the equation $\mathbf{r} \cdot \mathbf{u} = p$.

EXAMPLE 1   *Find a Cartesian equation for the plane Π, through A, $(5, -2, 4)$. with normal vector* $\begin{pmatrix} -1 \\ 2 \\ 3 \end{pmatrix}$.

The point P, (x, y, z), lies in $\Pi \Leftrightarrow \overrightarrow{AP}$ is perpendicular to $\begin{pmatrix} -1 \\ 2 \\ 3 \end{pmatrix}$

$$\Leftrightarrow \begin{pmatrix} x - 5 \\ y + 2 \\ z - 4 \end{pmatrix} \cdot \begin{pmatrix} -1 \\ 2 \\ 3 \end{pmatrix} = 0$$

$$\Leftrightarrow \begin{pmatrix} x \\ y \\ z \end{pmatrix} \cdot \begin{pmatrix} -1 \\ 2 \\ 3 \end{pmatrix} = \begin{pmatrix} 5 \\ -2 \\ 4 \end{pmatrix} \cdot \begin{pmatrix} -1 \\ 2 \\ 3 \end{pmatrix} = 3.$$

So the equation of Π is $-x + 2y + 3z = 3$.

EXAMPLE 2 *Find the coordinates of the point, P, of intersection of the line*
$\mathbf{r} = \begin{pmatrix} 2 \\ 1 \\ 3 \end{pmatrix} + t \begin{pmatrix} -1 \\ 2 \\ 1 \end{pmatrix}$, *and the plane* $\mathbf{r} \cdot \begin{pmatrix} 1 \\ -1 \\ 1 \end{pmatrix} = -6.$

Let *P* be the point on the line corresponding to $t = p$. Then

$$\overrightarrow{OP} = \begin{pmatrix} x \\ y \\ z \end{pmatrix} = \mathbf{r} = \begin{pmatrix} 2-p \\ 1+2p \\ 3+p \end{pmatrix},$$

and this vector satisfies the equation of the plane, if *P* is the point of intersection. Hence

$$-6 = \begin{pmatrix} 2-p \\ 1+2p \\ 3+p \end{pmatrix} \cdot \begin{pmatrix} 1 \\ -1 \\ 1 \end{pmatrix} = 2-p-(1+2p)+3+p = 4-2p,$$

so $p = 5$ and the coordinates of *P* are **(−3, 11, 8)**.

EXAMPLE 3 *Find the coordinates of N, the foot of the perpendicular from the origin O on to the plane* $\mathbf{r} \cdot \begin{pmatrix} 1 \\ -5 \\ 3 \end{pmatrix} = 5.$ *Deduce the coordinates of the reflection Q of O in this plane.*

The normal to the plane is $\begin{pmatrix} 1 \\ -5 \\ 3 \end{pmatrix}$ ($= \mathbf{n}$) and so the equation of *ON* is $\mathbf{r} = t\mathbf{n}$. This line meets the plane at *N* where $(t\mathbf{n}) \cdot \mathbf{n} = 5$, that is,

$$t = \frac{5}{\mathbf{n} \cdot \mathbf{n}} = \frac{5}{1+25+9} = \frac{1}{7}.$$

Thus *N* is the point $(\frac{1}{7}, -\frac{5}{7}, \frac{3}{7})$.

Since $\overrightarrow{ON} = t\mathbf{n}$ and *Q* is the reflection of *O* in the plane, $\overrightarrow{OQ} = 2\overrightarrow{ON} = 2t\mathbf{n}$ and *Q* is the point $(\frac{2}{7}, -\frac{10}{7}, \frac{6}{7})$ (see Fig. 12.18).

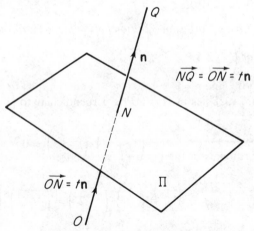

Fig. 12.18

EXERCISE 12.8

1   Find a Cartesian equation of the plane:

   (i) through $(1, 2, 3)$ with normal vector $\begin{pmatrix} 1 \\ 0 \\ 1 \end{pmatrix}$,

   (ii) through $(3, -4, 0)$ with normal vector $\begin{pmatrix} 2 \\ -1 \\ 3 \end{pmatrix}$,

   (iii) through $(-5, 2, 1)$ with normal vector $3\mathbf{i} - 2\mathbf{j} + \mathbf{k}$.

2   Find a vector equation of the plane in the form $\mathbf{r} \cdot \mathbf{u} = p$, where $|\mathbf{u}| = 1$, given that the Cartesian equation of the plane is:
   (i) $2x + y + 2z = 6$, (ii) $3x + 2y = 4$, (iii) $x - y + z = 1$,
   (iv) $y - 4x + 5z + 7 = 0$.

3   Prove that the line $\mathbf{r} = \begin{pmatrix} 3 \\ 2 \\ -1 \end{pmatrix} + t \begin{pmatrix} 1 \\ 2 \\ -3 \end{pmatrix}$ is parallel to the plane

   $x + y - z = 2$.

4   Find a vector equation of the plane described and its Cartesian equation:
   (i) through the point $(2, 1, 0)$ parallel to the plane $x = y$,
   (ii) through the point $(-1, 3, 1)$ parallel to the plane $2x - y + 3z = 6$,
   (iii) through the point $(1, -2, 3)$ perpendicular to the vector $\mathbf{i} - \mathbf{j} + 2\mathbf{k}$,

   (iv) through the point $(2, 1, 0)$ and perpendicular to the line

   $\mathbf{r} = \begin{pmatrix} 1 \\ 3 \\ 2 \end{pmatrix} + t \begin{pmatrix} 4 \\ 2 \\ 3 \end{pmatrix}$,

   (v) through the point $(3, -2, 1)$ and parallel to both the vectors $2\mathbf{i} + \mathbf{j} - 3\mathbf{k}$ and $3\mathbf{i} - \mathbf{j}$,

   (vi) through the point $(1, 2, 1)$ and parallel to both the lines

   $\mathbf{r} = \begin{pmatrix} 1 \\ 2 \\ 3 \end{pmatrix} + t \begin{pmatrix} 1 \\ -1 \\ -1 \end{pmatrix}$ and $\mathbf{r} = \begin{pmatrix} 4 \\ 3 \\ 2 \end{pmatrix} + t \begin{pmatrix} 0 \\ -1 \\ 1 \end{pmatrix}$,

   (vii) through the line $\mathbf{r} = \begin{pmatrix} 0 \\ 1 \\ 4 \end{pmatrix} + t \begin{pmatrix} -1 \\ 2 \\ 1 \end{pmatrix}$ and parallel to the line

   $\mathbf{r} = \begin{pmatrix} 2 \\ 1 \\ 4 \end{pmatrix} + t \begin{pmatrix} 1 \\ -2 \\ 1 \end{pmatrix}$.

5   Find the coordinates of the point of intersection of:

   (i) the line $\mathbf{r} = \begin{pmatrix} 2 \\ 0 \\ 1 \end{pmatrix} + t \begin{pmatrix} 1 \\ 1 \\ 0 \end{pmatrix}$ and the plane $\mathbf{r} \cdot \begin{pmatrix} 1 \\ -1 \\ -1 \end{pmatrix} = 1$,

   (ii) the line $\mathbf{r} = (1 + t)\mathbf{i} + (2 - t)\mathbf{j} + t\mathbf{k}$ and the plane $\mathbf{r} \cdot (\mathbf{i} - 3\mathbf{k}) = 5$.

6   Find the perpendicular distance from the origin to the plane:
(i) $x + 2y + 2z = 5$,   (ii) $3x + 4 = 2y - 6z$,   (iii) $\mathbf{r}.(3\mathbf{i} - 4\mathbf{j}) = 7$.

12.9 Parametric equation of a plane

Let $\mathbf{p}$ and $\mathbf{q}$ be non-parallel vectors, and let A be the point with position vector $\mathbf{a} = \overrightarrow{OA}$. Then there is one plane, Π, through A and parallel to both $\mathbf{p}$ and $\mathbf{q}$, Fig. 12.19.

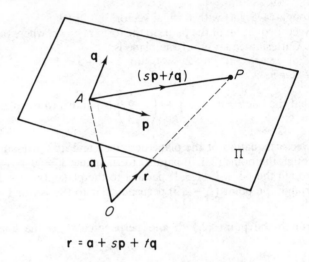

$$\mathbf{r} = \mathbf{a} + s\mathbf{p} + t\mathbf{q}$$

Fig. 12.19 Parametric equation of a plane.

A general point P in Π is given by $\overrightarrow{AP} = s\mathbf{p} + t\mathbf{q}$. The values of s and t are found by drawing a parallelogram with sides parallel to $\mathbf{p}$ and $\mathbf{q}$ and diagonal $\overrightarrow{AP}$ (theorem 5.5). Then

$$P \text{ lies in } \Pi \Leftrightarrow \mathbf{r} = \overrightarrow{OP} = \overrightarrow{OA} + \overrightarrow{AP} = \mathbf{a} + s\mathbf{p} + t\mathbf{q}$$

and so the equation of Π, in terms of two parameters s and t, is

$$\mathbf{r} = \mathbf{a} + s\mathbf{p} + t\mathbf{q}.$$

EXAMPLE 1   *Find a parametric equation of the plane* $2x + 3y - z = 4$.

In this case, we can use x and y as the parameters. Let $x = s$, $y = t$, then $2s + 3t - z = 4$ and so $z = 2s + 3t - 4$, and the equation of the plane is

$$\mathbf{r} = \begin{pmatrix} x \\ y \\ z \end{pmatrix} = \begin{pmatrix} s \\ t \\ 2s + 3t - 4 \end{pmatrix} = s\begin{pmatrix} 1 \\ 0 \\ 2 \end{pmatrix} + t\begin{pmatrix} 0 \\ 1 \\ 3 \end{pmatrix} + \begin{pmatrix} 0 \\ 0 \\ -4 \end{pmatrix}.$$

EXAMPLE 2   *Find a Cartesian equation of the plane, given by the parametric equation* $\mathbf{r} = \begin{pmatrix} 2 \\ 1 \\ 3 \end{pmatrix} + s\begin{pmatrix} 4 \\ -1 \\ 2 \end{pmatrix} + t\begin{pmatrix} 1 \\ -1 \\ 3 \end{pmatrix}$.

The vector equation is equivalent to three scalar equations,

$$x = 2 + 4s + t, \quad y = 1 - s - t \quad \text{and} \quad z = 3 + 2s + 3t.$$

We eliminate s and t from these equations. Add the first two equations to give

$$x + y = 3 + 3s, \quad \text{so} \quad s = \tfrac{1}{3}(x + y - 3).$$

Add four times the second equation to the first to give

$$x + 4y = 6 - 3t, \quad \text{so} \quad t = \tfrac{1}{3}(-x - 4y + 6).$$

Substitute these expressions for s and t into the third equation,

$$z = 3 + \tfrac{2}{3}(x + y - 3) + (-x - 4y + 6) = -\tfrac{1}{3}x - \tfrac{10}{3}y + 7,$$

giving as a Cartesian equation for the plane,

$$x + 10y + 3z = 21.$$

EXERCISE 12.9

1 Find a Cartesian equation and a vector equation for the plane given by the parametric equation:

   (i) $\mathbf{r} = \begin{pmatrix} 4 \\ -1 \\ 2 \end{pmatrix} + s \begin{pmatrix} 1 \\ 2 \\ -3 \end{pmatrix} + t \begin{pmatrix} -1 \\ 2 \\ -1 \end{pmatrix},$

   (ii) $\mathbf{r} = \mathbf{i} - 2\mathbf{j} + s(2\mathbf{i} + \mathbf{j} - 3\mathbf{k}) + t(4\mathbf{j} - \mathbf{i} - \mathbf{k}),$

   (iii) $\mathbf{r} = (2s + 3)\mathbf{i} + s\mathbf{j} + (t - s)\mathbf{k}.$

2 Find a parametric equation for the plane:

   (i) $x = 0,$ (ii) $x = z,$ (iii) $2x - y + z = 3,$ (iv) $y + 4 = 3z,$

   (v) $2z = x + 2y + 3.$

3 Show that the vector $\begin{pmatrix} 2 \\ 6 \\ 5 \end{pmatrix}$ is normal to the plane

$$\mathbf{r} = \begin{pmatrix} 1 \\ 1 \\ 0 \end{pmatrix} + s \begin{pmatrix} 2 \\ 1 \\ -2 \end{pmatrix} + t \begin{pmatrix} 1 \\ -2 \\ 2 \end{pmatrix}.$$ Hence, write down an equation of the plane

   in the form $\mathbf{r} \cdot \mathbf{n} = p$ and deduce a Cartesian equation for the plane. Verify that, for all values of s and t, the coordinates given by the parametric equation of the plane satisfy the Cartesian equation.

4 A plane passes through a fixed point with position vector $\mathbf{k}$ relative to a given origin, and contains lines in the directions of two given non-parallel vectors $\mathbf{a}$ and $\mathbf{b}$. Write down an expression for the position vector, $\mathbf{r}$, of any point of the plane in terms of $\mathbf{k}$, $\mathbf{a}$, $\mathbf{b}$ and two scalar parameters s and t.

   If $\mathbf{c}$ is a fixed vector at right angles to both $\mathbf{a}$ and $\mathbf{b}$, prove that the product $\mathbf{r} \cdot \mathbf{c}$ has a constant value at all points of the plane. (*SMP*)

12.10 Angles between lines and planes

Definition The angle between a line and a plane is the angle between the line and the projection of the line on to the plane.

Let $\overrightarrow{AB}$ be a line in the direction of a vector **u**. Suppose that AB intersects a plane Π at A. Let **n** be a normal vector of Π. If $\overrightarrow{AB}$ is not perpendicular to Π, then **n** and **u** define a plane ABC, which intersects Π in the line AC. AC is the projection of AB on to Π, (Fig. 12.20).

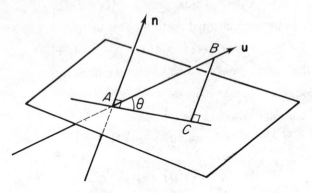

Fig. 12.20

Then the angle θ between AB and Π is the angle BAC. Therefore, θ is the complement of the angle between **n** and **u**. So

$$\mathbf{n} \cdot \mathbf{u} = |\mathbf{n}||\mathbf{u}| \cos(90° - \theta) = |\mathbf{n}||\mathbf{u}| \sin\theta,$$

and θ can be found from this equation.

EXAMPLE 1   *Find the angle between the line* $\mathbf{r} = \begin{pmatrix} 4 \\ 3 \\ 1 \end{pmatrix} + t\begin{pmatrix} 1 \\ -1 \\ 2 \end{pmatrix}$ *and the*

plane $\mathbf{r} \cdot \begin{pmatrix} 1 \\ 0 \\ 1 \end{pmatrix} = 5.$

The line is in the direction $\mathbf{u} = \begin{pmatrix} 1 \\ -1 \\ 2 \end{pmatrix}$ and the normal vector to the plane is

$\mathbf{n} = \begin{pmatrix} 1 \\ 0 \\ 1 \end{pmatrix}$. Hence, the angle θ between the line and the plane is given by

$$\sin\theta = \frac{|\mathbf{n} \cdot \mathbf{u}|}{|\mathbf{n}||\mathbf{u}|} = \frac{1+0+2}{\sqrt{2}\sqrt{6}} = \frac{\sqrt{3}}{2}, \text{ so } \theta = 60°.$$

The angle between two planes

If the planes are parallel, the angle between them is zero. If they are not parallel, the two planes, say Π_1 and Π_2 will meet in a line L. The angle between Π_1 and Π_2 is then defined as the angle between two lines, one in each plane, each perpendicular to L.

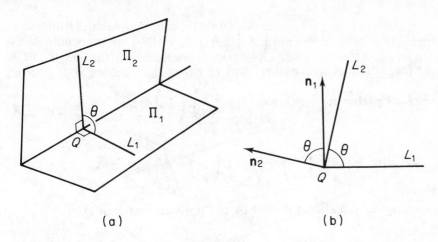

Fig. 12.21

Suppose that two such lines are L_1 and L_2 meeting at a point Q on L, (see Fig. 12.21(a)). Then L_1 and L_2 define a plane Π which is perpendicular to both Π_1 and Π_2. In Fig. 12.21(b) we show the cross-section of the three-dimensional figure of (a), in the plane Π. The line L is shown now as a single point Q and the planes Π_1 and Π_2 are shown by the lines L_1 and L_2. The normals $\mathbf{n}_1$ and $\mathbf{n}_2$ to Π_1 and Π_2, are perpendicular to L_1 and L_2 respectively. Therefore, the required angle θ between the planes Π_1 and Π_2, which is the angle between L_1 and L_2, is also the angle between $\mathbf{n}_1$ and $\mathbf{n}_2$. Therefore, θ may be found from the equation

$$|\mathbf{n}_1.\mathbf{n}_2| = |\mathbf{n}_1||\mathbf{n}_2|\cos\theta.$$

EXAMPLE 2   *Find the angle between the two planes* $\mathbf{r}.\begin{pmatrix} -2 \\ 3 \\ 1 \end{pmatrix} = 5$ *and*

$\mathbf{r}.\begin{pmatrix} 4 \\ 2 \\ -1 \end{pmatrix} = 3.$

In this case $\mathbf{n}_1.\mathbf{n}_2 = -8 + 6 - 1 = -3$, $|\mathbf{n}_1| = \sqrt{(14)}$, $|\mathbf{n}_2| = \sqrt{(21)}$, so $\cos\theta = \dfrac{3}{7\sqrt{6}}$, which gives $\theta \approx \mathbf{79 \cdot 9°}$.

12.11   Distance of a point from a plane

The shortest distance from a point $P(x_0, y_0, z_0)$ to a plane Π, $\mathbf{r}.\begin{pmatrix} a \\ b \\ c \end{pmatrix} = d$,

is the length of the perpendicular PN from P on to the plane. Let Q, (x, y, z)

be any point on Π so that $\begin{pmatrix} x \\ y \\ z \end{pmatrix} \cdot \begin{pmatrix} a \\ b \\ c \end{pmatrix} = d$. The required length PN is

the magnitude of the resolved part of $\overrightarrow{PQ}$ in the direction $\mathbf{n}$. If p is this

resolved part, $p\,|\mathbf{n}| = \overrightarrow{PQ} \cdot \mathbf{n} = \begin{pmatrix} x - x_0 \\ y - y_0 \\ z - z_0 \end{pmatrix} \cdot \begin{pmatrix} a \\ b \\ c \end{pmatrix}$. Thus

$$p\,|\mathbf{n}| = \begin{pmatrix} x \\ y \\ z \end{pmatrix} \cdot \begin{pmatrix} a \\ b \\ c \end{pmatrix} - \begin{pmatrix} x_0 \\ y_0 \\ z_0 \end{pmatrix} \cdot \begin{pmatrix} a \\ b \\ c \end{pmatrix} = d - (ax_0 + by_0 + cz_0),$$

on using the equation satisfied by the position vector of Q.

Thus
$$p = \frac{d - (ax_0 + by_0 + cz_0)}{\sqrt{(a^2 + b^2 + c^2)}}.$$

If $p > 0$, then P is on the opposite side of the plane Π from the direction of its normal vector $\mathbf{n}$, as shown in Fig. 12.22. If $p < 0$, then P is on the same side of Π as $\mathbf{n}$. The required perpendicular distance PN is $|p|$.

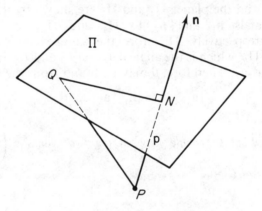

Fig. 12.22

EXAMPLE 1   *Find the perpendicular distance from the point* $(-2, 1, -5)$ *to the*

plane $\mathbf{r} \cdot \begin{pmatrix} 2 \\ 6 \\ -3 \end{pmatrix} = 4$.

Using the notation of the theory above, $\mathbf{n} = \begin{pmatrix} 2 \\ 6 \\ -3 \end{pmatrix}$, and so $|\mathbf{n}| = 7$. Then

$7p = 4 - (-4 + 6 + 15) = -13$, so P is on the same side of the plane as the vector $\mathbf{n}$, and the perpendicular distance is $\frac{13}{7}$.

EXERCISE 12.11

1 Find, in surd form, the perpendicular distance of the point A from the plane Π, and state whether or not the point A is on the same side of Π as is the origin:

(i) A is the point $(1, 2, -3)$ and Π is the plane $\mathbf{r} . \begin{pmatrix} 3 \\ -1 \\ 4 \end{pmatrix} = -5$,

(ii) A is the point $(5, -3, 2)$ and Π is the plane $\mathbf{r} . \begin{pmatrix} 2 \\ 3 \\ 4 \end{pmatrix} = 7$.

2 The point B has coordinates $(-2, 3, -5)$. In each of the two cases in question 1 above, find (a) a vector equation, (b) Cartesian equations, of the line AB, and find also the angle between AB and the given plane Π.

3 For each of the following, find the angle between the two planes with the given equations and find also a vector equation of their common line:

(i) $\mathbf{r} . \begin{pmatrix} -2 \\ 3 \\ 0 \end{pmatrix} = 4$ and $\mathbf{r} . \begin{pmatrix} 3 \\ -2 \\ 1 \end{pmatrix} = 2$,

(ii) $\mathbf{r} . \begin{pmatrix} -4 \\ 1 \\ 5 \end{pmatrix} = 5$ and $\mathbf{r} . \begin{pmatrix} 3 \\ 5 \\ -2 \end{pmatrix} = 2$.

4 Show that

$$\mathbf{r} = \mathbf{a} + s(\mathbf{b} - \mathbf{a}) + t(\mathbf{c} - \mathbf{a})$$

is an equation of the plane which passes through the noncollinear points whose position vectors are $\mathbf{a}$, $\mathbf{b}$, $\mathbf{c}$, where $\mathbf{r}$ is the position vector of a general point on the plane and s and t are scalars. Find a Cartesian equation of the plane containing the points $(1, 1, -1)$, $(2, 0, 1)$ and $(3, 2, 1)$, and show that the points $(2, 1, 2)$ and $(0, -2, -2)$ are equidistant from, and on opposite sides of, this plane. (L)

5 Obtain equations of the straight line through the origin parallel to the straight line through the points $P(3, 2, 3)$ and $Q(-1, -2, 1)$. Find an equation of the plane which passes through both these two lines.

The line of intersection of a plane through P and a plane through Q is the y-axis. Find the angle between these two planes. (L)

6 The point O is the origin and points A, B, C, D have position vectors

$$\begin{pmatrix} 4 \\ 3 \\ 4 \end{pmatrix}, \begin{pmatrix} 6 \\ 1 \\ 2 \end{pmatrix}, \begin{pmatrix} 0 \\ 9 \\ -6 \end{pmatrix}, \begin{pmatrix} -1 \\ 1 \\ 1 \end{pmatrix},$$

respectively.
Prove that
  (i) the triangle OAB is isosceles,
  (ii) D lies in the plane OAB,
  (iii) CD is perpendicular to the plane OAB,
  (iv) AC is inclined at an angle of $60°$ to the plane OAB. (C)

MISCELLANEOUS EXERCISE 12

1 Find the angle between the vectors $\mathbf{a}$ and $\mathbf{b}$ given that $|\mathbf{a}| = 3$, $|\mathbf{b}| = 5$, $|\mathbf{a} - \mathbf{b}| = 7$. $\hspace{2cm}$ (L)

2 State a relation which exists between the vectors $\mathbf{p}$ and $\mathbf{q}$ when these vectors are (a) parallel, (b) perpendicular.

   The position vectors of the vertices of a tetrahedron $ABCD$ are
   $A: -5\mathbf{i} + 22\mathbf{j} + 5\mathbf{k}$, $\quad B: \mathbf{i} + 2\mathbf{j} + 3\mathbf{k}$, $\quad C: 4\mathbf{i} + 3\mathbf{j} + 2\mathbf{k}$, $\quad D: -\mathbf{i} + 2\mathbf{j} - 3\mathbf{k}$.
   Find the angle CBD and show that AB is perpendicular to both BC and BD.
   Calculate the volume of the tetrahedron. If $ABDE$ is a parallelogram, find the position vector of E. $\hspace{2cm}$ (L)

3 Prove that if two pairs of opposite edges of a tetrahedron are perpendicular then the third pair of edges are also perpendicular.

4 $OABC$ is a tetrahedron such that $\overrightarrow{OA} = \mathbf{a}$, $\overrightarrow{OB} = \mathbf{b}$ and $\overrightarrow{OC} = \mathbf{c}$ where $\mathbf{a}$, $\mathbf{b}$ and $\mathbf{c}$ are unit vectors. If angle $BOA = \cos^{-1}(2/5)$, angle $COB = \cos^{-1}(9/10)$ and OA is perpendicular to OC, find p and q such that the vector $\mathbf{r} = \mathbf{a} + p\mathbf{b} + q\mathbf{c}$ is perpendicular to the plane ABC. $\hspace{2cm}$ (JMB)

5 The position vectors of two points A and B relative to an origin O are $\mathbf{a}$ and $\mathbf{b}$. You are given that $\mathbf{a}$ and $\mathbf{b}$ have unit length, and that the angle between these vectors is $60°$. Write down the values of the products $\mathbf{a.a}$, $\mathbf{b.b}$ and $\mathbf{a.b}$.

   The point C on the line-segment AB is such that $AC = 2CB$. If $\mathbf{c}$ is the position vector of C, express $\mathbf{c}$ in terms of $\mathbf{a}$ and $\mathbf{b}$. Hence calculate (i) the length of $\mathbf{c}$, (ii) the cosine of the angle between $\mathbf{a}$ and $\mathbf{c}$. $\hspace{0.5cm}$ (SMP)

6 The position vectors of the points A, B, C relative to the origin O are $\mathbf{a}$, $\mathbf{b}$, $\mathbf{c}$ respectively. The point P is on BC such that $BP:PC = 2:3$; the point Q is on CA such that $CQ:QA = 1:4$. Find the position vector of the common point R of AP and BQ. $\hspace{2cm}$ (JMB)

7 The position vectors of points A, B, C relative to a fixed origin O are $\mathbf{a}$, $\mathbf{b}$, $\mathbf{c}$ respectively. If D is the mid-point of AB, and if E is the point which divides CB internally in the ratio $1:2$, write down the position vectors of D and E in terms of $\mathbf{a}$, $\mathbf{b}$, $\mathbf{c}$.

   Show that the mid-point F of CD is on AE and find the ratio $AF:FE$. $\hspace{2cm}$ (JMB)

8 A tetrahedron $OABC$ with vertex O at the origin is such that $\overrightarrow{OA} = \mathbf{a}$, $\overrightarrow{OB} = \mathbf{b}$ and $\overrightarrow{OC} = \mathbf{c}$. Show that the line segments joining the mid-points of opposite edges bisect one another. Given that two pairs of opposite edges are perpendicular prove that $\mathbf{a.b} = \mathbf{b.c} = \mathbf{c.a}$ and show that the third pair of opposite edges is also perpendicular.

   Prove also that, in this case, $OA^2 + BC^2 = OB^2 + AC^2$. $\hspace{2cm}$ (L)

9 A plane passes through A, $(2,2,1)$, and is perpendicular to the line joining the origin to A. Write down a vector equation of the plane in the form $\mathbf{r.n} = p$, where $\mathbf{n}$ is a unit vector. $\hspace{2cm}$ (L)

10 Find the position vector of the point of intersection of the line

$$\mathbf{r} = \begin{pmatrix} 3 \\ 2 \\ 3 \end{pmatrix} + t \begin{pmatrix} 3 \\ 1 \\ 2 \end{pmatrix} \text{ and the plane } \mathbf{r}.\begin{pmatrix} 4 \\ -3 \\ -2 \end{pmatrix} = 10.$$

11 The position vectors of A and C with respect to the origin O are $(4\mathbf{i} + 3\mathbf{j})$ and $(-3\mathbf{i} + 4\mathbf{j})$ respectively. Show that $|\overrightarrow{OA}| = |\overrightarrow{OC}|$ and that OA is perpendicular to OC.

Find the vector $\overrightarrow{OB}$ given that $OABC$ is a square. The mid-point of OB is M and the mid-point of CM is L. Show that the position vector of L with respect to the origin is $-(5/4)(\mathbf{i}-3\mathbf{j})$. $\hspace{1cm}$ (L)

12 (i) In a triangle ABC the altitudes through A and B meet in a point O. Let $\mathbf{a}$, $\mathbf{b}$, $\mathbf{c}$ be the position vectors of A, B, C relative to O as origin. Show that $\mathbf{a}.\mathbf{b} = \mathbf{a}.\mathbf{c}$ and $\mathbf{b}.\mathbf{a} = \mathbf{b}.\mathbf{c}$. Deduce that the altitudes of a triangle are concurrent.

(ii) Find the point of intersection of the line through the points $(2, 0, 1)$ and $(-1, 3, 4)$ and the line through the points $(-1, 3, 0)$ and $(4, -2, 5)$. Calculate the acute angle between the two lines. $\hspace{1cm}$ (L)

13 The vectors $\mathbf{a}$ and $\mathbf{b}$ are $(2\mathbf{i}-\mathbf{j}+3\mathbf{k})$ and $(3\mathbf{i}+2\mathbf{j}-2\mathbf{k})$ respectively. The vector $\mathbf{a}$ is resolved into two components, one in the direction of $\mathbf{b}$ and the other perpendicular to $\mathbf{b}$. Find these components. $\hspace{1cm}$ (L)

14 Given that the points A and B have position vectors $(2\mathbf{i}+4\mathbf{j}+7\mathbf{k})$ and $(-4\mathbf{i}+\mathbf{j}+\mathbf{k})$ respectively, find the position vector of the point P on AB which is such that $\overrightarrow{AP} = 2\overrightarrow{PB}$. $\hspace{1cm}$ (L)

15 Given that $\mathbf{a} = -\mathbf{i}+\mathbf{j}+2\mathbf{k}$ and $\mathbf{b} = \mathbf{j}+\mathbf{k}$, find $|\mathbf{a}|$, $|\mathbf{b}|$ and the acute angle between $\mathbf{a}$ and $\mathbf{b}$. $\hspace{1cm}$ (JMB)

16 Given the vectors $\mathbf{u} = 3\mathbf{i}+2\mathbf{j}$ and $\mathbf{v} = 2\mathbf{i}+\lambda\mathbf{j}$, determine the values of λ so that

(a) $\mathbf{u}$ and $\mathbf{v}$ are at right angles,

(b) $\mathbf{u}$ and $\mathbf{v}$ are parallel,

(c) the acute angle between $\mathbf{u}$ and $\mathbf{v}$ is $\pi/4$. $\hspace{1cm}$ (L)

17 The angle between the vectors $(\mathbf{i}+\mathbf{j})$ and $(\mathbf{i}+\mathbf{j}+p\mathbf{k})$ is $\pi/4$. Find the possible values of p. $\hspace{1cm}$ (L)

18 Given that A and B are the points whose position vectors referred to the origin O are $(\mathbf{i}+2\mathbf{j}+\mathbf{k})$ and $(5\mathbf{i}-2\mathbf{j}+3\mathbf{k})$ respectively, determine $|\overrightarrow{AB}|$ and direction cosines for the line AB.

19 Find the resolved part of the vector $3\mathbf{i}-\mathbf{j}+3\mathbf{k}$ in the direction of the vector $6\mathbf{i}+3\mathbf{j}+2\mathbf{k}$.

20 The line l passes through the points $A(-1, -3, -3)$ and $B(5, 0, 6)$. Write down, in terms of a single parameter, the position vector of a general point P on l relative to the origin O.

Find the position of P for which OP is perpendicular to l. Hence find the minimum distance from the origin to the line l. $\hspace{1cm}$ (SMP)

21 Find the vector equation of the line through the origin and parallel to the line of intersection of the planes $\mathbf{r}.\begin{pmatrix} 2 \\ 1 \\ 3 \end{pmatrix} = 6$ and $\mathbf{r}.\begin{pmatrix} 1 \\ -1 \\ 2 \end{pmatrix} = -2$.

22 Find the coordinates of the point of intersection of:

(i) the line $\mathbf{r} = \mathbf{i}-\mathbf{j}+2\mathbf{k}+t(3\mathbf{j}+4\mathbf{k})$ and the plane $\mathbf{r}.(\mathbf{i}+\mathbf{j}-\mathbf{k}) = 6$,

(ii) the line $\mathbf{r} = \begin{pmatrix} 1 \\ 2 \\ 3 \end{pmatrix} + t\begin{pmatrix} 4 \\ 2 \\ -1 \end{pmatrix}$ and the plane $\mathbf{r}.\begin{pmatrix} -2 \\ 0 \\ 1 \end{pmatrix} = 4$.

23 The point $P(14+2\lambda, 5+2\lambda, 2-\lambda)$ lies on a fixed straight line for all values of λ. Find Cartesian equations for this line and find the cosine of the acute angle between this line and the line $x = z = 0$.

Show that the line $2x = -y = -z$ is perpendicular to the locus of P.

Hence, or otherwise, find the equation of the plane containing the origin and all possible positions of P. (L)

24 Find a vector equation for the plane passing through the points A, B, C with position vectors $(\mathbf{i} - \mathbf{k})$, $(2\mathbf{i} + \mathbf{j} + 3\mathbf{k})$, $(3\mathbf{i} + 2\mathbf{j} + \mathbf{k})$ respectively. By finding the scalar product $\overrightarrow{AC}.\overrightarrow{BC}$, or otherwise, show that angle ACB is a right angle. (L)

25 Find a unit vector which is perpendicular to the vector $(4\mathbf{i} + 4\mathbf{j} - 7\mathbf{k})$ and to the vector $(2\mathbf{i} + 2\mathbf{j} + \mathbf{k})$. (L)

26 Points O, A, B in space have coordinates $(0,0,0)$, $(-2,1,2)$ and $(-1,0,1)$ respectively. Show that the cosine of angle AOB is equal to $\frac{2}{3}\sqrt{2}$.

   If C divides AB in the ratio $k:1-k$, find the coordinates of C in terms of k. Find k if OC is perpendicular to AB. (SMP)

27 Prove that $\frac{1}{7}\begin{pmatrix} 2 \\ -3 \\ 6 \end{pmatrix}$ is a unit vector. Hence find the Cartesian equation of one of the two planes perpendicular to this vector whose distance from the origin is 10 units. (SMP)

28 The point $A(5, 1, 5)$ has position vector $\mathbf{a}$ relative to the origin O and the point $B(4, 2, 6)$ has position vector $\mathbf{b}$. The line l is given by $\mathbf{r} = \mathbf{a} + s(5\mathbf{i} - 2\mathbf{j} + 5\mathbf{k})$; the line m is given by $\mathbf{r} = \mathbf{b} + t(4\mathbf{i} - \mathbf{j} + 6\mathbf{k})$. Show that the lines l and m intersect and state the coordinates of the common point. Prove that AB is perpendicular to m and, hence, write down the coordinates of the reflection of A in m.

29 A plane passes through the three points A, B, C whose position vectors, referred to an origin O, are $(\mathbf{i} + 3\mathbf{j} + 3\mathbf{k})$, $(3\mathbf{i} + \mathbf{j} + 4\mathbf{k})$, $(2\mathbf{i} + 4\mathbf{j} + \mathbf{k})$ respectively. Find, in the form $(l\mathbf{i} + m\mathbf{j} + n\mathbf{k})$, a unit vector normal to this plane.
   Find also the Cartesian equation of the plane, and the perpendicular distance from the origin to this plane. (L)

30 Of the following equations, which represent lines and which represent planes?

   (i) $\dfrac{x-2}{1} = \dfrac{y-1}{2} = \dfrac{z-3}{-1}$;

   (ii) $x + 2y - z = 1$;

   (iii) $\begin{pmatrix} x \\ y \\ z \end{pmatrix} = \begin{pmatrix} 2 \\ 1 \\ 3 \end{pmatrix} + t\begin{pmatrix} 1 \\ -1 \\ -1 \end{pmatrix}$.

   Describe, or show in a clear diagram, how these lines and planes are related to each other. (SMP)

31 The angles between the non-zero vectors $\mathbf{b}$ and $\mathbf{c}$, $\mathbf{c}$ and $\mathbf{a}$, $\mathbf{a}$ and $\mathbf{b}$ are α, β, γ respectively. The vectors $\mathbf{u}$ and $\mathbf{v}$ are defined as

$$\mathbf{u} = (\mathbf{a}.\mathbf{c})\mathbf{b} - (\mathbf{a}.\mathbf{b})\mathbf{c}, \quad \mathbf{v} = (\mathbf{a}.\mathbf{c})\mathbf{b} - (\mathbf{b}.\mathbf{c})\mathbf{a}.$$

   Given that $\mathbf{u}$ and $\mathbf{v}$ are at right angles, show that either $\cos\beta = \cos\alpha\cos\gamma$, or $\mathbf{a}$ is perpendicular to $\mathbf{c}$. (L)

32 The lines m and n are given by

$$\mathbf{r} = \begin{pmatrix} 3 \\ 4 \\ 1 \end{pmatrix} + s\begin{pmatrix} 2 \\ -1 \\ 1 \end{pmatrix} \quad \text{and} \quad \mathbf{r} = \begin{pmatrix} 1 \\ 5 \\ 7 \end{pmatrix} + t\begin{pmatrix} 1 \\ 0 \\ 1 \end{pmatrix},$$

respectively. Show that the lines m and n are skew and find the acute angle between them.

33 The position vectors of the points A and B are given by

$$\mathbf{a} = 5\mathbf{i} + \mathbf{j} + 2\mathbf{k} \quad \text{and} \quad \mathbf{b} = -\mathbf{i} + 7\mathbf{j} + 8\mathbf{k}$$

respectively. Show that the two lines

$$\mathbf{r} = \mathbf{a} + \lambda\mathbf{l} \quad \text{and} \quad \mathbf{r} = \mathbf{b} + \mu\mathbf{m},$$

where $\mathbf{l} = -4\mathbf{i} + \mathbf{j} - \mathbf{k}$ and $\mathbf{m} = 2\mathbf{i} - 5\mathbf{j} - 7\mathbf{k}$, intersect, and find the position vector of the point of intersection C.

If D has position vector $3\mathbf{i} + 7\mathbf{j} - 2\mathbf{k}$ show that the line CD is perpendicular to the line AB. *(JMB)*

34 Find an equation, in both Cartesian and vector form, of the plane containing the point $(1, 1, 1)$ and the line defined by the equations

$$x - 2y + 3z = 1, \quad 2x + y + z = -3.$$

State direction cosines of a vector which is normal to this plane, and write down a unit normal vector to this plane.

Find the lengths of the perpendiculars from (*a*) the origin, (*b*) the point $(2, 2, 2)$ to the plane. *(L)*

35 The equations of two lines, l_1 and l_2, are

$$\frac{x-3}{2} = \frac{y}{-2} = \frac{z}{3} \quad \text{and} \quad x = \frac{y-6}{-4} = \frac{z}{-3}$$

respectively. Show that l_1 and l_2 intersect at a point, P, and find the coordinates of P. Find also the equation of the plane, Π, in which both l_1 and l_2 lie.

Show that the line

$$x + 1 = \frac{y+4}{-2}, z = 1,$$

is parallel to the plane Π, and find its distance from Π. *(JMB)*

36 A pyramid has a square base $OABC$ and vertex V. The position vectors of A, B, C, V referred to O as origin are given by $\overrightarrow{OA} = 2\mathbf{i}$, $\overrightarrow{OB} = 2\mathbf{i} + 2\mathbf{j}$, $\overrightarrow{OC} = 2\mathbf{j}$, $\overrightarrow{OV} = \mathbf{i} + \mathbf{j} + 3\mathbf{k}$.

(i) Express $\overrightarrow{AV}$ in terms of $\mathbf{i}$, $\mathbf{j}$ and $\mathbf{k}$.

(ii) Using scalar products, or otherwise, find a vector $\mathbf{x}$ which is perpendicular to both $\overrightarrow{OV}$ and $\overrightarrow{AV}$.

(iii) Calculate the angle between the vector $\mathbf{x}$, found in (ii), and $\overrightarrow{VB}$, giving your answer to the nearest degree.

(iv) Write down the acute angle between VB and the plane OVA. *(C)*

37 In three dimensions, O is the origin of position vectors and points A and B are given by $\overrightarrow{OA} = \mathbf{i} + \mathbf{k}$, $\overrightarrow{OB} = \mathbf{j} + \mathbf{k}$ respectively. The position vector of a general point P in the plane OAB is given by $\overrightarrow{OP} = \lambda\overrightarrow{OA} + \mu\overrightarrow{OB}$, where λ and μ are variable scalars. The position vector of the fixed point Q (which is not in the plane OAB) is given by $\overrightarrow{OQ} = \mathbf{i} + \mathbf{j} + \mathbf{k}$. Determine λ and μ such that QP is perpendicular to both OA and OB (that is, such that QP is perpendicular to the plane OAB). Hence, or otherwise, find the acute angle between OQ and the plane OAB to the nearest half degree.   (C)

38   The points O, P, Q, R have coordinates $(0, 0, 0)$, $(2, 6, -2)$, $(2, 4, -1)$, $(4, 6, 0)$ respectively. Find a vector perpendicular to both PQ and QR and hence, or otherwise, determine an equation for the plane PQR.

The line $$\frac{x-2}{1} = \frac{y}{1} = \frac{z-1}{3}$$

meets the plane PQR at the point T, and the point S on this line is such that T is the mid-point of OS. Find
(a) the cosine of the acute angle between the line and the plane PQR,
(b) the perpendicular distance from S to the plane PQR.   (L)

39   The coordinates of a point A are $(3, -1, 0)$, and the equation of a plane Π is

$$x - y + 2z = -2.$$

(i) Find, in parametric form, the equation of the line through A perpendicular to Π.
(ii) Find the coordinates of the foot of the perpendicular from A to Π.
(iii) Find the coordinates of the reflection A' of A in Π.

The line through A parallel to the vector $\begin{pmatrix} 1 \\ -2 \\ 3 \end{pmatrix}$ is denoted by l.

(iv) Find the coordinates of the point where l meets Π.
(v) Using your answers to (iii) and (iv), show that the reflection l' of l in Π has

direction $\begin{pmatrix} 2 \\ -1 \\ 3 \end{pmatrix}$, and find the equation of l' in parametric form.   (SMP)

40   The points A, B, C, D have coordinates $(1, 0, -1)$, $(2, -1, 2)$, $(0, 1, 0)$,

$(-1, 0, 1)$ respectively. Find the values of λ and μ if the vector $\begin{pmatrix} \lambda \\ \mu \\ 1 \end{pmatrix}$ is

perpendicular to both AB and CD. Hence find in the form $\mathbf{r.n} = k$ the equation of the plane through CD parallel to AB.

Write down the vector parametric equation of the line through A normal to this plane, and hence find the perpendicular distance from A to the plane.

Explain why your answer gives the length of the common perpendicular to the lines AB and CD.   (SMP)

41   Two points A and B have position vectors $\mathbf{a}$ and $\mathbf{b}$ respectively. Show that the point which divides AB internally in the ratio $m:n$ has position vector

$$\frac{n\mathbf{a} + m\mathbf{b}}{n + m}.$$

Three non-collinear points A, B and C have position vectors $\mathbf{a}$, $\mathbf{b}$ and $\mathbf{c}$ respectively. The point D divides AB internally in the ratio $2:1$. The point E divides BC internally in the ratio $2:1$. Show that DE produced meets AC

produced at the point with position vector

$$\frac{4}{3}\mathbf{c} - \frac{1}{3}\mathbf{a}.$$

(L)

42 The plane Π_1 is normal to the vector $\begin{pmatrix} 1 \\ -2 \\ 1 \end{pmatrix}$ and passes through the point

$A(3, 1, -1)$. The plane Π_2 has equation $x - y = 1$. Find the acute angle between the planes Π_1 and Π_2.

If Π_1 and Π_2 meet in the line l, verify that l passes through $(1, 0, -1)$ and is

parallel to the vector $\begin{pmatrix} 1 \\ 1 \\ 1 \end{pmatrix}$. Write down a parametric equation for l, and

hence find the point C on l such that AC is perpendicular to l.

If the normal through A to Π_2 meets Π_2 in B, show that the area of the triangle ACB is $\frac{1}{4}\sqrt{3}$ square units. (*SMP*)

43 If $\mathbf{a}$ is a fixed non-zero vector in 3-dimensional space, describe geometrically the set S of points whose position vectors $\mathbf{r}$ are given by $\mathbf{r} = t\mathbf{a}$, where t is a variable real number.

The position vectors $\mathbf{r}$ of the points of a plane Z satisfy the equation $\mathbf{r} \cdot \mathbf{n} = k$, where $\mathbf{n}$ is a fixed non-zero vector and $k \neq 0$. State (i) how the direction of $\mathbf{n}$ is related to Z, (ii) the geometrical significance of the condition $k \neq 0$.

Find the value of t which gives a point of S which lies in Z. How are S and Z related if $\mathbf{a} \cdot \mathbf{n} = 0$? (*SMP*)

44 A, B, C, D are points with position vectors $\mathbf{j} + 2\mathbf{k}$, $-\mathbf{i} - \mathbf{j}$, $4\mathbf{i} + \mathbf{k}$ and $3\mathbf{i} + \mathbf{j} + 2\mathbf{k}$ respectively. Prove that triangle ABC is right-angled and that triangle ABD is isosceles.

Prove that the vector $\mathbf{j} - \mathbf{k}$ is perpendicular to the plane of triangle ABC and that A, B, C and D are coplanar.

Show that BD passes through the mid-point, E, of AC and find the ratio $BE:ED$. (*C*)

45 The points A, B, C, D have position vectors $\mathbf{p} + \mathbf{q}$, $2\mathbf{p} + 3\mathbf{q}$, $4\mathbf{p} + 7\mathbf{q}$, $5\mathbf{p} - 3\mathbf{q}$ respectively. Given that $|\mathbf{p}| = 2|\mathbf{q}|$, show that
(a) A, B, C are collinear, (b) AC is perpendicular to BD.
Given also that $\mathbf{p} = 8\mathbf{i} - 6\mathbf{j}$ and $\mathbf{q} = 3\mathbf{i} + 4\mathbf{j}$, find
(c) the cosine of the angle BAD, (d) the area of the triangle BAD. (*L*)

46 The lines L_1 and L_2 are given by the equations

$$L_1 : \mathbf{r} = \begin{pmatrix} 1 \\ 6 \\ 3 \end{pmatrix} + t\begin{pmatrix} 2 \\ -1 \\ 1 \end{pmatrix}, \qquad L_2 : \mathbf{r} = \begin{pmatrix} 3 \\ 3 \\ 8 \end{pmatrix} + s\begin{pmatrix} 1 \\ 0 \\ 1 \end{pmatrix}.$$

(i) Calculate the angle between the directions of L_1, L_2.
(ii) Show that the lines do not intersect.

(iii) Verify that the vector $\mathbf{a} = \begin{pmatrix} 1 \\ 1 \\ -1 \end{pmatrix}$ is perpendicular to each of the lines.

The point P on L_1 is given by $t = p$; the point Q on L_2 is given by $s = q$. Write down the column vector representing $\overrightarrow{PQ}$. Hence calculate p and q so that the vectors $\overrightarrow{PQ}$ and $\mathbf{a}$ are parallel. (*JMB*)

47 Prove that the two lines with equations

$$\mathbf{r} = \begin{pmatrix} 0 \\ 2 \\ -3 \end{pmatrix} + s \begin{pmatrix} 1 \\ -1 \\ -1 \end{pmatrix} \quad \text{and} \quad \mathbf{r} = \begin{pmatrix} -1 \\ +6 \\ -1 \end{pmatrix} + t \begin{pmatrix} 2 \\ 1 \\ -1 \end{pmatrix}$$

have a point in common.

Find, in parametric form, an equation for the plane containing the two lines.   (*SMP*)

13 Inequalities

13.1 Ordering the real numbers

Consider the real line as the x-axis, with the number zero represented by the origin O, with any positive number x represented by a point to the right of O distant x from O, and with any negative number x represented by a point to the left of O distant $(-x)$ from O (see Fig. 13.1). For any real number x, just one of these three statements is true:

        (i) x is positive, that is, x is greater than 0, written '$x > 0$';

or   (ii) x is negative, that is, x is less than 0,   written '$x < 0$';

or   (iii) x is equal to 0,   written '$x = 0$'.

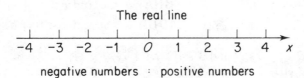

The real line

negative numbers · positive numbers

Fig. 13.1

Notation

$<$ means 'is less than'	$\leqslant$ means 'is less than or equal to'
$\not<$ means 'is not less than'	$\not\leqslant$ means 'is not less than or equal to'
$>$ means 'is greater than'	$\geqslant$ means 'is greater than or equal to'
$\not>$ means 'is not greater than'	$\not\geqslant$ means 'is not greater than or equal to'

The notations $<$ and $>$ are called strict inequalities, since equality is excluded. For example: $2 < 5$, $-3 \leqslant 2$, $-6 \not\geqslant -3$, $4 \leqslant 4$, $3 > -5$, $3 \not< 3$.

We now define an order relation on the real numbers in terms of positive numbers.

Definition   Given two real numbers x and y:

        x is less than $y \Leftrightarrow (y - x)$ is positive $\Leftrightarrow y$ is greater than x

that is, $x < y \Leftrightarrow (y - x) > 0 \Leftrightarrow y > x$.

It then follows that if x is positive, $0 < x = 0 - (-x)$ so $-x$ is negative. Similarly, if x is negative then $-x$ is positive.

Thus $x < y$ means that x is to the left of y on the real axis,

and $x > y$ means that x is to the right of y on the real axis.

Similar results to the above will hold for $<$ and $>$ replaced by $\leqslant$ and $\geqslant$, with the addition that for any two real numbers x, y either $x \leqslant y$ or $y \leqslant x$, and if $x \leqslant y$ and $y \leqslant x$ then $x = y$.

Transitivity

The inequality relation is transitive. Namely, for all x, y, z

$$x < y \text{ and } y < z \Rightarrow x < z.$$

Proof If $x < y$ and $y < z$ then $y - x > 0$ and $z - y > 0$, so $z - x = (z - y) + (y - x) > 0$ and hence $x < z$.

Some of the processes which are used in manipulation of equations can also be used for inequalities, but not all. We list these results, and then give examples and proofs. Let a, b, c, d be real numbers.

Equations	Inequalities
(i) $a = b \Rightarrow a + c = b + c$	(I) $a < b \Rightarrow a + c < b + c$
(ii) $a = b \Rightarrow a - c = b - c$	(II) $a < b \Rightarrow a - c < b - c$
(iii) $a = b \Rightarrow ac = bc$	(IIIA) $a < b$ and $c > 0 \Rightarrow ac < bc$
	(IIIB) $a < b$ and $c < 0 \Rightarrow bc < ac$

EXAMPLES: (I) $3 < 7$ and $3 + 4 < 7 + 4$, i.e. $7 < 11$;
(II) $3 < 7$ and $3 - 9 < 7 - 9$, i.e. $-6 < -2$;
(IIIA) $3 < 7$ and $3 \times 4 < 7 \times 4$, i.e. $12 < 28$;
(IIIB) $3 < 7$ and $3 \times (-2) > 7 \times (-2)$, i.e. $-6 > -14$.

Proofs:
(I) $a < b \Rightarrow 0 < b - a \Rightarrow 0 < (b + c) - (a + c) \Rightarrow a + c < b + c$,
(II) $a < b \Rightarrow 0 < b - a \Rightarrow 0 < (b - c) - (a - c) \Rightarrow a - c < b - c$,
(IIIA) $a < b$ and $c > 0 \Rightarrow 0 < b - a$ and $0 < c \Rightarrow 0 < (b - a)c = bc - ac$
$$\Rightarrow ac < bc,$$
(IIIB) $a < b$ and $c < 0 \Rightarrow 0 < b - a$ and $0 < -c \Rightarrow 0 < (b - a)(-c)$
$$\Rightarrow 0 < ac - bc \Rightarrow bc < ac.$$

If $d \neq 0$, then in (III) c can be replaced by $1/d$, which is of the same sign as d. Then we see that an inequality can also be divided by a positive number without reversing the inequality, while division by a negative number reverses the inequality, just as occurs with multiplication.

Intervals

If x satisfies both inequalities $2 < x$ and $x < 5$, then these may be written as one expression $2 < x < 5$. The set S_1 of all x which satisfy both these inequalities is a finite interval on the real line. This is shown by a thick line in Fig. 13.2(a). In set notation,

$$S_1 = \{x : 2 < x\} \cap \{x : x < 5\} = \{x : 2 < x < 5\}.$$

The set of all x satisfying the inequality $5 < x$ is S_2, where $S_2 = \{x : 5 < x\}$. This is an infinite interval on the real line, shown in Fig. 13.2(b).

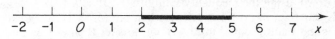

(a) The real interval: $S_1 = \{x : 2 < x < 5\}$

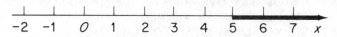

(b) The infinite interval: $S_2 = \{x : 5 < x\}$

(c) The two intervals: $S_3 = \{x : 5 < x\} \cup \{x : x < 2\}$

Fig. 13.2

If the set S_3 consists of all x satisfying either $5 < x$ or $x < 2$, then S_3 is the union of two infinite intervals, Fig. 13.2(c), so that

$$S_3 = \{x : 5 < x \text{ or } x < 2\} = \{x : 5 < x\} \cup \{x : x < 2\}.$$

Note that, in this case it is wrong to write $5 < x < 2$, since the two inequalities are not both true at once. If they were, then the transitivity of the order relation would imply that $5 < 2$, which is nonsense.

Note In these examples, the strict inequalities have been used, so that the end points 2 and 5 are excluded from the intervals. If any of the inequalities are replaced by non-strict inequalities, then the corresponding end points of the intervals would be included.

EXAMPLE 1 *Solve the inequality* $3x - 4 < 5x + 6$.

Add $(-3x - 6)$ to both sides of the inequality, to give $-10 < 2x$. Multiply by $\frac{1}{2}$, which is positive and so preserves the inequality, $-5 < x$, and the solution set is $\{x : x > -5\}$.

EXAMPLE 2 *Solve* $2 \leqslant 5 - 3x \leqslant 8$.

Subtract 5 from each side of the inequalities, $-3 \leqslant -3x \leqslant 3$. Divide by (-3), which is negative and so reverses the inequalities, $1 \geqslant x \geqslant -1$, giving the solution set $\{x: -1 \leqslant x \leqslant 1\}$.

EXERCISE 13.1

1 Write down, using the inequality symbols:
  (i) a is less than b, (ii) c is greater than b, (iii) x is not greater than y,
  (iv) y is less than or equal to z, (v) x is greater than -1 and less than 5.
2 Write down, using the inequality symbols, for points on the real line:
  (i) x is to the left of 2, (ii) y is to the right of 4, (iii) z lies between -4 and 2,
  (iv) w does not lie between 5 and 9, inclusive.
  In each case, give one value of the variable which satisfies the conditions.
3 Describe in words the subsets of the real numbers:
  (i) $\{x: x < 4\}$, (ii) $\{x: x - 3 < 4\}$, (iii) $\{x: 3x + 5 \leqslant -4\}$,
  (iv) $\{x: -2 \leqslant x < 3\}$, (v) $\{x: x \leqslant -2\} \cup \{x: 2 < x\}$.
4 Solve the inequality:
  (i) $2x - 3 < 5x + 3$, (ii) $3x + 4 \leqslant 6x - 5$, (iii) $6x - 5 < 2x + 1$,
  (iv) $2x + \dfrac{x - 2}{3} < 4x - \frac{3}{4}$, (v) $(2x + 1)^2 \leqslant 9$.
5 Solve the following so that x lies alone between the inequality signs:
  (i) $2 < x - 3 < 4$, (ii) $-2 \leqslant 3x + 4 \leqslant 1$, (iii) $4 < 2x - 5 \leqslant 7$,
  (iv) $3 < 5 - 2x < 7$.
6 Given that $x > 0$ prove that $x + \dfrac{1}{x} \geqslant 2$. (Hint: $\left(a - \dfrac{1}{a}\right)^2$ is positive.)
7 If a and b are both positive, prove that $a < b \Leftrightarrow a^2 < b^2$. (Remember that the implication has to be proved in each direction.)

13.2 Use of the modulus sign

The *absolute value*, or the *modulus*, of a real number a is $|a|$, given by
$$|a| = \sqrt{(a^2)}.$$
Thus $|a| = a$ if $a \geqslant 0$ and $|a| = -a$ if $a < 0$.

For example: $|7| = 7$, $|-2| = 2$, $|\sqrt{2}| = \sqrt{2}$, $|-\pi| = \pi$.
  For all a, $0 \leqslant |a|$, and $0 = |a| \Leftrightarrow a = 0$.
  On the real line, $|x|$ is the distance of the number x from O, and similarly $|a - b| = |b - a|$ is the distance from the point a to the point b (see Fig. 13.3). This means that a single inequality involving a modulus is

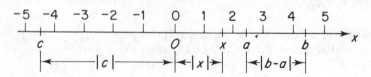

Fig. 13.3 Moduli of real numbers.

equivalent to two inequalities. Thus $|x| < 3$ means that the distance of x from O is less than 3 units, so that both $x < 3$ and $-x < 3$, that is, $x > -3$ or $-3 < x$. Hence $|x| < 3 \Leftrightarrow -3 < x < 3$.

On the other hand, $5 < |x|$ means that 5 is nearer to the origin than is x, and so either $x < -5$ or $x > 5$.

EXAMPLE 1   *Solve the inequality $|2x-3| < 7$.*

This means that $2x - 3$ is nearer to the origin than is 7 and so the inequality is equivalent to the two inequalities $-7 < 2x - 3 < 7$. Add 3 to each term to give $-4 < 2x < 10$, and on dividing by 2, which is positive, $-2 < x < 5$.

EXAMPLE 2   *The subset S of the real numbers is the interval on the real line which is centred at the point a and is of length 2b, b > 0. Describe S in terms of the modulus sign.*

S is the set of real numbers whose distance from a is less than or equal to b and so $S = \{x: a - b \leqslant x \leqslant a + b\}$. The distance of x from a is $|x - a|$ and so $a - b \leqslant x \leqslant a + b$ provided $|x - a| \leqslant b$. Therefore, $S = \{x: |x-a| \leqslant b\}$.

EXERCISE 13.2

1   Rewrite, without using the modulus sign:
(i) $|x| < 3$,   (ii) $|x - 3| < 4$,   (iii) $|3x - 1| \leqslant 7$,   (iv) $|x| \geqslant 5$,
(v) $|2x + 4| > 6$.
2   Describe the set in terms of a single inequality involving the modulus sign:
(i) $\{x: -2 < x < 2\}$,   (ii) $\{x: x \geqslant 7 \text{ or } x \leqslant -7\}$,   (iii) $\{x: -1 < x < 3\}$,
(iv) $\{x: x \geqslant 2 \text{ or } x \leqslant -4\}$,
(v) $\{x: x \text{ is a real number less than 5 units distant from the origin}\}$,
(vi) the set of all real numbers more than 2 units away from 6.
3   Prove that, for any real number a, $-|a| \leqslant a \leqslant |a|$ and $-|a| \leqslant -a \leqslant |a|$.
4   Prove that, for real numbers a and b,
(i) $|a||b| = |ab|$,   (ii) $|a| - |b| \leqslant |a + b| \leqslant |a| + |b|$.
5   On the same diagram, draw the graph of
(i) $y = x$,   (ii) $y = |x|$,   (iii) $y = x + |x|$,   (iv) $y = x - |x|$,   (v) $y = 3|x|$,
(vi) $y = |x - 2|$,   (vii) $y = |3(x - 2)|$,   (viii) $y = |3x + 4|$.

13.3 Solution of inequalities involving linear quotients

So far, we have dealt with linear inequalities and the solution has been obtained by means of addition and multiplication by a constant. When the inequality involves the quotient of one linear term by another we can use both graphical and algebraic methods of solution.

EXAMPLE 1   *Solve the inequality $(3x + 4)/x > 3$.*

The left-hand side may be simplified by separation into two terms, thus

$3 + 4/x > 3$, which is the same as $4/x > 0$, so the solution is $x > 0$, since the quotient is positive when numerator and denominator are the same sign.

EXAMPLE 2   *Solve the inequality* $\dfrac{(2x-3)}{x} \leqslant 5$.

Method 1   Simplify the left-hand side, as in Example 1. Then $2 - \dfrac{3}{x} \leqslant 5$. Adding $(3/x - 5)$ to both sides:

$$2 - 5 \leqslant 5 + \left(\frac{3}{x} - 5\right) \quad \text{or} \quad -3 \leqslant \frac{3}{x}.$$

Dividing by 3, $-1 \leqslant 1/x$.

Consider the graph $y = 1/x$ in Fig. 13.4(a) and observe that $1/x \geqslant -1$ for two parts of the graph, namely for $x > 0$ and for $x \leqslant -1$. The solution is, therefore, $\{x: x \leqslant -1\} \cup \{x: x > 0\}$.

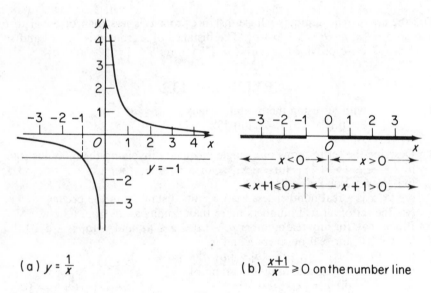

(a) $y = \dfrac{1}{x}$ (b) $\dfrac{x+1}{x} \geqslant 0$ on the number line

Fig. 13.4

Method 2   Subtract $(2x - 3)/x$ from both sides of the inequality to bring all the terms to one side,

$$0 \leqslant 5 - \frac{2x-3}{x} = \frac{5x - (2x-3)}{x} = \frac{3x+3}{x} = 3\frac{x+1}{x}.$$

Now $(x+1)/x$ is positive when both numerator and denominator are the same sign. This occurs when $(x+1)$ and x are both positive or when they are both negative. By considering the signs of $(x+1)$ and of x on the number line in Fig. 13.4(b), we find the solution set is $\{x: x \leqslant -1\} \cup \{x: x > 0\}$, indicated on the number line by heavy shading.

EXAMPLE 3 *Solve the inequality* $\dfrac{3x+1}{x+1} < 1$.

Method 1 $\dfrac{3x+1}{x+1} = \dfrac{3(x+1)-2}{x+1} = 3 - \dfrac{2}{x+1}$, so the inequality becomes

$3 - \dfrac{2}{x+1} < 1$ or $3 - 1 < \dfrac{2}{x+1}$, i.e. $1 < \dfrac{1}{x+1}$. Now the graph of $y = 1/(x+1)$ is
the graph of $y = 1/x$ translated a distance -1 in the direction Ox, and this graph is
shown in Fig. 13.5(a). From the graph, we see that $1/(x+1) > 1$ for $-1 < x < 0$,
so the solution set is $\{x: -1 < x < 0\}$.

Method 2 We bring all the terms to one side,

$$0 < 1 - \frac{3x+1}{x+1} = \frac{-2x}{x+1}, \text{ i.e. } \frac{x}{x+1} < 0.$$

This requires x and $(x+1)$ to be of opposite sign and, by reference to the number
line in Fig. 13.5(b), we find the solution $\{x: -1 < x < 0\}$.

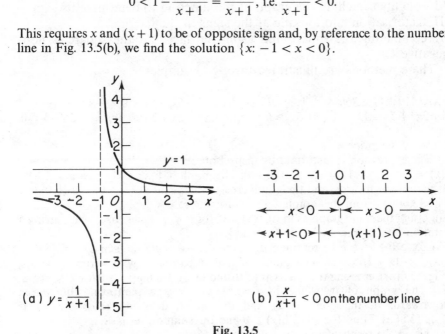

(a) $y = \dfrac{1}{x+1}$

(b) $\dfrac{x}{x+1} < 0$ on the number line

Fig. 13.5

EXERCISE 13.3

1 Solve the inequalities:

(i) $\dfrac{1}{x} < 5$, (ii) $2 \leqslant \dfrac{x}{3}$, (iii) $\dfrac{-2}{x} > 4$, (iv) $2 + \dfrac{5}{x} \leqslant 7$, (v) $\dfrac{2x-3}{x} > 5$,

(vi) $\dfrac{2-x}{x} \geqslant 3$, (vii) $\dfrac{2x+1}{x} < 4$.

2 For what values of x are the inequalities satisfied:

(i) $\dfrac{5x}{x+3} < 2$, (ii) $7 \leqslant \dfrac{2}{x-1}$, (iii) $\dfrac{1-x}{x+2} < -2$, (iv) $\dfrac{3x+4}{2x-1} \geqslant 7$.

3 Obtain the set of values of x for which $\dfrac{1}{x+3} < \dfrac{1}{x-3}$.

4 Find the set of values of x for which $\dfrac{1}{x-2} < \dfrac{1}{x+2}$.   (L)

5 Find the set of values of x for which $\dfrac{4x}{(x+2)} > 1$.   (L)

13.4 Quadratic inequalities

Inequalities which involve quadratic functions arise because the square of any real number is positive or zero, giving an inequality of the form $a^2 \geq 0$. There are three methods of dealing with a quadratic inequality:

I roughly sketching the graph of the function to determine its sign;
II factorisation and looking at the signs of the factors;
III completing the square and using the fact that the square is non-negative.

These methods are illustrated through examples.

EXAMPLE 1   *Solve the inequalities:* (i) $2x^2 + 5x - 12 < 0$,
(ii) $2x^2 + 5x - 12 > 0$, (iii) $5x \geq 6 + x^2$, (iv) $x^2 - 2x + 3 \geq 0$, (v) $x^2 - 2x + 3 \leq 0$.

I Use of the graph
(i) The expression is factorised by inspection,
$f(x) = 2x^2 + 5x - 12 = (2x - 3)(x + 4)$. The graph of f is clearly a parabola with its vertex downwards, and the graph crosses the x-axis at $(-4, 0)$ and $(\frac{3}{2}, 0)$ (see Fig. 13.6(a)). From this graph, $f(x) < 0$ for $-4 < x < \frac{3}{2}$.
(ii) Using the same graph as in (i), $f(x) > 0$, for $x < -4$ and for $x > 1\frac{1}{2}$, giving a solution set $\{x: x < -4\} \cup \{x: x > 1\frac{1}{2}\}$.
(iii) $5x \geq 6 + x^2$ can be rearranged, $-x^2 + 5x - 6 = g(x) \geq 0$, and,
$g(x) = -(x-2)(x-3)$. The graph of g is a parabola, with vertex upwards, (Fig. 13.6(b)), crossing the x-axis at $(2, 0)$ and $(3, 0)$. Then $g(x) \geq 0$ for $2 \leq x \leq 3$.
(iv) The graph of h, where $h(x) = x^2 - 2x + 3$ is a parabola, with vertex downwards at the point $(1, 2)$, and this graph does not cross the x-axis (Fig. 13.6(c)). Thus, for all x, $h(x) \geq 0$ and the solution set is $\mathbb{R}$.
(v) Using the result of (iv), the solution set for $h(x) \leq 0$ is ϕ.

II Using factors
(i) and (ii) The signs of the factors are studied for various subsets of the domain and these are used to obtain the signs of $f(x)$. In §13.3, this was done by marking intervals on the real line. Here, we shall use a table.

	$2x - 3$	$x + 4$	$f(x) = (2x - 3)(x + 4)$
$x < -4$	$-$ve	$-$ve	$+$ve
$-4 < x < 1\frac{1}{2}$	$-$ve	$+$ve	$-$ve
$1\frac{1}{2} < x$	$+$ve	$+$ve	$+$ve

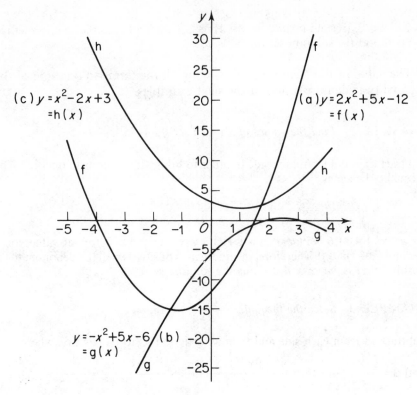

Fig. 13.6

Then the solution for (i), $f(x) < 0$, is the set $\{x: -4 < x < 1\frac{1}{2}\}$;
the solution set for (ii), $f(x) > 0$, is the set $\{x: x < -4\} \cup \{x: x > 1\frac{1}{2}\}$.
(iii) Similarly, $g(x) = -x^2 + 5x - 6 = -(x-2)(x-3)$ and we tabulate the signs
of the factors of $g(x)$.

	-1	$x-2$	$x-3$	$g(x) = -(x-2)(x-3)$
$x < 2$	$-$ve	$-$ve	$-$ve	$-$ve
$2 < x < 3$	$-$ve	$+$ve	$-$ve	$+$ve
$3 < x$	$-$ve	$+$ve	$+$ve	$-$ve

The solution set for $g(x) \geqslant 0$ is then $\{x: 2 \leqslant x \leqslant 3\}$.
(iv) and (v) Since $h(x)$ does not factorise, the method cannot be used.

III Completing the square
(i) and (ii) $f(x) = 2(x+5/4)^2 - 121/8$, so $f(x) < 0$ when $(x+5/4)^2 < 121/16$,
which occurs for $-11/4 < x + 5/4 < 11/4$, which gives the solution for (i).
  Similarly, $f(x) > 0$ for $(x+5/4)^2 > 121/16$, that is, for either $x + 5/4 < -11/4$
or $11/4 < x + 5/4$, giving the solution for (ii), $\{x: x < -4\} \cup \{x: x > 1\frac{1}{2}\}$.

(iii) $g(x) = -(x-5/2)^2 + \frac{1}{4} \geqslant 0$ for $(x-5/2)^2 \leqslant \frac{1}{4}$, i.e. $-\frac{1}{2} \leqslant x - 5/2 \leqslant \frac{1}{2}$ or $2 \leqslant x \leqslant 3$. (iv) and (v) $h(x) = (x-1)^2 + 2 \geqslant 2$ for all x, so the solution set for (iv) is $\mathbb{R}$ and the solution set for (v) is ϕ.

The value of the technique of bringing all the terms to one side of the inequality is demonstrated in the next examples.

EXAMPLE 2   *Find the values of x which satisfy* $\dfrac{(x+2)(x+1)}{x} > \dfrac{x+5}{3}$.

Subtract $(x+5)/3$ from each side in order to bring all the terms to one side. The inequality becomes

$$\frac{(x+2)(x+1)}{x} - \frac{x+5}{3} = \frac{3(x^2+3x+2) - x(x+5)}{3x} = \frac{2x^2+4x+6}{3x} > 0.$$

Now $2x^2 + 4x + 6 = 2(x^2 + 2x + 1 + 2) = 2(x+1)^2 + 4 > 0$ for all values of x because $(x+1)^2 \geqslant 0$. Therefore, the quotient is positive when the denominator is positive, that is, when $x > 0$. Thus the solution set is $\mathbb{R}^+$.

EXAMPLE 3   *Solve the inequality* $\dfrac{x(x+9)}{x-1} > 18$.

Subtract 18 from each side and then we require f(x) to be positive, where

$$\text{f}(x) = \frac{x(x+9)}{x-1} - 18 = \frac{x^2+9x-18x+18}{x-1} = \frac{x^2-9x+18}{x-1} = \frac{(x-3)(x-6)}{x-1}.$$

Of the three terms in this expression, either three or one must be positive in order for f(x) to be positive. We look at the signs of the terms for various values of x;

	$x < 1$	$1 < x < 3$	$3 < x < 6$	$x > 6$
$x-1$	$-$ve	$+$ve	$+$ve	$+$ve
$x-3$	$-$ve	$-$ve	$+$ve	$+$ve
$x-6$	$-$ve	$-$ve	$-$ve	$+$ve
f(x)	$-$ve	$+$ve	$-$ve	$+$ve

and pick out the solution set $\{x: 1 < x < 3\} \cup \{x: x > 6\}$.

EXERCISE 13.4

1 Solve the inequality:

(i) $(x-1)(x+2)(x-3) > 0$,   (ii) $\dfrac{(x+2)(x-3)}{x-1} < 0$,   (iii) $x^2 + x \leqslant 12$,

(iv) $x^2 + x + 1 > 0$.

2 Find the set of real values of x for which $\dfrac{x(x-5)}{(x-3)} > 6$.   (L)

3 Given that $px^2 + 2px - 3 < 0$ for all real values of x, determine the set of possible values of p.   (L)

4 Find the set of values of x for which $2x^2 + 3x + 2 < 4$.   (*L*)

5 Find the set of values of x for which $\dfrac{(2x-1)(x+4)}{(x-5)} > 0$.   (*L*)

6 Solve the inequality $(x+1)(x-2)(x+3)(x-4) > 0$.   (*L*)

13.5 Inequalities involving the modulus sign

We illustrate an algebraic method and a graphical method of solving an inequality which includes the modulus symbol.

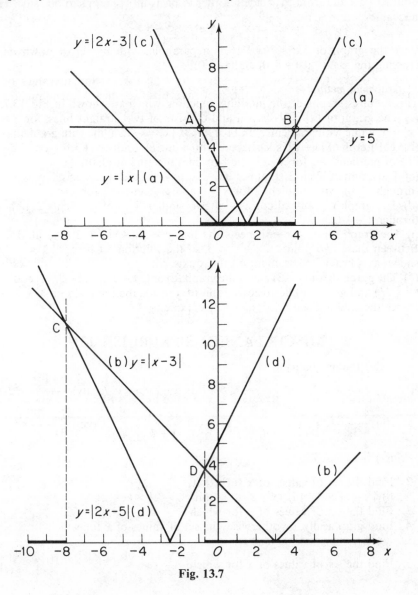

Fig. 13.7

EXAMPLE　*Solve the inequality:* (i) $|2x-3| < 5$,　(ii) $|x-3| < |2x+5|$.

I Algebraic method

(i) $|2x-3| < 5 \Leftrightarrow (2x-3)^2 < 25 \Leftrightarrow 4x^2 - 12x + 9 - 25 < 0 \Leftrightarrow x^2 - 3x - 4 < 0$

$$\Leftrightarrow (x+1)(x-4) < 0 \Leftrightarrow -1 < x < 4,$$

since the quadratic function $(x+1)(x-4)$ has a graph which is a parabola with its vertex downwards, crossing the x-axis at $(-1, 0)$ and $(4, 0)$.

(ii) $|x-3| < |2x+5| \Leftrightarrow (x-3)^2 < (2x+5)^2 \Leftrightarrow x^2 - 6x + 9 < 4x^2 + 20x + 25$

$$\Leftrightarrow 3x^2 + 26x + 16 > 0 \Leftrightarrow (3x+2)(x+8) > 0$$

$$\Leftrightarrow \text{either } x > -\tfrac{2}{3} \text{ or } x < -8$$

since the graph of $3x^2 + 26x + 16$ is a parabola, with its vertex downwards, crossing the x-axis at $(-8, 0)$ and $(-\tfrac{2}{3}, 0)$.

II Graphical method

This requires graphs of the modulus function, which are shown in Fig. 13.7.
(a) The graph of $|x|$ is V-shaped and consists of two straight lines; for $x > 0$, $|x| = x$, a line with gradient 1, and for $x < 0$, $|x| = -x$, a line with gradient -1.
(b) The graph of $|x-3|$ is V-shaped, with a line of gradient $+1$ for $x > 3$ and a line of gradient -1 for $x < 3$, the two lines meeting at $(3, 0)$.
(c) The graph of $|2x-3|$, that is, $|2(x-\tfrac{3}{2})|$, is V-shaped and consists of two lines through $(\tfrac{3}{2}, 0)$, with gradient 2 for $x > \tfrac{3}{2}$ and gradient -2 for $x < \tfrac{3}{2}$.
(d) The graph of $|2x+5|$ consists of two similar lines of gradients 2 and -2 through $(-\tfrac{5}{2}, 0)$.
(i) The graph (c) of $|2x-3|$ meets the line $y = 5$ at $A(-1, 5)$ and at $B(4, 5)$. Between these two points $|2x-3| < 5$, so the solution set is $\{x: -1 < x < 4\}$, shown as a heavy line segment on the x-axis.
(ii) The graph (b) of $|x-3|$ meets the graph (d) of $|2x-5|$ at $C(-8, 11)$ and at D $(-\tfrac{2}{3}, \tfrac{11}{3})$ and graph (b) lies below graph (d) outside the interval $-8 \leqslant x \leqslant -\tfrac{2}{3}$. Hence the solution set is $\{x: x < -8\} \cup \{x: x > -\tfrac{2}{3}\}$.

MISCELLANEOUS EXERCISE 13

1　Solve the inequality:

(i) $5x - 4 < 7$,　(ii) $\dfrac{5x-4}{x} \leqslant 5$,　(iii) $5 - 2x > x - 7$,　(iv) $\dfrac{2x+3}{x} \geqslant 2$,

(v) $\dfrac{4-2x}{3x} > 1$,　(vi) $\dfrac{8}{x+3} < 1$,　(vii) $\dfrac{3x-4}{x+2} < 1$,　(viii) $\dfrac{2x-3}{3x-4} \leqslant 1$,

(ix) $\dfrac{5-2x}{4-3x} < 3$.

2　Find the set of values of x such that:
(i) $|x-2| > 3$, (ii) $|x+2| \leqslant 3$, (iii) $|x-2| > |x+2|$.

3　Find the set of values of x for which $x^2 - 4x + 3 > 0$.　　　　　　　(L)

4　Find graphically, or otherwise, the set of values of x for which
$\sqrt{(16-x^2)} > |x|$.　　　　　　　(L)

5　Find the set of values of x for which $\dfrac{x}{x-2} > \dfrac{1}{x+1}$.　　　　　　　(L)

6 Find the set of values of x for which $\dfrac{(x+2)}{(x+1)(x-2)} > 0$.   (L)

7 Find the set of values of x for which $|x-2| > 5|x-3|$.   (L)

8 Find the set of values of x for which $|x-1| > 2|x-2|$.   (L)

9 Show that, if x is real, $2x^2 + 6x + 9$ is always positive. Hence, or otherwise, solve the inequality

$$\frac{(x+1)(x+3)}{x} > \frac{(x+6)}{3}. \qquad (JMB)$$

10 Find a quadratic inequality for which the solution set is:
(i) $\{x: -1 < x < 1\}$,   (ii) $\{x: a < x < b\}$,   (iii) $\{x: x < 0\} \cup \{x: 4 < x\}$.

11 The curve $y = \dfrac{2x+3}{x-1}$ has asymptotes parallel to the coordinate axes. State the equations of these asymptotes. Draw a sketch of this curve, showing the asymptotes and giving the coordinates of the points where the curve meets the axes.

    Solve the inequalities

(a) $2x + 3 \leqslant 3(x-1)$,   (b) $\dfrac{2x+3}{x-1} \leqslant 3$.   (L)

12 Given that $3 - t \leqslant 4 + t \leqslant 13 - 2t$, find the least and greatest values of t.   (L)

13 Given that $y = ax^2 + bx + c$, that $y = 8$ when $x = 1$ and that $y = 2$ when $x = -1$, show that $b = 3$ and find a in terms of c.

    Find the range of values of c for which $y > 0$ for all real values of x.

14 Prove that, for all real values of a and b: (JMB)
(i) $a^2 + 2ab + b^2 \geqslant 0$,   (ii) $a^2 - ab + b^2 \geqslant 0$.
    Deduce that $a^4 + a^3 b + ab^3 + b^4 \geqslant 0$.

15 Given that $x \in \mathbb{R}$, find the solution set of $x^2 - 1 < 2(x+1)$.   (L)

16 Find the set of values of θ, lying in the interval $-\pi \leqslant \theta \leqslant \pi$, for which

$$2\cos^2\theta + 3\sin\theta < 0. \qquad (JMB)$$

17 Solve the inequality $x^2 - |x| - 12 < 0$.   (JMB)

18 Find the sets of values of x for which
(a) $0 < \dfrac{(x-1)(x+3)}{(3x-1)} < 1$,   (b) $|x^2 + 1| < |x^2 - 4|$.   (L)

19 Find the set of values of θ, where $|\theta| < \pi/2$, for which

$$\frac{1}{1 + \cos 2\theta} < 1. \qquad (L)$$

20 Solve the inequalities $(x+3)^2 < (x-2)^2 < 9$.

21 Find the equation of the tangent to the curve $y = x(x-3)$ at the point $P(2, -2)$. Using the same axes sketch the curve and the tangent at P.

    Using sketches, or otherwise, find the values of x for which

(a) $x(x-3) \leqslant x - 4$,   (b) $x - 3 < \dfrac{x-4}{x}$,   (c) $\dfrac{1}{x-3} < \dfrac{x}{x-4}$.   (L)

14 Numerical Methods for Solving Equations

14.1 Graphical methods

Consider the problem of solving the equation $x^3 + x - 4 = 0$. As a first step, some idea of the behaviour of the function f, where $f(x) = x^3 + x - 4$, can be obtained from a sketch graph by plotting a few points and joining them by a smooth curve (Fig. 14.1).

x :	-2	-1	0	1	2	3
$f(x)$:	-14	-6	-4	-2	6	26

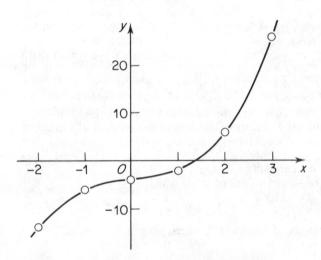

Fig. 14.1 Graph of $x^3 + x - 4 = 0$.

The sketch suggests that the graph of f crosses the x-axis between $x = 1$ and $x = 2$, in fact, somewhere near $x = 1\cdot4$. This assumes that the graph of the cubic function f is continuous, so that, because $f(1) < 0$ and $f(2) > 0$, there must be a root of the equation $f(x) = 0$ between $x = 1$ and $x = 2$. The graph must cross the x-axis somewhere between these two points. The gradient of $f(x)$, given by $f'(x)$, is always positive so $f(x)$ can only intercept the x-axis in one point.

It is important to remember that the function must be continuous for this method to give an approximate root. For example, if $g(x) = \dfrac{1}{x}$, then $g(1) = 1$ and $g(-1) = -1$, but there is no solution to $g(x) = 0$.

EXERCISE 14.1

1 On the same axes draw the graphs of $y = x^3$ and $y = -x + 4$ and find the value of x where these two graphs meet. Explain how this answer can be used to find the zero of $f(x)$ where

$$f(x) = x^3 + x - 4.$$

In the rest of this chapter, we demonstrate some methods for obtaining better approximations to the root than can be obtained from the graph. The reader should repeat the working on a calculator. Indeed, much of the chapter involves repeated calculation and, if available, a microcomputer could well be used to demonstrate the methods.

14.2 Decimal search

The first method involves sandwiching the root between decimal numbers, with an increasing number of decimal places. For the function f of §14.1, the root lies between 1 and 2. One could plot more points, draw the graph between $x = 1$ and $x = 2$ with a larger scale and so obtain a more accurate approximation to the root. The same result can be obtained numerically by using a decimal approximation method. First calculate $f(x)$ at the nine points between 1 and 2, with interval 0·1, to obtain two decimals between which the root lies. Then repeat the procedure for the nine points at an interval of 0·01 between the two points just found. The calculations are shown in tabular form:

		Stage 1		Stage 2		Stage 3	
x	$f(x)$	x	$f(x)$	x	$f(x)$	x	$f(x)$
1	-2	1·1	$-1·569$	1·31	$-0·4419$	1·371	...
2	6	1·2	$-1·072$	1·32	$-0·3800$	1·372	...
		1·3	$-0·503$	1·33	$-0·3173$	1·373	...
		1·4	$+0·144$	1·34	$-0·2539$	1·374	etc
		1·5	0·875	1·35	$-0·1896$	1·375	...
		1·6	1·696	1·36	$-0·1245$	1·376	...
		1·7	2·613	1·37	$-0·0586$	1·377	...
		1·8	3·632	1·38	$+0·0081$	1·378	...
		1·9	4·759	1·39	0·0756	1·379	...

The change in sign of $f(x)$ between $x = 1·3$ and $1·4$ indicates that these bracket the root. It takes nine calculations to reduce the interval from $2 - 1 = 1$ to $1 \cdot 4 - 1 \cdot 3 = 0 \cdot 1$. Repeating the calculations for stage 2 reduces the interval to $0 \cdot 01$ and the table shows that the root lies between $1 \cdot 37$ and $1 \cdot 38$. If we need more accuracy, we then move to stage 3 and reduce the interval to $0 \cdot 001$. If we do not wish to go to stage 3, we can note that $f(1 \cdot 37)$ is roughly 7 times $f(1 \cdot 38)$ and so it is likely that the root lies nearer to $1 \cdot 38$ than to $1 \cdot 37$. Thus, after 18 calculations, we have achieved an accuracy of two decimal places. This is not particularly impressive although it might be a satisfactory way to solve the problem by machine.

Comments on the method

(i) Quite a large number of the calculations could have been omitted. For instance, in stage 1, it would have been sensible to have started with $f(1 \cdot 5)$ and, since this turns out to be positive, there would be little point in calculating $f(1 \cdot 6)$, $f(1 \cdot 7)$, $f(1 \cdot 8)$ and $f(1 \cdot 9)$, since the root lies between 1 and $1 \cdot 5$. This idea of halving the interval, from 1 to $0 \cdot 5$, and so on, is used in §14.3. Another way of reducing the number of calculations would be to stop during each stage as soon as $f(x)$ changes sign and there are two numbers bracketing the root.

(ii) It was found above that an accuracy of two decimal places is obtained after 18 calculations. However, if the values at the change of sign of $f(x)$ had been more evenly spaced, say if $f(1 \cdot 37) = -0 \cdot 06$ and $f(1 \cdot 38) = 0 \cdot 06$, a further calculation would have been needed, namely to calculate the sign of $f(1 \cdot 375)$ in order to be sure whether the root is $1 \cdot 37$ or $1 \cdot 38$ to two decimal places of accuracy. So 19 calculations are needed for the stated level of accuracy.

EXERCISE 14.2

1 Continue the approximation process of §14.2 to find a three-decimal approximation to the root of the equation $x^3 + x - 4 = 0$.
2 Sketch a graph to find an approximate root of the equation $x^3 + 3x - 1 = 0$. Use a decimal-search method to improve your answer to an accuracy of two decimal places.
3 Show that the equation $x^2 + 2x - 7 = 0$ has a root between 1 and 2.
   (a) Find the root, accurate to three decimal places, by using a decimal search.
   (b) Use the formula for solution of a quadratic equation to find, in surd form, the exact value of this root.
     State the error in your decimal search approximation.
4 Solve the equation $x + 2/x - 5 = 0$, giving your answers to a suitable level of accuracy, using the decimal-search method. (Note that there are two solutions.)
5 (For those with access to a computer.) Write a computer program to solve an equation $f(x) = 0$ by the decimal-search method. Remember to allow for the user to enter the first two values and the number of stages to be calculated. The flow chart in Fig. 14.2 may help you.

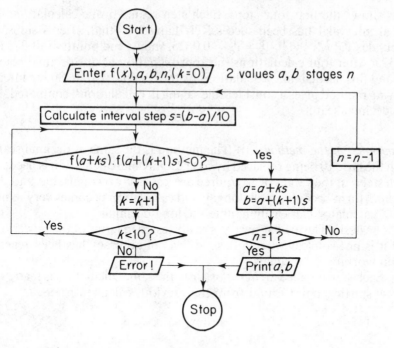

Fig. 14.2

14.3 The midpoint method (interval bisection)

An alternative to the decimal search is to use the suggested idea of halving the interval. For example, given $f(1).f(2) < 0$, calculate $f(1.5)$, the sign of which will show on which side of 1.5 the root lies. The interval halving is then repeated until the desired accuracy is obtained. The same equation, $f(x) = x^3 + x - 4 = 0$, is used to demonstrate the method.

Step 1: $f(1) = -2$ and $f(2) = 6$, so calculate $f(1.5) = 0.875$ which is greater than 0 so the root lies between 1 and 1.5. Now repeat the process for step 2. The work may be laid out as follows:

$f(1) = -2$

$f(1.25) = -0.8$

$f(1.375) = -0.03$

$f(1.4375) = 0.41 \ldots$

$f(1.5) = 0.9$

$f(2) = 6$

step 1   step 2   step 3   step 4

This shows the first four steps. Each step required one calculation. The initial interval 1 has been successively halved, so that, after 4 steps, the interval is $1 \times \frac{1}{2} \times \frac{1}{2} \times \frac{1}{2} \times \frac{1}{2} = \frac{1}{16} = 0.0625$, and its end points are 1·375 and 1·4375. After four calculations the approximation to one decimal place is 1·4 and this needed nine calculations by the decimal search. So the interval halving method gives a considerable saving in calculations, compared with the decimal search.

Comments on the method   (i) The method is the same as decimal search, with the base 10 being replaced by base 2. Only one calculation is needed at each step but the values of x required are less easily recognisable, and their decimal form becomes quite long. Hand calculation becomes very tedious but a calculator can easily handle the long decimals.
(ii) The desired level of accuracy is easily tested in terms of the powers of $\frac{1}{2}$, but it is not so clear when a desired level of accuracy has been reached, when working in decimals.
(iii) Each step is called an *iteration* since the same calculation is made with a new starting point found from the previous calculation.

EXERCISE 14.3

1 Continue the calculations of the above example for nine steps and compare your answer with the answer given by question 1 of Exercise 14.2 after nine steps.
2 Find the positive root of the equation $x^4 + 3x^2 - x - 8 = 0$ by the midpoint method, using eight steps. Give your answer to the best level of accuracy that these calculations allow.
3 Find the complete solution of the equation $x^2 - 7 = 0$ by the midpoint method, giving your answer accurate to three decimal places. Use a calculator or tables to find $\sqrt{7}$ and explain why the answer is similar to one of your answers.
4 The solution of $4x - x^3 - 6 = 0$ is -2.525, accurate to three decimal places. Find how many iterations are needed to produce this answer by interval halving, if the first two values taken are -3 and -2.
5 Find $\sqrt{(99)}$ accurate to one decimal place, using the midpoint method.
6 Use both a decimal search and the midpoint method to solve the equation $2x^3 + 4x^2 - x + 8 = 0$, accurate to two decimal places. Write down how many calculations were needed in each method.
7 Fig. 14.3 indicates another technique for improving an approximation to the root of an equation, this time, for the equation $x^3 - 3 = 0$. Given two points on the graph on opposite sides of the x-axis, $(1, -2)$ and $(2, 5)$, join them by a chord and find the point of intersection of the chord with the x-axis; in this case $x = 9/7$. Then calculate the value of $x^3 - 3$ at this value of x and repeat the process. Use the process a total of three times to find more accurate solutions. Compare the speed of improving the accuracy with the two previous methods by going through three steps of each method.

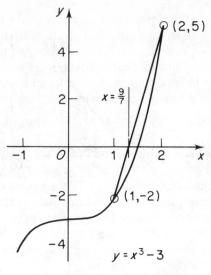

Fig. 14.3

14.4 Linear interpolation, the chord method

The technique described in question **7** of Exercise 14.3 uses the idea of approximating to the shape of a curve by a straight line joining two of its points, that is by a chord. This is a *linear approximation* to the curve. If the chord is then used to calculate an approximate root, the process is called *linear interpolation*.

Suppose that two points, A $(a, f(a))$ and B $(b, f(b))$, are taken on the graph of f, which are near to a point where the graph of f crosses the x-axis. This is shown in Fig. 14.4. In (a) the points A and B are on opposite sides of the x-axis but in (b) and (c) they are on the same side.

The chord AB crosses the x-axis at $(c, 0)$ and in case (a) it is clear that $x = c$ will be a better approximation to the root than $x = a$ and $x = b$. In (b) it is still true that c is a better approximation, but this may no longer be true in case (c). Much will depend upon the shape of the graph and, in particular, its gradient.

Let M, N, P be the points $(a, 0)$, $(b, 0)$ $(c, 0)$ respectively. Then triangles AMP and BNP are similar and $\dfrac{AM}{MP} = \dfrac{BN}{NP}$, so $AM \cdot NP = BN \cdot MP$.

Case (a), $-f(a)(b-c) = f(b)(c-a)$ or $c(f(a)-f(b)) = bf(a)-af(b)$.
Case (b), $f(a)(b-c) = f(b)(a-c)$; case (c), $-f(a)(c-b) = -f(b)(c-a)$; and again $c(f(a)-f(b)) = bf(a)-af(b)$.

So, in all three cases, provided that $f(a) \neq f(b)$, we find that

$$c = \frac{bf(a) - af(b)}{f(a) - f(b)}.$$

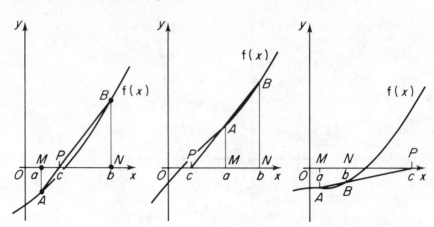

Fig. 14.4

Clearly we can expect that this will not give a good approximation when f(a) and f(b) are nearly equal. This is the case when the chord AB is nearly parallel to the x-axis.

EXAMPLE   *Use three steps of linear interpolation to obtain an approximate solution to the equation* $x^3 - 3 = 0$.

In Fig. 14.3, we see that two starting values are $x = 1$ and $x = 2$. So we use in the first step $a = 1, f(a) = -2, b = 2, f(b) = 5$. Then

$$c = \frac{2(-2) - 5}{-2 - 5} = \frac{9}{7} = 1\cdot2857.$$

Now $f(1\cdot2857) = -0\cdot8747$, so, for step **2** we use $a = 1\cdot2857$ and $b = 2$. Then

$$c = \frac{2(-0\cdot8747) - 1\cdot2857(5)}{-0\cdot8747 - 5} = 1\cdot3920.$$

Then $f(c) = -0\cdot3028$, so, for step **3** we use $a = 1\cdot2857$ and $b = 1\cdot3920$. Then

$$c = \frac{1\cdot3920(-0\cdot8747) - 1\cdot2857(-0\cdot3028)}{-0\cdot8747 + 0\cdot3028} = \mathbf{1\cdot448},$$

which may be compared with $\sqrt[3]{3}$ to three decimal places, which is $1\cdot442$.

EXERCISE 14.4

1   Use the chord method again on the above example to find the next approximation to $\sqrt[3]{3}$. How many more steps are needed to ensure an accuracy to three decimal places?

2 Use the chord method to find an improved approximation to the root of the equation in the given interval:

(i) $2x^3 - 3x + 6 = 0$, $(-2, -1)$, (ii) $3x^2 - 6x + 1 = 0$, $(0, 1)$,

(iii) $x - \dfrac{4}{x} + 2 = 0$, $(1, 2)$.

3 (i) Draw a large accurate graph of the function $x^3 + x - 4$ between $x = 1$ and $x = 2$;

(ii) apply the chord method once to find c, if $a = 1$ and $b = 2$;

(iii) apply the chord method again with $a = 1$ and $b = c$;

(iv) apply the chord method again with $a = c$ and $b = 2$;

(v) discuss which of (iii) and (iv) gives the better second approximation to the correct solution $1\cdot3788$, and suggest guidelines to decide which interval to use in the chord method.

4 The points $A(a, f(a))$ and $B(b, f(b))$ lie on the graph of the function f. Find the Cartesian equation of the line AB. Find the value of c such that the line AB meets the x-axis at $(c, 0)$ and prove that the expression obtained for c is the same as that obtained in §14.4.

5 On plotting values on a graph, a student thought that the equation $x^3 - 4x^2 + x + 7 = 0$ probably had a root between 2 and 3. Apply the chord method to this interval and explain by a diagram what happens. Solve the equation by the chord method giving your answer accurate to three decimal places. State whether the student was right or wrong.

14.5 The use of tangents

The three iterative methods considered in the last three sections have each required 2 values of x, bracketing the desired root, in order to produce a better approximation. There is another method which only requires one starting value and which converges faster for many functions. The method uses the tangent to the function at a point as a linear approximation and is demonstrated in Fig. 14.5.

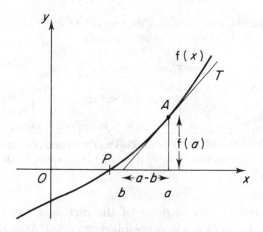

Fig. 14.5

Let the function f have a zero, given by the point P where its graph crosses the x-axis. Let a be a first approximation to the root of $f(x) = 0$, with a corresponding point $A(a, f(a))$ on the graph. The gradient of the tangent AT to the graph at A is $f'(a)$, the derivative of f evaluated at $x = a$. Taking the tangent as a linear approximation to the graph, a new approximation to the root is given by $x = b$, where the tangent AT crosses the x-axis. Then

$$\frac{f(a)}{a-b} = f'(a) \text{ and, by rearranging, } b = a - \frac{f(a)}{f'(a)}.$$

When the function is decreasing, with $f'(a)$ negative, instead of increasing as in Fig. 14.5, it can be shown that the same formula holds. This formula is called the *Newton-Raphson* formula,

$$b = a - \frac{f(a)}{f'(a)}.$$

Starting with a single initial value $a = p_1$, we may apply the formula repeatedly to produce a sequence of approximations to the root, namely

$$p_1, p_2, p_3, \ldots \text{ with } p_{n+1} = p_n - \frac{f(p_n)}{f'(p_n)}.$$

For many functions, the process, called the *Newton-Raphson process*, converges to the required root of the equation $f(x) = 0$.

Note The process requires the function to be differentiated and so we restrict its use to the set of differentiable functions. Since, in the formula, we divide by $f'(a)$, the process will fail if the graph is parallel to the x-axis at $x = a$, and it will not give a good approximation if $f'(a)$ is small numerically.

EXAMPLE *Use the Newton-Raphson process to improve the approximation $x = 4$ to the solution of the equation $x^3 - 50 = 0$.*

Since $f(x) = x^3 - 50$, $f'(x) = 3x^2$, so

$$p_{n+1} = p_n - \frac{p_n^3 - 50}{3p_n^2}, \text{ and so } p_{n+1} = \frac{2p_n^3 + 50}{3p_n^2}.$$

Taking $p_1 = 4$, $p_2 = (2 \times 4^3 + 50)/3 \times 4^2 = (128 + 50)/48 = 3.7083$ and $p_3 = (2(3.7083)^3 + 50)/3(3.7083)^2 = \mathbf{3.68419}$. This result of only two iterations gives a result which is correct to three decimal places since, to six decimal places, $\sqrt[3]{50} = 3.6840315$.

Comments on method As in most of the methods, the better the first approximation the quicker a good result is obtained. Also, if the graph of f is fairly steep near the root, then f' is numerically large and this speeds up the convergence. Also, it can normally be assumed that the error in the final approximation is less than the correction $-\dfrac{f(x)}{f'(x)}$ evaluated at the previous approximation.

EXERCISE 14.5

1 Use the Newton-Raphson method, twice in each case, to find improved approximations to the root of the equation, using the given starting value:
(i) $x^3 + x - 4 = 0$, $x = 2$;   (ii) $x^3 + 3x - 1 = 0$, $x = 1$;
(iii) $x^2 + 2x - 7 = 0$, $x = -4$;   (iv) $x^4 + 3x^2 - x - 8 = 0$, $x = -2$;
(v) $x^3 - 25 = 0$, $x = 3$.

2 Show that $x - \dfrac{f(x)}{f'(x)} = \dfrac{3x^3 - 1}{3x^2 - 3}$, when $f(x) = x^3 - 3x + 1$.

   Use the Newton-Raphson method to find the roots of the equation $x^3 - 3x + 1 = 0$, accurate to three decimal places. (Hint: there are three answers).

3 Show that the Newton-Raphson method gives this iteration:

$$p_{n+1} = \frac{6p_n^3 - 1}{9p_n^2 - 1}, \text{ when } f(x) = 3x^3 - x - 1$$

   and find the solution of the equation $f(x) = 0$.

4 Use the function $f(x) = x^4 - 17$ to find $\sqrt[4]{(17)}$ accurate to five decimal places.

5 Find the negative roots of the equation $x^4 - 7x^2 + 1 = 0$ and check your answers by solving for x^2 directly.

6 Draw a flow chart to solve the equation $f(x) = 0$, using the Newton-Raphson method. Arrange to input the first value p_1, and to halt when the consecutive iterations differ by less than 10^{-5}, to obtain an answer correct to four decimal places.

7 Draw an alternative to Fig. 14.5 in which the function f is shown to be decreasing, instead of increasing. Prove that, in this case, the Newton-Raphson formula still holds.

MISCELLANEOUS EXERCISE 14

1 Show that the equation $x^3 - 4x^2 + 2 = 0$ has a root α between -1 and 0, a root β between 0 and 1, and that its third root γ is between 1 and 4.

   Use the Newton-Raphson method to find a second approximation to α with a first approximation of -1, and also a second approximation to β taking 1 as first approximation. Determine and explain what happens if 0 is used as the first approximation to either root.

   State, with reasons, a suitable first approximation to γ, and use this to find a second approximation to γ.   (*C*)

2 The equation $f(x) = 0$ is known to have precisely one root in the interval $a \leqslant x \leqslant b$. Construct a flow diagram giving an algorithm to estimate this root, using the method of interval bisection. The following are given: the numbers a and b (with $a < b$), the function f, and the positive number δ which is the maximum permissible magnitude of the error in the root.

Demonstrate the working of your flow diagram by applying it to the equation $\cos x - x = 0$ with $a = 0$, $b = 1$ and $\delta = 0 \cdot 1$, giving your answer correct to one decimal place.   (C)

3   By applying Newton's method to the function f defined by

$$f(x) = 1 - \frac{5}{x^2}$$

develop an iterative formula for calculating $\sqrt{5}$, and show that if x_n is in error by a small quantity d_n then the next approximation x_{n+1} is in error by about $0 \cdot 7 \, d_n^2$.   (JMB)

4   Show that there is precisely one real root of the equation

$$x^3 + 2x - 1 = 0$$

and that it lies in the interval $0 < x < 1$. Use Newton's method to find the root to three decimal places.   (JMB)

5   Prove that the equation $e^{-x} - x + 2 = 0$ has a root between 0 and 3, and that the equation $e^x + x - 2 = 0$ has a root between 0 and $0 \cdot 5$.

By using the iterations $x_{n+1} = 2 + e^{-x_n}$ and $x_{n+1} = 2 - e^{x_n}$ respectively and taking, in each case, $x_1 = 0$ find, for each equation, values for x_2, x_3 and x_4 to three significant figures.

Show, by means of clear sketches with these values entered in, the manner in which the iteration for the first equation is converging and that for the second is diverging.

Find the root of $e^{-x} - x + 2 = 0$ correct to two decimal places, making it clear from your working why your answer is correct to that degree of accuracy.   (C)

6   Show that the equation $x^3 - 12x - 7 \cdot 2 = 0$ has one positive and two negative roots. Obtain the positive root to three significant figures by the Newton-Raphson method.   (L)

7   Find, correct to one decimal place, the real root of the equation $x^3 + 2x - 1 = 0$.   (L)

8   Use the Newton-Raphson method to find the real root of the equation $x^3 + 2x^2 + 4x - 6 = 0$ taking $x = 0 \cdot 9$ as the first approximation and carrying out one iteration.

Draw a flow diagram designed to evaluate this root using this method to within an accuracy of $0 \cdot 001 \%$, starting with $x = 0 \cdot 9$.   (L)

9   Hero's method for finding the square root of the number k, starting with the value a, is given by the iterative formula:

$$p_1 = a, \quad p_{n+1} = \frac{1}{2}\left(p_n + \frac{k}{p_n}\right).$$

(i) Take $k = 5$, $a = 2$, and show how to find $\sqrt{5}$ to three decimal places;
(ii) take $k = 5$ and $a = 3$, and show how this process converges;
(iii) explain why the method works. (Hint: use Newton-Raphson method on the function $f(x) = x^2 - k$.)

10   Using the idea of question **9**, generate an iterative process for finding the cube root of any given number.

11   Repeat question **9** to find the nth root of any positive number.

12 Show by means of a sketch how the Newton-Raphson iterative formula

$$x_{n+1} = x_n - \frac{f(x_n)}{f'(x_n)}$$

will, under suitable conditions, give a sequence $\{x_n\}$ converging to a root of the equation $f(x) = 0$.

The equation $\tan x = 2x$ has one positive root α in the interval $(0, \frac{1}{2}\pi)$. Taking 1 as a first approximation x_1, apply the Newton-Raphson formula once to obtain a second approximation x_2, giving three significant figures in your result.

By considering the graph of $y = \tan x - 2x$ for $0 < x < \frac{1}{2}\pi$, or otherwise, show that any initial approximation x_1 such that $\alpha < x_1 < \frac{1}{2}\pi$ will produce a sequence converging to α, whereas not all initial approximations less than α will do so. *(C)*

13 Sketch the graph of $y = \cos x$ for $0 \leqslant x \leqslant 4\pi$. (An accurately plotted curve on graph paper is *not* necessary.) Use your sketch to find the *number* of positive roots of the equation $\cos x = \frac{1}{10}x$.

Prove that the equation of the tangent to the curve $y = \cos x$ at $(\frac{1}{2}\pi, 0)$ is $y = \frac{1}{2}\pi - x$.

By using $\frac{1}{2}\pi - x$ as an approximation for $\cos x$ (valid for x near $\frac{1}{2}\pi$), calculate an estimate of the smallest positive root of the equation $\cos x = \frac{1}{10}x$, giving your answer to two decimal places.

Show on a new sketch the part of the curve $y = \cos x$ near $x = \frac{1}{2}\pi$ and the tangent at $x = \frac{1}{2}\pi$. Hence determine whether the estimate of the root calculated above is larger or smaller than the correct value. *(C)*

14 Show that the equation $x^3 - x - 2 = 0$ has a root between 1 and 2. Using Newton's approximation with starting point 1·5 (and showing all relevant working) determine, by means of two iterations, an approximation to this root, giving your answer to two decimal places. *(JMB)*

15 Complex Numbers

15.1 Extensions of number systems

As was demonstrated in Chapters 2 and 7, one technique for solving algebraic equations is to spot factors of quadratic and other polynomial functions. The factor theorem states that

$$f(\alpha) = 0 \Leftrightarrow (x - \alpha) \text{ is a factor of } f(x)$$

and this is used to find the roots of a polynomial equation, since

$$f(\alpha) = 0 \Leftrightarrow \alpha \text{ is a root of the equation } f(x) = 0.$$

For quadratic equations, the method of completing the square leads to a formula for the solution. The solution of the equation $ax^2 + bx + c = 0$ is given by

$$x = \frac{-b \pm \sqrt{(b^2 - 4ac)}}{2a}.$$

The existence of one root or of two distinct roots of the equation depends upon the sign of the discriminant $\Delta = b^2 - 4ac$.

If $\Delta > 0$, then $\sqrt{\Delta}$ is a real number and the equation has two roots

$$x = \frac{-b + \sqrt{\Delta}}{2a} = \frac{-b + \sqrt{(b^2 - 4ac)}}{2a} \quad \text{and} \quad x = \frac{-b - \sqrt{\Delta}}{2a}.$$

If $\Delta = 0$, then we have the situation of a repeated root, $x = \dfrac{-b}{2a}$, and the graph of the quadratic touches the x-axis at the point $\left(\dfrac{-b}{2a}, 0\right)$.

If $\Delta < 0$, then the equation has no real solution, since no real number can be the square root of a negative number. This corresponds to the graph of the quadratic lying either totally above, or totally below, the x-axis and never meeting it.

In many situations, although there is no real solution to the quadratic equation when $\Delta < 0$, it is very convenient to deal with the roots of the equation as if they were numbers, even if not real numbers. This means that we have to enlarge our definition of number. We have done this before in extending the natural numbers to the integers, then to the rationals and then to the real numbers.

Consider the natural numbers $\mathbb{N}$. In this system of numbers, we can add and multiply numbers but we cannot always subtract and divide and still obtain a natural number. We cannot solve the equation $x + 5 = 2$ in the set of natural numbers. We have to extend $\mathbb{N}$ to the set $\mathbb{Z}$, the set of integers, by introducing negative numbers. Then $x + 5 = 2$ has a solution, $x = -3$, in $\mathbb{Z}$.

In the set $\mathbb{Z}$, we can add, subtract and multiply and still obtain an integer. We say that $\mathbb{Z}$ is closed under the three operations $+, -, \times$. However, we cannot always divide and obtain another integer. For instance, the equation $5x = 2$ has no solution in the set of integers. It is necessary to extend the set of integers to the set of rational numbers, that is, the set $\mathbb{Q}$, in order to be able to solve this equation. The solution in $\mathbb{Q}$ of the equation is $x = \frac{2}{5}$. In the set $\mathbb{Q}$, we can add, subtract, multiply and divide by a non-zero, so we say that $\mathbb{Q}$ is closed under the four operations $+, -, \times, \div$.

However, in the set $\mathbb{Q}$, we are still unable to solve the equation $5x^2 = 2$, because the real number $\sqrt{\frac{2}{5}}$ is irrational. In order to have a number system in which we can take roots of all positive numbers, and also to include all numbers which are represented by infinite decimals, we need to extend $\mathbb{Q}$ to the set of all real numbers, that is, $\mathbb{R}$. We can represent every number in $\mathbb{R}$ by a length along a line and so all real numbers are represented by all the points along an axis, Ox, which we refer to as the real line.

We, therefore, have a chain of number systems, each contained in the next, and each with the property that, in it, more equations can be solved than in its predecessor; $\mathbb{N} \subset \mathbb{Z} \subset \mathbb{Q} \subset \mathbb{R}$. It is possible to insert other number systems in between two members of this chain. For example,

$$\text{let } \mathbb{Q}(\sqrt{2}) = \{a + b\sqrt{2} : a, b \in \mathbb{Q}\}.$$

Then $\mathbb{Q}(\sqrt{2})$ is formed by adding $\sqrt{2}$ to the rationals and forming all possible real numbers by using the four processes of arithmetic. Under these four operations, $\mathbb{Q}(\sqrt{2})$ is closed, for, suppose that a, b, c, d are all rational numbers,

then
$$(a + b\sqrt{2}) + (c + d\sqrt{2}) = (a + c) + (b + d)\sqrt{2},$$
$$(a + b\sqrt{2}) - (c + d\sqrt{2}) = (a - c) + (b - d)\sqrt{2},$$
$$(a + b\sqrt{2})(c + d\sqrt{2}) = (ac + 2bd) + (ad + bc)\sqrt{2},$$
and
$$(a + b\sqrt{2}) \div (c + d\sqrt{2}) = \frac{(a + b\sqrt{2})(c - d\sqrt{2})}{(c + d\sqrt{2})(c - d\sqrt{2})}$$
$$= \frac{ac - 2bd}{c^2 - 2d^2} + \frac{bc - ad}{c^2 - 2d^2}\sqrt{2}, \quad c + d\sqrt{2} \neq 0.$$

In each of the above four equations, the right hand side is of the form $x + y\sqrt{2}$, where x and y are rational numbers, so all these numbers lie in $\mathbb{Q}(\sqrt{2})$. Thus $\mathbb{Q}(\sqrt{2})$ is closed under the four arithmetical operations.

The above process of forming the set of numbers $\mathbb{Q}(\sqrt{2})$ can be described as extending the set $\mathbb{Q}$ in order to include the numbers which are solutions of the equation $x^2 = 2$. We now do the same thing for the equation $x^2 + 1 = 0$, or $x^2 = -1$, and extend the real numbers, $\mathbb{R}$, by adding a new number $\sqrt{(-1)}$, which is a solution of the equation $x^2 = -1$. We define this new number by the notation

$$i = \sqrt{(-1)}, \text{ so that } i^2 = -1.$$

Then the equation $x^2 + 1 = 0$ will have two solutions, $x = i$ and $x = -i$. Also, we can factorise the quadratic function $x^2 + 1$, since $x^2 + 1 = (x - i)(x + i)$.

Note   The notation $i = \sqrt{(-1)}$ is not universal. Many engineers use $j = \sqrt{(-1)}$ instead, because they reserve the symbol i for the rate of flow of current in electrical theory. However, the notation i is more common in mathematical texts.

The new number, i, is not a real number so it does not appear as a point on the real line Ox. It is called an *imaginary* number because it is not real. This is unfortunate since i is no more imaginary than $\sqrt{2}$, if we use the everyday meaning of the word imaginary. We must regard the words *real* and *imaginary* as mathematical words having a clearly defined mathematical meaning, quite distinct from their everyday use. We now have the real numbers and the new number i, and we then form a set of numbers, which is an extension of $\mathbb{R}$, by using all the processes of algebra. The new set of numbers is called the set of complex numbers and is denoted by $\mathbb{C}$. Thus $\mathbb{C} = \{a + ib: a, b \in \mathbb{R}\}$. In $\mathbb{C}$ we replace i^2 by -1, wherever it occurs.

The four arithmetical operations are defined on the complex numbers in a natural way, by extending the corresponding operations on the real numbers. Thus, for all real numbers a, b, c, d,

$$(a + ib) + (c + id) = (a + c) + i(b + d),$$

$$(a + ib) - (c + id) = (a - c) + i(b - d),$$

$$(a + ib) \times (c + id) = (ac + ibid) + (aid + ibc)$$

$$= (ac - bd) + i(ad + bc),$$

$$(a + ib) \div (c + id) = \frac{(a + ib)(c - id)}{(c + id)(c - id)} = \frac{ac - ibid + ibc - aid}{cc - idid + idc - cid}$$

$$= \frac{ac + bd}{c^2 + d^2} + i\frac{bc - ad}{c^2 + d^2}, \text{ if } c^2 + d^2 \neq 0.$$

The right-hand sides of the above equations are all complex numbers so that $\mathbb{C}$ is closed under the four operations of arithmetic. Also, all the properties of arithmetical and algebraic manipulation are still valid in $\mathbb{C}$. For example, addition and multiplication are both commutative (i.e. the order does not matter), and we identify

$$a + ib = a + bi = ib + a = bi + a.$$

Compare the above rules with those we had for $\mathbb{Q}(\sqrt{2})$.

Equality

Consider two equal complex numbers, $a + ib = c + id$, then

$$a - c = id - ib = i(d - b)$$

and, on squaring both sides,

$$(a - c)^2 = -(d - b)^2.$$

Since each square is positive, the left side of the equation is positive and the right side is negative. Since the only number which is positive and negative is zero, we conclude that $a - c = 0 = d - b$, and so $a = c$ and $b = d$. In the complex number $a + ib$, with a and b real, a is called the *real* part and b is called the *imaginary* part. Thus, when we equate two complex numbers, we put their real parts equal and also their imaginary parts equal, giving two real equations for one complex one.

This process is called *equating of real and imaginary parts* of two complex numbers which are equal. It is a very useful process since, when we have one equation in $\mathbb{C}$, we can then obtain two equations in $\mathbb{R}$.

Notation   If $z = x + iy$, with $x, y \in \mathbb{R}$, then $\operatorname{Re}(z) = x$ and $\operatorname{Im}(z) = y$.

A real number a is a complex number, since $a = a + i0$, and it has zero imaginary part. A complex number ib, which is of the form, $0 + ib$ and which has zero real part is called a pure imaginary number. Thus, the equation $x^2 = -2 = -1 \times 2$ has two pure imaginary roots, namely $x = i\sqrt{2}$ and $x = -i\sqrt{2}$. Similarly, the equation $x^2 = -9$ has two pure imaginary roots $3i$ and $-3i$.

EXAMPLE   *Express* $\dfrac{2 - i}{3 + 4i}$ *in the form* $a + ib$, *where* a *and* b *are real.*

$$\frac{2 - i}{3 + 4i} = \frac{(2 - i)(3 - 4i)}{(3 + 4i)(3 - 4i)} = \frac{6 - 8i - 3i + 4i^2}{3^2 - 4^2 i^2} = \frac{6 - 4 - (8 + 3)i}{9 + 16}$$

$$= \frac{2}{25} - \frac{11}{25}i.$$

EXERCISE 15.1

1   Express in the form $a + ib$, $a, b \in \mathbb{R}$:
   (i) $(2 + 3i) + (5 + 7i)$,   (ii) $(3 - 4i) + (4 - 3i)$,   (iii) $(5 - 6i) + (2 - 9i)$,
   (iv) $(4 - i) - (3 + 2i)$,   (v) $(14 - 3i) - (3 - 7i)$,   (vi) $(1 - i) - (1 + i)$.
2   Simplify:
   (i) $3(4 + 2i) - 2(5 + i)$,   (ii) $6(3 + i) + 2(5 + 3i)$,   (iii) $2(7 - i) - 7(3 - 2i)$.
3   Write as a single complex number:
   (i) $(2 + 3i)(4 + i)$,   (ii) $(3 - 4i)(4 - 3i)$,   (iii) $(5 + 2i)(4 - 3i)$,
   (iv) $(1 + i)(1 - i)$,   (v) $(2 + 3i)^2$,   (vi) $3i(2 + 5i)$,   (vii) $(3 - i)(2 + i)(3 + 2i)$,
   (viii) $(4 - 2i)^2 - (3 + i)^2$,   (ix) $(2 - i)(3i)(4 + i)$,   (x) $\left(-\dfrac{1}{2} + \dfrac{\sqrt{3}}{2}i \right)^3$.

4   Express the quotient in the form $a + ib$, with a and b real:

(i) $\dfrac{(2+i)}{(3+2i)}$,   (ii) $\dfrac{5-3i}{3-4i}$,   (iii) $\dfrac{4i}{2-5i}$,   (iv) $\dfrac{1}{1-i}$,   (v) $\dfrac{5+4i}{5-4i}$,   (vi) $\dfrac{(2+3i)}{(5+i)^2}$,

(vii) $\dfrac{(3+i)^2}{(4-i)^2}$.

5   Consider the alternative way of writing a complex number, if $a, b \in \mathbb{R}$, $a + ib \equiv (a, b)$, as an *ordered pair*, or pair of coordinates.
(a) Write the following using this new notation:
(i) $(3+2i) + (5+i) = (8+3i)$,   (ii) $(4-2i) + (6i) = (4+4i)$,
(iii) $(3+4i) - (8-2i) = (-5+6i)$,   (iv) $(-4-2i) - (7) = (-11-2i)$,
(v) $3(2-5i) = (6-15i)$,   (vi) $2(3i) - 3(1+i) = -3+3i$,
(vii) $(5+2i)(3-4i) = (23-14i)$,   (viii) $(1-2i) \div (2-i) = (\frac{4}{5} - \frac{3}{5}i)$.
(b) Using the ordered pair notation for complex numbers, complete the equation:
(i) $(2,4) + (-2,3) = (\ ,\ )$,   (ii) $-(4,-6) + (-3,5) = (\ ,\ )$,
(iii) $(4,7) - (\ ,\ ) = (7,2)$,   (iv) $(5,6) \times (1,3) = (\ ,\ )$,   (v) $(1,2) \times (0,1) = (\ ,\ )$,
(vi) $(2,3) \times (\ ,\ ) = (-5,-1)$,   (vii) $(\ ,\ ) \times (4,1) = (3,22)$,
(viii) $(5,6) \div (1,3) = (\ ,\ )$,   (ix) $(\ ,\ ) \div (2,4) = (14,8)$,
(x) $(a,b) + (c,d) = (\ ,\ )$,   (xi) $(a,b) \times (c,d) = (\ ,\ )$,
(xii) $(a,b) \div (c,d) = (\ ,\ )$.

6   Given that $z = -1 + 3i$, express $z + 2/z$ in the form $a + ib$, where a and b are real.   (L)

7   Given that $\dfrac{1}{x+iy} + \dfrac{1}{1+2i} = 1$, where x, y are real, find x and y.   (L)

8   Find the real numbers x and y given that $\dfrac{1}{x+iy} = 2 - 3i$.   (L)

9   Write in as simple a form as possible the complex number $\dfrac{2+5i}{5-2i}$.

15.2 Solution of polynomial equations

In the set of complex numbers, a quadratic equation, with real coefficients, will have two roots. These may be two different real roots, or two equal real roots (that is, a double root), or two complex roots. The three cases correspond to the parabola with the quadratic equation meeting the x-axis in two points, or touching it at one (double) point, or not meeting it at all. The graph of a cubic polynomial will always cross the x-axis at at least one point, corresponding to the fact that a cubic equation always has one real root. If the cubic function is factorised into the product of a linear factor corresponding to this real root, the other factor will be quadratic. Thus the cubic equation with real coefficients will always have three roots, of which one will be real and the other two may be real or complex. Two or three of the real roots may be equal.

There is a fundamental theorem of algebra, which cannot be proved here, stating that a polynomial equation of degree n, with real coefficients, has n roots in the set of complex numbers, provided that repeated roots are counted as many times as they are repeated. In fact, the same is true of a polynomial equation with complex coefficients, and so no further extension of the system of numbers, beyond the complex numbers, is needed to ensure that every polynomial can be factorised into linear factors. In this sense, the set $\mathbb{C}$, of complex numbers, is as large a number set as we need, whereas the set $\mathbb{R}$, of real numbers, is not large enough.

These ideas are applied now to the solution of polynomial equations. One difficulty should be mentioned. For a polynomial equation of degree higher than two, there is no formula for its solution, in the same way that there is for the solution of a quadratic equation. The only tool at our disposal is the factor theorem, as far as finding exact roots is concerned. Otherwise, if we can locate a real root approximately, we can use numerical methods to improve the approximation.

EXAMPLE 1   *Find the solution, in the set of complex numbers, of the quadratic equation* $x^2 + x + 1 = 0$.

Use the formula for the roots of a quadratic equation and replace $\sqrt{(-1)}$ by i wherever it occurs. The solution of the equation is

$$x = \frac{-1 \pm \sqrt{(1-4)}}{2} = \frac{-1 \pm \sqrt{(-3)}}{2} = \frac{-1 \pm i\sqrt{3}}{2}$$

so the two roots are $\frac{1}{2}(-1+i\sqrt{3})$ **and** $\frac{1}{2}(-1-i\sqrt{3})$.

EXAMPLE 2   *Find the real and complex roots of the cubic equation*

$$x^3 - 11x - 20 = 0.$$

Let $f(x) = x^3 - 11x - 20$, then we use the factor theorem to investigate as possible roots the factors of -20, that is, ± 1, ± 2, ± 4, ± 5, ± 10, ± 20.
$f(1) = -30$, $f(2) = -34$, $f(4) = 0$, $f(5) = 50$, $f(10) = 870$, $f(20) = 7760$,
$f(-1) = -10$, $f(-2) = -6$, $f(-4) = -40$, $f(-5) = -90$, $f(-10) = -910$,
$f(-20) = -7800$
Therefore, there is a real root $x = 4$ with a corresponding factor $(x-4)$. Now

$$f(x) = (x-4)(x^2 + 4x + 5),$$

so we investigate the roots of $x^2 + 4x + 5 = 0$. The discriminant $\Delta = 4^2 - 4 \times 1 \times 5 = -4$, so there are no further real factors of $f(x)$. The formula for the roots of a quadratic equation gives the solutions

$$x = \frac{-4 \pm \sqrt{(16-20)}}{2} = \frac{-4 \pm 2i}{2}, \quad \text{that is,} \quad x = -2+i \text{ and } -2-i,$$

with the corresponding linear factors $x-(-2+i)$ and $x-(-2-i)$. Finally, the cubic function $f(x)$ factorises into one real and two complex factors,

$$f(x) = (x-4)(x+2-i)(x+2+i)$$

and the equation $f(x) = 0$ has the solution set $\{4, -2+i, -2-i\}$.

EXERCISE 15.2

1   Find the roots of the equation, in the set of real numbers:
    (i) $x^2 + 4x + 3 = 0$ (ii) $2x^2 - 7x + 5 = 0$ (iii) $x^2 - 6x + 6 = 0$
    (iv) $x^2 + 6x + 9 = 0$ (v) $3x^2 + 5x - 1 = 0$ (vi) $4x^2 - 7x + 2 = 0$.

2   Find the roots of the equation, in the set of complex numbers:
    (i) $x^2 - 4x + 5 = 0$ (ii) $x^2 - 6x + 10 = 0$ (iii) $x^2 - 8x + 17 = 0$
    (iv) $x^2 + 12x + 40 = 0$ (v) $3x^2 + x + 1 = 0$ (vi) $x^2 - x + 1 = 0$.

3   Verify that the given linear polynomial is a factor of $f(x)$ and factorise $f(x)$ into a product of real polynomials, as far as possible:
    (i) $f(x) = 3x^2 - 4x + 1,\ x - 1$ (ii) $f(x) = 2x^2 + 5x + 2,\ x + 2$
    (iii) $f(x) = x^3 + 3x^2 + 2x,\ x + 1$ (iv) $f(x) = x^3 - 4x,\ x - 2$.

4   Factorise $f(x)$ into a product of real polynomials, as far as possible:
    (i) $f(x) = 2x^2 + 3x + 2$ (ii) $f(x) = x^2 + 8x + 17$
    (iii) $f(x) = x^2 - 9x + 20$ (iv) $f(x) = 3x^2 - x + 6$
    (v) $f(x) = x^3 - x^2 + x - 1$ (vi) $f(x) = x^4 + 7x^2 + 10$
    (vii) $f(x) = x^4 - x^3 - 6x^2$ (viii) $f(x) = x^3 + 2x^2 - 4x - 8$.

5   Draw a sketch of a polynomial function with the properties:
    (i) a quadratic with two real roots, with positive coefficient of x^2;
    (ii) a quadratic with one repeated root, with negative coefficient of x^2;
    (iii) a quadratic with two complex roots, with positive constant coefficient;
    (iv) a cubic with three distinct real roots, with positive coefficient of x^3;
    (v) a cubic with one repeated real root and one distinct real root, with coefficient of x^3 negative;
    (vi) a cubic with one real root and two complex roots, with positive coefficient of x^3;
    (vii) a quartic with two real and two complex roots, with negative coefficient of x^4;
    (viii) a quartic with one repeated real root and two complex roots, with positive coefficient of x^4.

6   Express in the form $a + ib$, where a and b are real, the complex number:
    (i) i^3, (ii) i^4, (iii) i^5, (iv) $\dfrac{1}{i}$, (v) $\dfrac{1}{i^2}$, (vi) $\dfrac{1}{i^3}$.

7   Write down the square of:
    (i) $2i$; (ii) $i\sqrt{5}$; (iii) $-7i$; (iv) $-i\sqrt{2}$.

8   Write down the two square roots of:
    (i) -5, (ii) -25, (iii) -8, (iv) -27.

9   Express the pair of zeros of the quadratic in the form $a + ib$ and $a - ib$:
    (i) $x^2 - 5x + 8$, (ii) $2x^2 + 3x + 4$, (iii) $3x^2 - x + 5$.
    (a) Write down the sum of the two complex zeros in each case.
    (b) Write down the product of the two complex zeros in each case.
    (c) Explain why the answers to (a) and (b) are all real.

15.3 Representation of complex numbers

You may have noticed a strong similarity between the addition and subtraction of complex numbers and the addition and subtraction of two dimensional column vectors. For example compare

$$(3 + 4i) + (-2 + i) = (1 + 5i) \quad \text{with} \quad \begin{pmatrix} 3 \\ 4 \end{pmatrix} + \begin{pmatrix} -2 \\ 1 \end{pmatrix} = \begin{pmatrix} 1 \\ 5 \end{pmatrix}.$$

A similar correspondence occurs when we consider the multiplication by a real number, for example compare

$$4(-2 + 3i) = (-8 + 12i) \quad \text{with} \quad 4\begin{pmatrix} -2 \\ 3 \end{pmatrix} = \begin{pmatrix} -8 \\ 12 \end{pmatrix}.$$

The structure of $\mathbb{C}$ under the operations of addition and of multiplication by a real number is precisely the same as the structure of all two-dimensional column vectors under the same operations. Also, the equality of two complex numbers corresponds to the equality of the corresponding vectors.* This means that we can use the geometrical properties of vectors, which we met in Chapter 5, to gain insight into the behaviour of complex numbers under addition and multiplication by real numbers. Such a use of an exact correspondence of structure of two apparently different areas of mathematics is an extremely valuable tool.

The representation of complex numbers as vectors in two dimensions is known as the *Argand diagram*, or sometimes as the *complex plane*. A complex number $3 + 4i$ is represented by the vector $\begin{pmatrix} 3 \\ 4 \end{pmatrix} = \overrightarrow{OP}$, where P is the point with coordinates $(3, 4)$ (see Fig. 15.1). In the vector, the top number corresponds to the *real* part of the complex number, and so this real part corresponds to the component of the vector parallel to Ox. A real number is represented by a vector parallel to Ox. Therefore, the axis Ox is the *real axis*, that is, *the real number line*. In the Argand diagram, the real numbers are represented by points on the real axis, Ox. A complex number, which is not real, such as $a + ib$, with $b \neq 0$, is represented by the point with coordinates (a, b) off the real axis. A purely imaginary number, such as bi, will be represented by a vector $\begin{pmatrix} 0 \\ b \end{pmatrix}$, parallel to the axis Oy, so this axis is called the *imaginary axis*, or the i *axis*. There is, therefore, a correspondence between complex numbers, vectors and coordinates:

the complex number $a + ib$ corresponds to the vector $\begin{pmatrix} a \\ b \end{pmatrix}$ which corresponds to the point (a, b).

In this way, the complex numbers, $\mathbb{C}$, are seen as an extension of the real numbers, $\mathbb{R}$, into a second dimension. The basis $\left\{ \begin{pmatrix} 1 \\ 0 \end{pmatrix}, \begin{pmatrix} 0 \\ 1 \end{pmatrix} \right\}$ of the two dimensional vector space $O(x, y)$ corresponds to a basis $\{1, i\}$ of $\mathbb{C}$.

---

* This complete correspondence one to one of the elements and the same correspondence of the structure under the given operations is called an *isomorphism* of the two systems.

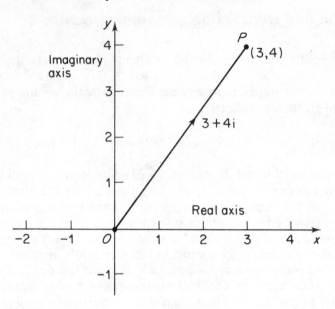

Fig. 15.1

EXERCISE 15.3A

1   Draw an Argand diagram and plot the points, draw in and label the vectors, which represent the following complex numbers:
(i) 1,   (ii) i,   (iii) -1,   (iv) $-i$,   (v) $1+i$,   (vi) $1-i$,   (vii) $-1+i$,
(viii) $-1-i$,   (ix) $3+4i$,   (x) $-4+3i$,   (xi) $-3-4i$,   (xii) $4-3i$.

2   Using the diagram you have drawn in question **1**, describe a geometrical transformation of the plane which transforms the first vector into the second vector, where the vector pair represent the given pair of complex numbers:
(i) 1, -1;   (ii) $3+4i$, $-3-4i$;   (iii) $1-i$, $-1+i$;   (iv) i, $-i$;
(v) $-1-i$, $-1+i$;   (vi) $1+i$, $1-i$;   (vii) 1, i;   (viii) $3+4i$, $-4+3i$;
(ix) $-4+3i$, $-3-4i$;   (x) 1, $-i$;   (xi) $3+4i$, $4-3i$;   (xii) -1, i.

3   Use the results of question **2** in order to pair off each of the following plane transformations with one of the operations on complex numbers:
(A) reflection in the axis Ox;   (a) multiply by -1;
(B) rotate (anticlockwise) 90°;   (b) multiply by i;
(C) rotate through 180°;   (c) multiply by $-i$;
(D) rotate through 270°;   (d) change the sign of i.

4   (a) Draw an Argand diagram to represent each of the complex numbers:
(i) $4+3i$,   (ii) $2-2i$,   (iii) $5+12i$,   (iv) $-6+8i$,   (v) $-3-i$.
(b) For each of the above, calculate the length of the vector and find the angle (in radians) between the vector and the positive real axis, measured positively anticlockwise.

5   Find the two square roots of $3-4i$ in the form $a+ib$, where a and b are real. Show, in an Argand diagram, the points P and Q representing these square roots.

Addition and subtraction of complex numbers

Suppose that a point $P(x, y)$ represents the complex number z in the Argand diagram, where $z = x + iy \in \mathbb{C}$. Then z is also represented by the vector $\overrightarrow{OP}$, and by any other vector equal to $\overrightarrow{OP}$. Consider two points $P_1(3, 4)$ and $P_2(-2, 1)$, representing z_1 and z_2, where $z_1 = 3 + 4i$ and $z_2 = -2 + i$ (see Fig. 15.2(a)).

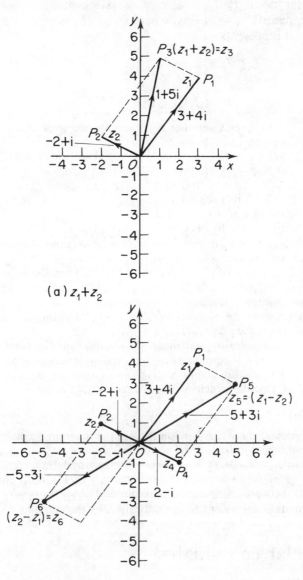

(a) $z_1 + z_2$

(b) $z_1 - z_2$ and $z_2 - z_1$

Fig. 15.2

Then $z_1 + z_2 = 1 + 5i = z_3$, the complex number represented by P_3, where $OP_1 P_3 P_2$ is a parallelogram. There is a correspondence between vector equations and complex number equations, thus

$$\overrightarrow{OP_3} = \overrightarrow{OP_1} + \overrightarrow{P_1 P_3} = \overrightarrow{OP_1} + \overrightarrow{OP_2} \quad \text{corresponds to } z_3 = z_1 + z_2, \text{ and}$$
$$\overrightarrow{OP_3} = \overrightarrow{OP_2} + \overrightarrow{P_2 P_3} = \overrightarrow{OP_2} + \overrightarrow{OP_1} \quad \text{corresponds to } z_3 = z_2 + z_1.$$

Similarly, the point P_4 $(2, -1)$ represents the number $z_4 = -z_2$, and P_5 $(5, 3)$ represents $z_5 = z_1 - z_2$, where $OP_4 P_5 P_1$ is a parallelogram, and P_6 $(-5, -3)$ represents

$$z_6 = z_2 - z_1 = -z_5 = -(z_1 - z_2) \text{ (Fig. 15.2(b)).}$$

EXERCISE 15.3B

1   Represent the complex number equation by means of vector triangles or polygons in the Argand diagram:
(i) $(2 + 3i) + (5 + i) = (7 + 4i)$,   (ii) $(3 - 2i) - (4 + 3i) = (-1 - 5i)$,
(iii) $2(4 - 3i) = (8 - 6i)$,   (iv) $-(3 - 2i) = (-3 + 2i)$,
(v) $(2 + 3i) + (2 - 3i) = 4$,   (vi) $(2 + 3i) - (2 - 3i) = 6i$,
(vii) $(1 + i) + (2 + 4i) + (-2 - 3i) = (1 + 2i)$,
(viii) $(3 - 2i) + (4 + 3i) + (-7 - i) = 0$,
(ix) $2(1) + 4(i) = (2 + 4i)$,   (x) $3(1 + i) + 6(1 - i) = (9 - 3i)$.
2   Given that $z_1 = (4 + 2i)$, $z_2 = (3 - 3i)$, on an Argand diagram, show the complex numbers $z_1, z_2, z_1 + z_2, z_1 - z_2, z_2 - z_1$.
3   Given that $z_1 = (3 + 4i)$, $z_2 = i$, on an Argand diagram show the numbers z_1, $z_2, z_1 z_2, z_1 z_2 z_2, z_1 z_2 z_2 z_2, z_1 z_2 z_2 z_2 z_2$. Explain, in geometrical terms, the effect that multiplication by i has on the vector representing z_1.
4   Given that $z_1 = (2 + i)$, $z_2 = 2i$, $z_3 = 3i$, $z_4 = 4i$, show, on an Argand diagram, the numbers $z_1, z_1 z_2, z_1 z_3, z_1 z_4, z_1 z_2 z_3$.
    Complete the following statement: "Multiplication of a complex number by ai (where a is real) is equivalent to the transformation, of the representative vector in the Argand diagram, given by a rotation of . . . . radians followed by an enlargement with scale factor . . . . ."
5   Given that $z_1 = (3 - 2i)$, $z_2 = i$, calculate $\dfrac{z_1}{z_2}, \dfrac{z_1}{z_2^2}, \dfrac{z_1}{z_2^3}$, and draw the three vectors which represent these numbers on the Argand diagram. Explain the geometrical significance of division by i.
6   Given that $z_1 = 4 + 3i$, $z_2 = 1 + i$, calculate $z_1 z_2$ and draw vectors representing $z_1, z_2, z_1 z_2$, on an Argand diagram. Measure the lengths of the vectors and the angle between each vector and the real axis. Propose a geometrical transformation equivalent to multiplication by the complex number z_2.

15.4 Polar coordinates

The position of a point P in a plane is known when its position vector $\overrightarrow{OP}$ is known, where O is some chosen origin in the plane. In order to fix the vector $\overrightarrow{OP}$, we need to know its length and its direction. Suppose that

$OP = r$, then the length of $\overrightarrow{OP}$ is its modulus $|\overrightarrow{OP}| = r$. The direction of $\overrightarrow{OP}$ may be fixed by means of the angle that $\overrightarrow{OP}$ makes with some fixed line. If this line, called the initial line, is the axis Ox, then the direction of $\overrightarrow{OP}$ is fixed by the angle θ between $\overrightarrow{OP}$ and Ox. The angle θ is measured in radians and the angle is measured positively in an anticlockwise direction from Ox to $\overrightarrow{OP}$ (see Fig. 15.3). We note that the addition of an amount 2π to θ does not change the position of P.

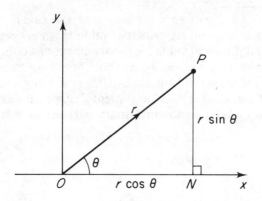

Polar coordinates (r, θ)

Fig. 15.3

Let PN be the perpendicular from P on to Ox, and let the coordinates of P, referred to Cartesian axes $O(x, y)$, be (x, y). The pair of numbers (r, θ) are called the *polar coordinates* of P and these are related to the Cartesian coordinates (x, y) by the equations

$$x = r \cos \theta, \quad r = \sqrt{(x^2 + y^2)},$$

$$y = r \sin \theta, \quad \theta \text{ is given by } \cos \theta = \frac{x}{\sqrt{(x^2 + y^2)}}, \sin \theta = \frac{y}{\sqrt{(x^2 + y^2)}}.$$

Note that $r > 0$ and that the addition of an integer multiple of 2π to θ does not affect the position of the point with polar coordinates (r, θ). The curves $r = c$, where c is a constant, are concentric circles with their centre at the origin. The curves $\theta = \alpha$, where α is a constant, are radial lines extending outwards from the origin O.

Modulus and argument of a complex number

The use of polar coordinates in the complex plane leads to another way of expressing a complex number, called its *polar representation*. The modulus of the complex number z, where $z = x + iy$, is defined to be the length, or modulus, of its representative vector $\begin{pmatrix} x \\ y \end{pmatrix}$ in the Argand diagram, and the argument of z is defined to be the angle θ which $\begin{pmatrix} x \\ y \end{pmatrix}$ makes with the real axis Ox, measured positively in the anticlockwise sense, as usual. Using the letter r for the modulus, (r, θ) are then the polar coordinates of the point P in the Argand diagram representing the complex number z.

Definition   Let $z = x + iy$ be a complex number, with $x,\ y \in \mathbb{R}$, then
the *modulus* of z is $|z| = r = \sqrt{(x^2 + y^2)}$;
the *argument* of z is θ, where $x = r \cos \theta$ and $y = r \sin \theta$;
θ is denoted by arg z, and its principal value is when $-\pi < \theta \leqslant \pi$.

The choice of a principal value for the argument of a complex number is a matter of convenience. Generally, we think of the argument of z, not as a single angle α, but as the set of angles $\{\theta : \theta = \alpha + 2n\pi,\ n \in \mathbb{Z}\}$.

In terms of the modulus r and argument θ of a complex number z, the real part of z is $r \cos \theta$ and the imaginary part is $r \sin \theta$ so that

$$z = r \cos \theta + r \sin \theta\, i = r(\cos \theta + i \sin \theta).$$

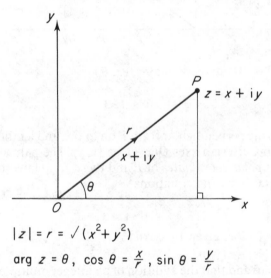

$$|z| = r = \sqrt{(x^2 + y^2)}$$

$$\arg z = \theta, \quad \cos \theta = \frac{x}{r}, \quad \sin \theta = \frac{y}{r}$$

Fig. 15.4

Referring to Fig. 15.4, we can describe a complex number z in various ways:
Cartesian form $z = x + iy$, real part x and imaginary part y;
polar form $z = (r, \theta)$, modulus r and argument θ;
trigonometric form $z = r(\cos \theta + i \sin \theta)$;

vector form $\mathbf{z} = \begin{pmatrix} x \\ y \end{pmatrix}$;

with the relations

$$x = r \cos \theta, \quad y = r \sin \theta, \quad r = \sqrt{(x^2 + y^2)}.$$

EXAMPLE 1 *Express the complex numbers $z = -2 + 2i$ and $w = 1 - i\sqrt{3}$ in polar form. Show these two polar forms on one Argand diagram.*

$$|z| = \sqrt{((-2)^2 + 2^2)} = \sqrt{8} = 2\sqrt{2},$$

$$\arg z = \theta, \sin \theta = \frac{2}{2\sqrt{2}}, \cos \theta = \frac{-2}{2\sqrt{2}}, \theta = \frac{3}{4}\pi,$$

$$|w| = \sqrt{(1+3)} = 2,$$

$$\arg w = \phi, \sin \phi = -\frac{1}{2}\sqrt{3}, \cos \phi = \frac{1}{2}, \phi = -\frac{\pi}{3}.$$

These results are shown on the Argand diagram in Fig. 15.5.

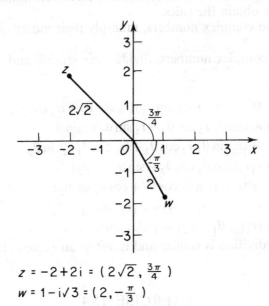

$$z = -2 + 2i = \left(2\sqrt{2}, \frac{3\pi}{4}\right)$$
$$w = 1 - i\sqrt{3} = \left(2, -\frac{\pi}{3}\right)$$

Fig. 15.5

EXAMPLE 2 *The complex numbers w and z are given in polar form $w = (2, \frac{\pi}{6})$ and $z = (3, \frac{\pi}{2})$. Calculate wz and express this complex number in polar form.*

$$w = \left(2, \frac{\pi}{6}\right) = 2\cos\frac{\pi}{6} + 2\sin\frac{\pi}{6}i = \sqrt{3} + i, \quad z = \left(3, \frac{\pi}{2}\right) = 0 + 3i = 3i.$$

$$wz = (\sqrt{3} + i)3i = 3\sqrt{3}i + 3i^2 = -3 + 3\sqrt{3}i,$$

so $|wz| = \sqrt{((-3)^2 + (3\sqrt{3})^2)}$, that is, $|wz| = \sqrt{(9 + 27)} = 6$.

$\text{Arg}(wz) = \theta$ where $\cos \theta = -\frac{3}{6} = -\frac{1}{2}$ and $\sin \theta = \frac{3\sqrt{3}}{6} = \frac{1}{2}\sqrt{3}$, so θ is in the

second quadrant and $\theta = \frac{2\pi}{3}$. Therefore, $wz = \left(6, \frac{2\pi}{3}\right) = \left(2, \frac{\pi}{6}\right) \times \left(3, \frac{\pi}{2}\right)$.

Multiplication and division of complex numbers

In example 2, notice that $|wz| = |w| \, |z|$ and $\arg(wz) = \arg(w) + \arg(z)$. This result is true in general. It means that, although it is easier to add and subtract complex numbers when they are expressed in Cartesian form, it is much easier to multiply and divide complex numbers in polar form. The results are stated as a theorem.

Theorem 15.4 Suppose that $z_1 = (r_1, \theta_1)$ and $z_2 = (r_2, \theta_2)$, then

$$z_1 z_2 = (r_1 r_2, \theta_1 + \theta_2)$$

and, if $z_2 \neq 0$, $z_1/z_2 = (r_1/r_2, \theta_1 - \theta_2)$.

Note that the arguments may be changed by an integer multiple of 2π. Put into words, we obtain the rules:

to multiply two complex numbers, multiply their moduli and add their arguments;

to divide two complex numbers, divide their moduli and subtract their arguments.

Proof $z_1 z_2 = (r_1 \cos\theta_1 + ir_1 \sin\theta_1)(r_2 \cos\theta_2 + ir_2 \sin\theta_2)$

$\qquad = r_1 \cos\theta_1 \, r_2 \cos\theta_2 - r_1 \sin\theta_1 \, r_2 \sin\theta_2$

$\qquad\quad + ir_1 \sin\theta_1 \, r_2 \cos\theta_2 + ir_1 \cos\theta_1 \, r_2 \sin\theta_2$

$\qquad = r_1 r_2 (\cos\theta_1 \cos\theta_2 - \sin\theta_1 \sin\theta_2)$

$\qquad\quad + ir_1 r_2 (\sin\theta_1 \cos\theta_2 + \cos\theta_1 \sin\theta_2)$

$\qquad = r_1 r_2 \cos(\theta_1 + \theta_2) + ir_1 r_2 \sin(\theta_1 + \theta_2)$

$\qquad = (r_1 r_2, \theta_1 + \theta_2)$ in polar form.

The proof for division is similar and is left as an exercise (Exercise 15.4, question 7).

EXERCISE 15.4

1 Draw the vectors, representing these complex numbers, on an Argand diagram and find the polar form of the numbers:
(i) $1 + i$, (ii) $4 + 3i$, (iii) $5 - 12i$, (iv) $-12 + 5i$, (v) $4 - 4i$,
(vi) $-6 - 8i$, (vii) $\sqrt{3} + i$, (viii) $-1 + i\sqrt{3}$, (ix) $2\sqrt{3} - 2i$,
(x) $7 + 24i$, (xi) $3 - 4i$.

2 Rewrite, in Cartesian form, the complex numbers which are given in polar form:

(i) $\left(2, \frac{1}{4}\pi\right)$, (ii) $\left(4, \frac{\pi}{6}\right)$, (iii) $\left(1, -\frac{1}{2}\pi\right)$, (iv) $\left(8, -\frac{\pi}{3}\right)$,

(v) $\left(3, -\frac{5\pi}{6}\right)$, (vi) $\left(12, \frac{1}{2}\pi\right)$, (vii) $\left(1, \frac{3}{4}\pi\right)$,

(viii) $\left(\sqrt{2}, -\frac{1}{4}\pi\right)$, (ix) $\left(2, \frac{\pi}{12}\right)$.

3 Find the product of the pair of complex numbers, (a) by direct multiplication, (b) by converting to polar form, multiplying by use of the rule, and then converting back:
(i) $(1+2i)(3+4i)$, (ii) $(5+12i)(2+2i)$, (iii) $(8-6i)(-1+i)$,
(iv) $(3+i)2i$, (v) $(4-5i)(7-8i)$, (vi) $(2+i)(3-5i)$.

4 (a) Write down the product $z_1 z_2$ and the quotient z_1/z_2, of the two numbers z_1 and z_2, given in polar form:

(i) $\left(2, \frac{1}{4}\pi\right), \left(3, \frac{1}{4}\pi\right)$; (ii) $\left(5, \frac{\pi}{3}\right), \left(2, \frac{\pi}{6}\right)$; (iii) $\left(1, \frac{\pi}{2}\right), \left(5, \frac{\pi}{12}\right)$;

(iv) $\left(1, \frac{3\pi}{4}\right), \left(7, -\frac{\pi}{3}\right)$; (v) $\left(2, -\frac{5\pi}{6}\right), \left(3, -\frac{3\pi}{4}\right)$;

(vi) $\left(2, \frac{\pi}{7}\right), \left(1, \frac{6\pi}{7}\right)$.

(b) For each of the six cases in part (a), draw the four vectors representing z_1, z_2, $z_1 z_2$, z_1/z_2.

5 Find the modulus and the argument of the complex number:
(i) $(1+i)^2$, (ii) $(1+i\sqrt{3})(\sqrt{3}+i)$, (iii) $(\frac{1}{2}+\frac{1}{2}\sqrt{3}i)^2$,

(iv) $\left(\frac{1}{\sqrt{2}}+\frac{i}{\sqrt{2}}\right)\bigg/(1-i)$, (v) $\left(\cos\frac{\pi}{6}+i\sin\frac{\pi}{6}\right)\left(\cos\frac{\pi}{4}+i\sin\frac{\pi}{4}\right)$,

(vi) $\left(\cos\frac{5\pi}{6}+i\sin\frac{5\pi}{6}\right)^3$, (vii) $\frac{(-1+i)}{(1-\sqrt{3}i)}$, (viii) $\frac{1+\sqrt{3}i}{-\sqrt{3}+i}$.

6 Given the complex numbers $z_1 = 2+i$ and $z_2 = 1+3i$. Find the vectors representing z_1+z_2, z_1-z_2, z_2-z_1, $z_1 z_2$, z_1/z_2, z_2/z_1. Draw these six vectors, together with z_1 and z_2, on an Argand diagram. List the eight vectors in increasing magnitude of their moduli.

7 The complex numbers z_1 and z_2 have polar form $z_1 = (r_1, \theta_1)$ and $z_2 = (r_2, \theta_2)$. By multiplying both the numerator and the denominator of

$$\frac{z_1}{z_2} = \frac{(r_1 \cos\theta_1 + r_1 \sin\theta_1 i)}{(r_2 \cos\theta_2 + r_2 \sin\theta_2 i)}$$

by $(r_2 \cos\theta_2 - r_2 \sin\theta_2 i)$, simplify this expression for $\frac{z_1}{z_2}$ and prove that

$$\frac{z_1}{z_2} = \left(\frac{r_1}{r_2}, \theta_1 - \theta_2\right).$$

8 Express $\dfrac{1}{1+i\sqrt{3}}$ in the form $r(\cos\theta + i\sin\theta)$ where $r > 0$ and $-\pi < \theta \leqslant \pi$.

(JMB)

9 (i) Find the modulus and argument of the complex number $\frac{1}{2}+i\frac{\sqrt{3}}{2}$.

(ii) Given that z_1 and z_2 are two complex numbers whose arguments are positive acute angles, express $|z_1 z_2|$ and $\arg(z_1 z_2)$ in terms of the moduli and arguments respectively of z_1 and z_2.

(iii) In an Argand diagram, O is the origin, and P represents $3 + i$. The point Q represents $a + bi$ (where a and b are both positive), and triangle OPQ is equilateral. Find a and b by calculation.   (C)

10   Find the modulus of each of the complex numbers

$$z_1 = 1 + 7i \text{ and } z_2 = -4 - 3i.$$

Hence, or otherwise, show that the modulus of z_1/z_2 is $\sqrt{2}$. Find also the argument of z_1/z_2.   (JMB)

11   Given that $\frac{1}{2}\pi < \theta < \pi$, find, in terms of θ, the modulus of $\cos^2 \theta + i \sin \theta \cos \theta$.   (L)

12   Given that $z = 4\left(\cos\frac{\pi}{3} + i \sin\frac{\pi}{3} \right)$ and $w = 2\left(\cos\frac{\pi}{6} - i \sin\frac{\pi}{6} \right)$, write down the modulus and the argument of each of the following:

(i) z,   (ii) w,   (iii) z^3,   (iv) $\dfrac{z^3}{w}$.

13   The complex numbers z_1, z_2, z_3, z_4 satisfy $\dfrac{z_1}{z_2} = \dfrac{z_3}{z_4}$. If $z_1 = 1 - 2i$, $z_2 = 4 + 3i$ and $z_3 = 10i$, find z_4, expressing your answer in the form $a + bi$.

Evaluate $|z_2 - z_1|^2$ and $|z_4 - z_3|^2$ and verify that $\dfrac{|z_4 - z_3|}{|z_2 - z_1|} = \dfrac{|z_3|}{|z_1|}$.   (C)

15.5 The triangle inequalities

In Fig. 15.6, the points P_1 and P_2 represent the complex numbers z_1 and z_2 in the Argand diagram. O is the midpoint of $P_2 P_4$ and $OP_1 P_3 P_2$ and $OP_1 P_5 P_4$ are parallelograms. Then P_3 and P_5 will represent the complex numbers $z_1 + z_2$ and $z_1 - z_2$, respectively.

Now $|z_1|$ is the length of $\overrightarrow{OP_1}$, $|z_2|$ is the length of $\overrightarrow{P_1 P_3}$ and $\overrightarrow{P_1 P_5}$,

$|z_1 + z_2|$ is the length of $\overrightarrow{OP_3}$, $|z_1 - z_2|$ is the length of $\overrightarrow{OP_5}$.

The length of one side of a triangle is always less than the sum of the lengths of the other two sides, because the shortest distance between two points is the straight line joining them. Therefore,

in the triangle $OP_1 P_3$, $|z_1 + z_2| \leqslant |z_1| + |z_2|$,

in the triangle $OP_1 P_5$, $|z_1| \leqslant |z_1 - z_2| + |z_2|$,

and $|z_2| \leqslant |z_1 - z_2| + |z_1|$.

The last two inequalities can be rearranged, to give

$$|z_1| - |z_2| \leqslant |z_1 - z_2| \text{ and } |z_2| - |z_1| \leqslant |z_1 - z_2|.$$

From the above results, we obtain the triangle inequalities, for any complex numbers z_1 and z_2,

$$|z_1 + z_2| \leqslant |z_1| + |z_2|,$$

and $$\big||z_1| - |z_2|\big| \leqslant |z_1 - z_2|.$$

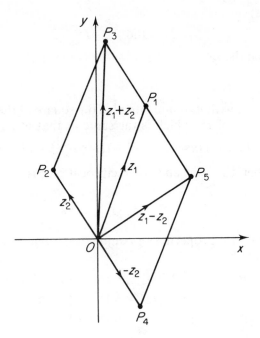

Fig. 15.6

EXERCISE 15.5

1 Let $\overrightarrow{OA}$ and $\overrightarrow{OC}$ be the vectors representing w and z, $w = 1 + 2i$ and $z = 4 + 3i$, respectively. Let B be the fourth vertex of the parallelogram $OABC$. Express $\overrightarrow{OB}$ in terms of w and z. Calculate $|w|$, $|z|$ and $|w + z|$, and show that $|w| + |z| > |w + z|$.

2 Find three complex numbers, z_1, z_2 and z_3, all distinct, such that

$$|z_1 + z_2| = |z_1| + |z_2|, \quad ||z_1| - |z_3|| = |z_1 - z_3|.$$

3 Given that $z_1 = 4 + i$ and $z_2 = 3 - 2i$, verify that:
   (i) $|z_1| + |z_2| > |z_1 + z_2|$, (ii) $|z_1|^2 + |z_2|^2 < |z_1 + z_2|^2$.

4 Show, geometrically or otherwise, that for any two complex numbers z_1, z_2,

$$|z_1 + z_2| \geqslant |z_1| - |z_2|. \tag{L}$$

15.6 Complex conjugates

Let z be a complex number, with $z = x + iy$, x, y real. Then

the *conjugate* of z is defined as z^*, where $z^* = x - iy$.

This will mean that z is the conjugate of z^* so we call z and z^* a *pair of conjugates*. If $P(x, y)$ represents z in the complex plane, then $P^*(x, -y)$ represents z^*, so that P and P^* are reflections of one another in the real axis Ox. Also

$$z + z^* = 2x \quad \text{and} \quad z - z^* = 2iy,$$

which means that the real part of z is $\frac{1}{2}(z + z^*)$ and the imaginary part of z is $\frac{1}{2i}(z - z^*)$.

The product of z with its conjugate z^* is the square of the modulus of z (or of the modulus of z^*, which is the same as that of z), for

$$zz^* = (x + iy)(x - iy) = x^2 - (iy)^2 = x^2 + y^2 = |z|^2.$$

So, if $|z| = r$, then $zz^* = r^2$ and the reciprocal of z is

$$z^{-1} = \frac{z^*}{r^2}.$$

We list some useful properties of conjugates, leaving the proofs as an exercise to the reader:

$$(z^*)^* = z; \; z + z^* \text{ and } zz^* \text{ are both real numbers;}$$
$$(z_1 + z_2)^* = z_1^* + z_2^*; \quad (z_1 - z_2)^* = z_1^* - z_2^*;$$
$$(z_1 z_2)^* = z_1^* z_2^*; \qquad (z_1/z_2)^* = z_1^*/z_2^*.$$

EXERCISE 15.6A

1 Write down the complex conjugate of:

(i) $2 + 3i$, (ii) $3 - 4i$, (iii) 4, (iv) $3i$, (v) $\left(2, \dfrac{\pi}{4}\right)$, (vi) $\left(3, -\dfrac{3\pi}{4}\right)$.

2 Given that $z_1 = 3 + 4i$ and $z_2 = 4 - 3i$, verify that:
   (i) $(z_1^*)^* = z_1$, (ii) $(z_1 + z_2)^* = z_1^* + z_2^*$, (iii) $(z_1 - z_2)^* = z_1^* - z_2^*$,
   (iv) $(z_1 z_2)^* = z_1^* z_2^*$, (v) $(z_1/z_2)^* = z_1^*/z_2^*$, (vi) $z_1^{-1} = \dfrac{z_1^*}{|z_1|^2}$.

3 Writing z_1 and z_2 in Cartesian form, $z_1 = x_1 + iy_1$, $z_2 = x_2 + iy_2$, prove that:
   (i) $(z_1 + z_2)^* = z_1^* + z_2^*$, (ii) $(z_1 - z_2)^* = z_1^* - z_2^*$.

4 Writing z_1 and z_2 in polar form, $z_1 = (r_1, \theta_1)$, $z_2 = (r_2, \theta_2)$, prove that
   $(z_1 z_2)^* = z_1^* z_2^*$ and that $(z_1/z_2)^* = z_1^*/z_2^*$.

5 Prove that, for a complex number z:
   (i) $|z| = |z^*|$, (ii) $\arg(z) + \arg(z^*) = 0$, (iii) $\arg(zz^*) = 0$,
   (iv) $\arg(z/z^*) = 2 \arg(z)$.

6 Given that $z = -5 + 12i$, draw an Argand diagram on which you represent the complex numbers:
   (i) z, (ii) z^*, (iii) zz^*, (iv) $z + z^*$, (v) $z - z^*$, (vi) z/z^*, (vii) z^2,
   (viii) $(z^*)^2$, (ix) $(z^2)^*$, (x) $(z + z^*)^*$.

Roots of real polynomial equations

Consider a quadratic equation, with real coefficients and complex roots, that is, $ax^2 + bx + c = 0$ with a, b, c real numbers and with $\Delta = b^2 - 4ac < 0$. Let $\Delta = -k^2$, so that $\sqrt{\Delta} = ik$, then the roots of the quadratic equation are

$$z_1 = \frac{-b}{2a} + \frac{k}{2a}i \quad \text{and} \quad z_2 = \frac{-b}{2a} - \frac{k}{2a}i = z_1^*.$$

So the roots of a real quadratic equation are complex conjugates, if they are not real.

Let $f(x) = ax^2 + bx + c$ be a real quadratic function, so that $a, b, c \in \mathbb{R}$. Then, if z is a complex number $f(z) = az^2 + bz + c$ and so

$$f(z^*) = a(z^*)^2 + bz^* + c = a(z^2)^* + bz^* + c = (az^2 + bz + c)^* = (f(z))^*.$$

Therefore, if z is a root of the quadratic equation $f(x) = 0$, we find that $f(z) = 0$ and so $f(z^*) = f(z)^* = 0^* = 0$ and so z^* is also a root of the equation $f(x) = 0$. If z is a real number then $z = z^*$ and we have no extra information. However, if z is complex (and not real) we have a *second* root of the equation, namely $x = z^*$.

The proof of the factor theorem (§7.5) involved the manipulation of polynomials by means of the four operations of arithmetic, finding quotients and remainders, and all of this applies equally well to complex numbers as to real numbers. Therefore the factor theorem is repeated for complex numbers.

Factor theorem   Let $f(x)$ be a polynomial in x, with complex or real coefficients, and let α be a complex number, then

$$(x - \alpha) \text{ is a factor of } f(x) \Leftrightarrow f(\alpha) = 0.$$

We apply this theorem to the quadratic function $f(x)$ with real coefficients, which has a complex (non-real) root z. Then $f(z) = 0$ and so $f(z^*) = f(z)^* = 0$, so that the roots are z and z^*. Then

$$f(x) = ax^2 + bx + c = a(x - z)(x - z^*).$$

Comparing coefficients of the powers of x,

$$\text{the sum of the roots} = z + z^* = -\frac{b}{a},$$

$$\text{the product of the roots} = zz^* = \frac{c}{a}.$$

These are the same results as were true for real roots. The results can be found by direct calculation from the form of the roots in terms of $k = -i\sqrt{\Delta}$, that is

$$z = \frac{-b + ki}{2a}, \quad \text{and} \quad z^* = \frac{-b - ki}{2a}.$$

More generally, suppose that $f(x)$ is a polynomial function, with real coefficients,

$$f(x) = a_n x^n + a_{n-1} x^{n-1} + \ldots + a_1 x + a_0, \; a_i \in \mathbb{R}, \; 0 \leqslant i \leqslant n.$$

Then, using the properties of conjugates, it will be seen that, for $z \in \mathbb{C}$,

$$f(z^*) \equiv a_n z^{*n} + a_{n-1} z^{*n-1} + \ldots + a_1 z^* + a_0 = (f(z))^*.$$

If $x = z$ is a non-real solution of $f(x) = 0$, then $f(z) = 0$, so $f(z^*) = f(z)^* = 0$ and $x = z^*$ is also a solution of $f(x) = 0$. Therefore, $f(x)$ has as factors $(x - z)$ and $(x - z^*)$. This means that $f(x)$ has a quadratic factor

$$(x - z)(x - z^*) = x^2 - (z + z^*)x + zz^*,$$

which is a quadratic with real coefficients, since $z + z^*$ and zz^* are both real.

Thus a polynomial equation with *real* coefficients will have roots, which may be real or may be complex. But the *non-real roots* will occur in *conjugate pairs* and they will give rise to *quadratic factors* with *real* coefficients.

EXAMPLE   *Given that the polynomial equation $x^4 + x^3 - 2x^2 + 2x + 4 = 0$ has a root $1 - i$, find all the roots of this equation.*

Since the coefficients of the quartic are all real and the equation has the root $1 - i$, it must also have a root $(1 - i)^* = 1 + i$. Thus the quadratic factor is

$$(x - 1 + i)(x - 1 - i) = x^2 - 2x + 2.$$

We now divide this quadratic into the quartic to give the other quadratic factor

$$x^4 + x^3 - 2x^2 + 2x + 4 = (x^2 - 2x + 2)(x^2 + 3x + 2).$$

This second quadratic factor can be factorised by inspection as $(x + 2)(x + 1)$. By use of the factor theorem, the roots of the quartic equation are $1 - i, 1 + i, -2, -1$.

EXERCISE 15.6B

1   The equation $x^3 - 7x^2 + 16x - 10 = 0$ has a root $3 + i$. Find the other two roots of this equation.

2   Given that $1 + 2i$ is one root of the equation

$$x^4 - 4x^3 - 15x^2 + 38x - 120 = 0$$

solve the equation.

3   Given that $2 + i$ is a root of the equation

$$z^3 = 11z + 20 = 0,$$

find the remaining roots.   (L)

4   Given that $z = 2 + 3i$ is a root of the equation

$$z^3 - 6z^2 + 21z - 26 = 0,$$

find the other two roots.   (L)

5   (i) Given that $3 + 4i = (x + iy)^2$, find x and y. Find also the square roots of i in the form $a + ib$, where a and b are real.
    (ii) Show that $1 + i$ is a root of the equation

$$z^3 - 4z^2 + 6z - 4 = 0$$

and find the other roots of this equation.   (L)

6 Obtain a quadratic function $f(z) = z^2 + az + b$, where a and b are real
   constants, such that $f(-1 - 2i) = 0$. (L)
7 (i) The equation $x^4 - 4x^3 + 3x^2 + 2x - 6 = 0$ has a root at $1 - i$. Find the three
   other roots.
   (ii) Given that $1, \omega_1, \omega_2$ are the roots of the equation $z^3 = 1$ express ω_1 and ω_2
   in the form $x + iy$ and hence, or otherwise, show that

   (a) $1 + \omega_1 + \omega_2 = 0$, (b) $\dfrac{1}{\omega_1} = \omega_2$. (L)

15.7 Loci in the complex plane

The modulus and argument of a complex number give rise to algebraic
descriptions of lines and circles in the complex plane. This is demonstrated
by means of examples.

EXAMPLE 1 *Sketch the loci given by each of the two equations; $|z| = 2$ and
$\arg(z) = \pi/3$. Find the complex number represented by the point of intersection of
these loci.*

Since $|z|$ is the distance of the point z in the Argand diagram from the origin, the
equation $|z| = 2$ gives the set of all points at a distance 2 units from O. The locus is
the circle, centre O and radius 2. If the argument of z is $\pi/3$, then the vector
representing z makes an angle $\pi/3$ with the x-axis. The locus is therefore a *half-line*,
starting from the origin O, at an angle of $\pi/3$ with Ox. The two loci are shown in
Fig. 15.7. They intersect at just one point, where $|z| = 2$ and $\arg(z) = \pi/3$. This is
the point $z = (2, \pi/3) = 1 + i\sqrt{3}$.

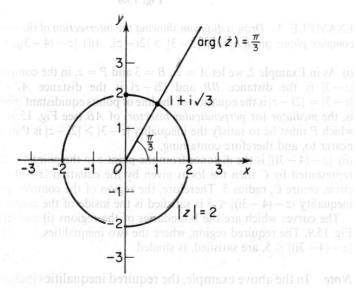

Fig. 15.7

EXAMPLE 2 *Describe the locus of z in the complex plane given by*
arg $(z+3) = -\pi/4$.

Since $z+3 = z-(-3)$, if the complex numbers z and -3 are represented by
P and A, then
$$\vec{AP} = \vec{OP} - \vec{OA} = z-(-3) = z+3.$$

Therefore, the locus is the set of points, P, such that $\vec{AP}$ makes an angle $-\pi/4$ with
the real axis, as shown in Fig. 15.8.

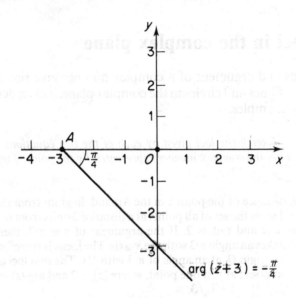

Fig. 15.8

EXAMPLE 3 *Draw a diagram showing the intersection of the two regions in the*
complex plane; given by: (i) $|z-3| \geqslant |2i-z|$, *(ii)* $|z-(4-3i)| \leqslant 5$.

(i) As in Example 2, we let $A = 2i$, $B = 3$ and $P = z$, in the complex plane. Then
$|z-3|$ is the distance BP and $|2i-z|$ is the distance AP. The equation
$|z-3| = |2i-z|$ is the equation of the line of points equidistant from A and B, that
is, the *mediator* (or *perpendicular bisector*) of AB (see Fig. 15.9). The region in
which P must lie to satisfy the inequality $|z-3| \geqslant |2i-z|$ is that side of the line
nearer to, and therefore containing, A.
(ii) $|z-(4-3i)|$ is the distance from the point z to the point $4-3i$. Let $4-3i$ be
represented by C then the locus given by the equation $|z-(4-3i)| = 5$ is the
circle, centre C, radius 5. Therefore, the region of the complex plane where the
inequality $|z-(4-3i)| \leqslant 5$ is satisfied is the inside of the above circle.
   The curves which are the boundaries of the regions (i) and (ii) are shown in
Fig. 15.9. The required region, where the two inequalities, $|z-3| \geqslant |2i-z|$ and
$|z-(4-3i)| \leqslant 5$, are satisfied, is shaded.

Note   In the above example, the required inequalities include *equality*, so
the required region of the plane includes the boundary of the region.

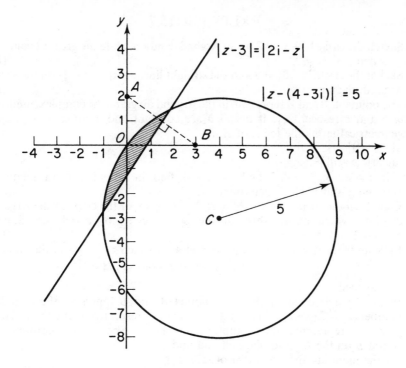

Shaded region: $|z-3| \geq |2i - z|$ and $|z - (4-3i)| \leq 5$

Fig. 15.9

Therefore, the boundary is shown as a solid curve. For a strict inequality, with equality excluded, the boundary could be dotted.

EXAMPLE 4 *The point P in an Argand diagram represents a complex number z. Given that z satisfies the equation $2|z - 2| = |z - 6i|$, show that P lies on a circle and find the radius and the centre of the circle.*

Let $z = x + iy$, then the equation can be manipulated as follows:

$$2|z - 2| = |z - 6i| \Leftrightarrow 2|(x - 2) + iy| = |x + i(y - 6)|$$
$$\Leftrightarrow 2\sqrt{\{(x-2)^2 + y^2\}} = \sqrt{\{x^2 + (y-6)^2\}}$$
$$\Leftrightarrow 4(x-2)^2 + 4y^2 = x^2 + (y-6)^2$$
$$\Leftrightarrow 4x^2 - 16x + 16 + 4y^2 = x^2 + y^2 - 12y + 36$$
$$\Leftrightarrow 3x^2 - 16x + 3y^2 + 12y = 20$$
$$\Leftrightarrow 3(x - 8/3)^2 + 3(y + 2)^2 = 20 + 64/3 + 12$$
$$\Leftrightarrow (x - 8/3)^2 + (y + 2)^2 = 160/9 = \left(\frac{4\sqrt{(10)}}{3}\right)^2.$$

The final equation is of a circle with **centre $\frac{8}{3} - 2i$** and **radius $(\frac{4}{3})\sqrt{(10)}$**.

EXERCISE 15.7

1 Sketch the circle $|z - 3| = 2$ on an Argand diagram. State the greatest value of $|z|$ when $|z - 3| = 2$.   (L)

2 Sketch the circle $|z - 2| = 2$ and the straight line $|z - i| = |z - 1|$ on the same Argand diagram.   (L)

3 The points A, B and Z represent, in the Argand diagram, the complex numbers a, b and z respectively. If a and b are constant and z varies, describe in geometrical terms the locus of Z if
  (i) $|z - a| = |z - b|$,
  (ii) $\arg(z - a) = \frac{1}{4}\pi$.
  If $a = 2 + 3i$, $b = -7i$ and $z = x + yi$, find, in simplified form, relations between x and y which determine each locus.   (C)

4 Express in the form $|z - a - ib| = R$ the equation of the circle in the Argand diagram which passes through the points given by $z = 4$, $z = 2i$ and $z = 4 + 2i$.   (L)

5 Indicate in an Argand diagram the region in which z lies, given that both
$$|z - 1| \leqslant 1 \quad \text{and} \quad 0 \leqslant \arg z \leqslant \pi/4$$
are satisfied.

6 In the Argand diagram the fixed points A and B represent the complex numbers $z_1 = 2(\cos \pi/3 + i \sin \pi/3)$ and $z_2 = -2$ respectively, and the variable point P represents the complex number z. Show the positions of A and B on the Argand diagram and find
  (i) the modulus and argument of $(z_1 - z_2)$,
  (ii) the Cartesian equation of the locus of P if $\arg(z - z_1) = 2\pi/3$, showing the locus on the Argand diagram.   (JMB)

MISCELLANEOUS EXERCISE 15

1 Given that $z_1 = 2 + i$, $z_2 = 1 - 2i$ and $\dfrac{1}{z_3} = \dfrac{1}{z_1} - \dfrac{1}{z_2}$, find z_3 in the form $a + ib$, where a and b are real.   (L)

2 Given that $z_1 = 2 + i$ and $z_2 = 1 + 2i$, express z_1/z_2 in the form $a + ib$, where a and b are real.   (L)

3 Find the complex numbers z and w satisfying the simultaneous equations
$$iz + (1 + 2i)w = 1,$$
$$(2 + i)z + (2 - i)w = -3.$$ (JMB)

4 Express $(6 + 5i)(7 + 2i)$ in the form $a + ib$. Write down $(6 - 5i)(7 - 2i)$ in a similar form. Hence find the prime factors of $32^2 + 47^2$.   (JMB)

5 Given that $(1 - i)$ is a root of the equation
$$z^3 - 4z^2 + 6z - 4 = 0,$$
find the other roots.   (L)

6 If $z^2 = 3 - i4$, find the two values of z in the form $a + ib$, where a and b are real.   (L)

7 Find, in a simple form not involving i the quadratic equation whose roots are $r(\cos \theta \pm i \sin \theta)$, where $r > 0$.
  If this equation is $x^2 + 2px + q = 0$ deduce that $r = \sqrt{q}$, and express $\cos \theta$ in terms of p and q.   (SMP)

8 The roots of the quadratic equation $z^2 + pz + q = 0$ are $1 + i$ and $4 + 3i$. Find the complex numbers p and q.

It is given that $1 + i$ is also a root of the equation $z^2 + (a + 2i)z + 5 + ib = 0$, where a and b are real. Determine the values of a and b. *(JMB)*

9 Verify that both $-\frac{1}{2}$ and i are roots of the equation

$$2z^5 - 7z^4 + 8z^3 - 2z^2 + 6z + 5 = 0.$$

Find all the roots of this equation.

Show that the quartic equation satisfied by the non-real roots can be written in the form

$$(z - p)^4 + q = 0,$$

where p and q are integers, and so find the four fourth roots of -4, expressing each root in the form $a + ib$. *(JMB)*

10 (i) Given that α and β are the roots of the quadratic equation

$$2x^2 - 3x + 4 = 0,$$

form an equation with integer coefficients whose roots are $\alpha(\alpha + \beta)$ and $\beta(\alpha + \beta)$.

(ii) Find the set of real values of k, $k \neq -1$, for which the roots of the equation

$$x^2 + 4x - 1 + k(x^2 + 2x + 1) = 0$$

are

(a) real and distinct,

(b) real and equal,

(c) complex with positive real part. *(L)*

11 Given that the roots of the equation $x^2 - 2x + 3 = 0$ are α and β, find a quadratic equation whose roots are α^2 and β^2. *(L)*

12 In an Argand diagram, the origin and the point representing the complex number $(1 + i)$ form two vertices of an equilateral triangle. Find, in any form, the complex number represented by the third vertex, given that its real part is positive. *(L)*

13 (i) Express in modulus-argument form the complex numbers

(a) $-1 + i\sqrt{3}$, (b) $\dfrac{(1 + i)}{(1 - i)}$.

(ii) Find the pairs of values of the real constants a and b such that $7 + 24i = -(a + ib)^2$.

(iii) Three complex numbers α, β and γ are represented in the Argand diagram by the three points A, B and C respectively. Find the complex number represented by D when $ABCD$ forms a parallelogram having BD as a diagonal. *(L)*

14 (i) Given that $z_1 = 3 + 4i$ and $z_2 = -1 + 2i$, represent z_1, z_2, $(z_1 + z_2)$ and $(z_2 - z_1)$ by vectors in the Argand diagram. Express $(z_1 + z_2)/(z_2 - z_1)$ in the form $a + ib$, where a and b are real. Find the magnitude of the angle between the vectors representing $(z_1 + z_2)$ and $(z_2 - z_1)$.

(ii) One root of the equation $z^3 - 6z^2 + 13z + k = 0$, where k is real, is $z = 2 + i$. Find the other roots and the value of k. *(L)*

15 Given that $z_1 = -\dfrac{1}{2} + \dfrac{i\sqrt{3}}{2}$, obtain $|z_1|$ and $\arg z_1$. Represent z_1, $1/z_1$ and $(z_1 - 1/z_1)$ by vectors in an Argand diagram.

Find in algebraic form all the roots of the equation $z^3 - 1 = 0$. Represent these roots by points on an Argand diagram, indicating their polar coordinates.   (L)

16 Two non-zero complex numbers z_1 and z_2 are such that $|z_1 + z_2| = |z_1 - z_2|$. Represent $z_1, z_2, z_1 + z_2$ and $z_1 - z_2$ by vectors on an Argand diagram. Hence, or otherwise, find the possible values of $\arg\left(\dfrac{z_1}{z_2}\right)$.   (JMB)

17 In the complex plane the points P_1, P_2 correspond to the numbers $\alpha_1 + \beta_1 i$, $\alpha_2 + \beta_2 i$. What complex number represents the displacement $\overrightarrow{P_1 P_2}$, and what complex number can be used by multiplication to rotate $\overrightarrow{P_1 P_2}$ anticlockwise through an angle θ and enlarge $|P_1 P_2|$ by a factor k?

The points A, B and C correspond respectively to the numbers $4\sqrt{3} + 2i$, $5\sqrt{3} + i$ and $6\sqrt{3} + 4i$. Write down the complex numbers z_1, z_2 representing the displacements $\overrightarrow{AB}, \overrightarrow{AC}$. Find z_3 such that $z_2 = z_3 z_1$. By writing z_3 in modulus-argument form, or otherwise, show that ABC is half an equilateral triangle ADC and give the complex number to which D corresponds.

The triangle ABC is now rotated through an angle of $\frac{1}{3}\pi$ anticlockwise about B. Find the new position of A.   (SMP)

18 Given that $z_1 = 1 + i\sqrt{3}, z_2 = 1 - i\sqrt{3}$, plot the points corresponding to z_1 and z_2 on an Argand diagram, and obtain the modulus and argument of each.

Show that $z_1^3 = z_2^3$, and indicate the points corresponding to $z_1 + z_2$, $z_1 z_2$ and z_1^3 on your diagram. Hence, or otherwise, obtain a quadratic equation with real coefficients which has z_1 as a root.

Find the values of the real constants a and b if the equation

$$z^3 + z^2 + az + b = 0$$

also has z_1 as a root.   (L)

19 If $a = \cos 3\theta + i \sin 3\theta$ and $b = \cos 7\theta + i \sin 7\theta$, find the modulus and argument of the complex number $c = a + b$. (You may assume that $0 < \theta < \frac{1}{10}\pi$.)   (SMP)

20 Sketch the locus in the Argand diagram of the point P representing z, where $|z - 1 + i| = 2$.   (L)

21 Find the modulus and argument of $z_1 = \sqrt{3} + i$. If $z_2 = \sqrt{3} - i$ express $q = z_1/z_2$ in the form $a + ib$ where a and b are real.

Plot z_1, z_2 and q on an Argand diagram and sketch the curve given by the equation

$$|z - z_2| = |q - z_1|.$$ (L)

22 Indicate on an Argand diagram the region in which z lies, given that both

$$|z - (3 + i)| \leqslant 3 \quad \text{and} \quad \frac{\pi}{4} \leqslant \arg[z - (1 + i)] \leqslant \frac{\pi}{2}$$

are satisfied.   (JMB)

23 (a) Express in the form $r(\cos\theta + i\sin\theta)$, where $r > 0$ and $180° < \theta \leqslant 180°$, each of the complex numbers $10i$, $-3 + 4i$, $\dfrac{10i}{-3 + 4i}$, giving each value of θ to the nearest $0\cdot1°$.

(b) In each of the following two cases, show, by a clear drawing in an Argand diagram, the set of points representing z, given that

(i) $\arg(z-1) = \frac{1}{4}\pi$,

(ii) $\arg(z+i) - \arg(z-i) = \frac{1}{2}\pi$.

Using Cartesian co-ordinates, describe each of these sets.   (C)

24   Given that z is a variable complex number such that

$$|z - 2 - 2i| = 2,$$

show, on an Argand diagram, the locus of the point P which represents z. Hence, or otherwise, find

(i) the greatest value of $|z|$,

(ii) z in the form $x + iy$ when $\arg(z - 2 - 2i) = 2\pi/3$.   (JMB)

25   The real number 5 is represented on the Argand diagram by the point A and the complex numbers u, v and w are represented by the points U, V and W respectively. Interpret geometrically the conditions that

(i) $|u - 5| = 3$,   (ii) $\arg\left(\dfrac{u-5}{v-5}\right) = \pi/2$,   (iii) $w - u = u - 5$.

Given that these three conditions are all satisfied and also that

$$\left|\frac{u-5}{v-5}\right| = \left|\frac{v-5}{w-5}\right|$$

show that the triangles UAV and VAW are similar and find the length of AV.

   (JMB)

26   (i) Find, without the use of tables, the two square roots of $5 - 12i$ in the form $x + iy$, where x and y are real.

(ii) Represent on an Argand diagram the loci $|z - 2| = 2$ and $|z - 4| = 2$. Calculate the complex numbers corresponding to the points of intersection of these loci.   (L)

27   The complex numbers $z_1 = \dfrac{a}{1+i}$, $z_2 = \dfrac{b}{1+2i}$, where a and b are real, are such that $z_1 + z_2 = 1$. Find a and b.

With these values of a and b, find the distance between the points which represent z_1 and z_2 in the Argand diagram.

28   Express the complex number $z_1 = \dfrac{11 + 2i}{3 - 4i}$ in the form $x + iy$ where x and y are real. Given that

$$z_2 = 2 - 5i,$$

find the distance between the points in the Argand diagram which represent z_1 and z_2.

Determine the real numbers α, β such that

$$\alpha z_1 + \beta z_2 = -4 + i.$$ (JMB)

29   Find the complex roots z_1, z_2 of the equation $z^2 - (\sqrt{3})z + 3 = 0$ and mark on an Argand diagram the points Z_1 and Z_2 representing them.

The points Z_1, Z_2, Z_3, Z_4 form a square $Z_1 Z_2 Z_3 Z_4$ which encloses the origin. Find the complex numbers z_3 and z_4 represented by the points Z_3 and Z_4.

Form a quadratic equation (with numerical coefficients) having roots z_3 and z_4.   (C)

30   Given that $z = 1 + i\sqrt{2}$, express in the form $a + ib$ each of the complex numbers

$$p = z + \frac{1}{z} \quad \text{and} \quad q = z - \frac{1}{z}.$$

In an Argand diagram, P and Q are the points which represent p and q respectively, O is the origin, M is the midpoint of PQ and G is the point on OM such that $OG = \frac{2}{3}OM$. Prove that the angle PGQ is a right angle.

(*JMB*)

31   If $z = \cos\theta + i\sin\theta$, prove that $\dfrac{1+z}{1-z} = i\cot\frac{1}{2}\theta$.

In an Argand diagram, the points O, A, Z, P, Q represent the complex numbers $0, 1, z, 1 + z$ and $1 - z$ respectively. Show these points on a diagram.

Prove that $P\hat{O}Q = \frac{1}{2}\pi$ and find, in terms of θ, the ratio $\left|\dfrac{OP}{OQ}\right|$.   (*C*)

32   (i)   Express the complex number $\dfrac{5 + 12i}{3 + 4i}$ in the form $a + ib$ and in the form $r(\cos\theta + i\sin\theta)$, giving the values of $a, b, r, \cos\theta, \sin\theta$.

(ii)   The points in the Argand diagram representing the complex numbers z_1, z_2, w_1, w_2 are Z_1, Z_2, W_1, W_2, and O is the origin. If the triangles $OZ_1 Z_2$ and $OW_1 W_2$ are similar (and in the same sense) with the angles O, Z_1, Z_2 of the first triangle corresponding respectively to the angles O, W_1, W_2 of the second triangle, find the relation connecting z_1, z_2, w_1, w_2.   (*L*)

33   Two complex numbers z_1 and z_2 are given by $z_1 = 10 - 2i$ and $z_2 = 2 - 3i$.

Show that $\dfrac{z_1}{z_2}$ is of the form $k(1 + i)$ where k is real.

If $z = x + yi$ is any non-zero complex number such that $\dfrac{z_1}{z} = a(1 + i)$ where a is real, prove that $3x + 2y = 0$. Hence find the possible values of $\arg z$, giving your answers in degrees correct to the nearest degree. (Take $\arg z$ to be in the interval $-180° < \arg z \leqslant 180°$.)   (*C*)

34   (a)   Express the complex number $z = \dfrac{1 + i}{3 - 2i}$ in the form $a + bi$ ($a, b \in \mathbb{R}$), and find the argument of z, giving your answer in radians correct to three significant figures. Write down the argument of z^2, and find the exact value of $|z^2|$.

(b)   The conjugate of the non-zero complex number z is denoted by z^*. In an Argand diagram with origin O, the point P represents z and Q represents $1/z^*$. Prove that O, P and Q are collinear, and find the ratio $OP:OQ$ in terms of $|z|$.   (*C*)

35   In an Argand diagram, the point P represents the complex number z, where $z = x + iy$. Given that

$$z + 2 = \lambda i(z + 8),$$

where λ is a real parameter, find the Cartesian equation of the locus of P as λ varies.

If also

$$z = \mu(4 + 3i),$$

where μ is real, prove that there is only one possible position for P.

(*JMB*)

36 Sketch the circle C with Cartesian equation $x^2 + (y-1)^2 = 1$. The point P, representing the non-zero complex number z, lies on C. Express $|z|$ in terms of θ, the argument of z. Given that $z' = 1/z$, find the modulus and argument of z' in terms of θ.

    Show that, whatever the position of P on the circle C, the point P' representing z' lies on a certain line, the equation of which is to be determined.

(JMB)

37 (i) Represent on the same Argand diagram the loci given by the equations $|z-3| = 3$ and $|z| = |z-2|$. Obtain the complex numbers corresponding to the points of intersection of these loci.

(ii) Find a complex number z whose argument is $\pi/4$ and which satisfies the equation

$$|z+2+i| = |z-4+i|.$$

(L)

38 (i) Given that $(1+5i)p - 2q = 3 + 7i$, find p and q when
(a) p and q are real, (b) p and q are conjugate complex numbers.

(ii) Shade on the Argand diagram the region for which $3\pi/4 < \arg z < \pi$ and $0 < |z| < 1$. Choose a point in the region and label it A. If A represents the complex number z, label clearly the points B, C, D and E which represent $-z$, iz, $z+1$ and z^2 respectively.

(L)

39 The complex number $z = x + iy$ is such that $\dfrac{z+3}{z-4i} = \lambda i$ where λ is real. Show that

$$x(x+3) + y(y-4) = 0.$$

    If z is represented in an Argand diagram by the point P, deduce that, when λ varies, P moves on a circle. Find the radius of this circle and the complex number corresponding to its centre.

(JMB)

40 (i) Given that x and y are real, find the values of x and y which satisfy the equation

$$\frac{2y+4i}{2x+y} - \frac{y}{x-i} = 0.$$

(ii) Given that $z = x + iy$, where x and y are real,

(a) show that, when $\text{Im}\left(\dfrac{z+i}{z+2}\right) = 0$, the point (x, y) lies on a straight line,

(b) show that, when $\text{Re}\left(\dfrac{z+i}{z+2}\right) = 0$, the point (x, y) lies on a circle with centre $(-1, -\frac{1}{2})$ and radius $\frac{1}{2}\sqrt{5}$.

(L)

41 Given that $(a+ib)^2 = 1 + i2\sqrt{2}$, where a and b are real, find the possible pairs of values of a and b.

(JMB)

42 Given that α is the complex number $1 - \sqrt{3}i$, write $(z-\alpha)(z-\alpha^*)$ as a quadratic expression in z with real coefficients (where α^* denotes the complex conjugate of α).

    Express α in modulus-argument form. Deduce the values of α^2 and α^3, and show that α is a root of

$$z^3 - z^2 + 2z + 4 = 0.$$

Find all three roots of this equation.

    Plot these roots on the complex plane as triangle ABC (where vertex A corresponds to the root in the first quadrant, vertex B to the real root).

What complex number represents the displacement $\overrightarrow{BA}$? What complex number can be used by multiplication to rotate $\overrightarrow{BA}$ through 30° anticlockwise?

If, in a 30° anticlockwise rotation about B, $A \rightarrow A'$, find the complex number that corresponds to the point A'. (*SMP*)

43 Using a scale of 1 cm to a unit, show on graph paper the point representing the complex number z such that $|z| = 3$ and $\arg z = \frac{1}{6}\pi$.

Show also on your diagram the points representing $2z$, iz and $w = (2+i)z$. (This should be done by construction, not calculation.)

Draw an arrow on your diagram to show a displacement corresponding to the complex number $w + 1$. Measure $|w + 1|$, giving your answer to one place of decimals. Measure $\arg (w + 1)$, and give its value in radians to one place of decimals. (*SMP*)

44 A regular hexagon $ABCDEF$ in the Argand diagram has its centre at the origin O and its vertex A at the point $z = 2$.

(a) Indicate in a diagram the region in the hexagon in which both the inequalities $|z| \geqslant 1$ and $-\pi/3 \leqslant \arg z \leqslant \pi/3$ are satisfied.

(b) Find in the form $|z - c| = R$ the equation of the circle through the points O, B, F.

(c) Find the values of z corresponding to the points C and E.

(d) The hexagon is now rotated clockwise about the origin through an angle of 45°. Express in the form $r (\cos \theta + i \sin \theta)$ the values of z corresponding to the new positions of the points C and E. (*L*)

45 (i) If $\operatorname{Im} \left(\dfrac{2z + 1}{iz + 1} \right) = -2$, show that the locus of the point representing z in the Argand diagram is a straight line.

(ii) If z_1 and z_2 represent points in the Argand diagram, show geometrically, or otherwise, that

$$|z_1| + |z_2| \geqslant |z_1 + z_2| \geqslant |z_1| - |z_2|.$$ (*L*)

Answers to Part A

Exercise 1.1A (p. 5)

1 (i) $\{a,b,c,d,e,f\}$ (ii) $\{a,c\}$ (iii) ϕ (iv) ϕ (v) $\{a,b,c,d\}$ (vi) $\{a,b,c,d,e\}$
(vii) $\{b,d\}$ (viii) $\{a,c\}$ (ix) $\{a,b,c,d\}$ (x) ϕ.
2 $\{a,b,c\},\{a,b,d\},\{a,b,e\},\{a,c,d\},\{a,\ c,e\},\{a,d,e\},\{b,c,d\},\{b,c,e\},\{b,d,e\},\{c,d,e\}$.
3 (i) T (ii) T (iii) I (iv) I (v) F (vi) T (vii) T (viii) F (ix) T (x) F.
4 (i) $\{-2,2\}$ (ii) ϕ (iii) $\{-1\}$ (iv) $\{0,1,2,3,4,5\}$ (v) $\{\pm 4, \pm 5, \pm 6, \pm 7\}$.
5 (i) ϕ, $\{1\},\{2\},\{1,2\}$ (ii) ϕ, $\{a\},\{b\},\{c\},\{d\}, \{a,b\}, \{a,c\}, \{a,d\}, \{b,c\}$,
$\{b,d\}, \{c,d\}, \{a,b,c\}, \{a,b,d\}, \{a,c,d\}, \{b,c,d\}, \{a,b,c,d\}$.
6 (i) $\{0,2\}$ (ii) $\{2n+2:n\in\mathbb{N}\}$ (iii) $\{x:x \geqslant 3\}$ (iv) $\{x:2 \neq x$ and $x < 3\}$ (v) $\{2\}$
(vi) $\{0\}$ (vii) $\{0\}$ (viii) $\{x:x < 3$ or $x\in\mathbb{Z}\}$.
7 $x\in A\cap(B\cup C)\Leftrightarrow x\in A$ and $(x\in B$ or $x\in C)\Leftrightarrow (x\in A$ and $x\in B)$ or $(x\in A$ and $x\in C)$
$\Leftrightarrow x\in(A\cap B)\cup(A\cap C)$. Hence $A\cap(B\cup C) = (A\cap B)\cup(A\cap C)$.

Exercise 1.1B (p. 5)

1 (i) T (ii) I (iii) T (iv) T (v) I (vi) T (vii) F (viii) F.
2 (i) T (ii) F (iii) T (iv) F (v) F (vi) T (vii) T (viii) F (ix) T.
3 (i) $\{x: x$ is a positive odd number less than $9\}$ (ii) $\{x: x = n^3, n\in\mathbb{N}, 1 \leqslant n \leqslant 4\}$
(iii) $\{x:x(x^2-4) = 0\}$. **4** $\{-5, -4, -3, -2, -1, 0, 1, 2, 3, 4, 5\}$.
5 (i) T (ii) T (iii) F, any set containing just one integer (iv) F, $\mathbb{Z}\backslash\{0\}$
(v) T (vi) F, $\{0\}$ (vii) T (viii) T (ix) T (x) F, $\{0,1\}$ (xi) F, $\{-1,2\}$
(xii) T (xiii) F, $\{1\}$ (xiv) F, $\{2\}$ (xv) F, $\{x:x\in\mathbb{Z}, x \leqslant 1\}$.
6 A non-empty set A contains 2 subsets ϕ and A. The empty set is the only set with just
one subset. **7** (i) $\{a\}, \{a,b\}, \{a,c\}, \{a,d\}, \{a,b,c\}, \{a,b,d\}, \{a,c,d\}, \{a,b,c,d\}$
(ii) $\{a,b\}, \{a,b,c\}, \{a,b,d\}, \{a,b,c,d\}$. **8** (i) T (ii) F (iii) T (iv) T (v) T
(vi) T (vii) T (viii) F (ix) F; $A = (0,4)$, $B = (2,5)$, $C = (1,3)$, $A\cap B = (2,4)$,
$B\cup C = (1,5)$, $(A\cap B)\cup C = (1,4) = A\cap(B\cup C)$.
9 (i) $P \not\Rightarrow R$, e.g. $A = \phi$, $R \not\Rightarrow P$ (ii) $Q \not\Rightarrow R$, $R \not\Rightarrow Q$
(iii) if $A\cap B \neq \phi$, then it contains an element of A which is in B, conversely a member of A
which is in B is in $A\cap B$, so $R\Leftrightarrow S$ (iv) $A\cup B \neq \phi \not\Rightarrow A\cap B \neq \phi$, $T \not\Rightarrow R$, $R \Rightarrow T$.

Exercise 1.2 (p. 10)

Throughout, $\mathbb{R}_0^+$ is used as defined in question **3**.
1 (i) $\{2,4,6,8\}$ (ii) $\{1,2,3,4\}$ (iii) $\{-3,0,5,12\}$ (iv) $\{1, \sqrt{2}, \sqrt{3}, 2\}$.
2 (i) $0\in\mathbb{Z}$, $f(0)\notin\mathbb{N}$ (ii) 4 elements in domain so not more than 4 in range
(iii) $2\in\mathbb{N}$, $f(x)^2 = x$ does not give unique $f(x)$, $\{\sqrt{2}, -\sqrt{2}\}$ (iv) $f(0)$ not defined.
3 Let $\mathbb{R}_0^+ = \mathbb{R}^+ \cup \{0\}$; (i) yes, $\mathbb{R}$, $\mathbb{R}$ (ii) yes, $\mathbb{R}$ (iii) yes, $\mathbb{R}$, $\mathbb{R}_0^+$ (iv) no
(v) yes, $\mathbb{R}$, $\mathbb{R}$ (vi) no (vii) yes, $\{0\}$, $\{0\}$ (viii) yes, $\mathbb{R}_0^+$, $\mathbb{R}_0^+$. **4** (i) $\mathbb{R}$, $\mathbb{R}$ (ii) $\mathbb{R}$, $\mathbb{R}_0^+$
(iii) $\mathbb{R}$, $\mathbb{R}$ (iv) $\mathbb{R}_0^+$, $\mathbb{R}_0^+$ (v) $\mathbb{R}$, $\{x:x \geqslant -1\}$ (vi) $\mathbb{R}\backslash\{2\}$, $\mathbb{R}\backslash\{0\}$ (vii) $\mathbb{R}^+$, $\mathbb{R}$ (viii) $\mathbb{R}$, $\mathbb{R}^+$
(ix) $\{x:x \leqslant -2\} \cup \{x:x \geqslant 2\}$, $\mathbb{R}_0^+$ (x) $\{x:-3 \leqslant x \leqslant 3\}$, $\{x:0 \leqslant x \leqslant 3\}$.
5 (i) x^2, $\mathbb{R}$, $\mathbb{R}_0^+$ (ii) $\sqrt{(-x)}$, $\{x:x \leqslant 0\}$, $\mathbb{R}_0^+$ (iii) $2x^2 + \sqrt{(-x)}$, $\{x:x \leqslant 0\}$, $\mathbb{R}_0^+$

(iv) $3x^2\sqrt{(-x)}, \{x : x \leqslant 0\}, \mathbb{R}_0^+$ (v) $x, \{x : x \leqslant 0\}, \{x : x \leqslant 0\}$
(vi) $x^2/\sqrt{(-x)}, \{x : x < 0\}, \mathbb{R}^+$ (vii) $\sqrt{(-x)}/x^2, \{x : x < 0\}, \mathbb{R}^+$.   **6** (i) 2
(ii) 4 (iii) 0.   **7** (a), (c), (f).   **8** (i) $\mathbb{R}_0^+$ (ii) $\mathbb{Z}$ (iii) $\mathbb{R}_0^+$ (iv) $\{x : x \leqslant 0\}$ (v) $\mathbb{R}$
(vi) $\{1, -1\}$ (vii) $\{x : 0 \leqslant x < 1\}$.

Exercise 1.3 (p. 12)

1 (i) $3x$ (ii) $-\frac{1}{2}x$ (iii) mx.   **2** (i) $x/3$, $1/3$ (ii) $-3x/5$, $-3/5$ (iii) $2x/3$, $2/3$
(iv) $3x/2$, $3/2$ (v) ax/b, a/b, $(b \neq 0)$ (vi) $2(2-x)/3$, $-2/3$ (vii) $5x+3$, 5
(viii) $-5/3$, 0 (ix) $-(ax+c)/b$, $-a/b$, $b \neq 0$.   **3** $\{2\}$, graphs cross at $(2, 4)$.

4 $f(x) = \dfrac{5-3x}{4}$, $g(y) = \dfrac{5-4y}{3}$, $g(x) = \dfrac{5-4x}{3}$.   **5** -1, $3-x$, $x = 3$.

6 $AB, 6 - \dfrac{3x}{2}$; $BC, \dfrac{x-4}{8}$; $AC, \dfrac{2x+5}{3}$; $90°$, -1, -1.   **7** $(-\frac{5}{2}, 3)$.

Exercise 2.1 (p. 15)

1 (i) $y-x$ (ii) $2x$ (iii) a (iv) $xy+2x-y$ (v) x^2+4 (vi) $5x^2+6x+1$
(vii) $3a^2-12b^2$ (viii) -6 (ix) $-4a+13b-3c$ (x) $r^2-s^2+2st-t^2$ (xi) $2ab-2b^2$
(xii) $x^3+3x^2y+3xy^2+y^3$ (xiii) $8x^3-36x^2+54x-27$ (xiv) a^3-b^3
(xv) $x^2+4y^2+9z^2-4xy+6xz-12yz$ (xvi) x^4-2x^2+1 (xvii) x^6+1
(xviii) $2abc+a^2b+bc^2-ac^2-a^2c-ab^2-b^2c$.   **2** (i) $\dfrac{x+y-1}{xy}$ (ii) $\dfrac{a^2-b^2}{ab}$

(iii) $\dfrac{15-2x}{12}$ (iv) $\dfrac{3}{x}$ (v) $\dfrac{5}{12x}$ (vi) $\dfrac{3y}{2x}$ (vii) $\dfrac{2x^2-3y^2}{xy}$ (viii) $\dfrac{3x-2y}{x^2y^2}$

(ix) $\dfrac{4x}{x^2-1}$ (x) $\dfrac{2x^2}{x-y}$ (xi) $\dfrac{x^3+xy^2+x^2-y^2}{(x^2-y^2)x}$ (xii) $\dfrac{2xy+y^2}{x^2-y^2}$.

3 (i) $2(a+b)$ (ii) $x(x+3)$ (iii) $x^2(x-1)$ (iv) $(a+c)b$ (v) $(a+b)(a-b)$
(vi) $(x+2y)(x-2y)$ (vii) $(6t+2)(6t-2)$ (viii) $b(a+c)(a-c)$ (ix) $(2x-1)(x+2)$
(x) $(2t-5)(t+6)$ (xi) $(x+y+z)(x+y-z)$ (xii) $(\sqrt{2}x+1)(\sqrt{2}x-1)(x+1)$
(xiii) $(a-b)(a^2+ab+b^2)$ (xiv) $(2x-3y)(4x^2+6xy+9y^2)$ (xv) $(r-st)(r^2+rst+s^2t^2)$
(xvi) $2(x+y)(x^2-xy+y^2)$ (xvii) $(a-b)(a+b)(a^2+b^2)$
(xviii) $(y+1)(y-1)(y^2+y+1)(y^2-y+1)$.   **4** (i) $\dfrac{-2b}{a^2-b^2}$ (ii) $\dfrac{2}{a^2+ab+b^2}$

(iii) $-\dfrac{x+2}{x+3}$ (iv) $\dfrac{y}{y^2-x^2}$ (v) q (vi) -2 (vii) $\dfrac{-1}{36}$ (viii) $\dfrac{5x-7}{x^3-4x^2+5x-2}$

(ix) $\dfrac{5a-8}{2a^2-8}$ (x) $\dfrac{a^2+b^2}{a^2-b^2}$ (xi) $\dfrac{4x}{1-x^4}$ (xii) $\dfrac{y}{x+y}$.

5 (i) $\sqrt{3}$ (ii) $4+\sqrt{3}$ (iii) $2+\sqrt{2}$ (iv) $\sqrt{6}$ (v) $9\sqrt{2}$ (vi) $\sqrt{6}-2\sqrt{3}$ (vii) $2\sqrt{2}$
(viii) $5\sqrt{2}+3\sqrt{5}-\sqrt{(10)}$ (ix) $6\sqrt{6}$ (x) 3 (xi) 12 (xii) $\frac{1}{2}$ (xiii) $3+2\sqrt{2}$
(xiv) $5-2\sqrt{6}$ (xv) $5\sqrt{6}$ (xvi) -1 (xvii) $a-b$ (xviii) $26+15\sqrt{3}$.

6 (i) $\dfrac{\sqrt{3}}{3}$ (ii) $\dfrac{2\sqrt{5}}{5}$ (iii) $\dfrac{\sqrt{6}}{2}$ (iv) $\dfrac{\sqrt{2}}{6}$ (v) $\sqrt{2}+1$ (vi) $\dfrac{\sqrt{5}-\sqrt{2}}{3}$

(vii) $\dfrac{\sqrt{6}-\sqrt{2}}{2}$ (viii) $\dfrac{10-2\sqrt{2}}{23}$ (ix) $3-2\sqrt{2}$ (x) $\dfrac{2\sqrt{3}-\sqrt{6}}{6}$ (xi) $7+5\sqrt{2}$

(xii) $\sqrt{(30)}+5-6\sqrt{2}-2\sqrt{(15)}$.   **7** (i) $3+2\sqrt{2}$ (ii) $2+2\sqrt{2}-\sqrt{3}-\sqrt{6}$
(iii) $2+2\sqrt{2}+\sqrt{3}+\sqrt{6}$ (iv) $3+\sqrt{2}+\sqrt{3}$ (v) $\dfrac{3\sqrt{2}+\sqrt{6}}{6}$ (vi) $1+\sqrt{2}+\sqrt{3}$

(vii) $4+2\sqrt{6}$ (viii) $\dfrac{2+\sqrt{2}-\sqrt{6}}{4}$ (ix) $4\sqrt{2}+2\sqrt{3}$.

Exercise 2.2 (p. 17)

1 (i) yes (ii) no (iii) no (iv) no (v) yes (vi) yes.   **2** (i) $1, -2$ (ii) $-4, -5$
(iii) $\frac{1}{2}, -3$ (iv) $0, -\frac{2}{3}$ (v) 4 (vi) $\frac{5}{3}$ (vii) $\frac{3}{2}, -\frac{2}{3}$ (viii) a, b (ix) $-1, 3$ (x) $-2, 3$

(xi) $-\frac{1}{3}, \frac{11}{3}$ (xii) $\dfrac{b+c}{a}, \dfrac{b-c}{a}$.   **3** (i) $(x-5)(x+3), 5, -3$ (ii) $(x-4)(x+3), 4, -3$

(iii) $(x-6)(x+9), 6, -9$ (iv) $(x+6)(x+7), -6, -7$ (v) $(x+24)(x-13), -24, 13$
(vi) $(3x-1)(x-1), \frac{1}{3}, 1$ (vii) $(2x+3)(x-4), -\frac{3}{2}, 4$ (viii) $(5x+2)(x-4), -\frac{2}{5}, 4$
(ix) $2(2x+5)(2x+1), -\frac{5}{2}, -\frac{1}{2}$ (x) $2(7x+3)(2x-1), -\frac{3}{7}, \frac{1}{2}$ (xi) $(3x-5)(3x-4), \frac{5}{3}, \frac{4}{3}$
(xii) $(x-17)(x-16), 17, 16$ (xiii) $2(x-5)(6x+1), 5, -\frac{1}{6}$ (xiv) $(3x-1)(2x+3), \frac{1}{3}, -\frac{3}{2}$
(xv) $(2x-7)(x+2), \frac{7}{2}, -2$ (xvi) $(x-2)^2, 2$ (xvii) $(x+2)(-2), -2$
(xviii) $(2x+1)(x-3), -\frac{1}{2}, 3$ (xix) as (xviii) (xx) $2(x-1), 1$.
4 (i) $x^2+4x+4 = (x+2)^2$ (ii) $x^2-10x+25 = (x-5)^2$ (iii) $s^2-3s+\frac{9}{4} = (s-\frac{3}{2})^2$
(iv) $x^2+2ax+a^2 = (x+a)^2$ (v) $x^2+x+\frac{1}{4} = (x+\frac{1}{2})^2$ (vi) $x^2+4xy+4y^2 = (x+2y)^2$.

5 (i) 36 (ii) 9 (iii) a^2 (iv) $\frac{9}{4}$ (v) 4 (vi) 1 (vii) $\frac{1}{2}$ (viii) $\frac{25}{12}$ (ix) $\dfrac{b^2}{4a}$.

6 (i) $(x-2)^2 = 3, 2+\sqrt{3}, 2-\sqrt{3}$ (ii) $(x-1)^2 = 5, 1+\sqrt{5}, 1-\sqrt{5}$
(iii) $(x-3)^2 = 2, 3+\sqrt{2}, 3-\sqrt{2}$ (iv) $(x-\frac{3}{2})^2 = \frac{5}{4}, \dfrac{3+\sqrt{5}}{2}, \dfrac{3-\sqrt{5}}{2}$

(v) $(2x-3)^2 = 2, \dfrac{3+\sqrt{2}}{2}, \dfrac{3-\sqrt{2}}{2}$ (vi) $(3x-1)^2 = 5, \dfrac{1+\sqrt{5}}{3}, \dfrac{1-\sqrt{5}}{3}$

(vii) $(x-a)^2 = a^2-b, a+\sqrt{(a^2-b)}, a-\sqrt{(a^2-b)}$ (viii) $(ax-1)^2 = 3, \dfrac{1+\sqrt{3}}{a}, \dfrac{1-\sqrt{3}}{a}$

(ix) $a\left(x+\dfrac{b}{2a}\right)^2 = \dfrac{b^2}{4a}-c, \dfrac{-b+\sqrt{(b^2-4ac)}}{2a}, \dfrac{-b-\sqrt{(b^2-4ac)}}{2a}$.

Exercise 2.3 (p. 21)

1 $(-1, 0), x = -1$ (ii) $(1, -4), x = 1$ (iii) $(-\frac{1}{2}, -\frac{9}{4}), x = -\frac{1}{2}$ (iv) $(\frac{3}{4}, \frac{121}{8}), x = \frac{3}{4}$
(v) $(\frac{1}{3}, \frac{1}{3}), x = \frac{1}{3}$ (vi) $(\frac{3}{4}, \frac{25}{8}), x = \frac{3}{4}$.   **2** (i) $y-3 = (x-2)^2$, vertex downwards
(ii) $y-4 = -(x+3)^2$, vertex upwards.

Exercise 2.4 (p. 22)

1 As for question **3** of Exercise 2.2.

Exercise 2.5 (p. 24)

1 (i) $0, 1$ (ii) $-8, 0$ (iii) $0, 1$ (iv) $1, 2$.   **2** (i) 2 (ii) 2 (iii) 0 (iv) 1
(v) 2 (vi) 1.   **3** $x^2-8x+18 = (x-4)^2+2, \{x:x \geqslant 2\}$.   **4** (i) $\{x:x \geqslant 8\}, 0$
(ii) $\{x:x \geqslant -\frac{25}{12}\}, 2$ (iii) $\{x:x \geqslant \frac{23}{16}\}, 0$ (iv) $\{x:x \leqslant 3\}, 2$ (v) $\{x:x \leqslant \frac{1}{4}\}, 2$
(vi) $\{x:x \leqslant -\frac{3}{4}\}, 0$.   **5** (i) $1, 2, \{y:y \geqslant -\frac{13}{2}\}$ (ii) $1, 2, \{y:y \geqslant -\frac{125}{4}\}$
(iii) $1, 2, \{y:y \leqslant \frac{31}{4}\}$ (iv) $1, 1, \{y:y \leqslant 0\}$ (v) $1, 2, \{y:y \geqslant -1\}$ (vi) $1, 2, \{y:y \leqslant 1\}$.
6 (i) $f(x) = -\frac{1}{10}x^2-\frac{9}{10}x+2$ (ii) $f(x) = \frac{7}{6}x^2-\frac{7}{6}x-7$ (iii) $f(x) = x^2-4x+3$
(iv) $f(x) = -3x^2+6x+2$.   **7** $x = 5, x = -3$.   **8** $3(x-2)^2+2$, x-shift 2, y-stretch 3,
y-shift 2.   **10** $f(0) = -8, x = \frac{4}{3}, x = -2, \{x:x < -2 \text{ or } x > \frac{4}{3}\}$.

Exercise 2.6 (p. 25)

1 Roots are real if $\Delta \geqslant 0$. We list Δ, sum, product of roots (i) $29, -3, -5$ (ii) $28, \frac{2}{3}, -\frac{2}{3}$
(iii) $52, -\frac{5}{2}, \frac{3}{4}$ (iv) $20d^2, -4d, -d^2$ (v) $12, 4, 1$ (vi) $41, \frac{3}{2}, -2$.
2 (i) $-9, 9, 29, 25, \frac{3}{2}, 13, -27, 45, -72$
(iii) $-15/2, 25/4, 13/4, 9/16, -10/3, 53/4, -125/8, -45/8, -10$
(v) $12, 16, 12, 1, 4, 33, 64, 12, 52$.

3 (i) $x^2 - 19x + 25, 5x^2 - 3x - 1, x^2 + 9x + 13, x^2 - 29, 5x^2 + 12x - 16, x^2 + 72x - 125$
(iii) $16x^2 - 76x + 9, 3x^2 + 10x + 4, 4x^2 + 30x + 53, 4x^2 - 13, 12x^2 + 70x + 49,$
$64x^2 + 640x + 27$ (v) $x^2 - 14x + 1, x^2 - 4x + 1, x^2 - 12x + 33, x^2 - 12, x^2 - 8x + 4,$
$x^2 - 52x + 1.$ **4** $p = -3, q = 19.$ **5** $\dfrac{a - b + c}{a}.$

Miscellaneous Exercise 2 (p. 26)

2 $q^2 x^2 - (p^2 - 2q)x + 1 = 0.$ **3** Hint: $x = 1$ is the common root.
5 (a) $p = \frac{1}{3}$ or -1, (b) $p > \frac{1}{3}$ or $p < -1$; $p = -2$ vertex $(-\frac{1}{2}, -\frac{7}{2})$ up; $p = 1$ vertex $(1, 4)$
down.   **6** $a = -1$ or $\frac{1}{4}$.   **7** $n > 3$ or $n < -3.$ **8** (b) $k \leqslant 0$ or $k \geqslant 3$ (c) $k \geqslant 3.$
10 $x^2 - 47x + 1 = 0.$ **11** $(-\frac{1}{4}, -\frac{121}{8}), (\frac{7}{4}, -\frac{185}{8}), (-\frac{9}{4}, -\frac{185}{8}).$
12 $k \leqslant 2$ or $k \geqslant 18$, $\{x : x \leqslant 2\} \cup \{x : x \geqslant 18\}.$
13 $9q - 2p^2.$ **14** $p = -\alpha - \beta, q = \alpha\beta, x^2 + 2px + 16q - 3p^2 = 0.$
15 $k = 7.$ **16** $3bc - b^3 - c - c^2.$ **17** $0, 0$ or $-1, 0$ or $1, 1$ or $-2, 1.$

Exercise 3.1 (p. 30)

2 (a) $A = 53\cdot1°, C = 36\cdot9°$ (b) $AB = 4\cdot6, BC = 3\cdot9, C = 50°$ (c) $BC = 1\cdot8, AC = 5\cdot3.$
3 $196\cdot7°, 599$ km.   **4** (a) $17\cdot3$ m (b) $15\cdot3$ m.   **5** $0\cdot0577$ m.   **7** $(1 - 0\cdot8\cos\theta)$ m.
10 $\tan\frac{1}{2}\theta$, if $\theta < 90°.$ **11** (i) $\sqrt{(r^2 + h^2)}$ (ii) $h/\sqrt{(r^2 + h^2)}$ (iii) $r/\sqrt{(r^2 + h^2)}$
(iv) $\pi r \sqrt{(r^2 + h^2)}.$

Exercise 3.2A (p. 32)

1 $(+, +)(-, -)(+, -)(-, +)(+, +)(+, -)$

Exercise 3.2B (p. 35)

1 Angle	$\theta°$	$-\theta°$	$(180 + \theta)°$	$(180 - \theta)°$	$(90 + \theta)°$	$(90 - \theta)°$
Quadrant	1	4	3	2	2	1
	2	3	4	1	3	4
	3	2	1	4	4	3
	4	1	2	3	1	2

2 Let $\cos\theta = c, \sin\theta = s$: (i) c (ii) $-s$ (iii) $-s/c$ (iv) $-c$ (v) $-s$ (vi) s/c (vii) $-c$
(viii) s (ix) $-s/c$ (x) $-s$ (xi) c (xii) $-c/s$ (xiii) s (xiv) c (xv) $c/s.$
3 (i) F (ii) T (iii) F (iv) F (v) F (vi) T.   **4** (a) $\cos\theta = 3/5$ and $\tan\theta = 4/3$ or
$\cos\theta = -3/5$ and $\tan\theta = -4/3$ (b) $\cos\theta = 12/13$ and $\sin\theta = -5/13$ or
$\cos\theta = -12/13$ and $\sin\theta = 5/13.$

Exercise 3.2C (p. 36)

1 (a) $\mathbb{R} \setminus \{(2n + 1) \times 90 : n \in \mathbb{Z}\}, \{x : x \geqslant 1\} \cup \{x : x \leqslant -1\}, 360$
(b) $\mathbb{R} \setminus \{n \times 180 : n \in \mathbb{Z}\}, \{x : x \geqslant 1\} \cup \{x : x \leqslant -1\}, 360$ (c) $\mathbb{R} \setminus \{n \times 180 : n \in \mathbb{Z}\}, \mathbb{R}, 180.$
4 (i) $\cos\theta$ (ii) 1 (iii) 1 (iv) 2 (v) $\sec\theta$ (vi) $\tan\theta + \sec\theta$

Exercise 3.3 (p. 39)

1 (i) $\{210, 330\}$ (ii) $\{0, 45, 135, 180, 225, 315\}$ (iii) $\{60, 300\}$ (iv) $\{199\cdot5, 340\cdot5\}.$
2 (i) $\{90n : n \in \mathbb{Z}\}$ (ii) $\{36 + 72n, 360n : n \in \mathbb{Z}\}$ (iii) $\{30 + 120n, 90 + 360n : n \in \mathbb{Z}\}$
(iv) $\{135 + 180n : n \in \mathbb{Z}\}.$ **3** (i) $\{0\}$ (ii) $\{0, 90\}$ (iii) $\{90, 180\}$ (iv) $\{-90, -60, 60\}.$
4 (i) $\{135, 153\cdot43\}$ (ii) $\{30, 150\}$ (iii) $\{0, 90\}$ (iv) $\{30, 150\}.$
5 (a) $\{150, 210\}$ (b) $\{135, 315\}.$ **6** $\{90 + \alpha + 360n, 90 - \alpha + 360n : n \in \mathbb{Z}\}.$ **7** $\{\frac{1}{2}, \frac{1}{3}\}.$

Exercise 3.4A (p. 41)

1

x	$\cos x$	$\sin x$	$\tan x$	$\sec x$	$\operatorname{cosec} x$	$\cot x$
$\dfrac{2\pi}{3}$	$-\dfrac{1}{2}$	$\dfrac{\sqrt{3}}{2}$	$-\sqrt{3}$	-2	$\dfrac{2}{\sqrt{3}}$	$-\dfrac{1}{\sqrt{3}}$
$\dfrac{3\pi}{4}$	$-\dfrac{1}{\sqrt{2}}$	$\dfrac{1}{\sqrt{2}}$	-1	$-\sqrt{2}$	$\sqrt{2}$	-1
$\dfrac{5\pi}{6}$	$-\dfrac{\sqrt{3}}{2}$	$\dfrac{1}{2}$	$-\dfrac{1}{\sqrt{3}}$	$-\dfrac{2}{\sqrt{3}}$	2	$-\sqrt{3}$
π	-1	0	0	-1	—	—

Exercise 3.4B (p. 43)

1 $\{0, \frac{\pi}{3}, \pi\}$.　**2** $\{\frac{\pi}{6}, \frac{5\pi}{6}, \frac{3\pi}{2}\}$.　**3** $\{(n+\frac{1}{3})\pi, (2n+\frac{1}{6})\pi, (2n+\frac{5}{6})\pi : n \in \mathbb{Z}\}$.
4 $\{n\pi, (2n+\frac{2}{3})\pi, (2n+\frac{4}{3})\pi : n \in \mathbb{Z}\}$.　**5** $\{\frac{\pi}{4}, \frac{\pi}{2}, \frac{3\pi}{4}\}$.　**6** $\{\frac{\pi}{4}, \frac{\pi}{2}, \frac{3\pi}{4}\}$.
7 $\{\frac{\pi}{6}, \frac{5\pi}{6}\}$.　**8** (a) $\{\frac{3\pi}{8}, \frac{7\pi}{8}, \frac{11\pi}{8}, \frac{15\pi}{8}\}$　(b) $\{0, \frac{2}{3}\pi, \frac{4}{3}\pi, 2\pi\}$.
9 $(x^2 - y^2)^2 = (x - 2y)^2 + (y - 2x)^2$.

Exercise 3.5 (p. 45)

1 (i) $12{\cdot}5\,\text{m}, 31{\cdot}25\,\text{m}^2$　(ii) $6\pi\,\text{cm}, 24\pi\,\text{cm}^2$　(iii) $\frac{5}{3}\pi\,\text{m}, \frac{25}{12}\pi\,\text{m}^2$
(iv) $0{\cdot}088\pi\,\text{m}, 0{\cdot}05808\pi\,\text{m}^2$.　**2** $28{\cdot}65°$.　**3** $6{\cdot}64\,\text{m}^2, 10{\cdot}71\,\text{m}$.

4 (i) $\dfrac{\pi}{6} - \dfrac{\sqrt{3}}{4} : \dfrac{5\pi}{6} + \dfrac{\sqrt{3}}{4}$　(ii) $\dfrac{4\pi}{3} - \sqrt{3} : \dfrac{2\pi}{3} + \sqrt{3}$　(iii) $1 : 6{\cdot}02$.

5 $\dfrac{a\pi}{a+b} - \dfrac{1}{2}\sin\left(\dfrac{2\pi a}{a+b}\right) : \dfrac{b\pi}{a+b} + \dfrac{1}{2}\sin\left(\dfrac{2\pi a}{a+b}\right)$.　**6** $\left(\sqrt{3} - \dfrac{\pi}{6}\right)a^2$.

Exercise 3.6A (p. 47)

1 (a) (i) 2 (ii) $\sqrt{3}$ (iii) $\sqrt{2}$ (iv) 1 (v) 0 (vi) -1 (vii) $-\sqrt{2}$ (viii) $-\sqrt{3}$
(ix) -2 (b) (i) 0 (ii) $\frac{1}{4}$ (iii) $\dfrac{1}{2\sqrt{2}}$ (iv) $\dfrac{\sqrt{3}}{4}$ (v) $\frac{1}{2}$ (vi) $\dfrac{\sqrt{3}}{4}$ (vii) $\dfrac{1}{2\sqrt{2}}$ (viii) $\frac{1}{4}$

(ix) 0 (c) (i) 3 (ii) $\dfrac{3\sqrt{3}+4}{2}$ (iii) $\dfrac{7}{\sqrt{2}}$ (iv) $\dfrac{3+4\sqrt{3}}{2}$

(v) 4 (vi) $\dfrac{4\sqrt{3}-3}{2}$ (vii) $\dfrac{1}{\sqrt{2}}$ (viii) $\dfrac{4-3\sqrt{3}}{3}$ (ix) -3.
2 cosine shape with amplitude 5.　**3** $\cos 2x$ (i) 1 (ii) $\frac{1}{2}$ (iii) 0 (iv) $-\frac{1}{2}$ (v) -1
(vi) $-\frac{1}{2}$ (vii) 0 (viii) $\frac{1}{2}$ (ix) $1, \sin 2x$ (i) 0 (ii) $\dfrac{\sqrt{3}}{2}$ (iii) 1 (iv) $\dfrac{\sqrt{3}}{2}$

(v) 0 (vi) $-\dfrac{\sqrt{3}}{2}$ (vii) -1 (viii) $-\dfrac{\sqrt{3}}{2}$ (ix) $0, \{\frac{\pi}{4} + \frac{1}{2}n\pi : n \in \mathbb{Z}\}$.　**4** a, b, a, b.

Exercise 3.6B (p. 49)

1 (i) $1, -1, \pi; (0, 1), (\frac{\pi}{4}, 0), (-\frac{\pi}{4}, 0)$　(ii) $1, -1, 2\pi, (0, -\sin 1), (1 - \pi, 0), (1, 0)$
(iii) $1, 1, 2, (0, \cos 2), (\frac{\pi}{2} - 2, 0), (\frac{3\pi}{2} - 2, 0)$　(iv) $2, -2, 6\pi, (0, 0), (3\pi, 0), (-3\pi, 0)$
(v) $6, -6, \frac{4\pi}{5}, (0, 6\cos 2), (\frac{4-\pi}{5}, 0), (\frac{4-3\pi}{5}, 0)$　(vi) $5, -5, 8, (0, 0), (4, 0), (-4, 0)$.
2 $3\cos 8x$.　**3** $a = 2, b = -2$ or $a = -2, b = 2$.　**4** (i) 5 (ii) 1 (iii) $\frac{2}{3}\pi$.

Exercise 3.7 (p. 52)

3 Hint: $\tan(A+B) = \dfrac{\sin(A+B)}{\cos(A+B)}$.   **4** (i) $2\sin A \cos B$ (ii) $2\cos A\cos B$ (iii) $\cos 3A$

(iv) $\cos A$ (v) $\cos B$ (vi) $\tan B$ (vii) $\cot B$ (viii) $\cos A\sin 2B$ (ix) $\tan 2A$ (x) $\cos A - \sin A$
(xi) $\tan A$ (xii) $\cot A$ (xiii) $\cos A$ (xiv) $\cos A$. **5** (i) $-\frac{7}{25}, \frac{24}{25}, -\frac{24}{7}$ (ii) $-\frac{7}{25}, -\frac{24}{25}, \frac{24}{7}$
(iii) $-\frac{119}{169}, \frac{120}{169}, -\frac{120}{119}$ (iv) $-\frac{119}{169}, -\frac{120}{169}, \frac{120}{119}$ (v) $\frac{1519}{1681}, \frac{720}{1681}, \frac{720}{1519}$.
7 $\{\frac{3}{4}, 0, -1\}, \{131\cdot4°, 180°, 270°\}, \{\theta: 131\cdot4° < \theta < 180°\}$. **9** (i) $\{30°, 90°, 150°\}$

(ii) $\{120°\}$. **10** 1. **11** $\frac{1}{2}$. **12** $\tan\theta$. **13** $\dfrac{1+\tan\theta}{1-\tan\theta}$ (i) $2+\sqrt{3}$ (ii) $2\sec 2\theta$. **14** $\frac{16}{33}$.

Exercise 3.8 (p. 58)

1 (i) $C = 50°, a = 11\cdot35, b = 9\cdot62$ (ii) $A = 30°, a = 7\cdot57, b = 14\cdot04$
(iii) $B = 46°, b = 2\cdot52, c = 0\cdot79$ (iv) $B = 61°, a = 13\cdot33, c = 21\cdot52$. **2** (i) $B = 22\cdot8°,$
$C = 142\cdot2°, c = 4\cdot74$ or $B = 157\cdot2°, C = 7\cdot8°, c = 1\cdot05$
(ii) $A = 45\cdot6°, C = 52\cdot4°, a = 3\cdot61$ (iii) $A = 55\cdot2°, C = 86\cdot8°, c = 14\cdot6$ or $A = 124\cdot8°,$
$C = 17\cdot2°, c = 4\cdot32$ (iv) $A = 17\cdot5°, B = 12\cdot5°, b = 4\cdot33$ (v) $B = 70\cdot4°, C = 16\cdot6°, b = 6\cdot6$
(vi) $B = 28\cdot3°, C = 110\cdot7°, c = 25\cdot66$. **3** (i) $B = 44\cdot4°, C = 101\cdot6°, a = 4\cdot0$
(ii) $A = 80\cdot8°, B = 41\cdot2°, c = 10\cdot31$ (iii) $A = 53\cdot5°, C = 15\cdot5°, b = 10\cdot46$
(iv) $A = 41\cdot4°, B = 55\cdot8°, C = 82\cdot8°$ (v) $A = 90°, C = 67\cdot4°, B = 22\cdot6$
(vi) $A = 21\cdot8°, B = 120°, C = 38\cdot2°$. **4** 5 m. **5** $-\frac{1}{8}$. **6** $46°34'$.
7 (a) $10/\sqrt{3}$ cm (b) $50/\sqrt{3}$ cm². **8** (i) (a) $\{36\cdot9, 143\cdot1\}$ (b) $\{228\cdot6, 311\cdot4\}$.
9 (a) 5 km (b) 330° (c) 7·05 km.

Miscellaneous Exercise 3 (p. 59)

1 (i) (a) $\{105°, 165°, 285°, 345°\}$ (b) $\{90°, 180°, 270°\}$ (ii) 38·36°.
2 120°, 60°. **3** 61·0°, 73·7°, 134·8°, 10·8 m. **4** $\{5, 6, 7\}$.
6 $0 < \theta < \frac{\pi}{3}, \cos\theta \equiv \dfrac{2p^2-1}{2p^2}, x = 0\cdot37.$
8 Either $A = (180n + 45)°, n \in \mathbb{Z}$ or $B = (180n - 45)°, n \in \mathbb{Z}$.
9 $\{\frac{3}{4}\pi, \frac{7}{4}\pi\}$. **10** $\{30°, 150°, 210°, 330°\}$. **11** 0·6, 0·8.
12 (i) $\{\frac{\pi}{3}, \frac{5\pi}{3}, \pi - \cos^{-1}\frac{1}{3}, \pi + \cos^{-1}\frac{1}{3}\}$ (ii) (a) 13 (b) 67·4°, 232·6°, 352·6°.
14 (a) $p = 5, q = 2$ (c) 2·35 hours. **15** $A = 2n\pi$ or $B = 2n\pi$ or $(A+B) = 2n\pi, n \in \mathbb{Z}$.

Exercise 4.1 (p. 62)

2 3/5, 8, −5/3. **3** 1/3, (3,4). **4** (i) 3, (ii) 18, (iii) −9, (iv) −4.
5

x	1	1·3	1·6	1·8	1·9	2	2·1	2·2	2·4	2·7	3
$f(x)$	1	2·38	4·12	5·48	6·22	7	7·82	8·68	10·52	13·58	17
grad. PQ	6	6·6	7·2	7·6	7·8	—	8·2	8·4	8·8	9·4	10

gradient of tangent is 8.

Exercise 4.2 (p. 64)

1 (d) $56\,\text{ms}^{-1}$

Exercise 4.7 (p. 71)

1 (i) $5x^4$ (ii) $-4x^{-5}$ (iii) $14x^6$ (iv) $-3/x^2$ (v) $-4/x^3$ (vi) $2x+2$ (vii) $2x$
(viii) $3+4x$ (ix) $9x^2+2$ (x) $4x^3 - 12x^2 + 4x$ (xi) $-3 - 8x$ (xii) $5x^4 - 20x^3$
(xiii) $2\cos x + 3\sin x$ (xiv) $-\sin x - \cos x$ (xv) $3\cos 3x$. **2** (i) $8x-6$ (ii) $2x-1$
(iii) $6x^2 + 6x$ (iv) $1 - \dfrac{1}{x^2}$ (v) $4x^3$ (vi) $2x - \dfrac{2}{x^3}$. **3** (i) $3x^2 - 2$ (ii) $1\sin 3x$ (iii) $1 - \dfrac{1}{x^2}$

(iv) $6x - 5$ (v) $4 - \dfrac{2}{x^3}$ (vi) $1 + 2\sin 2x$ (vii) $6x - 8x^3$ (viii) $24x^2 + 24x + 6$ (ix) $2 - \dfrac{2}{x^2}$

(x) $3x^2 - 3 - \dfrac{3}{x^2} + \dfrac{3}{x^4}$.   **4** (i) $1 - 6t$ (ii) $4\pi r^2$ (iii) $u + at$ (iv) $-\dfrac{25}{V^2}$.

5 (i) 12, $y = 12x - 46$ (ii) 0, $y = -1$.   **6** (i) 2, $y = 2x - 3$ (ii) 6, $y = 6x - 2$
(iii) $4/3$, $3y = 4x + 20$.   **7** -10, -6, -2, 2, $6\,\mathrm{ms}^{-1}$; $\tfrac{5}{2}\,\mathrm{s}$, $5\,\mathrm{s}$, $10\,\mathrm{ms}^{-1}$.
8 $-\tfrac{5\pi}{12}\sin\tfrac{\pi}{12}(x+3)$, $13\,\mathrm{m}$.   **9** (a) 2 (b) $24\,\mathrm{ms}^{-1}$.

Exercise 5.2 (p. 77)

2 (i) $\overrightarrow{OD}$ (ii) $\overrightarrow{OJ}$ (iii) $\overrightarrow{OE}$ (iv) $\overrightarrow{OK}$ (v) $\overrightarrow{OM}$ (vi) $\overrightarrow{OQ}$ (vii) $\overrightarrow{ON}$ (viii) $\overrightarrow{OL}$ (ix) $\overrightarrow{OP}$.
3 (i) $2\mathbf{a}$ (ii) $3\mathbf{a}$ (iii) $2\mathbf{b}$ (iv) $2\mathbf{a} + \mathbf{b}$ (v) $\mathbf{a} + 2\mathbf{b}$ (vi) $2\mathbf{a} - \mathbf{b}$ (vii) $3\mathbf{b} - 2\mathbf{a}$ (viii) $-\mathbf{a} - 2\mathbf{b}$
(ix) $-4\mathbf{b}$ (x) $-4\mathbf{a} - \mathbf{b}$.   **4** (i) $\overrightarrow{OC}, \overrightarrow{AE}, \overrightarrow{PF}$ (ii) $\overrightarrow{BP}, \overrightarrow{HG}, \overrightarrow{ID}$ (iii) $\overrightarrow{EO}, \overrightarrow{HB}, \overrightarrow{GP}$
(iv) $\overrightarrow{GO}, \overrightarrow{KD}, \overrightarrow{HA}$.   **5** (i) $\mathbf{t} - \mathbf{s}$ (ii) $\mathbf{s} - \mathbf{t}$ (iii) $2\mathbf{t} - \mathbf{s}$ (iv) $2\mathbf{s} - 2\mathbf{t}$ (v) $\mathbf{t} - \mathbf{s}$.

Exercise 5.3 (p. 79)

1 (i) $\mathbf{r} = \mathbf{c} + \lambda(\mathbf{c} + \mathbf{d})$ (ii) $\mathbf{r} = 2\mathbf{d} + \lambda\mathbf{c}$ (iii) $\mathbf{r} = 3\mathbf{c} + \lambda\mathbf{d}$ (iv) $\mathbf{r} = 2\mathbf{d} + \mathbf{c} + \lambda(2\mathbf{c} + \mathbf{d})$
(v) $\mathbf{r} = 3\mathbf{c} + \lambda(\mathbf{d} - \mathbf{c})$ (vi) $\mathbf{r} = 2\mathbf{c} + \lambda(2\mathbf{c} + \mathbf{d})$.   **2** Line through (i) $\{N, F, D\}$
(ii) $\{H, D, K\}$ (iii) $\{O, G\}$ (iv) $\{C, I\}$ (v) $\{D, C\}$.   **3** (i) $\mathbf{r} = \mathbf{a} + \lambda\mathbf{b}$ (ii) $\mathbf{r} = \lambda(\mathbf{a} - \mathbf{b})$
(iii) $\mathbf{r} = \tfrac{1}{2}\mathbf{b} + \lambda\mathbf{a}$ (iv) $\mathbf{r} = (1 - \lambda)\mathbf{a} + \lambda\mathbf{b}$.   **4** (i) Bisector of angle POQ
(ii) line through P parallel to $\overrightarrow{OQ}$ (iii) line through O parallel to $\overrightarrow{PQ}$ (iv) line PQ
(v) plane OPQ.   **5** (i) (a) $k = 1$ (b) $k = \tfrac{1}{2}$ (c) $k = \tfrac{1}{3}$ (ii) $\overrightarrow{OD} = \mathbf{b} + \mathbf{c}$, $\overrightarrow{OE} = \mathbf{c}$.
7 $\overrightarrow{BA} = \mathbf{a} - \mathbf{b}$, $\overrightarrow{OD} = \mathbf{a} - \mathbf{b} + \mathbf{c}$.

Exercise 5.4 (p. 84)

1 $\mathbf{b} + \mathbf{d}, \tfrac{1}{2}\mathbf{b} + \mathbf{d}, \tfrac{1}{3}\mathbf{b} + \tfrac{2}{3}\mathbf{d}$.   **2** Hint: prove $\overrightarrow{EF} = \overrightarrow{HG}$.
4 $\mathbf{q} - \mathbf{p}, -\mathbf{p}, -\mathbf{q}, \mathbf{p} - \mathbf{q}, 2\mathbf{p} - 2\mathbf{q}, \mathbf{p} - 2\mathbf{q}$.   **5** $\tfrac{1}{2}\mathbf{a} - \tfrac{1}{2}\mathbf{b}, \tfrac{1}{2}\mathbf{a} + \tfrac{1}{2}\mathbf{b}, \tfrac{1}{2}\mathbf{b} - \tfrac{1}{2}\mathbf{a}, -\tfrac{1}{2}\mathbf{a} - \tfrac{1}{2}\mathbf{b}$.
6 $\tfrac{1}{2}\mathbf{b} + \tfrac{1}{2}\mathbf{c}, \tfrac{1}{2}\mathbf{c} + \tfrac{1}{2}\mathbf{a}, \tfrac{1}{2}\mathbf{a} + \tfrac{1}{2}\mathbf{b}$.   **8** $\tfrac{1}{2}\mathbf{b} + \tfrac{1}{2}\mathbf{c}, \tfrac{1}{3}(\mathbf{a} + \mathbf{b} + \mathbf{c})$.   **9** $\tfrac{1}{2}\mathbf{a} + \tfrac{1}{2}\mathbf{b}, \tfrac{1}{8}\mathbf{a} + \tfrac{1}{8}\mathbf{b} + \tfrac{3}{4}\mathbf{c}$.
10 8.   **12** (ii) $\tfrac{1}{3}\overrightarrow{PQ} + \tfrac{2}{3}\overrightarrow{PS}$, $-\tfrac{2}{3}\overrightarrow{PQ} + \tfrac{2}{3}\overrightarrow{PS}$, $-\tfrac{1}{6}\overrightarrow{PQ} + \tfrac{1}{6}\overrightarrow{PS}$.

Exercise 5.5 (p. 87)

1 $\mathbf{b} + \mathbf{d}, \tfrac{1}{2}\mathbf{b} + \mathbf{d}, \tfrac{1}{3}\mathbf{b} + \tfrac{2}{3}\mathbf{d}$.   **2** $\tfrac{3}{4}\mathbf{a} + \tfrac{3}{4}\mathbf{b}, 3\mathbf{a}, 3\mathbf{b}$.   **3** $\tfrac{1}{2}\mathbf{c} - \tfrac{1}{2}\mathbf{b}, 2\mathbf{b}, \tfrac{3}{4}\mathbf{b} + \tfrac{1}{4}\mathbf{c}$.
4 $\tfrac{1}{3}\overrightarrow{AB} + \tfrac{2}{3}\overrightarrow{AC}$.

Miscellaneous Exercise 5 (p. 89)

1 (a) $\tfrac{1}{2}\mathbf{b} + \lambda(\mathbf{a} - \tfrac{1}{2}\mathbf{b})$ (b) $\tfrac{1}{2}\mathbf{a} + \mu(\mathbf{b} - \tfrac{1}{2}\mathbf{a}), \tfrac{1}{3}\mathbf{a} + \tfrac{1}{3}\mathbf{b}$.   **2** (i) $\tfrac{1}{3}\mathbf{a} + \tfrac{2}{3}\mathbf{b}$
(ii) $\tfrac{1}{6}\mathbf{a} + \tfrac{1}{3}\mathbf{b}, \tfrac{2}{3}, -6$.   **3** $2\mathbf{c} - \mathbf{b}, \tfrac{1}{3}\mathbf{a} + \tfrac{2}{3}\mathbf{b}, \tfrac{1}{5}\mathbf{a} + \tfrac{4}{5}\mathbf{c}$.   **4** $\mathbf{b} - \mathbf{a}, \mathbf{b} - 2\mathbf{a}, 2\mathbf{a} - 2\mathbf{b}$.
5 $\tfrac{1}{2}\mathbf{a} + \tfrac{1}{2}\mathbf{b}, \tfrac{1}{8}\mathbf{a} + \tfrac{1}{8}\mathbf{b} + \tfrac{3}{4}\mathbf{c}$.   **6** R is midpoint of CU, where U divides AB in the ratio $1:2$, R divides AT in the ratio $2:1$, where T divides BC in the ratio $3:1$.
7 $\mathbf{a} + \lambda(\tfrac{1}{2}\mathbf{b} + \tfrac{1}{2}\mathbf{c} - \mathbf{a}), \mathbf{b} + \mu(\tfrac{4}{3}\mathbf{c} - \tfrac{1}{3}\mathbf{a} - \mathbf{b}), \tfrac{1}{7}(-\mathbf{a} + 4\mathbf{b} + 4\mathbf{c})$.   **8** $\tfrac{1}{4}\mathbf{a} + \tfrac{1}{4}\mathbf{b} + \tfrac{1}{4}\mathbf{c}$.

Exercise 6.1 (p. 92)

1 $3x^2 + 2, x^3 + c, x^2 + x + c, c \in \mathbb{R}$.   **2** (i) $2x$ (ii) x^2 (iii) 1 (iv) $3x^2$ (v) $3x^2$
(vi) 0 (vii) $4x^3$ (viii) $4x^3$ (ix) $\tfrac{3}{2}x^{\frac{1}{2}}$ (x) $2x + 2$ (xi) $x^{-\frac{1}{2}} + x^{-\frac{3}{2}}$ (xii) $\tfrac{1}{4}x^{-\frac{3}{4}} - \tfrac{3}{4}x^{-\frac{7}{4}}$.
3 (i) $\tfrac{1}{3}x^3 + c$ (ii) $\tfrac{1}{4}x^4 + c$ (iii) $\tfrac{2}{3}x^{\frac{3}{2}} + c$ (iv) $\tfrac{1}{2}x^2 - \tfrac{2}{3}x^{\frac{3}{2}} + c$ (v) $\tfrac{1}{3}x^3 + \tfrac{7}{2}x^2 + 7x + c$
(vi) $\tfrac{1}{3}x^3 + \tfrac{7}{2}x^2 + c$ (vii) $7x + c$ (viii) $\dfrac{x^4}{16} + cx + d$ (ix) $\tfrac{4}{7}x^{\frac{7}{4}} + \tfrac{4}{3}x^{\frac{3}{4}} + c, c \in \mathbb{R}$.

Exercise 6.3 (p. 94)

2 (i) $2x^3 + c$ (ii) $2x^2 - 3x + c$ (iii) $\tfrac{1}{5}x^5 + c$ (iv) $\tfrac{1}{2}\sin 2x + c$ (v) $\tfrac{1}{4}x^4 - x^2 + c$
(vi) $\tfrac{1}{7}x^7 + c$.   **3** (i) $\tfrac{3}{4}x^2 + c$ (ii) $\tfrac{2}{x} + c$ (iii) $\dfrac{-5}{x} + x + c$ (iv) $x + \tfrac{1}{x} + c$

(v) $\frac{1}{3}x^3 + x^2 + x + c$ (vi) $\frac{1}{3}x^3 - \frac{1}{2}x^2 - 6x + c.$ **4**   (i) $x^3 + x^2 + x + c$

(ii) $2x^3 - \frac{11}{2}x^2 + 4x + c$ (iii) $x - \frac{1}{2}\cos 2x + c$ (iv) $\frac{1}{5}x^5 + 2x - \dfrac{1}{3x^3} + c$

(v) $\frac{1}{3}ax^3 + \frac{1}{2}bx^2 + cx + d$ (vi) $4x^{\frac{1}{2}} + 2x^{\frac{3}{2}} + c.$ **5**   (i) $\frac{1}{3}x^3 + x^2 + c$ (ii) $\frac{1}{3}t^3 + t^2 + c$

(iii) $-\frac{1}{2}\cos 2\theta + c$ (iv) $u^2 + \frac{2}{3}u^{\frac{3}{2}} + c$ (v) $\frac{1}{4}s^4 + 4s^3 + 24s^2 + 64s + c$

(vi) $\frac{1}{3}u^3 + 2u - \frac{1}{u} + c.$

Exercise 6.4 (p. 95)

1 $2x^2 + c, c = -14.$ **2**   (i) $2x^2 + 4$ (ii) $x^3 - 3x^2 + 3x - 1$ (iii) $x^3 + 6x^2 + 12x + 8$
(iv) $\sin 2x + \cos 2x.$ **3** $s = 6.$ **4**   (i) $44\frac{3}{4}$ (ii) $11\frac{1}{4}.$ **5**   (i) $s = ut - \frac{1}{2}gt^2 + c$

(ii) $h = \dfrac{u^2}{2g}$ (iii) time taken is $\dfrac{2u}{g}.$

Exercise 7.1 (p. 97)

1   (i) $1\,3\,2\,3\,7$ (ii) $2\,0\,1\,0 - 4$ (iii) $3 - 11 - 19$ (iv) $0\,4\,4\,4\,4$ (v) $1 - 21 - 20$
(vi) $2 - 41 - 40$ (vii) $15 - 251$ (viii) $2 - 7 - 3 - 7 - 11$ (ix) $50 - 108.$
2   The constant term of p(x) is p(0).

Exercise 7.2 (p. 99)

1   (i) $x + 1$ (ii) $3x^2 + 2x + 4$ (iii) $-8x - 2$ (iv) $-3x^2 + 10x - 23$
(v) $5x^3 - 6x^2 - 11x - 3$ (vi) $41x^2 + 43x - 12.$ **2**   (i) $x^3 - x^2 + x - 6$
(ii) $x^5 - 2x^4 - 4x^3 - 2x^2 - 5x$ (iii) $4x^4 + 4x^3 + 5x^2 + 2x + 1$ (iv) $x^3 + x^2 - 8x - 12$
(v) $x^3 + 12x^2 + 48x + 64$ (vi) $x^5 - x^3 - 6x.$ **3**   (i) $3x^3 - x^2 + 6x - 2$
(ii) $6x^3 - 5x^2 - 2x + 1$ (iii) $12x^5 - 6x^4 - 4x^3 + 2x^2$ (iv) $25x^3 + 5x^2 - 17x - 6$
(v) $-8x^4 + 16x^3 - 4x^2 + x + 1$ (vi) $30x^5 + 3x^4 - 17x^3 - x^2 + x$
(vii) $3x^5 - 7x^4 + 11x^3 - 15x^2 + 19x - 5$ (viii) $3x^5 - 11x^4 + 16x^3 - 19x^2 + 10x - 4.$
4   (i) $3x^2 - 3x + 5$ (ii) $7x^2 - 6x + 11$ (iii) $2x^2 - 9x + 8$ (iv) $2x^4 - 3x^3 + 6x^2 - 3x + 4$
(v) $4x^4 - 12x^3 + 25x^2 - 24x + 16$ (vi) $x^4 + 2x^2 + 1$ (vii) $3x^4 - 12x^3 + 23x^2 - 24x + 15.$
5 $4x^5 - 5x^4 + 1.$ **6**   (i) 5   (ii) 7.   **7**   (i) $1 - 2x + 5x^2 - 6x^3$ (ii) $2 + 9x + 7x^2 - 4x^3$
(iii) $32 - 32x - 38x^2 + 41x^3$ (iv) $8x - 28x^2 + 34x^3.$

Exercise 7.3 (p. 100)

(i) $\dfrac{x - 1}{x^2}$ (ii) $\dfrac{1 + 2x - x^2}{1 - x^2}$ (iii) $\dfrac{2x^2 + 2x + 1}{x + 1}$ (iv) $\dfrac{2x - 4}{3 - x}$ (v) $\dfrac{x^3 + 4x^2 + 2x - 1}{x + 3}$

(vi) $\dfrac{x^3 - x^2 + 3x - 1}{x^2 + 1}$ (vii) $\dfrac{2(x - 4)}{(x - 5)}, x \neq -3$ (viii) $\dfrac{2x^4 + 4x^3 + 16x^2 + 16x + 32}{x^5 + 2x^4 + 4x^3 - 8x^2 - 16x - 32}.$

Exercise 7.4 (p. 101)

1   (i) $x + 2, -9$ (ii) $x^2 - x + 1, 0$ (iii) $x + 2, 0$ (iv) $3x - 2, 2$ (v) $x^2 - 4x + 11, -28$
(vi) $2x^2 - 2x + 6, -6$ (vii) $x^2 - 2x - 3, 18$ (viii) $x^3 + 2x^2 + x + 2, 4$ (ix) $x^2 + x - 3, 0$
(x) $x^2 + \frac{1}{2}x - \frac{1}{4}, \frac{31}{4}$ (xi) $3x^3 - 7x^2 + 20x - 56, 163$ (xii) $\frac{1}{2}x^3 + \frac{3}{4}x^2 - \frac{3}{8}x + \frac{7}{16}, \frac{21}{16}.$

Exercise 7.5 (p. 103)

1   (i) -1 (ii) 3   (iii) 15   (iv) 0   (v) -1 (vi) 21.   **2**   (i) yes   (ii) yes   (iii) no
(iv) no   (v) yes   (vi) yes   (vii) no   (viii) yes.   **3**   (i) $(x - 1)(x + 1)(2x + 1)$
(ii) $(x + 1)(x + 1)(2x - 1)$ (iii) $(x + 1)(x + 1)(x + 1)$ (iv) $(x - 1)(x + 1)(x^2 + 1)$
(v) $(x - 1)(x^2 + x + 1)$ (vi) $(x - 1)(x - 1)(x + 1)$ (vii) $(x + 2)(x^2 - x + 1)$
(viii) $(x - 1)(x^2 + 1)$ (ix) $(2x - 1)(2x^2 + x + 2)$ (x) $(3x + 1)(2x^2 - 5x + 6)$
(xi) $(x - 1)(3x + 2)(x - 2)$
(xii) $(x + 1)(x - 2)(x - 3)$ (xiii) $(x - 1)(x - 4)(x + 5)$
(xiv) $(x - 1)(2x + 1)(2x - 3)$

(xv) $(2x-1)(3x+1)(x+3)$.   **4**   (i) $a = 6$
(ii) $a = -10$ (iii) $a = \frac{3}{4}$ (iv) $a = \frac{7}{2}$ (v) $a = \frac{23}{4}$.
5 (i) $a = 3, b = -3$ (ii) $a = -5, b = 8$ (iii) $a = -13, b = 6$.   **6**   (i) n is odd
(ii) n is even (iii) never true.   **7**   (i) $(x-2)$ (ii) no linear factor (iii) $(x+3)$
(iv) $(2x-5)$.   **8** 95.   **9** $a = 14, 2(x+3)(x-1)(x-2)$.   **10** $a = 1, b = -37$.
11 6.   **12** $(x-3)(x+4)(x+5)$.   **13** $\sqrt{7}, \dfrac{\sqrt{5}-\sqrt{7}}{2}, \dfrac{-\sqrt{5}-\sqrt{7}}{2}$.

Exercise 7.6 (p. 105)

2 (i) $0\,1\,2$ (ii) $1\,1\,-1\,-1$ (iii) $0\,1\,-1$ (iv) $-1\frac{1}{2}\,2$. **3** $p(x) = 2x^3 - 6x^2 - 12x + 16$.
4 (a) $x^2 - 2x - 8$ (b) $x^3 + 2x^2 - 11x - 12$ (c) $x^4 - 4x^3 - 11x^2 + 30x$.

Miscellaneous Exercise 7 (p. 105)

1 $f(2) = f(-1) = 0, f(x) = (x-2)(x+1)(x^2+4x+1), 2, -1, -2+\sqrt{3}, -2-\sqrt{3}$.
2 $a = -2, b = 2, c = 1$.   **3** $p = -1, q = 5$.
4 $a = 1, b = 1, (x+1)(x^2-3x+3), 9 < 12$ so no other real roots.
5 quotient $(2x-1)$, remainder $(1-5x)$.   **7** $(2x-1)(x+1), 25x-23$.

Exercise 8.1 (p. 108)

1 (i) John Smith's school; (ii) your paternal grandmother; (iii) the colour of Manchester
United's strip. **2** your grandparents. **3** (i) 'the class containing the school' has no
meaning; (iii) 'the strip of the colour' has no meaning. fg has a meaning when C is a subset
of A. **4** (i) -1 (ii) 1 (iii) 13 (iv) 31 (v) 97 (vi) $2x^2 - 1$ (vii) $(2x-1)^2$
(viii) $2t^2 - 1$.

Exercise 8.2 (p. 110) Let $\mathbb{R}^+ \cup \{0\} = \mathbb{R}_0^+$.

1 (i) (a) $\mathbb{R}^+, \mathbb{R}^+$ (b) $\mathbb{R}, \mathbb{R}_0^+$ (ii) (a) $\mathbb{R}_0^+, \mathbb{R}_0^+$ (b) $\mathbb{R}, \mathbb{R}_0^+$ (iii) (a) $\mathbb{R}\backslash\{1, -1\}, \mathbb{R}\backslash\{3\}$
(iv) (a) $\{x: -2 \leqslant x \leqslant 0\}\backslash\{-1\}, \mathbb{R}_0^+$ (b) $\mathbb{R}\backslash\{x: -2 < x < 2\}, \{x: 0 < x \leqslant 2\}$.

Exercise 8.6 (p. 116)

1 (a) (i) $f(x-1)$, translation $+1$ along Ox (ii) $f(1-x)$, reflection in $x = \frac{1}{2}$
(iii) $f(3x)$, stretch along Ox S.F. $\frac{1}{3}$ (iv) $f(\frac{1}{2}x)$, stretch along Ox S.F.2
(v) $f(-2x)$, stretch along Ox S.F. $-\frac{1}{2}$ (vi) $f(\frac{1-x}{2})$, stretch along Ox S.F. 2 followed by
reflection in $x = \frac{1}{2}$; (b) (i) $f(x) - 1$, translation -1 along Oy
(ii) $1 - f(x)$, reflection in $y = \frac{1}{2}$ (iii) $3f(x)$, stretch along Oy S.F. 3
(iv) $\frac{1}{2}f(x)$, stretch along Oy S.F. $\frac{1}{2}$ (v) $-2f(x)$, stretch along Oy S.F. -2
(vi) $\frac{1}{2} - \frac{1}{2}f(x)$, reflection in $y = \frac{1}{2}$ followed by stretch along Oy S.F. $\frac{1}{2}$.
2 (i) $g(x) = 3x, h(x) = \frac{1}{3}x$ (ii) $g(x) = -x, h(x) = x - 1$
(iii) $g(x) = \frac{1}{2}x, h(x) = -x$ (iv) $g(x) = 4 - x, h(x) = x$.

Exercise 8.7 (p. 118)

1 (i) even (ii) none (iii) odd (iv) even and periodic (v) odd and periodic
(vi) odd and periodic (vii) even (viii) odd (ix) even.   **2** (i) 4 (ii) $4\frac{7}{8}$ (iii) 4 (iv) -5.
3 (i) even (ii) none (iii) odd (iv) periodic, period 1 (v) even (vi) none (vii) even.
5 (a) not periodic (b) periodic, period π.   **6** 3π.   **10** $\pi, \pi, \pi, \{x: 0 < x < 2\}$.

Exercise 8.8 (p. 120)

1 (i) range $A = \{1, 0, -1, -2, -3, \ldots\}$, $f^{-1}(x) = 1 - x$ for $x \in A$
(ii) range $B = \{0, 1, 4, 9, 16, \ldots\}$, $f^{-1}(x) = \sqrt{x}$ for $x \in B$
(iii) range $C = \{1, \frac{1}{2}, \frac{1}{3}, \frac{1}{4}, \ldots\}$, $f^{-1}(x) = (1-x)/x, x \in C$
(iv) range $D = \{0, 1, 8, 27, 64, \ldots\}$, $f^{-1}(x) = \sqrt[3]{x}$ for $x \in D$.

2   (i) domain $\mathbb{R}$, range $\mathbb{R}$, inverse $f^{-1}(x) = 1 - x$
(ii) domain $\mathbb{R}$, range $\mathbb{R}^+ \cup \{0\}$, no inverse because $f(2) = f(-2)$
(iii) domain $\mathbb{R} \backslash \{-1\}$, range $\mathbb{R} \backslash \{0\}$, inverse $f^{-1}(x) = (1-x)/x$
(iv) domain $\mathbb{R}$, range $\mathbb{R}$, inverse $f^{-1}(x) = \sqrt[3]{x}$.
3 $a = 0$, range $\{b\}$, no inverse since $f(1) = b = f(2)$
$a \neq 0$, range $\mathbb{R}$, inverse $f^{-1}(x) = (x-b)/a$

4   (i) $f(1) = f(-1)$ so no inverse, restrict domain to $\mathbb{R}^+$, $f^{-1}(x) = \left(\dfrac{x+1}{2}\right)$

(ii) $f(2) = f(-2)$ so no inverse, restrict domain to $\{x : x > 1\}$, $f^{-1}(x) = \left(1 + \dfrac{1}{x}\right)$

(iii) $f(0) = f(\pi)$ so no inverse, restrict domain to $\left\{x : -\dfrac{\pi}{2} \leqslant x \leqslant \dfrac{\pi}{2}\right\}$, $f^{-1}(x) = \sin^{-1} x$

(iv) $f(0) = f(\pi)$ so no inverse, restrict domain to $\left\{x : 0 \leqslant x \leqslant \dfrac{\pi}{4}\right\}$, $f^{-1}(x) = \tfrac{1}{4}\cos^{-1} x$

(v) $f(1) = f(-1)$ so no inverse, restrict domain to $\mathbb{R}^+ \cup \{0\}$, $f^{-1}(x) = x$

(vi) $f(0) = f(\tfrac{1}{2})$ so no inverse, restrict domain to $\{x : 0 \leqslant x < 1\}$, $f^{-1}(x) = -x$.

Exercise 8.9 (p. 122)

1   (i) $f^{-1}(x) = \tfrac{1}{2}x$, domain $\mathbb{R}$ (ii) $f^{-1}(x) = x + 3$, domain $\mathbb{Z}$
(iii) $f^{-1}(x) = (x-2)/3$, domain $\mathbb{R}$ (iv) $f^{-1}(x) = \sqrt{x} - 2$, domain $\{x : x > 4\}$
(v) $f^{-1}(x) = \sqrt[3]{x}$, domain $\mathbb{R}$ (vi) $f^{-1}(x) = x^2$, domain $\mathbb{R}^+$
(vii) $f^{-1}(x) = \sqrt{x}$ for $x \in \mathbb{R}^+$, $f^{-1}(x) = -\sqrt{(-x)}$ for $x \in \mathbb{R} \backslash \mathbb{R}^+$, domain $\mathbb{R}$
(viii) $f^{-1}(x) = 2 - x$, domain $\{x : -1 < x \leqslant 0\}$
(ix) $f^{-1}(x) = \sqrt{(1 - x^2)}$, domain $\{x : 0 \leqslant x \leqslant 1\}$.
2   Range of f is A where $A = \{x : x \geqslant 1\}$, $d = 2$, g^{-1} and h^{-1} have domain A,
$g^{-1}(x) = 2 + \sqrt{(x-1)}$, $h^{-1}(x) = 2 - \sqrt{(x-1)}$.
3   For all x, $g(x) = eg(x) = hfg(x) = he(x) = h(x)$, so $g = h$.
4 $P(1)$ and $P(4)$ are each both necessary and sufficient. $P(2)$ and $P(3)$ are always true so they are necessary but not sufficient.
5   For all a in A, $gf(a) = a$ and so $g = f^{-1}$. For all b in B, since B is the range of f, there is some a in A with $b = f(a)$, then $fg(b) = fgf(a) = f(a) = b$.
6   Let $f^{-1}(x) = f^{-1}(y)$, with x, $y \in \mathbb{R}$. Then $x = ff^{-1}(x) = ff^{-1}(y) = y$, so f^{-1} is $1-1$.
7   Since gf is the identity function on the domain D, $gf(d) = d$ for every d in D, and so $f(d)$ lies in the domain of g, and the range of f is a subset of the domain of g. Let $f(a) = f(b)$ for $a, b \in D$, then $a = gf(a) = gf(b) = b$, and hence f is $1-1$. When $f(x) = \sqrt{x}$, $g(x) = x^2$, with the domains of f and g being $\mathbb{R}^+$ and $\mathbb{R}$ respectively, $gf(x) = (\sqrt{x})^2 = x$, so gf is the identity function on $\mathbb{R}^+$. The inverse of f is the function h with $h(x) = x^2$, with domain $\mathbb{R}^+$, and so $h \neq g$.

Miscellaneous Exercise 8 (p. 123)

1 $\dfrac{x+2}{x-1}$, $\mathbb{R} \backslash \{1\}$.   **2**   Both exist, $f^{-1}g = \cos^{-1} \tan$, domain $A = \left\{x : 0 \leqslant x \leqslant \dfrac{\pi}{4}\right\}$, range

$B = \left\{x : 0 \leqslant x \leqslant \dfrac{\pi}{2}\right\}$, $g^{-1}f = \tan^{-1} \cos = (f^{-1}g)^{-1}$, domain B, range A; solution to

$f(x) = g(x)$ is $x = 0.67$.   **3**   (i) $\{x : x \geqslant -\tfrac{1}{4}\}$,   (ii) $\{x : x \leqslant -4\} \cup \{x : x > 0\}$.

4   (a) $gf(x) = \dfrac{2}{4x^2 - 1}$, domain $\mathbb{R} \backslash \{\tfrac{1}{2}, -\tfrac{1}{2}\}$ (b) $g^{-1}(x) = \dfrac{3x - 2}{4x}$.

5 $f(x) = 3(x - \tfrac{7}{6})^2 + \tfrac{11}{12} > 0$, $g(-2) < 0 < g(0)$, $h(0) < 0 < h(3)$, $g(x) > 0$ for
$-1 < x < 5$, $h(x) > 0$ for $x < -2$ or $x > \tfrac{5}{2}$.   **6**   (i) $\{x : x \leqslant -\tfrac{3}{4}\}$,   (ii) $4, -4$.

7 (a) even, odd and periodic period 2π, even (b) $\frac{1}{2}$.
8 (i) $y = x + 1$ (ii) $y = 3 - x$.   **9** (c).

Exercise 9.2 (p. 126)

2 (i) $\dfrac{1}{a}$ (ii) $\dfrac{1}{a^3}$ (iii) $\dfrac{1}{a}$ (iv) $\dfrac{1}{a^4}$.

Exercise 9.3A (p. 127)

3 $2^{\frac{1}{2}} = 1{\cdot}41$, $2^{-\frac{1}{2}} = 0{\cdot}71$, $2^{\frac{3}{8}} = 1{\cdot}30$, $2^{-\frac{7}{8}} = 0{\cdot}55$; $2^x > 0$.   **4** (i) $0{\cdot}1$ (ii) 100 (iii) x
(iv) x^9 (v) $0{\cdot}001$ (vi) 4 (vii) a^4 (viii) x^6 (ix) $\frac{64}{27}$ (x) 2^{12} (xi) a^7 (xii) $\dfrac{1}{2^{12}}$
(xiii) 2^{12} (xiv) 1 (xv) $\frac{3}{4}$ (xvi) $\dfrac{3^{10}}{9^4}$ (xvii) 1 (xviii) $a^4 b^6$ (xix) a^{12}/b^6 (xx) $1/a^4 b^6$
(xxi) $a^{12} b^8$.

Exercise 9.3B (p. 128)

1 (i) $2{\cdot}7 \times 10^1$ (ii) $2{\cdot}7 \times 10^2$ (iii) $2{\cdot}7 \times 10^{-1}$ (iv) $2{\cdot}7 \times 10^{-2}$ (v) $3{\cdot}456 \times 10^3$
(vi) $3{\cdot}4 \times 10^4$ (vii) $3{\cdot}456 \times 10^1$ (viii) $3{\cdot}456 \times 10^6$ (ix) $3{\cdot}456 \times 10^{-5}$
2 (i) $2\,700$ (ii) $27\,890$ (iii) $0{\cdot}001\,234$ (iv) $1\,200\,000\,000\,000\,000\,000\,000$
(v) $0{\cdot}000\,000\,000\,25$ (vi) $3\,450\,000$.

Exercise 9.4 (p. 129)

1 (i) 27 (ii) 3 (iii) 2 (iv) $2/3$ (v) $3/2$ (vi) 1 (vii) $1/3$ (viii) $\frac{1}{36}$ (ix) $\frac{1}{2}$ (x) 2 (xi) $\frac{3}{2}$
(xii) $\frac{3}{2}$.   **2** (i) 4 (ii) $\frac{1}{8}$ (iii) 125 (iv) $5^{\frac{13}{6}}$ (v) 2 (vi) 12 (vii) $6\sqrt[3]{2}$ (viii) 3 (ix) 2
(x) $\frac{1}{3^n}$ (xi) 2.   **4** (i) 3^3 (ii) $4^{\frac{3}{4}}$ (iii) $5^{\frac{3}{2}}$ (iv) $5^{\frac{5}{2}}$ (v) $a^{\frac{1}{2}}$ (vi) $x^{\frac{3}{2}}$ (vii) $a^{\frac{41}{20}}$ (viii) x^2
(ix) a^2 (x) $a^{\frac{1}{4}}$ (xi) 2^2 (xii) a^2 (xiii) $x^{-\frac{7}{12}}$ (xiv) $(\frac{8}{9})^n$ (xv) 2^4 (xvi) 2^{-3}.

5

x	-2	$-1\frac{1}{2}$	-1	$-\frac{1}{2}$	$-\frac{1}{4}$	0	$\frac{1}{4}$	$\frac{1}{2}$	$\frac{3}{4}$	1	$1\frac{1}{2}$	2
16^x	$0{\cdot}004$	$0{\cdot}016$	$0{\cdot}063$	$0{\cdot}25$	$0{\cdot}5$	1	2	4	8	16	64	256

(i) true (ii) true.

Exercise 9.6 (p. 132)

1 (i) 2 (ii) 3 (iii) 5 (iv) 1 (v) 0 (vi) -1 (vii) -2 (viii) $\frac{1}{2}$ (ix) $\frac{1}{3}$.   **2** Reflection
in $y = x$ followed by reflection in $y = 0$, i.e. rotation $90°$ about O.

Exercise 9.7 (p. 133)

1 (i) 1 (ii) 2 (iii) 3 (iv) 3 (v) 4 (vi) 3 (vii) 10 (viii) $\frac{1}{2}$ (ix) $\frac{3}{2}$ (x) $\frac{5}{2}$ (xi) $\frac{5}{3}$
(xii) -4 (xiii) $\frac{1}{2}$ (xiv) $\frac{1}{2}$ (xv) 3 (xvi) $\frac{1}{4}$ (xvii) $\frac{2}{3}$ (xviii) 0 (xix) 10.   **2** (i) $s^2 = t^2$
(ii) $\frac{1}{9} = 3^{-2}$ (iii) $y = x^z$.   **3** (i) $\log_{10} 1\,000\,000 = 6$ (ii) $\log_5 125 = 3$
(iii) $\log_2 1024 = 10$ (iv) $\log_{16} \frac{1}{2} = -\frac{1}{4}$ (v) $\log_9 27 = \frac{3}{2}$ (vi) $\log_a b = c$.
4 $n = \dfrac{\log_{10} A - \log_{10} P}{\log_{10}\left(1 + \frac{x}{100}\right)}$.

Exercise 9.8 (p. 134)

1 (i) $\log_a x + \log_a y - \log_a z$ (ii) $\log_a x + \log_a y + \log_a z$
(iii) $2\log_a x + 3\log_a y + 4\log_a z$ (iv) $\frac{1}{2}\log_a x + \frac{1}{2}\log_a z$ (v) $-\frac{1}{3}\log_a y$.   **2** (i) $\log_a 2$
(ii) $\log_a\left(\dfrac{xy}{z}\right)$ (iii) $\log_a\left(\dfrac{y^3}{z}\right)$ (iv) $\log_a(3a)$ (v) $\log_a\left(\dfrac{a^5}{5}\right)$.

3 (i) 2 (ii) $-\frac{3}{2}$ (iii) $-\frac{1}{2}$ (iv) $\frac{5}{2}$ (v) 2
(vi) $-\frac{1}{6}$ (vii) 1 (viii) $\log_a b$ (ix) 0.
4 (i) 4 (ii) 27 (iii) 4 (iv) $\sqrt{2}$ (v) $\frac{125}{9}$
(vi) $\frac{864}{25}$. **5** (i) $\log_3 7$ (ii) $\log_{10} 4$
(iii) $\log_{0.1} 7$ (iv) $1 - \log_5 3$ (v) $\log_2 24$. **6** (i) -2 (ii) $-\frac{1}{2}$ (iii) 1.

Exercise 9.9 (p. 135)

1 (i) 1 (ii) $3\log_c b$. **2** $\log_3 \sqrt{2}$. **3** 2·322.

Miscellaneous Exercise 9 (p. 136)

1 $\log_5 2$, $\log_5 3$. **2** (i) 1·58 (ii) 0·63 (iii) 2·32 (iv) 2·58 (v) 0·38. **3** $\frac{3}{4}$.
4 $\{2, 16\}$. **5** $\{4\}$. **6** $y = 1000x^{-\frac{3}{4}}$. **7** $x = -\frac{1}{2}$. **8** $2\sqrt{2} - 2$. **9** $e(1 - e^x)$.
10 4. **11** $\frac{1}{2}\ln 3$. **12** $\{1, 3\}$. **13** $y = 6/\sqrt{x}$. **14** $y = ex^{-3}$.
15 $\ln 2$. **16** 8·6. **17** (a) $\frac{2}{5}$ (b) 1·48. **18** $x = 27 = y$ or $x = 9$, $y = 81$.
19 $x = 3$, $y = 9$ or $x = -4$, $y = 16$. **20** $x = 3$, $y = 9$ or $x = 9$, $y = 3$.
21 (a) (i) 1·2041200 (ii) $-0·9542426$ (iii) 0·3597271, (b) (i) 3
(ii) 2, (c) 13/5, 3/5, (d) $-0·30$.

Exercise 10.1 (p. 139)

1 (i) $y = 0·29x + 3·4$ (ii) $y = -0·004x + 0·23$ (iii) $y = 133x - 319$.

Exercise 10.4 (p. 143)

1 $a = 1·3$, $b = 2·1$. **2** $c = 10$, $d = 0·5$. **3** $a = -23$, $b = 42$. **4** $u = 1$, $f = 7$.
5 $n = 2$, $k = 0·0002$. **6** $a = 22·9$, $b = 0·03$. **7** $R = 8·3$.
8 $n = 3$, $a = 2·5$, $r = 2·5s^3$. **9** $a = 25$, $b = 1·6$.

Miscellaneous Exercise 10 (p. 145)

1 $a = 7$, $b = 5$. **2** $a = 2$, $x = 3·1$. **3** $a = 10$, $K = 7·4$. **4** $a = -0·5$, $b = 2·3$.
5 $n = 3$, $a = 50$. **6** $a = 0·4$, $b = 1·5$. **7** $f = 12$.
8 (a) $2y + 3x = 4$ (b) $\ln 5 + \ln x = y \ln 6$. **9** $a = -12$, $b = 100$.
10 $a = 1·5$, $b = 1·1$. **11** $a = 2$, $b = 3$, 202. **12** $a = -5$, $b = 8$.
13 $a = 0·4$, $b = 2·7$, 0·9. **14** $a = 0·5$, $b = 12$.

Exercise 11.4 (p. 154)

1 (i) 2, 1 max. and 1 min. (ii) 3, 2 max. and 1 min. or 1 max. and 2 min. (iii) $n - 1$, if n is
odd the graph can have $\frac{1}{2}(n - 1)$ max. and $\frac{1}{2}(n - 1)$ min., if n is even the graph can have $\frac{1}{2}n$
max. and $\frac{1}{2}n - 1$ min. or $\frac{1}{2}n - 1$ max. and $\frac{1}{2}n$ min. **3** Use $f(x) = :$ (i) $x(1 - x)(1 + x)$
(v) $x^3(1 - x)(1 + x)$ (vi) $x^2(x - 1)(x + 1)$ (vii) and (viii) $f(x) > 0$ for all x.

Exercise 11.6 (p. 156)

2 (i) 1 way stretch parallel to Oy factor 2 (ii) translation -2 parallel to Oy
(iii) translation $+3$ parallel to Oy (iv) reflection in Ox followed by translation $+3$
parallel to Oy (v) translation $+3$ parallel to Ox (vi) translation -2 parallel to Ox
(vii) 1 way stretch parallel to Oy factor 2 followed by translation $+5$ parallel to Oy
(viii) 1 way stretch parallel to Oy factor -1 followed by translation $+4$ parallel to Oy
(ix) translation $+3$ parallel to Ox followed by 1 way stretch parallel to Oy factor 4
(x) translation -1 parallel to Ox followed by 1 way stretch parallel to Oy factor 4
(xi) translation -1 parallel to Ox followed by translation $+4$ parallel to Oy.

Exercise 12.1 (p. 162)

1

$$\binom{x_1}{y_1} = \binom{x_2}{y_2} \Leftrightarrow x_1\mathbf{i}+y_1\mathbf{j} = x_2\mathbf{i}+y_2\mathbf{j} \Leftrightarrow (x_1-x_2)\mathbf{i} = (y_1-y_2)\mathbf{j} \Leftrightarrow x_1 = x_2,\ y_1 = y_2$$

$$\binom{x_1}{y_1} + \binom{x_2}{y_2} = x_1\mathbf{i}+y_1\mathbf{j}+x_2\mathbf{i}+y_2\mathbf{j} = (x_1+x_2)\mathbf{i}+(y_1+y_2)\mathbf{j} = \binom{x_1+x_2}{y_1+y_2}$$

$$\lambda\binom{x}{y} = \lambda(x\mathbf{i}+y\mathbf{j}) = \lambda x\mathbf{i}+\lambda y\mathbf{j} = \binom{\lambda x}{\lambda y}.$$

2 (i) $\binom{3}{-3}$ (ii) $\binom{-4}{3}$ (iii) $\binom{3}{-11}$ (iv) $\binom{a-2b}{2a-b}$ (v) $\binom{7x}{-y}$ (vi) $\binom{ax+bu}{ay+bv}$.

Exercise 12.2 (p. 165)

1 (i) 5 (ii) 13 (iii) $\sqrt{2}$ (iv) 10 (v) 5 (vi) $\sqrt{5}$ (vii) 3. **2** (i) $\binom{-\frac{3}{5}}{-\frac{4}{5}}$

(ii) $\binom{\frac{5}{13}}{\frac{12}{13}}$ (iii) $\binom{\frac{1}{\sqrt{2}}}{\frac{1}{\sqrt{2}}}$ (iv) $\binom{-\frac{4}{5}}{\frac{3}{5}}$ (v) $\frac{3}{5}\mathbf{i}-\frac{4}{5}\mathbf{j}$ (vi) $\frac{1}{\sqrt{5}}\mathbf{i}+\frac{2}{\sqrt{5}}\mathbf{j}$ (vii) $\mathbf{j}$.

3 $\overrightarrow{AB} = \binom{3}{2},\ \overrightarrow{CD} = \binom{-3}{-2}$, so $\overrightarrow{AB} = \overrightarrow{DC}$ and $ABCD$ is a parallelogram.

4 (i) $\mathbf{a}+\mathbf{b}$ (ii) $3\mathbf{a}$ (iii) $\frac{3}{11}\mathbf{a}+\frac{2}{11}\mathbf{b}$ (iv) $\frac{4}{11}\mathbf{a}-\frac{1}{11}\mathbf{b}$ (v) $\frac{4x+3y}{11}\mathbf{a}+\frac{2y-x}{11}\mathbf{b}$

(vi) $\frac{10}{11}\mathbf{a}+\frac{3}{11}\mathbf{b}$ (vii) $3\mathbf{a}+2\mathbf{b}$. **5** (i) $-\frac{1}{5}\mathbf{a}+\frac{6}{5}\mathbf{b}$ (ii) $\frac{1}{5}\mathbf{a}+\frac{4}{5}\mathbf{b}$ (iii) $-\frac{2}{5}\mathbf{a}+\frac{1}{5}\mathbf{b}$

(iv) $-\frac{1}{5}\mathbf{a}+\frac{3}{5}\mathbf{b}$ (v) $\frac{3}{5}\mathbf{a}+\frac{1}{5}\mathbf{b}$ (vi) $\frac{x-2y}{5}\mathbf{a}+\frac{2x+y}{5}\mathbf{b}$ (vii) $-\frac{8}{5}\mathbf{a}+\frac{9}{5}\mathbf{b}$

(viii) $\frac{s-2t}{5}\mathbf{a}+\frac{2s+t}{5}\mathbf{b}$ **6** The point with position vector $\binom{x}{y}$ lies on the line joining the

points with position vectors $\binom{p}{q}$ and $\binom{s}{t}$ and divides the line in the ratio $\mu:\lambda$ if and only if

$\lambda\binom{p}{q}+\mu\binom{s}{t} = (\lambda+\mu)\binom{x}{y}$. The point (x, y) lies on the line joining (p, q) and (s, t) and

divides this line in the ratio $\mu:\lambda$ if and only if $\lambda p+\mu s = (\lambda+\mu)x$ and $\lambda q+\mu t = (\lambda+\mu)y$.

7 $\left(\frac{11}{2}, \frac{77}{18}\right)$. **8** $A\binom{1}{3}, B\binom{3}{-1}, C\binom{5}{7}, D\binom{4}{3}, E\binom{3}{5}, F\binom{2}{1}, G\binom{3}{3}$,

$3\overrightarrow{OG} = 2\overrightarrow{OD}+\overrightarrow{OA} = 2\overrightarrow{OE}+\overrightarrow{OB} = 2\overrightarrow{OF}+\overrightarrow{OC}$.

9 $\mathbf{u} = \binom{\frac{4}{5}}{\frac{3}{5}}, \mathbf{v} = \binom{\frac{3}{5}}{-\frac{4}{5}}, \mathbf{u}+\mathbf{v} = \binom{\frac{7}{5}}{-\frac{1}{5}}, \mathbf{u}-\mathbf{v} = \binom{\frac{1}{5}}{\frac{7}{5}}, \left(\frac{7}{5\sqrt{2}}, \frac{-1}{5\sqrt{2}}\right), \left(\frac{1}{5\sqrt{2}}, \frac{7}{5\sqrt{2}}\right)$.

10 $m = 3,\ n = -2$. **11** $-15/2$.

12 $|\overrightarrow{PQ}| = \sqrt{(2\lambda^2+2\mu^2)},\ \overrightarrow{OR} = \frac{1}{2}(\lambda-\mu)\mathbf{i}+\frac{1}{2}(\lambda+\mu)\mathbf{j}$.

Exercise 12.3 (p. 169)

1 (b) (a) (a).

2 $3, 5\sqrt{2}, \sqrt{(29)}, 3\sqrt{(74)}, \frac{1}{3}\begin{pmatrix}1\\2\\-2\end{pmatrix}, \frac{1}{5\sqrt{2}}\begin{pmatrix}3\\-4\\5\end{pmatrix}, \frac{1}{\sqrt{29}}\begin{pmatrix}4\\-2\\3\end{pmatrix}, \frac{1}{3\sqrt{74}}\begin{pmatrix}-7\\16\\-19\end{pmatrix}$.

3 $\begin{pmatrix}-2\\5\\1\end{pmatrix}, \sqrt{30}$. **4** (i) $\begin{pmatrix}-3\\1\\1\end{pmatrix}$ (ii) $\begin{pmatrix}0\\7\\-1\end{pmatrix}$ (iii) $\begin{pmatrix}14\\5\\-2\end{pmatrix}$ (iv) $\begin{pmatrix}9\\10\\0\end{pmatrix}$.

5 (i) $\begin{pmatrix} 1 \\ 2 \\ 3 \end{pmatrix}, \begin{pmatrix} 1 \\ 2 \\ 3 \end{pmatrix}$, yes (ii) $\begin{pmatrix} -2 \\ 2 \\ -3 \end{pmatrix}, \begin{pmatrix} -2 \\ 1 \\ -3 \end{pmatrix}$, no (iii) $\begin{pmatrix} 2 \\ 1 \\ -4 \end{pmatrix}, \begin{pmatrix} -2 \\ -1 \\ 4 \end{pmatrix}$, yes.

6 $\begin{pmatrix} 1 \\ 2 \\ 3 \end{pmatrix}, \begin{pmatrix} -2 \\ 0 \\ 2 \end{pmatrix}, \begin{pmatrix} -1 \\ 1 \\ 3 \end{pmatrix}, \begin{pmatrix} -4 \\ 4 \\ 12 \end{pmatrix}$, $2\overrightarrow{PQ} + 3\overrightarrow{PR} = 4\overrightarrow{PS}$, hence coplanar.

7 $(-\frac{7}{3}, -\frac{2}{3}, -\frac{5}{3})$.

Exercise 12.4 (p. 171)

1 (i) $x = 2 + 5t$, $y = 3 + t$, $5y = x + 13$ (ii) $x = 1 - 2t$, $y = -3 + 5t$, $2y + 5x + 1 = 0$
(iii) $x = 1 - 2t$, $y = t$, $2y + x = 1$.

2 (i) $r = \begin{pmatrix} 4 \\ -3 \end{pmatrix} + t\begin{pmatrix} -1 \\ 8 \end{pmatrix}$, $y + 8x = 29$ (ii) $r = 2\mathbf{i} + 3\mathbf{j} + t(3\mathbf{i} + 2\mathbf{j})$, $3y = 2x + 5$.

3 $r = \begin{pmatrix} \frac{5}{8} \\ \frac{3}{4} \end{pmatrix} + t\begin{pmatrix} 2 \\ 1 \end{pmatrix}$. 4 $\begin{pmatrix} -\frac{4}{5} \\ \frac{3}{5} \end{pmatrix}$, $(-9, 10)$, $(15, -8)$. 5 $r = t\begin{pmatrix} 1 \\ 3 \end{pmatrix}$.

6 (i) $(3, 6)$ (ii) $r = \begin{pmatrix} 2 \\ 3 \end{pmatrix} + t\begin{pmatrix} -1 \\ 1 \end{pmatrix}$, $r = \begin{pmatrix} -1 \\ 2 \end{pmatrix} + t\begin{pmatrix} 1 \\ 1 \end{pmatrix}$ (iii) $\begin{pmatrix} 1 \\ 4 \end{pmatrix}$.

Exercise 12.5 (p. 175)

1 (i) $x = 2 - 3t$, $y = 5 + t$, $z = 4 - 2t$, $\dfrac{x-2}{-3} = \dfrac{y-5}{1} = \dfrac{z-4}{-2}$

(ii) $x = 2t$, $y = 2 + t$, $z = -1 + 4t$, $\dfrac{x}{2} = \dfrac{y-2}{1} = \dfrac{z+1}{4}$

(iii) $x = 4 - 3t$, $y = 1 + 12t$, $z = 2 - 5t$, $\dfrac{x-4}{-3} = \dfrac{y-1}{12} = \dfrac{z-2}{-5}$

(iv) $x = 2 + s$, $y = -2 + s$, $z = 3 + s$, $x - 2 = y + 2 = z - 3$.

2 (i) $r = \begin{pmatrix} 3 \\ -1 \\ 4 \end{pmatrix} + t\begin{pmatrix} -1 \\ 3 \\ 2 \end{pmatrix}$ (ii) $r = \begin{pmatrix} 1 \\ 1 \\ -1 \end{pmatrix} + t\begin{pmatrix} 4 \\ 3 \\ 5 \end{pmatrix}$ (iii) $r = \begin{pmatrix} 2 \\ -3 \\ 4 \end{pmatrix} + t\begin{pmatrix} 1 \\ 0 \\ 2 \end{pmatrix}$

(iv) $r = \begin{pmatrix} 2 \\ -4 \\ 1 \end{pmatrix} + t\begin{pmatrix} -1 \\ -1 \\ 6 \end{pmatrix}$ (v) $r = \begin{pmatrix} -7 \\ 3 \\ -1 \end{pmatrix} + t\begin{pmatrix} 3 \\ -2 \\ 2 \end{pmatrix}$.

3 (i) $r = \begin{pmatrix} 2 \\ -1 \\ 3 \end{pmatrix} + t\begin{pmatrix} 1 \\ 3 \\ -4 \end{pmatrix}$ (ii) $r = \begin{pmatrix} 0 \\ 2 \\ 0 \end{pmatrix} + t\begin{pmatrix} 1 \\ -3 \\ 4 \end{pmatrix}$ (iii) $r = \begin{pmatrix} 4 \\ 0 \\ 0 \end{pmatrix} + t\begin{pmatrix} -2 \\ 1 \\ 0 \end{pmatrix}$

(iv) $r = \begin{pmatrix} 3 \\ 2 \\ 0 \end{pmatrix} + t\begin{pmatrix} 0 \\ -1 \\ 1 \end{pmatrix}$. 4 (i) $r = \begin{pmatrix} 0 \\ 3 \\ -2 \end{pmatrix} + x\begin{pmatrix} 1 \\ 2 \\ -4 \end{pmatrix}$ (ii) $r = \begin{pmatrix} -7 \\ 0 \\ -1 \end{pmatrix} + y\begin{pmatrix} 3 \\ 1 \\ 2 \end{pmatrix}$

(iii) $r = \begin{pmatrix} -4 \\ 0 \\ 0 \end{pmatrix} + z\begin{pmatrix} 3 \\ 1 \\ 1 \end{pmatrix}$ (iv) $r = \begin{pmatrix} 6 \\ 0 \\ 2 \end{pmatrix} + y\begin{pmatrix} 2 \\ 1 \\ 0 \end{pmatrix}$ (v) $r = \begin{pmatrix} 0 \\ 3 \\ 2 \end{pmatrix} + x\begin{pmatrix} 1 \\ -1 \\ 1 \end{pmatrix}$

(vi) $r = \begin{pmatrix} 0 \\ 2 \\ \frac{4}{5} \end{pmatrix} + x\begin{pmatrix} 1 \\ \frac{3}{2} \\ -\frac{1}{5} \end{pmatrix}$. 5 (i) $\begin{pmatrix} 4 \\ 2 \\ 3 \end{pmatrix}$ (ii) $\begin{pmatrix} 2 \\ \frac{3}{2} \\ 1 \end{pmatrix}$

(iii) $\begin{pmatrix} -\frac{7}{3} \\ -\frac{5}{3} \\ \frac{3}{2} \end{pmatrix}$ (iv) $\begin{pmatrix} 2 \\ 1 \\ -3 \end{pmatrix}$ (v) $\begin{pmatrix} 2 \\ -1 \\ -2 \end{pmatrix}$ (vi) $\begin{pmatrix} 3 \\ 2 \\ 1 \end{pmatrix}$.

6 (i) $\dfrac{x-2}{-3} = \dfrac{y-4}{-14} = \dfrac{z-3}{9}$ (ii) $3x = -2y = z$ (iii) $\dfrac{x-3}{-1} = \dfrac{y-5}{2}, z = 7$

(iv) $\dfrac{x-5}{3} = \dfrac{y-7}{3} = \dfrac{z+3}{-8}$ (v) $x = 3, y = \dfrac{z-1}{-2}$.   **7** No.

8 $t = 0, t = 3, t = -4, AB:AC = 3:4$.   **9** $(0, 1, -9)$.   **10** (i) no (ii) no.
11 $\mathbf{r} = \mathbf{i} - \mathbf{j} + 3\mathbf{k} + t(3\mathbf{j} - \mathbf{k}), \mathbf{i} + \mathbf{j} + \tfrac{7}{3}\mathbf{k}$.   **12** (ii) is a line and (i) is a plane. The point
$(6, 3, 2)$ lies on both loci and is the point of intersection of the line and the plane.

Exercise 12.6A (p. 179)

1 $\mathbf{a} . \mathbf{b} = ab \cos\theta = b(a \cos\theta) = a(b \cos\theta)$.
2 $1, 25, 4, 5/2, \sqrt{3}, 0, 5/2 + \sqrt{3}$, all in km^2.
3 $-\tfrac{1}{2}15\sqrt{3}, -10, 0, -\tfrac{1}{2}15\sqrt{3}, -10 - \tfrac{1}{2}15\sqrt{3}$.   **5** 17, 17.   **7** 7.

Exercise 12.6B (p. 184)

1 Approximately, in radians, (i) 1·38 (ii) 2·47 (iii) 0·28 (iv) 1·03 (v) 1·41.

2 0·79.   **4** 7.   **5** $\dfrac{1}{\sqrt{14}}(2\mathbf{i} - \mathbf{j} - 3\mathbf{k}), 4/\sqrt{14}$.   **6** (i) no (ii) yes $(6, -7, -8)$

(iii) no.   **7** $3, (\tfrac{2}{3}, \tfrac{2}{3}, \tfrac{1}{3})$.   **8** (i) 5·34, $(4, -\tfrac{1}{2}, \tfrac{7}{2})$ (ii) 3·94, $(1, -\tfrac{7}{2}, \tfrac{3}{2})$.

9 $(2, 6, 8), 4\sqrt{(247)}$.   **11** $(1\tfrac{9}{19}, 2\tfrac{13}{19}, 1\tfrac{4}{19})$.   **12** $1:6, \mathbf{r} = \mathbf{a} + t(\mathbf{a} + 6\mathbf{b}), \begin{pmatrix} -12 \\ -6 \\ 6 \end{pmatrix}$.

13 $3, (4, -1, 2)$.   **14** $\mathbf{r} = \begin{pmatrix} 3 + 5t \\ 2 - 7t \\ 1 + t \end{pmatrix}, (13, -12, 3), 10\sqrt{3}$.

15 $\overrightarrow{PQ} = \begin{pmatrix} 3q - 2p - 2 \\ 2p + 4q - 1 \\ 7q + 1 \end{pmatrix}, k = -4/3, 4\sqrt{3}/3$.   **16** $1/\sqrt{15}$.

Exercise 12.7 (p. 188)

1 (i) $y = 0$ (ii) $2x + 3y + 4z = 0$ (iii) $2x + 3z = y$ (iv) $x = y$ (v) $5x + 3y = 2z$.
2 (i) $\mathbf{i} + 2\mathbf{j}$ (ii) $\mathbf{i} + \mathbf{j} + \mathbf{k}$ (iii) $3\mathbf{i} - 4\mathbf{j} + 2\mathbf{k}$.   **3** $x + 3z = 2y$.
4 $2\mathbf{i} + \mathbf{j} - 3\mathbf{k}, -2\mathbf{i} + 3\mathbf{j} + \mathbf{k}, 3x = 6y = -2z, -3x = 2y = 6z, 1·86$.

Exercise 12.8 (p. 193)

1 (i) $x + z = 4$ (ii) $2x - y + 3z = 10$ (iii) $3x - 2y + z = -18$.

2 (i) $\mathbf{r} . \begin{pmatrix} 2/3 \\ 1/3 \\ 2/3 \end{pmatrix} = 2$ (ii) $\mathbf{r} . \begin{pmatrix} 3/\sqrt{10} \\ 2/\sqrt{10} \\ 0 \end{pmatrix} = 4/\sqrt{10}$ (iii) $\mathbf{r} . \begin{pmatrix} 1/\sqrt{3} \\ -1/\sqrt{3} \\ 1/\sqrt{3} \end{pmatrix} = 1/\sqrt{3}$

(iv) $\mathbf{r} . \begin{pmatrix} -4/\sqrt{42} \\ 1/\sqrt{42} \\ 5/\sqrt{42} \end{pmatrix} = -7/\sqrt{42}$.   **4** (i) $\mathbf{r} . \begin{pmatrix} 1 \\ -1 \\ 0 \end{pmatrix} = 1, x - y = 1$

(ii) $\mathbf{r} . \begin{pmatrix} 2 \\ -1 \\ 3 \end{pmatrix} = -2, 2x - y + 3z = -2$ (iii) $\mathbf{r} . \begin{pmatrix} 1 \\ -1 \\ 2 \end{pmatrix} = 9, x - y + 2z = 9$

(iv) $\mathbf{r} . \begin{pmatrix} 4 \\ 2 \\ 3 \end{pmatrix} = 10, 4x + 2y + 3z = 10$ (v) $\mathbf{r} . \begin{pmatrix} 3 \\ 9 \\ 5 \end{pmatrix} = -4, 3x + 9y + 5z = -4$

(vi) $\mathbf{r} . \begin{pmatrix} 2 \\ 1 \\ 1 \end{pmatrix} = 5, 2x + y + z = 5$ (vii) $\mathbf{r} . \begin{pmatrix} 2 \\ 1 \\ 0 \end{pmatrix} = 1, 2x + y = 1$.
5 (i) $(2, 0, 1)$ (ii) $(-1, 4, -2)$.   **6** (i) 5/3 (ii) 4/7 (iii) 7/5.

Exercise 12.9 (p. 195)

1 (i) $x+y+z=5$, $\mathbf{r}.\begin{pmatrix}1\\1\\1\end{pmatrix}=5$ (ii) $2x+5y+3z=-8$, $\mathbf{r}.\begin{pmatrix}2\\5\\3\end{pmatrix}=-8$

(iii) $x-2y=3$, $\mathbf{r}.\begin{pmatrix}1\\-2\\0\end{pmatrix}=3$. **2** (i) $\mathbf{r}=s\mathbf{j}+t\mathbf{k}$ (ii) $\mathbf{r}=s\mathbf{j}+t(\mathbf{i}-\mathbf{k})$

(iii) $\mathbf{r}=3\mathbf{k}+s(\mathbf{i}+2\mathbf{j})+t(\mathbf{j}+\mathbf{k})$ (iv) $\mathbf{r}=-4\mathbf{j}+s\mathbf{i}+t(3\mathbf{j}+\mathbf{k})$

(v) $\mathbf{r}=-3\mathbf{i}+s(2\mathbf{i}+\mathbf{k})+t(\mathbf{j}+\mathbf{k})$. **3** $\mathbf{r}.\begin{pmatrix}2\\6\\5\end{pmatrix}=8$, $2x+6y+5z=8$.

4 $\mathbf{r}=\mathbf{k}+s\mathbf{a}+t\mathbf{b}$, $\mathbf{r}.\mathbf{c}=\mathbf{k}.\mathbf{c}$, for all s, t.

Exercise 12.11 (p. 199)

1 (i) $6/\sqrt{26}$, opposite side (ii) $2/\sqrt{29}$, same side.

2 (i) $\mathbf{r}=\begin{pmatrix}1\\2\\-3\end{pmatrix}+t\begin{pmatrix}3\\-1\\2\end{pmatrix}$, $\dfrac{x-1}{3}=\dfrac{y-2}{-1}=\dfrac{z+3}{2}$, $70\cdot6°$

(ii) $\mathbf{r}=\begin{pmatrix}5\\-3\\2\end{pmatrix}+t\begin{pmatrix}7\\-6\\7\end{pmatrix}$, $\dfrac{x-5}{7}=\dfrac{y+3}{-6}=\dfrac{z-2}{7}$, $33\cdot1°$.

3 (i) $27\cdot2°$, $\mathbf{r}=\begin{pmatrix}-2\\0\\8\end{pmatrix}+t\begin{pmatrix}3\\2\\5\end{pmatrix}$ (ii) $64\cdot8°$, $\mathbf{r}=\begin{pmatrix}-1\\1\\0\end{pmatrix}+t\begin{pmatrix}27\\-7\\23\end{pmatrix}$.

4 $4x=2y+3z+5$. **5** $\dfrac{x}{2}=\dfrac{y}{2}=z$, $4x-3y-2z=0$, $\tfrac{1}{2}\pi$.

Miscellaneous Exercise 12 (p. 200)

1 $\dfrac{2\pi}{3}$. **2** $64\cdot6°$, $198\cdot7$. **4** $p=-2$, $q=2$. **5** $1,1,\tfrac{1}{2}$, $\dfrac{\sqrt{7}}{3}$, $\dfrac{2}{\sqrt{7}}$.

6 $\dfrac{1}{11}(a+6b+4c)$. **7** $\tfrac{1}{2}(a+b)$, $\tfrac{1}{3}(b+2c)$, $1:3$. **9** $\mathbf{r}.\begin{pmatrix}2/3\\2/3\\1/3\end{pmatrix}=3$. **10** $\begin{pmatrix}9\\4\\7\end{pmatrix}$.

11 $\mathbf{i}+7\mathbf{j}$. **12** (ii) $(1,-1,2)$, $70\cdot5°$. **13** $\tfrac{1}{17}(-6,-4,4)$, $\tfrac{1}{17}(40,-13,47)$.

14 $-2\mathbf{i}+2\mathbf{j}+3\mathbf{k}$. **15** $\sqrt{6}$, $\sqrt{2}$, $\tfrac{\pi}{6}$. **16** (a) -3. (b) $4/3$ (c) 10, -0.4.

17 $\sqrt{2}$, $-\sqrt{2}$. **18** 6, $(\tfrac{2}{3},-\tfrac{2}{3},\tfrac{1}{3})$. **19** $\tfrac{3}{7}(6\mathbf{i}+3\mathbf{j}+2\mathbf{k})$.

20 $\begin{pmatrix}-1\\-3\\-3\end{pmatrix}+t\begin{pmatrix}2\\1\\3\end{pmatrix}=\mathbf{r}$, $(1,-2,0)$, $\sqrt{5}$. **21** $\mathbf{r}=t\begin{pmatrix}5\\-1\\-3\end{pmatrix}$. **22** (i) $(1,-7,-6)$

(ii) $(-\tfrac{1}{3},\tfrac{4}{3},\tfrac{10}{3})$. **23** $x-14=y-5=4-2z$, $2/3$, $x=2y+2z$. **24** $\mathbf{r}.(\mathbf{i}-\mathbf{j})=1$.

25 $\dfrac{1}{\sqrt{2}}(\mathbf{i}-\mathbf{j})$. **26** $(-1-k,k,1+k)$, $-2/3$.

27 $2x-3y+6z=70$. **28** $(0,-3,0)$, $(3,-1,7)$.

29 $(3\mathbf{i}+5\mathbf{j}+4\mathbf{k})\dfrac{1}{5\sqrt{2}}$, $3x+5y+4z=30$, $3\sqrt{2}$.

30 (i) line, (ii) plane, (iii) line; line (iii) lies in plane (ii), line (i) is at right angles to plane (ii) and meets it and line (iii) at $(2,1,3)$. **32** $\dfrac{\pi}{6}$. **33** $\begin{pmatrix}1\\2\\1\end{pmatrix}$.

34 $x - 3y + 4z = 2, \dfrac{1}{\sqrt{26}}(1, -3, 4), \dfrac{1}{\sqrt{26}}(\mathbf{i} - 3\mathbf{j} + 4\mathbf{k}), \dfrac{2}{\sqrt{26}}, \dfrac{2}{\sqrt{26}}.$

35 $(1, 2, -3), 6x = 3y - 2z = 18, \frac{38}{7}.$ **36** (i) $-\mathbf{i} + \mathbf{j} + 3\mathbf{k}$, (ii) $3\mathbf{j} - \mathbf{k}$, (iii) $55 \cdot 1°$

(iv) $34 \cdot 9°$. **37** $\lambda = \mu = \frac{2}{3}, 19 \cdot 5°$. **38** $\begin{pmatrix} -2 \\ 1 \\ 2 \end{pmatrix}, 2x = y + 2z + 2, \dfrac{\sqrt{74}}{11}, \frac{2}{3}.$

39 (i) $x = 3 + t, y = -1 - t, z = 2t$ (ii) $(2, 0, -2)$ (iii) $(1, 1, -4)$

(iv) $(\frac{7}{3}, \frac{1}{3}, -2)$ (v) $\mathbf{r} = \begin{pmatrix} 3 \\ 0 \\ -1 \end{pmatrix} + t\begin{pmatrix} 2 \\ -1 \\ 3 \end{pmatrix}.$

40 $\lambda = -1, \mu = 2, \mathbf{r} \cdot \begin{pmatrix} -1 \\ 2 \\ 1 \end{pmatrix} = 2, \mathbf{r} = \begin{pmatrix} 1 \\ 0 \\ -1 \end{pmatrix} + t\begin{pmatrix} -1 \\ 2 \\ 1 \end{pmatrix}, \dfrac{4}{\sqrt{6}}.$

42 $\dfrac{\pi}{6}, \mathbf{r} = \begin{pmatrix} 1 \\ 0 \\ -1 \end{pmatrix} + t\begin{pmatrix} 1 \\ 1 \\ 1 \end{pmatrix}, (2, 1, 0).$ **43** S is line OA, where $\overrightarrow{OA} = \mathbf{a}$,

(i) $\mathbf{n}$ is normal to Z (ii) O is not in $Z, t = \dfrac{k}{\mathbf{a} \cdot \mathbf{n}}$ if $\mathbf{a} \cdot \mathbf{n} \neq 0$, S is parallel to Z if $\mathbf{a} \cdot \mathbf{n} = 0$.

45 (c) $\dfrac{1}{\sqrt{10}}$ (d) 300. **46** (i) $\dfrac{\pi}{6}, \begin{pmatrix} 2 + q - 2p \\ -3 + p \\ 5 + q - p \end{pmatrix}, p = 1, q = -2.$

47 $\mathbf{r} = \begin{pmatrix} -3 \\ 5 \\ 0 \end{pmatrix} + s\begin{pmatrix} 1 \\ -1 \\ -1 \end{pmatrix} + t\begin{pmatrix} 2 \\ 1 \\ -1 \end{pmatrix}.$

Exercise 13.1 (p. 210)

1 (i) $a < b$ (ii) $c > b$ (iii) $x \not> y$ (iv) $y \leqslant z$ (v) $-1 < x < 5$. **2** (i) $x < 2$, 0
(ii) $y > 4$, 5 (iii) $-4 < z < 2$, 1 (iv) $w < 5$ or $w > 9$, 4. **3** The subset of real
numbers: (i) less than 4 (ii) less than 7 (iii) less than or equal to -3 (iv) greater than
or equal to -2 and less than 3 (v) either less than or equal to -2 or greater than 2.
4 (i) $\{x : x > -2\}$ (ii) $\{x : x \geqslant 3\}$ (iii) $\{x : x < \frac{3}{2}\}$ (iv) $\{x : x > \frac{1}{20}\}$
(v) $\{x : -2 \leqslant x \leqslant 1\}$. **5** (i) $\{x : 5 < x < 7\}$ (ii) $\{x : -2 \leqslant x \leqslant -1\}$
(iii) $\{x : \frac{9}{2} < x \leqslant 6\}$ (iv) $\{x : -1 < x < 1\}$.

Exercise 13.2 (p. 211)

1 (i) $-3 < x < 3$ (ii) $-1 < x < 7$ (iii) $-2 \leqslant x \leqslant \frac{8}{3}$ (iv) $x \leqslant -5$ or $x \geqslant 5$
(v) $x \leqslant -5$ or $x \geqslant 1$. **2** (i) $\{x : |x| < 2\}$ (ii) $\{x : |x| \geqslant 7\}$ (iii) $\{x : |x - 1| < 2\}$
(iv) $\{x : |x + 1| \geqslant 3\}$ (v) $\{x : |x| < 5\}$ (vi) $\{x : |x - 6| > 2\}$.

Exercise 13.3 (p. 213)

1 (i) $\{x : x < 0\} \cup \{x : x > \frac{1}{5}\}$ (ii) $\{x : x \geqslant 6\}$ (iii) $\{x : -\frac{1}{2} < x < 0\}$
(iv) $\{x : x < 0\} \cup \{x : x \geqslant 1\}$ (v) $\{x : -1 < x < 0\}$ (vi) $\{x : 0 < x \leqslant \frac{1}{2}\}$
(vii) $\{x : x < 0\} \cup \{x : x > \frac{1}{2}\}$. **2** (i) $\{x : -3 < x < 2\}$ (ii) $\{x : 1 < x \leqslant \frac{9}{7}\}$
(iii) $\{x : -5 < x < -2\}$ (iv) $\{x : \frac{1}{2} < x \leqslant 1\}$. **3** $\{x : |x| > 3\}$.
4 $\{x : |x| < 2\}$. **5** $\{x : x < -2\} \cup \{x : x > \frac{2}{3}\}$.

Exercise 13.4 (p. 216)

1 (i) $\{x : -2 < x < 1\} \cup \{x : x > 3\}$ (ii) $\{x : x < -2\} \cup \{x : 1 < x < 3\}$
(iii) $\{x : -4 \leqslant x \leqslant 3\}$ (iv) $\mathbb{R}$. **2** $\{x : 2 < x < 3\} \cup \{x : x > 9\}$.

3 $\{p:-3 < p < 0\}$.   **4** $\{x:-2 < x < \frac{1}{2}\}$ **5** $\{x:-4 < x < \frac{1}{2}\} \cup \{x:x > 5\}$.
6 $\{x:x < -3\} \cup \{x:-1 < x < 2\} \cup \{x:x > 4\}$.

Miscellaneous Exercise 13 (p. 218)

1 (i) $\{x:x < \frac{11}{5}\}$ (ii) $\mathbb{R}^+$ (iii) $\{x:x < 4\}$ (iv) $\mathbb{R}^+$ (v) $\{x:0 < x < \frac{4}{5}\}$
(vi) $\{x:x < -3\} \cup \{x:x > 5\}$ (vii) $\{x:-2 < x < 3\}$ (viii) $\{x:x > \frac{4}{3}\} \cup \{x:x \leqslant 1\}$
(ix) $\{x:x > \frac{4}{3}\} \cup \{x:x < 1\}$.   **2** (i) $\{x:x < -1\} \cup \{x:x > 5\}$
(ii) $\{x:-5 \leqslant x \leqslant 1\}$ (iii) $\{x:x < 0\}$.   **3** $\{x:x < 1\} \cup \{x:x > 3\}$.
4 $\{x:-2\sqrt{2} < x < 2\sqrt{2}\}$.   **5** $\{x:x < -1\} \cup \{x:x > 2\}$
6 $\{x:-2 < x < -1\} \cup \{x:x > 2\}$.   **7** $\{x:\frac{17}{6} < x < \frac{13}{4}\}$.   **8** $\{x:\frac{5}{3} < x < 3\}$.
9 $\mathbb{R}^+$.   **10** (i) $x^2 - 1 < 0$ (ii) $(x-a)(x-b) < 0$ (iii) $x^2 > 4x$.
11 (a) $\{x:x \geqslant 6\}$ (b) $\{x:x < 1\} \cup \{x:x \geqslant 6\}$.   **12** $-\frac{1}{2}, 3$.
13 $a = 5 - c$, $\{c:\frac{1}{2} < c < 4\frac{1}{2}\}$.   **15** $\{x:-1 < x < 3\}$.
16 $\{\theta:-\frac{5}{6}\pi < \theta < -\frac{\pi}{6}\}$.   **17** $\{x:|x| < 4\}$.
18 (a) $\{x:-3 < x < -1\} \cup \{x:1 < x < 2\}$ (b) $\{x:|x| < \sqrt{(\frac{3}{2})}\}$.
19 $\{\theta:-\frac{\pi}{4} < \theta < \frac{\pi}{4}\}$.   **20** $\{x:-1 < x < -\frac{1}{2}\}$.
21 $y = x - 4$, (a) $\{2\}$ (b) $\{x:x < 0\}$ (c) $\{x:x < 3\} \cup \{x:x > 4\}$.

Exercise 14.1 (p. 221)

1 1·4.

Exercise 14.2 (p. 222)

1 1·379.   **2** 0·32.   **3** (a) 1·828 (b) $2\sqrt{2} - 1$, error is 0·0004 to 1 S.F.
4 4·56, 0·44 to 2 D.P.

Exercise 14.3 (p. 224)

1 Answer between 1·3769 and 1·3789.   **2** 1·38.   **3** 2·646, $-2\cdot646$, 13 steps
needed.   **4** 12 iterations.   **5** 9·9.   **6** 2·72, 20 calculations, 2·72, 10 calculations.
7 $f(1\cdot286) = -0\cdot873$, $f(1\cdot392) = -0\cdot303$, $f(1\cdot427) = -0\cdot094$.

Exercise 14.4 (p. 226)

1 1·44203, 1·44224, 2 steps.   **2** (i) $-1\cdot784$ (ii) 0·1835 (iii) 1·236.
3 (ii) $c = 1\cdot25$ (iii) $c = 1\cdot4156$ (iv) $c = 1\cdot3379$.
5 no root between 2 and 3, $-4\cdot164$, wrong.

Exercise 14.5 (p. 229)

1 (i) 1·54, 1·39 (ii) 0·5, 0·33 (iii) $-3\cdot83$, $-3\cdot828$ (iv) $-1\cdot5$, $-1\cdot28$
(v) 2·925, 2·924.   **2** 1·532, $-1\cdot879$, 0·347.   **3** 0·8514.   **4** 2·03054.
5 $-2\cdot61803$, $-0\cdot38197$.

Miscellaneous Exercise 14 (p. 229)

1 $\alpha \approx -0\cdot727$, $\beta \approx 0\cdot8$, $f'(0) = 0$ gives division by zero, start $\gamma = 4$, $\gamma \approx 3\cdot875$.
2 $f(0) = 1$, $f(1) = -0\cdot45$, $f(0\cdot5) = 0\cdot37$, $f(0\cdot75) = -0\cdot01$, $f(0\cdot625) = 0\cdot18$,
$f(0\cdot6875) = 0\cdot08$, so the answer is 0·7 to 1 D.P.   **3** $x_{n+1} = 1\cdot5x_n - 0\cdot1x_n^3$.   **4** 0·453.
5 (i) 0, 3, 2·05, 2·13 (ii) 0, 1, $-0\cdot72$, 1·51, 2·12.   **6** 3·73.   **7** 0·5.   **8** 0·9051.
9 (i) 2·236 (ii) 3, 2·33, 2·238, 2·236.   **10** $p_{n+1} = \frac{1}{3}\left(2p_n + \frac{k}{p_n^2}\right)$.

11 $p_{m+1} = \frac{1}{n}\left((n-1)p_m + \frac{k}{p_m^{n-1}}\right)$.   **12** 1·31, α is the root, approximately 1·166.
13 3, 1·43, larger.   **14** 1·52.

Exercise 15.1 (p. 235)

1 (i) $7+10i$ (ii) $7-7i$ (iii) $7-15i$ (iv) $1-3i$ (v) $11+4i$ (vi) $-2i$.
2 (i) $2+4i$ (ii) $28+12i$ (iii) $-7+12i$. **3** (i) $5+14i$ (ii) $-25i$ (iii) $26-7i$
(iv) 2 (v) $-5+12i$ (vi) $-15+6i$ (vii) $19+17i$ (viii) $4-22i$ (ix) $6+27i$ (x) 1.

4 (i) $\dfrac{8}{13}-\dfrac{i}{13}$ (ii) $\dfrac{27}{25}+\dfrac{11}{25}i$ (iii) $\dfrac{-20}{29}+\dfrac{8}{29}i$ (iv) $\frac{1}{2}+\frac{1}{2}i$ (v) $\dfrac{9}{41}+\dfrac{40}{41}i$

(vi) $\dfrac{78}{676}+\dfrac{52}{676}i$ (vii) $\dfrac{72}{289}+\dfrac{154}{289}i$. **5(a)** (i) $(3,2)+(5,1)=(8,3)$

(ii) $(4,-2)+(0,6)=(4,4)$ (iii) $(3,4)-(8,-2)=(-5,6)$
(iv) $(-4,-2)-(7,0)=(-11,-2)$ (v) $3(2,-5)=(6,-15)$
(vi) $2(0,3)-3(1,1)=(-3,3)$ (vii) $(5,2)(3,-4)=(23,-14)$

(viii) $\dfrac{(1,-2)}{(2,-1)}=\left(\dfrac{4}{5},\dfrac{-3}{5}\right)$ **(b)** (i) $(0,7)$ (ii) $(-7,11)$ (iii) $(-3,5)$

(iv) $(-13,21)$ (v) $(-2,1)$ (vi) $(-1,1)$ (vii) $(2,5)$ (viii) $(2\cdot3,-0\cdot9)$
(ix) $(-4,72)$ (x) $(a+c,b+d)$ (xi) $(ac-bd,ad+bc)$ (xii) $\left(\dfrac{ac+bd}{c^2+d^2},\dfrac{-ad+bc}{c^2+d^2}\right)$.

6 $-1\cdot2+2\cdot4i$. **7** $x=1,\ y=-\frac{1}{2}$. **8** $x=\dfrac{2}{13},\ y=\dfrac{3}{13}$. **9** i.

Exercise 15.2 (p. 238)

1 (i) $-1,-3$ (ii) $1,\frac{5}{2}$ (iii) $3+\sqrt{3},3-\sqrt{3}$ (iv) -3

(v) $\dfrac{-5+\sqrt{37}}{6},\dfrac{-5-\sqrt{37}}{6}$ (vi) $\dfrac{7+\sqrt{17}}{8},\dfrac{7-\sqrt{17}}{8}$. **2** (i) $2+i,2-i$

(ii) $3+i,3-i$ (iii) $4+i,4-i$ (iv) $-6-2i,-6+2i$ (v) $-\dfrac{1}{6}+\dfrac{\sqrt{11}}{6}i,-\dfrac{1}{6}-\dfrac{\sqrt{11}}{6}i$

(vi) $\dfrac{1}{2}+\dfrac{\sqrt{3}}{2}i,\dfrac{1}{2}-\dfrac{\sqrt{3}}{2}i$. **3** (i) $(x-1)(3x-1)$ (ii) $(x+2)(2x+1)$

(iii) $(x+1)(x+2)x$ (iv) $(x-2)(x+2)x$. **4** (i) $2x^2+3x+2$ (ii) $x^2+8x+17$
(iii) $(x-4)(x-5)$ (iv) $3x^2-x+6$ (v) $(x-1)(x^2+1)$ (vi) $(x^2+5)(x^2+2)$
(vii) $x^2(x-3)(x+2)$ (viii) $(x-2)(x+2)^2$. **6** (i) $-i$ (ii) 1 (iii) i (iv) $-i$
(v) -1 (vi) i. **7** (i) -4 (ii) -5 (iii) -49 (iv) -2. **8** (i) $i\sqrt{5},-i\sqrt{5}$
(ii) $5i,-5i$ (iii) $2i\sqrt{2},-2i\sqrt{2}$ (iv) $3i\sqrt{3},-3i\sqrt{3}$.

9 (i) $\dfrac{5}{2}+\dfrac{\sqrt{7}}{2}i,\dfrac{5}{2}-\dfrac{\sqrt{7}}{2}i,5,8$ (ii) $\dfrac{-3}{4}+\dfrac{\sqrt{23}}{4}i,\dfrac{-3}{4}-\dfrac{\sqrt{23}}{4}i,\dfrac{-3}{2},2$

(iii) $\dfrac{1}{6}+\dfrac{\sqrt{59}}{6}i,\dfrac{1}{6}-\dfrac{\sqrt{59}}{6}i,\dfrac{1}{3},\dfrac{5}{3}$.

Exercise 15.3A (p. 240)

2 (i), (ii) & (iii) rotation $180°$ about O (iv), (v) & (vi) reflection in $\mathbb{R}$-axis
(vii), (viii) & (ix) rotation $90°$ about O (x), (xi) & (xii) rotation $270°$ about O.

3 $A=d,B=b,C=a,D=c$. **4** (i) $5,0\cdot64$ (ii) $2\sqrt{2},\dfrac{\pi}{4}$ (iii) $13,1\cdot18$

(iv) $10,2\cdot21$ (v) $\sqrt{10},3\cdot46$. **5** $2-i,-2+i$.

Exercise 15.3B (p. 242)

3 rotation $90°$ about O. **4** $\dfrac{\pi}{2},a$. **5** rotation $-90°$ about O.

6 $5,\sqrt{2},5\sqrt{2},0\cdot64,0\cdot79,1\cdot43$, rotation $\dfrac{\pi}{4}$ about O, enlargement S.F. $\sqrt{2}$.

Exercise 15.4 (p. 246)

1 (i) $\left(\sqrt{2},\dfrac{\pi}{4}\right)$ (ii) (5, 0·64)   (iii) (13, −1·18)   (iv) (13, 2·75)   (v) $\left(4\sqrt{2},-\dfrac{\pi}{4}\right)$

(vi) (10, −2·21)   (vii) $\left(2,\dfrac{\pi}{6}\right)$ (viii) $\left(2,\dfrac{2\pi}{3}\right)$ (ix) $\left(4,-\dfrac{\pi}{6}\right)$ (x) (25, 1·29)

(xi) (5, −0·93).   2 (i) $\sqrt{2}+i\sqrt{2}$ (ii) $2\sqrt{3}+2i$ (iii) $-i$ (iv) $4-4i\sqrt{3}$

(v) $\dfrac{-3\sqrt{3}}{2}-\dfrac{3}{2}i$ (vi) 12i   (vii) $\dfrac{-1}{\sqrt{2}}+\dfrac{1}{\sqrt{2}}i$ (viii) $1-i$ (ix) $1·93+0·52i$.

3 (i) $-5+10i$ (ii) $-14+34i$ (iii) $-2+14i$ (iv) $-2+6i$ (v) $68-67i$

(vi) $11-7i$.   4 (a) (i) $\left(6,\dfrac{\pi}{2}\right)\left(\tfrac{1}{3},0\right)$ (ii) $\left(10,\dfrac{\pi}{2}\right)\left(\dfrac{5}{2},\dfrac{\pi}{6}\right)$ (iii) $\left(5,\dfrac{7\pi}{12}\right)\left(\dfrac{1}{5},\dfrac{5\pi}{12}\right)$

(iv) $\left(7,\dfrac{5\pi}{12}\right)\left(\dfrac{1}{7},\dfrac{-11\pi}{12}\right)$ (v) $\left(6,\dfrac{5\pi}{12}\right)\left(\dfrac{2}{3},\dfrac{-\pi}{12}\right)$ (vi) $(2,\pi)\left(2,\dfrac{-5\pi}{7}\right)$

5 (i) $\left(2,\dfrac{\pi}{2}\right)$ (ii) $\left(4,\dfrac{\pi}{2}\right)$ (iii) $\left(1,\dfrac{2\pi}{3}\right)$ (iv) $\left(\dfrac{1}{\sqrt{2}},\dfrac{\pi}{2}\right)$ (v) $\left(1,\dfrac{5\pi}{12}\right)$

(vi) $\left(1,\dfrac{\pi}{2}\right)$ (vii) $\left(\dfrac{1}{\sqrt{2}},\dfrac{-11\pi}{12}\right)$ (viii) $\left(1,\dfrac{-\pi}{2}\right)$.

6 $\dfrac{z_1}{z_2}=\tfrac{1}{2}-\tfrac{1}{2}i$, $\dfrac{z_2}{z_1}=1+i$, $z_1-z_2=1-2i$, $z_2-z_1=-1+2i$, $z_1+z_2=3+4i$,

$z_1z_2=-1+7i$.   8 $\dfrac{1}{2}\left(\cos\dfrac{-\pi}{3}+i\sin\dfrac{-\pi}{3}\right)$.

9 (i) $1,\dfrac{\pi}{3}$ (ii) $|z_1z_2|=|z_1||z_2|$, $\arg(z_1z_2)=\arg(z_1)+\arg(z_2)$

(iii) $a=0·63$, $b=3·1$.   10 $\sqrt{50},5,-2·4$.   11 $-\cos\theta$.   12 (i) $\left(4,\dfrac{\pi}{3}\right)$

(ii) $\left(2,-\dfrac{\pi}{6}\right)$ (iii) $(64,\pi)$ (iv) $\left(32,\dfrac{-5\pi}{6}\right)$.   13 $-22-4i,34,680$.

Exercise 15.5 (p. 249)

1 $\overrightarrow{OB}=w+z$, $\sqrt{5},5,\sqrt{50}$.   2 $z_1=1+2i$, $z_2=2+4i$, $z_3=3+6i$.

Exercise 15.6A (p. 250)

1 (i) $2-3i$ (ii) $3+4i$ (iii) 4   (iv) $-3i$ (v) $\left(2,\dfrac{-\pi}{4}\right)$ (vi) $\left(3,\dfrac{3\pi}{4}\right)$.

Exercise 15.6B (p. 252)

1 (i) $3-i,1$.   2 $x=1+2i, 1-2i, 1+i\sqrt{23}, 1-i\sqrt{23}$.   3 $2-i,-4$.

4 $2-3i,2$.   5 (i) $x=2, y=1, \dfrac{1}{\sqrt{2}}+\dfrac{i}{\sqrt{2}}, \dfrac{-1}{\sqrt{2}}-\dfrac{i}{\sqrt{2}}$ (ii) $1-i,2$.

6 $a=2, b=5$.   7 (i) $1+i, 3, -1$ (ii) $\dfrac{-1}{2}+\dfrac{\sqrt{3}}{2}i, \dfrac{-1}{2}-\dfrac{\sqrt{3}}{2}i$.

Exercise 15.7 (p. 256)

1 5.   3 (i) mediator of A and B (ii) half line from A parallel to $1+i$

$\{(x,y):x+5y+9=0\}$, $\{(x,y):y-x=1, x\geqslant 2\}$.   4 $|z-2-i|=\sqrt{5}$.   6 (i) $2\sqrt{3},\dfrac{\pi}{6}$

(ii) $y+\sqrt{3}x=2\sqrt{3}, y\geqslant\sqrt{3}$.

Miscellaneous Exercise 15 (p. 256)

1 $-\frac{1}{2}+\frac{7}{2}i$.   **2** $\frac{4}{5}-\frac{3}{5}i$.   **3** $z=-1+2i,\ w=1-i$.   **4** $32+47i,\ 32-47i,\ 61,\ 53$.

5 $1+i,\ 2$.   **6** $2-i,\ -2+i$.   **7** $x^2-2r\cos\theta x+r^2=0,\ \dfrac{-p}{\sqrt{q}}$.

8 $p=-5-4i,\ q=1+7i,\ a=-3,\ b=-1$.

9 $-\frac{1}{2},\ i,\ -i,\ 2+i,\ 2-i,\ p=1,\ q=4,\ -1+i,\ -1-i,\ 1+i,\ 1-i$.

10   (i) $4x^2-9x+18=0$ (ii) (a) $k>-\frac{5}{4}$ (b) $k=-\frac{5}{4}$ (c) $-\frac{5}{3}<k<-\frac{5}{4}$.

11 $x^2+2x+9=0$.   **12** $\left(\sqrt{2},\dfrac{-\pi}{12}\right)$.   **13**   (i) (a) $\left(2,\dfrac{2\pi}{3}\right)$ (b) $\left(1,\dfrac{\pi}{2}\right)$

(ii) $a=-3,\ b=4$ or $a=3,\ b=-4$ (iii) $\alpha-\beta+\gamma$.   **14**   (i) $-1-i,\ \dfrac{3\pi}{4}$

(ii) $2-i,\ 2,\ k=-10$.   **15** $z_1=\left(1,\dfrac{2\pi}{3}\right)z_2=\left(1,\dfrac{-2\pi}{3}\right)z_3=(1,0)$.   **16** $\dfrac{\pi}{4},\ -\dfrac{\pi}{4}$.

17 $(\alpha_2-\alpha_1)+(\beta_2-\beta_1)i,\ k(\cos\theta+i\sin\theta),\ \overrightarrow{AB}=\sqrt{3}-i,\ \overrightarrow{AC}=2\sqrt{3}+2i,\ z_3=\left(2,\dfrac{\pi}{3}\right)$,

$D=6\sqrt{3},\ 4\sqrt{3}$.   **18** $\left(2,\dfrac{\pi}{3}\right)\left(2,-\dfrac{\pi}{3}\right),\ z^2-2z+4=0,\ a=-2,\ b=12$.

19 $(2\cos 2\theta,\ 5\theta)$.   **21** $\left(2,\dfrac{\pi}{6}\right),\left(2,\dfrac{-\pi}{6}\right),\dfrac{1}{2}+\dfrac{\sqrt{3}}{2}i$.

23   (a) $(10,90°),\ (5,126\cdot9°),\ (2,-36\cdot9°)$ (b) $x-y=1,\ x^2+y^2=1$.

24   (i) $2(1+\sqrt{2})$ (ii) $1+(2+\sqrt{3})i$.   **25**   (i) U lies on a circle centre A radius 3

(ii) UAV is a right angle

(iii) WUA is a straight line with $WU=UA,\ |AV|=\sqrt{18}$.   **26**   (i) $3\cdot54-2\cdot55i$

(ii) $3+\sqrt{3}i,\ 3-\sqrt{3}i$.   **27** $a=4,\ b=-5,\ 5$.   **28** $1+2i,\ \sqrt{50},\ \alpha=-2,\ \beta=-1$.

29 $\frac{1}{2}(\sqrt{3}+3i),\ \frac{1}{2}(\sqrt{3}-3i),\ \frac{1}{2}(\sqrt{3}-6+3i),\ \frac{1}{2}(\sqrt{3}-6-3i)$,

$z^2+(6-\sqrt{3})z+12-3\sqrt{3}=0$.   **30** $\dfrac{4}{3}+\dfrac{2\sqrt{2}}{3}i,\dfrac{2}{3}+\dfrac{4\sqrt{2}}{3}i$.

31 $\cot\frac{1}{2}\theta$.   **32**   (i) $a=\dfrac{63}{25},\ b=\dfrac{16}{25},\ r=\dfrac{13}{5},\ \cos\theta=\dfrac{63}{65},\ \sin\theta=\dfrac{16}{65}$ (ii) $\dfrac{z_2}{z_1}=\dfrac{w_2}{w_1}$.

33 $k=2,\ 56°,\ -124°$.   **34**   (a) $\dfrac{1}{13}+\dfrac{5}{13}i,\ 1\cdot37,\ 2\cdot75,\ \dfrac{2}{13}$ (b) $|z|^2:1$.

35 $y^2+(x+5)^2=9$.   **36** $4\sin^2\theta,\ \dfrac{1}{4\sin^2\theta},\ -\theta,\ y=-\frac{1}{2}$.

37   (i) $1+\sqrt{5}i,\ 1-\sqrt{5}i$ (ii) $1+i$.   **38**   (i) (a) $p=\dfrac{7}{5},\ q=\dfrac{-4}{5}$

(b) $p=2-i,\ q=2+i$.   **39** $\frac{5}{2},\ -1\frac{1}{2}+2i$.

40   (i) $x=1,\ y=2$ or $x=-1,\ y=-2$.

41 $a=\sqrt{2},\ b=1$ or $a=-\sqrt{2},\ b=-1$.

42 $z^2-2z+4,\ \alpha=\left(2,\dfrac{-\pi}{3}\right),\ \alpha^2=\left(4,\dfrac{-2\pi}{3}\right),\ \alpha^3=(8,\pi),\ \alpha,\ \alpha^*,\ -2,\ 3+\sqrt{3}i,\ \left(1,\dfrac{\pi}{6}\right)$,

$\sqrt{3}-2+3i$.   **43** $8,\ 0\cdot8$.   **44**   (b) $|z-2|=2$ (c) $-1+\sqrt{3}i,\ -1-\sqrt{3}i$

(d) $\left(2,\dfrac{11\pi}{12}\right)\left(2,\dfrac{-5\pi}{12}\right)$.

16 Lines and Circles in the Plane

16.1 The equation of a line (revision)

The equation of a line has been obtained in various forms, (§12.4):

in vector form: $\quad \mathbf{r} = \mathbf{a} + t\mathbf{b}, \quad$ e.g. $\mathbf{r} = \begin{pmatrix} x \\ y \end{pmatrix} = \begin{pmatrix} 1 \\ 2 \end{pmatrix} + t\begin{pmatrix} 3 \\ -4 \end{pmatrix};$

in parametric form: $x = a + pt, \quad$ e.g. $x = 1 + 3t;$

$\qquad\qquad\qquad\quad y = b + qt \qquad\qquad\quad y = 2 - 4t$

in Cartesian form: $\quad y = mx + c, \quad$ e.g. $y = -\dfrac{4}{3}x + \dfrac{10}{3}.$

EXERCISE 16.1

1   Verify that the three example equations, given above, all represent the same line.
2   The line L passes through $A(1, 1)$ and $B(-3, 4)$. Find the equation of L:
(i) in vector form,   (ii) in parametric form,   (iii) in Cartesian form.

16.2 The intersection of two lines

Given the equations of two lines in any one of the forms, we can first check whether or not the lines are parallel and, if they are not parallel, we can find the coordinates of their point of intersection. The vector and the parametric forms of the equation of a line are clearly similar. There is either one vector equation or two parametric equations, but, in each case, just one parameter occurs. When dealing with two different lines, the two lines *must* be given in terms of *different* parameters, say s and t, because the required point of intersection is given by different values of the parameter on each line.

EXAMPLE 1   *Find the point of intersection of the two lines:*

(i) $\begin{pmatrix} x \\ y \end{pmatrix} = \begin{pmatrix} 1 \\ 2 \end{pmatrix} + t\begin{pmatrix} 3 \\ -4 \end{pmatrix}$ *and* $\begin{pmatrix} x \\ y \end{pmatrix} = \begin{pmatrix} 2 \\ -4 \end{pmatrix} + t\begin{pmatrix} 1 \\ 1 \end{pmatrix},$

(ii) $\begin{pmatrix} x \\ y \end{pmatrix} = \begin{pmatrix} 1 \\ 2 \end{pmatrix} + t\begin{pmatrix} 3 \\ -4 \end{pmatrix}$ *and* $\begin{pmatrix} x \\ y \end{pmatrix} = \begin{pmatrix} 3 \\ 0 \end{pmatrix} + t\begin{pmatrix} -3 \\ 4 \end{pmatrix}.$

(i) We must use different parameters for the two lines, so let the lines be given by $x = 1 + 3t$, $y = 2 - 4t$ and by $x = 2 + s$, $y = -4 + s$. At the common point P, (x, y), $x = 1 + 3t = 2 + s$ and $y = 2 - 4t = -4 + s$. To solve the two simultaneous equations in s and t, bring the variables to the same side of the equations

$$3t - s = 1,$$
$$4t + s = 6.$$

Eliminate s by adding these equations to give $7t = 7$ or $t = 1$. Then $s = 3t - 1 = 2$, and **P is the point $(4, -2)$.**

(ii) The direction vectors along the two lines are $\begin{pmatrix} 3 \\ -4 \end{pmatrix}$ and $\begin{pmatrix} -3 \\ 4 \end{pmatrix}$, $= -\begin{pmatrix} 3 \\ -4 \end{pmatrix}$, so the lines are parallel. The displacement vector from the point $(1, 2)$ on one line to $(3, 0)$ on the other line is $\begin{pmatrix} 3 \\ 0 \end{pmatrix} - \begin{pmatrix} 1 \\ 2 \end{pmatrix}$, which is not parallel to $\begin{pmatrix} 3 \\ -4 \end{pmatrix}$. Hence, **the lines are distinct parallel lines and do not meet**.

When the equations of the lines are given in Cartesian form, we again have to solve two simultaneous equations, this time in x and y.

EXAMPLE 2   *Find the point of intersection of the two lines:*

(i) $3y - 2x = 5$ and $4x = 6y - 10$,   (ii) $2y - x = 1$ and $y + 3x = 11$.

(i) We have to solve $3y - 2x = 5$, $4x = 6y - 10$, and the second equation can be written $6y - 4x = 10$. Both lines have the same gradient $\frac{2}{3}$, so they are parallel. Also $(-1, 1)$ lies on each line, so they are the same line. Alternatively, by dividing the second equation by 2, we obtain the first equation, so **the two lines are identical**.
(ii) We solve the pair of equations $2y - x = 1$, $y + 3x = 11$ by eliminating y. From the second equation $y = 11 - 3x$ and this is substituted into the first equation giving $2(11 - 3x) - x = 1$, that is, $22 - 6x - x = 1$ or $7x = 21$, whence $x = 3$. Then $y = 2$ and **the lines intersect at the point $(3, 2)$**.

EXERCISE 16.2

1   Find the point of intersection of the two lines:

(i) $\begin{pmatrix} x \\ y \end{pmatrix} = \begin{pmatrix} -1 \\ 3 \end{pmatrix} + t\begin{pmatrix} 2 \\ -1 \end{pmatrix}$ and $\begin{pmatrix} x \\ y \end{pmatrix} = \begin{pmatrix} 5 \\ 7 \end{pmatrix} + s\begin{pmatrix} 1 \\ 3 \end{pmatrix}$;

(ii) $\begin{pmatrix} x \\ y \end{pmatrix} = \begin{pmatrix} 3 \\ -5 \end{pmatrix} + t\begin{pmatrix} -1 \\ 2 \end{pmatrix}$ and $\begin{pmatrix} x \\ y \end{pmatrix} = \begin{pmatrix} -3 \\ 3 \end{pmatrix} + t\begin{pmatrix} 1 \\ -1 \end{pmatrix}$;

(iii) $x = 3 + 2t$, $y = 2 - 3t$ and $x = -1 + 4t$, $y = -2t$;

(iv) $x = 5 + 3t$, $y = 2 - t$ and $x = 4 - 6t$, $y = 1 + 2t$;

(v) $3x + 4y = 5$ and $2y - x = 5$; (vi) $4x - 2y = 7$, $-3x + 4y = 1$.

2 Find the equations of the two lines described and the coordinates of their common point:

(i) through $(-2, -2)$ with gradient 2 and through $(0, 3)$ with gradient $-6/5$,

(ii) through $(-3, 3)$ and $(5, 0)$ and through $(-2, -3)$ and $(4, 6)$,

(iii) through $(-3, -5)$ and $(1, 1)$ and through $(-1, -2)$ and $(5, 7)$.

16.3 Distance of a point from a line

If the equation of a line is given in vector form as $\mathbf{r} = \mathbf{w} + t\mathbf{v}$, then the vector $\mathbf{v}$ is a vector in the direction of the line. If the line is given by a Cartesian equation $ax + by + d = 0$ and if $b \neq 0$, then on using t as a parameter with $x = -bt, y = -\dfrac{d}{b} + at$, the vector equation of the line is

$$\mathbf{r} = \begin{pmatrix} x \\ y \end{pmatrix} = \begin{pmatrix} 0 \\ -d/b \end{pmatrix} + t\begin{pmatrix} -b \\ a \end{pmatrix}.$$

If $b = 0$, the line is parallel to $0y$, and hence parallel to $\begin{pmatrix} 0 \\ a \end{pmatrix}$.

In either case, the vector $\begin{pmatrix} -b \\ a \end{pmatrix}$ is in the direction of the line, and since $\begin{pmatrix} a \\ b \end{pmatrix} \cdot \begin{pmatrix} -b \\ a \end{pmatrix} = -ab + ab = 0$, the vector $\mathbf{u}$, where

$$\mathbf{u} = \frac{1}{\sqrt{(a^2 + b^2)}}\begin{pmatrix} a \\ b \end{pmatrix},$$

is a unit vector perpendicular to the line.

Let P be the point (x_0, y_0) and let PN be the perpendicular from P on to the line $ax + by + d = 0$. Then $\overrightarrow{NP} = p\mathbf{u}$, where the length of the perpendicular PN is $|p|$, and $p > 0$ when P is on the same side of the line as $\mathbf{u}$, and $p < 0$ when P is on the opposite side of the line from $\mathbf{u}$. The two possibilities are shown in Fig. 16.1 (a) and (b). This two-dimensional problem of finding the distance of a point from a line may be compared with the three-dimensional problem of finding the distance of a point from a plane, considered in §12.11.

Now let the coordinates of N be (x, y). Then $ax + by + d = 0$, and

$$p\mathbf{u} = \overrightarrow{NP} = \overrightarrow{OP} - \overrightarrow{ON} = \begin{pmatrix} x_0 - x \\ y_0 - y \end{pmatrix}.$$

Also $p = p\mathbf{u} \cdot \mathbf{u}$ and so

$$p = \begin{pmatrix} x_0 - x \\ y_0 - y \end{pmatrix} \cdot \begin{pmatrix} a \\ b \end{pmatrix}\frac{1}{\sqrt{(a^2 + b^2)}} = \frac{ax_0 + by_0 - ax - by}{\sqrt{(a^2 + b^2)}} = \frac{ax_0 + by_0 + d}{\sqrt{(a^2 + b^2)}},$$

since $ax + by = -d$.

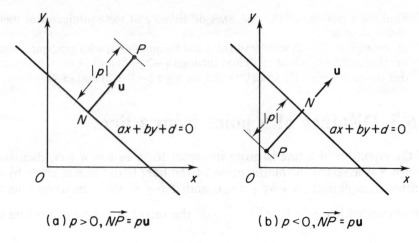

$$(a)\,p>0,\,\overrightarrow{NP}=p\mathbf{u} \qquad (b)\,p<0,\,\overrightarrow{NP}=p\mathbf{u}$$

Fig. 16.1

When $x_0 = 0, y_0 = 0$, the distance from the origin O to the line is $|p|$, where $p = \dfrac{d}{\sqrt{(a^2 + b^2)}}$. If $p > 0$ then O is on the same side of the line as $\mathbf{u}$ and if $p < 0$ then O is on the opposite side.

EXAMPLE 1 *Find the distances from the line $x + 2y = 3$ of the points $O(0,0)$, $A(4,2), B(1,1)$. Determine whether the points A and B are on the same side of the line as O or on the opposite side.*

A direction vector along the line is $\begin{pmatrix} -2 \\ 1 \end{pmatrix}$ and a unit vector perpendicular to the line is $\mathbf{u}$, where $\mathbf{u} = \dfrac{1}{\sqrt{5}}\begin{pmatrix} 1 \\ 2 \end{pmatrix}$. Let $\overrightarrow{LO} = p\mathbf{u}$, $\overrightarrow{MA} = q\mathbf{u}$, $\overrightarrow{NB} = r\mathbf{u}$, where OL, AM, BN are the perpendiculars from O, A, B, on to the line. Then

$$p = \frac{0+0-3}{\sqrt{5}} = \frac{-3}{\sqrt{5}}, q = \frac{4+4-3}{\sqrt{5}} = \sqrt{5}, r = \frac{1+2-3}{\sqrt{5}} = 0.$$

Thus the **point B lies on the line, the points O and A are on opposite sides of the line and are distant** $\dfrac{3}{\sqrt{5}}$ **and** $\sqrt{5}$ **respectively, from the line.**

EXAMPLE 2 *Find the equations of the lines which bisect the angles between the lines with equations $2x + 9y = 12$ and $7x + 6y = -9$.*

The point (x, y) lies on a bisector of the angle between the two lines if it is equidistant from the two lines. Therefore, we write down the condition that the distances of (x, y) from the two lines are equal,

$$\frac{|2x + 9y - 12|}{\sqrt{(2^2 + 9^2)}} = \frac{|7x + 6y + 9|}{\sqrt{(7^2 + 6^2)}}.$$

Since $2^2+9^2 = 85 = 7^2+6^2$, we equate the numerators and obtain two equations $2x+9y-12 = 7x+6y+9$ and $2x+9y-12 = -7x-6y-9$. These equations give us the equations of the two bisectors and may be simplified to $5x-3y+21=0$ **and** $3x+5y-1=0$. Note that the gradients of these bisectors are $5/3$ and $-3/5$, with a product -1, which shows that the two bisectors are perpendicular.

EXAMPLE 3 *Find the equation of the perpendicular bisector of the line segment joining the two points* $(1,4)$ *and* $(5,-2)$.

The point (x, y) lies on the perpendicular bisector of the segment if it is equidistant from the end points. This condition can be written

$$(x-1)^2 + (y-4)^2 = (x-5)^2 + (y+2)^2,$$

that is, $\quad x^2 - 2x+1+y^2 - 8y+16 = x^2 - 10x+25 + y^2 + 4y+4,$

which is simplified to give the equation of the perpendicular bisector of the line segment, namely $2x = 3y + 3$.

EXERCISE 16.3

1 Find the distance of the point with the given coordinates from the line with the given equation:

(i) $(1,2), 3x+4y+5 = 0$; (ii) $(2, 2), 4x-3y-2 = 0$;
(iii) $(1, -1), 5x+12y = 0$; (iv) $(2, 1) 15y = 8x+4$; (v) $(2, 3), x+y = 3$;
(vi) $(-3, 4), x+2 = 3y$; (vii) $(2, 1), 2x = 3y+4$;
(viii) $(a, b), y = 4x - 7$.

2 Show that the parametric equations of the line through $(1, 2)$ with gradient $\tan\theta$ are
$$x = 1+t\cos\theta,\ y = 2+t\sin\theta,$$
and that $|t|$ is the distance of the point on the line with parameter t from the point $(1, 2)$. Show that the intersection of this line with the line $L, 2x+3y = 4$, is given by
$$t = \frac{-4}{2\cos\theta+3\sin\theta}.$$

Show that the first line is perpendicular to L if $\tan\theta = 3/2$, and deduce that the distance of $(1, 2)$ from L is $\dfrac{4}{\sqrt{13}}$.

3 Show that the line through (a, b) with gradient $\tan\theta$ is given by
$$x = a+t\cos\theta,\ y = b+t\sin\theta,$$
and that the distance from (a, b) to the point on the line with parameter t is $|t|$.

The line L has equation $px+qy = r$. Show that this line meets the first line at the point with parameter
$$\frac{pa+qb-r}{p\cos\theta+q\sin\theta}.$$

Deduce that the distance of (a, b) from L is $\left|\dfrac{pa+qb-r}{\sqrt{(p^2+q^2)}}\right|.$

4 Show that the point (x, y) is equidistant from the two lines

$$3x - 4y = 2 \quad \text{and} \quad 4x + 3y = 11, \quad \text{if} \quad |3x - 4y - 2| = |4x + 3y - 11|.$$

Deduce that the equations of the lines which bisect the angles between the two given lines are $x + 7y = 9$ and $7x = y + 13$.

5 Find the equations of the lines which bisect the angles between the lines with equations:
(i) $y + x = 2$ and $y = x + 5$, (ii) $y = 1 + x\sqrt{3}$ and $y = 1$,
(iii) $3y = 4x + 2$ and $5y = 12x - 2$.

6 Show that $P(x, y)$ is equidistant from $A(5, 2)$ and $B(2, -4)$ if $4y + 2x = 3$.
Deduce a Cartesian equation for the perpendicular bisector of AB.

7 Find an equation of the perpendicular bisector of:
(i) $(2, 2)$ and $(6, 4)$, (ii) $(3, -5)$ and $(6, 1)$, (iii) $(-2, 0)$ and $(0, -6)$,
(iv) $(-3, 1)$ and $(2, -3)$.

16.4 The equation of a circle

The circle of radius a, with centre at the origin O, consists of those points whose position vectors satisfy the equation

$$|\mathbf{r}| = \left| \begin{pmatrix} x \\ y \end{pmatrix} \right| = \sqrt{(x^2 + y^2)} = a,$$

that is
$$r^2 = x^2 + y^2 = a^2.$$

If we move the centre of the circle to the point $C(h, k)$, this transformation is achieved by replacing x and y by $(x - h)$ and $(y - k)$. A point $P(x, y)$ lies on the circle, centre $C(h, k)$ and radius a, if

$$|\overrightarrow{CP}| = \left| \begin{pmatrix} x \\ y \end{pmatrix} - \begin{pmatrix} h \\ k \end{pmatrix} \right| = \left| \begin{pmatrix} x - h \\ y - k \end{pmatrix} \right| = \sqrt{[(x - h)^2 + (y - k)^2]} = a.$$

So the equation of the circle, centre (h, k) and radius a, is

$$(x - h)^2 + (y - k)^2 = a^2,$$

or
$$x^2 + y^2 - 2hx - 2ky = a^2 - h^2 - k^2.$$

Note that, in the equation for a circle, there is a constant term, there are terms in x and in y, terms in x^2 and y^2 with the same coefficient, but there is no term in xy.

Conversely, suppose that we begin with an equation of this kind and find out if it is the equation of a circle. For example, consider the equation

$$x^2 + y^2 + fx + gy + c = 0.$$

To transform this equation into the previous form, we complete the squares of the terms in x and also of the terms in y. Thus

$$x^2 + fx + \tfrac{1}{4}f^2 = (x + \tfrac{1}{2}f)^2, \quad y^2 + gy + \tfrac{1}{4}g^2 = (y + \tfrac{1}{2}g)^2,$$

so
$$(x + \tfrac{1}{2}f)^2 + (y + \tfrac{1}{2}g)^2 = x^2 + xf + y^2 + gy + \tfrac{1}{4}f^2 + \tfrac{1}{4}g^2$$

so the original equation becomes

$$(x + \tfrac{1}{2}f)^2 + (y + \tfrac{1}{2}g)^2 = \tfrac{1}{4}f^2 + \tfrac{1}{4}g^2 - c.$$

Compare this equation with the equation of a circle, centre (h, k), radius a, namely

$$(x - h)^2 + (y - k)^2 = a^2,$$

and we see that we have an equation of the circle, centre $(-\tfrac{1}{2}f, -\tfrac{1}{2}g)$ and radius $a = \sqrt{(\tfrac{1}{4}f^2 + \tfrac{1}{4}g^2 - c)}$, provided that $\tfrac{1}{4}f^2 + \tfrac{1}{4}g^2 - c > 0$.

If $f^2 + g^2 = 4c$, the equation becomes $(x + \tfrac{1}{2}f)^2 + (y + \tfrac{1}{2}g)^2 = 0$, and this is satisfied by only one point $(-\tfrac{1}{2}f, -\tfrac{1}{2}g)$.

If $f^2 + g^2 < 4c$, the equation means that $(x + \tfrac{1}{2}f)^2 + (y + \tfrac{1}{2}g)^2$ must be negative, and so there are no points (x, y) satisfying the equation.

Definition   The *locus* given by an equation is the set of points in the plane whose coordinates satisfy the equation.

EXAMPLE 1   *Find the locus of points satisfying the equation:*
(i) $x^2 + y^2 + 2x + 6y = 6$,   *(ii)* $x^2 + y^2 + 2x + 6y = -10$,
(iii) $x^2 + y^2 + 2x + 6y = -12$.

The left-hand side of all three equations is the same, so we complete the squares for these terms.

$$x^2 + 2x + y^2 + 6y = (x + 1)^2 + (y + 3)^2 - 1 - 9$$

so $$(x + 1)^2 + (y + 3)^2 = x^2 + 2x + y^2 + 6y + 10.$$

We then substitute in this equation for the value given for $x^2 + y^2 + 2x + 6y$ in each of the three cases:

(i) $(x + 1)^2 + (y + 3)^2 = x^2 + 2x + y^2 + 6y + 10 = 6 + 10 = 16 = 4^2$, so the locus is **a circle of radius 4 with centre $(-1, -3)$**;
(ii) $(x + 1)^2 + (y + 3)^2 = x^2 + 2x + y^2 + 6y + 10 = -10 + 10 = 0$, and the locus is **the single point $(-1, -3)$**;
(iii) $(x + 1)^2 + (y + 3)^2 = x^2 + 2x + y^2 + 3y + 10 = -12 + 10 = -2$,   and   the locus is **the empty set**.

EXAMPLE 2   *Find the equation of the circle which touches the y-axis at $(0, 4)$ and passes through the point $(1, 1)$.*

For the circle to touch $x = 0$ at $(0, 4)$, the centre must lie on the line $y = 4$, so let the centre be at the point $(a, 4)$. Then the radius is the distance between $(0, 4)$ and $(a, 4)$, which is $|a|$. For the circle to pass through $(1, 1)$ we need $(1 - a)^2 + (1 - 4)^2 = a^2$, that is, $1 - 2a + 9 = 0$, or $a = 5$. Therefore, the equation of the circle is $(x - 5)^2 + (y - 4)^2 = 25$.

EXAMPLE 3   *Find the equation of the circle passing through the three points $A(1, -2)$, $B(6, 3)$ and $C(1, 4)$.*

One method of solution is to find the equations of the perpendicular bisectors of AB and BC, and solve two simultaneous equations to find the coordinates of the centre of the circle, which is where these two lines intersect. Then the radius is the distance from the centre to any one of the given points. We shall work through another method of solution. Suppose that the equation of the circle is $x^2 + y^2 + fx + gy + c = 0$. Then this equation is satisfied by the coordinates of A, B and C. This gives three equations

$$f - 2g + c + \ 5 = 0,$$
$$6f + 3g + c + 45 = 0,$$
$$f + 4g + c + 17 = 0.$$

Subtract the third equation from the first, $-6g - 12 = 0$, so $g = -2$. Then $f + c = -9$ and $6f + c = -39$, so on subtracting $5f = -30$, $f = -6$ and then $c = -3$. Therefore, the equation of the circle is $x^2 + y^2 - 6x - 2y - 3 = 0$. As a check, substitute the coordinates of A, B and C into the equation.

EXERCISE 16.4

1 Roughly sketch the circle with the given centre and radius. Find the equation of the circle:
   (i) centre $(0, 0)$, radius 4;   (ii) centre $(0, 3)$, radius 3;
   (iii) centre $(-2, 3)$, radius 2;   (iv) centre $(-3, -4)$, radius 5;
   (v) centre $(2, -2)$, radius $\sqrt{7}$.

2 Find the locus of the given equation; if it is a circle, state the radius and coordinates of its centre; if it is a single point, state its coordinates:
   (i) $x^2 + y^2 - 4x - 6y + 9 = 0$, (ii) $x^2 + y^2 - 4x + 2y = 4$,
   (iii) $x^2 + y^2 + 6x + 8y = 0$,   (iv) $x^2 + y^2 + 2ax = 0$,
   (v) $x^2 + y^2 + 2ax + 2by + a^2 + b^2 = 0$,
   (vi) $x^2 + y^2 + a^2 + b^2 = c^2 + 2ax + 2by$,   (vii) $x^2 + y^2 + 2ax + 2by = 0$,
   (viii) $9x^2 + 9y^2 = 4$,   (ix) $4x^2 + 4y^2 - 4x + 12y + 9 = 0$.

3 For the given equation, describe the locus of points (x, y) satisfying it. If the locus is a circle, state its radius and the coordinates of its centre. If the locus is a line, state the coordinates of two points on the line. If the locus is a parabola, state the coordinates of its vertex and whether the vertex is upwards or downwards.
   (i) $x^2 + y^2 = 16$,   (ii) $x + y = 16$,   (iii) $x^2 + y = 16$,
   (iv) $x^2 + 6x + y^2 - 4y = 14$,   (v) $x^2 + 6x + y = 14$,   (vi) $x^2 + 6x - 4y = 14$,
   (vii) $x^2 - y + 2ax = 0$,   (viii) $x^2 + y^2 + 2ax = 0$,
   (ix) $x^2 + y^2 + 10 = 2x + 6y$,   (x) $x^2 + 10 = 2x + 6y$,
   (xi) $x^2 + y^2 + 11 = 2x + 6y$,   (xii) $10 = 2x + 6y$.

4 Find the equation of the circle:
   (i) touching Ox at $(3, 0)$ and Oy at $(0, 3)$;
   (ii) touching Ox at $(2, 0)$ and passing through the point $(0, 3)$;
   (iii) passing through the three points $(0, 0)$, $(1, 0)$, $(0, 1)$;
   (iv) passing through $(0, 0)$, $(3, 0)$, $(0, 4)$;
   (v) passing through $(2, 1)$, $(-4, 3)$, $(-6, 5)$.

16.5 Intersection of a line and a circle

Given a line and a circle in the plane, there are three possibilities. These are, as shown in Fig. 16.2, (a) the line meets the circle in two points; (b) the line meets the circle in one point, in which case the line is a tangent to the circle: (c) the line does not meet the circle at all.

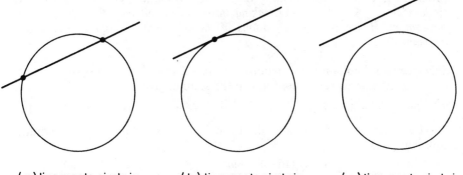

(a) line meets circle in   (b) line meets circle in   (c) line meets circle in
     2 points   1 point, so is a tangent   0 points

Fig. 16.2

In Cartesian coordinates, the line is given by a linear equation and the circle is given by a quadratic equation. The problem of finding points of intersection of the line and the circle is reduced to finding the solution of a pair of simultaneous equations in x and y, one equation being linear and one quadratic. One variable may be eliminated by expressing it in terms of the other variable, using the linear equation, and then substituting in the quadratic equation.

There are then three possibilities concerning the number of roots of the quadratic equation (in one variable) which is obtained:

(a) two roots : the line meets the circle in two points;
(b) one root : the line meets the circle at one point, where it is the tangent;
(c) no roots : the line does not meet the circle at all.

EXAMPLE 1   *Find the point of intersection of the line* $3x - 2y = 5$ *and the circle* $x^2 - 2x + y^2 = 7$.

Use the linear equation to express y in terms of x, that is, $y = \frac{1}{2}(3x - 5)$. Substitute in the quadratic equation,

$$x^2 - 2x + \tfrac{1}{4}(3x - 5)^2 = 7.$$

Multiply by 4 and expand the bracket,

$$4x^2 - 8x + 9x^2 - 30x + 25 = 28,$$

rearrange and factorise,

$$13x^2 - 38x - 3 = (13x + 1)(x - 3) = 0.$$

There are two solutions to the pair of simultaneous equations, either $x = -\frac{1}{13}$ and $y = \frac{1}{2}(-\frac{3}{13} - 5) = -\frac{34}{13}$, or $x = 3$ and $y = \frac{1}{2}(9 - 5) = 2$. Therefore, the line meets the circle at $(-\frac{1}{13}, -\frac{34}{13})$ **and at (3, 2).**

EXAMPLE 2   *Find the points of intersection of the line $y = mx + 2$ with the circle $x^2 + y^2 - 8x - 4y = -12$, taking into account all possible values of the constant m. Hence, find the equations of the two tangents to the circle from the point (0, 2).*

Solve the two simultaneous equations $y = mx + 2$ and $x^2 + y^2 - 8x - 4y = -12$, by using the first to substitute for y in the second, giving

$$x^2 + (mx + 2)^2 - 8x - 4(mx + 2) = x^2 + m^2x^2 + 4mx + 4 - 8x - 4mx - 8 = -12.$$

So $(1 + m^2)x^2 - 8x + 8 = 0$, and the formula for the solution of a quadratic equation gives

$$x = \frac{4 \pm \sqrt{(16 - 8 - 8m^2)}}{1 + m^2} = \frac{4 \pm \sqrt{(8 - 8m^2)}}{1 + m^2}.$$

There are three cases which are shown graphically in Fig. 16.3.

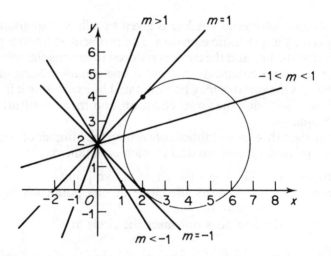

Intersection of line $y = mx + 2$ with the circle
$x^2 + y^2 - 8x - 4y = -12$
(a) $-1 < m < 1$: two points of intersection,
(b) $-1 = m$ and $m = 1$: line touches circle
(c) $m < -1$ and $m > 1$: line does not meet circle

Fig. 16.3

(a) If $-1 < m < 1$, then the line meets the circle in two points,

$$\left(\frac{4+\sqrt{(8-8m^2)}}{1+m^2}, 2+\frac{m[4+\sqrt{(8-8m^2)}]}{1+m^2}\right)$$

$$\text{and } \left(\frac{4-\sqrt{(8-8m^2)}}{1+m^2}, 2+\frac{m[4-\sqrt{(8-8m^2)}]}{1+m^2}\right).$$

(b) If $m = 1$, the line touches the circle at $(2, 4)$,
 if $m = -1$, the line touches the circle at $(2, 0)$.
(c) if $m < -1$, the line does not meet the circle at all,
 if $m > 1$, the line does not meet the circle at all.

EXAMPLE 3 *Find the equation: (i) of the tangent L to the circle $x^2 + y^2 = 6x + 8y$, at the point $A(6,8)$; (ii) of the other tangent to the circle parallel to L.*

The equation of the circle may be written $(x-3)^2 + (y-4)^2 = 25$, so it has centre $C(3, 4)$ and radius 5. The line CA has gradient $(8-4)/(6-3) = 4/3$, so the tangent at A, which is perpendicular to CA, has gradient $-\frac{3}{4}$, and hence equation $y-8 = -\frac{3}{4}(x-6)$ or $4y+3x=50$. The equation of AC is $y-4 = \frac{4}{3}(x-3)$, that is, $3y = 4x$. This line meets the circle at points where x satisfies

$$x^2 + \tfrac{16}{9}x^2 = 6x + \tfrac{32}{3}x,$$

that is, $x^2 = 6x$, $x = 6$, (corresponding to A), or $x = 0$. If $x = 0$ then $y = 0$, so CA meets the circle again at $(0, 0)$ which is the point of contact of the tangent parallel to L. The gradient of this tangent is also $-\frac{3}{4}$, so its equation is $4y+3x=0$.

EXAMPLE 4 *Find the equation of the circle with diameter AB, where the coordinates of A and B are (p, q) and (r, s).*

The point $P(x, y)$ lies on the circle, diameter AB, if AP and BP are perpendicular, that is, if the product of their gradients is -1. The gradients of AP and BP are $\dfrac{y-q}{x-p}$ and $\dfrac{y-s}{x-r}$, respectively, so the equation of the circle is

$$\frac{y-q}{x-p}\cdot\frac{y-s}{x-r} = -1, \quad \text{or} \quad (y-q)(y-s)+(x-p)(x-r)=0.$$

EXAMPLE 5 *Find the equation of the common chord of the two circles with equations $x^2 + y^2 - 2x - 2y - 6 = 0$, $x^2 + y^2 - 8x + 3y = 0$.*

If the point (x, y) satisfies both given equations then it will satisfy the linear equation

$$(-2x-2y-6)-(-8x+3y) = 6x-5y-6=0,$$

obtained by subtracting them. The common points of the two circles, therefore, lie on the line with this equation, so this equation must be the equation of the common chord of the two circles.

EXERCISE 16.5

1   Find the length of the tangents:
   (i) from $(0, 0)$ to the circle $x^2 + y^2 - 2x - 4y + 1 = 0$,
   (ii) from $(2, 8)$ to the circle $x^2 + y^2 - 8x - 4y - 5 = 0$.
2   Find the points of intersection:
   (i) of the line $y = x$ with the circle $x^2 - 10x + y^2 - 4y + 24 = 0$,
   (ii) of the line $x + y = 2$ with the circle $x^2 + y^2 - 8x + 4y + 16 = 0$.
3   Find the equation of the tangent:
   (i) at $(3, 4)$ to the circle $x^2 + y^2 - 8x - 2y + 7 = 0$,
   (ii) at $(0, 0)$ to the circle $x^2 + y^2 = x + 3y$,
   (iii) at $(-1, -5)$ to the circle $x^2 + y^2 - 2x - 4y = 48$.
4   Find the equation of the circle with a diameter whose end points have coordinates:
   (i) $(0, 0)$ and $(8, -6)$, (ii) $(2, 4)$ and $(6, 2)$, (iii) $(4, 1)$ and $(-3, -2)$,
   (iv) $(-2, 7)$ and $(4, 5)$.
5   Find the length of the tangents from the point $(5, -1)$ to the circle $x^2 + y^2 = 13$. Find the equation of the line joining the points of contact of these tangents with the circle.
6   Show that the equation

$$x^2 + y^2 + 2x - 4y + \lambda(x + y) = 0,$$

where λ is a constant, represents a circle which passes through the points of intersection of the circle

$$x^2 + y^2 + 2x - 4y = 0$$

and the straight line

$$x + y = 0.$$

   Find the coordinates of the centres of the two circles, and hence, or otherwise, show that if the circles intersect at right angles, $\lambda = 10$. For this value of λ, find the radii of the circles, and write down (or determine) the area of the quadrilateral which has its vertices at the points of intersection and the centres of the two circles.

MISCELLANEOUS EXERCISE 16

1   Find the distance from the origin of the point on the x-axis which is equidistant from the points $P(-2, 2)$ and $Q(4, 4)$.   (L)
2   A straight line parallel to the line $2x + y = 0$ intersects the x-axis at A and the y-axis at B. The perpendicular bisector of AB cuts the y-axis at C. Prove that the gradient of the line AC is $-\frac{3}{4}$.
   Find also the tangent of the acute angle between the line AC and the bisector of the angle AOB, where O is the origin.   (JMB)
3   Find the equation of the circle which has its centre on the line $x - 2y + 2 = 0$ and touches the y-axis at the point $(0, 3)$.   (L)
4   Find the centre and radius of each of the circles

$$x^2 + y^2 - 8x - 6y = 0,$$
$$x^2 + y^2 - 24x - 18y + 200 = 0.$$

   Deduce that the circles touch each other externally.   (L)

5 State the equation of the straight line which has gradient m and which passes through the point $P(0, 18)$.

Show that this line is a tangent to the circle, centre $C(4, 6)$ and radius 10, provided that m satisfies the equation $21m^2 - 24m - 11 = 0$.

Find the product of the gradients of the tangents from P to this circle. *(AEB)*

6 Write down the equation of the tangent at the origin to the curve with equation

$$x^2 + y^2 + 2x - y = 0.$$

Sketch the tangent and the curve. *(L)*

7 Find the distance from the origin of the point on the circle

$$x^2 + y^2 - 8x + 6y = 75$$

which is nearest to the origin. *(L)*

8 Find the centre and the radius of the circle C which passes through the points $(4, 2)$, $(2, 4)$ and $(2, 6)$. If the line $y = mx$ is a tangent to C, obtain the quadratic equation satisfied by m. Hence or otherwise find the equations of the tangents to C which pass through the origin O. Find also

(i) the angle between the two tangents,

(ii) the equation of the circle which is the reflection of C in the line $y = 3x$. *(AEB)*

9 The point $M(2, 3)$ is the mid-point of the chord AB of the circle whose equation is $x^2 + y^2 = 26$ and O is the origin.

(i) Calculate the coordinates of A and B.

(ii) Find the equation of the circle which has AB as a diameter and show that this circle passes through O. *(AEB)*

10 Find the equation of the perpendicular bisector of the line segment joining the points $Q(-5, 0)$ and $R(7, 6)$. This bisector meets the y-axis at S. Find the coordinates of the point P in which the line through the point Q perpendicular to QS meets the bisector.

Show that the circle through Q, R, and S passes through P, and find the co-ordinates of its centre. *(L)*

11 The circles whose equations are $x^2 + y^2 - x + 6y + 7 = 0$ and $x^2 + y^2 + 2x + 2y - 2 = 0$ intersect at the points A and B.

Find

(i) the equation of the line AB,   (ii) the coordinates of A and B.

Show that the two given circles intersect at right angles and obtain the equation of the circle which passes through A and B and which also passes through the centres of the two circles. *(AEB)*

12 Two circles, C_1 and C_2, have equations

$$x^2 + y^2 - 4x - 8y - 5 = 0$$

and

$$x^2 + y^2 - 6x - 10y + 9 = 0,$$

respectively. Find the x-coordinates of the points P and Q at which the line $y = 0$ cuts C_1, and show that this line touches C_2.

Find the tangent of the acute angle made by the line $y = 0$ with the tangents to C_1 at P and Q.

Show that, for all values of the constant λ, the circle C_3 whose equation is

$$\lambda(x^2 + y^2 - 4x - 8y - 5) + x^2 + y^2 - 6x - 10y + 9 = 0$$

passes through the points of intersection of C_1 and C_2. Find the two possible values of λ for which the line $y = 0$ is a tangent to C_3.   (*JMB*)

13 On the same diagram sketch the circles $x^2 + y^2 = 4$ and $x^2 + y^2 - 10x = 0$. The line $ax + by + 1 = 0$ is a tangent to both these circles. State the distances of the centres of the circles from this tangent. Hence, or otherwise, find the possible values of a and b and show that, if 2ϕ is the angle between the common tangents, then $\tan \phi = \frac{3}{4}$.   (*L*)

14 Find the points of intersection of the circle

$$x^2 + y^2 - 6x + 2y - 17 = 0$$

and the line $x - y + 2 = 0$. Show that an equation of the circle which has these points as the ends of a diameter is

$$x^2 + y^2 - 4y - 5 = 0.$$

Show also that this circle and the circle

$$x^2 + y^2 - 8x + 2y + 13 = 0$$

touch externally.   (*L*)

15 Prove that the point $B(1, 0)$ is the mirror image of the point $A(5, 6)$ in the line $2x + 3y = 15$.

Find the equation of
(a) the circle on AB as diameter, (b) the circle which passes through A and B and touches the x-axis.   (*L*)

16 Given that $a^2 + b^2 = c^2$, show that the two circles $x^2 + y^2 + ax + by = 0$ and $x^2 + y^2 = c^2$ touch each other and find the coordinates of the point of contact.

Two circles, which pass through the origin and the point $(1, 0)$ touch the circle $x^2 + y^2 = 4$. Find the coordinates of the points of contact. Find also the equation of the circle which has these points of contact as the ends of a diameter.   (*L*)

17 Finite Series I

17.1 Sequences

In the definition of a set, such as $\{1, 3, 5, 7\}$, the order of the elements does not matter and $\{1, 3, 5, 7\} = \{5, 3, 7, 1\}$. If we wish to consider a list of numbers in which the order does matter, then we use round brackets and the list is called a *sequence*. Thus $(1, 5, 3, 7)$ is a sequence in which the first element (or term) is 1, the second term is 5 and the fourth term is 7.

Definition A *sequence* of numbers is an ordered list of numbers. In the sequence $(u_1, u_2, u_3, \ldots u_n)$, $u_r \in \mathbb{R}$, u_r is the rth *term*, u_1 is the first term and u_n is the nth term, and n is the *length* of the sequence.

It is possible to consider a sequence such as $\mathbb{N} = (0, 1, 2, 3, \ldots)$ which has infinitely many terms, but we shall not consider such infinite sequences here. We shall restrict our work to finite sequences whose number of terms is a natural number.

In order to define a sequence it is necessary to know the number n of terms in the sequence and also the value of the term u_r for each r, $1 \leqslant r \leqslant n$. This may be given by a formula, but it may not; for example;
in the sequence $(1, 2, 3, 4, 5)$, $n = 5$, the rth term is $u_r = r$;
in the sequence $(2, 5, 6, 2)$, $n = 4$, there is no formula for u_r;
in the sequence $(-2, 4, -8)$, $n = 3$, the rth term is $u_r = (-2)^r$.

Note that in the definition of the rth term u_r of a sequence, the symbol r is a dummy, in that it may be replaced by any other letter. For example, $u_r = (-2)^r$ could be written $u_i = (-2)^i$ or $u_p = (-2)^p$.

EXAMPLE 1 *Write down the first, fourth and last term of the sequence of 10 terms whose rth term is $u_r = 2r(r + 1)$.*

The first term is $u_1 = 2 \times 1(1 + 1) = 4$, the fourth term is $u_4 = 2 \times 4(4 + 1) = 40$, the last term is $u_{10} = 2 \times 10(10 + 1) = 220$.

EXAMPLE 2 *The sequence $(u_1, u_2, u_3, \ldots, u_n)$ of n terms has its rth term given by $u_r = ar + b$, where a and b are constants. The first term is $u_1 = 3$ and the third term is $u_3 = -1$. Find a and b and the last term.*

Substituting $r = 1$ and $r = 3$ in $u_r = ar + b$ gives $3 = a + b$ and $-1 = 3a + b$. Subtracting, $2a = -4$, so $a = -2$ and then $b = 3 - (-2) = 5$, and $u_n = -2n + 5$.

Being given the first few terms of a sequence, gives no knowledge of the values of later terms. For example, if asked what is the sixth term of the sequence $(5, 6, 5, 6, 5, \ldots)$, the answer might be (a) 6, to continue the pattern of the first five terms, or (b) there is no sixth term because the sequence has only 5 terms, or (c) 5, because u_r is the length of the English word indicating the position in the sequence (there are 5, 6, 5, letters in 'first', 'second', 'third', respectively).

EXERCISE 17.1

1   Write down the sequence $(u_1, u_2, u_3, \ldots, u_n)$, given that:
    (i) $u_r = 2r, n = 3$; (ii) $u_r = 2 - r, n = 4$; (iii) $u_r = 2, n = 6$;
    (iv) $u_r = r^2, n = 5$; (v) $u_r = (r - 1)(r + 2), n = 3$; (vi) $u_r = (-1)^r, n = 5$;
    (vii) $u_r = \dfrac{1}{r(r + 1)}, n = 5$; (viii) $u_r = \dfrac{(-x)^r}{r}, n = 4$.

2   The sequence (u_r) of ten terms has its rth term given by $u_r = ar + b$, where a and b are constants. Find a and b and u_8, given that:
    (i) $u_1 = 10, u_{10} = 100$; (ii) $u_3 = 7, u_5 = 11$; (iii) $u_2 = 6, u_6 = -10$.

3   Find a formula for the nth term of a sequence, which has the given first four terms:
    (i) $(1, 2, 3, 4, \ldots)$, (ii) $(2, 4, 6, 8, \ldots)$, (iii) $(-1, -2, -3, -4, \ldots)$,
    (iv) $(1, \frac{1}{2}, \frac{1}{3}, \frac{1}{4}, \ldots)$, (v) $(1, \frac{1}{2}, \frac{1}{4}, \frac{1}{8}, \ldots)$, (vi) $(7, 10, 13, 16, \ldots)$,
    (vii) $(10, 5, 0, -5, \ldots)$, (viii) $(1, -1, 1, -1, \ldots)$, (ix) $(1, 2, 6, 24, \ldots)$.

4   Write down the sequence of length n, satisfying the conditions:
    (i) $n = 5, u_1 = 1, u_{r+1} = 2 \times u_r$; (ii) $n = 6, u_1 = -5, u_{r+1} = 3 + u_r$;
    (iii) $n = 4, u_1 = 1, u_2 = 2, u_{r+2} = u_{r+1} + u_r$;
    (iv) $u_1 = 2, u_2 = 3, u_{r+2} = u_{r+1} \times u_r, n = 5$.

5   Which term is equal to 64 in the sequence in (a) question **3** (ii); (b) question **3** (vi); (c) question **4** (i); (d) question **4** (ii)? In (c) and (d), assume that the sequence has 100 terms, instead of the number of terms given in question **4**.

17.2 Series

A series is obtained by adding the terms of a sequence. The result is called the sum of the series. Thus, $1 + 2 + 3 + 4$, $1 + \frac{1}{2} + \frac{1}{4} + \frac{1}{8}$, $1 + 1 - 1 - 1 + 1$, $a_1 + a_2 + a_3$ are all series and the sum of the first of these series is 10. We shall only consider finite series here, that is, those obtained by adding the terms of a finite sequence. For an infinite sequence $(u_1, u_2, u_3, \ldots)$, the infinite series $u_1 + u_2 + u_3 + \ldots$ will be considered in Chapter 26.

Notation   A shorthand notation for a series is to use the Σ sign (the Greek capital letter sigma), meaning 'sum'. We write

$$\sum_{r=1}^{n} u_r = u_1 + u_2 + u_3 + \ldots + u_n,$$

so
$$\sum_{r=1}^{4} 2r = 2+4+6+8 = 20.$$

When there is no ambiguity, we also write $\sum_1^n u_r = \sum_{r=1}^n u_r$.

If we wish to sum a sequence of terms which does not begin with the first term of a given sequence, then we write

$$\sum_p^q u_r = \sum_{r=p}^q u_r = u_p + u_{p+1} + u_{p+2} + \ldots + u_{q-1} + u_q, \; p,q \in \mathbb{N}, p \leqslant q,$$

so that
$$\sum_5^7 r^2 = 5^2 + 6^2 + 7^2 = 110, \; \sum_{i=5}^6 (-1)^i = (-1)^5 + (-1)^6 = 0.$$

In the sum $\sum_p^q u_r$, p and q are called the *limits* of the sum and the summation is made over all the natural numbers between p and q inclusive.

EXERCISE 17.2

1 Write down the sum in full and evaluate it:

(i) $\displaystyle\sum_{i=1}^{5} i$, (ii) $\displaystyle\sum_{i=1}^{4} i(i+1)$, (iii) $\displaystyle\sum_{4}^{9} (-1)^r$, (iv) $\displaystyle\sum_{1}^{6} r^2$, (v) $\displaystyle\sum_{1}^{3} (-r)^3$,

(vi) $\displaystyle\sum_{2}^{6} \frac{1}{r(r-1)}$.

2 Express in the $\sum$ notation, the sums:

(i) $1+3+5+7+9$, (ii) $\dfrac{1}{1\times 2} + \dfrac{1}{2\times 3} + \dfrac{1}{3\times 4} + \dfrac{1}{4\times 5}$,

(iii) $1\times 2\times 3 + 2\times 3\times 4 + 3\times 4\times 5$, (iv) $1+3+5+ \ldots$ to 27 terms,

(v) $\frac{1}{2} - \frac{2}{3} + \frac{3}{4} - \frac{4}{5} + \ldots$ to 101 terms, (vi) $1+x+x^2+ \ldots$ to n terms.

3 Write down the first three terms of the series $\displaystyle\sum_1^{20} u_r$, where:

(i) $u_r = r^3$, (ii) $u_r = \dfrac{r(r+2)}{r+1}$, (iii) $u_r = \dfrac{(-x)^r}{r}$, (iv) $u_r = \cos r\theta$.

17.3 Arithmetic progression

A particular simple series is given by $\displaystyle\sum_1^n a_r$, where $a_r - a_{r-1} = d$, a constant for all $r, 2 \leqslant r \leqslant n$. If the first term $a_1 = a$ and the last term $a_n = l$, then the series is

$$\sum_1^n a_r = a + (a+d) + (a+2d) + (a+3d) + \ldots + (l-d) + l.$$

The rth term is $a_r = a + (r-1)d$ and the series $\sum_{1}^{n}[a+(r-1)d]$ is called an *arithmetic progression*, or A.P., with first term a, last term l and *common difference d*.

The following are examples of arithmetic progressions:

$$
\begin{array}{lllll}
1+2+3+4+5, & a=1, & n=5, & l=5, & d=1; \\
10+8+6+4, & a=10, & n=4, & l=4, & d=-2; \\
x+2x+3x+4x+5x, & a=x, & n=5, & l=5x, & d=x.
\end{array}
$$

The sum of an arithmetic progression

Let the sum of the first 100 natural numbers be S,

$$S = 1 + 2 + 3 + \ldots + 98 + 99 + 100,$$
$$S = 100 + 99 + 98 + \ldots + 3 + 2 + 1.$$

So $\quad 2S = 101 + 101 + 101 + \ldots + 101 + 101 + 101$, a sum of 100 terms,

hence $\quad 2S = 100 \times 101 = 10100$ and $S = 5050$.

A similar method may be used to sum a general A.P. Let S_n be the sum of the A.P. $\sum_{1}^{n}[a+(r-1)d]$, then

$$S_n = a + (a+d) + (a+2d) + \ldots + (l-2d) + (l-d) + l$$

and $\quad S_n = l + (l-d) + (l-2d) + \ldots + (a+2d) + (a+d) + a.$

There are n terms in each sum, so on adding the two lines

$$2S_n = (a+l) + (a+l) + (a+l) + \ldots + (a+l) + (a+l) + (a+l)$$
$$= n(a+l),$$

and $\quad S_n = \tfrac{1}{2}n(a+l) = \tfrac{1}{2}n(a+a+(n-1)d) = \tfrac{1}{2}n[2a+(n-1)d].$

Expressing this result in words: the sum of an arithmetic progression is one half of the sum of the first and last terms multiplied by the number of terms, or it is the average term multiplied by the number of terms.

EXERCISE 17.3

1 Find the 11th term and the sum of the first 11 terms of the A.P. whose first three terms are:
   (i) $2+5+8+\ldots$, (ii) $4+-1+-6+\ldots$, (iii) $27+22+17+\ldots$.
2 Find the number of terms and the sum of the A.P.
   (i) $3+7+\ldots+43$, (ii) $8+3+\ldots+-22$, (iii) $4+\ldots+28+34$.
3 The third term of an A.P. is 6 and the seventh term is 12. Find the common difference and the sum of the first nine terms.
4 An A.P. has 20 terms. Its eighth term is -20 and its common difference is 5. Find the first and the last terms and the sum of the A.P.
5 Find the number of terms in the A.P., and the fifth term, given that the first term is 8, the last term is 12 and the sum is 100.

6 Find the number of terms in an A.P. given that it has first term 3, common difference 4 and sum 1378.
7 Mary's pocket money started at 50p in the first week and increased each week by 5p. Calculate her total pocket money for one year.
8 The sum of the first 20 terms of an arithmetic progression is 45, and the sum of the first 40 terms is 290. Find the first term and the common difference. Find the number of terms in the progression which are less than 100. (*JMB*)

17.4 Geometric progression

A *geometric progression*, or G.P., or a geometric series, is a series in which each term, other than the first term, is a constant multiple r of the previous term. The constant ratio r of each term to its predecessor is called the *common ratio*. For example:

$1 + \frac{1}{2} + \frac{1}{4} + \frac{1}{8}$ is a G.P. with common ratio $\frac{1}{2}$;

$2 - 6 + 18 - 54 + 162$ is a G.P. with common ratio -3.

Let S_n be the sum of the geometric series with n terms, with the first term equal to a and the common ratio r, so that

$$S_n = a + ar + ar^2 + ar^3 + \ldots + ar^{n-1}.$$

Multiply the equation by r and subtract the new equation from the old,

$$rS_n = ar + ar^2 + ar^3 + \ldots + ar^{n-1} + ar^n,$$
$$S_n - rS_n = (1 - r)S_n = a - ar^n = a(1 - r^n),$$

and, providing that $r \neq 1$, we can divide by $(1 - r)$ and obtain

$$S_n = a\frac{(1 - r^n)}{(1 - r)}.$$

Note that if $r = 1$ we have an A.P. with sum equal to na. Note also that, in a G.P. with n terms, the last term is ar^{n-1}.

EXAMPLE 1 *The third term of a G.P. is 2 and the sixth term is 54. Find (i) the common ratio, (ii) the first term and (iii) the sum of seven terms.*

Let the first term be a and the common ratio r. Then $ar^2 = 2$, $ar^5 = 54$, so

$$r^3 = \frac{ar^5}{ar^2} = \frac{54}{2} = 27$$

and $r = 3$, and then $a = \frac{2}{3^2} = \frac{2}{9}$.

The sum of seven terms is then

$$a\frac{(1 - r^7)}{(1 - r)} = \frac{2(-2186)}{9(-2)} = \frac{2186}{9} = 240\frac{8}{9}.$$

EXAMPLE 2   *A geometric progression has six terms. The sum of the first three terms is eight times the sum of the last three. Find the common ratio. If the sum of the G.P. is $15\frac{3}{4}$, calculate the first term.*

Let the first term be a and the common ratio r. Then

$$a(1+r+r^2) = 8a(r^3+r^4+r^5) = 8a(1+r+r^2)r^3,$$

so $8r^3 = 1$ and $r = \frac{1}{2}$. $S_6 = a\dfrac{(1-r^6)}{(1-r)}$, and so

$$15\tfrac{3}{4} = \frac{63}{4} = a\,\frac{1-\frac{1}{64}}{1-\frac{1}{2}} = \frac{63a}{32},$$

and $a = 8$.

EXAMPLE 3   *Find the smallest integer n such that the G.P.* $1+\frac{1}{2}+\frac{1}{4}+ \ldots$ *has a sum to n terms which is equal to* 2·000 *to four significant figures.*

$S_n = \dfrac{1-(\frac{1}{2})^n}{1-\frac{1}{2}} = 2-(\frac{1}{2})^{n-1}$. This sum is equal to 2·000 to four significant figures if it is larger that 1·9995. The value of n must satisfy

$$(\tfrac{1}{2})^{n-1} < 0.0005, \text{ that is, } 2^n > (0.001)^{-1} = 1000.$$

We, therefore, require $n \log_{10} 2 > 3$ or $n > \dfrac{3}{0.3010} = 9.9$, and the least number of terms required is **10**.

EXAMPLE 4   *A man borrows £3000 in order to purchase a car. He agrees to repay in equal instalments at the end of each month, over a total period of 24 months. The interest each month is 2% of the current outstanding loan. Find (i) his monthly payment and (ii) the total interest paid over the two years.*

Suppose that he pays £x per month. Then if he owes £y at the beginning of any month, just after making a repayment, the interest during that month is $0.02y$ pounds. He then repays £x and so the amount owing at the beginning of the next month is $y+0.02y-x = 1.02y-x$ pounds. Calculating the amount still owing at the end of each month, in pounds:

month 1: $3000(1.02)-x$;
month 2: $[3000(1.02)-x](1.02)-x = 3000(1.02)^2 - x(1+1.02)$;
month 3: $3000(1.02)^3 - x(1+1.02)(1.02)-x$
$\quad = 3000(1.02)^3 - x[1+1.02+(1.02)^2]$.
If we put $1.02 = r$, the amount owing at the end of the 24th month is

$$3000r^{24} - x(1+r+r^2+r^3+ \ldots +r^{23}) = 3000r^{24} - x\,\frac{1-r^{24}}{1-r} = 0,$$

on using the formula for the sum of the G.P. On rearranging this equation

$$x = \frac{3000r^{24}(1-r)}{1-r^{24}} = 3000 \times 0.02\,\frac{r^{24}}{r^{24}-1}, r = 1.02.$$

Hence, $x = \mathbf{156.21}$ and the total interest paid, in pounds, is $24 \times 156.21 - 3000$ $= \mathbf{749}$.

EXERCISE 17.4

1 Find the common ratio, the sixth term, and the sum to six terms, of the G.P.,
   whose first two terms are:
   (i) $\frac{1}{3} + 3 + \ldots$, (ii) $1 - 1 + \ldots$, (iii) $12 + 6 + \ldots$,
   (iv) $-1 + 2 + \ldots$, (v) $1 + \frac{1}{2}x + \ldots$.
2 Find the number of terms, and the sum, of the G.P.:
   (i) $54 + 18 + \ldots + \frac{2}{3}$, (ii) $3 + 6 + \ldots + 384$, (iii) $1 - a + \ldots + a^{2k}$,
   (iv) $1 - 2 + \ldots + 256$.
3 Prove that the series whose terms are the squares of the terms of a G.P. is also
   a G.P.
4 The fourth term of a G.P., containing eight terms, is 8·1 and the seventh term
   is 0·3. Find the common ratio and the sum of the G.P.
5 The third term of a G.P., containing eight terms, is 1, and the seventh term is
   16. Find two possible values for the common ratio and the sum of each
   corresponding G.P.
6 A G.P. has four terms. The sum of the first two terms is 20 and the sum of the
   second and third terms is 30. Find the sum of all four terms.
7 A house is mortgaged for £20 000. This is to be paid back in 20 equal
   instalments, paid at the end of each year. Each year, the interest is 15% of the
   amount owing during that year, and this is added to the amount owing. Find
   the annual repayment. Find that fraction of the annual repayment which is
   interest
   (i) in the first repayment, (ii) in the last repayment.
8 Prove that the sum of the geometric series

$$\sum_{k=1}^{n} ar^{k-1} \quad (r \neq 1) \text{ is } a(r^n - 1)/(r-1).$$

   Find the least number of terms of the geometric series
   $1\cdot2 + (1\cdot2)^2 + (1\cdot2)^3 + \ldots$ for its sum to exceed 594.   *(JMB)*
9 (a) The first and last terms of an arithmetic progression are -2 and 73
   respectively, and the sum of all the terms is 923. Calculate the number of
   terms and the common difference of the progression.
   (b) The three positive numbers $x - 2$, x, $2x - 3$ are successive terms of a
   geometric progression. Calculate the value of x. Given that x is the second
   term of the progression, calculate the value of the seventh term.   *(JMB)*
10 (a) The sum of seven consecutive terms of an arithmetic progression of
   positive terms is 147 and the product of the first and last of these terms is 297.
   Calculate the common difference of the arithmetic progression.
   (b) A geometric progression has first term a and common ratio r. If S denotes
   the sum of the first n terms of this progression and if R denotes the sum of the
   reciprocals of these terms, show that $S = a^2 r^{n-1} R$.   *(JMB)*
11 The first term of a progression is 24 and the fourth term is 81. Calculate the
   second and third terms
   (i) if the progression is arithmetic, (ii) if the progression is geometric.
   (JMB)
12 The first, second and third terms of an arithmetic series are a, b and c
   respectively. Prove that the sum of the first ten terms can be expressed as
   $\frac{5}{2}(9c - 5a)$.
   These numbers a, b and c are *also* the first, third and fourth terms,
   respectively, of a geometric series. Prove that $(2b - c)c^2 = b^3$.   *(AEB)*

13 Evaluate the following, giving each answer correct to two significant figures:

(i) $\sum_{n=1}^{20} (1\cdot1)^n$,   (ii) $\sum_{n=1}^{20} \log_{10}(1\cdot1)^n$.   *(JMB)*

14 Show that $\sum_{k=7}^{18} \left(\frac{1}{2}\right)^k = \left(\frac{1}{2}\right)^6 - \left(\frac{1}{2}\right)^{18}$.   *(L)*

17.5 Arithmetic and geometric means

If three numbers a, b, c, are adjacent terms in an arithmetic progression, then b is said to be the *arithmetic mean* of a and c. If the common difference of the A.P. is d, then $b = a + d$ and $c = b + d$. Hence

$$a + c = (b - d) + (b + d) = 2b, \quad \text{and} \quad b = \frac{a+c}{2}.$$

Thus the arithmetic mean of two numbers is their average. The arithmetic mean of n numbers $a_1, a_2, a_3, \ldots, a_n$ is similarly defined as their average

$$\frac{1}{n}(a_1 + a_2 + a_3 + \ldots + a_n).$$

In a similar manner, if three positive numbers a, b, c, are adjacent terms in a geometric progression, then b is said to be the *geometric mean* of a and c. If the common ratio is r, then $b = ra$ and $c = rb$, so $ac = b^2$ and $b = \sqrt{(ac)}$. The geometric mean of n numbers is the nth root of their product.

EXERCISE 17.5

1 Find (a) the arithmetic mean and (b) the geometric mean of:
(i) 1 and 1·21; (ii) 5/8 and $2\frac{1}{2}$; (iii) a/b and b/a; (iv) 3, 6, 9 and 12;
(v) $-2, 4, -8$ and 16; (vi) a, a^2, a^3, a^4 and a^5.
2 Use the fact that $(a - b)^2 \geqslant 0$ to show that the arithmetic mean of two positive numbers, a and b, is greater than, or equal to, their geometric mean. Find the condition on a and b required to make the two means equal.
3 The harmonic mean of two positive numbers is the reciprocal of the arithmetic mean of their reciprocals. Prove that the harmonic mean of a and b is $2ab/(a + b)$. Find the arithmetic mean, the geometric mean and the harmonic mean of the numbers 16 and 25.

17.6 How many choices?

There are four different ways in which a choice can be made from a set of items. These are illustrated in the following examples.

EXAMPLE 1 *There are four runners, A, B, C, D, in a race. In how many ways can the first two places be filled, allowing no ties?*

We list the possible results for the first two places, namely: A, B; A, C; A, D; B, A; B, C; B, D; C, A; C, B; C, D; D, A; D, B; D, C; making a total of 12 possibilities. A systematic way of calculating the number of ways is to note that the first place can be filled in four ways, either A or B or C or D will be first. If A is first, then the second place can be filled in three ways, either B or C or D. In fact, whichever of the four is first, there are three left who may take the second place. Thus the total number of ways of filling the first two places is $4 \times 3 = \mathbf{12}$.

EXAMPLE 2 *Four schools, A, B, C, D, each enter 50 runners for a cross-country race. In how many ways can the schools take the first two places, allowing no ties?*

In this case, we do not distinguish between individual runners, but only between the schools that they represent. Therefore, the first place can be filled by any one of the four schools A, B, C or D, making four ways of filling the first place. But, whichever school takes the first place, it is still possible for any one of the four schools to take the second place. Therefore, the total number of ways is $4 \times 4 = \mathbf{16}$. The 16 possible results are: $AA, AB, AC, AD, BA, BB, BC, BD, CA, CB, CC, CD, DA, DB, DC, DD$.

EXAMPLE 3 *How many two-element subsets are there of the four-element set $\{A, B, C, D\}$?*

As in Examples 1 and 2, we have to choose two letters from the four as the elements of a subset. But in a set, there are no repetitions and the order of the elements is unimportant. The possible subsets are $\{A, B\}, \{A, C\}, \{A, D\}, \{B, C\}, \{B, D\}, \{C, D\}$, so the answer is that there are six two-element subsets of a four-element set. This could have been obtained by first noting that we can choose 12 ordered pairs, as in Example 1, and then counting the number of repetitions of an ordered pair giving the same set. In this case each subset arises from two pairs, for example, $\{A, B\}$ arises from the two pairs (A, B) and (B, A). Hence, this gives the number of different subsets as $12 \div 2 = \mathbf{6}$.

EXAMPLE 4 *In how many ways can two identical balls be painted using four colours A, B, C and D?*

This is similar to Example 3 in that the order of the two balls does not matter. If two colours are used, then, as in Example 3, there are six ways of painting the two balls, namely A, B or A, C or A, D or B, C or B, D or C, D. But it is also possible to paint the two balls the same colour, giving four cases A, A or B, B or C, C or D, D. Hence, the total number of ways is $6 + 4 = \mathbf{10}$.

These four examples show that, when choosing r objects from n objects, the different ways in which one can go about this are:
Example 1: with order and without repetition,
Example 2: with order and with repetition,
Example 3: without order and without repetition,

Example 4: without order and with repetition.
We shall consider the first three in the general case, but the fourth is too difficult for the general case to be considered here.

17.7 Permutations

A *permutation* of a sequence is a rearrangement of the terms of the sequence to form another sequence. Thus, $(2, 4, 1, 3)$ is a permutation of $(1, 2, 3, 4)$. In order to count how many permutations of $(1, 2, 3, 4)$ there are, we may proceed as in example 1 of §17.6, in four steps.

Step 1: Choose the first term in 4 possible ways. It can be 1 or 2 or 3 or 4.

Step 2: If the first term is 1, then the second can be 2 or 3 or 4. Similarly, whichever term is chosen as the first term, there are three possible choices for the second term. Therefore, the first two terms can be chosen in 4×3 ways.

Step 3: If the first two terms are $(1, 4)$ then the third term can be either 2 or 3. Similarly, for each possible choice of the first two terms, there are 2 possibilities for the third term. This means that the first three terms can be chosen in $4 \times 3 \times 2$ ways.

Step 4: After three terms have been chosen, there is only one term left, so the fourth term can only be chosen in one way. This means that the four terms of the sequence can be chosen in $4 \times 3 \times 2 \times 1 = 24$ ways, and this is the number of permutations of four terms.

Notation   The number $4 \times 3 \times 2 \times 1$ is called 4 factorial, and written 4! Generally, $n! = n(n-1)(n-2)(n-3)\ldots 3.2.1$ is called n factorial. Conventionally, we also define $0! = 1$.

   Thus $0! = 1$, $1! = 1$, $2! = 2.1 = 2$, $3! = 3.2.1 = 6$, $4! = 24$. Generally $n! = n \times (n-1)!$ for $n \in \mathbb{Z}^+$.

Theorem 17.7A   The number of permutations of a sequence of length n is $n!$

Proof   We use the same method as we used above for the case $n = 4$. In order to permute, or rearrange, a sequence of length n, we can choose the first term of the permutation in one of n ways. After the first term has been chosen, there are $(n-1)$ terms left and so the second term can be chosen in $(n-1)$ ways. This means that the first two terms can be chosen in $n(n-1)$ ways. Continuing in this manner, the third term can be chosen in $(n-2)$ ways, the fourth in $(n-3)$ ways, so that the first three terms can be chosen in $n(n-1)(n-2)$ ways and the first four terms can be chosen in $n(n-1)(n-2)(n-3)$ ways. For $1 \leqslant r \leqslant n$, the rth term can be chosen in $(n-r+1)$ ways and the first r terms can be chosen in $n(n-1)(n-2)\ldots(n-r+1)$ ways. Therefore, on putting $r = n$, the total number of permutations of n terms is $n(n-1)\ldots 3.2.1 = n!$

Permutation of *r* from *n*

A sequence of *r* terms, chosen from *n* terms, is called a *permutation of r from n*. The total number of permutations of *r* from *n* is denoted by $_nP_r$, which stands for *n* perm. *r*. Let us again consider the sequence (1, 2, 3, 4), so that *n* = 4. The previous four steps correspond to taking *r* equal to 1, 2, 3 and 4 in turn. Thus

$_4P_1 = 4$, since we can choose one term from 4 in four ways;

$_4P_2 = 4 \times 3 = 12$, since we can choose the first term in 4 ways and the second in 3 ways so that a sequence of length two can be chosen in 4×3 ways;

$_4P_3 = 4 \times 3 \times 2$, since having chosen two terms, we can chose the third in two ways, so we can choose the sequence of three terms in $4 \times 3 \times 2 = 24$ ways;

$_4P_4 = 4 \times 3 \times 2 \times 1 = 4!$, since to choose a sequence of length four from the four terms is the same as permuting the four terms.

The number $_4P_0$ can also be given a meaning. It must be the number of ways of choosing no terms from 4 terms. This can be done in just one way, namely by not choosing any terms, and so $_4P_0 = 1$.

Thus, $_4P_0 = 1$, $_4P_1 = 4$, $_4P_2 = 4 \times 3$, $_4P_3 = 4 \times 3 \times 2$ and $_4P_4 = 4 \times 3 \times 2 \times 1$. These results suggest that a sequence of length *r* can be chosen from a sequence of length *n* in $n(n-1)(n-2) \ldots (n-r+1)$ ways, which we now prove.

Theorem **17.7B**   The number of permutations of *r* from *n* is $_nP_r$, given by

$$_nP_r = n(n-1)(n-2)(n-3) \ldots (n-r+1) = \frac{n!}{(n-r)!}, \quad 0 \leqslant r \leqslant n.$$

Proof   If *r* = 0, we can choose no terms from *n* in one way, by choosing none, and so $_nP_0 = 1$. The number of ways of choosing a sequence of *r* terms from *n* terms was found during the proof of theorem 17.7A. One term can be chosen in *n* ways. A sequence of two terms can be chosen in $n(n-1)$ ways. A sequence of three can be chosen in $n(n-1)(n-2)$ ways, and so on. A sequence of *r* terms can be chosen in $n(n-1)(n-2) \ldots (n-r+1)$ ways. Therefore,

$$_nP_r = \frac{n(n-1) \ldots (n-r+1)(n-r)(n-r-1) \ldots 3.2.1}{(n-r)(n-r-1) \ldots 3.2.1} = \frac{n!}{(n-r)!}.$$

EXAMPLE 1   *Find in how many ways the first three places (with no ties) can be filled in a race with 15 contestants.*

The first place can be filled in 15 ways, since any contestant can come first. When the first place has been filled, there are 14 more contestants to choose from for the second place. Hence the first two places can be filled in 15.14 ways. Finally, for each of these ways, the third place can be filled by any of the remaining 13 contestants, and the total number of ways is 15.14.13. Alternatively, from the formula, $_{15}P_3 = \dfrac{15!}{(15-3)!} = \dfrac{15!}{12!} = \mathbf{15.14.13}$.

EXAMPLE 2   *Eight people are to sit down at a round table. Determine the number of ways in which they can sit relative to one another.*

Suppose that one person sits down first. Then there are seven places to be filled by a permutation of the other seven. Thus this can be done in $_7P_7 = \mathbf{7!}$ ways.

EXAMPLE 3   *I wish to arrange six books on my bookshelf, but I do not want to have the two maths books together because their colours clash. Find the number of ways in which I can arrange the books, with this restriction.*

The six books can be arranged on the shelf in 6! ways. For the moment, regard the two maths books as being stuck together. Then the **5** 'books' can be arranged on the shelf in 5! ways. However, the maths books can be stuck together in two ways, taking order into account. So, out of the 6! ways of arranging the 6 books, the two maths books will be together in 2(5!) ways. Therefore the required number of arrangements is $6! - 2(5!) = (6-2)5! = \mathbf{4(5!)}$.

This method of subtracting, from the total number of ways, the number of ways which are not required, in order to obtain the number of ways required, is a very useful technique.

Counting sequences with repetition

The problem of deciding how many sequences exist when the terms can be repeated is somewhat easier. Suppose that there are n different terms to choose from in order to form a sequence of length r, and that each term can be used any number of times. Then, in each place of the sequence any of the n terms can be chosen, so that the total number of sequences is n^r.

EXAMPLE 4   *Find how many three-letter codes can be formed from an alphabet of 26 letters.*

The first letter of the code may be any of the 26 in the alphabet. The same is true for the second letter and for the third letter. Therefore, the total number of codes is $26.26.26 = \mathbf{(26)^3}$.

Permutations of a sequence containing repeated terms

If a word contains n letters, all of them different, then the number of permutations of the letters of the word is $n!$ Sometimes a word may contain repeated letters. If so, the method used to count the number of permutations is:
(i) assume that all the letters are distinct and find the number of permutations in this case;
(ii) if a letter is repeated k times, then any given permutation of the word will occur $k!$ times because the k identical letters can themselves be

permuted $k!$ times. Therefore, divide the number of permutations by $k!$; (iii) repeat step (ii) for each letter that is repeated, using the appropriate number k.

EXAMPLE 5   *Find the number of permutations of the nine letters of the word 'sequences'.*

If all the letters were distinct, the nine letters could be permuted 9! times. In the 9! permutations, every permutation will occur twice because of the repetition of the letter 's' and for each of the permutations, where we have allowed for the repeated 's', it will occur $6 = 3!$ times because of the three 'e's in the word. We must, therefore, divide the number of permutations by $2(3!) = 12$, and the number of permutations of the given word is **9!/12**.

EXERCISE 17.7

1   Find the number of permutations of the letters of the word 'number'.
2   A family of six, including two brothers, sit in six adjacent seats at the cinema. If the two brothers do not sit together, in how many different ways can the family be seated?   (L)
3   Find the number of permutations of the seven letters of the word 'maximum'. Find also how many of these permutations begin with a consonant and end with a vowel.   (L)
4   Ten different books (four green, four blue and two red) are arranged on a shelf. The green books are always placed together and in the same order, but the red books are always separated. Calculate the number of ways in which the books can be arranged.   (L)
5   Define a 'word' to be a sequence of four letters from an alphabet of 26 letters, of which five are vowels. Determine the number of words:
(i) with all the letters different, (ii) with repetitions of the letters allowed, (iii) with all the letters different and containing one vowel, (iv) with all the letters different and containing at least one vowel, (v) with repetitions allowed and containing at least one vowel. (Hint: The set of words containing at least one vowel is the set of all words less the set of words containing no vowels.)
6   A four-figure number is to be made using the digits $2, 3, 4, 6, 8$, without any repetitions. Find
(i) how many such numbers can be formed, (ii) how many odd numbers can be formed, (iii) how many even numbers can be formed.

17.8 Combinations

A combination is a selection of r objects from n objects, all distinct, without paying any attention to the order. Therefore, it is an r-element subset of an n-element set. The number of such combinations is denoted by $_nC_r$.

Notation $_nC_r$ = the number of r-element subsets of an n-element set.

EXAMPLE 1   *Find the number of ways of forming a class committee of three members from a class of* 20.

Begin by choosing the three members of the committee in order. The first can be chosen in 20 ways, the second in 19 ways, and the third in 18 ways. The sequence of three can be chosen from 20 in $_{20}P_3 = \dfrac{20!}{17!} = 20.19.18$ ways. Now a given sequence of 3 members can be permuted in $3! = 3.2.1$ ways. Therefore, every set of 3 members will occur $3!$ times in the $_{20}P_3$ ordered sets of 3. So, the required number of committees, which is the number of 3 member sets, unordered, is

$$\dfrac{_{20}P_3}{3!} = \dfrac{20.19.18}{3.2.1} = 1140 = {_{20}C_3}.$$

Example 1 can be generalised by replacing 20 by n and 3 by r, in order to determine $_nC_r$. The proof may be more easily understood by comparing it with the special case in Example 1.

***Theorem* 17.8**   The number of combinations of r from n, that is, the number of r-element subsets of an n-element set is $_nC_r$, where

$$_nC_r = (_nP_r)/r! = \dfrac{n!}{r!(n-r)!}.$$

Proof   We first count the number of sequences of length r which can be formed from a set of n elements. By theorem 17.7B, this is $_nP_r = \dfrac{n!}{(n-r)!}$. Now the number of permutations of each sequence of length r is $r!$ (theorem 17.7A). Therefore, in the $_nP_r$ sequences of r elements, each set of r elements occurs permuted $r!$ times. Hence the number of distinct subsets, which is the number of combinations of r from n, is

$$_nC_r = \dfrac{_nP_r}{r!} = \dfrac{n!}{r!(n-r)!} = \dfrac{n(n-1)(n-2)\ldots(n-r+1)}{r!}.$$

EXAMPLE 2   *Find the number of ways in which a team of* 11 *players can be chosen from a group of* 20 *players. Find in how many ways a cricket team of* 11 *can be chosen, with one wicket-keeper, five bowlers and five batsmen, from* 20 *players of whom two are wicket-keepers, eight are bowlers and* 10 *are batsmen.*

A selection of 11 from 20 can be made in

$$_{20}C_{11} = \dfrac{20!}{9!11!} = \dfrac{20.19.18.17.16.15.14.13.12}{9.8.7.6.5.4.3.2.1}$$

ways. With the restrictions on the players, one wicket-keeper can be chosen from two in two ways, five bowlers can be chosen from eight in $_8C_5$ ways, five batsmen can be chosen from ten in $_{10}C_5$ ways. These choices are all independent, so that the total numbers of ways of selecting a team will be

$$2 \times {_8C_5} \times {_{10}C_5} = 2 \cdot \dfrac{8.7.6.5.4}{5.4.3.2.1}\dfrac{10.9.8.7.6}{5.4.3.2.1} = \dfrac{2.8.7.6.10.9.8.7}{5.4.3.2} = \mathbf{9.8.8.7.7.}$$

Notation   We shall use the following notation

$$\binom{n}{r} = {}_nC_r = \frac{n(n-1)(n-2)\ldots(n-r+1)}{r!},$$

and, more generally, even if n is not a natural number, so that ${}_nC_r$ has no meaning, we shall define

$$\binom{n}{r} = n(n-1)(n-2)\ldots(n-r+1)/r!, \quad r \in \mathbb{N}, n \in \mathbb{R}.$$

EXERCISE 17.8

1   Express in factorial notation:

(i) $1 \times 2 \times 3 \times 4 \times 5$, (ii) $5 \times 6 \times 7$, (iii) $\binom{10}{4}$, (iv) $\binom{9}{5}$.

2   Prove that: (i) $\binom{5}{3} = \binom{5}{2}$, (ii) $\binom{n}{r} = \binom{n}{n-r}$, $r, n \in \mathbb{N}$.

3   Determine the number of: (i) one-, (ii) two-, (iii) three-, (iv) four-element subsets there are of the set $\{a, b, c, d\}$. Show that the total number of subsets of this set is $2^4 = 16$.

   Generalise this result to find the number of subsets of an n-element set.

4   In a tennis club of 68 members, 30 are women. Find the number of ways of forming a committee of four members:

   (i) with no restrictions, (ii) with at least one woman on the committee.

5   Find the number of different ways in which a committee of two men and two women students can be selected from five men and four women students.   (*L*)

6   A water-polo team consisting of a goalkeeper and six other players is to be selected from 11 players. Just two of the 11 players are goalkeepers. Find the number of ways in which the team may be selected.   (*L*)

17.9 The binomial series

Consider the following expansions, known as binomial expansions:

$$(1+x)^2 = 1 + 2x + x^2,$$
$$(1+x)^3 = (1+x)(1+x)^2 = (1+x)(1+2x+x^2)$$
$$= 1 + 2x + x^2$$
$$ + x + 2x^2 + x^3$$
$$= 1 + 3x + 3x^2 + x^3,$$
$$(1+x)^4 = (1+x)(1+x)^3 = (1+x)(1+3x+3x^2+x^3)$$
$$= 1 + 3x + 3x^2 + x^3$$
$$ + x + 3x^2 + 3x^3 + x^4$$
$$= 1 + 4x + 6x^2 + 4x^3 + x^4.$$

Note that the coefficient of x^{r+1} in the expansion of $(1+x)^{n+1}$ is the sum of the coefficients of x^r and x^{r+1} in $(1+x)^n$. This is made clear by displaying the coefficients by themselves.

Polynomial	Coefficients
$1 = (1+x)^0$	1
$1 + x = (1+x)^1$	1 1
$(1+x)^2$	1 2 1
$(1+x)^3$	1 3 3 1
$(1+x)^4$	1 4 6 4 1

This is known as Pascal's triangle and the entry in any place in one line is the sum of the entries in the previous line to either side of the required entry.

EXERCISE 17.9

1   Expand $(1+x)^5$ and $(1+x)^6$, using the identity $(1+x)^n \equiv (1+x)(1+x)^{n-1}$. Verify your results by using the previous expansions and expanding $(1+x)^2 (1+x)^3$ and $(1+x)^2 (1+x)^4$.

2   Work out Pascal's triangle, as explained above, repeating the first five lines and extending to a further two lines. Verify that the numbers in these last two lines are the coefficients in the expansions of $(1+x)^5$ and $(1+x)^6$, which you obtained in question **1**.

3   Verify that, in the Pascal's triangle which you have produced in question **2**, if the rows are labelled $0, 1, 2, 3, 4, 5, 6$, then the entry in the $(r+1)$th place in the row labelled n is $\dbinom{n}{r} = \dfrac{n(n-1)(n-2)\ldots(n-r+1)}{r!}$.

17.10   The binomial theorem

We shall show that the result you proved in question **3** of Exercise 17.9 is true for all natural numbers n and r; that is, that the coefficient of x^r in the expansion of $(1+x)^n$ is $\dbinom{n}{r}$. Because of this result, the numbers $\dbinom{n}{r}$ are called the *binomial coefficients*.

***Theorem* 17.10** $(1+x)^n = \displaystyle\sum_{r=0}^{n} \binom{n}{r} x^r$,   for $n, r \in \mathbb{N}$.

Proof   In the expansions

$$(a+b)(c+d) = ac + bc + ad + bd \quad \text{and}$$
$$(a+b)(c+d)(e+f) = ace + acf + ade + adf + bce + bcf + bde + bdf,$$

there are 2^n terms in the case of a product of n brackets and each term is obtained by choosing one element from each of the brackets. Similarly,

$$(a+b)^2 = aa + ba + ab + bb = a^2 + 2ab + b^2, \quad \text{and}$$

$$(a+b)^3 = aaa + aab + aba + abb + baa + bab + bba + bbb$$
$$= a^3 + 3a^2b + 3ab^2 + b^3.$$

These equations are obtained from the lines above by putting $a = c = e$, $b = d = f$. Note that the coefficients are again the binomial coefficients. Using the same idea, if we wish to find the term in $(a+b)^5$ containing a multiple of $a^2b^3 = aabbb$, we shall get a contribution to this term by choosing a from two of the brackets and b from three of the brackets in $(a+b)^5 = (a+b)(a+b)(a+b)(a+b)(a+b)$.

Repeating the argument for $a = 1$ and $b = x$, a contribution to the term in x^r in the expansion of $(1+x)^n = (1+x)(1+x)(1+x) \ldots (1+x)$ is obtained by choosing x from r of the brackets and 1 from the other $(n-r)$ brackets. Thus the coefficient of x^r is the number of ways of choosing r brackets from the n brackets, and this, of course, is $_nC_r = \binom{n}{r}$.

Q.E.D.

Corollary 17.10 $(a+b)^n = \sum_{r=0}^{n} \binom{n}{r} a^{n-r} b^r$, for $n, r \in \mathbb{N}$.

Proof $(a+b)^n = \left[a\left(1 + \dfrac{b}{a}\right) \right]^n = a^n \left(1 + \dfrac{b}{a}\right)^n$

$$= a^n \sum_{r=0}^{n} \binom{n}{r} \left(\dfrac{b}{a}\right)^r = \sum_{r=0}^{n} \binom{n}{r} a^{n-r} b^r.$$

In expanded form, this corollary means that

$$(a+b)^n = a^n + na^{n-1}b + \frac{n(n-1)}{2} a^{n-2}b^2 + \frac{n(n-1)(n-2)}{3!} \times a^{n-3}b^3 +$$

$$+ \ldots + b^n.$$

EXAMPLE 1 *Expand* $(3x-2)^5$.

Using the corollary,

$$(3x-2)^5 = (3x)^5 + 5(3x)^4(-2) + 10(3x)^3(-2)^2 + 10(3x)^2(-2)^3 + 5(3x)^1(-2)^4$$
$$+ (-2)^5$$

$$= 243x^5 + 5.81(-2)x^4 + 10.27(+4)x^3 + 10.9(-8)x^2 + 5.3(+16)x$$
$$+ (-32)$$

$$= 243x^5 - 810x^4 + 1080x^3 - 720x^2 + 240x - 32.$$

EXAMPLE 2   *Find the term independent of x in the expansion of* $\left(2x + \dfrac{1}{x}\right)^6$.

The term required is $\dbinom{6}{3}(2x)^3\left(\dfrac{1}{x}\right)^3 = \dfrac{6.5.4}{3.2.1}2^3 = 5.4.8 = \mathbf{160}$.

EXAMPLE 3   *Find the value of* $(1\cdot99)^{10}$, *accurate to two decimal places, without using tables or calculator.*

$$(1\cdot99)^{10} = (2 - 0\cdot01)^{10} = 2^{10} - 10.2^9(0\cdot01) + \frac{10.9}{2}2^8(0\cdot01)^2 - \frac{10.9.8}{3.2.1}2^7(0\cdot01)^3$$

$$+\frac{10.9.8.7}{4.3.2.1}2^6(0\cdot01)^4 - \dots$$

The last term shown is $13\,440\ 10^{-8}$, so this, and further terms, will not affect the result to two decimal places. Therefore, the result is

$$(1\cdot99)^{10} = 2^7(2^3 - 0\cdot4 + 0\cdot009 - 0\cdot00012) = 128(7\cdot59888) = \mathbf{972\cdot66}.$$

EXERCISE 17.10

1   Use the binomial series to expand:
(i) $(1+x)^6$, (ii) $(1+a)^7$, (iii) $(1-x)^5$, (iv) $(1+2x)^3$, (v) $(1-\frac{1}{4}x)^4$,
(vi) $(a+b)^5$, (vii) $(2+3x)^3$, (viii) $(1+x^2)^4$, (ix) $(x-2)^3$, (x) $(x+1/x)^5$.

2   Write down the fifth term in the series, in ascending powers of x, of
(i) $(1+x)^8$, (ii) $(1-2x)^{12}$, (iii) $(3x-2)^9$, (iv) $(2+\frac{1}{2}x)^8$.

3   Find the coefficient of x^3 in the expansion of
(i) $(1+2x)^9$, (ii) $(3-5x)^6$, (iii) $(1+x)(1-x)^4$, (iv) $(1+\frac{1}{2}x)^3(1-x)^5$.

4   Use the substitution $y = x + x^2$ to expand $(1+x+x^2)^5$.

5   Find the term which is independent of y in the expansion of $(2y - 1/y)^{10}$.

6   Without using tables or calculator, evaluate:
(i) $(1-\sqrt{2})^2 + (1+\sqrt{2})^2$, (ii) $(2+\sqrt{3})^5 + (2-\sqrt{3})^5$,
(iii) $(\sqrt{2}+\sqrt{5})^8 + (\sqrt{2}-\sqrt{5})^8$.

7   Using the expansion of $(1+x)^7$, evaluate, correct to three decimal places:
(i) $(1\cdot02)^7$, (ii) $(0\cdot97)^7$.

8   Find the numerical value of the term independent of x in the expansion of $(2x+1/x^2)^6$.   (*L*)

9   Find the coefficient of x^6 in the expansion in powers of x of
$(1+x-2x^2)(1+3x^2)^5$.   (*JMB*)

10   Use the binomial theorem to express $(1+x)^5 + (1-x)^5$ in powers of x.
Hence find the real roots of the equation $(1+x)^5 + (1-x)^5 = 242$.
                                                                                     (*JMB*)

MISCELLANEOUS EXERCISE 17

1   The sum of the first six terms of a geometric series of positive terms is 1, and the sum of the first 12 terms is 65. Find the first term.   (*L*)

2   Find the sum of all the positive integers less than 1000 which are divisible by 7.   (*L*)

3 The coefficients of x^2 and x^3 are equal in the expansion of $(1+x)^n$, where $n > 1$. Find the value of n. *(L)*

4 It is given that $\dfrac{1}{b+c}, \dfrac{1}{c+a}, \dfrac{1}{a+b}$ are three consecutive terms of an arithmetic series. Show that a^2, b^2 and c^2 are also three consecutive terms of an arithmetic series. *(JMB)*

5 A rod 1 metre in length is divided into ten pieces whose lengths are in geometrical progression. The length of the longest piece is eight times the length of the shortest piece. Find, to the nearest millimetre, the length of the shortest piece. *(JMB)*

6 Find the number of permutations of the letters of the word 'topology'. In how many of these permutations are no two o's together? *(L)*

7 Ten lines are drawn parallel to the x-axis and eight lines parallel to the y-axis. Find the total number of rectangles (of all sizes) formed by these lines. *(L)*

8 The second and fourth terms of an arithmetic progression are -3 and 7 respectively. Find the 100th term. *(L)*

9 Show that if p is an integer, $p(1 + 3p)$ is an even integer. Write out the expansion of $(1+x)^4$ in ascending powers of x. By putting $x = 2p$, or otherwise, show that the fourth power of an odd integer is of the form $16k + 1$, where k is an integer. *(JMB)*

10 Find
(a) the sum of the first 20 terms of the arithmetic progression $1, 3, \ldots$,
(b) an expression for the 17th term of the geometric progression $1, 3, \ldots$. *(L)*

11 The numbers $-1, 1, 3$ are the first three terms of either an arithmetic or a geometric progression. State which of these alternatives is true. Find the tenth term and the sum of the first 20 terms of this progression. *(L)*

12 One sequence of alternating terms of the series

$$1 + 2 + 3 + 4 + 5 + 8 + \ldots$$

forms an arithmetic progression, while the other sequence of alternating terms forms a geometric progression. Sum the first ten terms of each progression and hence find the sum of the first 20 terms of the series. *(L)*

13 (a) The first term of a geometric progression is 7, its last term is 448 and its sum is 889. Find the common ratio.
(b) Given that $\dfrac{1}{y-x}, \dfrac{1}{2y}$ and $\dfrac{1}{y-z}$ are consecutive terms of an arithmetic progression, prove that x, y and z are consecutive terms of a geometric progression. *(C)*

14 The sum to n terms of a series is $2n^2 - n$. Find the first term. By subtracting the sum to $(n-1)$ terms from the sum to n terms, find the nth term. Prove that the series is an A.P. and find its common difference.

15 Prove that the sum of the first n odd numbers is a square. Find p and q such that the sum of consecutive odd numbers between the pth and the qth odd number, inclusive, is $43^2 - 23^2$.

16 The first term of a G.P. is 25 and the sum of the first three terms is 61. Find the possible values of the second term.

17 Given the digits $1, 2, 3, 4, 6, 8$, and using each digit only once, determine how many numbers can be formed, satisfying the condition:
(i) they are less than 100,   (ii) they lie between 100 and 1 000,
(iii) they lie between 1 000 and 10 000,   (iv) they are below 10 000,
(v) they are odd numbers below 1 000.

18 Find how many different four-digit numbers can be formed from the seven digits $1, 2, 4, 5, 6, 8, 9$.
(a) if each digit may be used as often as desired,
(b) if no digit is repeated.
   In case (b), find how many of the numbers include two odd digits and two even digits and state how many such numbers are even.   *(L)*

19 Seven people are to be seated in a row, on the platform at the school speech day. The chairman of the governors must sit in the centre and the visiting speaker and the headteacher must sit next to the chairman. Find the number of ways in which the seating plan can be arranged.

20 The first and last terms of an arithmetic progression are a and l, respectively. If the progression has n terms, prove from first principles that its sum is $\frac{1}{2}n(a + l)$. An array consists of 100 rows of numbers, in which the rth row contains $r + 1$ numbers. In each row, the numbers form an arithmetic progression whose first and last terms are 1 and 99, respectively. Calculate the sum of the numbers in the rth row, and hence determine the sum of all the numbers in the array.   *(JMB)*

21 Show that the number of permutations of n things, of which p are alike of one kind, q are alike of a different kind, and the remainder are different from these and from each other, is

$$\frac{n!}{(p!)(q!)}$$

   Naval signals are made by arranging coloured flags in a vertical line and the flags are then read from top to bottom. How many different signals can be made from one green, three identical red and two identical blue flags, if
(a) all six of the flags are used, (b) at least five of them are used?   *(L)*

22 Calculate the number of ways in which six people can form
(a) a queue (that is, a single file) of six people,
(b) a queue of two people and another queue of four people,
(c) a group of two people and another group of four people,
(d) first, second and third pairs,
(e) three pairs.
Assume that the order within a group or a pair is *not* significant.   *(L)*

23 The first and last terms of an arithmetic progression are a and l respectively. If the progression has n terms, prove from first principles that its sum is $\frac{1}{2}n(a + l)$.
   A circular disc is cut into twelve sectors whose areas are in arithmetic progression. The area of the largest sector is twice that of the smallest. Find the angle (in degrees) between the straight edges of the smallest sector. *(JMB)*

24 Given that n is a positive integer and that the coefficient of x^2 in the binomial expansion of $(3 + 2x)^n$ is twice the coefficient of x, find n.   *(L)*

25 Three Englishmen, four Frenchmen and five Germans are available for selection to a European committee of four, on which each nation has to be represented. In how many different ways can the committee be selected?   *(L)*

26 (a) Obtain the first three terms in the expansion in ascending powers of x of $(1-x)^6 (1+5x)^4$.

(b) Given that $\dfrac{1}{b+c} + \dfrac{1}{a+b} = \dfrac{2}{c+a}$, prove that a^2, b^2, c^2 are in arithmetic progression. Prove that the sum of the first n terms of the arithmetic progression $a^2, b^2, c^2, \ldots$ is $\frac{1}{2}(n^2-n)b^2 - \frac{1}{2}(n^2-3n)a^2$.   (*JMB*)

27 Seven students are eligible for selection to a delegation of four students from a school to attend a conference. Two of them will not attend together but each is prepared to attend in the absence of the other. In how many different ways can the delegation be chosen?   (*L*)

18 Differentiation II

18.1 Limits of functions

In the study of calculus, we need to consider the behaviour of a function near to some point, which may not be in its domain. That is, we need to know how $f(x)$ behaves near to $x = a$. We shall not be concerned with the value of $f(a)$; in fact, this will not exist when the point $x = a$ does not lie in the domain of f. By means of some examples, we illustrate the limiting process which describes the behaviour of $f(x)$ as x tends to a.

EXAMPLE 1 *How does the function* f, *given by* $f(x) = \dfrac{x^2 - 9}{x - 3}$, *behave near to the point* $x = 3$?

Note that, in the expression for $f(x)$, the denominator vanishes when $x = 3$, so $f(3)$ is not defined, that is, the point $x = 3$ is not in the domain of f. We tabulate some values of x and $f(x)$ near $x = 3$.

x	3·5	3·1	3·01	3·001	2·9	2·999
$f(x)$	6·5	6·1	6·01	6·001	5·9	5·999.

These results suggest that $f(x)$ tends to 6 as x tends to 3. In fact, for $x \neq 3$,
$$f(x) = \frac{(x+3)(x-3)}{x-3} = x+3,$$ which clearly tends to 6 as x tends to 3. This is expressed in two (equivalent) ways:

$f(x)$ *tends to* 6 *as* x *tends to* 3, (in symbols) $\mathbf{f(x) \to 6}$ **as** $x \to 3$,

or   *the limit of* $f(x)$ *as* $x \to 3$ *is* 6, (in symbols) $\displaystyle\lim_{x \to 3} \mathbf{f(x) = 6}$.

EXAMPLE 2 *Given that* $f(x) = 2x^2 - 4$, *find the limit* $\displaystyle\lim_{h \to 0} \dfrac{f(2+h) - f(2)}{h}$.

We calculate the numerator,
$$\begin{aligned} f(2+h) - f(2) &= 2(2+h)^2 - 4 - (8-4) \\ &= 8 + 8h + 2h^2 - 4 - 4 \\ &= 8h + 2h^2, \end{aligned}$$

so $\dfrac{f(2+h) - f(2)}{h} = 8 + 2h$, which clearly tends to 8 as h tends to 0. Hence

$$\lim_{h \to 0} \frac{f(2+h) - f(2)}{h} = \mathbf{8}.$$

EXAMPLE 3 *Given that* $f(x) = 2x^2 - 4$, *find the limit* $\lim\limits_{h \to 0} \dfrac{f(x+h) - f(x)}{h}$.

In this case, we keep x fixed and consider the limit as h tends to 0.

$$\frac{f(x+h) - f(x)}{h} = \frac{2(x^2 + 2hx + h^2) - 4 - (2x^2 - 4)}{h}$$

$$= 4x + 2h,$$

which tends to $4x$, so $\lim\limits_{h \to 0} \dfrac{f(x+h) - f(x)}{h} = 4x$.

It is useful to use some rules for calculating limits, and these are listed without proof. The results are quite plausible but their proofs cannot be dealt with in an elementary treatment.

Rules for limits

The same limiting process is used in each of the limits.
Sum of functions: $\qquad\qquad\qquad \lim (f + g) = \lim f + \lim g$.
Multiplication by a constant k: $\quad \lim (kf) \quad = k \lim f$.
Product of functions: $\qquad\qquad \lim (f.g) \quad = (\lim f)(\lim g)$.
Quotient of functions: $\qquad\qquad \lim (f/g) \quad = (\lim f)/(\lim g)$, if $\lim g \neq 0$.
The use of these rules is illustrated in the following example.

EXAMPLE 4 *Find the following limits, when they exist:*

(i) $\lim\limits_{x \to 2} (x^2 - 3x)$, (ii) $\lim\limits_{x \to 5} \dfrac{x^2 - 6x + 5}{x^2 - 3x + 2}$, (iii) $\lim\limits_{x \to 1} \dfrac{x^2 - 6x + 5}{x^2 - 3x + 2}$,

(iv) $\lim\limits_{x \to 2} \dfrac{x^2 - 6x + 5}{x^2 - 3x + 2}$.

(i) $\lim\limits_{x \to 2} (x^2 - 3x) = \lim\limits_{x \to 2} x^2 - 3 \lim\limits_{x \to 2} x = 4 - 6 = -\mathbf{2}$.

(ii) $\lim\limits_{x \to 5} \dfrac{x^2 - 6x + 5}{x^2 - 3x + 2} = \left(\lim\limits_{x \to 5} x^2 - 6x + 5 \right) \Big/ \left(\lim\limits_{x \to 5} x^2 - 3x + 2 \right) = \dfrac{0}{12} = \mathbf{0}$.

(iii) $\dfrac{x^2 - 6x + 5}{x^2 - 3x + 2} = \dfrac{(x-1)(x-5)}{(x-1)(x-2)} = \dfrac{x-5}{x-2}$ as long as $x \neq 1$, and so

$$\lim\limits_{x \to 1} \frac{x^2 - 6x + 5}{x^2 - 3x + 2} = \lim\limits_{x \to 1} \frac{x-5}{x-2} = \mathbf{4}.$$

(iv) As in (iii) above, $\dfrac{x^2 - 6x + 5}{x^2 - 3x + 2} = \dfrac{x-5}{x-2}$, for $x \neq 1$, and as x tends to 2 the numerator tends to -3, whereas the denominator tends to zero. Thus, the fraction will be very large and negative when x is just greater than 2 and will be very large and positive when x is just less than 2. Therefore, **no limit exists**.

EXERCISE 18.1

1   For each of the following functions, find how the value of the function
    behaves near to the point $x = 2$ by calculating the value for $x = 2\cdot1$, $2\cdot01$,
    $2\cdot001$, $1\cdot9$, $1\cdot99$, $1\cdot999$. Deduce whether the function has a limit as x tends to 2
    and, if it has, what this limit is:

    (i) $f(x) = 2x^2 - 3x + 2$,   (ii) $f(x) = \dfrac{x^2 + 6x + 5}{x^2 - 2x - 8}$,   (iii) $f(x) = \dfrac{x + 4}{x^2 - 4}$,

    (iv) $f(x) = \dfrac{x^4 - 16}{x^2 - 4}$.

    Note: to calculate the value of a polynomial using a calculator, write it in the
    form:
    $$ax^2 + bx + c = (ax + b)x + c$$
    or $$ax^4 + bx^3 + cx^2 + dx + e = (((ax + b)x + c)x + d)x + e.$$

2   Find the limit, or show that no limit exists:

    (i) $\lim\limits_{x \to 2} (x + 4)(x^2 - 4)$,   (ii) $\lim\limits_{x \to 3} \dfrac{x^2 + 6x + 5}{x^2 - 2x - 3}$,   (iii) $\lim\limits_{x \to -1} \dfrac{x^2 + 6x + 5}{x^2 - 2x - 3}$,

    (iv) $\lim\limits_{x \to 1} \dfrac{x^2 - 1}{x^2 - 1}$,   (v) $\lim\limits_{x \to 0} \dfrac{x^2 + 6x}{3x}$,   (vi) $\lim\limits_{x \to 0} \dfrac{x + \dfrac{1}{x}}{x - \dfrac{1}{x}}$.

3   Given that, as x tends to 1, the limit of $f(x)$ is 3 and the limit of $g(x)$ is 5, find the
    limit of:

    (i) $3f(x)$,   (ii) $2f.g(x)$,   (iii) $(4f - g)(x)$,   (iv) $f^2(x)$,   (v) $\dfrac{f}{g}(x)$.

4   Evaluate: (i) $\lim\limits_{x \to 0} \dfrac{2x^2 - x}{5x}$,   (ii) $\lim\limits_{x \to 0} \dfrac{4x^2 + 3x}{x^2 - x}$,   (iii) $\lim\limits_{x \to 1} \dfrac{x^2 - 1}{x^3 - x^2}$.

5   Show that the limit of $\left(\dfrac{1}{x - 1} - \dfrac{2}{x^2 - 1} \right)$ as x tends to 1 is $\frac{1}{2}$.

18.2 Continuous functions

A continuous function may be thought of as a function whose graph may
be drawn without raising the pen from the paper. For example, the
functions x^2, $3x^2 - 4x + 2$ are both continuous at all real values of x.
However, consider the function f given by

$$f(x) = \frac{x - 1}{|x - 1|},$$

for $x \neq 1$, and $f(1) = 0$. For $x > 1$, $f(x) = 1$, and for $x < 1$, $f(x) = -1$, so
clearly $f(x)$ cannot be continuous at $x = 1$. Also, we can not make the
function continuous at $x = 1$ by changing the value of $f(1)$, because $f(x)$
has no limit as x tends to 1 (Fig. 18.1).

Now, consider the function g given by $g(x) = \dfrac{x^2 - 1}{x - 1}$, for $x \neq 1$, and

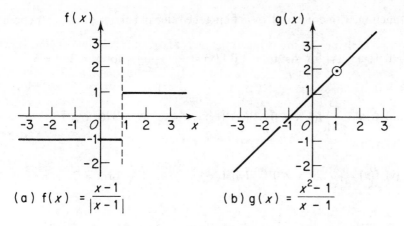

(a) $f(x) = \dfrac{x-1}{|x-1|}$

(b) $g(x) = \dfrac{x^2-1}{x-1}$

Fig. 18.1

$g(1) = 2$. If $x \neq 1$, then $g(x) = x+1$, and this is also true for $x = 1$. So, for all values of x, $g(x) = x+1$, the limit of $g(x)$ as x tends to 1 is equal to $g(1)$, so g is *continuous* for all x. However, suppose that a new function h is formed by changing the value of $g(1)$, say $h(x) = g(x)$ for $x \neq 1$ and $h(1) = 0$. Then h will not be continuous at $x = 1$.

In order that the function f be continuous at $x = a$, the point $x = a$ must lie in the domain of f and the limit of $f(x)$ as x tends to a must be the value $f(a)$ of f at $x = a$. We use this to make a formal definition.

Definition   The function f is *continuous at* $x = a$ if and only if

$$\lim_{x \to a} f(x) = f(a).$$

The function f is *continuous* if and only if it is continuous at all points of its domain.

EXAMPLE   *Given that* $f(x) = \dfrac{x^2 + 4x + 3}{x^2 - 1}$ *for* $x \neq \pm 1$, *find the value of* $f(-1)$ *required to make f a continuous function with domain* $\mathbb{R} \backslash \{1\}$.

For

$$x \neq \pm 1, \; f(x) = \frac{(x+1)(x+3)}{(x+1)(x-1)} = \frac{x+3}{x-1}, \quad \text{so} \quad \lim_{x \to -1} f(x) = \frac{-1+3}{-1-1} = -1.$$

Thus, by adding the point -1 to the domain of f and defining $f(-1) = -1$, we make f into a continuous function with domain $\mathbb{R} \backslash \{1\}$.

EXERCISE 18.2

1   The function f is defined as indicated for all but one or two real values of x. Determine whether or not the function can be made into a continuous

function, with domain $\mathbb{R}$, by defining the value of f at one or two extra points:

(i) $f(x) = |x|$ for $x \neq 0$ (ii) $f(x) = \dfrac{x^2 - 4x + 3}{x^2 - 5x + 6}$ for $x \neq 2$, $x \neq 3$

(iii) $f(x) = \dfrac{x^3}{|x|}$ for $x \neq 0$ (iv) $f(x) = \dfrac{x^2 + \dfrac{1}{x}}{x^2 - \dfrac{1}{x}}$ for $x \neq 0$, $x \neq 1$

(v) $f(x) = \dfrac{x}{|x|}$ for $x \neq 0$ (vi) $f(x) = \dfrac{x^2 - 4x + 3}{x^2 - 6x + 9}$ for $x \neq 3$.

18.3 The tangent to the graph of a function

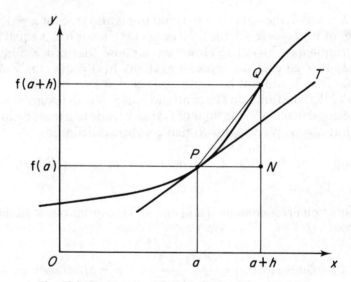

Fig. 18.2 The tangent PT as the limit of the chord PQ.

Consider a fixed point P on the graph $y = f(x)$ of the function f. Let Q be a neighbouring point to P on the graph, so that the coordinates of P and Q are $(a, f(a))$ and $(a+h,\ f(a+h))$ respectively (Fig. 18.2). Thus, a is considered fixed whilst h can vary but remains small and non-zero. Let N be the point $(a+h,\ f(a))$, then the gradient of the chord PQ is

$$\frac{QN}{PN} = \frac{f(a+h) - f(a)}{h} = g(h), \text{ say, for } h \neq 0.$$

Now, as h tends to zero, Q will approach P along the graph, and the chord PQ will approach the tangent PT at P. We assume that the graph is smooth near P, having no abrupt change of direction or kink in it, so that a tangent PT exists and $g(h)$ has a limit as h tends to zero.

Then, as h tends to zero, the limit of the chord PQ is the tangent PT and the limit of the gradient $g(h)$ of PQ is the gradient of the tangent PT. Thus, the equation of the tangent PT is

$$y = f(a) + m(x - a)$$

where
$$m = \lim_{h \to 0} g(h) = \lim_{h \to 0} \frac{f(a+h) - f(a)}{h}.$$
18.3

EXAMPLE *Find the equation of the tangent to the curve $y = x^3$ at (i) the point $(1, 1)$, (ii) the point (a, a^3).*

(i) Let P be the point $(1, 1)$ and Q the point $(1 + h, (1 + h)^3)$. Then the gradient of PQ is

$$g(h) = \frac{(1+h)^3 - 1^3}{h} = \frac{1 + 3h + 3h^2 + h^3 - 1}{h} = 3 + 3h + 3h^2.$$

The gradient of the tangent PT is $m = \lim_{h \to 0} g(h) = 3$, so the equation of PT is
$y - 1 = 3(x - 1)$ or $y = 3x - 2$.
(ii) Similarly, the gradient of the tangent at (a, a^3) is

$$\lim_{h \to 0} \frac{(a+h)^3 - a^3}{h} = \lim_{h \to 0} \frac{3a^2 h + 3ah^2 + h^3}{h} = \lim_{h \to 0} (3a^2 + 3ah + h^2) = 3a^2.$$

Thus, the equation of the tangent is $y = a^3 + 3a^2(x - a)$ or $y = 3a^2 x - 2a^3$. (This was the method used in §4.5.)

EXERCISE 18.3

1   The function f is given by $f(x) = x^3 - 4x^2 + 3x + 5$. Express this in the form $f(x) = ((x - 4)x + 3)x + 5$ and hence calculate $f(x)$ for various values of x in order to sketch the graph of $f(x)$ between $x = 1$ and $x = 3$. Complete the following table in order to estimate the gradient of the tangent to the graph at $x = 2$; find the equation of this tangent and draw the tangent on your graph.

Coordinates of P		Coordinates of Q		Gradient of PQ
x	$f(x)$	x	$f(x)$	NQ/PN
2	3	3	5	2
		2·5	3·125	0·5
		2·1		
		2·01	2·990201	−0·9799
		2·001		
		2·0001		

2 Use two methods, that of the example above and that of question **1**, to find the equation of the tangent to the curve $y = x^2 - 1$ at $(2, 3)$.

3 Find the equation of the tangent to the curve $y = f(x)$ at P, where:
   (i) $f(x) = 2x(x - 1)$ and P is the point $(1, 0)$;
   (ii) $f(x) = 3x^2 - x^3$, and P is the point $(a, 3a^2 - a^3)$.

18.4 Linear approximation to a function

In §18.3, the gradient m of the tangent to $y = f(x)$ at $x = a$ is given by

$$m = \lim_{h \to 0} g(h), \quad \text{where} \quad g(h) = \frac{f(a+h) - f(a)}{h}, \quad h \neq 0.$$

If we now extend the domain of g by defining $g(0) = m$, then g becomes a *continuous* function at $h = 0$, and $f(a+h) = f(a) + h\,g(h)$. We now put $a + h = x$ and $f(x) = f(a+h) = y$, and compare the equation of the graph of f with the equation of the tangent PT at P, $(a, f(a))$:

equation of the graph of f $y = f(x) = f(a) + (x - a)g(h), \quad h = x - a,$

equation of the tangent at P, $y = f(a) + m(x - a), \quad m = g(0).$

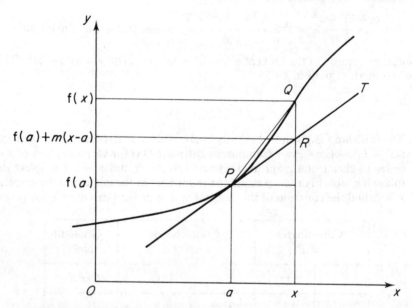

Fig. 18.3 The tangent as a linear approximation to a function at P.

Let P and Q be the points $(a, f(a))$ and $(x, f(x))$ on the graph of f and let R be the point $(x, f(a) + m(x - a))$ on the tangent PT with the same x-coordinate as Q (Fig. 18.4). The difference between the y-coordinates of Q

and R is

$$QR = f(x) - f(a) - m(x-a) = (g(h) - m)(x-a) = h(g(h) - m).$$

As h tends to zero, $g(h) - m$ also tends to zero, since $g(h)$ tends to m, so QR tends to zero even more rapidly. This indicates that the function

$$f(a) + m(x-a),$$

whose graph is the tangent PT, is a good linear approximation to the function f near to the point $x = a$. We shall refer to this function as *the linear approximation* to f near $x = a$.

EXAMPLE   *Find the linear approximation to the function f near the point $x = 1$, where (i) $f(x) = x(x+2)$,   (ii) $f(x) = x^4 - 2x^2$.*

(i) $f(x) = x(x+2)$, $f(1) = 3$. We manipulate $f(1+h)$ into the desired form $f(1) + h\, g(h)$, where $g(h)$ is continuous at $h = 0$.

$$f(1+h) = (1+h)(3+h) = 3 + 4h + h^2 = f(1) + h(4+h),$$

so that $g(h) = 4 + h$. Thus $m = g(0) = 4$ and the linear approximation to f near $x = 1$ is

$$f(1) + m(x-1) = 3 + 4(x-1) = \mathbf{4x-1}.$$

(ii) $f(x) = x^4 - 2x^2$, $f(1) = -1$. Similarly,

$$f(1+h) = (1+h)^4 - 2(1+h)^2 = 1 + 4h + 6h^2 + 4h^3 + h^4 - 2(1 + 2h + h^2)$$
$$= -1 + 4h^2 + 4h^3 + h^4 = f(1) + h g(h)$$

where $g(h) = 4h + 4h^2 + h^3$, so that $m = g(0) = 0$. The linear approximation to f near $x = 1$ is thus the constant function taking the value $\mathbf{f(1) = -1}$.

EXERCISE 18.4

1   Find the linear approximation to the function f given by
   (i) $f(x) = x^2 - 1$, near to the point $x = 2$,
   (ii) $f(x) = 2x(x-1)$, near to the point $x = 1$,
   (iii) $f(x) = 3x^2 - x^3$, near to the point $x = a$.
   Compare your answers with those you obtained in Exercise 18.3, questions **2** and **3**.

18.5 The gradient of a function

The gradient of the function f at the point $x = a$ is the gradient of the graph of f at P, $(a, f(a))$ (see §4.3), that is, the gradient m of the tangent PT to the graph at P.

   In §18.4 we saw that this gradient m can be derived as a limit in two ways:

$$m = \lim_{h \to 0} \frac{f(a+h) - f(a)}{h} \qquad \text{18.5A}$$

or $m = g(0)$, where $g(h)$ is the continuous function given by equation

$$f(a+h) = f(a) + h\,g(h). \qquad\qquad \text{18.5B}$$

These two forms for the gradient m of f at a are clearly equivalent. The first is useful in some calculations of m for particular functions, and it relates the gradient m of f to the limit of the gradient of a chord PQ of the graph of f. The second form has great advantages in some proofs of future results, see §§18.6–18.10, both in the calculation of m for some simple functions and in the proof of theorems.

Equation 18.5B also has the advantage that, as the definition of the gradient of a function, it generalises to functions of more than one variable, whereas 18.5A does not.

EXERCISE 18.5

1   Find the gradient of the function f at the point $(a, f(a))$, where:
  (i) $f(x) = 3x$ and $a = 3$ (ii) $f(x) = x + 5$ and $a = -2$
  (iii) $f(x) = x^2 - 1$ and $a = 2$ (iv) $f(x) = (x+1)^2$ and $a = 3$
  (v) $f(x) = (x+1)^3$ and $a = -1$ (vi) $f(x) = x^3 - 3x^2$ and $a = 1$.

18.6 The derived function

Some of the material of Chapter 4 is now dealt with more formally and, for completeness, the definitions are repeated.

When the graph of a function f is continuous and smooth, a *tangent* exists at every point on the graph, so that the gradient is defined for each value a in the domain of f. This means that we have a function, with the same domain as f, which takes as its value at a point a in the domain, the *gradient* of f at a. This function is called the *derived function* of f and is denoted by f′, so that

$$f'(a) = m, \text{ the gradient of f at the point } x = a,$$

and f′(x) is called the *derivative* of $f(x)$.

If we replace a by x in 18.5 A and B, we see that f′(x) may be defined in two equivalent ways:

$$f'(x) = \lim_{h \to 0} \frac{f(x+h) - f(x)}{h} \qquad\qquad \text{18.6A}$$

or f′$(x) = g(0)$, where $g(h)$ is continuous at $h = 0$ and

$$\text{18.6B}$$
$$f(x+h) = f(x) + h\,g(h).$$

There is another notation for the derivative, due to Leibnitz, which is in common use, and this is now defined, repeating the comments of 4.3.

Suppose that f′ is the derived function of the function f and that

$y = f(x)$, then we write

$$f'(x) = \frac{dy}{dx}.$$

Here, the three letters $\frac{d}{dx}$ (in words: dee by dee x) form a single mathematical symbol, which is a differential operator, operating upon y. They cannot be separated and $\frac{dy}{dx}$ must *not* be regarded as an algebraic fraction.

The above is now illustrated by finding the derivatives of some simple functions.

(a) *The constant function.* Let $f(x) = c$, a constant. Then

$$f(x+h) = c = f(x) = f(x) + hg(h),$$

where $g(h) = 0$ for all h, so $f'(x) = 0$, and f' is the zero function, that is, **if $y = c$, $\frac{dy}{dx} = 0$.**

(b) *The identity function.* Let $f(x) = x$. Then

$$f(x+h) - f(x) = x + h - x = h = hg(h),$$

where $g(h) = 1$. So $g(0) = f'(x) = 1$, that is **if $y = x$ then $\frac{dy}{dx} = 1$.**

(c) *The square function*, given by $f(x) = x^2$. In this case

$$f(x+h) - f(x) = (x+h)^2 - x^2 = x^2 + 2xh + h^2 - x^2$$
$$= 2xh + h^2 = h(2x+h) = hg(h)$$

where $g(h) = 2x + h$. Thus $f'(x) = g(0) = 2x$, that is,
if $y = x^2$ then $\frac{dy}{dx} = 2x$.

(d) Similarly, let $f(x) = x^4$ and proceed as above.

$$f(x+h) - f(x) = 4x^3h + 6x^2h^2 + 4xh^3 + h^4 = h(4x^3 + 6x^2h + 4xh^2 + h^3)$$
$$= hg(h),$$

where $g(h) = 4x^3 + 6x^2h + 4xh^2 + h^3$ and $g(0) = f'(x) = 4x^3$. In the Leibnitz notation, **if $y = x^4$ then $\frac{dy}{dx} = 4x^3$.**

We now consider some negative powers of x and in the next two examples we assume that both x and $x + h$ are non-zero.

(e) Let $f(x) = x^{-1}$, then $f(x+h) - f(x) = \dfrac{1}{x+h} - \dfrac{1}{x} = \dfrac{-h}{x(x+h)} = hg(h),$

where $g(h) = \dfrac{-1}{x(x+h)}$ and $g(0) = f'(x) = \dfrac{-1}{x^2}$, that is,

if $y = x^{-1}$ then $\frac{dy}{dx} = -x^{-2}$.

(f) Similarly, let $f(x) = x^{-2}$, then $f(x+h) - f(x) = \dfrac{-2xh - h^2}{x^2(x+h)^2} = hg(h)$,

with $g(h) = \dfrac{-2x - h}{x^2(x+h)^2}$ so that $g(0) = f'(x) = \dfrac{-2x}{x^4} = \dfrac{-2}{x^3}$, that is,

if $y = x^{-2}$ then $\dfrac{dy}{dx} = -2x^{-3}$.

EXERCISE 18.6

1   Using the above method, find $f'(x)$ in the cases
   (i) $f(x) = x^3$,   (ii) $f(x) = x^5$,   (iii) $f(x) = x^{-3}$.
   Do these results suggest to you what function is the derivative of x^n, for any integer n?

18.7 The derivative of x^n

The derivative of x^n is nx^{n-1}, for any integer n. This result was assumed in rule 1 of §4.7, and we now prove it, using the definition 18.6B. We need to quote an algebraic identity, which is easily verified by multiplication, namely for a positive integer n,

$$a^n - b^n = (a-b)(a^{n-1} + a^{n-2}b + \ldots + b^{n-1}).$$

Now let $f(x) = x^n$, then on putting $a = x+h$, $b = x$,

$$f(x+h) - f(x) = (x+h)^n - x^n$$
$$= h[(x+h)^{n-1} + x(x+h)^{n-2} + \ldots + x^{n-2}(x+h) + x^{n-1}]$$
$$= h\,g(h),$$

where $g(h) = (x+h)^{n-1} + x(x+h)^{n-2} + \ldots + x^{n-2}(x+h) + x^{n-1}$, which is a sum of n terms.

Now $g(h)$ is a polynomial function of h and so is continuous, and as $h \to 0$, $g(h) \to g(0) = nx^{n-1}$. Thus, by 18.5B

$$f'(x) = nx^{n-1}.$$

This gives the required result for any positive integer n, and for $n = 0$ we already know that $\dfrac{d}{dx}(1) = 0$.

For a negative integer n, let $m = -n$, so that $m > 0$, and let $f(x) = x^n = 1/x^m$. Then

$$f(x+h) - f(x) = \frac{1}{(x+h)^m} - \frac{1}{x^m}.$$

$$= \frac{1}{(x+h)^m x^m}(x^m - (x+h)^m)$$

$$= \frac{-h}{(x+h)^m x^m}(x^{m-1} + x^{m-2}(x+h) + \ldots + (x+h)^{m-1})$$

on putting $a = x$ and $b = x+h$.

Thus
$$f(x+h)-f(x) = hg(h)$$

where $g(h) = \dfrac{-(x^{m-1}+x^{m-2}(x+h)+ \ldots +(x+h)^{m-1})}{(x+h)^m x^m}$,

assuming that $x \neq 0$, $x+h \neq 0$.

Then $g(h)$ is continuous at $h = 0$, $g(0) = \dfrac{-mx^{m-1}}{x^{2m}}$ and on putting
$n = -m$, $f'(x) = g(0) = nx^{n-1}$.

We have, therefore, proved that

$$\frac{d}{dx}x^n = nx^{n-1} \qquad 18.7$$

for any integer n. Note that for negative values of n, the point $x = 0$ is not in the domain of the function x^n and so the condition $x \neq 0$ applies, and also, for small enough values of h, $x+h \neq 0$.

18.8 Derivative of a linear combination of two functions

We now prove rule 1 of §4.7, beginning with the derivative of the sum of two functions.

Suppose that g and G are continuous functions and that

$$f(x+h)-f(x) = hg(h), \quad F(x+h)-F(x) = hG(h),$$

so that $f'(x) = g(0)$ and $F'(x) = G(0)$. Let the sum of the functions f and F be $k = f+F$. Then

$$\begin{aligned}
k(x+h)-k(x) &= f(x+h)+F(x+h)-f(x)-F(x)\\
&= f(x+h)-f(x)+F(x+h)-F(x)\\
&= hg(h)+hG(h) = h(g(h)+G(h)),
\end{aligned}$$

where $g(h)+G(h) = (g+G)(h)$ and $g+G$ is continuous. Therefore,

$$k'(x) = (g+G)(0) = g(0)+G(0) = f'(x)+F'(x),$$

so the derivative of the sum of two functions is the sum of their derivatives.

For a constant c,

$$\begin{aligned}
(cf)(x+h)-(cf)(x) &= cf(x+h)-cf(x)\\
&= chg(h) = h(cg(h))
\end{aligned}$$

and cg is a continuous function. Therefore,

$$(cf)'(x) = cg(0) = cf'(x)$$

so the derivative of a constant multiple of a function is that same multiple of its derivative.

These results are combined to give rule 1. Given constants a, b and functions f, F with $u = f(x)$, $v = F(x)$,

$$(af + bF)' = af' + bF'$$

or $$\frac{d}{dx}(au + bv) = a\frac{du}{dx} + b\frac{dv}{dx}.$$ 18.8

This result, together with equation 18.7, enables us to differentiate polynomial functions and functions involving negative powers.

EXAMPLE 1   *Find* $\dfrac{d}{dx}\left(x + \dfrac{1}{x}\right)^3.$

Expanding the cube,

$$\frac{d}{dx}\left(x + \frac{1}{x}\right)^3 = \frac{d}{dx}\left(x^3 + 3x + \frac{3}{x} + \frac{1}{x^3}\right)$$

$$= \frac{d}{dx}x^3 + 3\frac{d}{dx}x + 3\frac{d}{dx}\frac{1}{x} + \frac{d}{dx}\frac{1}{x^3}$$

$$= 3x^2 + 3 - \frac{3}{x^2} - \frac{3}{x^4}.$$

EXAMPLE 2   *Given that* $f(t) = t^3 - t + \dfrac{1}{t}$, *find the derivative of* f *and the gradient of* f *when* $t = 2$.

Let $y = f(t)$ then $f'(t) = \dfrac{dy}{dt} = 3t^2 - 1 - \dfrac{1}{t^2}$, since we have to differentiate with respect to t. Then $f'(2) = 12 - 1 - \frac{1}{4} = 10\frac{3}{4}$.

EXERCISE 18.8

1   Differentiate with respect to x:

(i) x^6,   (ii) x^{-5},   (iii) $3x^4$,   (iv) $\dfrac{5}{x}$,   (v) $\dfrac{3}{x^3}$,   (vi) $x^2 - 5x + 1$,

(vii) $3x^4 - 2x^3 + 5x - 1$,   (viii) $(x - 1)^3$,   (ix) $(3x + 4)^2$,   (x) $\left(x - \dfrac{1}{x}\right)^2.$

2   Find the derivative of: (i) $x^4 - \frac{1}{2}x^3 + 5x$,   (ii) $2x^2 + 4x^5$,   (iii) $(3x + 2)^2$,

(iv) $(x + 1)^3 - (x - 1)^3$,   (v) $\left(x + \dfrac{1}{x}\right)^2.$

3   Find the gradient of the tangent to the curve, with equation $y = 2(x - 1)^4$ at the point $(2, 2)$ and write down the equation of the tangent to the curve at this point.

4   The displacement of a particle P from O, which is moving on the straight line OQ, is x. At time t, $x = t^3 - 12t^2$. Find the speed of P at time t and the distances from O at which the particle is at rest.

18.9 Derivative of the composition of functions

It often occurs that more complicated functions are built up from simpler functions by means of composition and it is then helpful to have a rule for the derivative of the composite function.

Suppose we have two functions f, F and wish to determine (Ff)′ in terms of f, F, f′ and F′.

Let
$$f(x+h) - f(x) = h\,g(h), \quad \text{with} \quad g(0) = f'(x)$$
$$F(y+k) - F(y) = k\,G(k), \quad \text{with} \quad G(0) = F'(y).$$

Then
$$Ff(x+h) - Ff(x) = F(f(x) + h\,g(h)) - F(f(x)),$$
and
$$F(f(x) + h\,g(h)) - F(f(x)) = h\,g(h)\,G(h\,g(h)),$$

on replacing y by f(x) and k by hg(h) in the equation two lines above. Hence
$$Ff(x+h) - Ff(x) = h\,g(h)\,G(h\,g(h))$$

and so
$$(Ff)'(x) = g(0)\,G(0) = f'(x)\,F'(y)$$
$$= f'(x)\,F'(f(x))$$

Thus
$$(Ff)' = f'.(F'f).$$

In the Leibnitz notation, this formula becomes the *chain rule*. Let $u = f(x)$, $y = F(u) = Ff(x)$. Then

$$\frac{du}{dx} = f'(x), \quad \frac{dy}{du} = F'(u) = F'(f(x)) = F'f(x)$$

so that
$$(Ff)'(x) = \frac{dy}{dx} = f'(x) \times F'(u) = \frac{du}{dx} \cdot \frac{dy}{du}$$

or
$$\frac{dy}{dx} = \frac{dy}{du} \cdot \frac{du}{dx}.$$

EXAMPLE 1   *Given that* $y = 3u^5$ *and* $u = x^2 + 2$, *find* $\dfrac{dy}{dx}$.

In this case $y = 3(x^2 + 2)^5$ and our previous method would involve expanding this expression as a polynomial. However, using the chain rule,

$$\frac{dy}{du} = 15u^4, \quad \frac{du}{dx} = 2x,$$

hence
$$\frac{dy}{dx} = \frac{dy}{du} \cdot \frac{du}{dx} = 15u^4 \cdot 2x = 30x(x^2 + 2)^4.$$

Note that, since we require $\dfrac{dy}{dx}$, we eliminate u from the answer.

In words, the chain rule means 'differentiate the bracket, treating it as a single term, and then multiply by the derivative of the function inside the bracket'.

EXAMPLE 2   *Find the derivative of the function* f, *where* $f(x) = (x^2 - 2x)^7$.

Let $u = x^2 - 2x$ and $y = f(x) = (x^2 - 2x)^7 = u^7$. Then

$$\frac{dy}{dx} = \frac{dy}{du} \cdot \frac{du}{dx} = 7u^6 \cdot (2x - 2) = \mathbf{14(x-1)(x^2-2x)^6}.$$

EXAMPLE 3   *Find the derivative of:* (i) $\sqrt{x}$, (ii) $x^{\frac{1}{n}}$, (iii) $x^{\frac{m}{n}}$, *where* m, n *are integers and* n > 0.

(i) In this case, one might be tempted to use equation 18.7 and say $\sqrt{x} = x^{\frac{1}{2}}$ and so putting $n = \frac{1}{2}$ in equation 18.7, $\dfrac{d}{dx}\sqrt{x} = \frac{1}{2}x^{-\frac{1}{2}}$. The answer is, in fact, correct, for $x \neq 0$, but the method is wrong because equation 18.7 only applied to integer values of n. So we must use another method.

Suppose that $u = \sqrt{x} = x^{\frac{1}{2}}$, let $y = u^2$, so $y = x$. By the chain rule $\dfrac{dy}{dx} = \dfrac{dy}{du} \cdot \dfrac{du}{dx}$ and so $1 = 2u \cdot \dfrac{du}{dx}$ which gives

$$\frac{du}{dx} = \frac{1}{2u} = \frac{1}{2\sqrt{x}} = \frac{1}{2}x^{-\frac{1}{2}}.$$

(ii) Using a method, similar to the method used in (i), let $y = u^n$, where $u = x^{\frac{1}{n}}$, so $y = x$. Then $\dfrac{dy}{dx} = \dfrac{dy}{du} \cdot \dfrac{du}{dx}$ so that $1 = n\,u^{n-1}\dfrac{du}{dx}$ and hence

$$\frac{du}{dx} = \frac{1}{n\,u^{n-1}} = \frac{1}{n} \cdot \frac{1}{(x^{\frac{1}{n}})^{n-1}} = \frac{1}{n}\frac{1}{x^{(1-\frac{1}{n})}} = \frac{1}{n}x^{\frac{1}{n}-1}.$$

(iii) Using the above result, let $y = u^m$ where $u = x^{\frac{1}{n}}$. Then

$$\frac{d}{dx}x^{\frac{m}{n}} = \frac{dy}{dx} = \frac{dy}{du}\frac{du}{dx} = mu^{m-1} \cdot \frac{1}{n}x^{\frac{1}{n}-1} = \frac{m}{n}x^{\frac{m-1}{n}} \cdot x^{\frac{1-n}{n}} = \frac{m}{n}x^{\frac{m}{n}-1}.$$

Note that we have now proved that

$$\frac{d}{dx}x^r = rx^{r-1} \qquad \text{for all rational numbers } r.$$

EXERCISE 18.9

1   Given that $y = u^2$ and $u = 1 - 2x$, find $\dfrac{dy}{du}$ and $\dfrac{du}{dx}$ and hence find the derivative of $(1 - 2x)^2$. Check your result by expanding $(1 - 2x)^2$ as a polynomial and then differentiating it.

2 (i) Given that $y = u^3$ and $u = 3x - 2$, find $\dfrac{dy}{dx}$.

(ii) Given that $z = 3v^2$ and $v = 2t - t^3$, find $\dfrac{dz}{dt}$.

3 Differentiate the following with respect to x:
(i) $(3x+4)^2$, (ii) $(2x-1)^3$, (iii) $(x^2-3x)^4$, (iv) $(1+2x^3)^2$,
(v) $(3x-2x^2)^5$, (vi) $(3x^4-1)^4$, (vii) $(2x^2+3x-1)^3$, (viii) $(x^2-1)^{-2}$,
(ix) $(3x^2+2x)^{-3}$, (x) $(3x^3-2x^2+5x)^{-4}$.

4 Differentiate the following with respect to x:

(i) $\dfrac{2}{(2x-1)}$, (ii) $\dfrac{4}{(2x^2+3)}$, (iii) $\dfrac{1}{(3x^3-2x)^2}$, (iv) $\dfrac{5}{(3+4x^2)^3}$,

(v) $\dfrac{6}{(3+2x^2-x^4)^2}$, (vi) $\dfrac{1}{(x^3+2x^5)^3}$.

5 Differentiate these functions:
(i) $x^{\frac14}$, (ii) $5x^{\frac25}$, (iii) $x^{0.3}$, (iv) $3x^{-\frac23}$, (v) $\sqrt{(1+x^2)}$, (vi) $(2x^3-1)^{\frac23}$.

18.10 Derivatives of products and quotients of functions

Another way in which more complicated functions may be built up from other functions is by multiplication and division of functions. Thus, it is necessary to have rules for finding the derivatives of products and of quotients.

Suppose that for two functions f and F,

$$f(x+h) - f(x) = h\,g(h), \quad \text{with } g(0) = f'(x) \quad \text{and}$$
$$F(x+h) - F(x) = h\,G(h), \quad \text{with } G(0) = F'(x),$$

$g(h)$ and $G(h)$ being continuous. Then for the product function f.F, $(f.F)(x) = f(x)\,F(x)$, so

$$(f.F)(x+h) - (f.F)(x) = f(x+h).F(x+h) - f(x).F(x)$$
$$= \{f(x) + h\,g(h)\}\{F(x) + h\,G(h)\} - f(x).F(x)$$
$$= h\{f(x)\,G(h) + g(h)F(x) + h\,g(h)\,G(h)\}$$

Then $k(h) = f(x)\,G(h) + g(h)\,F(x) + h\,g(h)\,G(h)$ is continuous and

$$k(0) = f(x)\,G(0) + g(0)\,F(x) = f(x)\,F'(x) + f'(x)\,F(x)$$

and so $\quad (f.F)'(x) = k(0) = f(x)\,F'(x) + F(x)\,f'(x).$

Also, if $F(x) \neq 0$, we may in addition assume that $F(x+h) \neq 0$ for such small h, and then

$$\left(\frac{f}{F}\right)(x+h) - \left(\frac{f}{F}\right)(x) = \frac{f(x+h)}{F(x+h)} - \frac{f(x)}{F(x)}$$

$$= \frac{F(x)\{f(x)+h\,g(h)\} - f(x)\{F(x)+h\,G(h)\}}{\{F(x)+h\,G(h)\}F(x)}$$

$$= h\,r(h)$$

where $r(h) = \dfrac{F(x)g(h) - f(x)G(h)}{\{F(x)+h\,G(h)\}F(x)}$, a continuous function of h. Then

$$\left(\frac{f}{F}\right)'(x) = r(0) = \frac{F(x)f'(x) - f(x)F'(x)}{F(x)^2}.$$

If we now put $u = f(x)$ and $v = F(x)$, we obtain the following two rules.

Product rule $(f.F)' = f.F' + f'.F$ or $\dfrac{d}{dx}(uv) = u\dfrac{dv}{dx} + v\dfrac{du}{dx}.$

Quotient rule $\left(\dfrac{f}{F}\right)' = \dfrac{F.f' - f.F'}{F^2}$ or $\dfrac{d}{dx}\left(\dfrac{u}{v}\right) = \dfrac{v\dfrac{du}{dx} - u\dfrac{dv}{dx}}{v^2}.$

EXAMPLE   *Differentiate* (i) $(2x+3)(x+1)^2$, (ii) $\dfrac{2x+3}{(x+1)^2}$.

Let $u = 2x+3$ and $v = (x+1)^2$, then $\dfrac{du}{dx} = 2$ and $\dfrac{dv}{dx} = 2(x+1)$.

(i) $\dfrac{d}{dx}(2x+3)(x+1)^2 = (2x+3)(2x+2) + 2(x+1)^2 = 6x^2 + 14x + 8.$

(ii) $\dfrac{d}{dx}\dfrac{2x+3}{(x+1)^2} = \dfrac{2(x+1)^2 - (2x+3)(2x+2)}{(x+1)^4} = \dfrac{2(x+1) - 2(2x+3)}{(x+1)^3}$

$$= \frac{-2x-4}{(x+1)^3}.$$

EXERCISE 18.10

1   Differentiate the following with respect to x:
   (i) $(x+2)(2x-1)$,   (ii) $(3x-1)(1+2x)$,   (iii) $(3x^2-1)(x-4)$,
   (iv) $(3+x)(5x^3-2x)$,   (v) $(2x^2-x)(3x-2)$,   (vi) $(4x^2+2x)^2$,
   (vii) $(2x^2+2)(x-1)^2$,   (viii) $(x-1)^3(2x+1)^2$,   (ix) $(x^3-2)^2(x^2+4x)^{\frac{1}{2}}$,
   (x) $(3x^2-2x)^{\frac{1}{3}}(2x^2+5)^{\frac{1}{2}}$.

2   Differentiate the following with respect to x:
   (i) $\dfrac{x}{x+1}$,   (ii) $\dfrac{2x-1}{2x+3}$,   (iii) $\dfrac{3x^2}{2x-1}$,   (iv) $\dfrac{x^2+2x}{5-2x^2}$,   (v) $\dfrac{(x+2)^2}{(2x+4)}$,

   (vi) $\dfrac{x^3}{(3x+1)^2}$,   (vii) $\dfrac{\sqrt{x}}{1-\sqrt{x}}$,   (viii) $\dfrac{1-x^3}{(1-x)^3}$,   (ix) $\dfrac{(x+3)}{(x^2-1)^2}$,

   (x) $\dfrac{(x^2+2)}{(2x^3-1)^{\frac{1}{2}}}$,   (xi) $\sqrt{\dfrac{(x+2)^3}{(x+3)}}$,   (xii) $\sqrt{\dfrac{(1+x^3)^3}{(x^2-3)}}$.

3 Write down the rule for evaluating $\dfrac{d}{dx}(uv)$ and $\dfrac{d}{dx}(zw)$. By substituting $z = uv$,

show that

$$\frac{d}{dx}(uvw) = uv\frac{dw}{dx} + uw\frac{dv}{dx} + vw\frac{du}{dx}.$$

Deduce that $\dfrac{d}{dx}u^3 = 3u^2\dfrac{du}{dx}$, by putting $v = w = u$, and find the derivative of

$\left(\dfrac{1-x}{1+x}\right)^3$.

4 Use the product rule derived in question **3** above to differentiate the following functions:
(i) $(x+1)(2x-1)(3x^2-2)$,   (ii) $(3x^2-1)(x^3+1)(1-3x)$,
(iii) $x(x+1)(x+2)^2$,   (iv) $3x^2(2x^3-x)(3x-1)^{\frac{1}{2}}$.

5 Find the gradient of the function $f(x)$, where $f(x) = (2x-1)(x+2x^2)$, for $x = -2, 0, 2$. For which values of x is the gradient of f zero?

MISCELLANEOUS EXERCISE 18

1 Determine the limits:
(i) $\lim\limits_{x\to 2} \dfrac{x^4-2x^3}{x-2}$,   (ii) $\lim\limits_{x\to -1} \dfrac{x^2+x^{-1}}{1+x}$,   (iii) $\lim\limits_{x\to a} \dfrac{a^3-x^3}{a-x}$.

2 The function f is defined by $f(x) = \dfrac{x^2-4}{x^3-8}$ with domain $\mathbb{R}\backslash\{2\}$. Find a suitable value for $f(2)$ in order to extend f to a continuous function with domain $\mathbb{R}$.

3 Find the derivatives of the functions:
(i) $\dfrac{x^2+1}{x+1}$,   (ii) $\dfrac{2x^3+3x}{3x^2+6}$,   (iii) $\dfrac{ax+b}{cx+d}$,   (iv) $(x^2+4)\sqrt{(x+1)}$,
(v) $(2x+x^2)^5$,   (vi) $(x^2+2)(x-1)^6$.

4 Find a linear approximation to the function f at the given point and hence determine its gradient at that point:
(i) $f(x) = x^3-x$, at $(0,0)$;   (ii) $f(x) = x^3-x$, at $(1,0)$;   (iii) $f(x) = (x+1)^2$, at $(0,1)$.

5 Find those points on the graph of f where the tangent is parallel to the axis Ox if $f(x)$ is
(i) $x(1-x)$,   (ii) $x^2(1-x)^2$,   (iii) $\dfrac{x^2}{1-x}$.

6 Determine the points where the tangent to the graph is parallel to the line $y = x$, if the equation of the graph is:
(i) $y = x(1-x)$,   (ii) $y = (1+x)^2(1-x)$,   (iii) $y = x^2(1-x)$.

7 Differentiate, with respect to x:
(i) $(3x-2)^3$,   (ii) $2(x+1)^4$,   (iii) $(2x-1)^{\frac{1}{2}}$,   (iv) $(2x+3)^{-1}$,
(v) $x(x^2+2x)^{\frac{3}{4}}$,   (vi) $\dfrac{1}{3x^2+2}$,   (vii) $\dfrac{1}{(3x+2)^2}$,   (viii) $\dfrac{1}{\sqrt{(3x+2)}}$,
(ix) $(x^3-2x)^{-\frac{1}{4}}$,   (x) $\dfrac{1}{x^2\sqrt{(x^2+2)}}$.

8 Find $\dfrac{dy}{dx}$ given that:

(i) $y = (x^2 + 4)(x^3 - 2x)$, (ii) $y = \dfrac{1 + x^2}{(1 + x)^2}$, (iii) $y = \dfrac{x^2 + 4x}{\sqrt{(x - 1)}}$,

(iv) $y = \left(\dfrac{2 + 3x}{4 - 3x}\right)^{\frac{1}{2}}$, (v) $y = x^5 \sqrt{(2 + x^3)}$.

9 Given that $y = 3x^4 - x^2$, $x = t^3 + 2t + 3$, find $\dfrac{dy}{dx}, \dfrac{dx}{dt}, \dfrac{dy}{dt}$.

10 Given that $y = 4u^{-6}$, $u = x^4 + x^5$, find $\dfrac{dy}{dx}$.

19 Partial Fractions

19.1 Algebraic identities

Two functions f and g, with the same domain A, are equal if they have the same value at each point x of A. The notation of an identity is used, $f(x) \equiv g(x)$ to be read 'f(x) *is identical to* g(x)', where

$$f = g \Leftrightarrow f(x) \equiv g(x) \Leftrightarrow \text{for each } x \text{ in } A, \, f(x) = g(x).$$

For example, $x^2 - 1 \equiv (x-1)(x+1)$ and $\dfrac{2}{x^2 - 1} \equiv \dfrac{1}{x-1} - \dfrac{1}{x+1}$.

When f and g are polynomials, they are equal if they have the same coefficient of any power of x. Therefore, we may equate the coefficients, term by term, of equal polynomials. We have often simplified a sum of polynomials and polynomial fractions, for example,

$$x + 1 - \frac{2}{x-1} + \frac{3x-2}{x^2+1} \equiv \frac{x^4 + x^2 - 5x - 1}{(x-1)(x^2+1)}.$$

We need to be able to perform the reverse process in applications to integration and series in Chapters 24, 26 and 28. In this chapter, we consider techniques for splitting up a polynomial fraction into the sum of a polynomial and certain simple proper polynomial fractions. A polynomial fraction g(x)/h(x) is said to be *proper* if the degree of g is less than the degree of h.

19.2 Fractions with prime power denominators

There is a parallel to the above in arithmetic. Rational numbers may be written as sums of other rational numbers whose denominators are either prime numbers or the power of a prime. The denominators of the component fractions are '*prime power factors*' of the denominator of the rational number. For example, $\frac{1}{6} = \frac{1}{2} - \frac{1}{3}$ and $\frac{5}{12} = \frac{1}{2} + \frac{1}{4} - \frac{1}{3}$. We shall not be concerned with this process for rational numbers, but the reader may like to try to split up other rational numbers in this way.

In the algebra of polynomials, the polynomials which correspond to the prime numbers in arithmetic are those polynomials which cannot be factorised into a product of polynomials of lower degree. These are called *irreducible* polynomials. They are either linear polynomials or quadratic with no linear factors, see Chapter 15.

19.3 Partial fractions

A *partial fraction* is a proper polynomial fraction in which either (i) the denominator is a linear polynomial, or the power of a linear polynomial, and the numerator is a constant; or (ii) the denominator is an irreducible quadratic polynomial, or a power of an irreducible quadratic polynomial, and the numerator is linear. Thus they are polynomial fractions with prime power denominators. Examples of partial fractions so defined are

$$\frac{2}{3x-1}, \quad \frac{4x+1}{x^2+4}, \quad \frac{-2x}{(x^2+x+1)^2}.$$

Our concern is to split up a polynomial fraction into the sum of a polynomial and partial fractions. The separate terms will have unknown constant multipliers. After multiplying by a common denominator, the unknown constants are found either by equating coefficients of powers of x or by giving x suitable values.

19.4 No repeated factors in the denominator

EXAMPLE 1   *Express in partial fractions* $\dfrac{3x-2}{(x+1)(2x-5)}$.

The partial fractions are of the form: constant divided by linear. Suppose that

$$f(x) = \frac{3x-2}{(x+1)(2x-5)} \equiv \frac{A}{x+1} + \frac{B}{2x-5}.$$

Multiplying by $(x+1)(2x-5)$,

$$3x-2 \equiv A(2x-5) + B(x+1).$$

Put $x = -1$, $-3-2 = A(-2-5)$ so $A = 5/7$. Put $x = 5/2$, $3 \times 5/2 - 2 = B((5/2)+1)$ or $15/2 - 2 = (7/2)B$ so $B = 11/7$. Hence

$$f(x) = \frac{1}{7}\left(\frac{5}{x+1} + \frac{11}{2x-5}\right).$$

Check: compare the coefficients of x and the constant term.

EXAMPLE 2   *Express in partial fractions* $\dfrac{x^2+2x+3}{(x^2+1)(x-1)}$.

The partial fraction, corresponding to the *irreducible* quadratic factor x^2+1 in the denominator must have a *linear* numerator, so we put

$$f(x) = \frac{x^2+2x+3}{(x^2+1)(x-1)} \equiv \frac{Ax+B}{x^2+1} + \frac{C}{x-1}.$$

Multiplying by the denominator of $f(x)$,
$$x^2 + 2x + 3 \equiv (Ax + B)(x - 1) + C(x^2 + 1).$$
Put $x = 1$, $6 = 2C$ so $C = 3$. Put $x = 0$, $3 = -B + C$ so $B = -3 + C = 0$.
Equating the coefficients of x^2, $1 = A + C$, so $A = 1 - C = -2$. Thus

$$f(x) = \frac{-2x}{x^2 + 1} + \frac{3}{x - 1}.$$

EXAMPLE 3   *Express in partial fractions* $\dfrac{2x^2 - x + 3}{(x + 1)(x - 2)(x + 3)}$.

Let $f(x) = \dfrac{2x^2 - x + 3}{(x + 1)(x - 2)(x + 3)} \equiv \dfrac{A}{x + 1} + \dfrac{B}{x - 2} + \dfrac{C}{x + 3}$.

Multiply by $(x + 1)(x - 2)(x + 3)$. Then

$$2x^2 - x + 3 \equiv A(x - 2)(x + 3) + B(x + 1)(x + 3) + C(x + 1)(x - 2).$$

Put $x = -1$, $6 = -6A$ so $A = -1$. Put $x = 2$, $9 = 15B$ so $B = \frac{3}{5}$. Put $x = -3$,
$24 = 10C$ so $C = \frac{12}{5}$. Thus

$$f(x) = \frac{-1}{x + 1} + \frac{3}{5(x - 2)} + \frac{12}{5(x + 3)}.$$

Note the use of making one of the factors vanish by the choice of x.

EXERCISE 19.4

1   Express as a single rational fraction:

(i) $\dfrac{1}{x + 2} - \dfrac{1}{x - 3}$,   (ii) $\dfrac{2}{x - 2} + \dfrac{3}{x + 3}$,   (iii) $\dfrac{4}{x + 2} - \dfrac{5}{x + 1}$,

(iv) $\dfrac{1}{x + 1} + \dfrac{2}{x^2 + 2}$,   (v) $\dfrac{2}{x^2 + x + 2} - \dfrac{3}{x + 3}$.

2   Express:

(i) $\dfrac{2x + 2}{(x - 1)(x + 3)}$ in the form $\dfrac{A}{x - 1} + \dfrac{B}{x + 3}$,

(ii) $\dfrac{1 - x}{(x - 2)(x + 5)}$ in the form $\dfrac{A}{x - 2} + \dfrac{B}{x + 5}$,

(iii) $\dfrac{-26}{(1 + x + x^2)(x - 3)}$ in the form $\dfrac{Ax + B}{1 + x + x^2} + \dfrac{C}{x - 3}$.

3   Express in partial fractions:

(i) $\dfrac{2x - 3}{(x - 1)(x - 2)}$,   (ii) $\dfrac{2}{x^2 - 1}$,   (iii) $\dfrac{x + 7}{(3 - 2x)(x^2 + 2)}$,   (iv) $\dfrac{x^2 - x - 2}{(x^2 + 1)(x - 1)}$,

(v) $\dfrac{x + 1}{(3x - 2)(2 - x)}$,   (vi) $\dfrac{3x^2 - 2}{(x + 1)(x^2 + x + 1)}$,   (vii) $\dfrac{1}{(x + 1)(x + 4)(x - 3)}$,

(viii) $\dfrac{2x^2 - x + 6}{(2x^2 + 3)(x^2 + 1)}$.

4   Check all your answers above by recombining the partial fractions.

19.5 Repeated factors in the denominator

The fraction $\dfrac{2x+3}{(x+1)^2}$ can be written

$$\frac{2x+3}{(x+1)^2} \equiv \frac{2(x+1)+1}{(x+1)^2} \equiv \frac{2}{x+1} + \frac{1}{(x+1)^2},$$

and this is the correct form as a sum of partial fractions. If the square $(ax+b)^2$ occurs in the denominator, then in the partial fractions we have two terms $\dfrac{A}{ax+b} + \dfrac{B}{(ax+b)^2}$. This can be extended to repeated quadratic factors and also to more than one repetition, but we shall only consider this case here.

EXAMPLE 1   *Express in partial fractions* $\dfrac{2x+3}{(x+2)^2}$.

Let $f(x) = \dfrac{2x+3}{(x+2)^2} \equiv \dfrac{A}{x+2} + \dfrac{B}{(x+2)^2} \equiv \dfrac{A(x+2)+B}{(x+2)^2}$.

Comparing the numerators, $2x+3 \equiv A(x+2)+B$. Put $x=-2$, $-4+3=B$, so $B=-1$. Put $x=0$, $3=2A+B=2A-1$ so $A=2$. Therefore,

$$f(x) = \frac{2}{x+2} - \frac{1}{(x+2)^2}.$$

EXAMPLE 2   *Express* $f(x)$ *in partial fractions where* $f(x) = \dfrac{-2x^2+14x-4}{(x^2+1)(x-3)^2}$.

The required form is $f(x) = \dfrac{-2x^2+14x-4}{(x^2+1)(x-3)^2} \equiv \dfrac{Ax+B}{x^2+1} + \dfrac{C}{x-3} + \dfrac{D}{(x-3)^2}$.

Multiply by the denominator of $f(x)$, to give the identity

$$-2x^2+14x-4 \equiv (Ax+B)(x-3)^2 + C(x^2+1)(x-3) + D(x^2+1).$$

Put $x=3$, $20=10D$, so $D=2$. Put $x=0$, $-4=9B-3C+D$, so $3B+2=C$. Put $x=1$, $8=4A+4B-4C+2D$ so $4A+4B-4C=4$, or $A+B-C=1$. Put $x=-1$, $-20=-16A+16B-8C+2D$ so $16A-16B+8C=24$, or $2A-2B+C=3$.

Eliminating A, $2(A+B-C)-(2A-2B+C)=4B-3C=2-3=-1$.

Substituting for C, $4B-3(3B+2)=-1$ so $-5B=5$ and $B=-1$.
Then $C=3B+2=-1$, and $A=1-B+C=1$. Finally,

$$f(x) = \frac{x-1}{x^2+1} - \frac{1}{x-3} + \frac{2}{(x-3)^2}.$$

EXERCISE 19.5

1 Write out as a single fraction:

(i) $\dfrac{2}{x-2} + \dfrac{3}{(x-2)^2}$, (ii) $\dfrac{4}{(x+1)^2} - \dfrac{2}{(x+1)}$, (iii) $\dfrac{2}{(2x-1)} - \dfrac{3}{(2x-1)^2}$,

(iv) $\dfrac{5}{(1-x)^2} + \dfrac{2}{(1-x)}$, (v) $\dfrac{3}{(x+2)} + \dfrac{1}{(x+2)^2} - \dfrac{2}{(x^2+1)}$.

2 Express in partial fractions:

(i) $\dfrac{x}{(x+1)^2}$, (ii) $\dfrac{x-1}{(2x-3)^2}$, (iii) $\dfrac{4x}{(2-x)^2}$, (iv) $\dfrac{3x+1}{(1-2x)^2}$, (v) $\dfrac{4-x}{(4x+1)^2}$,

(vi) $\dfrac{x^2-4x+18}{(4+x)(x-1)^2}$, (vii) $\dfrac{x^3-12x+17}{(x-2)^2(x^2-x+1)}$, (viii) $\dfrac{1-12x}{(x+2)(x-3)^2}$,

(ix) $\dfrac{5x^2+6x+7}{(1+x)^2(1-2x)}$.

19.6 Improper polynomial fractions

If we are given the polynomial fraction $g(x)/h(x)$, such that the degrees of the polynomials g and h are respectively m and n with $m \geqslant n$, then we must first divide g by h and obtain a quotient q and a remainder r of lower degree than h. This means that we have a polynomial identity

$$g(x) \equiv q(x)h(x) + r(x),$$

degree of $r(x)$ less than degree of $h(x)$. Then

$$\frac{g(x)}{h(x)} \equiv q(x) + \frac{r(x)}{h(x)},$$

where $r(x)/h(x)$ is proper and can be split into partial fractions.

The calculation of q and r can be done either by long division of the polynomials g and h or by finding the unknown coefficients of q and r. Both methods are shown.

EXAMPLE 1 *Express* f(x) *as the sum of a polynomial and a proper polynomial fraction where* $f(x) = \dfrac{x^3+2}{x^2-x}$.

Method 1: long division. We write the polynomial long division sum and include all the terms of each polynomial by inserting zero coefficients where necessary.

```
                       x + 1          quotient
        x² − x + 0 ) x³ + 0x² + 0x + 2
                     x³ −  x² + 0x
                     ─────────────
                          x² + 0x + 2
                          x² −  x + 0
                          ──────────
                                x + 2    remainder
```

Thus, $\dfrac{x^3+2}{x^2-x} \equiv x+1+\dfrac{x+2}{x^2-x}$.

Method 2: unknown coefficients.   Suppose that

$$x^3+2 \equiv (Ax+B)(x^2-x)+Cx+D,$$

since we know that the quotient must have degree 1 ($=3-2$) and that the remainder must be linear, that is, degree one less than the degree of the denominator. Put $x = 0$, $2 = D$. Put $x = 1$, $1+2 = C+D$ so $C = 3-D = 1$. Equating the coefficients of x^3 and of x^2, $1 = A$ and $0 = B-A$, so $B = 1$. Hence

$$\frac{x^3+2}{x^2-x}=x+1+\frac{x+2}{x^2-x}.$$

EXAMPLE 2   *Express* $\dfrac{x^3+2}{x^2-x}$ *in partial fractions.*

Using Example 1 to express the fraction as a polynomial and a proper fraction,

$$\frac{x^3+2}{x^2-x} \equiv x+1+\frac{x+2}{x^2-x}.$$

We next factorise the denominator, $x^2-x \equiv x(x-1)$, and then write

$$\frac{x+2}{x^2-x} \equiv \frac{A}{x}+\frac{B}{x-1}.$$

Multiply by x^2-x, then $x+2 \equiv A(x+1)+Bx$.
Put $x = 1$, $B = 3$. Put $x = 0$, $A = -2$. Finally

$$\frac{x^3+2}{x^2-x} \equiv x+1-\frac{2}{x}+\frac{3}{x-1}.$$

19.7 Summary of the procedure

In order to express the polynomial fraction $f(x) \equiv g(x)/h(x)$ in partial fractions:
A   if the degree of $g(x)$ is greater than or equal to the degree of $h(x)$, divide the polynomials to give a quotient $q(x)$ and a remainder $g_1(x)$ of degree less than the degree of $h(x)$, so that $g(x) \equiv q(x)h(x)+g_1(x)$ and $f(x) \equiv q(x)+\dfrac{g_1(x)}{h(x)}$, the sum of a polynomial and a proper fraction;
B   now assume that the degree of $g_1(x)$ is less than the degree of $h(x)$, factorise $h(x)$ into irreducible factors, and

(i) for each linear factor $ax+b$ use a partial fraction $\dfrac{A}{ax+b}$,

(ii) for each repeated linear factor $(ax+b)^2$ use partial fractions
$$\frac{A}{ax+b}+\frac{B}{(ax+b)^2},$$

(iii) for each irreducible quadratic factor ax^2+bx+c, with $b^2 < 4ac$, use
a partial fraction $\dfrac{Ax+B}{ax^2+bx+c}$;

C express f(x) as the sum of the partial fractions (using different constants for the different terms), multiply by h(x) to give a polynomial identity;

D solve for the unknown constants, either by giving x a number of suitable values or by comparing the coefficients of powers of x.

Note that the procedure can be extended to include repeated quadratic factors and factors repeated more than twice.

EXERCISE 19.7

1 Express as the sum of a polynomial and a proper polynomial fraction:

(i) $\dfrac{x^3+1}{x-1}$, (ii) $\dfrac{x^2}{x^2-1}$, (iii) $\dfrac{x^3-5x^2+3x}{(x+1)(x-2)}$,

(iv) $\dfrac{x^4-2x^3-10x^2+39x-18}{(x+4)(x-1)^2}$, (v) $\dfrac{x^3+x^2}{(x^2+x+2)(x-3)}$, (vi) $\dfrac{2x^3+5x^2}{(x+2)^2}$.

2 Express the polynomial fractions of question **1** in partial fractions. Check your results by recombining the fractions.

3 Express in partial fractions:

(i) $\dfrac{x}{(x+2)(x+3)}$, (ii) $\dfrac{5x+1}{(x-4)(x+3)}$, (iii) $\dfrac{3}{(2x+1)(x+1)}$,

(iv) $\dfrac{x^3-4x+5}{(x-2)(x+3)}$, (v) $\dfrac{x^3+6}{(x-1)(x+2)(x-3)}$.

4 Express in partial fractions:

(i) $\dfrac{x-1}{x(x+1)^2}$, (ii) $\dfrac{x^5}{x^4-x^2}$, (iii) $\dfrac{x+7}{2x^2+3x-2}$, (iv) $\dfrac{7x+4}{(x-3)(x+2)^2}$,

(v) $\dfrac{2x-2}{(x+1)^2(x^2+1)}$, (vi) $\dfrac{x}{x^3-19x-30}$.

5 Express $\dfrac{(x-1)}{x(x+1)}$ in partial fractions. (L)

6 Express $\dfrac{8-x}{(3x-2)(4x+1)}$ in partial fractions. (L)

7 Express $\dfrac{3x+2}{(x-3)(x^2+2)}$ in partial fractions. (L)

20 Integration II

20.1 Revision of Chapter 6

In Chapter 6, we met the inverse operation of differentiation, called integration. In this chapter, we consider integration from the point of view of the area under a graph. We start with a reminder of the results of chapter 6 and a revision exercise.

If the function $f(x)$ has an antiderivative function $F(x)$, then $F'(x) = f(x)$ and the *indefinite integral* of f is written

$$\int f(x)\,dx = F(x) + c,$$

where c is a constant of integration. In certain situations, this constant can be determined from a given condition to be satisfied by $F(x)$.

EXERCISE 20.1

1   Find an antiderivative function for the function:
(i) $4x^3 + 2x$, (ii) $3x^2 - 2$, (iii) $(x+2)^2$, (iv) $\sin x$, (v) $3\cos 2x$,
(vi) $\sqrt{x} - \dfrac{1}{\sqrt{x}}$, (vii) $\sqrt[3]{x} + \sqrt{(x^3)}$, (viii) $\left(x - \dfrac{1}{x}\right)^2$, (ix) $\sin(3x+2)$.

2   Find:
(i) $\int (4 - 2x + x^2)\,dx$, (ii) $\int (3 - 2x)(4x+2)\,dx$,
(iii) $\int (x^3 + 3x^2 - 2x + 5)\,dx$, (iv) $\int (\cos x + \cos 2x - 1)\,dx$,
(v) $\int (\sin x + \cos x)\,dx$, (vi) $\int \sqrt[4]{x}\,dx$, (vii) $\int (t^2 - 2t)\,dt$,
(viii) $\int (p^3 - 3p^2 + 8)\,dp$.

3   Find the indefinite integral $F(x)$, satisfying the given condition:
(i) $\int (2x^2 - 3x^3)\,dx$, $F(0) = 3$, (ii) $\int \cos 2x\,dx$, $F(\frac{\pi}{4}) = 1$,
(iii) $\int (2\sin 3x + 4)\,dx$, $F(0) = 2\pi$, (iv) $\int \left(\sqrt{x} + \dfrac{1}{\sqrt{x}}\right)dx$, $F(4) = 10$.

20.2 The definite integral

We begin by taking a further look at the constant of integration. Suppose that the velocity of a particle P is $v\,\mathrm{cm\,s^{-1}}$, given by $v = v(t) = 3t^2 + 2$ when the time is t seconds. Then, by means of integration, we can find a function $s(t)$ describing the distance in centimetres travelled by P in t

seconds, namely

$$s(t) = \int v \, dt = \int (3t^2 + 2) \, dt = t^3 + 2t + c,$$

where c is a constant of integration. This constant plays the part of fixing the starting point of the motion. For instance, if P is assumed to start from the origin when $t = 0$, then $s(0) = 0$ so that $c = 0$. However, if motion starts from a point $10\,\text{cm}$ away from the origin, then $s(0) = 10$ and so $c = 10$. In fact, in this type of problem, we are usually looking for the distance travelled by P during some interval of time, for example between $t = 3$ and $t = 5$. This distance is given by

$$s(5) - s(3) = (5^3 + 2 \times 5 + c) - (3^3 + 2 \times 3 + c)$$
$$= 135 + c - 33 - c = 102.$$

The constant c has been cancelled out from the answer by the subtraction. Therefore, we would have obtained the same answer if we had omitted the constant c from the expression for $s(t)$. This process of evaluating the *change* in the value of the antiderivative function is used so often that a notation is used, namely $s(5) - s(3) = [s(t)]_3^5$ and the result is known as a *definite integral*. Thus, the distance travelled between $t = 3$ and $t = 5$ is written as

$$\int_3^5 v(t) \, dt = \int_3^5 (3t^2 + 2) \, dt = \left[t^3 + 2t \right]_3^5$$
$$= [5^3 + 2 \times 5] - [3^3 + 2 \times 3] = 102.$$

Notice that the constant c of integration is omitted, since it would only be cancelled out in the subtraction. Also the two values of t (5 and 3) appear in the initial notation for the integral and also in the next step after integration before they are substituted in the antiderivative. These values of t correspond to the beginning and end of the time interval concerned and are called the *limits* of the integral.

Definition The definite integral of $f(x)$ between the limits $x = a$ and $x = b$ is

$$\int_a^b f(x) \, dx = \left[F(x) \right]_a^b = F(b) - F(a),$$

where $F(x)$ is the antiderivative of $f(x)$, that is, $F'(x) = f(x)$.

EXAMPLE *Find* (i) $\displaystyle\int_1^3 (2t + 4) \, dt$, (ii) $\displaystyle\int_0^{\frac{\pi}{2}} \sin 2\theta \, d\theta$.

(i) $\displaystyle\int_1^3 (2t + 4) \, dt = \left[t^2 + 4t \right]_1^3 = [3^2 + 4 \times 3] - [1^2 + 4 \times 1] = 21 - 5 = \mathbf{16}.$

(ii) $\displaystyle\int_0^{\frac{\pi}{2}} \sin 2\theta \, d\theta = \left[-\tfrac{1}{2} \cos 2\theta \right]_0^{\frac{\pi}{2}} = [-\tfrac{1}{2} \cos \pi] - [-\tfrac{1}{2} \cos 0] = \tfrac{1}{2} + \tfrac{1}{2} = \mathbf{1}.$

EXERCISE 20.2

1   Evaluate the following, remembering that $\left[F(x)\right]_a^b = F(b) - F(a)$:

(i) $\left[2x^2 + 3x\right]_1^3$,   (ii) $\left[4 - 2x\right]_1^2$,   (iii) $\left[x^3\right]_2^5$,   (iv) $\left[3x^3 - 2x\right]_0^3$,

(v) $\left[\dfrac{2}{x^2} - \dfrac{3}{x}\right]_1^6$,   (vi) $\left[\sin x\right]_0^{\frac{\pi}{2}}$,   (vii) $\left[(3x - 2)^2\right]_1^4$,   (viii) $\left[2 - x^2\right]_{-1}^1$,

(ix) $\left[x^3 + 3x - 4\right]_1^4$,   (x) $\left[\sqrt{(x^2 - 3)}\right]_2^5$.

2   Write down the antiderivative functions of the following, and check your results by differentiation:
(i) $x - 3$,   (ii) $2x + 6$,   (iii) $x^2 - 1$,   (iv) $3x^3$,   (v) $x^2 + x + 1$,   (vi) x^{-2},
(vii) $\dfrac{3}{x^2}$,   (viii) $\dfrac{4}{x^3}$,   (ix) $x^{\frac{1}{2}}$,   (x) $3\sqrt{x} - \sqrt[3]{x}$,   (xi) $\dfrac{1}{x\sqrt{x}}$,   (xii) $\cos x$.

3   Complete the following definite integral evaluations:

(i) $\displaystyle\int_1^2 (2x + 4)\,dx = \left[x^2 + 4x\right]_1^2 = [\qquad] - [\qquad] = \qquad$,

(ii) $\displaystyle\int_0^3 (3 - x)\,dx = \left[3x - \tfrac{1}{2}x^2\right]_0^3 = [\qquad] - [\qquad] = \qquad$,

(iii) $\displaystyle\int_1^4 (3x^2 - 2x)\,dx = \left[\qquad\right]_1^4 = [\qquad] - [\qquad] = \qquad$,

(iv) $\displaystyle\int_{-2}^2 (2 + x^2)\,dx = \left[\qquad\right]_{-2}^2 = [\qquad] - [\qquad] = \qquad$.

4   Evaluate:

(i) $\displaystyle\int_0^5 x^2\,dx$,   (ii) $\displaystyle\int_{-1}^1 (4 - x^2)\,dx$,   (iii) $\displaystyle\int_1^4 \sqrt{x}\,dx$,   (iv) $\displaystyle\int_{-2}^{-1} (2x^2 - x^3)\,dx$,

(v) $\displaystyle\int_{-4}^{-2} (4 + 2x)\,dx$,   (vi) $\displaystyle\int_0^{\frac{\pi}{4}} \sin x\,dx$,   (vii) $\displaystyle\int_{\frac{\pi}{6}}^{\frac{\pi}{3}} \cos x\,dx$,

(viii) $\displaystyle\int_{-8}^{-1} \dfrac{1}{\sqrt[3]{x}}\,dx$,   (ix) $\displaystyle\int_2^3 x^{-2}\,dx$,   (x) $\displaystyle\int_1^2 \left(1 - \dfrac{8}{x^3}\right)\,dx$.

5   Given that $f(x) = 3(x^2 - 4)$, find:
(i) $\displaystyle\int_{-4}^{-2} f(x)\,dx$,   (ii) $\displaystyle\int_{-2}^2 f(x)\,dx$,   (iii) $\displaystyle\int_2^4 f(x)\,dx$.

20.3 The area under a curve

Consider the velocity-time graph for a particle moving in a straight line with velocity v given, as a function of the time t, by $v = v(t) = at + u$, where

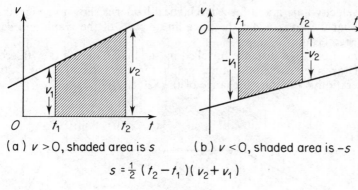

(a) $v > 0$, shaded area is s (b) $v < 0$, shaded area is $-s$

$$s = \tfrac{1}{2}(t_2 - t_1)(v_2 + v_1)$$

Fig. 20.1

a and u are constant. See Fig. 20.1(a) for $v > 0$, (b) for $v < 0$. If $s(t)$ is the distance moved in time t, and $s(0) = 0$, then $s(t) = \tfrac{1}{2}at^2 + ut$. Let s be the distance moved (with sign taken into account) in the time interval $t_1 < t < t_2$, then

$$
\begin{aligned}
s &= s(t_2) - s(t_1) = (\tfrac{1}{2}at_2^2 + ut_2) - (\tfrac{1}{2}at_1^2 + ut_1) \\
&= \tfrac{1}{2}a(t_2^2 - t_1^2) + u(t_2 - t_1) = \tfrac{1}{2}(t_2 - t_1)(at_2 + u + at_1 + u) \\
&= \tfrac{1}{2}(t_2 - t_1)(v_1 + v_2), \quad \text{where} \quad v_1 = v(t_1),\ v_2 = v(t_2).
\end{aligned}
$$

The shaded area in each of the two cases (a) and (b), bounded by the curve $v = v(t)$, the lines $t = t_1$ and $t = t_2$ and the line $v = 0$, is a trapezium with area S, given by

$$S = \tfrac{1}{2}(t_1 - t_2)(|v_1| + |v_2|) = |s|.$$

Thus, for $v > 0$, $s = S$ and, for $v < 0$, $s = -S$. The situation will be extended to the case when $v(t)$ is any function. We, therefore, define an area to be positive when above the t-axis and negative when below the t-axis. Then, in both cases, the distance travelled is equal to the area below the curve $v = v(t)$ between the times t_1 and t_2.

Definition   The *area under the curve* $y = f(x)$ between $x = a$ and $x = b$ is defined to be the area between the curve $y = f(x)$, the x-axis, and the lines $x = a$ and $x = b$, measured positively for areas above the axis Ox and negatively for areas below the axis Ox.

EXERCISE 20.3

1   A train accelerates steadily from rest for 10 seconds. During that time, the velocity v is given by $v = 2t$. The time t is measured in seconds and the velocity v is measured in metres per second.
(i) Draw a velocity-time graph for this motion.
(ii) Find how far the train travels during the time intervals (a) $0 < t < 10$, (b) $2 < t < 3$, (c) $5 < t < 8$.
(iii) Show on your graph, by shading, the areas which correspond to these distances.

2   The velocity-time graph for a mechanical loader is shown in Fig. 20.2.
   (i) Find the areas under the three linear parts of the graph and describe their significance.
   (ii) Find the total distance travelled by the loader during the 20 seconds.
   (iii) Draw the acceleration-time graph for the interval $0 \leqslant t \leqslant 20$, (the acceleration is the rate of change of the velocity).

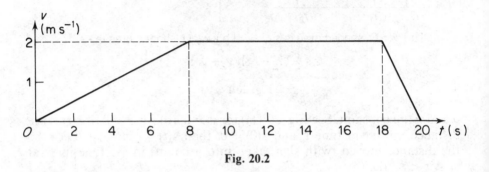

Fig. 20.2

20.4 The definite integral as an area

By extending the idea that the area under the velocity-time graph is the distance travelled, we can give a meaning to the definite integral in terms of the area under the graph of the function being integrated. We need to link these ideas with the use of the antiderivative as a means of calculating definite integrals. The important fact used is known as the *fundamental theorem of calculus*, which is now stated and proved.

***Theorem* 20.4**   Let the real function f have domain $\{x : a \leqslant x \leqslant b\}$. Let $F(x)$ denote the area under the graph of f between $A(a, f(a))$ and $P(x, f(x))$. Then $F' = f$ and the area under the graph of f between $x = a$ and $x = b$ is $\int_a^b f(x)\,dx$, provided that F has a derivative.

Proof   Let A, B, P and Q be the points on the graph of f where x takes the values a, b, x and $(x + h)$ respectively, see Fig. 20.3. The area under the curve $y = f(x)$ between A and P is $F(x)$ and is shown shaded in (a). The area under the curve between A and Q is $F(x + h)$ and the area between P and Q is $F(x + h) - F(x)$. This area is shown in (b) and lies between the areas mh and Mh of two rectangles, where m and M are the minimum and maximum values of $f(x)$ between P and Q. Thus, for $h > 0$,

$$mh < F(x + h) - F(x) < Mh \quad \text{or} \quad m < \frac{F(x + h) - F(x)}{h} < M.$$

Provided that f is a continuous function, as h tends to zero both m and M

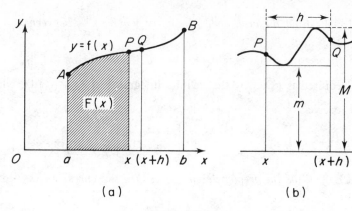

Fig. 20.3

will tend to $f(x)$, the value of f at P. Therefore, in the limit,

$$f(x) = \lim_{h \to 0} \frac{F(x+h)-F(x)}{h} = F'(x).$$

This means that F is an antiderivative of f and the area under the curve between $x = a$ and $x = b$ is

$$F(b)-F(a) = \int_a^b f(x)\,\mathrm{d}x.$$

In the proof of theorem 20.4, we assumed that $f(x)$ is positive for $a \leqslant x \leqslant b$ and, in this case, the integral $\int_a^b f(x)\,\mathrm{d}x$ is equal to the area between the graph of f and the lines $x = a$, $x = b$ and $y = 0$. In the case when $f(x)$ is negative, the theorem still holds because of our definition of the area under a curve, but the integral is now the *negative* of the area between the graph and the three lines.

Now suppose that $F(x)$ is any antiderivative of $f(x)$. Then the indefinite integral $\int f(x)\,\mathrm{d}x = F(x)+c$ can itself be interpreted as an area. Let $c = -F(x_0)$, then

$$F(x)+c = F(x)-F(x_0) = \int_{x_0}^x f(x)\,\mathrm{d}x = \int f(x)\,\mathrm{d}x$$

and the indefinite integral is the area under the graph of f between $x = x_0$ and $x = x$. There is confusion in the meaning of x in the above. This can be avoided by using another variable in drawing the graph of f. For example, the area under the graph $y = f(u)$ between $u = x_0$ and $u = x$ is

$$F(x)-F(x_0) = \int_{x_0}^x f(u)\,\mathrm{d}u = \int f(x)\,\mathrm{d}x.$$

The value of the constant c determines the left-hand edge of the area under consideration.

EXAMPLE 1   *Find the area under the curve* $y = 3x + \dfrac{2}{x^2}$ *between* $x = 2$ *and* $x = 5$.

The area required is given by the definite integral $\displaystyle\int_2^5 \left(3x + \dfrac{2}{x^2}\right) dx$, which is evaluated to give:

$$\left[\frac{3x^2}{2} - \frac{2}{x}\right]_2^5 = \left[\frac{3 \times 25}{2} - \frac{2}{5}\right] - \left[\frac{3 \times 4}{2} - \frac{2}{2}\right] = \frac{63}{2} + \frac{3}{5} = \mathbf{32 \cdot 1}.$$

EXAMPLE 2   *Find the area enclosed between the two curves* $y = x^2 - 6x + 10$ *and* $y = 10 - x$.

The first step is to find the points of intersection of the two curves by solving the equations simultaneously. Eliminating y between the equations,
$x^2 - 6x + 10 = 10 - x$ so $x^2 - 5x = 0$ or $x(x - 5) = 0$. The solutions are, therefore, $x = 0$ and $y = 10$ or $x = 5$ and $y = 5$, both pairs of values satisfying both equations. The situation is shown in a sketch in Fig. 20.4. Since

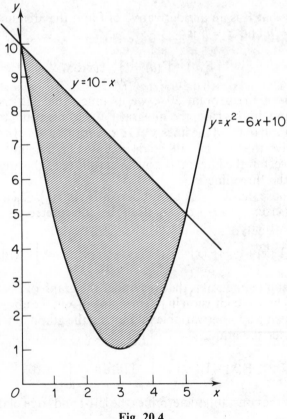

Fig. 20.4

$x^2 - 6x + 10 = (x - 3)^2 + 1$, the graph of the quadratic function is a parabola with vertex (downwards) at $(3, 1)$. The required shaded area is the difference between the area under the line $y = 10 - x$ between $x = 0$ and $x = 5$ and the area under the parabola between the same points. So the area is

$$\int_0^5 (10 - x)\,dx - \int_0^5 (x^2 - 6x + 10)\,dx = \left[10x - \tfrac{1}{2}x^2\right]_0^5 - \left[\frac{x^3}{3} - 3x^2 + 10x\right]_0^5$$

$$= [50 - 12\tfrac{1}{2} - 0] - [41\tfrac{2}{3} - 75 + 50 - 0] = 20\tfrac{5}{6}.$$

EXERCISE 20.4

1 Draw a graph to illustrate the area given by the definite integral and calculate the integral:

(i) $\displaystyle\int_1^3 (2x + 4)\,dx$, (ii) $\displaystyle\int_1^2 (3 - x)\,dx$, (iii) $\displaystyle\int_0^5 x^2\,dx$, (iv) $\displaystyle\int_{-1}^1 (4 - 2x^2)\,dx$,

(v) $\displaystyle\int_2^4 \sqrt{x}\,dx$.

2 Find the area under the graph of $f(x)$ between the given limits:

(i) $f(x) = x^2 + 3$, -2, 3; (ii) $f(x) = \dfrac{3}{x^2} - \dfrac{2}{x^4}$, 2, 5; (iii) $f(x) = \cos x$, $\tfrac{1}{4}\pi$, $\tfrac{1}{2}\pi$;

(iv) $f(x) = (x + 2)(x - 4)$, 0, 3.

3 Given that $f(x) = x^3 - 3x^2 - 6x + 8$, evaluate:

(i) $\displaystyle\int_{-2}^1 f(x)\,dx$, (ii) $\displaystyle\int_{-5}^{-2} f(x)\,dx$, (iii) $\displaystyle\int_1^4 f(x)\,dx$, (iv) $\displaystyle\int_4^6 f(x)\,dx$.

Factorise $f(x)$ and sketch the graph of $f(x)$. Explain why your answers to parts (ii) and (iii) are negative.

4 Find the area bounded by the x-axis and the graph of the function:

(i) $x^2 - 2x$, (ii) $(4 - x)(3 - x)$, (iii) $(x + 2)(7 - x)$, (iv) $x^2 - 5x + 6$,
(v) $x^3 - 6x^2$, (vi) $3 - x^2$.

5 Find the area bounded by the curves with equations:

(i) $y = x + 3$ and $y = 12 + x - x^2$, (ii) $y = 3x + 5$ and $y = x^2 + 1$,
(iii) $y = 3 - x^2$ and $y = 2x^2$, (iv) $y = x^2$ and $y = 3x$.

6 Verify that each of the curves $y^2 = 3x$ and $x^2 = 3y$ passes through the point $(3, 3)$ and find the area of the finite region bounded by these curves. (L)

7 Calculate the area of the finite region bounded by the curve, $y = x(4 - 3x)$ and the straight line $y = x$. (L)

20.5 Limits of integration

We prove some results concerning definite integrals and interpret the results in terms of the corresponding areas under the graph.

Theorem 20.5

A $\displaystyle\int_a^c f(x)\,dx = \int_a^b f(x)\,dx + \int_b^c f(x)\,dx.$

B $\displaystyle\int_a^b f(x)\,dx - \int_a^b g(x)\,dx = \int_a^b \{f(x) - g(x)\}\,dx.$

C $\displaystyle\int_a^b f(x)\,dx = -\int_b^a f(x)\,dx.$

Proof   These results all follow from the definition of a definite integral. Suppose that $F(x)$ and $G(x)$ are antiderivatives of $f(x)$ and $g(x)$. Then

$$\int_a^c f(x)\,dx = F(c) - F(a) = F(c) - F(b) + F(b) - F(a)$$

$$= F(b) - F(a) + F(c) - F(b) = \int_a^b f(x)\,dx + \int_b^c f(x)\,dx.$$

$$\int_a^b f(x)\,dx - \int_a^b g(x)\,dx = F(b) - F(a) - \{G(b) - G(a)\}$$

$$= F(b) - G(b) - \{F(a) - G(a)\} = (F - G)(b) - (F - G)(a)$$

$$= \int_a^b \{f(x) - g(x)\}\,dx.$$

$$\int_a^b f(x)\,dx = F(b) - F(a) = -\{F(a) - F(b)\} = -\int_b^a f(x)\,dx.$$

Interpretation   Regarding the definite integrals as the areas under the graph of f, between the given limits, result **A** shows that the area of the union of two regions which do not overlap is the sum of their areas. In Fig. 20.5(a), if $a < b < c$,

$$S_1 = \int_a^b f(x)\,dx, \quad S_2 = \int_b^c f(x)\,dx \quad \text{and} \quad S_3 = \int_a^c f(x)\,dx,$$

and clearly $S_1 + S_2 = S_3$. Result **B** is the one we used in §20.4, Example 2 to find the area between two curves and this is shown in Fig. 20.5(b). Result **C** shows that if the limits of integration are interchanged, then the integral changes sign. This enables us to interpret the integral $\int_a^b f(x)\,dx$ when $a > b$, since it is then the negative of the area under the graph of f between b and a. The change of sign is because the integration is 'backwards', in the direction of negative x.

EXAMPLE 1   *Find the area bounded by the curve $y = x(x^2 - 4)$ and the x-axis.*

We start by sketching the curve, which is the cubic curve $y = x(x - 2)(x + 2)$. It crosses the x-axis at the three points where x takes the values $0, 2$ and -2. Therefore, by symmetry, the area which we have to find is split into two regions which are equal in area, one above and one below the x-axis. This is shown in Fig. 20.6. Suppose that we first try to find the area by integrating y between the

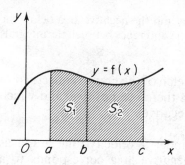

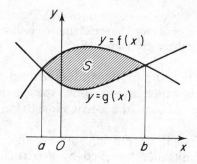

(a) Shaded area $S_3 = S_1 + S_2$

(b) $S = \int_a^b f(x)\,dx - \int_a^b g(x)\,dx$

Fig. 20.5

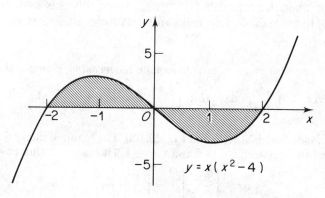

Fig. 20.6

limits -2 and 2. This will give us

$$\int_{-2}^{2} x(x^2 - 4)\,dx = \int_{-2}^{2} (x^3 - 4x)\,dx = \left[\tfrac{1}{4}x^4 - 2x^2\right]_{-2}^{2} = [\tfrac{16}{4} - 8] - [\tfrac{16}{4} - 8]$$
$$= 0.$$

The answer is zero, which can not be the area shaded, since that is clearly non-zero. So, we now try splitting up the range of integration into two parts, that where $y > 0$ and that where $y < 0$.

$$\int_{-2}^{0} (x^3 - 4x)\,dx = \left[\tfrac{1}{4}x^4 - 2x^2\right]_{-2}^{0} = [\tfrac{0}{4} - 2 \times 0] - [\tfrac{16}{4} - 8] = 4.$$

$$\int_{0}^{2} (x^3 - 4x)\,dx = \left[\tfrac{1}{4}x^4 - 2x^2\right]_{0}^{2} = [\tfrac{16}{4} - 8] - [\tfrac{0}{4} - 2 \times 0] = -4.$$

It is now clear why the first integral was zero. The integral between $x = -2$ and $x = 0$ is 4, which is the area of the first shaded area above the x-axis. The integral between 0 and 2 is -4, which is the negative of the second shaded area below the x-axis. The sum of these two integrals is zero which is the integral from $x = -2$ to

$x = 2$, because the positive area above the axis and the negative area below the axis cancel out. The area required is therefore the difference between the integrals last calculated, namely **8**.

This example illustrates the need of a sketch of the graphs concerned to show which area is above the x-axis (and therefore positive) and which area is below the x-axis (and therefore negative).

Note   If we are interpreting the area under a velocity-time graph as the distance travelled by a particle, a negative area corresponds to a distance travelled backwards. If this distance is the same as the distance previously travelled forward, the final displacement is zero but the total distance travelled (forwards and backwards) is non-zero.

Another need for a sketch graph is that we must ensure that $f(x)$ is continuous over the intervals concerned, since this was assumed in the proof of Theorem 20.4. The following example illustrates cases where trouble arises.

EXAMPLE 2   *Use sketch graphs to explain the meaning, if any, of the integral:*

(i) $\displaystyle\int_{-2}^{-1} \frac{1}{x^2}\, dx$,   (ii) $\displaystyle\int_{0}^{2} \frac{1}{x^2}\, dx$,   (iii) $\displaystyle\int_{-2}^{2} \frac{1}{x^2}\, dx$,   (iv) $\displaystyle\int_{-2}^{-1} \frac{1}{\sqrt{x}}\, dx$.

(i) The graph of $1/x^2$ is sketched in Fig. 20.7(a). The required integral is the left-hand shaded area between $x = -2$ and $x = -1$. This can be evaluated as follows,

$$\int_{-2}^{-1} \frac{1}{x^2}\, dx = \left[-\frac{1}{x} \right]_{-2}^{-1} = \left[\frac{-1}{-1} \right] - \left[\frac{-1}{-2} \right] = \tfrac{1}{2}.$$

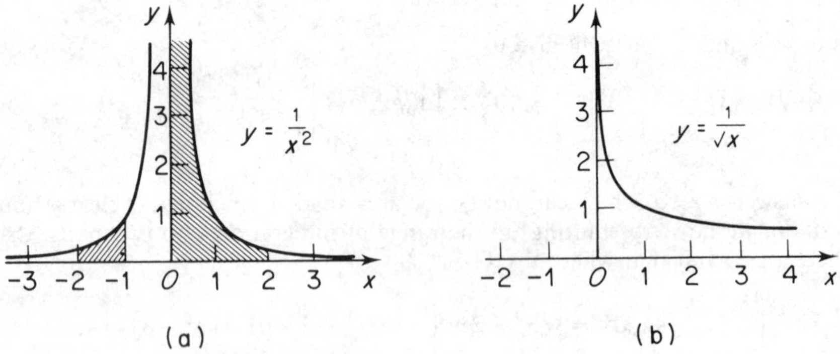

Fig. 20.7

(ii) The integral $\displaystyle\int_{0}^{2} \frac{1}{x^2}\, dx$ is the area between the graph of $\dfrac{1}{x^2}$, the x-axis and the lines $x = 0$ and $x = 2$, which is the second shaded region in Fig. 20.7(a). However, $f(x)$, given by $f(x) = 1/x^2$, is not defined for $x = 0$ and $f(x) \to \infty$ as $x \to 0$. In order to decide if the area concerned is bounded, so that some meaning may be attached to the integral, we evaluate the integral with the limits e and 2 and investigate its behaviour as $e \to 0$.

$$I(e) = \int_e^2 \frac{1}{x^2}\,dx = \left[-\frac{1}{x} \right]_e^2 = -\tfrac{1}{2} + \frac{1}{e}$$

and this tends to infinity as e tends to zero. This means that the area is unbounded (or infinite) and the integral has **no meaning**.

(iii) Suppose that we try to evaluate the integral using the antiderivative. Then

$$\int_{-2}^2 \frac{1}{x^2}\,dx = \left[-\frac{1}{x} \right]_{-2}^2 = \left[\frac{-1}{-2} \right] - \left[\frac{-1}{2} \right] = 1.$$

But this is nonsense, because we know from (ii) that the area between the graph, the x-axis and the lines $x = -2$ and $x = 2$ is infinite. What has happened here is that neither $1/x^2$ nor $1/x$ are defined at $x = 0$ and tend to infinity as x tends to zero. In relating a definite integral to the area under a curve we used the condition that $f(x)$ is continuous. This is not true in this case and so the theorem does not hold. Again, we can attach **no meaning** to the definite integral.

(iv) The graph of the function $1/\sqrt{x}$ is drawn in Fig. 20.7(b). Its domain is $\mathbb{R}^+$ so it is not defined for $-2 < x < -1$. Therefore, the integral has **no meaning**.

These examples show that care must be taken to check that the function being integrated is defined for the whole of the range of integration.

Summary Points to keep in mind when evaluating the integral $\int_a^b f(x)\,d(x)\,(= I)$ in order to find the area S of the region between the graph of f, the x-axis and the lines $x = a$ and $x = b$:

A make sure that $f(x)$ is defined at every point in the domain $\{x : a \leqslant x \leqslant b\}$;

B if $f(x)$ is positive in the domain, then $I = S$;

C if $f(x)$ is negative in the domain then $-I = S$;

D if $b < a$, then the domain is $\{x : b \leqslant x \leqslant a\}$ and the sign of I is changed, so that if $f(x) > 0$ then $-I = S$ and if $f(x) < 0$ then $I = S$.

EXERCISE 20.5

1 Evaluate the following definite integral and interpret the meaning of your result by sketching a graph:

(i) $\displaystyle\int_{-\frac{\pi}{2}}^{\frac{\pi}{2}} \sin x\,dx$,   (ii) $\displaystyle\int_{-3}^1 (x+1)\,dx$,   (iii) $\displaystyle\int_0^3 (x^2 - 2x)\,dx$,   (iv) $\displaystyle\int_{-2}^{-3} x^2\,dx$,

(v) $\displaystyle\int_4^2 \sqrt{x}\,dx$,   (vi) $\displaystyle\int_2^2 \frac{1}{x^2}\,dx$,   (vii) $\displaystyle\int_{\frac{\pi}{2}}^{\frac{3\pi}{2}} \cos x\,dx$,   (viii) $\displaystyle\int_{-2}^2 \frac{1}{x^3}\,dx$,

(ix) $\displaystyle\int_0^5 \frac{1}{\sqrt[3]{x}}\,dx$,   (x) $\displaystyle\int_{\frac{\pi}{3}}^{\frac{2\pi}{3}} \sin 3x\,dx$.

2 Find the area bounded by the curve $y = x^3 - 6x^2 + 11x - 6$ and the x-axis.

3 Find, as a function of a, with domain $\{a : a \geqslant 1\}$, $I(a) = \displaystyle\int_1^a \frac{1}{x^2}\,dx$. Evaluate

$I(2)$, $I(20)$, $I(200)$. Determine whether it is possible to give a meaning to $\displaystyle\lim_{a \to \infty} I(a)$ and, if so, state the value of this limit.

4 Find, as a function of k, with domain $\mathbb{R}^+$, the integral $I(k) = \displaystyle\int_0^k \sin x\,dx$.

Evaluate $I(\frac{\pi}{2})$, $I(\pi)$, $I(\frac{3\pi}{2})$, $I(2\pi)$, $I(4\pi)$, $I(5\pi)$. Determine whether any meaning can be attached to $\displaystyle\lim_{k \to \infty} I(k)$.

20.6 Numerical integration

Using the basic idea of an antiderivative, we can already integrate a number of different types of functions. Further techniques are given in Chapters 23 and 24 but this will still leave many functions which can not be integrated by any known method. In this situation, our only recourse is to find an approximate value to a definite integral by means of a numerical approximation to the area under a curve. Two techniques are dealt with here, of which the first is revision of work done in elementary mathematics and is included here for completeness.

The trapezium rule

In order to find an approximation to $\displaystyle\int_a^b f(x)\,dx$, we begin by choosing a number n and split up the required area under the curve into n equal strips. Each strip is of width h, where $h = \dfrac{b-a}{n}$. This is shown in Fig. 20.8(a), and

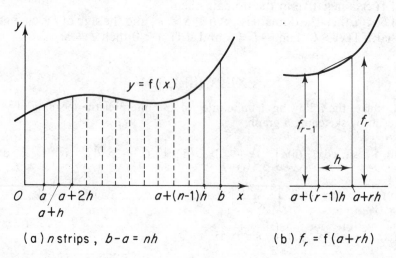

(a) n strips, $b-a = nh$ (b) $f_r = f(a+rh)$

Fig. 20.8

in Fig. 20.8(b) we show an enlarged picture of the rth strip, which is bounded by the lines $x = a + (r-1)h$ and $x = a + rh$. If the top of the strip is closed by a chord of the graph of f, the area under this chord is a trapezium of width h and sides of height f_{r-1} and f_r, where $f_r = f(a + rh)$. Note that $f_0 = f(a)$ and $f_n = f(b)$. The area of the trapezium is $\frac{1}{2}h(f_{r-1} + f_r)$ and this is an approximation to the area under the graph between $x = a + (r-1)h$ and $x = a + rh$. If we sum the areas of the n trapezia, we obtain an approximation to the integral;

$$\tfrac{1}{2}h(f_0 + f_1) + \tfrac{1}{2}h(f_1 + f_2) + \ldots + \tfrac{1}{2}h(f_{n-1} + f_n),$$

which may be written

$$\tfrac{1}{2}h(f_0 + 2f_1 + 2f_2 + \ldots + 2f_{n-1} + f_n)$$

Quite a good approximation can often be obtained by a few strips. It may be possible to improve the accuracy of the approximation by increasing the number of strips, but for a very large number of strips there will be a loss of accuracy due to rounding-off errors in the arithmetic. This can be overcome by doing the calculation to a larger number of significant figures but this is limited by the calculator used.

EXAMPLE 1   *Find an approximation to* $\displaystyle\int_0^1 \frac{1}{1+x^2}\,dx$ *by using the trapezium rule with:* (i) *four strips,* (ii) *eight strips.*

(i) $a = 0$, $b = 1$, $n = 4$, $h = \dfrac{1-0}{4} = 0.25$.   Using $f_r = f(a + rh) = f(0.25r)$,

$f_0 = f(0) = 1$, $f_1 = f(0.25) = 0.9412$, $f_2 = f(0.5) = 0.8$, $f_3 = f(0.75) = 0.64$,
$f_4 = f(1) = 0.5$. The approximation to the area (and, therefore, to the integral) is

$$\tfrac{1}{2} \times 0.25 \times (1 + 2 \times 0.9412 + 2 \times 0.8 + 2 \times 0.64 + 0.5) = 0.7828 = \mathbf{0.783}$$

to three decimal places.

(ii) $n = 8$, $h = \dfrac{1-0}{8} = 0.125$, and the values of f_r, $0 \leqslant r \leqslant 8$ are: $f_0 = 1.0000$,

$f_1 = 0.9846$, $f_2 = 0.9412$, $f_3 = 0.8767$, $f_4 = 0.8000$, $f_5 = 0.7191$, $f_6 = 0.6400$,
$f_7 = 0.5664$, $f_8 = 0.5000$. The approximation to the area, and the integral, is, therefore, $\tfrac{1}{2} \times 0.125 \times \{1 + 0.5 + 2(0.9846 + 0.9412 + 0.8767 + 0.8000 + 0.7191 + 0.6400 + 0.5664)\} = 0.7848 = \mathbf{0.785}$ to three decimal places.

Note A   We write the sum as the sum of f_0 and f_n and twice the sum of the intermediate values of the function.
B   It is necessary to work to four decimal places in order to obtain accuracy to three decimal places.

In Example 1, the exact value of the definite integral is $\pi/4$, that is, 0.7854 to four decimal places, so the trapezium rule is accurate to three decimal places with eight strips and is only in error by 2 in the third decimal place with four strips. This approximation to the integral could be used to calculate π to any desired accuracy by using enough strips, but allowance must be made for rounding errors in the calculation and the working must be made to many more significant figures than are required in the answer.

Simpson's rule

The trapezium rule approximates to the graph by chords so it uses a linear approximation to the function over many equal intervals. The second method is more sophisticated and approximates to the function by means of quadratic functions over pairs of strips (see Fig. 20.9). It is called *Simpson's rule*.

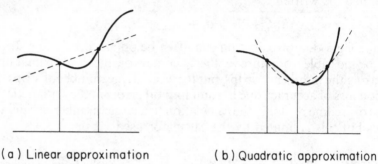

(a) Linear approximation   (b) Quadratic approximation

Fig. 20.9

Three points on the graph of f are used and a parabola is made to fit them. This compares with the use of two points, to which a line is fitted, in the trapezium rule. For ease of calculation, the origin is moved so that the y-axis passes through the centre point. Consider the three points $(-h, f(-h))$, $(0, f(0))$ and $(h, f(h))$ on the graph of the function f. We approximate to the graph of f by a parabola passing through these points, that is, we approximate to f by a quadratic function g, given by $g(x) = ax^2 + bx + c$. The graphs of f and g are shown in Fig. 20.10. We require that:

$$f(-h) = g(-h) = ah^2 - bh + c, \tag{1}$$
$$f(0) = g(0) = c, \tag{2}$$
$$f(h) = g(h) = ah^2 + bh + c. \tag{3}$$

The area under the graph of the quadratic function g is given by

$$S = \int_{-h}^{h} (ax^2 + bx + c)\,dx = \left[\frac{ax^3}{3} + \frac{bx^2}{2} + cx \right]_{-h}^{h}$$
$$= \frac{2}{3}ah^3 + 2ch = \frac{h}{3}(2ah^2 + 6c).$$

From equations 1 and 3, $f(-h) + f(h) = 2ah^2 + 2c$, and on using equation 2,

$$2ah^2 + 6c = f(-h) + 4f(0) + f(h).$$

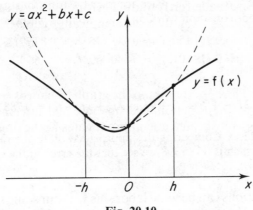

Fig. 20.10

The approximation to the area under the graph of f between $x = -h$ and $x = h$ is, therefore, S, given by

$$S = \frac{h}{3}[f(-h) + 4f(0) + f(h)].$$

We now divide the interval between $x = a$ and $x = b$ into n strips, where n is *even*. Use the previous notation that $f(a + rh) = f_r$, $h = \dfrac{b-a}{n}$, and apply the above argument repeatedly for pairs of strips. This then gives Simpson's rule for the approximation to the area under the graph of f between $x = a$ and $x = b$. That is, an approximation to $\displaystyle\int_a^b f(x)\,dx$ is:

$$\frac{h}{3}(f_0 + 4f_1 + f_2 + f_2 + 4f_3 + f_4 + \ldots + f_{n-2} + 4f_{n-1} + f_n)$$

$$= \frac{h}{3}(f_0 + 4f_1 + 2f_2 + 4f_3 + 2f_4 + \ldots + 2f_{n-2} + 4f_{n-1} + f_n).$$

Note The pattern of the coefficients of f_r in Simpson's rule is $1, 4, 2, 4, 2, \ldots, 2, 4, 1$ compared with the pattern $1, 2, 2, 2, \ldots, 2, 1$ in the trapezium rule and the factor outside the bracket is $h/3$ instead of $h/2$.

EXAMPLE 2 *Use Simpson's rule to find an approximation to* $\displaystyle\int_0^1 \frac{1}{1+x^2}\,dx$, *with: (i) four strips; (ii) eight strips.*

We can use the same data as we used in Example 1.

(i) The approximation is $\dfrac{0 \cdot 25}{3}(1 + 4 \times 0 \cdot 9412 + 2 \times 0 \cdot 8 + 4 \times 0 \cdot 64 + 0 \cdot 5) = 0 \cdot 7854$ or **0·785** to three decimal places.

(ii) Since we have three decimal place accuracy by using four strips, we calculate the eight strip approximation working to six decimal places, with the values:

$$f_0 = 1 \cdot 000\,000, f_1 = 0 \cdot 984\,150, f_2 = 0 \cdot 941\,176, f_3 = 0 \cdot 876\,712,$$
$$f_4 = 0 \cdot 800\,000, f_5 = 0 \cdot 719\,101, f_6 = 0 \cdot 640\,000, f_7 = 0 \cdot 566\,372, f_8 = 0 \cdot 500\,000.$$

Then
$$\frac{0 \cdot 125}{3}(f_0 + 4f_1 + 2f_2 + 4f_3 + 2f_4 + 4f_5 + 2f_6 + 4f_7 + f_8) = 0 \cdot 785\,3205 = \mathbf{0 \cdot 785\,32}$$

to five decimal places. Comparing this approximation to the value of $\pi/4$ to six decimal places, which is $0 \cdot 785\,398$ we see that the error in the estimate is of the order of 8×10^{-5}.

These calculations can become laborious without some sort of calculating device but they are quite easily programmed on a computer, if available. The need to retain sufficient figures during the calculation to ensure accuracy to the desired level cannot be overstressed, especially when using a large number of strips.

EXERCISE 20.6

1 Find approximations to $\displaystyle\int_1^5 (3x^3 + 2x - 5)\,dx$ using: (i) the trapezium rule with eight strips, (ii) Simpson's rule with eight strips. Compare the two answers with the exact answer obtained by integration.

2 Find an approximation to the integral, accurate to two decimal places, by the use of the trapezium rule:

(i) $\displaystyle\int_1^3 \frac{1}{2+x^2}\,dx$,   (ii) $\displaystyle\int_3^5 \frac{3x}{x^3-2}\,dx$,   (iii) $\displaystyle\int_0^{0 \cdot 8} \sin(x^2)\,dx$.

3 Repeat question 2 using Simpson's rule, increasing the accuracy to four decimal places.

4 Given that $\displaystyle\int_0^1 \frac{4}{(1+x^2)}\,dx = \pi$, use Simpson's rule with eight strips to find an approximate value of π.

5 Use Simpson's rule, with eight strips, to evaluate:

(i) $\displaystyle\int_1^3 \frac{1}{x}\,dx$,   (ii) $\displaystyle\int_1^2 \frac{1}{x}\,dx$,   (iii) $\displaystyle\int_1^6 \frac{1}{x}\,dx$.

Show that $\displaystyle\int_1^2 \frac{1}{x}\,dx + \int_1^3 \frac{1}{x}\,dx \approx \int_1^6 \frac{1}{x}\,dx$.

6 Use the trapezium rule with five strips of equal width to estimate, to three significant figures, the value of $\displaystyle\int_0^1 10^x\,dx$. Show all your working.

7 Estimate the value of $\displaystyle\int_1^{49} \frac{1}{3+\sqrt{x}}\,dx$ by applying the trapezium rule using three ordinates.

(L)

8 Use Simpson's rule with five ordinates to find an approximate value of

$$\int_2^4 \log_{10} x \, dx,$$

giving sufficient details of your working to indicate how your result has been obtained. Give your answer to two decimal places.   (*JMB*)

9 Tabulate, to three decimal places, the values of the function $f(x) = \sqrt{(1 + x^2)}$ for values of x from 0 to 0·8 at intervals of 0·1.

Use these values to estimate $\int_0^{0\cdot8} f(x) \, dx$

(a) by the trapezium rule, using all the ordinates,
(b) by Simpson's rule, using only ordinates at intervals of 0·2.   (*L*)

10 A tree trunk is of length 8 m. At a distance x m from one end, its cross-sectional area A m^2 is given by the following table:

x	0	2·0	4·0	6·0	8·0
A	0·6	0·8	1·1	1·5	2·0

Using the trapezium rule with five ordinates, estimate the volume, in m^3, of the tree trunk to one decimal place.   (*L*)

11 Estimate $\int_{\pi/6}^{\pi/2} \sqrt{(\sin x)} \, dx$ by Simpson's rule, using five ordinates and giving your answer to two decimal places.   (*L*)

12 The function g is an even function whose graph is a smooth curve. Use Simpson's rule to prove that

$$\int_{-0\cdot1}^{0\cdot1} g(x) \, dx \approx \tfrac{1}{15}\{g(0\cdot1) + 2g(0)\},$$

and deduce an approximation for $\int_0^{0\cdot1} g(x) \, dx.$ (*SMP*)

20.7 The definite integral as a limit of a sum

In the last section, we gave two procedures for numerical approximations to a definite integral, that is, to the area under a curve, by splitting up the area under the curve into a number of strips and calculating the total area of all these strips. This method predates the antiderivative method historically, and it provides an interesting and useful alternative definition for the definite integral.

Let f be a real function, defined on the domain $\{x : a \leqslant x \leqslant b\}$, and let

$$\int_a^b f(x) \, dx = I = \text{the area under the graph of f between } x = a \text{ and } x = b.$$

We split up the area under the graph into n strips, each of width δx, so that $\delta x = (b - a)/n$. Consider one strip bounded by $x = x_{r-1} = a + (r-1)\delta x$ and $x = x_r = a + r\delta x$, (Fig. 20.11). Let the maximum and minimum values

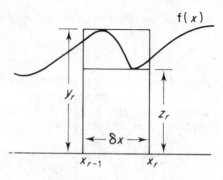

Fig. 20.11

of $f(x)$ for $x_{r-1} \leqslant x \leqslant x_r$ be y_r and z_r. Then the area which this strip contributes to I lies between the areas of the two rectangles of heights y_r and z_r and of width δx, summing over all the strips, we obtain two inequalities:

$$\sum_{r=1}^{n} z_r \, \delta x \leqslant I \leqslant \sum_{r=1}^{n} y_r \, \delta x.$$

When f is a continuous function, the difference $(y_r - z_r)$ between the heights of the two rectangles tends to zero as $n \to \infty$ and $\delta x \to 0$, and the two sums tend to the same limit, which must be I, since one sum is less than or equal to I and one is greater than or equal to I. The value of the function f at x_r may be written $f_r = f(x_r)$ and $z_r \leqslant f_r \leqslant y_r$ and we can regard I as the limit of either of the two sums above or of $\sum_{r=0}^{n} f_r \, \delta x$. In this way it is possible to define the definite integral as the limit of a sum:

$$\int_a^b f(x) \, dx = \lim_{n \to \infty} \sum_{r=1}^{n} f_r \, \delta x.$$

The full analysis of these limiting procedures is complicated and it has only been possible to outline the theory here. The concept of the integral as a limit of a sum is useful in many applications of integration such as in the calculation of means and volumes (§20.8, §20.9, §20.10). It is not very useful as a technique for calculating a definite integral because of the difficulty of finding the limit of the sum. Evaluation of the integral is best done by using the antiderivative. The following is an example of the use of the limit of the sum to evaluate the integral.

EXAMPLE   *Calculate* $\int_0^5 3x^2 \, dx$ *using rectangle sums.*

We try five strips, each of width 1, showing the rectangles in Fig. 20.12. We

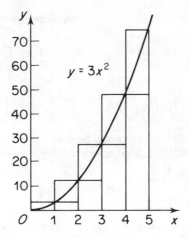

Fig. 20.12

calculate the upper and lower rectangle sums:

$$3 \times 1^2 \times 1 + 3 \times 2^2 \times 1 + 3 \times 3^2 \times 1 + 3 \times 4^2 \times 1 + 3 \times 5^2 \times 1 = 165$$
$$3 \times 0^2 \times 1 + 3 \times 1^2 \times 1 + 3 \times 2^2 \times 1 + 3 \times 3^2 \times 1 + 3 \times 4^2 \times 1 = 90.$$

Hence, the area under the curve lies between these sums, $90 < \displaystyle\int_0^5 3x^2 \, dx < 165$.

This does not give much information and the process needs to be repeated with a larger number of strips to give some closer bounds. Some results are:

> for ten strips the integral lies between 106·875 and 144·375
> for 20 strips the integral lies between 115·781 and 134·531.

This method only gives a rough idea of the value of the integral; it gives the limits between which the integral must lie. We would need 188 strips in order to obtain an accuracy of 1 per cent.

However, an algebraic method will give us the value of the integral by calculating the limits of the upper and lower sums. Divide the interval $0 \leqslant x \leqslant 5$ into n strips of width δx, with $\delta x = 5/n$. The rth strip has sides $x = (r-1)\delta x$ and $x = r\delta x$, so the heights of the upper and lower rectangles are $3r^2\delta x^2$ and $3(r-1)^2\delta x^2$ and the areas of the two rectangles are given by $3r^2\delta x^3 = 3r^2 5^3/n^3$ and $3(r-1)^2\delta x^3 = 3(r-1)^2 5^3/n^3$. Removing the common factor $3 \times 5^3/n^3$ from the two sums we obtain the upper rectangle sum $\dfrac{375}{n^3} \displaystyle\sum_{r=1}^{n} r^2$ and the lower rectangle

sum $\dfrac{375}{n^3} \displaystyle\sum_{r=1}^{n} (r-1)^2$. It is shown in §30.1 that

$$\sum_{r=1}^{n} r^2 = n(n+1)(2n+1)/6$$

and, from this result,

$$\sum_{r=1}^{n} (r-1)^2 = \sum_{r=0}^{n-1} r^2 = (n-1)n(2n-1)/6.$$

Therefore, the upper and lower rectangle sums can be written

$$375n(n+1)(2n+1)/6n^3 = \frac{125}{2}\left(1+\frac{1}{n}\right)\left(2+\frac{1}{n}\right)$$

and

$$375(n-1)n(2n-1)/6n^3 = \frac{125}{2}\left(1-\frac{1}{n}\right)\left(2-\frac{1}{n}\right).$$

As n tends to infinity, both these sums tend to 125, so this is the value of the integral. Check this result using an antiderivative.

EXERCISE 20.7

1   Use the result $\sum\limits_{r=1}^{n} r^3 = n^2(n+1)^2/4$ and the method of rectangle sums to
prove that $\int_0^3 4x^3\,dx = 81$.

2   Use the result $\sum\limits_{r=1}^{n} r(r+1) = n(n+1)(n+2)/3$ and the method of rectangle
sums to prove that $\int_0^4 (x^2+x)\,dx = 88/3$.

20.8 Mean values

The concept of the mean value of a function has important applications in mechanics and statistics. Consider the linear function f, given by $y = f(x) = \frac{1}{2}x+1$, with domain $\{x : a \leqslant x \leqslant b\}$, where $a < b$ (Fig. 20.13). The graph of f is the line segment AB, where A and B have coordinates $(a, f(a))$ and $(b, f(b))$. Since the graph is linear, the average value of f is $f\left(\dfrac{a+b}{2}\right)$, that is,

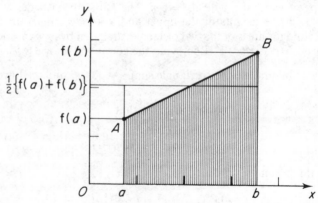

Fig. 20.13

$\frac{1}{2}\{f(a)+f(b)\}$. This is the average height of the graph between A and B and is equal to the height of a rectangle, with base $(b-a)$, which has the same area as the area under the graph between $x=a$ and $x=b$. Thus the average height is $\dfrac{1}{b-a}\displaystyle\int_a^b f(x)\,dx$. This is used as our definition of mean value.

Definition The *mean value* of the function f over the interval $\{x:a\leqslant x\leqslant b\}$, when $a<b$, is

$$\frac{1}{b-a}\int_a^b f(x)\,dx.$$

Notation If $y=f(x)$, the mean value of y is denoted by $\bar{y}$, so that

$$\bar{y}=\frac{1}{b-a}\int_a^b f(x)\,dx$$

The mean value of f can be thought of as the limit of a sum. Suppose that the area S under the curve $y=f(x)$ between $x=a$ and $x=b$ is divided into small strips, of width δx, and area δS. The area δS is given approximately by $\delta S\approx f(x)\,\delta x$, since the height of the strip is approximately $f(x)$, Fig. 20.11. Then the mean value of f between $x=a$ and $x=b$ is $\bar{y}$, given by

$$(b-a)\bar{y}=\lim\Sigma\,\delta S=\lim\Sigma f(x)\,\delta x=\int_a^b f(x)\,dx,$$

and $\bar{y}$ is the height of a rectangle on base $(b-a)$ and equal in area to S.

EXAMPLE *Find the mean value of (i)* $\cos 2x$, *(ii)* $\cos^2 2x$, *over one period.*

The period of each function is π, so we calculate their mean values over the interval $\{x:0\leqslant x\leqslant\pi\}$.

(i) $\dfrac{1}{\pi}\displaystyle\int_0^\pi\cos 2x\,dx=\dfrac{1}{2\pi}\left[\sin 2x\right]_0^\pi=\dfrac{1}{2\pi}[0-0]=\mathbf{0}.$

(ii) $\dfrac{1}{\pi}\displaystyle\int_0^\pi\cos^2 2x\,dx=\dfrac{1}{2\pi}\int_0^\pi(1+\cos 4x)\,dx=\dfrac{1}{2\pi}\left[x+\tfrac{1}{4}\sin 4x\right]_0^\pi=\tfrac{1}{2}.$

In case (i), the mean value is zero because the positive values of $\cos 2x$ for $0\leqslant x<\tfrac{1}{2}\pi$ are cancelled out by the negative values for $\tfrac{1}{2}\pi<x\leqslant\pi$.

In case (ii), the function $\cos^2 2x$ is never negative and the integrals over the two half periods are the same and add together. This particular result is important in electricity theory. If an alternating supply has voltage V, $V=v\cos nt$ at time t, where v is the peak voltage, then a voltmeter reads the 'root mean square' voltage, that is, the square root of the mean value of the square of the voltage. The time for one period is $2\pi/n$, and the measured voltage is V_1, where $V_1^2 = v^2 \dfrac{n}{2\pi} \displaystyle\int_0^{2\pi/n} \cos^2 nt \, dt$. Then

$$V_1^2 = \frac{v^2 n}{4\pi} \int_0^{2\pi/n} (1 + \cos 2nt) \, dt = \tfrac{1}{2}v^2, \text{ so } V_1 = v/\sqrt{2}.$$

The measured voltage is the peak voltage divided by $\sqrt{2}$. Thus, a mains voltage of 240 volts has a peak voltage of $240\sqrt{2} = 339$ volts.

EXERCISE 20.8

1   Over the interval $\{x : -1 \leqslant x \leqslant 1\}$, find the mean value of:
   (i) x^3,   (ii) x^4,   (iii) $|x|$,   (iv) $\sin x$,   (v) $\cos x$,   (vi) $\sin^2 x$.
2   Find the mean value of:
   (i) $3x^2 + 2x + 1$ over the interval $\{x : 0 \leqslant x \leqslant 1\}$,
   (ii) $\sin x + \cos x$ over the interval $\{x : 0 \leqslant x \leqslant \tfrac{1}{2}\pi\}$,
   (iii) $\sin x + \cos x$ over the interval $\{x : 0 \leqslant x \leqslant 2\pi\}$,
   (iv) $(\sin x + \cos x)^2$ over the interval $\{x : 0 \leqslant x \leqslant 2\pi\}$,
   (v) $(\sin x + \cos x)^2$ over the interval $\{x : 0 \leqslant x \leqslant \pi\}$.
3   Find the mean value of:

   (i) $x^2 - 4$,   (ii) $(x-1)(x+3)$,   (iii) $2x^3 - 4x$,   (iv) $\sqrt{x}$,   (v) $\dfrac{1}{x^2} - \dfrac{4}{x^3}$,

   (vi) $x^3 - 3x^2 + 2x - 6$,   (vii) $\cos \dfrac{\pi x}{2}$,   (viii) $\sin^2 \dfrac{\pi x}{2}$,

   over the interval $\{x : 1 \leqslant x \leqslant 3\}$, in each case.
4   Show that the mean value of x^{2n}, $n \in \mathbb{N}$, over the interval between $x = -1$ and $x = 1$, tends to zero as n tends to infinity. This is so in spite of the fact that the graph of x^{2n} always passes through $(-1, 1)$ and $(1, 1)$. Explain why this is so, using a sketch graph.

20.9 Centroids

Another important concept, used in mechanics and in statistics, is that of the first moment of an area about an axis. Before we consider this idea, we will look again at area as a limit of a sum.

Consider some area S in the Cartesian plane. Let δS be a small element of the area S, situated at the point (x, y) (Fig. 20.14 (a)). Then the area S can be regarded as the limit of a sum of elementary areas, thus,

$$S = \lim \Sigma \, \delta S,$$

the limiting process being performed by letting δS tend to zero while the number of elementary area in the sum tends to infinity. This limit of a sum can be expressed as an integral, and two ways of doing this are shown.

First, consider an element of area δS as a strip of width δx, situated at a distance x from the y-axis and parallel to this axis, and bounded by $y = y_1(x)$ and $y = y_2(x)$ (Fig. 20.14 (b)). The limits of the strip are

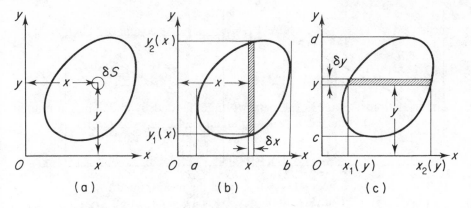

Fig. 20.14

determined by the particular boundary curve of the area S. Then

$$\delta S = \{y_2(x) - y_1(x)\}\,\delta x \quad \text{and} \quad \Sigma \delta S = \Sigma \{y_2(x) - y_1(x)\}\,\delta x,$$

and in the limit $\qquad S = \int_a^b \{y_2(x) - y_1(x)\}\,\mathrm{d}x,$

where, for the whole of S, $a < x < b$. This is the way in which we calculated areas in §20.4. In the case of the area under the curve $y = \mathrm{f}(x)$, $y_2(x) = \mathrm{f}(x)$ and $y_1(x) = 0$.

An alternative procedure is to choose, as the element of area δS, a strip of width δy parallel to Ox and distance y from Ox (Fig. 20.14(c)). If the length of the strip is $x_2(y) - x_1(y)$ then its area δS is given by

$$\delta S = \{x_2(y) - x_1(y)\}\delta y \quad \text{and} \quad \Sigma \delta S = \Sigma\{x_2(y) - x_1(y)\}\delta y.$$

In the limit, if $c < y < d$, for the whole of S,

$$S = \int_c^d \{x_2(y) - x_1(y)\}\,\mathrm{d}y.$$

This gives the area as an integral with respect to y, instead of x. In order to evaluate these integrals, it may be necessary to split up the range of integration, as shown in the following example.

EXAMPLE 1 *Verify that the above two methods give the area under the line segment AB of Fig. 20.13.*

Method 1   Use a strip of width δx, parallel to the y-axis as the element δS of area (Fig. 20.15 (a)). The required area S is given by

$$S = \int_a^b y\,\mathrm{d}x = \int_a^b (1 + \tfrac{1}{2}x)\,\mathrm{d}x = \left[x + \tfrac{1}{4}x^2\right]_a^b = b + \tfrac{1}{4}b^2 - a - \tfrac{1}{4}a^2.$$

This agrees with the known area of the trapezium, which can be written

$$(b - a)\tfrac{1}{2}\{\mathrm{f}(a) + \mathrm{f}(b)\} = \tfrac{1}{2}(b - a)\{1 + \tfrac{1}{2}a + 1 + \tfrac{1}{2}b\} = b - a + \tfrac{1}{4}b^2 - \tfrac{1}{4}a^2.$$

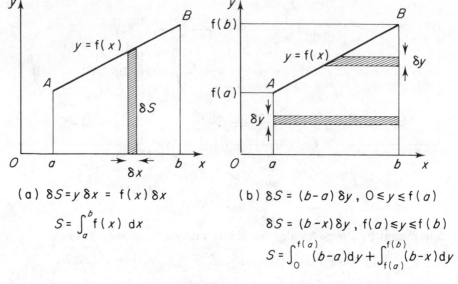

(a) $\delta S = y\,\delta x = f(x)\,\delta x$

$$S = \int_a^b f(x)\,dx$$

(b) $\delta S = (b-a)\,\delta y,\ 0 \le y \le f(a)$

$\delta S = (b-x)\,\delta y,\ f(a) \le y \le f(b)$

$$S = \int_0^{f(a)}(b-a)\,dy + \int_{f(a)}^{f(b)}(b-x)\,dy$$

Fig. 20.15

Method 2   This time, use a strip of width δy, parallel to Ox as δS (Fig. 20.15 (b)). For $0 \le y \le f(a)$, the length of the strip is $(b-a)$. For $f(a) \le y \le f(b)$, the length of the strip is $(b-x)$, where $y = f(x)$, that is, $y = 1 + \frac{1}{2}x$, so $x = 2(y-1)$, so the length of the strip is $b - 2y + 2$. We calculate the area by dividing the integral into two parts:

$$S = \int_0^\alpha (b-a)\,dy + \int_\alpha^\beta (b-x)\,dy,$$

where $\alpha = 1 + \frac{1}{2}a$, $\beta = 1 + \frac{1}{2}b$. Now

$$\int_0^\alpha (b-a)\,dy = [(b-a)y]_0^\alpha = (b-a)\alpha = (b-a)(1 + \tfrac{1}{2}a),$$

and

$$\int_\alpha^\beta (b-x)\,dy = \int_\alpha^\beta (b - 2y + 2)\,dy = \left[(b+2)y - y^2\right]_\alpha^\beta$$
$$= (b+2)(\beta - \alpha) - \beta^2 + \alpha^2$$
$$= (\beta - \alpha)(b + 2 - \beta - \alpha) = \tfrac{1}{2}(b-a)(b - \tfrac{1}{2}b - \tfrac{1}{2}a) = \tfrac{1}{4}(b-a)^2.$$

Therefore,
$$S = (b-a)(1 + \tfrac{1}{2}a) + \tfrac{1}{4}(b-a)^2 = \tfrac{1}{4}(b-a)(4 + a + b),$$

as before. In this example, the first method is the easier, but this may not always be so.

We are now in a position to define the first moment of an area about a line and also the centroid of an area relative to axes $O(x, y)$. Referring to the notation of Fig. 20.14:
the first moment about Oy of an element δS distant x from Oy is $x\delta S$;

the first moment about Ox of an element δS distant y from Ox is $y\delta S$. For the whole area S, we sum the elementary moments and take the limit to obtain an integral. In each case, we take as the element of area a strip parallel to the axis about which we are taking moments.

Definition The *first moment about Oy* of the plane area S is given by

$$\lim \Sigma x\delta S = \lim \Sigma x(y_2 - y_1)\delta x = \int_a^b x\{y_2(x) - y_1(x)\}\,dx = \bar{x}S.$$

The *first moment about Ox* of the plane area S is given by

$$\lim \Sigma y\delta S = \lim \Sigma y(x_2 - x_1)\delta y = \int_c^d y\{x_2(y) - x_1(y)\}\,dy = \bar{y}S.$$

The point $(\bar{x}, \bar{y})$ is *the centroid* of the area S.

In many cases, the area S has an axis of symmetry. In this case, if we take this axis as the axis Oy, then, because of the symmetry, $\bar{x}S = 0$ and so $\bar{x} = 0$, and the centroid lies on the axis of symmetry. The centroid is, in fact, the geometric centre of the area S. The position of the centroid of S has been defined with reference to a particular set of axes $O(x, y)$ but the centroid is independent of the axes used and depends only on the shape of the area S.

EXAMPLE 2 *Find the coordinates $(\bar{x}, \bar{y})$ of the centroid of the region below the curve $y = 2\sqrt{x}$ between $x = 0$ and $x = 4$.*

The curve $y = 2\sqrt{x}$ is a part of a parabola lying on its side, with the axis of symmetry $y = 0$ (Fig. 20.16). We begin by finding the x coordinate $\bar{x}$ of the

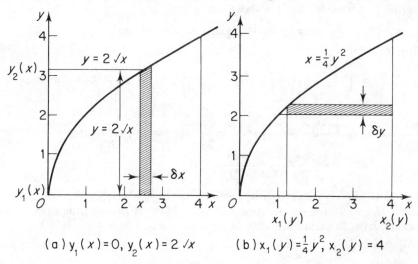

(a) $y_1(x) = 0$, $y_2(x) = 2\sqrt{x}$ (b) $x_1(y) = \frac{1}{4}y^2$, $x_2(y) = 4$

Fig. 20.16

centroid, using Fig. 20.16(a). The area S of the region is given by the integral

$$S = \int_0^4 y \, dx = \int_0^4 2\sqrt{x} \, dx = \left[\frac{4}{3}(x)^{\frac{3}{2}} \right]_0^4 = \frac{32}{3}.$$

Then

$$\bar{x}S = \frac{32}{3}\bar{x} = \int_0^4 xy \, dx = \int_0^4 x.2\sqrt{x} \, dx = \int_0^4 2x^{\frac{3}{2}} \, dx = \left[\frac{4}{5}x^{\frac{5}{2}} \right]_0^4$$

$$= \frac{4.32}{5},$$

and so $\qquad \bar{x} = \frac{4.3}{5} = 2{\cdot}4.$

To find $\bar{y}$, we refer to Fig. 20.16 (b), and note that $x_2(y) = 4$ and $x_1(y) = \frac{1}{4}y^2$. Then

$$\bar{y}\frac{32}{3} = \int_0^4 y(x_2(y) - x_1(y)) \, dy = \int_0^4 y(4 - \frac{1}{4}y^2) \, dy = \left[2y^2 - \frac{y^4}{16} \right]_0^4$$

$$= 32 - 16 = 16, \quad \text{and so} \quad \bar{y} = \frac{3}{2} = 1{\cdot}5.$$

Therefore, the centroid of the area is at the point **(2·4, 1·5)**.

EXAMPLE 3　*Find the coordinates $(\bar{x}, \bar{y})$ of the centroid of the trapezium bounded by the lines $y = 1 + \frac{1}{2}x$, $y = 0$, $x = a$ and $x = b$, where $a < b$.*

This is the figure whose area S we calculated in Example 1, where we found that $S = b - a + \frac{1}{4}b^2 - \frac{1}{4}a^2$. To find $\bar{x}$, we refer to Fig. 20.15 (a), since we require an element of area δS of constant distance x from Oy.

$$\delta S = y\delta x = (1 + \frac{1}{2}x)\delta x,$$

and the first moment of δS about Oy is

$$x\delta S = x(1 + \frac{1}{2}x)\delta x = (x + \frac{1}{2}x^2)\delta x.$$

Therefore,

$$\bar{x}S = \int_a^b (x + \frac{1}{2}x^2) \, dx = \left[\frac{1}{2}x^2 + \frac{1}{6}x^3 \right]_a^b = \frac{1}{6}(3b^2 - 3a^2 + b^3 - a^3),$$

and $\qquad \bar{x} = \frac{1}{6}\frac{3b^2 - 3a^2 + b^3 - a^3}{b - a + \frac{1}{4}b^2 - \frac{1}{4}a^2} = \frac{1}{6}\frac{(b-a)(3b + 3a + b^2 + ab + a^2)}{(b-a)(1 + \frac{1}{4}b + \frac{1}{4}a)}$

$$= \frac{2}{3}\frac{3b + 3a + a^2 + ab + b^2}{4 + b + a}.$$

In order to find $\bar{y}$, we need to use an element of area S which is a constant distance y from Ox, so we use the notation of Fig. 20.15 (b). Because of the shape of the area, we have to divide the integral into two parts. For $0 < y < \alpha = 1 + \frac{1}{2}a$,

$$y\delta S = y(b - a)\delta y;$$

for $1 + \frac{1}{2}a = \alpha < y < \beta = 1 + \frac{1}{2}b$,

$$y\delta S = y(b - x)\delta y.$$

Then
$$\bar{y}S = \int_0^\alpha y(b-a)\,dy + \int_\alpha^\beta y(b-2y+2)\,dy$$

$$= \tfrac{1}{2}\alpha^2(b-a) + \tfrac{1}{2}(\beta^2-\alpha^2)(b+2) - \tfrac{2}{3}(\beta^3-\alpha^3).$$

Now $\beta^2-\alpha^2 = (\beta-\alpha)(\beta+\alpha) = \tfrac{1}{4}(b-a)(4+a+b),$

and $\beta^3-\alpha^3 = (\beta-\alpha)(\beta^2+\alpha\beta+\alpha^2) = \tfrac{1}{8}(b-a)(12+6a+6b+a^2+ab+b^2).$

Also
$$\tfrac{1}{2}\alpha^2(b-a) + \tfrac{1}{2}(\beta^2-\alpha^2)(b+2) = (b-a)[\tfrac{1}{2}(1+\tfrac{1}{2}a)^2 + \tfrac{1}{8}(b+2)(4+a+b)]$$

$$= \tfrac{1}{8}(b-a)(4+4a+a^2+4b+ab+b^2+8+2a+2b)$$

$$= \tfrac{1}{8}(b-a)(12+6a+6b+a^2+ab+b^2),$$

therefore, $\bar{y}S = \tfrac{1}{8}(b-a)(12+6a+6b+a^2+ab+b^2)(1-\tfrac{2}{3})$

and so $$\bar{y} = \frac{12+6a+6b+a^2+ab+b^2}{6(4+a+b)}.$$

EXERCISE 20.9

1 Use the formula for $\bar{x}$ and $\bar{y}$ to prove that the centroid of the rectangle bounded by the four lines $x=1$, $x=4$, $y=2$ and $y=6$, lies at $(2\tfrac{1}{2},4)$.
2 The triangle ABC has vertices A $(1,0)$, B $(5,0)$, C $(3,6)$. Find Cartesian equations for the lines AC and BC. Prove that the y coordinate of the centroid of the triangle lies at a distance 2 from the x-axis. Show that the three medians of the triangle ABC intersect at the centroid.
3 Prove that the derivative of the function f, given by $f(x)=(a^2-x^2)^{\frac{3}{2}}$ is given by $f'(x)=-3x\sqrt{(a^2-x^2)}$. Find the centroid of the half circle
$$y=\sqrt{(a^2-x^2)}, \quad -a\leqslant x\leqslant a.$$
4 Prove that the two parabolas $y^2=16x$ and $y=\tfrac{1}{2}x^2$ meet at the points $(0,0)$ and $(4,8)$. Prove that the centroid of the region enclosed between these two parabolas has coordinates $(1\cdot8, 3\cdot6)$.

20.10 Volumes of revolution

The integral, as a limit of a sum, can also be used to calculate the volume of a solid of revolution, formed by rotating an area S about an axis in its plane. As an example, suppose that the area S is the area under the curve $y=f(x)$ between $x=a$ and $x=b$. The volume V is formed by one full revolution of S about Ox (Fig. 20.17). An element of volume δV is formed by rotating about Ox an element of area δS, consisting of a strip of width δx and height y, where $y=f(x)$. This element of volume is approximately a circular disc, of thickness δx and radius y, and so $\delta V \approx \pi y^2 \delta x$. Summing for

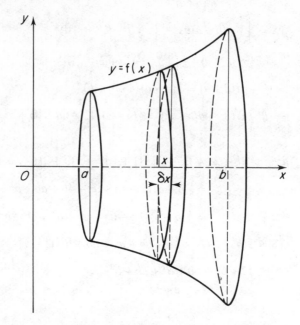

Fig. 20.17

the whole volume and taking the usual limit,

$$V = \lim \Sigma \delta V = \lim \Sigma \pi y^2 \delta x = \int_a^b \pi y^2 \, dx.$$

If the area S is not bounded by the x-axis, as is the case in Fig. 20.14 (b), then the volume of revolution of the area S about the x-axis is given by

$$V = \int_a^b \pi (y_2^2 - y_1^2) \, dx,$$

since the element of volume δV will be in the shape of a washer, of thickness δx, lying between two circles of radii $y_1(x)$ and $y_2(x)$. For a corresponding volume of revolution about the axis Oy, the volume is given by

$$V = \int_c^d \pi (x_2^2 - x_1^2) \, dy.$$

EXAMPLE 1   *Find the volume of a sphere of radius a.*

We regard the sphere as the volume of revolution through one complete turn about the axis Ox of the area below the curve $y = \sqrt{(a^2 - x^2)}$, between $x = -a$ and $x = a$. Therefore, the volume of the sphere is V, given by

$$V = \int_{-a}^a \pi y^2 \, dx = \int_{-a}^a \pi (a^2 - x^2) \, dx = \pi \left[(a^2 x - \tfrac{1}{3} x^3) \right]_{-a}^a = \tfrac{4}{3} \pi a^3.$$

EXAMPLE 2   *A bowl is formed by rotating the curve $y = x^4$ from $(0, 0)$ to $(1, 1)$ through one complete revolution about the y-axis. Find the volume of liquid which the bowl can hold. If liquid is poured into the bowl at a rate K units of volume per second, find the rate of increase of the depth of liquid in the bowl when the depth is $\frac{1}{4}$.*

The volume V of the bowl is given by

$$V = \int_0^1 \pi x^2 \, dy = \int_0^1 \pi y^{\frac{1}{2}} \, dy = \left[\tfrac{2}{3} \pi y^{\frac{3}{2}} \right]_0^1 = \tfrac{2}{3} \pi.$$

Approximately, $V \approx \pi x^2 \delta y$ so $\dfrac{dV}{dt} = \pi x^2 \dfrac{dy}{dt} = \pi y^{\frac{1}{2}} \dfrac{dy}{dt} = K.$ Hence $\dfrac{dy}{dt} = \dfrac{K}{\pi y^{\frac{1}{2}}}$ and

when $y = \frac{1}{4}$, $\dfrac{dy}{dt} = \dfrac{2K}{\pi}.$

EXERCISE 20.10

1   Find the volume of revolution formed when the portion of the curve $y = 9 - x^2$ between $x = -3$ and $x = 3$ is rotated through 2π about the x-axis.
2   A wine glass is formed by rotating the curve $y = x^3$, $0 \leqslant x \leqslant 2$, through 2π about the y-axis. Find the volume of the wine glass. Given that 10π cubic units of liquid are poured into the glass, find the depth of the liquid in the glass.
3   A pulley is formed by rotating the curve $y = 5 + 2x^2 - x^4$, between $x = \sqrt{3}$ and $x = -\sqrt{3}$, about the x-axis through one whole turn. Find the volume of the pulley.
4   A hemispherical bowl of radius 50 cm is filled with water to a depth of 30 cm. Find, by integration, the volume, in cm^3, of water in the bowl. (Leave your answer in terms of π.)   (L)
5   Find the volume generated when the finite region bounded by the curve $y = 1 + \sqrt{x}$, the x-axis and the lines $x = 1$ and $x = 4$, is rotated through 2π radians about the x-axis.   (L)

MISCELLANEOUS EXERCISE 20

1   Evaluate: (i) $\displaystyle\int_0^\pi \sin\tfrac{1}{4}x \, dx$,   (ii) $\displaystyle\int_2^3 (1 - x)(x^2 - 3x + 4) \, dx$.
2   Find the area enclosed by the graph of the function $x^3 - 5x^2 + 3x + 9$ and the x-axis.
3   Find the area of the region of the plane enclosed by the two curves with equations $y = x$ and $y = 8x - 12 - x^2$.
4   Find the area of the triangle enclosed between the lines $3y = x$, $x = a$, and the x-axis.
    Find also the volume of revolution formed when this triangle is rotated through one revolution about the x-axis to form a cone.
5   The region S in the first quadrant of the x-y plane is bounded by the axes, the line $x = 3$ and the curve $y = \sqrt{(1 + x^2)}$.
    (i) Use the trapezium rule with ordinates at $x = 0, 1, 2, 3$ to estimate the area of S.
    (ii) Show that the volume of the solid formed when S is rotated through one revolution about the y-axis is $\frac{2}{3}\pi(\sqrt{1000} - 1)$.   (C)

6 Given that $f(x) = a_0 + a_1x + a_2x^2 + a_3x^3 + a_4x^4 + a_5x^5$, prove that when

$$\int_{-h}^{h} f(x)\,dx$$

is evaluated by the trapezoidal rule using only the three ordinates at $x = -h$, 0 and h, then the error is $\frac{1}{15}h^3(5a_2 + 9a_4h^2)$.

Find an expression for the error when the same integral is evaluated by Simpson's rule using the same three ordinates.

Show that this latter error is reduced by over 93 % if Simpson's rule is used with ordinates at $x = -h$, $-\frac{1}{2}h$, 0, $\frac{1}{2}h$ and h.   (C)

7 Use the binomial theorem to expand $(1 + x^3)^{10}$ in ascending powers of x up to and including the term in x^9. Hence estimate I, where

$$I = \int_{0}^{0 \cdot 2} (1 + x^3)^{10}\,dx,$$

to three decimal places.

Make another estimate of I, again to three decimal places, by using Simpson's rule with three ordinates, showing all your working.   (L)

8 An arc PQ of a circle with centre O and radius r subtends an angle θ radians at O. Show that the area of the segment between the arc PQ and the line PQ is $\frac{1}{2}r^2(\theta - \sin\theta)$.

Find the coordinates of the points of intersection of the circle $x^2 + y^2 = 4a^2$ and the curve $y^2 = 3ax$, $(a > 0)$.

Find also the area of the smaller of the finite regions enclosed by the curves.   (JMB)

9 Find by integration the volume of the sphere formed by rotating the circle $x^2 + y^2 = a^2$ through π radians about the x-axis.   (L)

10 The region defined by the inequalities

$$0 \leqslant x \leqslant \pi, \quad 0 \leqslant y \leqslant \log_{10}(1 + \sin x)$$

is rotated completely about the x-axis. Using any appropriate rule for approximate integration with five ordinates, find the volume of the solid of revolution formed, giving your answer to three significant figures. (AEB)

11 The triangle ABC has area $\triangle$ and the perpendicular from O to the plane ABC is of length h. Prove by integration that the volume of the tetrahedron $OABC$ is $\frac{1}{3}h\triangle$.   (L)

12 Values of a continuous function f were found experimentally as given below.

t	0	0·3	0·6	0·9	1·2	1·5	1·8
f(t)	2·72	3·00	3·32		4·06	4·48	4·95

Use linear interpolation to estimate f$(0 \cdot 9)$. Then use Simpson's rule with seven ordinates to estimate $\int_{0}^{1 \cdot 8} f(t)\,dt$, tabulating your working and giving your answer to two places of decimals.   (JMB)

13 Find the mean value of $\sin t$ with respect to t over the interval $0 \leqslant t \leqslant \pi/2$.   (L)

14 Find the coordinates of the centroid of the finite region bounded by the curve $y = x^2$, the lines $x = 2$, $x = 3$ and the x-axis.   (L)

15   The curves $cy^2 = x^3$ and $y^2 = ax$ (where $a > 0$ and $c > 0$) intersect at the origin O and at a point P in the first quadrant. The areas of the regions enclosed by the arcs OP, the x-axis and the ordinate through P are A_1 and A_2 for the two curves; the volumes of the two solids formed by rotating these regions through four right angles about the x-axis are V_1 and V_2 respectively. Prove that $A_1/A_2 = \frac{3}{5}$ and $V_1/V_2 = \frac{1}{2}$.

16   Write down the area of the circle in which a sphere of radius r is cut by a plane at a distance y from the centre, where $y < r$.

By integration prove that the volume of a 'cap' of height $\frac{1}{4}r$ cut off the top of the sphere, as shown in the diagram, is $11\pi r^3/192$.   *(SMP)*

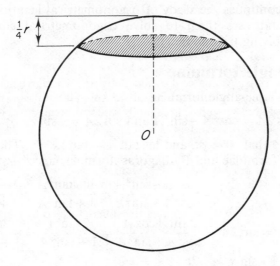

21 Trigonometry II

21.1 Further trigonometrical identities

This section continues the study of trigonometrical identities, begun in Chapter 3, and uses them to solve trigonometrical equations and problems involving trigonometry.

The half-angle formulae

We use the double-angle formulae of §3.7:

$$\cos 2A = \cos^2 A - \sin^2 A \text{ and } \sin 2A = 2 \sin A \cos A.$$

Let $x = 2A$ so that $A = \frac{1}{2}x$, and let $\tan A = \tan \frac{1}{2}x = t$. Then, using the double-angle formulae and Pythagoras' formula, $\cos^2 A + \sin^2 A = 1$,

$$\cos x = \cos 2A = \frac{\cos^2 A - \sin^2 A}{\cos^2 A + \sin^2 A} = \frac{1 - \tan^2 A}{1 + \tan^2 A} = \frac{1 - t^2}{1 + t^2},$$

$$\sin x = \sin 2A = \frac{2 \sin A \cos A}{\cos^2 A + \sin^2 A} = \frac{2 \tan A}{1 + \tan^2 A} = \frac{2t}{1 + t^2},$$

$$\tan x = \frac{\sin x}{\cos x} = \frac{2t}{1 - t^2}.$$

These formulae are useful in solving equations, since an equation involving any of the trigonometric functions can be rewritten as an equation in the one variable t.

EXAMPLE 1 *Find the general solution of the equation* $\cos x + 2 \sin x = 1$.

Substitute for $\cos x$ and $\sin x$ in terms of $t = \tan \frac{1}{2}x$,

$$\frac{1 - t^2}{1 + t^2} + 2 \frac{2t}{1 + t^2} = 1$$

and so $1 - t^2 + 4t = 1 + t^2$, on multiplying by $(1 + t^2)$. This gives a quadratic equation

$$2t^2 - 4t = 2t(t - 2) = 0,$$

with solution $t = 0$ or $t = 2$.
If $t = \tan \frac{1}{2}x = 0$, then $\frac{1}{2}x = n\pi$, $n \in \mathbb{Z}$, so $x = 2n\pi$, $n \in \mathbb{Z}$.
If $t = \tan \frac{1}{2}x = 2$, then let $\tan \alpha = 2$ so that $\alpha = 1\cdot11$ radians, then $\frac{1}{2}x = \alpha + n\pi$ so $x = 2\alpha + 2n\pi$, $n \in \mathbb{Z}$.
The solution set is $\{2n\pi,\ 2\alpha + 2n\pi : n \in \mathbb{Z}\}$, $\boldsymbol{\alpha = 1\cdot11}$ **radians**.

The function $a \cos x + b \sin x$

The function f, with real domain, given by $f(x) = a \cos x + b \sin x$, occurs frequently in applications of mathematics and it is useful to transform $f(x)$ into a single cosine or sine function.

Let $r = \sqrt{(a^2 + b^2)}$, so that $(a/r)^2 + (b/r)^2 = 1$, and the point P, with coordinates $(a/r, b/r)$, lies on the unit circle, centre the origin. The signs of a and b will determine in which quadrant the point P lies. Let α (in radians) be the angle between Ox and OP, measured positively in the anticlockwise sense, as usual. Then

$$\cos \alpha = \frac{a}{r} \quad \text{and} \quad \sin \alpha = \frac{b}{r}.$$

If P lies in the 1st quadrant, $0 < \alpha < \frac{1}{2}\pi$, that is, $a > 0$, $b > 0$;
if P lies in the 2nd quadrant, $\frac{1}{2}\pi < \alpha < \pi$, that is, $a < 0$, $b > 0$;
if P lies in the 3rd quadrant, $\pi < \alpha < \frac{3}{2}\pi$, that is, $a < 0$, $b < 0$;
if P lies in the 4th quadrant, $\frac{3}{2}\pi < \alpha < 2\pi$, that is, $a > 0$, $b < 0$.

Then $\quad f(x) = r\left(\frac{a}{r}\cos x + \frac{b}{r}\sin x\right) = r(\cos x \cos \alpha + \sin x \sin \alpha)$

$$= r\cos(x - \alpha),$$

using the addition formula of §3.7. An immediate conclusion is that $f(x)$ is a cosine function with maximum and minimum values r and $-r$, occurring when $x = \alpha + 2n\pi$ and $x = \alpha + (2n+1)\pi$, respectively, for all integers n.

Other forms of f(x)

If $a < 0, b > 0$, so that $\frac{1}{2}\pi < \alpha < \pi$, then let $\beta = \alpha - \frac{1}{2}\pi, \alpha = \frac{1}{2}\pi + \beta$, so that β is an acute angle. Then

$$f(x) = b \sin x + a \cos x = r(\sin x \cos \beta - \cos x \sin \beta) = r \sin(x - \beta).$$

Similarly, if $a > 0$ and $b < 0$, let $\gamma = -\alpha$, and then γ is acute and

$$f(x) = r(\cos x \cos \gamma - \sin x \sin \gamma) = r \cos(x + \gamma).$$

Note that since $\cos \alpha = a/r$ and $\sin \alpha = b/r$, then $\tan \alpha = b/a$. However, the one equation $\tan \alpha = b/a$ is not enough to define α, since it leaves an ambiguity between α and $\alpha + \pi$.

EXAMPLE 2 *Express* $f(x) = 3 \cos x + 4 \sin x$ *in the form* $r \cos(x - \alpha)$.

Since $3^2 + 4^2 = 5^2$, let $\cos \alpha = 3/5$ and $\sin \alpha = 4/5$, so that $\alpha = 0.93$. Then

$$f(x) = 5(\cos x \cos \alpha + \sin x \sin \alpha) = \mathbf{5 \cos(x - \alpha)}.$$

EXAMPLE 3 *Express* $f(x) = 4 \sin x - 3 \cos x$ *in the form* $r \sin(x - \beta)$, *where* $0 < \beta < \frac{1}{2}\pi$.

Let $\cos \beta = 4/5$ and $\sin \beta = 3/5$, then $\beta = 0.64$.

$$f(x) = 5(4/5 \sin x - 3/5 \cos x) = 5(\sin x \cos \beta - \cos x \sin \beta) = \mathbf{5 \sin(x - \beta)}.$$

EXAMPLE 4   *Find the maximum and the minimum values of*
$f(x) = 12 \cos x - 5 \sin x$.

Since $12^2 + 5^2 = 13^2$, let $\cos \alpha = 12/13$, $\sin \alpha = 5/13$, then $f(x) = 3 \cos(x + \alpha)$ so that $f(x)$ has maximum and minimum values **13** and **−13**.

EXAMPLE 5   *Solve the equation* $\cos \theta^\circ - \sin \theta^\circ = 1, 0 \leqslant \theta < 360$.

In Example 6 of §3.3, we solved this equation by squaring to eliminate one of the circular functions. Now we transform the left-hand side to a single cosine. Let $r = \sqrt{2}$, then

$$1/\sqrt{2} = (1/\sqrt{2}) \cos \theta^\circ - (1/\sqrt{2}) \sin \theta^\circ$$

so $\cos 45^\circ = \cos \theta^\circ \cos 45^\circ - \sin \theta^\circ \sin 45^\circ = \cos(\theta + 45)^\circ.$

For the required range of values of θ, $\theta + 45 = 45$ or $\theta + 45 = 360 - 45$. Hence, either $\theta = 0$ or $\theta = 270$, giving a solution set $\{\mathbf{0, 270}\}$.

   The equation can also be solved using the half-angle formulae. Let $t = \tan \frac{1}{2} \theta^\circ$, then $\dfrac{1 - t^2}{1 + t^2} - \dfrac{2t}{1 + t^2} = 1$, so $1 - t^2 - 2t = 1 + t^2$, $2t(t + 1) = 0$, giving a solution $t = 0$ or $t = -1$. Therefore, either $\frac{1}{2}\theta = 0$ or $\frac{1}{2}\theta = 135$, and either $\theta = 0$ or $\theta = 270$ as before.

The factor formulae

We rearrange the addition formulae of §3.7 to derive four more useful identities. Consider the formulae for the sine:

$$\sin(A + B) = \sin A \cos B + \cos A \sin B,$$
$$\sin(A - B) = \sin A \cos B - \cos A \sin B.$$

On adding and subtracting these equations,

$$\sin(A + B) + \sin(A - B) = 2 \sin A \cos B,$$
$$\sin(A + B) - \sin(A - B) = 2 \cos A \sin B.$$

Likewise, by adding and subtracting the formulae for the cosine of a sum and a difference,

$$\cos(A + B) = \cos A \cos B - \sin A \sin B,$$
$$\cos(A - B) = \cos A \cos B + \sin A \sin B,$$

give

$$\cos(A + B) + \cos(A - B) = 2 \cos A \cos B,$$
$$\cos(A + B) - \cos(A - B) = -2 \sin A \sin B.$$

These formulae are now rearranged by putting $x = A + B$, $y = A - B$, so that $A = \dfrac{x+y}{2}$, $B = \dfrac{x-y}{2}$. This gives the *factor formulae*:

$$\sin x + \sin y = 2 \sin \frac{x+y}{2} \cos \frac{x-y}{2},$$

$$\sin x - \sin y = 2 \cos \frac{x+y}{2} \sin \frac{x-y}{2},$$

$$\cos x + \cos y = 2 \cos \frac{x+y}{2} \cos \frac{x-y}{2},$$

$$\cos x - \cos y = -2 \sin \frac{x+y}{2} \sin \frac{x-y}{2}.$$

Note carefully the presence of the negative sign in the last equation.

EXAMPLE 6   *Solve the equation* $\cos x + \cos 3x = 0$, *for* $-\pi < x \leqslant \pi$.

$$\cos x + \cos 3x = 2 \cos \frac{x+3x}{2} \cos \frac{x-3x}{2} = 2 \cos 2x \cos x,$$

since $\cos(-x) = \cos x$. The equation becomes $2 \cos 2x \cos x = 0$, so either $\cos x = 0$ and $x = \frac{1}{2}\pi$ or $x = -\frac{1}{2}\pi$, or $\cos 2x = 0$ and

$$2x \in \{\tfrac{1}{2}\pi, \tfrac{3}{2}\pi, -\tfrac{1}{2}\pi, -\tfrac{3}{2}\pi\}.$$

Hence the solution set for the equation is

$$\{-\tfrac{3}{4}\pi, -\tfrac{1}{2}\pi, -\tfrac{1}{4}\pi, \tfrac{1}{4}\pi, \tfrac{1}{2}\pi, \tfrac{3}{4}\pi\}.$$

EXAMPLE 7   *Solve the equation* $\sin x + \sin 2x + \sin 3x = 0$.

Using the factor formula, $\sin x + \sin 3x = 2 \sin 2x \cos x$, and so the equation becomes $\sin 2x(1 + 2\cos x) = 0$. Therefore, either $\sin 2x = 0$, so $2x = n\pi$ and $x = \frac{1}{2}n\pi$ for any integer n, or $\cos x = -\frac{1}{2}$, so $x = (2n+1)\pi \pm \frac{1}{3}\pi$. A check shows that all these values of x satisfy the equation, so the solution set is the set

$$\{\tfrac{1}{2}n\pi, (2n+1)\pi + \tfrac{1}{3}\pi, (2n+1)\pi - \tfrac{1}{3}\pi : n \in \mathbb{Z}\}.$$

EXERCISE 21.1

1   Use the half-angle formulae to solve the equation, for $0 \leqslant x < 2\pi$:
   (i) $\cos x + 2 \sin x = 2$,   (ii) $2 \cos x - \sin x = 1$,
   (iii) $2 \cos x + 3 \sin x + 3 = 0$,   (iv) $4 \cos x = 3 \sin x + 3$.
2   Find a positive number r and an acute angle α, such that:
   (i) $2 \cos \theta + 3 \sin \theta = r \cos(\theta - \alpha)$,   (ii) $2 \cos \theta + 3 \sin \theta = r \sin(\theta + \alpha)$,
   (iii) $3 \cos \theta - 4 \sin \theta = r \cos(\theta + \alpha)$,   (iv) $\sin \theta - \cos \theta = r \sin(\theta - \alpha)$.
3   Find the maximum and the minimum values of $f(x)$ where:
   (i) $f(x) = 6 \cos x - 8 \sin x$,   (ii) $f(x) = 12 \sin x + 5 \cos x$,
   (iii) $f(x) = 3 \sin x + 4 \cos x$,   (iv) $f(x) = 2 \sin x - \cos x$.

4 Solve the equation, for $0 \leqslant \theta < 360$, (a) by using the $r \cos(\theta - \alpha)$ formula, and (b) by using the half-angle formulae:
(i) $4 \sin \theta + 3 \cos \theta = 2$, (ii) $\cos \theta + \sin \theta = \frac{1}{2}$, (iii) $2 \cos \theta + 5 \sin \theta = 4$.

5 Find the general solution of the equation:
(i) $\cos \theta + \sin \theta = 2$, (ii) $3 \cos \theta - 4 \sin \theta = 5$, (iii) $3 \cos \theta - 4 \sin \theta = 2$,
(iv) $3 \cos \theta - 4 \sin \theta = 6$, (v) $3 \sin \theta - 4 \cos \theta = 1$.

6 Expand, as a product of factors:
(i) $\sin x + \sin 3x$, (ii) $\cos x + \cos 5x$, (iii) $\sin 3\theta + \sin 5\theta$,
(iv) $\cos 2x - \cos 2y$, (v) $\sin 3\theta - \sin 2\theta$, (vi) $\sin(x + \frac{1}{6}\pi) + \sin(x - \frac{1}{6}\pi)$,
(vii) $\cos(\frac{1}{2}\pi - x) + \cos 3x$, (viii) $\sin 3x + \cos x$, (ix) $\sin 2x + \cos 2y$.

7 Prove the following trigonometrical identities:
(i) $\dfrac{\sin A + \sin B}{\cos A + \cos B} = \tan\frac{1}{2}(A + B)$, (ii) $\dfrac{\sin A - \sin B}{\cos A + \cos B} = \tan\frac{1}{2}(A - B)$,

(iii) $\dfrac{\sin(2A + B) + \sin B}{\cos(2A + B) + \cos B} = \tan(A + B)$, (iv) $\dfrac{\sin 3x - \sin x}{\cos 3x + \cos x} = \tan x$,

(v) $\cos x + \cos 2x + \cos 3x = \cos 2x(1 + 2\cos x)$,
(vi) $\cos x + 2\cos 3x + \cos 5x = 4\cos^2 x \cos 3x$.

8 Find the general solution of the equation:
(i) $\cos x - \cos 5x = 0$, (ii) $\sin 2x + \sin 4x = 0$, (iii) $\cos x - \sin 3x = 0$,
(iv) $\cos x + \cos 2x + \cos 3x = 0$, (v) $\sin x + 2 \sin 3x + \sin 5x = 0$.

9 Given that $a \cos \theta - b \sin \theta \equiv 2 \cos(\pi/6 + \theta)$, find the constants a and b.
   Write down the general solution of the equation $\sqrt{6} \cos \theta - \sqrt{2} \sin \theta = 2$ giving your answer in radians. *(L)*

10 Express $5 \cos \theta + 12 \sin \theta$ in the form $r \cos(\theta - \alpha)$, where $r > 0$ and $0 < \alpha < \pi/2$, giving the value of α to the nearest 0.01 of a radian. *(L)*

11 Given that $\sin(A - B) \neq 0$, prove that

$$\frac{\sin A + \sin B}{\sin(A - B)} = \sin\left(\frac{A + B}{2}\right)\operatorname{cosec}\left(\frac{A - B}{2}\right).$$

Use this result to find cosec $15°$ in surd form. *(L)*

12 (a) Find all angles x between $0°$ and $360°$ inclusive for which

$$\sin^2 x + 2 \cos 2x = 2 \cos x$$

(b) Prove that $\sin \theta + \sin 2\theta + \cos \theta + \cos 2\theta = 2\sqrt{2} \cos \frac{1}{2}\theta \sin(\frac{3}{2}\theta + \frac{1}{4}\pi)$. Hence, or otherwise, determine all values of θ between 0 and 2π for which

$$\sin \theta + \sin 2\theta + \cos \theta + \cos 2\theta = 0. \qquad (C)$$

21.2 The derivatives of $\cos x$ and $\sin x$

In §4.6, we estimated the values of the derivatives of $\cos x$ and $\sin x$, and in §4.7 we used, without proof, the derived functions of these circular functions. We now justify the results and begin by finding the value of the limit, $\displaystyle\lim_{\theta \to 0} \frac{\sin \theta}{\theta}$, repeating example 3 of §3.5.

On the unit circle in the Cartesian plane, let A be the point $(1, 0)$ and let P be the point $(\cos \theta, \sin \theta)$. Let the tangent to the circle at P meet Ox at T

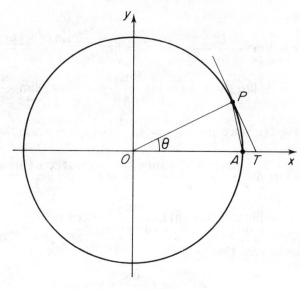

Fig. 21.1

(Fig. 21.1). The angle AOP is θ, so, by §3.5, the area of the sector OAP is $\frac{1}{2}\theta$. Using the sine formula, the area of triangle OAP is $\frac{1}{2}\sin\theta$ and this is less than the area of the sector. The triangle OPT has area $\frac{1}{2}PT = \frac{1}{2}\tan\theta$ and this is larger than the area of the sector. We are assuming that $0 < \theta < \frac{1}{2}\pi$. Thus

$$\tfrac{1}{2}\sin\theta < \tfrac{1}{2}\theta < \tfrac{1}{2}\tan\theta$$

so $$\sin\theta < \theta < \tan\theta.$$

For $0 < \theta < \frac{1}{2}\pi$, $\cos\theta > 0$, so we may divide the left inequality by θ to give $\dfrac{\sin\theta}{\theta} < 1$, and multiply the right inequality by $\dfrac{\cos\theta}{\theta}$ to give $\cos\theta < \dfrac{\sin\theta}{\theta}$, and so

$$\cos\theta < \frac{\sin\theta}{\theta} < 1.$$

Now as θ tends to zero, $\cos\theta$ tends to 1, and then, since $\dfrac{\sin\theta}{\theta}$ is sandwiched between $\cos\theta$ and 1, it also must tend to 1. We have proved that

$$\lim_{\theta\to 0} \frac{\sin\theta}{\theta} = 1.$$

Note that in the above, θ must be measured in *radians*, of course.
    We are now in a position to determine the derivatives of $\cos x$ and $\sin x$.

Using §18.6A,

$$\frac{d}{dx}\cos x = \lim_{h \to 0} \frac{\cos(x+h) - \cos x}{h} = \lim_{h \to 0} \frac{-2\sin(x+\frac{1}{2}h)\sin\frac{1}{2}h}{h}$$

$$= -\lim_{h \to 0} \sin(x+\tfrac{1}{2}h) \lim_{h \to 0} \frac{\sin\frac{1}{2}h}{\frac{1}{2}h} = -\sin x \lim_{\theta \to 0} \frac{\sin\theta}{\theta},$$

on putting $h = 2\theta$. Since $\lim\limits_{\theta \to 0} \dfrac{\sin\theta}{\theta} = 1$, $\dfrac{d}{dx}\cos x = -\sin x$. Similarly,

$$\frac{d}{dx}\sin x = \lim_{h \to 0} \frac{\sin(x+h) - \sin x}{h} = \lim_{h \to 0} \frac{2\cos(x+\frac{1}{2}h)\sin\frac{1}{2}h}{h}$$

$$= \lim_{h \to 0} \cos(x+\tfrac{1}{2}h) \lim_{h \to 0} \frac{\sin\frac{1}{2}h}{\frac{1}{2}h} = \cos x.$$

We have thus proved that

$$\frac{d}{dx}\cos x = -\sin x, \quad \frac{d}{dx}\sin x = \cos x.$$

EXAMPLE   *Find the derivative of the function (i)* $3\cos x - \sin^2 x$,
(ii) $2\sin x \cos(2x + \pi/2)$.

(i) $\dfrac{d}{dx}3\cos x = -3\sin x$. Treating $\sin^2 x$ as a composition of the function
$u = \sin x$ and the function u^2 we apply the chain rule as follows

$$\frac{d}{dx}\sin^2 x = 2\sin x \frac{d}{dx}\sin x = 2\sin x \cos x.$$

So $\qquad\qquad \dfrac{d}{dx}3\cos x - \sin^2 x = \mathbf{\sin x(2\cos x - 3)}.$

(ii) Applying the product rule for differentiation,

$$\frac{d}{dx}2\sin x \cos(2x + \pi/2) = 2\sin x (-\sin(2x + \pi/2)2) + 2\cos x \cos(2x + \pi/2),$$

$$= \mathbf{-4\sin x \sin(2x + \pi/2) + 2\cos x \cos(2x + \pi/2)}.$$

EXERCISE 21.2

1   Using a calculator, evaluate $\sin\theta$, $\cos\theta$, $\tan\theta$, $(\sin\theta)/\theta$, for θ taking values
between $0\cdot1$ and 1, inclusive, at intervals of $0\cdot1$. Verify that, over this range,

$$\sin\theta < \theta < \tan\theta, \; \cos\theta < (\sin\theta)/\theta < 1.$$

2   Use the chain rule for finding the derivative of a composition of functions
(§18.9), to find the derivatives of:
(i) $\cos 5x$,   (ii) $\sin(x+1)$,   (iii) $\cos(ax+b)$,   (iv) $\sin(ax+b)$,   (v) $\cos^2 x$,
(vi) $\cos^2 x + \sin^2 x$,   (vii) $\cos^3 x$,   (viii) $\sin^4(ax+b)$.

3 Write the sum as a product, using the factor theorem. Differentiate the sum and the product separately, and prove that the derivatives are equal:
(i) $\cos 2x + \cos 4x$, (ii) $\sin 2x - \sin x$.

4 Find the derivative of the function;
(i) $2 \sin x + 3 \cos x$, (ii) $4 \sin 2x - 2 \cos 4x$, (iii) $3 \sin x \cos 2x$,
(iv) $5 \cos 3x \sin^2 x$, (v) $2 \sin^2 3x + 3 \cos^2 2x$, (vi) $(4 \sin x + \cos x)^2$,
(vii) $3 \cos x \sin x (2 \sin x - \cos x)$, (viii) $(2 \sin x - 3 \cos x)(\cos 2x + 1)$,
(ix) $\dfrac{2 \sin x}{\cos x}$, (x) $\dfrac{3 \sin^2 x + \cos x}{(2 \sin x - 1)^2}$.

5 Differentiate with respect to x
(a) $\sin^2 x$, (b) $\sin (x^2)$, (c) $\sin 2x$. *(L)*

6 Differentiate $x(\sin x)^{\frac{1}{2}}$ with respect to x. *(L)*

7 Given that $y = \dfrac{1}{\sin x + \cos x}$, find $\dfrac{dy}{dx}$. *(L)*

8 Differentiate $(x^2 + \cos 3x)^4$ with respect to x. *(L)*

9 Differentiate with respect to x

(a) $\dfrac{x-1}{x+1}$, (b) $\cos^2 (3x + 4)$. *(L)*

21.3 Derivatives of circular functions

The results $\dfrac{d}{dx} \cos x = -\sin x$ and $\dfrac{d}{dx} \sin x = \cos x$ are used in order to obtain the derivatives of other circular functions.

$$\frac{d}{dx} \tan x = \frac{d}{dx} \frac{\sin x}{\cos x} = \left(\cos x \frac{d}{dx} \sin x - \sin x \frac{d}{dx} \cos x \right) \Big/ \cos^2 x.$$

$$= \frac{\cos x \cos x - \sin x (-\sin x)}{\cos^2 x} = \frac{\cos^2 x + \sin^2 x}{\cos^2 x}$$

$$= \frac{1}{\cos^2 x} = \sec^2 x.$$

The derivatives of the other circular functions are obtained in a similar manner and the details are left as an exercise. The results for the six circular functions are:

$f(x)$	$f'(x)$
$\cos x$	$-\sin x$
$\sin x$	$\cos x$
$\tan x$	$\sec^2 x$
$\sec x$	$\sec x \tan x$
$\operatorname{cosec} x$	$-\operatorname{cosec} x \cot x$
$\cot x$	$-\operatorname{cosec}^2 x$

EXAMPLE *Differentiate with respect to x:* (i) $\tan 5x$, (ii) $\cos^3 x$, (iii) $\cos x^3$, (iv) $\tan x^5$, (v) $\sec^2 (x^3 - 1)$, (vi) $x^2 \sec x \tan x$.

(i) $\dfrac{d}{dx} \tan 5x = \sec^2 5x \dfrac{d}{dx} 5x = \mathbf{5\sec^2 5x,}$

(ii) $\dfrac{d}{dx} \cos^3 x = 3\cos^2 x \dfrac{d}{dx} \cos x = \mathbf{-3\cos^2 x \sin x,}$

(iii) $\dfrac{d}{dx} \cos x^3 = -\sin x^3 \dfrac{d}{dx} x^3 = \mathbf{-3x^2 \sin x^3,}$

(iv) $\dfrac{d}{dx} \tan x^5 = \sec^2 x^5 \dfrac{d}{dx} x^5 = \mathbf{5x^4 \sec^2 x^5,}$

(v) $\dfrac{d}{dx} \sec^2 (x^3 - 1) = 2\sec (x^3 - 1) \dfrac{d}{dx} \sec (x^3 - 1)$

$\qquad = 2\sec (x^3 - 1) \sec (x^3 - 1) \tan (x^3 - 1) \dfrac{d}{dx} (x^3 - 1)$

$\qquad = \mathbf{6x^2 \sec^2 (x^3 - 1) \tan(x^3 - 1),}$

(vi) $\dfrac{d}{dx} x^2 \sec x \tan x = x^2 \sec x \dfrac{d}{dx} \tan x + \tan x \dfrac{d}{dx} x^2 \sec x$

$\qquad = x^2 \sec x \sec^2 x + \tan x (2x \sec x + x^2 \sec x \tan x)$

$\qquad = \mathbf{x^2 \sec^3 x + 2x \tan x \sec x + x^2 \sec x \tan^2 x.}$

EXERCISE 21.3

1 Prove the results, given in the table, for the derivatives of $\sec x$, $\csc x$ and $\cot x$.

2 Differentiate with respect to x:
(i) $x \sin x$, (ii) $x^2 \cos x$, (iii) $\cos^3 x$, (iv) $3 \sin^4 x$, (v) $\sqrt{(\cos x)}$,
(vi) $x \sec 3x$, (vii) $\sec x \tan x$, (viii) $\dfrac{\sin x}{x}$, (ix) $\cos (3x + 2)$,
(x) $\sin (x^2 + 1)$.

3 Find the derivatives of:
(i) $\cos (x + 1) \sin x$, (ii) $\sin (x + 1)^2$, (iii) $\sec^2 3x$, (iv) $x^2 \sec x - \tan x$,
(v) $x^3 \sin (4x + 3)$, (vi) $(1 + \sin 2x)^2$.

4 Find $\dfrac{dy}{dx}$, given that:
(i) $y = \tan (2x - 1)$, (ii) $y = x^2 \cos x^3$, (iii) $y = x^2 \cos^3 x$,
(iv) $y = \cos^2 2x \sin 3x$, (v) $y = x^2 \cot x - x \csc x$, (vi) $(\cos x + \sec x)^3$,
(vii) $(1 + \sin 2x)^2 + (1 - \sin 2x)^2$.

5 A point moves on a straight line Ox and, at time t, its distance from O is x. Find the speed of the point at time t, the time at which it is next at rest after $t = 0$ and the distance moved in that time, given that

$$x = t \sin t + \cos t.$$

6   Find those points on the graph, with equation $y = \sin(x^2 - 1)$, where the tangent is parallel to the axis Ox.

7   Find the equation of the tangent to the curve $y = x \tan 2x$ at the point $(\pi/8, \pi/8)$.

8   A curve has the equation $y = x \sin 2x$. Find the gradient of the curve at $x = \pi/3$.

21.4 Inverse circular functions

Fig. 3.9 shows the graphs of $\cos x$, $\sin x$, $\tan x$, and these are repeated in Fig. 21.2. These functions are all periodic, so they are not one-one

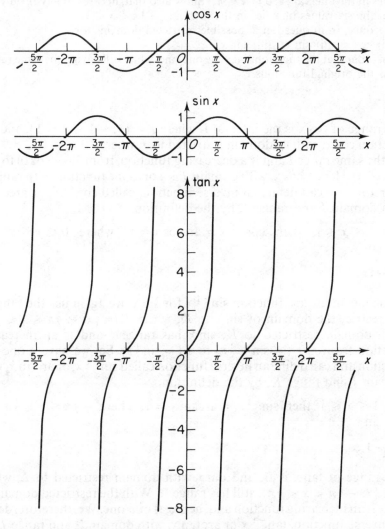

Fig. 21.2 Graphs of $\cos x$, $\sin x$ and $\tan x$.

functions and do not have inverse functions. However, we can define inverse functions if we restrict the domains of the circular functions so that they become one-one and yet retain their full ranges.

EXERCISE 21.4A

1   Using Fig. 21.2, and your knowledge of the trigonometric ratios of $\pi/3$ and $\pi/6$, find six values of x such that:
(i) $\cos x = \frac{1}{2}$,   (ii) $\cos x = \frac{1}{2}\sqrt{3}$,   (iii) $\cos x = -\frac{1}{2}$,   (iv) $\sin x = \frac{1}{2}$,
(v) $\sin x = \frac{1}{2}\sqrt{3}$,   (vi) $\tan x = \sqrt{3}$,   (vii) $\tan x = -\sqrt{3}$.

2   Show that in question **1**, in cases (i), (ii) and (iii) only one of the six values of x lies in the interval $\{x: 0 \leqslant x \leqslant \pi\}$. Show also that, in cases (iv)–(vii), only one of the six values of x lies in the interval $\{-\frac{1}{2}\pi \leqslant x \leqslant \frac{1}{2}\pi\}$.

3   Deduce, from question **2**, possible restricted domains for
(i) $\cos x$,   (ii) $\sin x$,   (iii) $\tan x$,
in order that the functions become one-one, while still having the same ranges as the original functions.

$\cos^{-1} x$

The range of $\cos x$ is the interval I, where $I = \{x: -1 \leqslant x \leqslant 1\}$, and the function $\cos x$, with its domain restricted to J, where $J = \{x: 0 \leqslant x \leqslant \pi\}$, has the same range I and is a decreasing function, from 1 ($= \cos 0$) to -1 ($= \cos \pi$). Hence, $\cos x$, with domain J is a one-one function with range I. Therefore, we can define an inverse function, called $\cos^{-1} x$, or $\arccos x$, with domain I and range J, by the definition:

if $-1 \leqslant x \leqslant 1$, then $\cos^{-1} x = \arccos x = y$, where $0 \leqslant y \leqslant \pi$ and $x = \cos y$.

$\sin^{-1} x$

To define an inverse function $\sin^{-1} x$ for $\sin x$, we again use the range I. We restrict the domain of $\sin x$ to K, where $K = \{x: -\frac{1}{2}\pi \leqslant x \leqslant \frac{1}{2}\pi\}$. With domain restricted to K, $\sin x$ has range I and is an increasing function from -1 ($= \sin -\frac{1}{2}\pi$) to 1 ($= \sin \frac{1}{2}\pi$). So, we have a one-one function and can define an inverse function called $\sin^{-1} x$, or $\arcsin x$, with domain I and range K, by the definition:

if $-1 \leqslant x \leqslant 1$, then $\sin^{-1} x = \arcsin x = y$, where $-\frac{1}{2}\pi \leqslant y \leqslant \frac{1}{2}\pi$ and $x = \sin y$.

$\tan^{-1} x$

The range of $\tan x$ is $\mathbb{R}$, and $\tan x$ with domain restricted to L, where $L = \{x: -\frac{1}{2}\pi < x < \frac{1}{2}\pi\}$, still has range $\mathbb{R}$. With the restricted domain L, $\tan x$ is an increasing function and, hence, is one-one. We, therefore, define an inverse function, $\tan^{-1} x$ or $\arctan x$, with domain $\mathbb{R}$ and range L, by the definition:

if $x \in \mathbb{R}$, then $\tan^{-1} x = \arctan x = y$, where $-\frac{1}{2}\pi < y < \frac{1}{2}\pi$ and $x = \tan y$.

The graphs of the three inverse circular functions are shown in Fig. 21.3. Each is the reflection in the line $y = x$ of its respective circular function with the domain appropriately restricted.

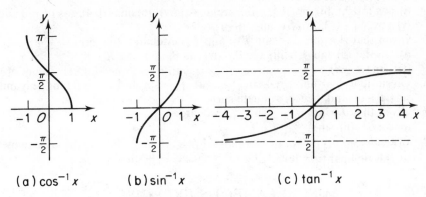

(a) $\cos^{-1} x$ (b) $\sin^{-1} x$ (c) $\tan^{-1} x$

Fig. 21.3 Inverse circular functions.

EXAMPLE 1   *Find*: (i) $\cos^{-1}(-\sqrt{3}/2)$,   (ii) $\sin^{-1}(-\frac{1}{2})$, (iii) $\tan^{-1}(-\sqrt{3})$.

(i) Since $\cos y = -\sqrt{3}/2$ for $y \in \{(2n+1)\pi \pm \pi/6 : n \in \mathbb{Z}\}$, in the interval $0 \leqslant y \leqslant \pi$, there is only one value of y satisfying $\cos y = -\sqrt{3}/2$, namely $y = 7\pi/6$, and so $\cos^{-1}(-\sqrt{3}/2) = \mathbf{7\pi/6}$.
(ii) Similarly, $\sin y = -\frac{1}{2}$ for $y \in \{(2n-\frac{1}{6})\pi, (n+\frac{1}{6})\pi : n \in \mathbb{Z}\}$, and if also $-\frac{1}{2}\pi \leqslant y \leqslant \frac{1}{2}\pi$, then $y = \sin^{-1}(-\frac{1}{2}) = \mathbf{-\pi/6}$.
(iii) If $\tan y = -\sqrt{3}$, then $y \in \{(n-\frac{1}{3})\pi : n \in \mathbb{Z}\}$, and if also $-\frac{1}{2}\pi < y < \frac{1}{2}\pi$, then $y = \tan^{-1}(-\sqrt{3}) = \mathbf{-\pi/3}$.

EXAMPLE 2   *Find* (i) $\sin(\cos^{-1} x)$,   (ii) $\tan(\sin^{-1} x)$.

(i) Let $y = \cos^{-1} x$ so that $x = \cos y$ and $\sin^2 y = 1 - \cos^2 y = 1 - x^2$. Now $\cos^{-1} x > 0$ and so $\sin y > 0$ and $\sin y = \sqrt{(1-x^2)}$. Thus,

$$\sin(\cos^{-1} x) = \sqrt{(1-x^2)}.$$

(ii) Let $z = \sin^{-1} x$, then $x = \sin z$ and $\cos^2 z = 1 - x^2$. Now x and z have the same sign and this is also the sign of $\tan z$, so

$$\tan(\sin^{-1} x) = \tan z = \frac{\sin z}{\cos z} = \frac{x}{\sqrt{(1-x^2)}}.$$

EXERCISE 21.4B

1   Verify that $\sin(\sin^{-1} x) = x$, $\cos(\cos^{-1} x) = x$, $\tan(\tan^{-1} x) = x$, but that, in general, $\sin^{-1}(\sin x) \neq x$, $\cos^{-1}(\cos x) \neq x$, $\tan^{-1}(\tan x) \neq x$.

2 Evaluate:
(i) $\cos^{-1} 0$, (ii) $\sin^{-1}\frac{1}{2}$, (iii) $\cos^{-1} 1$, (iv) $\tan^{-1} 1$, (v) $\cos^{-1}\frac{1}{2}$,
(vi) $\cos^{-1}(-\frac{1}{2})$, (vii) $\tan^{-1}(-1)$, (viii) $\sin^{-1}(-\frac{1}{2})$.

3 Express as a rational number:
(i) $\sin(\cos^{-1}\frac{3}{5})$, (ii) $\tan(\arcsin -\frac{3}{5})$, (iii) $\cos(\tan^{-1} -\frac{3}{4})$,
(iv) $\cos(\sin^{-1}\frac{12}{13})$, (v) $\sin(\tan^{-1} -\frac{5}{12})$.

4 Evaluate to one decimal place:
(i) $\cos^{-1} 0\cdot 9$, (ii) $\sin^{-1}\frac{2}{3}$, (iii) $\arctan 4$, (iv) $\arcsin(-0\cdot 4)$,
(v) $\cos^{-1}(-0\cdot 35)$, (vi) $\tan^{-1}(-1)$.

5 Given that $A = \sin^{-1}\frac{12}{13}$ and $B = \tan^{-1}\frac{3}{4}$, evaluate:
(i) $\cos A$, (ii) $\tan A$, (iii) $\cos B$, (iv) $\sin B$, (v) $\cos(A+B)$,
(vi) $\sin(A-B)$.

6 Given that $\theta = \sin^{-1}\frac{5}{13} + \sin^{-1}\frac{4}{5}$, put $\frac{5}{13} = \sin A$, $\frac{4}{5} = \sin B$ and expand $\sin(A+B)$, and, hence, evaluate $\sin\theta$.

7 Given that $\theta = \tan^{-1}\frac{1}{2} - \tan^{-1}\frac{3}{4}$, evaluate
(i) $\cos\theta$, (ii) $\sin\theta$.

8 Evaluate:
(i) $\tan(\tan^{-1}\dfrac{1}{\sqrt{3}} + \tan^{-1}\frac{3}{4})$, (ii) $\sin(\cos^{-1}\frac{1}{3} + \sin^{-1}\frac{1}{4})$.

MISCELLANEOUS EXERCISE 21

1 Obtain all the values of x for which $0 \leqslant x < 2\pi$ and $\sin x + \sin 3x = 0$.

2 Given that X is the acute angle such that $\sin X = 4/5$ and Y is the obtuse angle such that $\sin Y = 12/13$, find the exact value of $\tan(X+Y)$.

3 (a) Find the values of θ lying between $0°$ and $360°$ for which
$\cos(\theta - 30°) = 2\sin\theta$.
(b) Solve the equation $\cos 3\theta + 2\cos\theta = 0$, giving all solutions lying between $0°$ and $360°$.   *(C)*

4 (a) Prove that $\cot A - \tan A = 2\cot 2A$.
(b) Prove the following identities:
(i) $\cos\theta + \cos(\theta + \frac{2}{3}\pi) + \cos(\theta + \frac{4}{3}\pi) = 0$,
(ii) $\cos^2\theta + \cos^2(\theta + \frac{2}{3}\pi) + \cos^2(\theta + \frac{4}{3}\pi) = \frac{3}{2}$.   *(C)*

5 (i) Use the formula for $\cos A + \cos B$ to express $\cos n\theta + \cos(n-2)\theta$ as a product.
Deduce that $\cos n\theta = 2\cos\theta\cos(n-1)\theta - \cos(n-2)\theta$, and express $\cos 3\theta$, $\cos 4\theta$ and $\cos 5\theta$ in terms of $\cos\theta$.
Find an expression for $\sin 5\theta$ in terms of $\sin\theta$.
(ii) Express $y = 2\sin x + \cos x + 2$ in terms of $\tan(x/2)$ and find the values of $\tan(x/2)$ for which $y = 0$.   *(L)*

6 (a) Find all values of x between $0°$ and $360°$ inclusive for which
$$3\tan 2x + 2\tan x = 0.$$
(b) Verify, without use of tables, calculator or slide rule, that $\theta = 18°$ satisfies the equation
$$\cos 3\theta = \sin 2\theta.$$
Deduce, or prove otherwise, that $\sin 18°$ is a root of the equation
$$4x^2 + 2x - 1 = 0.$$
Solve this equation, giving the roots in surd form. State which of the roots is $\sin 18°$, giving a brief reason for your answer.   *(C)*

7   (a) Given that $\sin \alpha = 3/5$ and $\cos \beta = 12/13$, prove that one possible value of $\cos (\alpha + \beta)$ is $33/65$, and find all the other possible values.
(b) By drawing appropriate sketch-graphs, find the number of solutions of the equation
$$\sin^{-1} x = \pi - 2x,$$
where $\sin^{-1} x$ is not restricted to the principal value.   (C)

8   Express the function $f(\theta) = \sin \theta \,(2 \sin \theta + \cos \theta)$ in terms of $\sin 2\theta$ and $\cos 2\theta$. Hence show that $f(\theta)$ can be written in the form
$$1 - R \cos (2\theta + \alpha)$$
where R is positive and α is acute. State the values of R and $\tan \alpha$.
Deduce the maximum value of $f(\theta)$ and the smallest positive value of θ for which this maximum value occurs.   (JMB)

9   Express $3 \cos x + 4 \sin x$ in the form $r \cos (x - \alpha)$, where r is a positive constant and α is acute. Hence, or otherwise,
(a) find the greatest and least values of $(3 \cos x + 4 \sin x)^2$,
(b) find, to the nearest tenth of a degree, all the solutions of the equation
$$3 \cos x + 4 \sin x = \frac{5\sqrt{3}}{2}$$
in the interval $0° \leqslant x \leqslant 360°$.   (L)

10   (i) By expressing the numerator as a difference of sines, or otherwise, show that, as θ varies, the greatest value of
$$\frac{2 \sin \theta \cos (\theta + \alpha)}{\cos^2 \alpha}$$
where $0 < \alpha < \pi/2$, is $1/(1 + \sin \alpha)$. Give in terms of α the smallest positive value of θ for which the expression has this greatest value.
(ii) By expressing $\operatorname{cosec} 2x$ and $\cot 2x$ in terms of $\tan x$, or otherwise, find the possible values of $\tan x$ for which
$$3 \cot 2x + 7 \tan x = 5 \operatorname{cosec} 2x.$$ (L)

11   (i) Find all the angles between 0 and 2π which satisfy the equation
$$\sin 3x - \sin x = \cos 2x.$$
(ii) Find, to the nearest tenth of a degree, the general solution of the equation
$$6 \sin x + 8 \cos x = 5.$$ (L)

12   Express $8 \cos \theta - 15 \sin \theta$ in the form $R \cos (\theta + \alpha)$ where R is positive and α is acute, giving the values of R and $\tan \alpha$. Hence or otherwise find the values of θ between $0°$ and $360°$ which satisfy the equation $8 \cos \theta - 15 \sin \theta = 3$.   (L)

13   For the function f given by
$$f(x) = \tan x - \cot x - 4x \quad (x \in \mathbb{R}; \; x \neq \tfrac{1}{2}n\pi, \, n \in \mathbb{Z}),$$
find the derivative, $f'(x)$, and prove that $f'(x)$ can be expressed as $(\tan x - \cot x)^2$.
Find the co-ordinates of any points of the graph of
$$y = \tan x - \cot x - 4x \quad (0 < x < \pi; \; x \neq \tfrac{1}{2}\pi)$$
for which $\dfrac{dy}{dx} = 0$. Sketch this graph.   (C)

14 (i) Using the substitution $t = \tan(x/2)$, or otherwise, find, without using tables, the two possible values of $\tan x$ when

$$\sin x - 7\cos x + 5 = 0.$$

(ii) Express $(\cos 3\theta + \sin 3\theta)$ in the form $R\cos(3\theta - \alpha)$, where $R > 0$ and $0 < \alpha < \pi/2$. Hence, or otherwise, find the general solution in radians of the equation

$$\cos 3\theta + \sin 3\theta = 1. \qquad (L)$$

15 (a) Prove that, if $\cos\theta\cos 2\theta \neq 0$, $\dfrac{\cos 3\theta}{\cos\theta} - \dfrac{\cos 6\theta}{\cos 2\theta} = 2(\cos 2\theta - \cos 4\theta)$.

(b) Solve the equation $10\sin^2(\tfrac{1}{2}\theta°) - 5\sin\theta° = 4$,

giving the values of θ between 0 and 360 to the nearest 0·1. $\qquad (C)$

16 (a) Find the angles between $-180°$ and $180°$ which satisfy the equation $3\sin\theta - \cos\theta = 2$.

(b) Write down expression(s) for all the angles having the same sine as the angle α. Hence, or otherwise, obtain the general solution, in radian measure, of the equation

$$\sin 3\theta = \sin 2\theta. \qquad (C)$$

17 Find the values of r and α when $3\cos x + 2\sin x$ is expressed in the form $r\cos(x - \alpha)$, where r is positive and α is an acute angle, giving the value of α to the nearest tenth of a degree. Hence, or otherwise,

(a) show that $(3\cos x + 2\sin x)^2 \leqslant 13$ for all values of x,

(b) find, to the nearest degree, the values of x between $0°$ and $360°$ which satisfy the equation

$$3\cos x + 2\sin x = 2·75. \qquad (L)$$

18 (a) If the angle x can vary, where $0 \leqslant x < 2\pi$, prove that the expression

$$\sin x + \sin(x + \tfrac{1}{3}\pi)$$

has its greatest value when $x = \tfrac{1}{3}\pi$, and find the value of x for which the given expression is least.

(b) Prove that

(i) $\sin 3A = 4\sin A\sin(\tfrac{1}{3}\pi + A)\sin(\tfrac{1}{3}\pi - A)$,

(ii) $\cos 3A = 4\cos A\cos(\tfrac{1}{3}\pi + A)\cos(\tfrac{1}{3}\pi - A)$,

(iii) $\tan 3A = \tan A\tan(\tfrac{1}{3}\pi + A)\tan(\tfrac{1}{3}\pi - A)$. $\qquad (C)$

19 Show that, provided x is not a multiple of $\pi/4$,

$$\frac{\sin x + \sin 3x}{\cos x + \cos 3x} = \tan 2x. \qquad (L)$$

20 Find the range of values of a for which the equation

$$\cos(x + 90°) + \cos x = a$$

has real solutions. For the case when $a = 0$, find all the solutions in the interval $0° \leqslant x \leqslant 360°$.

Sketch the graph of $y = \cos(x + 90°) + \cos x$ for $0° \leqslant x \leqslant 360°$. $\quad (JMB)$

21 Given that $y = \sqrt{3}\cos x + \sin x$, where $0 \leqslant x \leqslant 2\pi$, find the maximum and minimum values of y.

Sketch the graph of y.

Calculate the values of x for which $\dfrac{\mathrm{d}y}{\mathrm{d}x} = -1$.

Show that the mean value of y^2 with respect to x in the interval $0 \leqslant x \leqslant \pi/2$ is $2(\pi + \sqrt{3})/\pi$.   (*JMB*)

22 (i) Given that $3\cos x + 2\sec x + 5 = 0$, find all the possible values of $\cos x$ and of $\tan^2 x$.

(ii) By putting $\tan(\theta/2) = t$, or otherwise, find the general solution of the equation $\sin\theta + 7\cos\theta = 5$, giving your answers to the nearest tenth of a degree.   (*L*)

23 Find all the values of x in the range $0 < x < \pi$ for which

$$\sin 5x + \sin 3x = 0.$$

24 By expressing $5\cos\theta + 12\sin\theta$ in the form $R\sin(\theta + \alpha)$ find the greatest and least values of

$$(5\cos\theta + 12\sin\theta)^2$$

as θ varies.   (*L*)

25 (i) Express $8\sin\theta + 15\cos\theta$ in the form $R\sin(\alpha + \theta)$, where R is positive and α is an acute angle. Hence, or otherwise, solve the equation

$$8\sin\theta + 15\cos\theta = 8{\cdot}5,$$

giving the general solution to the nearest tenth of a degree.

(ii) The sides AC, AB, BC of an acute-angled triangle ABC are of length x, $x+1$, $x+2$ respectively and BD is an altitude of the triangle. Show that

$$\cos A = \frac{x-3}{2x}$$

and find expressions for AD and BD in terms of x.   (*L*)

26 Express $7\cos\theta - 24\sin\theta$ in the form $R\cos(\theta + \alpha)$ where R is positive and α is acute, giving the values of R and $\tan\alpha$. Hence, or otherwise, find the values of θ between $0°$ and $360°$ which satisfy the equation

$$7\cos\theta - 24\sin\theta = 20$$

27 Express $\sin\theta + \sqrt{3}\cos\theta$ in the form $r\sin(\theta + \alpha)$, where $r > 0$ and $0 < \alpha < \pi/2$, giving the value of α in radians. Hence find in radians the general solution of the equation

$$\sin\theta + \sqrt{3}\cos\theta = 1. \qquad (L)$$

28 (i) Find, in radians, the general solution of the equation

$$\sin x + \sin 2x = \sin 3x.$$

(ii) By expressing $\sec 2x$ and $\tan 2x$ in terms of $\tan x$, or otherwise, solve the equation

$$2\tan x + \sec 2x = 2\tan 2x,$$

giving all solutions between $-180°$ and $+180°$.   (*L*)

29   Express $3 \cos 2t + 4 \sin 2t$ in the form $r \cos (2t - \alpha)$, where $r > 0$ and $0 < \alpha < 90°$, giving the value of α to the nearest degree.   (L)

30   (a) Express $\cos \theta - \sqrt{3} \sin \theta$ in the form $R \cos (\theta + \lambda)$, where R is positive and $0 < \lambda < \frac{1}{2}\pi$. Hence prove that the solutions of the equation $\cos \theta - \sqrt{3} \sin \theta = \sqrt{2}$ which lie in the interval $0 \leqslant \theta \leqslant 2\pi$ are $\theta = (17/12)\pi$ and $\theta = (23/12)\pi$.

(b) Find all values of x in the interval $0° \leqslant x \leqslant 90°$ for which $\sin 5x = \sin x$.
    (C)

31   (i) Prove that $\cos (A + B) \cos (A - B) = \cos^2 A - \sin^2 B$, and use this result to show that

$$\cos 15° \cos 105° = -\tfrac{1}{4}.$$

(ii) By expressing $\sin \theta$ and $\cos \theta$ in terms of t, where $t = \tan (\theta/2)$, in the equation

$$5 \sin \theta + 2 \cos \theta = 5,$$

form a quadratic equation in t and solve this equation to find θ, correct to the nearest $0 \cdot 1°$, in the range $40° < \theta < 50°$.   (L)

32   Given that

$$x = \sin (\theta + 105°) + \sin (\theta - 15°) + \sin (\theta + 45°)$$

and

$$y = \cos (\theta + 105°) + \cos (\theta - 15°) + \cos (\theta + 45°)$$

show that

$$\frac{x}{y} = \tan (\theta + 45°).$$

Express

$$\frac{x}{y} - \frac{y}{x}$$

in terms of $\tan 2\theta$.   (JMB)

33   Assuming that $\sin x \approx a + bx$ and $\cos x \approx c + dx$, for small values of x, use the method of §18.4 to prove that $a = 0 = d$ and $b = 1 = c$. Assuming quadratic approximations for $\sin x$ and $\cos x$ which give rise to the above linear approximations by differentiation, show that, for small x,

$$\sin x \approx x,$$
$$\cos x \approx 1 - \tfrac{1}{2}x^2,$$
$$\tan x \approx x.$$

22 Differentiation Techniques and Applications

In the study of differentiation in Chapter 18, we considered the gradient of the graph of a function f, that is, the gradient of f, and this was the value of the derived function f', also called the derivative of f. The functions concerned had values given explicitly. This means that the value y of $f(x)$ is given explicitly by an equation $y = f(x)$. Then the derivative $\dfrac{dy}{dx} = f'(x)$ was determined for a number of types of functions: polynomials, circular functions, sums, products, quotients and composition of functions.

However, it often happens that the function f may not be given explicitly by an equation $y = f(x)$. Rather, the value of y is connected to x by means of some equation or equations which are sufficient to determine the function. We shall consider some possibilities where this situation can arise before we move on to the applications of differentiation. We shall use the idea that, given an equation relating functions, that is, an identity in x, we may differentiate both sides of the equation to obtain another equation relating the derivatives of the functions. Thus, if $F(x) = G(x)$ then $F'(x) = G'(x)$.

22.1 Derivative of an inverse function

Suppose that we have a one-one function f, with domain A and range B. Then f has an inverse function f^{-1}, with domain B and range A. For convenience, write $f^{-1} = g$ so that, for all x in A, $gf(x) = x$. Then, if $y = f(x)$, $x = f^{-1}(y) = g(y)$ so

$$\frac{dx}{dy} = g'(y) = g'(f(x)) = g'f(x).$$

Using the chain rule for the derivative of a composite function:

$$1 = \frac{dx}{dx} = \frac{dg}{dx} = \frac{dg}{dy} \cdot \frac{dy}{dx} = g'(y)f'(x).$$

This is usually used when we are given f as an inverse function and know the value of g'(y), so the equation is rearranged:

$$f'(x) = \frac{1}{g'(y)} = \frac{1}{g'f(x)} \quad \text{so} \quad f' = \frac{1}{g'f},$$

or

$$\frac{dy}{dx} = \frac{1}{\frac{dx}{dy}}.$$

EXAMPLE 1   *Differentiate with respect to x the function* f, *given by* f $(x) = \sqrt[3]{x}$.

The function f has an inverse function g, where g(y) = y^3. Therefore, g'(y) = $3y^2$, so

$$f'(x) = \frac{1}{g'f(x)} = \frac{1}{3y^2} = \frac{1}{3(\sqrt[3]{x})^2} = \tfrac{1}{3}x^{-\tfrac{2}{3}}.$$

This is the same result as arises from the formula $\dfrac{d}{dx}x^n = nx^{n-1}$, $n = \tfrac{1}{3}$.

EXAMPLE 2   *Find the derivative of* $\tan^{-1}x$.

Let $\tan^{-1}x = y$, then $x = \tan y$, and we differentiate this equation with respect to x.

$$1 = \frac{dx}{dx} = \frac{d}{dx}\tan y = \frac{d}{dy}\tan y\,\frac{dy}{dx} = \sec^2 y\,\frac{dy}{dx}.$$

Therefore,

$$\frac{dy}{dx} = \frac{1}{\sec^2 y} = \frac{1}{1+\tan^2 y} = \frac{1}{1+x^2}.$$

EXAMPLE 3   *Find* $\dfrac{dy}{dx}$, *given that* $y = \cos^{-1}x$.

Since $y = \cos^{-1}x$, $x = \cos y$ and so

$$1 = -\sin y\,\frac{dy}{dx} = -\sqrt{(1-\cos^2 y)}\,\frac{dy}{dx}$$

and $\dfrac{dy}{dx} = \dfrac{-1}{\sqrt{(1-x^2)}}$, the negative sign arising from the fact that $\cos^{-1}x$ is a decreasing function.

EXERCISE 22.1

1   Find $\dfrac{d}{dx}\sin^{-1}x$, and verify that $\dfrac{d}{dx}(\sin^{-1}x + \cos^{-1}x) = 0$. Explain this result.

2   Find the derivative of:
   (i) $\tan^{-1}2x$,   (ii) $\sin^{-1}(3x-1)$,   (iii) $\arccos x^2$,   (iv) $(\arccos x)^2$.

3   Find $\dfrac{dy}{dx}$, given that:
   (i) $y = x - \tan^{-1}x$,   (ii) $y = \sin^{-1}(x^2+2x)$,   (iii) $y = \cos(\sin^{-1}x)$.

22.2 Implicit differentiation

In cases when the function is defined implicitly by means of some equation relating x and $y = f(x)$, we obtain the derivative of f by differentiating the equation and using the chain rule.

EXAMPLE 1 *The function f, with domain $\{x: -a \leqslant x \leqslant a\}$, is defined by the equations $y = f(x)$, $x^2 + y^2 = a^2$, $y \geqslant 0$. Find $f'(x)$.*

Since $x^2 + y^2 = a^2$, $2x + 2y\dfrac{dy}{dx} = 0$, so $\dfrac{dy}{dx} = -\dfrac{x}{y}$. Therefore,

$$f'(x) = -\frac{x}{\sqrt{(a^2 - x^2)}}.$$

Note that in this case, the graph of f is the upper half circle, centre the origin, radius a. Had f been defined so that $f(x)$ was the negative value of y satisfying the equation, then we should still have had $x^2 + y^2 = a^2$, and $2x + 2y\dfrac{dy}{dx} = 0$, so that $\dfrac{dy}{dx} = -\dfrac{x}{y}$. However, since y is now negative,

$f'(x) = \dfrac{x}{\sqrt{(a^2 - x^2)}}$. The difference in signs between the derivatives of the

two functions reflects the fact that, for the same value of x, the gradients of the upper and lower half circles are of opposite sign.

EXAMPLE 2 *Find the equation of the tangent to the curve, with equation $y^3 - 2x^2y + x^3 = 0$, at the point $(1, 1)$.*

Differentiating with respect to x, $3y^2\dfrac{dy}{dx} - 2\left(2xy + x^2\dfrac{dy}{dx}\right) + 3x^2 = 0$. When

$$x = 1 = y, \ 3\frac{dy}{dx} - 2\left(2 + \frac{dy}{dx}\right) + 3 = 0 \quad \text{so} \quad \frac{dy}{dx} = \frac{4-3}{3-2} = 1,$$

and this is the gradient of the tangent. So, the equation of the tangent is $y - 1 = 1(x - 1)$, that is, $y = x$.

EXERCISE 22.2

1 Find $\dfrac{dy}{dx}$ in terms of x and y, given that:
   (i) $2x^2 + 3y^2 = 6$, (ii) $x \sin y + y \cos x = 0$, (iii) $x^3y + y^3x = a^4$.
2 Find the equation of the tangent at the given point to the curve, with the given equation:
   (i) $(1, 1)$, $x^3 + 2y^3 = 3$; (ii) $(1, \frac{1}{2}\pi)$, $xy + \cos(xy) = \frac{1}{2}\pi$;
   (iii) $(1, 2)$ $xy^3 = 4 + x^2y^2$.

22.3 Parametric differentiation

Suppose that the function f, with $y = f(x)$, is defined indirectly, with x and y each being given in terms of a parameter t. Thus

$$x = x(t) \text{ and } y = y(t).$$

Then we can differentiate the equation $y = f(x)$ with respect to t.

$$\frac{dy}{dt} = f'(x)\frac{dx}{dt} = \frac{dy}{dx}\frac{dx}{dt},$$

by the chain rule. Therefore, provided that $\frac{dx}{dt}$ is not zero, we can divide by it, to give

$$\frac{dy}{dx} = \frac{dy}{dt} \bigg/ \frac{dx}{dt}.$$

EXAMPLE 1 *Find the tangent to the curve given by the equations $y = 2t^3$, $x = t^2$ at the point where $t = 2$.*

We first find the gradient of the curve, $\frac{dy}{dt} = 6t^2$, $\frac{dx}{dt} = 2t$, so $\frac{dy}{dx} = \frac{6t^2}{2t} = 3t$, and at the point $(4, 16)$ the gradient is 6. Therefore, the equation of the tangent is $y - 16 = 6(x - 4)$ or $\mathbf{y = 6x - 8}$.

EXAMPLE 2 *Find $\frac{dy}{dx}$, given that $y = a\sin\theta$, $x = a\cos\theta$.*

$\frac{dy}{d\theta} = a\cos\theta$, $\frac{dx}{d\theta} = -a\sin\theta$, so $\frac{dy}{dx} = \frac{a\cos\theta}{-a\sin\theta} = -\cot\theta$.

Note that this means that $\frac{dy}{dx} = -\frac{x}{y}$, agreeing with Example 1 of §22.2.

EXERCISE 22.3

1   Find the equation of the tangent to the curve given by $y = \tan t$, $x = \sec t$, at the point where $t = \frac{1}{4}\pi$.

2   A curve is given parametrically by the equations $y = t\cos t - \sin t$, $x = t\sin t + \cos t$. Prove that the tangent to the curve at the point P, where $t = p$, has equation $y\cos p + x\sin p = p$.

3   A point $P(x, y)$ moves in a plane so that, at time t, its coordinates are given by $x = a\sin t$, $y = a\cos t$. Find the equation of the tangent to the path at time t. Find also the values of t for which the particle is travelling parallel to the line $y = x$.

4   The parametric equations of a curve are $y = a\sin\theta$, $x = b\cos\theta$. Find the gradient of the curve at the point (p, q) of parameter θ:
(i) in terms of θ, (ii) in terms of p and q. Find the Cartesian equation of the

. curve and, by differentiating this equation, find $\dfrac{dy}{dx}$. Verify that this answer is the same as the answers to (i) and (ii) above.

22.4 Higher derivatives

The velocity of a point P, moving along the axis Ox, when its distance from O is $x = f(t)$ in terms of the time t, is $v = f'(t) = \dfrac{dx}{dt}$. The rate of change of the velocity is the acceleration a of the particle. Thus

$$a = \frac{dv}{dt} = (f'(t))' = \frac{d}{dt}\left(\frac{dx}{dt}\right).$$

The derivative of the derivative of f is $(f')'$ and this is called the *second derivative* of f.

Notation $(f')' = f'', f''(t) = \dfrac{d}{dt}\left(\dfrac{dx}{dt}\right) = \dfrac{d^2x}{dt^2}$, where $f(t) = x$.

As with the differential operator $\dfrac{d}{dt}$, the second differential operator $\dfrac{d^2}{dt^2}$ is a single symbol and not a fraction.

Similarly, if $y = f(x)$, with a derivative $\dfrac{dy}{dx} = f'(x)$, then the second derivative is

$$\frac{d^2y}{dx^2} = f''(x) = \frac{d}{dx}\left(\frac{dy}{dx}\right).$$

Higher derivatives can also be defined, thus

$$\frac{d}{dx}\left(\frac{d^2y}{dx^2}\right) = \frac{d^3y}{dx^3} = f'''(x),$$

and so on. For example,

$$\frac{d}{dx}x^4 = 4x^3, \frac{d^2}{dx^2}x^4 = \frac{d}{dx}4x^3 = 12x^2, \frac{d^3}{dx^3}x^4 = \frac{d}{dx}12x^2 = 24x.$$

EXAMPLE 1   *Find the second derivative of* $3x^2 + x \sin x$.

Let $y = 3x^2 + x \sin x$, then $\dfrac{dy}{dx} = 3(2x) + \sin x + x \cos x$, differentiating the product in the usual manner. Differentiating again,

$$\frac{d^2y}{dx^2} = 6 + \cos x + \cos x - x \sin x = \mathbf{6 + 2\cos x - x\sin x}.$$

In determining higher derivatives of a function which is not defined explicitly, care must be taken to use the chain rule correctly and also to differentiate correctly any products that occur. For example, if $y = f(x)$ is given by an inverse function $x = g(y)$, then $\dfrac{dy}{dx} = \dfrac{1}{g'(y)} = \dfrac{1}{\dfrac{dy}{dx}}$.

Differentiating again

$$\frac{d^2y}{dx^2} = \frac{d}{dx}\left(\frac{1}{g'(y)}\right) = \frac{d}{dy}\left(\frac{1}{g'(y)}\right)\frac{dy}{dx}.$$

If $u = g'(y)$, $\dfrac{d}{dy}\dfrac{1}{u} = \dfrac{-1}{u^2}\dfrac{du}{dy} = \dfrac{-1}{(g'(y))^2}g''(y)$

and so $\dfrac{d^2y}{dx^2} = \dfrac{-g''(y)}{(g'(y))^2}\dfrac{1}{g'(y)} = \dfrac{-g''(y)}{(g'(y))^3} = \dfrac{\dfrac{d^2x}{dy^2}}{\left(\dfrac{dx}{dy}\right)^3}.$

EXAMPLE 2   *Given that* $x^3 - 3x = y^3$, *find* $\dfrac{d^2y}{dx^2}$.

Differentiating the equation with respect to x,

$$3x^2 - 3 = 3y^2\frac{dy}{dx}, \text{ and so } x^2 - 1 = y^2\frac{dy}{dx}.$$

Differentiating again,

$$2x = y^2\frac{d^2y}{dx^2} + \left(2y\frac{dy}{dx}\right)\frac{dy}{dx} = y^2\frac{d^2y}{dx^2} + 2y\left(\frac{dy}{dx}\right)^2.$$

Thus $\dfrac{d^2y}{dx^2} = \dfrac{1}{y^2}\left(2x - 2y\left(\dfrac{dy}{dx}\right)^2\right) = 2\dfrac{x}{y^2} - 2\dfrac{1}{y}\left(\dfrac{x^2-1}{y^2}\right)^2.$

EXAMPLE 3   *Given that* $x = t^2, y = t^3$, *are the parametric equations of a curve, find the gradient of the curve at the point with parameter t, and find also the rate of change of the gradient with respect to x*

Differentiate the parametric equations to give $\dfrac{dx}{dt} = 2t, \dfrac{dy}{dt} = 3t^2$, so

$$\frac{dy}{dx} = \frac{\dfrac{dy}{dt}}{\dfrac{dx}{dt}} = \frac{3t^2}{2t} = \tfrac{3}{2}t,$$

and this is the gradient of the curve. The rate of change of the gradient with respect to x is

$$\frac{d^2y}{dx^2} = \frac{d}{dt}\left(\frac{dy}{dx}\right)\frac{dt}{dx}$$

that is,

$$\frac{d^2y}{dx^2} = \frac{\frac{3}{2}}{\dfrac{dx}{dt}} = \frac{\frac{3}{2}}{2t} = \frac{3}{4t}.$$

EXERCISE 22.4

1 Find $\dfrac{d^2x}{dt^2}$, given that:

   (i) $x = t^5$, (ii) $x = t^4 + t^2$, (iii) $x = \dfrac{1}{t^2}$, (iv) $x = t\cos t$, (v) $x = \sin(t^2)$,

   (vi) $x = \tan t$, (vii) $x = \tan^{-1} t$.

2 Determine $\dfrac{dy}{dx}, \dfrac{dx}{dy}, \dfrac{d^2y}{dx^2}, \dfrac{d^2x}{dy^2}$, given that:

   (i) $y = 2t - 1$, $x = \dfrac{1}{t}$; (ii) $y = t^2$, $x = t^3$; (iii) $y = \sin t$, $x = \cos t$;

   (iv) $y = 2at$, $x = at^2$; (v) $y = t\sin t$, $x = t\cos t$; (vi) $y = \sec t$, $x = \tan t$.

3 Find the equation of the tangent at the point $(1, 2)$ to the curve with equation $4x^2 + 5y^2 = 24$.

4 Find $\dfrac{dy}{dx}$ and $\dfrac{d^2y}{dx^2}$, in terms of x and y, given that:

   (i) $x^2 + xy + y^2 = 3$, (ii) $\sin x + \cos y = 1$, (iii) $\sin y = \tan x$.

5 Given that $y = \dfrac{1 - \sin x}{\cos x}$, show that $\dfrac{dy}{dx} = -\dfrac{1}{1 + \sin x}$ and express $\dfrac{d^3y}{dx^3}$ in terms of $\sin x$ only, in its simplest form.

6 Prove that if $f'(a) > 0$, then $f(x)$ is increasing, as x increases, at $x = a$, and if $f'(a) < 0$ then $f(x)$ is decreasing at $x = a$.

7 Given that $f'(a) = 0$, and that $f''(a) < 0$, prove that near to the point $(a, f(a))$ the gradient of the graph of f is changing from positive to negative. Use a sketch to indicate the general shape of the graph near to the point where $x = a$.

22.5 Rates of change

Velocity and acceleration

We have already defined the velocity and acceleration of a point moving in a straight line as the first and second derivatives of the displacement with respect to time. A similar definition may be made for motion in two or three dimensions. We shall deal with the two-dimensional case, which is easily extended to three dimensions.

Referred to Cartesian axes in the plane, the position of a point P is given by its coordinates (x, y). If P is moving, then x and y will be functions of the time t, $x = x(t)$ and $y = y(t)$. The position vector of P is

$$\mathbf{r} = \begin{pmatrix} x(t) \\ y(t) \end{pmatrix} = \mathbf{r}(t),$$

so that this vector is a function of the time too. We define the derivative with respect to t of $\mathbf{r}$ to be the vector whose components are the time derivatives of the components of $\mathbf{r}$, so

$$\frac{d\mathbf{r}}{dt} = \begin{pmatrix} \dfrac{dx}{dt} \\[2mm] \dfrac{dy}{dt} \end{pmatrix}.$$

Definition $\dfrac{d\mathbf{r}}{dt} = \mathbf{v}(t) =$ the velocity of P at time t,

$$\frac{d^2\mathbf{r}}{dt^2} = \frac{d\mathbf{v}}{dt} = \mathbf{a}(t) = \text{the acceleration of } P \text{ at time } t.$$

Thus, the velocity is a vector which is the first time derivative of the position vector, and the acceleration is a vector which is the time derivative of the velocity and is, therefore, the second derivative of the position vector.

It is common in kinematics (the study of motion) to denote differentiation with respect to time t by placing a dot over the scalar or vector concerned, thus $\dfrac{dx}{dt} = \dot{x}$ and $\dfrac{d\mathbf{r}}{dt} = \dot{\mathbf{r}}$. Then the second derivative is denoted by two dots over the symbol, thus $\dfrac{d^2x}{dt^2} = \ddot{x}$.

EXAMPLE 1   *A particle P moves in a vertical plane and its acceleration is* $\begin{pmatrix} 0 \\ -g \end{pmatrix}$, *referred to Cartesian axes, Ox horizontal and Oy vertical upwards. If the particle is projected from O at time $t = 0$ with an initial velocity* $\begin{pmatrix} u \\ v \end{pmatrix}$, *find the position vector of P at time t.*

We may integrate the equation $\dfrac{d^2\mathbf{r}}{dt^2} = \begin{pmatrix} 0 \\ -g \end{pmatrix}$, with initial conditions $\dfrac{d\mathbf{r}}{dt} = \begin{pmatrix} u \\ v \end{pmatrix}$ when $t = 0$, to give $\dfrac{d\mathbf{r}}{dt} = \begin{pmatrix} u \\ v - gt \end{pmatrix}$, and this is the velocity of P at time t. A further integration, with the initial condition $\mathbf{r} = \begin{pmatrix} 0 \\ 0 \end{pmatrix}$ when $t = 0$, gives $\mathbf{r} = \begin{pmatrix} ut \\ vt - \frac{1}{2}gt^2 \end{pmatrix}$. The path of P is a parabola, with vertex upwards, the path of a projectile moving under gravity.

Angular velocity and acceleration

Suppose that the line $\overrightarrow{OP}$ is rotating in the plane about O. For example, let P be the point $(a \cos \theta, a \sin \theta)$ relative to axes $O(x, y)$ so that OP makes a (positive) angle θ with the axis Ox. Then we define

$$\frac{d\theta}{dt} = \text{the angular velocity of } \overrightarrow{OP} \text{ about } O,$$

$$\frac{d^2\theta}{dt^2} = \text{the angular acceleration of } \overrightarrow{OP} \text{ about } O.$$

EXAMPLE 2 *The coordinates (x, y) of a point P, moving in a plane, are given in terms of the time t by $x = a \cos pt$, $y = b \sin pt$, where a, b and p are constants. Show that the position vector of P satisfies the equation $\ddot{\mathbf{r}} = -p^2 \mathbf{r}$. Find the angular velocity of $\overrightarrow{OP}$ at time t.*

$$\dot{\mathbf{r}} = \begin{pmatrix} \dot{x} \\ \dot{y} \end{pmatrix} = \begin{pmatrix} -ap \sin pt \\ bp \cos pt \end{pmatrix},$$

and so

$$\ddot{\mathbf{r}} = \begin{pmatrix} \ddot{x} \\ \ddot{y} \end{pmatrix} = \begin{pmatrix} -ap^2 \cos pt \\ -bp^2 \sin pt \end{pmatrix} = -p^2 \mathbf{r}.$$

Let OP make an angle θ with Ox so that $x = r \cos \theta$ and $y = r \sin \theta$. Then $\tan \theta = \dfrac{y}{x} = \dfrac{b}{a} \tan pt$ and so $\sec^2 \theta \dfrac{d\theta}{dt} = p \dfrac{b}{a} \sec^2 pt$. The angular velocity of $\overrightarrow{OP}$ is

$$\frac{d\theta}{dt} = \frac{pb}{a} \frac{\sec^2 pt}{\sec^2 \theta} = \frac{pab}{a^2 \cos^2 pt (1 + \tan^2 \theta)} = \frac{pab}{x^2 (1 + y^2/x^2)}$$

$$= \frac{pab}{x^2 + y^2} = \frac{pab}{a^2 \cos^2 pt + b^2 \sin^2 pt}.$$

Related rates of change

When two quantities are related by a functional relationship, possibly given by an equation, then, if both are varying, there will be a connection between their rates of change. For example, if $y = f(x)$ and both x and y vary with time t, then their rates of change $\dfrac{dy}{dt}$ and $\dfrac{dx}{dt}$ are connected by the chain rule, $\dfrac{dy}{dt} = \dfrac{dy}{dx} \dfrac{dx}{dt}$, and $\dfrac{dy}{dx}$ is the rate of change of y with respect to x.

EXAMPLE 3 *Water is dropping into a circular puddle at a uniform rate of $1 \, cm^3 \, s^{-1}$. The depth of water in the puddle remains constant at $0.1 \, cm$. When the diameter of the puddle is $10 \, cm$, find: (i) the rate of increase of the area, (ii) the rate of increase of the circumference, (iii) the rate of increase of the diameter, of the puddle.*

Suppose that, at time t, the volume of water in the puddle is V, the area of the puddle is A, the circumference is C, the radius is R and the depth is D. Then we are told that $D = 0.1$ cm, a constant, and that $\dfrac{dV}{dt} = 1 \text{ cm}^3 \text{ s}^{-1}$, when t is measured in seconds.

(i) $\dfrac{dV}{dt} = \dfrac{d}{dt} DA = D\dfrac{dA}{dt}$, since D is constant, so

$$\dfrac{dA}{dt} = \dfrac{1}{D}\dfrac{dV}{dt} = \dfrac{1}{0.1}.1 = \mathbf{10\,cm^2\,s^{-1}}.$$

(ii) $\dfrac{dC}{dt} = \dfrac{d}{dt} 2\pi R = 2\pi \dfrac{dR}{dt}$ and $\dfrac{dA}{dt} = \dfrac{d}{dt} \pi R^2 = 2\pi R\dfrac{dR}{dt}$, so

$$\dfrac{dC}{dt} = 2\pi \dfrac{\dfrac{dA}{dt}}{2\pi R} = \dfrac{1}{R}\dfrac{dA}{dt} = \dfrac{1}{5}.10 = \mathbf{2\,cm\,s^{-1}}.$$

(iii) The diameter is $2R$ and its rate of change is $2\dfrac{dR}{dt} = \dfrac{1}{\pi}\dfrac{dC}{dt}$ which is $\mathbf{2/\pi\,cm\,s^{-1}}$.

EXAMPLE 4   *Find the rate of increase of the surface area of a spherical balloon, when its radius is r, if the volume of the balloon is increasing at a rate q.*

Let the volume be V and the surface area S. Then $V = \frac{4}{3}\pi r^3$ and $S = 4\pi r^2$. Hence

$$\dfrac{dV}{dt} = \dfrac{dV}{dr}\dfrac{dr}{dt} = 4\pi r^2 \dfrac{dr}{dt} \quad \text{and} \quad \dfrac{dS}{dt} = \dfrac{dS}{dr}\dfrac{dr}{dt} = 8\pi r \dfrac{dr}{dt}.$$

Therefore, $\qquad \dfrac{dS}{dt} = 8\pi r \dfrac{dV}{dt}\dfrac{1}{4\pi r^2} = \dfrac{2}{r}\dfrac{dV}{dt} = \dfrac{2q}{r}.$

EXERCISE 22.5

1   At time t, the distance from O of a point P moving along the x-axis, is given by $x = t^3 - 2t^2 + t$. Find the velocity and the acceleration of P in terms of t. State the interval of time during which the point P is moving towards O.

2   Find the velocity and acceleration, at time t, of the point P, whose coordinates are (t^3, t^4), relative to Cartesian axes $O(x, y)$. Find also the angular velocity of OP.

3   The ends A and B of a rod of length L slide on the axes Ox and Oy, respectively and $OA = a$, $OB = b$. The angle BAO is θ. Express a and b in terms of θ and, given that $\theta = \pi/4$ and $\dfrac{da}{dt} = 2$, find $\dfrac{db}{dt}$.

4   The product PV of the pressure, P, and the volume, V, of a gas is a constant. Given that $\dfrac{dV}{dt} = -2$, express $\dfrac{dP}{dt}$ in terms of P and V.

5   A cube, of side 10 cm, is being heated so that its volume is changing. (i) Find the rate of change in the volume and the rate of change of the surface area given that the rate of increase of the length of each edge is 1 mm per second.

(ii) Find the rate of increase of the length of each edge, and the rate of increase of the surface area, if the volume is increasing at the rate $8\,\text{cm}^3$ per second.

6 At time t, the position vector $\mathbf{r}$ of a point P is given by $\mathbf{r} = t\sin t\mathbf{i} + t\cos t\mathbf{j}$, where $(\mathbf{i}, \mathbf{j})$ are Cartesian unit vectors. Find the velocity and the acceleration of P and the angular velocity of $\overline{OP}$.

7 The volume of a circular cylinder, of height h and radius r, is constant and equal to V. If h is changing at a rate $p = \dfrac{\mathrm{d}h}{\mathrm{d}t}$, find the rate of change of r and of the total surface area of the cylinder.

22.6 Approximations to small changes

In our development of the theory of differentiation in Chapter 18, we showed that the derivative $\mathrm{f}'(a)$ of the function f is given by $\mathrm{f}'(a) = m$, where the line $y = \mathrm{f}(a) + m(x - a)$ is a linear approximation to the graph of f near to $x = a$. In Fig 22.1 we show the graph of f near to the point $P\,(a, \mathrm{f}(a))$. PT is the tangent to the graph at P and PT has gradient $m = \mathrm{f}'(a)$. $P'\,(a + \delta x, \mathrm{f}(a + \delta x))$ is a point on the graph near to P and P' is approximated to by the point $Q\,(a + \delta x, \mathrm{f}(a) + m\delta x)$ on PT. As x changes from a to $a + \delta x$, the value of f changes from y to $y + \delta y$, where $y = \mathrm{f}(a)$ and $y + \delta y = \mathrm{f}(a + \delta x)$, while the corresponding change in the linear approximation to y is $m\delta x = \mathrm{f}'(a)\,\delta x$. The linear approximation, therefore, gives

$$\delta y \approx \mathrm{f}'(a)\,\delta x = \frac{\mathrm{d}y}{\mathrm{d}x}\,\delta x.$$

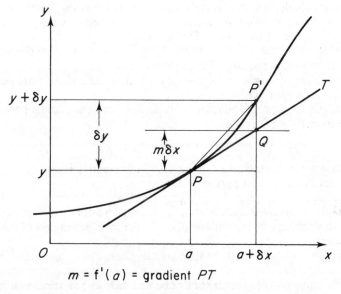

$$m = \mathrm{f}'(a) = \text{gradient } PT$$

Fig. 22.1

EXAMPLE 1   *Given that $y^3 = 2x^2$ and that x increases by 1%, find the percentage increase in y.*

Differentiating implicity, $3y^2 \dfrac{dy}{dx} = 4x$. Let the small change in x be δx, so that $\delta x = x/100$, then the change in y is δy which is given approximately by $\delta y \approx \dfrac{dy}{dx}\delta x$, so the percentage change in y is

$$\frac{100\,\delta y}{y} \approx \frac{100}{y}\frac{dy}{dx}\delta x = \frac{100}{y}\frac{4x}{3y^2}\delta x = \frac{400x\delta x}{3y^3}$$

$$= \frac{400x\,\delta x}{6x^2} = \frac{2\delta x}{3x}100 = \frac{2}{3}.$$

So the increase in y is $\frac{2}{3}\%$.

EXAMPLE 2   *Find an approximate value of $(0\cdot99)^5$.*

Let $f(x) = x^5$, so that $f'(x) = 5x^4$. Since $0\cdot99 = 1 - 0\cdot01$, we put $a = 1$, $\delta x = -0\cdot01$, so that $(a + \delta x) = 0\cdot99$. Then

$$f(0\cdot99) = (0\cdot99)^5 = f(a + \delta x) \approx f(a) + f'(a)\delta x = 1 - 5(0\cdot01),$$

so the approximation to $(0\cdot99)^5$ is **$0\cdot95$**.

EXERCISE 22.6

1   Given that $y = 4x^3 + 2x^2$ and that x increases from 100 to 101, find the approximate increase in y.

2   You intend to measure the length of a piece of string wrapped ten times round a circular cylinder and then calculate the radius of the cylinder. If you make an error of $1\,mm$ in your measurement, find the approximate error in the calculated value of the radius. Explain why the error in the radius does not depend upon the size of the cylinder being measured, but that the percentage error decreases as the size of the cylinder increases.

3   If the volume of a sphere increases by 6% find the approximate percentage increase in the surface area.

4   The length l of a simple pendulum, its period T and the constant acceleration g due to gravity are related by

$$T = 2\pi\sqrt{(l/g)}.$$

Evaluate the percentage error in the length of a simple pendulum which would cause a 1% error in its period.   (*L*)

5   The volume V of a sphere is to be obtained from the formula $V = \pi d^3/6$ after the diameter d is measured. Without using a calculator or tables, estimate the percentage change in the volume due to a percentage increase of $0\cdot5\%$ in the diameter.   (*L*)

6   Given that $s = t^5 - t^3$, find the approximate change in s when t changes from 2 to $2\cdot01$.

7   Find the approximate volume of a cube with side $2\cdot98\,m$, treating it as a small change from a cube of side $3\,m$.

22.7 Stationary points

The shape of the graph of a function f, with equation $y = f(x)$, can be studied with the help of the derivative f' of f. Consider the shape of the graph near to the point where $x = a$, say the point $A(a, f(a))$ on the graph. When $f'(a)$ is positive then $f(x)$ is increasing as x increases at A, and when $f'(a)$ is negative $f(x)$ is decreasing at A. We shall omit the words 'as x increases at A' since we shall always consider what happens to $f(x)$ as x increases. When $f'(a) = 0$ then the tangent to the graph at A is parallel to Ox and we do not know if $f(x)$ is increasing or decreasing. In fact, the linear approximation to $f(x)$ at A is the constant function $y = f(a)$, which is neither increasing nor decreasing but is stationary. So, in this case, the point A is called a stationary point of f. Clearly the stationary points of f can be found by solving the equation $f'(x) = 0$.

To understand in more detail what the graph looks like near to a stationary point, we study the signs of $f'(x)$ near to the stationary point. Suppose that near to A, $f'(x) > 0$ if $x < a$, and $f'(x) < 0$ if $x > a$, as shown in Fig 22.2. Then, as x increases through A, $f'(x)$ changes from positive to negative and the graph changes from increasing to decreasing. Thus $f(a)$ is greater than $f(x)$ for values of x near to a. We say that the point A is a *maximum*, or f has a maximum at $x = a$. This does not mean that $f(a)$ is greater than $f(x)$ for all other values of x. There could be a point B, where $x = b$, and $f(b) > f(a)$, as shown in Fig. 22.2. To be correct, we should say that A is a local maximum, meaning that we are only considering the graph near to A, but, for convenience, the word 'local' is usually omitted.

Similarly, if $f'(c) = 0$ and C is the point on the graph where $x = c$, and, near to C, if $x < c$ then $f'(x) < 0$, and if $x > c$ then $f'(x) > 0$, then at C $f'(x)$ changes from negative to positive, the graph changes from decreasing to increasing, and $f(c)$ is less than neighbouring values of $f(x)$. Therefore, in this case, we call C a (local) *minimum* (see Fig. 22.2).

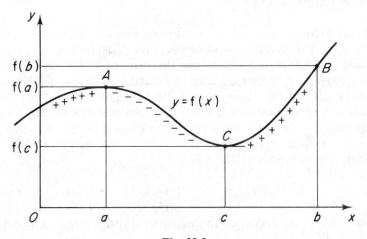

Fig. 22.2

We, therefore, need to take three steps in the investigation.

Step 1 Find $f'(x)$ and solve the equation $f'(x) = 0$. Let $x = a$ be a solution, so $f'(a) = 0$. Calculate $f(a)$, then the point $(a, f(a))$ is a stationary point.

Step 2 Consider values of x near to $x = a$ and determine the signs of $f'(x)$ near $x = a$.

Step 3 If $f'(x)$ changes from positive to negative at $x = a$ then the stationary point is a maximum.

If $f'(x)$ changes from negative to positive at $x = a$ then the stationary point is a minimum.

EXAMPLE 1 *Find the stationary values of the function* f, *given by* $f(x) = x^3 - 3x^2 - 24x + 20$, *and determine their nature.*

Step 1 $f'(x) = 3x^2 - 6x - 24 = 3(x^2 - 2x - 8) = 3(x - 4)(x + 2)$, so the stationary points of f occur when $x = 4$ and $x = -2$, and at these points

$$f(4) = 64 - 48 - 96 + 20 = -60, \quad f(-2) = -8 - 12 + 48 + 20 = 60.$$

Step 2 Consider values of $f'(x)$ near the stationary values.

Let e be a small positive number, then list the sign of the factors of $f'(x)$:

x	sign of $(x - 4)$	sign of $(x + 2)$	sign of $f'(x)$
$4 - e$	$-$ ve	$+$ ve	$-$ ve
$4 + e$	$+$ ve	$+$ ve	$+$ ve
$-2 - e$	$-$ ve	$-$ ve	$+$ ve
$-2 + e$	$-$ ve	$+$ ve	$-$ ve

Step 3 At $x = 4$, $f'(x)$ is increasing from negative to positive and so the point **(4, −60) is a minimum** of f.

At $x = -2$, $f'(x)$ is decreasing from positive to negative and so the point **(−2, 60) is a maximum** of f.

Another method of deciding whether or not the stationary point is a maximum or a minimum is to use the second derivative of f. If, at the stationary point, $f''(a)$ is positive, then this means that $f'(x)$ is increasing at that point so it must be increasing from negative to positive and the value is a minimum. If, on the other hand, $f''(a)$ is negative, then $f'(x)$ must be decreasing, and so must be decreasing from positive to negative, and the stationary value is a maximum. This argument relies on the fact that $f''(x) = (f')'(x)$ and so its sign determines the sign of the gradient of f'. The following is a summary of the method.

Step 1 Find f', solve $f'(x) = 0$ giving $x = a$, and find $f(a)$.

Step 2 Find f'', test its sign,   if $f''(a) < 0$ the point is a maximum,
if $f''(a) > 0$ the point is a minimum.

Note that it is not necessary to calculate $f''(a)$ accurately, but only to be sure of its sign.

EXAMPLE 2 *Find the maximum and minimum points of* f *where*
$f(x) = x^3 - 3x + 2.$

Step 1 $f'(x) = 3x^2 - 3 = 3(x-1)(x+1)$, so $f'(x) = 0$ at $x = 1$, $x = -1$. The stationary values are $f(1) = 0$ and $f(-1) = 4$.
Step 2 $f''(x) = 6x$, so $f''(1) > 0$ and f has a minimum at **(1, 0)**,
$\qquad\qquad$ also $f''(-1) < 0$ and f has a maximum at **(-1, 4)**.

Sometimes this method will not give the answer because the second derivative vanishes at the stationary point. Then the first method is used, as shown in the next example.

EXAMPLE 3 *Find, and determine the nature of, the stationary points of the graph with equation* $y = x^4$.

$\dfrac{dy}{dx} = 4x^3$, which vanishes at the one point, $x = 0$, so there is just one stationary

point at (0, 0). The second derivative $\dfrac{d^2y}{dx^2} = 12x^2$ and this vanishes when $x = 0$. So

we have to look at the sign of $\dfrac{dy}{dx}$ at values near $x = 0$. Suppose that x is negative,

then $\dfrac{dy}{dx} = 4x^3$ is also negative, and, if x is positive, then $\dfrac{dy}{dx}$ is positive. So, at the

stationary point **(0, 0)**, $\dfrac{dy}{dx}$ is increasing from negative to positive and, therefore,

the stationary value is **a minimum**.

The one case which we have not dealt with is when $f'(a) = 0$ but $f'(x)$ is the same sign at points near to $x = a$. If $f'(x)$ is positive at all points near $x = a$, then the stationary point is a minimum point of the derived function $f'(x)$. Similarly, when $f'(x)$ is negative near $x = a$ on both sides, then the point is a maximum point for f'. In both these cases, the second derivative f'' will also vanish at $x = a$, because we have a stationary value of f'. A point on the graph where f'' vanishes and changes sign as x increases is called a *point of inflexion*.

At a point of inflexion, $A(a, f(a))$, $f''(a) = 0$, but it is not necessary that $f'(a)$ should also be zero. Since, as x increases through a, $f''(x)$ changes sign, the graph changes from curving one way to curving the other way and so, at A, the graph crosses the tangent at A. This is another way of defining a point of inflexion, and various cases are shown in Fig 22.3.

At a point of inflexion where $f'(a) = 0$ as well as $f''(a)$, the tangent is horizontal, that is, parallel to Ox, and, since $f'(x)$ does not change sign, either the graph continues increasing each side of the inflexion or it continues to decrease each side. The two cases may be referred to as increasing point of inflexion and decreasing point of inflexion.

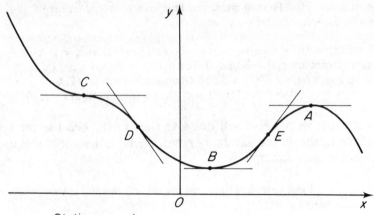

Stationary values
  A : maximum
  B : minimum
  C : decreasing inflexion
Points of inflexion : *C,D,E*, where
    curve crosses tangent

Fig. 22.3

EXAMPLE 4   *The function* f *is given by* $f(x) = 3x^4 + 4x^3 - 6x^2 - 12x + 5$. *Find the stationary values of* f *and determine their nature. Find also any non-stationary points of inflexion on the graph of* f, *and sketch this graph.*

We first find the first and second derivatives of f and the values of x where they vanish.

$$f'(x) = 12x^3 + 12x^2 - 12x - 12 = 12(x^3 + x^2 - x - 1)$$
$$= 12(x+1)(x^2 - 1) = 12(x+1)^2(x-1),$$
$$f''(x) = 12(3x^2 + 2x - 1) = 12(x+1)(3x-1).$$

Therefore, $f'(x)$ vanishes at A, where $x = -1$ and at B, where $x = 1$, so these are the two stationary points. Also $f(-1) = 10$ and $f(1) = -6$. At A, $f''(x) = f''(-1) = 0$, so this is a case where we must test whether or not $f'(x)$ changes sign at A. Let e be a small positive number, then

$$f'(-1-e) = 12e^2(-2-e) < 0 \quad \text{and} \quad f'(-1+e) = 12e^2(-2+e) < 0,$$

and so the stationary point A is a decreasing point of inflexion.
  At B, $f'(x) = 0$, $f''(x) = f''(1) = 12 \times 2 \times 2 > 0$ and so B is a minimum point.
  The second derivative $f''(x)$ also vanishes at $x = \frac{1}{3}$, so let C be the point $(\frac{1}{3}, 16/27)$, and then, at C, f' takes the value $-96/27$.

$$f''(\tfrac{1}{3} - e) = 12(\tfrac{1}{3} - e)(-3e) < 0 \quad \text{and} \quad f''(\tfrac{1}{3} + e) = 12(\tfrac{1}{3} + e)(3e) > 0,$$

so at C f'' changes sign and C is a point of inflexion. Thus:
f has a **decreasing point of inflexion at** $A(-1, 10)$,
f has a **minimum at** $B(1, -6)$,

f has a **point of inflexion at** $C(\frac{1}{3}, 16/27)$ where the gradient is $-96/27$, the graph of f crosses the y-axis at $(0, 5)$,
for large values of x, both positive and negative, $f(x)$ is large and positive since x^4 is the dominant term in $f(x)$.

From all this information, the sketch of the graph of f can be made. This is shown in Fig. 22.4.

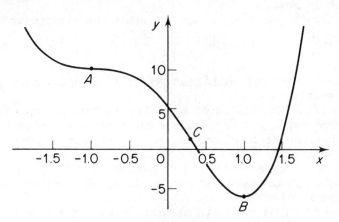

Fig. 22.4 Graph of $3x^4 + 4x^3 - 6x^2 - 12x + 5$.

EXAMPLE 5 *An open box is to be made with a square base of side x, and with height h. If the volume of the box is to be 32 cm³ and the area of the material forming the sides and base of the box is to be a minimum, find the dimensions of the box.*

The volume of the box is $x^2 h = 32$, and so $h = 32/x^2$. The area of the surface of the box is $S = x^2 + 4xh$ and we require S to be a minimum. Substituting for h,

$$S = x^2 + 4x(32/x^2) = x^2 + (128/x).$$

So $\dfrac{dS}{dx} = 2x - \dfrac{128}{x^2}$ and, for S to be a minimum, $\dfrac{dS}{dx} = 0$, that is, $2x = 128/x^2$, or

$x^3 = 64$, so $x = 4$. Now $\dfrac{d^2 S}{dx^2} = 2 + \dfrac{256}{x^3} > 0$, for $x = 4$ and so the value $x = 4$ will

give a minimum for S equal to $16 + (128/4) = 48 \, \text{cm}^2$. Then $h = 32/16 = 2$, and the dimensions of the box are **height 2 cm and square base with edge 4 cm.**

EXERCISE 22.7

1　Find at which part of its domain the function f is (a) increasing, (b) decreasing, where f is given by:
(i) $f(x) = x^2 + 4x$, (ii) $f(x) = x^4$, (iii) $f(x) = x^2 - 6x + 2$,
(iv) $f(x) = x^3 - 12x + 4$, (v) $f(x) = x^5 - 5x$,
(vi) $f(x) = -x^3 + 3x^2 + 24x - 16$, (vii) $f(x) = x^3 + 5x + 2$,
(viii) $f(x) = (x + 1)^3 (x + 2)$, (ix) $f(x) = x - \dfrac{1}{x^2}$, (x) $f(x) = \cos 3x$.

2　Find the maximum and the minimum points for the functions given in question **1**, above.

3 Find the stationary values, and determine their nature, for the curve with equation:
(i) $y = x^2 - 6x + 7$, (ii) $y = 4x - x^2 + 3$, (iii) $y = 2x^2 - 8x + 5$,
(iv) $y = x^3 + 3x^2$, (v) $y = (x-1)^3$, (vi) $y = 3x^3 - 4x + 5$,
(vii) $y = 8x - x^3 - x^2$, (viii) $y = 9(x+2)^4$.
   Find any points of inflexion of the curve, in each of the above cases, and sketch the curve.

4 Find the maximum and the minimum values, (if any) of the function:
(i) $x^2 - \dfrac{1}{x}$, (ii) $x - \dfrac{1}{x}$, (iii) $\dfrac{x}{x^2+1}$, (iv) $x^2 + \dfrac{a}{x}$, (v) $(x+2)(x-1)^3$,
(vi) $x^4 - 4x^3$.

5 Prove that a rectangle, with a perimeter of a given length, has the maximum area when it is a square.

6 A manufacturer wishes to make a can, which is to be in the shape of a circular cylinder, that holds 1 litre. Find the dimensions of the cylinder with the condition that the area of metal required to form the can is to be a minimum, (i) if the can is to be closed, (ii) if the can is to have no lid.

7 Find the minimum value of $x^2 + y^2$, given that $2x + 3y = 6$.

8 By finding the least value of $x^2 + y^2$, find the least distance from the origin to the line $3y = 4x + 12$.

9 A closed hollow right-circular cone has internal height a and the internal radius of its base is also a. A solid circular cylinder of height h just fits inside the cone with the axis of the cylinder lying along the axis of the cone. Show that the volume of the cylinder is

$$V = \pi h(a - h)^2.$$

If a is fixed, but h may vary, find h in terms of a when V is maximum.   (*JMB*)

10 Find the dimensions of an open rectangular box, of largest volume, if it is to be made from a sheet of metal, which is a rectangle with sides 1 m and 2 m, by cutting four equal squares from the corners, and bending up the sides.

11 A hollow cone is to be made from a sector of a circle, of radius a, by joining the bounding radii. Find the angle of the sector so that the volume of the cone is a maximum.

22.8 Tangents and normals

The gradient of the tangent at A, where $x = a$, to the curve $y = \mathrm{f}(x)$ is $\mathrm{f}'(a)$ and the equation of the tangent is

$$y - \mathrm{f}(a) = \mathrm{f}'(a)(x - a).$$

The *normal* to the curve at A is the line through A perpendicular to the tangent at A. Therefore, the equation of the normal at A is

$$y - \mathrm{f}(a) = -\frac{(x - a)}{\mathrm{f}'(a)},$$

because the product of the gradients of perpendicular lines is -1, and so the gradient of the normal is $-\dfrac{1}{f'(a)}$.

EXAMPLE *Find the equation of the tangent and the equation of the normal to the curve* $y = x^3 - 2x^2 + 2x$, *at the point where* $x = 2$.

The gradient of the tangent at $x = 2$ is $\dfrac{dy}{dx} = 3x^2 - 4x + 2$, evaluated at $x = 2$, giving $12 - 8 + 2 = 6$, so the gradient of the normal is $-\tfrac{1}{6}$. When $x = 2$, $y = 8 - 8 + 4 = 4$, so both lines pass through $(2, 4)$. Thus, the equation of the tangent is

$$y - 4 = 6(x - 2) \quad \text{or} \quad y + 8 = 6x,$$

and the equation of the normal is

$$-6(y - 4) = x - 2 \text{ or } x + 6y = 26.$$

EXERCISE 22.8

1 Find the equations of the tangent and the normal to the curve $y^2 = 2x^2 + 1$ at the point $(2, 3)$.
2 Find the coordinates of the point where the normal to the curve $y^2 = 4ax$ at the point $(a, 2a)$ meets the curve for the second time.
3 Find the coordinates of the point of intersection of (i) the tangents, (ii) the normals, to the curve $y = x^2 - 3x + 2$ at the points where the curve crosses the axis Ox.
4 The tangent and the normal, at the point $P(3, 2)$ on the curve $4x^2 = 9y$, meet the y-axis at points A and B. Find the coordinates of the mid-point C of AB and show that $CP = \tfrac{1}{2}AB$.

MISCELLANEOUS EXERCISE 22

1 Given that $y = \sin^{-1}\sqrt{x}$, prove that $\dfrac{dy}{dx} = \dfrac{1}{\sin 2y}$.

2 If $x^2 = \tan y$, find $\dfrac{dy}{dx}$ in terms of x. *(L)*

3 Given that x and y are related by $x^3 + y^3 = 3xy$, find $\dfrac{dy}{dx}$ in terms of x and y. *(L)*

4 Given that $y = \dfrac{\sin x - \cos x}{\sin x + \cos x}$, show that

$$\dfrac{dy}{dx} = 1 + y^2.$$

Prove that $\dfrac{d^2y}{dx^2}$ is zero only when $y = 0$. *(JMB)*

5 If $x = ct$ and $y = c/t$, find $\dfrac{dy}{dx}$ in terms of t and deduce the equation of the tangent to the curve $xy = c^2$ at the point where $t = 2$.   (L)

6 Given that $x = 2\theta + \sin 2\theta$, $y = \cos 2\theta$, prove that

$$\frac{dy}{dx} = -\tan\theta$$

and find the value of $\dfrac{d^2 y}{dx^2}$ when $\theta = \dfrac{\pi}{4}$.   (L)

7 Given that $x = at^2$ and $y = 2at$, where a is a constant, find $\dfrac{d^2 y}{dx^2}$ in terms of a and t.   (L)

8 If $x = \cot\theta$ and $y = \sin^2\theta$, find $\dfrac{dy}{dx}$ in terms of θ and show that

$$\frac{d^2 y}{dx^2} = 2\sin^3\theta \sin 3\theta.$$ (JMB)

9 Given that $y = \dfrac{x^4}{144} + x^3 + 54x^2 + ax + b$, where a and b are constants, find a if $\left(\dfrac{d^2 y}{dx^2}\right)^2 = y$ for all values of x.   (L)

10 Given that $y = \dfrac{1 - x^3}{x}$, find the values of $\dfrac{dy}{dx}$ and $\dfrac{d^2 y}{dx^2}$ and determine the nature of the stationary point on the graph of y.   (L)

11 The surface area of a sphere is decreasing at a rate of $0.4\,\text{m}^2/\text{s}$ when the radius is $0.5\,\text{m}$. Calculate the rate of decrease of the volume of the sphere at this instant.   (L)

12 Find the maximum and minimum values of the expression $\dfrac{x}{1 + x^2}$. Sketch the curve $y = \dfrac{x}{1 + x^2}$.   (L)

13 The area of the region enclosed between two concentric circles of radii x and $y\,(x > y)$ is denoted by A. Given that

x is increasing at the rate of $2\,\text{m s}^{-1}$,
y is increasing at the rate of $3\,\text{m s}^{-1}$ and,
when $t = 0$, $x = 4$ metres and $y = 1$ metre,

find

(i) the rate of increase of A when $t = 0$,
(ii) the ratio of x to y when A begins to decrease,
(iii) the time at which A is zero.   (JMB)

14 The intensity of illumination I cd, (candela), on a surface is inversely proportional to the square of the distance x m from the surface to the source of light. The intensity is 1000 cd when the distance is 0.5 m. Find, when $x = 1$, both the intensity and its rate of change with respect to the distance.   (L)

15 A particle moves on the x-axis so that its displacement x from O is related to the time t by

$$t = x^2 + 2x.$$

Find the speed of the particle when $x = 1$. *(L)*

16 Apply the small increment formula $f(x + \delta x) - f(x) \approx \delta x f'(x)$, to $\tan x$ to find an approximate value of

$$\tan\left(\frac{100\pi + 4}{400}\right) - \tan\frac{\pi}{4}.$$

17 For a certain gas at a constant temperature, the pressure p and the volume V are related by the formula $pV^{1.4} = C$, where C is a constant. Estimate the percentage change in the pressure due to a percentage increase of 0.5% in the volume.

18 The height h and the base radius r of a right circular cone vary in such a way that the volume remains constant. Find the rate of change of h with respect to r at the instant when h and r are equal. *(L)*

19 If $V = r(150 - \frac{1}{2}r^2)$, find the maximum value of V as r varies $(r > 0)$. *(L)*

20 Given that $T = k\sqrt{x}$ where k is a positive constant, find dT/dx.

Calculate the approximate percentage change in the value of T when x increases by 0.2 per cent. *(JMB)*

21 A horizontal circular disc of radius a is fixed at a height a above a horizontal floor. A small lamp vertically above the centre of the disc is moving downward with speed v. Find the rate of increase with respect to time of the area of the shadow of the disc at the instant when the height of the lamp above the floor is $2a$. *(L)*

22 The fixed points A, O, B and C are on a straight line such that $AO = OB = BC = 1$ unit. The points A and B are also joined by a semicircle of radius 1 unit, and P is a variable point on this semicircle such that the angle POC is θ. Calculate the value of θ for which the area of the region R bounded by the arc AP of the semicircle and the straight lines PC and AC is a maximum.

Show that the perimeter of R is of length

$$L = 3 + \pi - \theta + (5 - 4\cos\theta)^{1/2}.$$

Prove that L has just one stationary point and that this occurs at the same value of θ for which the area of R is a maximum.

Find the greatest value and the least value of L in the interval $0 \leqslant \theta \leqslant \pi$. *(JMB)*

23 (a) The function f is given by $f : x \mapsto x^2 + \dfrac{1}{x}\ (x \in \mathbb{R},\ x \neq 0)$. Prove that the graph of f has one turning point, and determine whether it is a maximum or a minimum. Sketch the graph of f.

(b) An isosceles triangle ABC, with $AB = AC$, is inscribed in a fixed circle, centre O, whose radius is 1 unit. Angle $BOC = 2\theta$, where θ is acute. Express the area of the triangle ABC in terms of θ, and hence show that, as θ varies, the area is a maximum when the triangle is equilateral. *(C)*

24 State the derivatives of $\sin x$ and $\cos x$, and use these results to show that the derivative of $\tan x$ is $\sec^2 x$. Show further that

$$\frac{d}{dx}(\tan^{-1}x) = \frac{1}{1 + x^2}.$$

A vertical rod AB of length 3 units is held with its lower end B at a distance 1 unit vertically above a point O. The angle subtended by AB at a variable point P on the horizontal plane through O is θ. Show that

$$\theta = \tan^{-1}x - \tan^{-1}\frac{x}{4},$$

where $x = OP$. Prove that, as x varies, θ is a maximum when $x = 2$, and that the maximum value of θ can be expressed as $\tan^{-1}\frac{3}{4}$.   (*JMB*)

25 A solid cylinder, of height h and base radius r, has a fixed volume. Find the ratio $r:h$ if the surface area of the cylinder is a minimum.   (*L*)

26 Find the coordinates of the maximum point T and the minimum point B of the curve

$$y = \frac{x^3}{3} - 2x^2 + 3x.$$

Find also the point of inflexion I and show that T, I, B are collinear. Calculate to the nearest $0.1°$ the acute angle between TIB and the normal to the curve at I.   (*L*)

27 (i) Given that $y = (k + x)\cos x$, where k is a constant, find $\dfrac{d^2 y}{dx^2} + y$ and show that it is independent of k.

(ii) Differentiate with respect to x $\left(\dfrac{1-x}{1+x}\right)^{\frac{1}{2}}$.

(iii) Given that x and y vary so that $ax + by = c$, where a, b, c are constants, show that the minimum value of $x^2 + y^2$ is

$$\frac{c^2}{(a^2 + b^2)}.$$ (*L*)

23 The Exponential and Logarithmic Functions

23.1 Indices and logarithms (revision)

In Chapter 9, we developed the power function, or exponential function, a^x, where a was a positive real number. We began with the obvious first idea of a power, $a^x = a \times a \times a \times \ldots \times a$, the product of x a's, where x is a natural number. Then we successively widened this definition to include first negative values of x, then rational values and finally all real values. Thus, we have the function a^x, with domain $\mathbb{R}$ and range $\mathbb{R}^+$. The graphs of a^x for $a \in \{\frac{1}{5}, \frac{1}{3}, \frac{1}{2}, 1, 2, 3, 5\}$ are shown in Fig. 23.1(a).

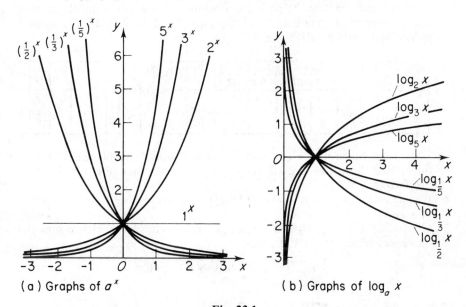

(a) Graphs of a^x

(b) Graphs of $\log_a x$

Fig. 23.1

The graph of a^x passes through $(0, 1)$ for all values of a. If $a > 1$, the graph is increasing, $a^x \to 0$ as $x \to -\infty$ and $a^x \to \infty$ as $x \to +\infty$. However, if $0 < a < 1$, the graph is decreasing, $a^x \to \infty$ as $x \to -\infty$ and $a^x \to 0$ as $x \to +\infty$. We defined the inverse function of a^x as $\log_a x$, the logarithm of x to the base a, which is the power to which a must be raised to obtain x. For example, $\log_2 8 = 3$ and $\log_{10}(0.01) = -2$. $\log_a x$ has domain $\mathbb{R}^+$ and range $\mathbb{R}$. Its graph for the same six values of a is shown in Fig. 23.1(b). The graphs of a^x and $\log_a x$ are reflections of one another in the line $y = x$ (see §9.6). The two functions are related by the equivalence $y = a^x \Leftrightarrow x = \log_a y$.

EXERCISE 23.1

1 Evaluate:
   (i) 3^4, (ii) 2^5, (iii) 4^{-2}, (iv) 3^{-1}, (v) $4^{\frac{1}{2}}$, (vi) $8^{-\frac{1}{3}}$, (vii) $25^{-\frac{1}{2}}$,
   (viii) $(\frac{1}{2})^4$, (ix) $(\frac{1}{4})^{-\frac{1}{2}}$, (x) $(\frac{1}{3})^{-3}$, (xi) $(\frac{1}{9})^{\frac{3}{2}}$, (xii) $16^{-\frac{3}{4}}$, (xiii) $\log_2 16$,
   (xiv) $\log_3 27$, (xv) $\log_5 0\cdot2$, (xvi) $\log_2 (\frac{1}{8})$, (xvii) $\log_{\frac{1}{2}} 4$,
   (xviii) $\log_{\frac{1}{3}} (\frac{1}{27})$.
2 For $-3 \leqslant x \leqslant 3$, draw graphs of the functions given by $y = 3^x$ and $y = \log_3 x$. State the domain and the range of each function.
   (a) Use your graphs to read off the following values: (i) $3^{0\cdot2}$, (ii) $3^{1\cdot6}$,
   (iii) $3^{2\cdot4}$, (iv) $3^{-0\cdot7}$. (v) $3^{-2\cdot4}$, (vi) $\log_3 2\cdot5$, (vii) $\log_3 2\cdot1$,
   (viii) $\log_3 1\cdot9$, (ix) $\log_3 0\cdot6$, (x) $\log_3 0\cdot1$.
   (b) Find the solution of the equation, from your graph, (i) $3^x = 1\cdot5$,
   (ii) $3^x = 0\cdot8$, (iii) $2.3^x = 7$, (iv) $3^x = -2$, (v) $\log_3 x = 1$,
   (vi) $\log_3 x = -0\cdot2$, (vii) $\log_3 x + 1\cdot2 = 0$, (viii) $3\log_3 x + 4 = 0$.
3 Calculate the values of 2^x from $x = -3$ to $x = 3$, in steps of $0\cdot5$ and draw the graph of the function 2^x with the domain $\{x: -3 \leqslant x \leqslant 3\}$. By careful measurement from your graph, find the gradient of 2^x at each of the 13 plotted points, and complete the table:

x	-3	$-2\cdot5$	-2	$-1\cdot5$	-1	$-0\cdot5$	0	0·5	1	1·5	2	2·5	3
$y = 2^x$ gradient of 2^x	$\frac{1}{8}$		$\frac{1}{4}$		$\frac{1}{2}$		1		2		4		8

Look for a functional relationship between 2^x and its gradient. Given that these two functions are directly proportional, find the constant of proportionality.
4 Given that $\log_2 (x - 5y + 4) = 0$ and $\log_2 (x + 1) - 1 = 2\log_2 y$, find the values of x and y. *(AEB)*
5 Given that $y = \log_a (x^3)$ and $z = \log_x a$, show that $yz = 3$. Hence find the numerical values of y and z when

$$\log_a (3\log_a x) - \log_a (\log_x a) = \log_a 27. \qquad (AEB)$$

23.2 The gradient of a^x

Question 3 of Exercise 23.1 investigated the gradient of 2^x graphically. We now wish to find a gradient function for the function f, where $f(x) = a^x$. Following the methods of §18.6:

$$g(h) = \frac{a^{x+h} - a^x}{h} = a^x \left(\frac{a^h - 1}{h} \right)$$

and $$f(x+h) = f(x) + h.g(h).$$

Therefore, the tangent to the graph of f at $x = b$ is given by

$$y = a^b + m(x - b),$$

where $m = \lim_{h \to 0} g(h)$, and m is the gradient at $x = b$. Now

$$g(h) = a^b \left(\frac{a^h - 1}{h} \right)$$

and so $m = a^b . \lim_{h \to 0} \left(\frac{a^h - 1}{h} \right) = k.a^b$, where $k = \lim_{h \to 0} \left(\frac{a^h - 1}{h} \right).$

Notice that the gradient m at $x = b$ is the constant factor k times a^b, for all b. The constant factor k is a limit which depends only upon a. In fact, k is the gradient of f at $x = 0$, because the gradient m at $x = 0$ is $k.a^0 = k$.

We can use k to find the gradient function f'. In the above, we replace b by x, then the gradient of f at x is ka^x. Therefore $f'(x) = ka^x$. This means that, if

$$y = a^x, \frac{dy}{dx} = ka^x = ky.$$

The limit k, which depends only on a, being the gradient of a^x at $(0, 1)$, can take any value, as can be seen from Fig. 23.1(a). In question **3** of Exercise 23.1, the approximate value of k, for $a = 2$, was 0·7.

To make progress, we choose one particular value of a. Of all the functions a^x, we choose that one which has gradient 1 at $(0, 1)$. From Fig. 23.1(a) this value of a will lie between 2 and 3. We call this value of a the number e.

Definition   The real number e is that number such that the function e^x has gradient equal to 1 at $x = 0$.

Thus, for $a = e$, $k = 1$, and so $\lim_{h \to 0} \left(\frac{e^h - 1}{h} \right) = 1$. The value of e is 2·71828, approximated to five decimal places. The derivative of the function e^x is

$$\frac{d}{dx} e^x = k.e^x = 1.e^x = e^x,$$

and so e^x is seen to be that exponential function whose gradient equals the value of the function. It is called the exponential function.

The exponential function is useful in modelling a situation where the rate of growth of some quantity is directly proportional to that quantity. For example, if the number of bacteria in a colony is $n(t)$ at time t and if $n(t) = me^t$, for some constant m, then m is the number of bacteria at time

$t = 0$, and the rate of growth of the colony is

$$\frac{d}{dt}n(t) = m\frac{d}{dt}e^t = me^t = n(t),$$

so the rate of growth equals the size of the colony.

EXAMPLE 1   *Differentiate: (i) e^{x^2}, (ii) $e^{-(3x+4)^2}$, (iii) $(x^2+1)e^x$.*

(i) Let $y = e^{x^2}$ and make the substitution $x^2 = u$, so $y = e^u$. Then $\dfrac{dy}{du} = e^u$, by the

definition of e. Also $\dfrac{du}{dx} = 2x$ and, by an application of the chain rule (§18.9),

$$\frac{dy}{dx} = \frac{dy}{du}\cdot\frac{du}{dx} = e^u.2x = \mathbf{2xe^{x^2}}.$$

(ii) Let $y = e^{-(3x+4)^2}$, which is sufficiently complicated to need two substitutions.

Let $v = 3x+4$, $u = -v^2 = -(3x+4)^2$ and $y = e^u$. Then $\dfrac{dy}{du} = e^u$, $\dfrac{du}{dv} = -2v$,

$\dfrac{dv}{dx} = 3$, and so

$$\frac{dy}{dx} = e^u(-2v)3 = \mathbf{-6(3x+4)e^{-(3x+4)^2}}.$$

(iii) Let $y = (x^2+1)e^x = uv$, where $u = x^2+1$ and $v = e^x$. Then $\dfrac{du}{dx} = 2x$ and

$\dfrac{dv}{dx} = e^x$, so, by the product rule

$$\frac{dy}{dx} = u\frac{dv}{dx}+v\frac{du}{dx} = (x^2+1)e^x+2xe^x = (x^2+2x+1)e^x = \mathbf{(x+1)^2\,e^x}.$$

EXAMPLE 2   *Find the turning points of the graph of f, where $f(x) = xe^{-x^2}$, and state whether they are maxima or minima.*

Let $y = xe^{-x^2}$, then $\dfrac{dy}{dx} = x(-2x)e^{-x^2}+e^{-x^2} = (1-2x^2)e^{-x^2}$. Then

$$\frac{dy}{dx} = 0 \Leftrightarrow e^{-x^2}(1-2x^2) = 0 \Leftrightarrow 1-2x^2 = 0,$$

since, for all real x, $e^{-x^2} > 0$. Solving the quadratic equation, we find either
$x = 1/\sqrt{2}$, $y = e^{-\frac{1}{2}}/\sqrt{2}$ or $x = -1/\sqrt{2}$, $y = -e^{-\frac{1}{2}}/\sqrt{2}$. Also

$$\frac{d^2y}{dx^2} = (1-2x^2)(-2x)e^{-x^2}+(-4x)e^{-x^2} = (4x^3-6x)e^{-x^2}.$$

For $x = 1/\sqrt{2}$, $\dfrac{d^2y}{dx^2} = \left(\dfrac{4}{2\sqrt{2}}-\dfrac{6}{\sqrt{2}}\right) < 0$,   giving   a   **maximum**   at
$\left(\dfrac{1}{\sqrt{2}}, \dfrac{1}{\sqrt{2}}e^{-\frac{1}{2}}\right)$.

For $x = -1/\sqrt{2}$, $\dfrac{d^2y}{dx^2} = \left(\dfrac{-4}{2\sqrt{2}} + \dfrac{6}{\sqrt{2}}\right) > 0$, giving a **minimum** at $\left(\dfrac{-1}{\sqrt{2}}, \dfrac{-1}{\sqrt{2}}e^{-\frac{1}{2}}\right)$.

EXERCISE 23.2

1 Differentiate with respect to x:
   (i) e^{x^3}, (ii) $e^{(-2x+4)}$, (iii) $e^{-\frac{1}{2}x^2}$, (iv) xe^{-x}, (v) $(x^3 - 2)e^x$,
   (vi) $(3 + x^2)e^x$, (vii) $(x^3 + 2x^2 + 1)e^{2x+1}$, (viii) $\dfrac{e^x}{4x}$,
   (ix) $\dfrac{1}{6x^2}e^{3x^2+1}$, (x) $e^x \sin x$, (xi) $(\cos^2 x)e^{x^2}$, (xii) $e^{(x+1)^2}$, (xiii) $e^{\cos x}$,
   (xiv) $e^{x \sin x}$, (xv) e^{ax+b}, (xvi) $(ax+b)e^{cx+d}$, (xvii) $e^{ax}(\cos bx + \sin bx)$.
2 Determine whether the statement is true or false, for all $a, b \in \mathbb{R}$. If it is false, give the largest subset of $\mathbb{R}$ in which it is true:
   (i) $a > 0 \Leftrightarrow e^a > 1$, (ii) $a < b \Leftrightarrow e^a < e^b$, (iii) $\dfrac{e^a}{e^b} > 1 \Leftrightarrow a < b$,
   (iv) $\dfrac{d}{dx}(e^{ax}) = e^{ax}$, (v) $\dfrac{d}{dx}(e^{bx^2}) = 2be^{bx^2}$.

3 Let f be any function, with derivative f'. Find an expression for $\dfrac{d}{dx}e^{f(x)}$.

   (a) Can you write down an expression for (i) $\int e^{f(x)} dx$, (ii) $\int f'(x)e^{f(x)} dx$?

   (b) Prove that: (i) if $f(x) = ax + b$, then $\int e^{f(x)} dx = \dfrac{1}{a}e^{f(x)} + c$,
       (ii) if $f(x) = x^2$, then $\int 2xe^{f(x)} dx = e^{f(x)} + c$.
4 Find the equation of the tangent to the curve $y = e^x$ at the point (a, e^a). Find the value of a such that this tangent passes through the origin, and state its gradient.
5 Find the maximum and minimum values of the real function f, given by $f(x) = x^2e^{-x}$. Assuming that $x^2e^{-x} \to 0$ as $x \to \infty$, sketch the graph of f.
6 Given that $f(x) \equiv \dfrac{x}{e^x - 1} + \dfrac{x}{2}$ for $x \neq 0$, show that $f(-x) \equiv f(x)$. (L)
7 Differentiate $e^{2x} \cos x$ with respect to x. (JMB)
8 Given that $y = (1+x)^2e^{-2x}$, find $\dfrac{dy}{dx}$. Find also the coordinates of the turning points of the graph of y. (L)
9 Given that $y = xe^y$ show that
   (a) $(1-y)\dfrac{dy}{dx} = e^y$, (b) $(1-y)\dfrac{d^2y}{dx^2} = (2-y)\left(\dfrac{dy}{dx}\right)^2$. (L)
10 Show that if $y = e^{4x} \cos 3x$ then, for a suitable constant α,
$$\dfrac{dy}{dx} = 5e^{4x} \cos(3x + \alpha)$$
    and find also $\dfrac{d^2y}{dx^2}$. (L)

11   Find the coordinates of any turning points and points of inflexion on the curve $y = xe^{-x}$. Sketch the curve.

Find the volume of the solid of revolution formed when the region enclosed by the curve, the x-axis and the line $x = 2$ is rotated completely about the x-axis. (Hint: use the derivative of $(2x^2 + 2x + 1)e^{-2x}$.)   (*AEB*)

12   Sketch the curve $y = 1 + 2e^{-x}$, showing clearly the behaviour of the curve as $x \to +\infty$.

Find the area of the finite region enclosed by the curve and the lines $x = 0$, $x = 1$, $y = 1$. Find also the volume formed when this region is rotated completely about the line $y = 1$.   (*L*)

13   Find by trial the two consecutive integers between which the solution of the equation

$$x + 2e^x = 0$$

lies. Of these two integers, take as first approximation the one which makes the absolute value of the left side smaller. Then find a second approximation to the solution by a single application of the Newton–Raphson method. Give two decimal places in your answer.   (*SMP*)

23.3 The natural logarithm function $\ln x$

The inverse of the exponential function e^x is $\log_e x$, the logarithm to the base e of x. This function is called the *natural logarithm function* and is denoted by $\ln x$.

Definition $y = \ln x \Leftrightarrow x = e^y$.

The domain of e^x is $\mathbb{R}$ and its range is $\mathbb{R}^+$. Therefore, the domain of $\ln x$ is $\mathbb{R}^+$ and its range is $\mathbb{R}$. This means that, in using the function $\ln x$, we must be careful to restrict the domain to positive real numbers. In this way, $\ln x$ is similar to the square root function $\sqrt{x}$. If either of these functions is called on an electronic calculator when the display entry is negative, you should obtain an error signal.

The derivative of $\ln x$

We use the derivative of the exponential function. Let $y = \ln x = \log_e x$ so that $e^y = x$. Differentiating this equation implicitly, $e^y \dfrac{dy}{dx} = 1$, so $\dfrac{dy}{dx} = \dfrac{1}{e^y} = \dfrac{1}{x}$. Therefore

$$\text{if}\quad f(x) = \ln x\quad\text{then}\quad f'(x) = \frac{1}{x}.$$

This result can also be obtained graphically. Consider the graphs of e^x and $\ln x$, which are reflections of each other in the line $y = x$ (Fig. 23.2). Let P be the point $(a, \ln a)$ on the graph $y = \ln x$, and let Q be the point given by

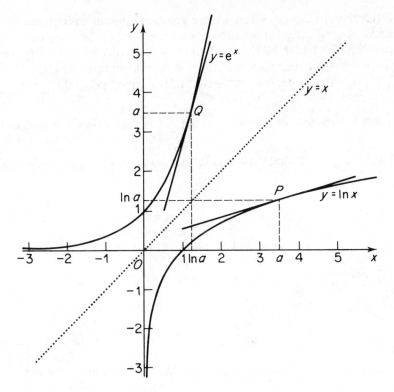

Fig. 23.2

$x = \ln a$ on the graph $y = e^x$. Since $e^{\ln a} = a$, the y coordinate of Q is a and Q is the point $(\ln a, a)$. This also follows from the fact that P and Q are reflections in $y = x$. The tangent at Q to $y = e^x$ has gradient $\dfrac{d}{dx} e^x$ evaluated at $x = \ln a$, that is, $e^{\ln a} = a$. By reflection in $y = x$, the gradient of the tangent at P to $y = \ln x$ is $1/a$, so $\dfrac{d}{dx}(\ln x)$ takes the value $1/a$ at $x = a$. Hence

$$\frac{dy}{dx} = \frac{1}{x}, \quad \text{for} \quad y = \ln x.$$

This result provides the answer to the missing link in the sequence of integrals of the functions x^n, for $n \in \mathbb{Z}$. In Chapter 6, we saw that, for $n \neq -1$,

$$\int x^n \, dx = x^{n+1}/(n+1) + c.$$

Now we can complete the list, for $n = -1$,

$$\int x^n \, dx = \int x^{-1} \, dx = \int \frac{1}{x} \, dx = \ln x + c.$$

Notice that, in this equation, we are assuming that $\ln x$ is defined. So the equation is true only for values of x in the domain of $\ln x$, that is, for positive values of x. In §23.6, we shall see what to do when we require the integral of $1/x$ for negative x. In·the rest of this section, and in §23.5, we shall assume that whenever we write $\ln f(x)$, then $f(x) > 0$.

EXAMPLE   *Find the derivative of*: (*i*) $\ln (x^2 + 2)$,   (*ii*) $x^{\frac{1}{2}} \ln x$,   (*iii*) $x(\ln x - 1)$.

(i) Let $y = \ln (x^2 + 2)$, and make the substitution $u = x^2 + 2$, so $y = \ln u$. Then $\dfrac{dy}{du} = \dfrac{1}{u}$, $\dfrac{du}{dx} = 2x$, and so

$$\frac{dy}{dx} = \frac{dy}{du}\frac{du}{dx} = \frac{1}{u}.2x = \frac{2x}{(x^2 + 2)}.$$

(ii) Let $y = x^{\frac{1}{2}} \ln x = uv$, where $u = x^{\frac{1}{2}}$ and $v = \ln x$; $\dfrac{du}{dx} = \frac{1}{2}x^{-\frac{1}{2}}$ and $\dfrac{dv}{dx} = \dfrac{1}{x}$.

Then

$$\frac{dy}{dx} = u\frac{dv}{dx} + v\frac{du}{dx} = x^{\frac{1}{2}}.\frac{1}{x} + (\ln x).\tfrac{1}{2}x^{-\frac{1}{2}} = \frac{2 + \ln x}{2\sqrt{x}}.$$

(iii) Let $y = x \ln x - x$, then $\dfrac{dy}{dx} = x.\dfrac{1}{x} + 1.\ln x - 1 = 1 + \ln x - 1 = \ln x$.

Note that a very useful result follows from (iii):

$$\int \ln x \, dx = x \ln x - x + c.$$

EXERCISE 23.3

1   Simplify, where possible:
   (i) $\ln e^3$,   (ii) $\ln\left(\dfrac{1}{e^2}\right)$,   (iii) $e^{\ln 2x}$,   (iv) $e^{2\ln x}$,   (v) $\ln (xe^{x^2})$,
   (vi) $\ln (x + e^x)$,   (vii) $e^{x + \ln x}$,   (viii) $e^{x \ln x}$.
2   Solve the equation for x:
   (i) $y = 3 + 2\ln x$,   (ii) $y = 5e^x$,   (iii) $y = \ln (e^x + 2)$,   (iv) $e^{xy} = 5$,
   (v) $\ln y - \ln x = 3$,   (vi) $\sin y + e^x = 3$.
3   Differentiate with respect to x:
   (i) $\ln 6x$,   (ii) $\ln (6x - 3)$,   (iii) $\ln\left(\dfrac{x}{5}\right)$,   (iv) $\ln x^3$,   (v) $3\ln x$,
   (vi) $\ln (x + 1)^2$,   (vii) $(\ln (x + 1))^2$,   (viii) $\ln \sqrt{x}$,   (ix) $\frac{1}{2}\ln \sqrt[3]{x}$,
   (x) $\ln (\sin x)$,   (xi) $\ln (\cos 3x + \sin 2x)$,   (xii) $\ln (2 \tan 3x)$,
   (xiii) $\ln (x\sqrt{(x^2 - 1)})$,   (xiv) $\ln (x + \sqrt{(x^2 - 1)})$,   (xv) $\ln \dfrac{\sqrt{(1 - x)}}{\sqrt{(1 + x)}}$.
4   Differentiate:
   (i) $x^2 \ln x^2$,   (ii) $(x^3 + 1) \ln x$,   (iii) $\sin x \ln x$,   (iv) $x(\ln x)^3$,

(v) $(3 \ln x - 2)^3$,   (vi) $\sin(\ln x - 2)$,   (vii) $\dfrac{x}{\ln x}$,   (viii) $\dfrac{x}{\ln \sin x}$,

(ix) $\dfrac{\ln x}{x^2}$,   (x) $\ln(3+x) + \ln(2-x)$,   (xi) $\ln(1+x^2) - \ln(1+x)$,

(xii) $\ln\left(\dfrac{2-x}{3+x}\right)$,   (xiii) $\ln\dfrac{(x+1)^2}{2-x}$,   (xiv) $\ln(x^2+1) \cdot e^{x^2}$,   (xv) $e^{2x} \cdot \ln x^2$.

5   Let f be a function with derivative f′. Find an expression for $\dfrac{d}{dx}(\ln f(x))$. Is

there an easy expression for $\displaystyle\int \dfrac{1}{f(x)}\,dx$, in general? Prove that:

(a) if $f(x) = x + 2$, then $\displaystyle\int \dfrac{1}{f(x)}\,dx = \ln f(x) + c$;

(b) if $f(x) = x^2 + 2$, then $\displaystyle\int \dfrac{x}{f(x)}\,dx = \tfrac{1}{2}\ln f(x) + c$.

6   (i) Find $\dfrac{d}{dx}\ln(2-x^3)$ and hence find $\displaystyle\int \dfrac{x^2}{2-x^3}\,dx$.

    (ii) Find $\dfrac{d}{dx}\ln(\sin x)$ and hence find $\displaystyle\int \cot x\,dx$.

7   Sketch on the same axes the graphs of the functions $\ln x$, $\ln(-x)$ and $1/x$. On a second set of axes, sketch the graphs of e^x, e^{-x} and $-e^{-x}$.

8   Differentiate with respect to x:
   (i) e^{-x}/x,   (ii) $\ln(\sin 2x)$,   (iii) $e^{2x}\ln(2x)$.   (L)

9   Show that the area of the finite region bounded by the x-axis, the line $x = e$ and the curve $y = \ln x$ is 1. (Hint: use Example (iii).)

10   Find the value, when $x = 1$, of the derivative of $\ln\left(\dfrac{4x^2 - 1}{3x + 1}\right)$.   (L)

11   Differentiate with respect to x
   (a) $\dfrac{\ln x}{\sin x}$,   (b) $e^{\tan x}$.   (L)

12   Differentiate $\ln\left(\dfrac{1+x}{1-x}\right)$ with respect to x.   (L)

13   Sketch the curve $y = \ln(1+x)$.   (L)

14   Differentiate with respect to x
   (i) $\log_e(3x-1)^3$,   (ii) xe^{x^2}.   (JMB)

15   A periodic function is defined by

$$\begin{cases} f(x) = \dfrac{1}{x}\ln(1+x) \cdot & \text{for } 0 < x \leqslant 1, \\[2mm] f(x+1) = f(x) & \text{for all } x. \end{cases}$$

Sketch the graph $y = f(x)$ for values of x from -2 to 2.

16   Sketch the curve whose equation is $y = \ln(1-3x)$, for $x < \tfrac{1}{3}$.   (L)

17   The successive stationary points of the curve

$$y = e^{-x}\sin x, \; (x \geqslant 0),$$

are denoted by (x_0, y_0), (x_1, y_1), (x_2, y_2), ...
Show that the values $x_0, x_1, x_2, \ldots$ form an arithmetic progression and that the values $y_0, y_1, y_2, \ldots$ form a geometric progression, and find the common difference of the arithmetic progression and the common ratio of the geometric progression.   (*JMB*)

18 Find the gradient of the tangent from the origin to the curve

$$y = \log_e x.$$

Hence, by graphical considerations determine the range of values of m for which the equation

$$\log_e x = mx$$

has two unequal real roots.   (*JMB*)

23.4 Derivative and integral of the function a^x

We now return to the original problem of finding the derivative of the general exponential function a^x, where $a > 0$. Let $y = a^x$, then

$$\ln y = \ln (a^x) = x \ln a.$$

Implicit differentiation of the equation will give us

$$\frac{1}{y}\frac{dy}{dx} = \ln a \quad \text{and so} \quad \frac{dy}{dx} = y . \ln a = a^x . \ln a.$$

Hence, the required derivative and integral are given by, if $y = a^x$,

$$\frac{dy}{dx} = a^x . \ln a,$$

$$\int a^x \, dx = \frac{a^x}{\ln a} + c.$$

Alternatively, we can use the result $\ln (a^x) = x \ln a$, so $a^x = e^{x \ln a}$, and

$$\frac{d}{dx} a^x = \frac{d}{dx} e^{x \ln a} = (\ln a) e^{x \ln a} = (\ln a) a^x.$$

Similarly

$$\int a^x \, dx = \int e^{x \ln a} \, dx = \frac{1}{\ln a} e^{x \ln a} + c = \frac{1}{\ln a} a^x + c.$$

We use a similar technique for $\log_a x$. Let $y = \log_a x$, then $a^y = x$ and, by implicit differentiation,

$$a^y . \ln a . \frac{dy}{dx} = 1, \quad \text{so} \quad \frac{dy}{dx} = \frac{1}{x \ln a}.$$

Alternatively, $y = \log_a x = (\ln x)/(\ln a)$, so $\dfrac{dy}{dx} = \dfrac{1}{x \ln a}$. Thus

$$\frac{d}{dx} \log_a x = \frac{1}{x} \cdot \frac{1}{\ln a}.$$

EXAMPLE 1 *Differentiate 2^x with respect to x.*

Let $y = 2^x$, then $\dfrac{dy}{dx} = 2^x . \ln 2$. Approximately, $\ln 2 = 0.693$ to three decimal places, and this is the answer that we approximated to in question **3** of Exercise 23.1.

EXAMPLE 2 *Evaluate $\displaystyle\int_2^3 5^x \, dx$.*

$$\int_2^3 5^x \, dx = \left[\frac{5^x}{\ln 5} \right]_2^3 = \frac{1}{\ln 5} [5^3 - 5^2] = \frac{100}{\ln 5}.$$

EXAMPLE 3 *Find expressions for $\dfrac{d}{dx}(3^{x^2})$ and $\int x 3^{x^2} \, dx$.*

Let $u = x^2$ so that $y = 3^{x^2} = 3^u$. Then $\dfrac{dy}{du} = 3^u . \ln 3$ and $\dfrac{du}{dx} = 2x$, so

$$\frac{dy}{dx} = 3^{x^2} . \ln 3.2x = 6x \ln 3.3^{x^2}$$

and

$$\int x 3^{x^2} \, dx = \frac{3^{x^2}}{6 \ln 3} + c.$$

EXERCISE 23.4

1 Find the derivative of the function:
   (i) 6^x, (ii) 2^{x^3}, (iii) $\log_2 (x + 4)$, (iv) $\log_5 (\sqrt{x})$, (v) 2^{3x-1}, (vi) $7^{\sqrt[3]{x}}$,
   (vii) $2^x + x^2$, (viii) $2x^3 \log_3 x$, (ix) $(3 \log_3 x)/(3^x)$.
2 Find the following indefinite integrals:
   (i) $\int 2^x \, dx$, (ii) $\int x^2 3^{x^3} \, dx$, (iii) $\int (x + 1) 5^{x^2 + 2x} \, dx$, (iv) $\int \sin x \, 2^{\cos x} \, dx$.
3 Find the equation of the tangent to the curve $y = 2^x$ at the point on the curve where $x = 3$. Show that this tangent meets the y-axis at $(0, 8 - 8 \ln 8)$.
4 Find the turning point on the graph $y = x/3^x$ and sketch the curve.
5 Differentiate $x 10^x$ with respect to x. (L)
6 (i) Find $\dfrac{d}{dx}(x \log_{10} x)$. (ii) Find the value of $\displaystyle\int_1^2 \log_{10} x \, dx$, leaving your answer in terms of logarithms.

23.5 Alternative definitions for e^x and $\ln x$.

In §23.2, we defined the function e^x to be that exponential function which has gradient 1 at $x = 0$. Then we developed $\ln x$ as the inverse function of e^x. From these definitions, we obtained the basic results

$$\frac{d}{dx}e^x = e^x \quad \text{and} \quad \frac{d}{dx}\ln x = \frac{1}{x}.$$

Alternatively, it is possible to start with the definition of a function F by the equation $F(a) = \int_1^a \frac{1}{x}dx$, and use the properties of the definite integral to show that F has all the properties of a logarithm function, in fact, that $F(a) = \log_e a$, where $F(e) = 1$ defines the number e. Then e^x is the inverse function of $F(x)$ and, if we put $F(x) = \ln x$ we can deduce all the properties of the exponential and logarithmic functions, finishing with the property $\frac{d}{dx}e^x = e^x$, which was the starting point of our original definition.

The following exercise is intended to guide the reader through the proofs of this alternative development. In doing so, it proves that the two definitions of the functions e^x and $\ln x$ are equivalent.

EXERCISE 23.5

Throughout this exercise none of the previous results of this chapter concerning e^x and $\ln x$ are used and it is assumed that these functions have not been defined. We define the function F, with domain $\mathbb{R}^+$, by

$$F(a) = \int_1^a \frac{1}{x}dx, \quad a > 0, \quad \text{or} \quad F(x) = \int_1^x \frac{1}{t}dt, \quad x > 0.$$

1   Draw the graph of $y = \frac{1}{x}$ for $-3 \leqslant x \leqslant 7$, $x \neq 0$. Use Simpson's rule with eight strips to calculate $F(4)$ to an accuracy of two decimal places.
Use the results of question 5 of Exercise 20.6 to find F(2), F(3), F(6).
Show that $F(2) + F(3) = F(6)$, $F(6) - F(2) = F(3)$, $F(4) = 2F(2)$.

2   Refer to the graph of $y = 1/x$ (Fig. 23.3), with points $P(1, 1)$ and $Q(2, \frac{1}{2})$ on the curve.

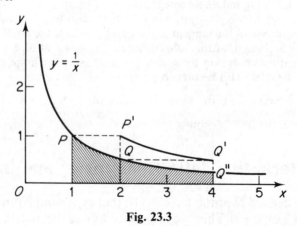

Fig. 23.3

(i) Stretch the graph, with scale factor 2 in the x-direction, shifting P to $P'(2, 1)$ and Q to $Q'(4, \frac{1}{2})$.

(ii) Stretch the new graph, with scale factor $\frac{1}{2}$ in the y-direction, shifting P' to P'' and Q' to Q''. Prove that P'' and Q'' lie on the original graph, because $P'' = Q$ and Q'' is the point $(4, \frac{1}{4})$.

(iii) Use the fact that the combination of the two one-way stretches leaves all areas unchanged, because $2 \times \frac{1}{2} = 1$, to prove that $F(2) = F(4) - F(2)$, and hence that $F(4) = 2F(2)$.

(iv) Use the same transformations to prove that $F(3) = F(6) - F(2)$.

(v) Use similar transformations, with scale factors 3 and $\frac{1}{3}$, to prove that $F(3) = F(9) - F(3)$ and hence that $F(9) = 2F(3)$.

3 Use the graph of $y = 1/x$, Fig. 23.4, to prove that:
(i) area A + area $C = F(2)$, (ii) area B = area C,
(iii) area A + area $B = F(2)$.
   Deduce that $F(\frac{1}{2}) = -F(2)$.
   Use similar diagrams to prove that $F(\frac{1}{3}) = -F(3)$ and $F(\frac{1}{4}) = -2F(2)$.

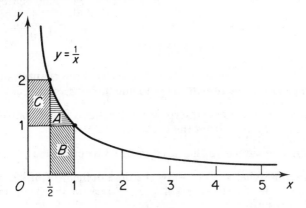

Fig. 23.4

4 Using the results of questions **1, 2** and **3**, write down whether the statement is true or false:
(i) $F(1) = 0$, (ii) $2F(3) = F(9)$, (iii) $2F(3) = F(6)$,
(iv) $F(2) + F(3) = F(5)$, (v) $F(2)F(3) = F(6)$, (vi) $F(6) - F(3) = F(3)$,
(vii) $F(\frac{1}{2}) + F(2) = 0$, (viii) $F(\frac{1}{2}) = -3F(1)$.

5 Generalise the above results to obtain the following:
(i) $F(x) + F(y) = F(x, y)$, (ii) $F\left(\dfrac{1}{x}\right) = -F(x)$, (iii) $F(x) - F(y) = F\left(\dfrac{x}{y}\right)$,
(iv) $xF(y) = F(y^x)$.

6 By use of the substitution $x = au$, prove that
$$\int_a^{ab} \frac{1}{x}\,dx = \int_1^b \frac{1}{u}\,du = F(b).$$
By use of the substitution $x^b = u$, prove that
$$\int_1^{a^b} \frac{1}{x}\,dx = b\int_1^a \frac{1}{u}\,du = bF(a).$$

Hence, prove that
$$\int_1^{ab} \frac{1}{x}\,dx = \int_1^{a} \frac{1}{x}\,dx + \int_1^{b} \frac{1}{x}\,dx$$

and that
$$\int_1^{a^b} \frac{1}{x}\,dx = b\int_1^{a} \frac{1}{x}\,dx.$$

Use these equations to prove the four results of question **5**.

The above properties of the function F show that it has all the properties of a logarithmic function. We now show that it is indeed a logarithmic function. Define the real number e such that $F(e) = 1$, that is:

Definition   e is the real number such that $F(e) = \displaystyle\int_1^{e} \frac{1}{x}\,dx = 1.$

Use this definition in the remaining questions of this exercise.

7   Using question **5** (iv), with $y = e$, show that $x = F(e^x)$ and hence deduce that
$$F(e^x) = \log_e e^x \quad \text{and} \quad F(x) = \log_e x = \int_1^{x} \frac{1}{u}\,du.$$
Prove that $\dfrac{d}{dx}\log_e x = \dfrac{1}{x}.$

8   Use Simpson's rule with eight strips to find $\displaystyle\int_1^{2.7} \frac{1}{x}\,dx$ and $\displaystyle\int_1^{2.75} \frac{1}{x}\,dx$ to three decimal places. Use linear interpolation (that is, a linear approximation) to find an estimate of e, from the equation $F(e) = 1$.

9   Use a method similar to that used in §23.3 to prove that $\dfrac{d}{dx}(e^x) = e^x.$

10   Draw two flowcharts: (i) to show the development of §23.2, §23.3, §23.4; (ii) to show the development of §23.5. Mark in the definitions and the important results.

11   Sketch on the same diagram the graphs of $1/x$ and $1/\sqrt{x}$ for $x \geqslant 1$, labelling the graphs clearly.
    By comparing the areas under the two graphs over the interval $1 \leqslant x \leqslant t$, show that, for $t \geqslant 1$,
$$\frac{\ln t}{t} \leqslant \frac{2}{\sqrt{t}} - \frac{2}{t}.$$

What can you deduce about the behaviour of the function $t \mapsto (\ln t)/t$ for large values of t? Sketch the graph of this function for $t > 0$, paying particular attention to the asymptotes and turning values.   (*SMP*)

23.6 Applications to integration

We noted in §23.3 that, provided that x is positive, $\displaystyle\int \frac{1}{x}\,dx = \ln x + c$. We need a similar result for negative x since we may wish to evaluate the integral of $1/x$ between negative limits.

Note   It is important to remember that the integral $\displaystyle\int_a^b \frac{1}{x}\,dx$ can only have a meaning when a and b are either both positive or both negative. This is because the integrand $1/x$ is not defined at the origin $x = 0$.

Now let $a < b < 0$, so that $0 < -b < -a$. Consider the complete graph of $y = 1/x$, Fig. 23.5. The function $1/x$ is an odd function and the graph has a symmetry of rotation through a half turn about the origin O. Therefore, the shaded areas on the diagram are equal. The one below the x-axis is the negative of the integral and so

$$-\int_a^b \frac{1}{x}\,dx = \int_{-b}^{-a} \frac{1}{x}\,dx = \ln\,(-a) - \ln\,(-b)$$

and so $$\int_a^b \frac{1}{x}\,dx = \ln\,(-b) - \ln\,(-a), \text{ when } a < b < 0.$$

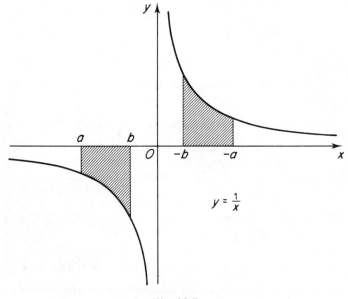

Fig. 23.5

This may be combined with the result for $0 < a < b$, namely that

$$\int_a^b \frac{1}{x}\,dx = \ln\,(b) - \ln\,(a), \quad \text{when} \quad 0 < a < b,$$

to give the result, provided that a and b are the same sign,

$$\int_a^b \frac{1}{x}\,dx = \ln\,|b| - \ln\,|a|.$$

Replacing b by x, we obtain as the indefinite integral of $1/x$, for any real x other than $x = 0$,

$$\int \frac{1}{x} dx = \ln|x| + c.$$

The modulus sign will then ensure that $\ln|x|$ is defined, since $|x| > 0$ for $x \neq 0$.

Further standard integral forms

Three standard forms for integrals are summarised here. They have arisen already in the Exercises 23.2 and 23.3 and they allow the integration of a wider class of functions.

$$\int \frac{1}{ax+b} dx = \frac{1}{a} \ln|ax+b| + c, \quad \text{for} \quad x \neq -\frac{b}{a},$$

$$\int \frac{f'(x)}{f(x)} dx = \ln f(x) + c,$$

$$\int f'(x) e^{f(x)} dx = e^{f(x)} + c.$$

Proof $\dfrac{d}{dx} \ln|ax+b| = \dfrac{a}{ax+b}$, $\dfrac{d}{dx} \ln f(x) = \dfrac{f'(x)}{f(x)}$, $\dfrac{d}{dx} e^{f(x)} = f'(x) e^{f(x)}$.

EXAMPLE   *Find* (i) $\displaystyle\int_{-3}^{-1} \frac{1}{2x-1} dx$, (ii) $\displaystyle\int_{1}^{2} \frac{1}{2x^2+3x+1} dx$,

(iii) $\displaystyle\int x e^{x^2} dx$, (iv) $\displaystyle\int_{-2}^{-\frac{3}{2}} \frac{x^2}{x^3+1} dx$.

(i) $\dfrac{1}{2x-1}$ is of the form $\dfrac{1}{ax+b}$, with $a = 2$ and $b = -1$. Therefore,

$$\int_{-3}^{-1} \frac{1}{2x-1} dx = \left[\tfrac{1}{2} \ln|2x-1| \right]_{-3}^{-1} = \tfrac{1}{2} \ln 3 - \tfrac{1}{2} \ln 7 = \tfrac{1}{2} \ln \tfrac{3}{7}.$$

(ii) The integrand can be converted by means of partial fractions into a sum of terms of the form $\dfrac{1}{ax+b}$. That is, $\dfrac{1}{2x^2+3x+1} = \dfrac{2}{2x+1} - \dfrac{1}{x+1}$,

$$\int_{1}^{2} \frac{1}{2x^2+3x+1} dx = \int_{1}^{2} \frac{2}{2x+1} dx - \int_{1}^{2} \frac{1}{x+1} dx = \left[\ln|2x+1| - \ln|x+1| \right]_{1}^{2}$$

$$= \left[\ln \left| \frac{2x+1}{x+1} \right| \right]_{1}^{2} = \ln \tfrac{5}{3} - \ln \tfrac{3}{2} = \mathbf{\ln \tfrac{10}{9}}.$$

(iii) $2xe^{x^2}$ is of the form $f'(x)e^{f(x)}$, with $f(x) = x^2$, $f'(x) = 2x$.

$$\int xe^{x^2}\, dx = \tfrac{1}{2}\int 2xe^{x^2}\, dx = \tfrac{1}{2}e^{x^2} + c.$$

(iv) $\dfrac{3x^2}{x^3+1}$ is of the form $\dfrac{f'(x)}{f(x)}$, with $f(x) = x^3 + 1$, $f'(x) = 3x^2$.

$$\int_{-2}^{-\frac{3}{2}} \frac{x^2}{x^3+1}\, dx = \tfrac{1}{3}\int_{-2}^{-\frac{3}{2}} \frac{3x^2}{x^3+1}\, dx = \left[\tfrac{1}{3}\ln|x^3+1|\right]_{-2}^{-\frac{3}{2}}$$

$$= \tfrac{1}{3}\ln|-\tfrac{27}{8}+1| - \tfrac{1}{3}\ln|-8+1| = \tfrac{1}{3}\ln\tfrac{19}{8} - \tfrac{1}{3}\ln 7 = \tfrac{1}{3}\ln\tfrac{19}{56}.$$

EXERCISE 23.6

1 Find the indefinite integral and state the value of x which is excluded from the domain:

(i) $\displaystyle\int \frac{1}{2-x}\, dx$, (ii) $\displaystyle\int \frac{3}{4x+1}\, dx$, (iii) $\displaystyle\int \frac{-2}{x+7}\, dx$.

2 Evaluate:

(i) $\displaystyle\int_0^1 \frac{1}{2-x}\, dx$, (ii) $\displaystyle\int_{-2}^{-1} \frac{3}{4x+1}\, dx$, (iii) $\displaystyle\int_{-10}^{-8} \frac{-2}{x+7}\, dx$.

3 Find:

(i) $\displaystyle\int \frac{3x}{2+x^2}\, dx$, (ii) $\displaystyle\int \frac{4x}{2x^2+1}\, dx$, (iii) $\displaystyle\int \frac{3}{x^2-9}\, dx$,

(iv) $\displaystyle\int \frac{2}{x^2+5x+6}\, dx$, (v) $\displaystyle\int \frac{1}{x^2+3x}\, dx$, (vi) $\displaystyle\int \frac{4}{(x+1)^2}\, dx$.

4 Find:

(i) $\displaystyle\int x^2 e^{2x^3}\, dx$, (ii) $\displaystyle\int \sec^2 x\, e^{\tan x}\, dx$, (iii) $\displaystyle\int x^{-2} e^{1/x}\, dx$,

(iv) $\displaystyle\int x^{-\frac{1}{2}} e^{\sqrt{x}}\, dx$, (v) $\displaystyle\int \frac{e^{1/x^2}}{x^3}\, dx$, (vi) $\displaystyle\int \frac{e^{\cot x}}{\sin^2 x}\, dx$.

5 Find:

(i) $\displaystyle\int_2^3 \frac{2x}{3-x^2}\, dx$, (ii) $\displaystyle\int_{-3}^{-2} \frac{x-1}{x^2-2x}\, dx$, (iii) $\displaystyle\int_1^3 e^{\frac{1}{2}(3x+1)}\, dx$,

(iv) $\displaystyle\int_1^{\frac{4}{3}} \frac{1}{3x-2}\, dx$, (v) $\displaystyle\int_{-\frac{1}{4}}^{\frac{1}{4}} \frac{1}{2x+1}\, dx$, (vi) $\displaystyle\int_1^2 3xe^{x^2-1}\, dx$,

(vii) $\displaystyle\int_{\frac{\pi}{3}}^{\frac{\pi}{2}} \cot x\, dx$, (viii) $\displaystyle\int_{\frac{\pi}{6}}^{\frac{\pi}{4}} \frac{\sec^2 x}{\tan x}\, dx$, (ix) $\displaystyle\int_0^3 \frac{2x-1}{x^2-x+1}\, dx$.

6 Evaluate $\displaystyle\int_0^1 xe^{-\frac{1}{2}x^2}\, dx$, leaving your answer in terms of e. (L)

7 By using partial fractions, show that $\displaystyle\int_9^{16} \frac{16-x}{(x-2)(x+5)}\, dx = 5\ln 2 - 3\ln 3$. (L).

8 Given that $f(x) = \dfrac{2}{1+2x} - \dfrac{x}{1+x^2}$, evaluate by direct integration $\displaystyle\int_0^{0\cdot1} f(x)\,dx$ correct to three decimal places.

9 Find the area of the region bounded by the curve $y = x + (1/x)$, the x-axis and the lines $x = 2$ and $x = 5$, leaving your answer in a form involving a natural logarithm. Find, to three decimal places, an approximation to this area by using the trapezium rule, dividing the area into three strips of unit width. Explain why the approximation is greater than the correct value. *(C)*

10 Evaluate $\displaystyle\int_0^1 \dfrac{e^x}{1+e^x}\,dx$. *(L)*

11 Find the coordinates and the nature of the turning point of the function $f(x) = 6/(x^2 - 9)$. Sketch the graph of f. Write $f(x)$ in partial fractions, and hence show that $\displaystyle\int_0^2 f(x)\,dx = -\ln 5$. Explain the negative sign. *(SMP)*

MISCELLANEOUS EXERCISE 23

1 The functions f and g are defined by $f(x) = e^x$, $g(x) = x - 1$. Let $F = fg$ and let $G = gf$, the two composite functions formed from f and g. Write down $F(x)$ and $G(x)$, state the domain and range of (i) F and (ii) G. Determine the inverse functions F^{-1} and G^{-1} and, using the same axes, sketch the graphs of F, G, F^{-1} and G^{-1}.

2 Differentiate with respect to x
(i) $x^2 \ln 6x$, (ii) $\cos^3 4x$. *(L)*

3 Determine which of the functions defined below are one-one functions and give the range of each function:
(i) $f: x \mapsto e^{-x}$ $(x \in \mathbb{R}, x \geqslant 0)$, (ii) $g: x \mapsto x^2 + 4x - 5$ $(x \in \mathbb{R}, x \geqslant 0)$,
(iii) $h: x \mapsto x + (1/x)$ $(x \in \mathbb{R}, x > 0)$.
  With the above definitions of f, g and h, determine which of the following functions exist:
(iv) fg, (v) hf, (vi) g^{-1}.
  In each of (iv) to (vi), if you assert that the function does not exist, give reasons for your assertion, and, if you assert that the function does exist, give its rule and its domain. *(C)*

4 Find $\dfrac{d}{dx} \ln\left(\dfrac{x}{1+x^2}\right)$. *(L)*

5 Given that $x = 2 + 2e^{-2t} - e^{-3t}$, find the maximum value of x as t varies. *(L)*

6 Given that $y = \log_e \sqrt{(1+x)}$, show that $(1+x)\dfrac{dy}{dx} = \dfrac{1}{2}$. *(JMB)*

7 Find the coordinates of any stationary points or points of inflexion on the curve given by

$$y = \frac{x}{\ln x} \quad (x > 0, \quad x \neq 1)$$

and sketch the curve, marking in these points. *(C)*

8   If $y = \ln(\sin^3 2x)$, find $\dfrac{dy}{dx}$ and prove that

$$3\frac{d^2y}{dx^2} + \left(\frac{dy}{dx}\right)^2 + 36 = 0. \qquad (C)$$

9   Given that $\lg y = 1 - 0.5 \lg x$, express y explicitly in terms of x and sketch the graph of y against x. Only the general shape of the graph and its position relative to the axes are required. ($\lg p$ means $\log_{10} p$.) $\qquad (L)$

10   Show that the curve $y = (1 + x)e^{-2x}$ has just one maximum and one point of inflexion.

Sketch the curve, marking the coordinates of the maximum point, the point of inflexion and the point where the curve crosses the x-axis.

(You may assume that $(1 + x)e^{-2x} \to 0$ as $x \to +\infty$.)

11   The function f is defined for $x \geq 0$ by $f : x \mapsto 1 + \log_e(2 + 3x)$. Find
(i) the range of f, (ii) an expression for $f^{-1}(x)$, (iii) the domain of f^{-1}.
On one diagram, sketch graphs of f and f^{-1}.

Show that the x-coordinate of the point of intersection of the graphs of f and f^{-1} satisfies the equation

$$\log_e(2 + 3x) = x - 1. \qquad (JMB)$$

12   By considering the area under the graph of $y = 1/u$ between $u = 1$ and $u = 1 + x$, or otherwise, show that for positive x

$$\frac{x}{1 + x} < \ln(1 + x) < x.$$

Deduce that, as x tends to zero, $(1/x) \ln(1 + x)$ tends to 1.

13   Evaluate:

(i) $\displaystyle\int_0^1 xe^{2x^2}\,dx$,   (ii) $\displaystyle\int_0^1 \frac{1 - 2x}{(x + 2)(x^2 + 1)}\,dx$.

14   It is given that the equation $10e^x - x = 70$ has a root near 2. Working to three places of decimals and using two applications of the iterative relation

$$x_{n+1} = \ln[(x_n + 70)/10]$$

obtain an approximation to the root, giving your answer to three significant figures.

15   A certain jet aircraft uses $10h$ tonne of fuel getting up to a height of h km, and $0.5e^{-0.2h}$ tonne per km cruising at this height. Write an expression for the total fuel consumed on a flight of distance a km (taking the horizontal distance covered going up and coming down to be negligible compared with a).

Find, in terms of the constant a (assumed to be greater than 100), the most economical cruising height, h km, for that distance. $\qquad (SMP)$

16   Sketch on the same diagram the graphs of (i) $y = \ln x$, (ii) $y = \alpha x$, where α is a small positive number. Explain why the equation

$$\ln x - \alpha x = 0$$

has a solution close to $x = 1$.

Using the Newton-Raphson process once, or otherwise, find a closer approximation to the solution in terms of α. $\qquad (SMP)$

17   The function f is defined for $-\pi \leqslant x \leqslant \pi$ by

$$f : x \mapsto e^{\frac{1}{2}\cos x} \sin x.$$

State the values of $f(0)$, $f(\frac{1}{2}\pi)$, $f(-\frac{1}{2}\pi)$, $f(\pi)$, $f(-\pi)$.
   Determine $f'(x)$, and show that $f'(\alpha) = 0$ if, and only if, $\cos \alpha = \sqrt{2} - 1$. For such a value α calculate $f(\alpha)$ and $f(-\alpha)$ correct to two significant figures. Sketch a graph of $f(x)$.
   Calculate the area bounded by the graph and the x-axis between $x = 0$ and $x = \pi$.   (*JMB*)

18   Simplify $\log_e \left[\dfrac{(1+x)e^{-2x}}{1-x} \right]^{\frac{1}{2}}$ and show that its derivative is $\dfrac{x^2}{1-x^2}$.

Hence, or otherwise, evaluate dy/dx at $x = 0$ for the function

$$y = \left[\frac{(1+x)e^{-2x}}{1-x} \right]^{\frac{1}{2}}.$$ (*JMB*)

19   Sketch the curve whose equation is $y = \dfrac{x+2}{x}$ and state the equations of its

asymptotes.
   By considering this sketch, or otherwise, show that the curve whose

equation is $y = \log_e \dfrac{x+2}{x}$ has no points whose x-coordinates lie in the

interval $-2 \leqslant x \leqslant 0$. Sketch this curve on a separate diagram.
   Prove that the area bounded by the *second* curve, the x-axis and the lines

$x = 1$ and $x = 2$ is $3 \log_e \dfrac{4}{3}$.   (*JMB*)

20   Given that $y = e^{-2x} \sin 3x$, write down $\log_e y$ as a sum of two terms. Hence, or otherwise, find

$$\frac{1}{y} \frac{dy}{dx} \quad \text{at } x = \pi/12.$$

Deduce to one significant figure the change in y as x increases from $\pi/12$ to $(\pi/12) + 0.02$.   (*JMB*)

21   Find the turning points on the curve $y = e^x \sin x$ for $0 \leqslant x \leqslant 2\pi$. Sketch the curve over this interval.   (*AEB*)

22   If $y = (x - 0.5)e^{2x}$, find $\dfrac{dy}{dx}$ and hence, or otherwise, calculate, correct to one

decimal place, the mean value of xe^{2x} in the interval $0 \leqslant x \leqslant 2.5$.   (*AEB*)

23   If $\tan y = \log_e x^2$, show that $x \dfrac{dy}{dx} = 2 \cos^2 y$. Hence show that

$$x^2 \frac{d^2 y}{dx^2} + 2(1 + 2\sin 2y)\cos^2 y = 0.$$ (*AEB*)

24   In the diagram of the graph of $y = 1/x$ for $x > 0$, the shaded area is equal to half the area of the rectangle. Obtain the equation $2 \ln z = z - 1$, and show that z lies between 3 and 4.
   Use the Newton–Raphson method to calculate an approximation to z, starting at $z = 3$ and continuing the iteration until successive values differ by less than 10^{-2}.   (*SMP*)

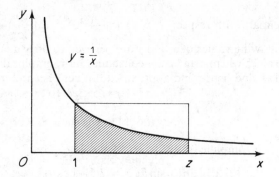

25 Solve, correct to two decimal places, $\log_e x + x - 2 = 0$.   (*JMB*)

26 The function g, with domain $x > -3$, is defined by $g(x) = \log_e(x+3)+1$. Find the inverse $g^{-1}(x)$ and state its domain and range. Explain how the graph of $y = g^{-1}(x)$ may be constructed from the graphs of $y = g(x)$ and $y = x$. Sketch the graphs of $y = g(x)$ and $y = g^{-1}(x)$ on the same axes.

Show that the x-coordinates of the points of intersection of the graphs satisfy the equation

$$\log_e(x+3) = x - 1.$$

Verify that this equation has a root between 2 and 3, and use linear interpolation to obtain this root correct to two decimal places.   (*JMB*)

27 Find the area of the finite region enclosed by the curve $xy = 6$ and the line $x + y = 5$ and calculate the coordinates of the centroid of this region, giving your answers to two significant figures. Show that the volume of the solid of revolution formed by rotating this region completely about the x-axis is $\pi/3$.

28 The function f is defined for $x > 0$ by $f : x \mapsto 2e^{-0.5x^2} - 1$. Find the positive solution a of $f(x) = 0$, and sketch the graph of the function.

A pile of gravel has a shape such that, at horizontal distance r from the central axis of the pile, the depth d of the pile is given by

$$\begin{cases} d = f(r), & r \leqslant a; \\ d = 0, & r > a. \end{cases}$$

Write down an approximation for the volume of gravel that is between r and $r + \delta r$ from the axis (assuming that both r and $r + \delta r$ are less than a). Hence obtain the total volume as an integral of form

$$V = k \int_0^\alpha r^\beta f(r) \, dr,$$

stating the values of the constants k, α and β.

Show that $V = 2\pi(1 - \ln 2)$.   (*SMP*)

29 A region of a plane is bounded by parts of the x-axis and of the line $x = k$, and by that part of the curve with equation

$$y = 1 - (1/\sqrt{x})$$

for which $1 < x < k$. Find, in terms of k, the volume formed when this region is rotated through a whole turn about the x-axis.   (*SMP*)

30 (a) Differentiate with respect to x, (i) $(x-1)^2 e^{x^3}$, (ii) $\dfrac{\cos^2 x}{\log_e x}$.

(b) The following values of x and y are believed to obey a law of the form $y - 3 = 20\, b^{x+a}$, where a and b are constants. Show graphically that they do obey this law and hence find approximate values of a and b.

x	0	2	5	7	
y	8233	744	23	5	(*AEB*)

31 (a) Find $\dfrac{dy}{dx}$ in simplified form when (i) $y = \log_e \sqrt{1 + 7x}$,

(ii) $y = 2(\theta - \sin\theta)$, $x = 2(1 - \cos\theta)$.

(b) Using the same axes sketch the curves given by $y = e^x$ and $y = 3 - 4x - x^2$. If t is the positive root of the equation $e^x = 3 - 4x - x^2$, prove that the area of the finite region in the first quadrant enclosed by the curves and the y-axis is given by $\frac{1}{3}(21t - 3t^2 - t^3 - 6)$.   (*AEB*)

32 (a) (i) Find $\dfrac{dy}{dx}$ when $y = e^{\tan x}$.

(ii) Evaluate $\dfrac{dy}{dx}$ when $x = 2$ if $y = \log_e \sqrt{\left[\dfrac{1 + x^2}{1 - 2x}\right]}$.

(b) Find the turning points on the curve $y = \cos x + 2\cos\frac{1}{2}x$ for $0 \leqslant x \leqslant 2\pi$. Hence sketch the curve over this range.   (*AEB*)

33 (a) Find $\dfrac{dy}{dx}$ when (i) $y = \dfrac{x^2 - 2}{(x+2)^2}$, (ii) $y = \log_e \sin 2x + e^{1/x}$,

(iii) $x = \sin^2\theta$, $y = \sin 2\theta$.

(b) Given that $y = e^{3x} \sin 4x$, express d^2y/dx^2 in the form $R\,e^{3x}\cos(4x + \alpha)$, giving values of R and $\tan\alpha$.   (*AEB*)

34 Find the minimum value of $x \ln x$, where x is positive. Given that $x \ln x$ tends to 0 as x tends to 0, sketch the curve $y = x \ln x$.

24 Integration Techniques

24.1 Indefinite integrals

In Chapter 6, we began the task of searching for antiderivatives, or indefinite integrals, and there we dealt with polynomials and the sine and cosine functions. We are now in a position to deal with many other functions. The problem of finding $\int f(x) \, dx$ is that of finding a function $F(x)$ such that $F'(x) = f(x)$. As mentioned in Chapter 6, a constant may be added to $F(x)$, so that in finding an indefinite integral, we include an arbitrary constant c of integration. Thus $\int f(x) \, dx = F(x) + c$. For convenience, in the present work we shall omit the constant c.

To some extent, integrating is a form of guesswork. From the form of the function $f(x)$, we guess what form the function $F(x)$ must take, and then we check that $F'(x) = f(x)$. To assist our guesses, there are a number of techniques, most of which involve transforming the function into a standard form of which we know the integral. We, therefore, begin with a list of standard forms of integrals, which we have found previously.

Standard integrals

Omitting the constant of integration:

$f(x)$	$\int f(x) \, dx$
$x^n, n \neq -1$	$\dfrac{1}{n+1} x^{n+1}$
x^{-1}	$\ln \lvert x \rvert$
e^x	e^x
$\cos x$	$\sin x$
$\sin x$	$-\cos x$
$\sec^2 x$	$\tan x$
$\dfrac{1}{1+x^2}$	$\tan^{-1} x$
$\dfrac{1}{\sqrt{(1-x^2)}}$	$\sin^{-1} x$

24.2 Integration by substitution

Suppose that we wish to find y, where $y = \int f(x) \, dx$. In this integral, we refer to x as the *independent variable*. It may be possible to manipulate the integral into a standard form, or a more recognisable form, by means of a change in the independent variable from x to some new variable u, where $u = u(x)$. Suppose that y is given as a function $F(x)$ of x and also as a function $G(u)$ of u, where $G'(u) = g(u)$. Then

$$y = \int f(x)\,dx = F(x), \quad \text{so that} \quad F'(x) = f(x),$$

and $\qquad y = \int g(u)\,du = G(u), \quad \text{so that} \quad G'(u) = g(u).$

The connection between the functions $f(x)$ and $g(u)$ are given by the chain rule for differentiation, since we are dealing with the composition of functions. Since $y = Gu(x)$, $\dfrac{dy}{dx} = \dfrac{dy}{du}\dfrac{du}{dx}$ so that $f(x) = g(u)\dfrac{du}{dx}$ and

$$y = \int f(x)\,dx = \int g(u)\frac{du}{dx}\,dx = \int g(u)\,du.$$

This means that we must find a function $u(x)$, such that $f(x) = g(u)\dfrac{du}{dx}$ and that $\int g(u)\,du$ is an easier integral to handle than $\int f(x)\,dx$. If possible we try and make the new integral a standard form. Some guesswork is needed to find $u(x)$ and if the first guess fails then we try again. The transformation of the integral changes the independent variable x to u, that is, we *substitute u* for x as independent variable, so the method is referred to as *substitution*.

EXAMPLE 1   Find $\int (3x+2)^3\,dx$.

The integral of u^n is a standard form so we try the substitution $u = 3x+2$. Then $\dfrac{du}{dx} = 3$,

$$f(x) = (3x+2)^3 = \frac{1}{3}(3x+2)^3\,\frac{du}{dx} = g(u)\frac{du}{dx}$$

where $g(u) = \frac{1}{3}(3x+2)^3 = \frac{1}{3}u^3$. Hence

$$\int f(x)\,dx = \int (3x+2)^3\,dx = \int \tfrac{1}{3}u^3\frac{du}{dx}\,dx = \int \tfrac{1}{3}u^3\,du = \tfrac{1}{12}u^4 = \tfrac{1}{12}(3x+2)^4.$$

If an arbitrary constant is included, the integral is $\tfrac{1}{12}(3x+2)^4 + c$.

A similar substitution may be used to find the integral of $(ax+b)^n$, as shown in the next two examples.

EXAMPLE 2   Find $F(x)$, where $F(x) = \displaystyle\int \frac{1}{(3x-4)^4}\,dx$.

Put $u = 3x-4$, then $\dfrac{du}{dx} = 3$ and $F(x) = \displaystyle\int \frac{1}{3}u^{-4}\frac{du}{dx}\,dx = \frac{1}{3}\int u^{-4}\,du$, so

$$F(x) = \frac{1}{3}\left(-\frac{1}{3}u^{-3}\right) + c = -\frac{1}{9}(3x-4)^{-3} + c = \frac{-1}{9(3x-4)^3} + c.$$

EXAMPLE 3 *Find* $\int \dfrac{1}{3x-4}\,dx$.

Again put $u = 3x - 4$, then $\dfrac{du}{dx} = 3$, $\dfrac{1}{3x-4} = \dfrac{1}{3}\dfrac{1}{u}\dfrac{du}{dx}$. Then

$$\int \frac{1}{3x-4}\,dx = \int \frac{1}{3}\frac{1}{u}\frac{du}{dx}\,dx = \int \frac{1}{3}u^{-1}\,du = \frac{1}{3}\ln|u| + c = \frac{1}{3}\ln|3x-4| + c.$$

Note A Remember the need for the modulus sign in the argument of the logarithm when integrating u^{-1}, unless it is known that u is positive.

B Since $\dfrac{du}{dx} = 1 \Big/ \left(\dfrac{dx}{du} \right)$ the transformation of $\int f(x)\,dx$ may be written

$$\int f(x)\,dx = \int g(u)\,du = \int \frac{f(x)}{\dfrac{du}{dx}}\,du = \int f(x)\frac{dx}{du}\,du,$$

so that, in terms of the inverse function $x = x(u)$ of the function $u = u(x)$, the transformation becomes $\int f(x)\,dx = \int fx(u)\dfrac{dx}{du}\,du$. Thus dx is replaced by $\dfrac{dx}{du}\,du$, which may be an aid to the memory.

EXAMPLE 4 *Find* $\int \sin(3x-4)\,dx$

This time we use the idea of the above note **B**. Put $u = 3x - 4$, so that $\dfrac{du}{dx} = 3$ and $\dfrac{dx}{du} = \dfrac{1}{3}$. Then

$$\int \sin(3x-4)\,dx = \int \sin u \frac{dx}{du}\,du = \int \frac{1}{3}\sin u\,du = -\frac{1}{3}\cos u + c$$

and the answer is $-\frac{1}{3}\cos(3x-4) + c$.

Note C Always remember that, in the final expression, u must again be replaced by $u(x)$ so that the integral is a function of x.

Definite integrals

The evaluation of definite integrals may be performed in one of two ways. The first method is to find the indefinite integral $F(x)$ and then evaluate $F(b) - F(a)$, where a and b are the limits of integration. Thus, if $\int f(x)\,dx = F(x) + c$, then $\displaystyle\int_a^b f(x)\,dx = F(b) - F(a)$.

The second method is to apply the method of substitution to the definite integral. This means that when the independent variable is changed, say from x to u, where $u = u(x)$, then the limits of integration must also be changed from a and b to $u(a)$ and $u(b)$. If $u = u(x)$,

$$\int_a^b f(x)\,dx = \int_{u(a)}^{u(b)} g(u)\,du, \quad \text{where} \quad f(x) = g(u)\frac{du}{dx}.$$

We shall illustrate this method at the same time as showing some further substitutions. Remember that we look for $u(x)$ and $g(u)$ such that $f(x) = g(u)\,u'(x)$.

EXAMPLE 5   *Evaluate* $\displaystyle\int_0^1 \frac{x}{1+x^2}\,dx.$

Put $u = 1 + x^2$, so that $u'(x) = 2x$, then $\dfrac{x}{1+x^2} = \dfrac{1}{2}\dfrac{1}{u}u'(x)$. Therefore

$$\int_0^1 \frac{x}{1+x^2}\,dx = \int_{u(0)}^{u(1)} \frac{1}{2u}\,du = \frac{1}{2}\int_1^2 u^{-1}\,du = \frac{1}{2}\Big[\ln|u|\Big]_1^2 = \frac{1}{2}\ln 2 = \mathbf{ln}\sqrt{2}.$$

The corresponding indefinite integral is $\ln\sqrt{(1+x^2)} + c$, where we may omit the modulus sign because the argument of the logarithm is positive.

When the integrand involves a square root, it may help to replace the square root by a new variable. We give two examples; one is a definite integral and one is indefinite.

EXAMPLE 6   *Evaluate* $\displaystyle\int_0^3 x\sqrt{(x+1)}\,dx.$

Try the substitution $u = \sqrt{(x+1)}$, so that $x = u^2 - 1$, $\dfrac{dx}{du} = 2u$. Then $u(0) = 1$ and $u(3) = 2$, so

$$\int_0^3 x\sqrt{(x+1)}\,dx = \int_1^2 (u^2 - 1)\,u\frac{dx}{du}\,du = \int_1^2 2u^2(u^2 - 1)\,du$$

$$= \int_1^2 (2u^4 - 2u^2)\,du = \Big[\tfrac{2}{5}u^5 - \tfrac{2}{3}u^3\Big]_1^2$$

$$= \tfrac{2}{5}(32 - 1) - \tfrac{2}{3}(8 - 1) = 7\tfrac{11}{15}.$$

EXAMPLE 7   *Find* $\displaystyle\int \frac{3x^2}{\sqrt{(1+x^3)}}\,dx.$

Put $u = \sqrt{(1+x^3)}$, so that $u^2 = 1+x^3$ and $2u = 3x^2 \dfrac{dx}{du}$. Then

$$\int \frac{3x^2}{\sqrt{(1+x^3)}}\,dx = \int \frac{3x^2}{u}\frac{dx}{du}\,du = \int \frac{3x^2}{u}\frac{2u}{3x^2}\,du = \int 2\,du = 2u+c$$

$$= 2\sqrt{(1+x^3)}+c.$$

Since $\dfrac{d}{dx}\ln|g(x)| = \dfrac{g'(x)}{g(x)}$, look out for an integrand of this form and make the substitution $u = g(x)$. Example 5 was of this type and we give two more examples.

EXAMPLE 8 *Find* $\displaystyle\int_0^1 \frac{2x+3}{x^2+3x+4}\,dx.$

The integrand is of the form $\dfrac{g'(x)}{g(x)}$, so substitute $g(x) = u$, then

$$\int_0^1 \frac{g'(x)}{g(x)}\,dx = \int_4^8 \frac{1}{u}\frac{du}{dx}\,dx = \int_4^8 \frac{1}{u}\,du = \Big[\ln|u|\Big]_4^8 = \ln 8 - \ln 4 = \textbf{ln 2}.$$

EXAMPLE 9 *Integrate* $\dfrac{1}{\sin x \cos x}$ *with respect to x.*

$$\frac{1}{\sin x \cos x} = \frac{\sec^2 x}{\tan x} = \frac{g'(x)}{g(x)},$$ so, on making the substitution $g(x) = u = \tan x$,

$$\int \frac{1}{\sin x \cos x}\,dx = \int \frac{\sec^2 x}{\tan x}\,dx = \int \frac{1}{u}\frac{du}{dx}\,dx = \int \frac{1}{u}\,du = \ln|u| = \textbf{ln}|\textbf{tan}\,x|+c.$$

So far, we have used substitutions of the form $u = g(x)$. In some cases, it is more straightforward to make the substitution by means of the inverse of the function g, that is, substitute $x = h(u)$, $h = g^{-1}$. Then $\dfrac{dx}{du} = h'(u)$, so

$$\int f(x)\,dx = \int f h(u) h'(u)\,du.$$

EXAMPLE 10 *Find* $\displaystyle\int \frac{1}{x^2+1}\,dx.$

Make the trigonometrical substitution, $x = \tan u$, so that $\dfrac{dx}{du} = \sec^2 u$. Then

$$\int \frac{1}{x^2+1}\,dx = \int \frac{1}{\tan^2 u+1}\frac{dx}{du}\,du = \int \frac{1}{\sec^2 u}\sec^2 u\,du = \int du = u.$$

Hence the integral is $\textbf{tan}^{-1}\,x$, as we had previously found.

EXAMPLE 11   *Find* $\int \dfrac{1}{\sqrt{(1-x^2)}}\,dx$.

Let $x = \sin\theta$, then $\sqrt{(1-x^2)} = \sqrt{(\cos^2\theta)} = \cos\theta$, and $\dfrac{dx}{d\theta} = \cos\theta$. Then

$$\int \frac{1}{\sqrt{(1-x^2)}}\,dx = \int \frac{1}{\cos\theta}\cos\theta\,d\theta = \int d\theta = \theta = \sin^{-1}x.$$

Examples 10 and 11 give results in our table of integrals, that is, they are standard forms, so we omit the arbitrary constants of integration.

A substitution, which may sometimes help in the integration of a function of circular functions is to use the half-angle formulae and replace $\sin\theta$ and $\cos\theta$ by $\dfrac{2t}{1+t^2}$ and $\dfrac{1-t^2}{1+t^2}$, respectively, where $t = \tan\tfrac{1}{2}\theta$.

EXAMPLE 12   *Find* $\displaystyle\int_0^{\frac{\pi}{2}} \dfrac{1}{1+\sin\theta}\,d\theta$.

Put $t = \tan\tfrac{1}{2}\theta$, then $\sin\theta = \dfrac{2t}{1+t^2}$ and $\dfrac{dt}{d\theta} = \tfrac{1}{2}\sec^2\tfrac{1}{2}\theta = \tfrac{1}{2}(1+t^2)$. For the change in the limits of the integral, $\tan 0 = 0$, $\tan(\tfrac{1}{2}.\tfrac{\pi}{2}) = 1$. Hence

$$\int_0^{\frac{\pi}{2}} \frac{1}{1+\sin\theta}\,d\theta = \int_0^1 \frac{1}{1+\dfrac{2t}{1+t^2}}\frac{d\theta}{dt}\,dt = \int_0^1 \frac{1+t^2}{1+t^2+2t}\frac{1}{\tfrac{1}{2}(1+t^2)}\,dt$$

$$= \int_0^1 \frac{2}{(1+t)^2}\,dt = \int_1^2 \frac{2}{u^2}\,du = \left[-\frac{2}{u}\right]_1^2 = -1+2 = \mathbf{1}.$$

In the last stage, we used a second substitution, $u = 1+t$.

EXERCISE 24.2

1   Use the given substitution to find the integral:
   (i) $\int(x+2)^2\,dx$, $u = x+2$;   (ii) $\int\sqrt{(3x-4)}\,dx$, $u = 3x-4$;
   (iii) $\int 2\sin(3x+4)\,dx$, $u = 3x+4$;   (iv) $\int x(1-x^2)^3\,dx$, $u = 1-x^2$;
   (v) $\int x\sqrt{(x^2+2)}\,dx$, $u^2 = x^2+2$;   (vi) $\int \sin\theta\cos^4\theta\,d\theta$, $u = \cos\theta$;
   (vii) $\int\tan\theta\,d\theta$, $u = \cos\theta$;   (viii) $\int\csc\theta\,d\theta$, $u = \tan\tfrac{1}{2}\theta$;
   (ix) $\int(x^2+1)^{-\frac{3}{2}}\,dx$; $x = \tan\theta$;   (x) $\displaystyle\int \frac{1}{4+x^2}\,dx$, $x = 2\tan\theta$.

2   Find the indefinite integral of:
   (i) $4(3-2x)^3$;   (ii) $(x-1)^{-\frac{1}{2}}$;   (iii) $\sin^5\theta\cos\theta$,   (iv) $\cot x$,
   (v) $\sec^2(2x-1)$,   (vi) $\tan x\sec^2 x$.

3   Evaluate:
   (i) $\displaystyle\int_0^{\frac{1}{2}\pi} \cos(\tfrac{1}{2}x+\tfrac{3}{4}\pi)\,dx$,   (ii) $\displaystyle\int_1^2 (4x-3)^3\,dx$,   (iii) $\displaystyle\int_0^4 x\sqrt{(2x+1)}\,dx$.

4 Evaluate $\int_0^4 3x\sqrt{(x^2+9)}\,dx$.   (*L*)

5 Using the substitution $2x = \sin\theta$, or otherwise, evaluate $\int_0^{\frac{1}{2}} \sqrt{(1-4x^2)}\,dx$.
                                                                                       (*L*)

6 Find $\int_1^e \dfrac{1+x^2}{x}\,dx$.   (*L*)

7 Using the substitution $u^2 = x^2 - 1$, or otherwise, evaluate

$$\int_{\sqrt{2}}^{\sqrt{5}} \frac{x^3}{\sqrt{(x^2-1)}}\,dx.$$ (*L*)

8 By substituting $x = \sin\theta$, or otherwise, evaluate $\int_0^{\frac{1}{\sqrt{2}}} \dfrac{x^2}{(1-x^2)^{3/2}}\,dx$, leaving
your answer in terms of π.   (*JMB*)

9 Evaluate $\int_0^2 \dfrac{x^2\,dx}{\sqrt{(x^3+1)}}$ using the substitution $u = x^3 + 1$ or otherwise.
                                                                                       (*L*)

10 Using the substitution $x = 3\tan\theta$, or otherwise, evaluate

$$\int_0^3 \frac{dx}{9+x^2}.$$ (*L*)

11 Using the substitution $u^2 = x^2 - 2x$, or otherwise, evaluate

$$\int_3^4 \frac{x-1}{\sqrt{(x^2-2x)}}\,dx.$$ (*L*)

12 Find $\int_0^1 xe^{-x^2}\,dx$.   (*L*)

13 (a) Find $\int \sin^5 x\,dx$.
   (b) Using the substitution $t = \tan\theta$, or otherwise, evaluate

$$\int_0^{\frac{1}{4}\pi} \frac{1}{4+5\cos 2\theta}\,d\theta,$$

   giving two significant figures in your answer.   (*C*)

14 Evaluate $\int_1^2 x(2-x)^7\,dx$.   (*L*)

15 Find $\int \tan^3 x\,dx$.   (*L*)

16 Evaluate $\int_0^1 xe^{2x^2}\,dx$, leaving your answer in terms of e.

24.3 Integration of rational functions

In order to integrate a rational function $f(x)$, given by $\dfrac{p(x)}{q(x)}$, where $p(x)$
and $q(x)$ are polynomials, it is necessary to factorise $q(x)$ and split $f(x)$ into

partial fractions. There is one case which can be integrated more directly, namely when p = q', and this we have met before:

$$\int \frac{q'(x)}{q(x)} \, dx = \ln |q(x)|.$$

The methods are illustrated by some examples and then the procedures are summarised. The first few examples provide some important results which are used throughout the methods.

EXAMPLE 1 *Prove that* $\int \dfrac{1}{ax+b} \, dx = \dfrac{1}{a} \ln |ax+b|.$

Let $u = ax + b$, then $1 = a \dfrac{dx}{du}$, so $\dfrac{dx}{du} = \dfrac{1}{a}$ and, by substituting in the integral,

$$\int \frac{1}{ax+b} \, dx = \int \frac{1}{u} \frac{dx}{du} \, du = \frac{1}{a} \int \frac{1}{u} \, du = \frac{1}{a} \ln |u| = \frac{1}{a} \ln |ax+b|.$$

EXAMPLE 2 *Prove that* $\int \dfrac{1}{(ax+b)^2} \, dx = \dfrac{-1}{a(ax+b)}.$

Let $u = ax + b$, then $\dfrac{dx}{du} = \dfrac{1}{a}$ and

$$\int \frac{1}{(ax+b)^2} \, dx = \frac{1}{a} \int \frac{1}{u^2} \, du = \frac{-1}{au} = \frac{-1}{a(ax+b)}.$$

EXAMPLE 3 *Prove that* $\int \dfrac{a}{x^2+a^2} \, dx = \tan^{-1} \dfrac{x}{a}.$

Let $x = au$, so that $\dfrac{dx}{du} = a$ and

$$\int \frac{a}{x^2+a^2} \, dx = \int \frac{a}{a^2 u^2 + a^2} \, a \, du = \int \frac{1}{u^2+1} \, du = \tan^{-1} u = \tan^{-1} \frac{x}{a}.$$

EXAMPLE 4 *Find* $\displaystyle\int_0^1 \dfrac{x^2}{x^2+3x+2} \, dx.$

Split the integrand into partial fractions, the first step being to find the quotient and remainder. Since $x^2 = x^2 + 3x + 2 - (3x + 2)$,

$$\frac{x^2}{x^2+3x+2} = \frac{x^2+3x+2-3x-2}{x^2+3x+2} = 1 - \frac{3x+2}{x^2+3x+2} = 1 + \frac{A}{x+2} + \frac{B}{x+1}.$$

Then $-3x - 2 = A(x+1) + B(x+2)$. Put $x = -1, 1 = B$. Put $x = -2, 4 = -A$.

So $\int_0^1 \frac{x^2}{x^2+3x+2}\,dx = \int_0^1 1\,dx - \int_0^1 \frac{4}{x+2}\,dx + \int_0^1 \frac{1}{x+1}\,dx$

$= \left[x - 4\ln|x+2| + \ln|x+1| \right]_0^1 = 1 + \ln\frac{32}{81}$ (using example 1).

EXAMPLE 5 *Evaluate* $\int_0^1 \frac{x+1}{x^2+x+1}\,dx$.

Since the denominator does not split into two linear factors, we complete the square, writing $x^2+x+1 = (x+\frac{1}{2})^2 + \frac{3}{4}$. Then the square term is used as the new variable. Let $u = x+\frac{1}{2}$, so $\frac{du}{dx} = 1$, $u(0) = \frac{1}{2}$, $u(1) = \frac{3}{2}$, and

$\int_0^1 \frac{x+\frac{1}{2}+\frac{1}{2}}{(x+\frac{1}{2})^2+\frac{3}{4}}\,dx = \int_{\frac{1}{2}}^{\frac{3}{2}} \frac{u+\frac{1}{2}}{u^2+\frac{3}{4}}\,du = \frac{1}{2}\int_{\frac{1}{2}}^{\frac{3}{2}} \frac{2u}{u^2+\frac{3}{4}}\,du + \frac{1}{2}\frac{2}{\sqrt{3}}\int_{\frac{1}{2}}^{\frac{3}{2}} \frac{\sqrt{3}/2}{u^2+\frac{3}{4}}\,du$

$= \left[\frac{1}{2}\ln|u^2+\frac{3}{4}| + \frac{1}{\sqrt{3}}\tan^{-1}\frac{2u}{\sqrt{3}} \right]_{\frac{1}{2}}^{\frac{3}{2}} = \ln\sqrt{3} + \frac{\pi}{6\sqrt{3}}$.

EXAMPLE 6 *Find* $\int \frac{x+2}{x^3-1}\,dx$.

Factorise the denominator and split the integrand into partial fractions.

$$\frac{x+2}{x^3-1} = \frac{x+2}{(x^2+x+1)(x-1)} = \frac{A}{x-1} + \frac{Bx+C}{x^2+x+1},$$

so we require

$$x+2 = A(x^2+x+1) + (Bx+C)(x-1).$$

Put $x=1$, then $3 = 3A$, $A = 1$. Put $x=0$, then $2 = A - C$, $C = -1$. Compare the coefficients of x^2, $0 = A+B$, $B = -1$. Therefore,

$\int \frac{x+2}{x^3-1}\,dx = \int \frac{1}{x-1}\,dx - \int \frac{x+1}{x^2+x+1}\,dx$

$= \ln|x-1| - \frac{1}{2}\ln(x^2+x+1) - \frac{1}{\sqrt{3}}\tan^{-1}\frac{2x+1}{\sqrt{3}}$,

on using Examples 1 and 5.

Summary To integrate a rational function:
  (i) split the function into partial fractions;
  (ii) for a term with linear denominator, use $\int \frac{1}{ax+b}\,dx = \frac{1}{a}\ln|ax+b|$;
  (iii) for a term with linear-squared denominator, use

$$\int \frac{1}{(ax+b)^2}\,dx = \frac{-1}{a(ax+b)};$$

(iv) for a term with quadratic denominator, complete the square of the denominator in the form $(x + a)^2 + b^2$, substitute $u = x + a$, and then use

$$\int \frac{cu + d}{u^2 + b^2}\, du = \tfrac{1}{2}c \ln (u^2 + b^2) + \frac{d}{b} \tan^{-1}\left(\frac{u}{b}\right),$$

and then substitute back again $u = x + a$.

EXERCISE 24.3

1 Find:

(i) $\displaystyle\int \frac{1}{x - 3}\, dx$,   (ii) $\displaystyle\int \frac{3}{x + 4}\, dx$,   (iii) $\displaystyle\int \frac{1}{2x - 1}\, dx$,   (iv) $\displaystyle\int \frac{x}{x - 1}\, dx$,

(v) $\displaystyle\int \frac{2x + 3}{x - 2}\, dx$,   (vi) $\displaystyle\int \frac{3x - 4}{2x - 5}\, dx$,   (vii) $\displaystyle\int \frac{x^2 + 1}{x - 1}\, dx$,

(viii) $\displaystyle\int \frac{3}{(x - 1)(x + 2)}\, dx$,   (ix) $\displaystyle\int \frac{1}{x^2 - x + 6}\, dx$,   (x) $\displaystyle\int \frac{x + 5}{(x + 1)(x + 3)}\, dx$,

(xi) $\displaystyle\int \frac{1}{(x - 1)^2}\, dx$,   (xii) $\displaystyle\int \frac{x^2 + 7}{(x + 1)(x - 1)^2}\, dx$.

2 Find the indefinite integrals of:

(i) $\dfrac{(x + 1)^2}{x + 2}$,   (ii) $\dfrac{1}{(x + 4)(x + 5)}$,   (iii) $\dfrac{6}{x^2 - 9}$,   (iv) $\dfrac{2x - 3}{x^2 - 5x + 6}$,

(v) $\dfrac{x}{(x - 1)^2}$,   (vi) $\dfrac{1}{x^2 + 4}$,   (vii) $\dfrac{1}{x^2 + 2x + 5}$,   (viii) $\dfrac{x + 1}{x^2 - 6x + 10}$,

(ix) $\dfrac{x - 4}{2x^2 + x - 3}$.

3 Use the substitution $t = \tan\tfrac{1}{2}x$ to find $\displaystyle\int \frac{1}{\sin x + \cos x}\, dx$.

4 Find the indefinite integral of $\dfrac{\sin x}{\cos x - 1}$.

5 Evaluate:

(i) $\displaystyle\int_1^2 \frac{1}{x(x + 1)}\, dx$,   (ii) $\displaystyle\int_0^1 \frac{1}{(x + 1)(x + 2)}\, dx$,   (iii) $\displaystyle\int_2^3 \frac{x + 2}{(x - 1)(x + 2)}\, dx$,

(iv) $\displaystyle\int_0^1 \frac{x^2}{(x + 1)^3}\, dx$.

6 Show that $\displaystyle\int_0^1 \frac{1}{(x + 1)(2x + 1)}\, dx = \ln \alpha$, where α is a rational number to be determined.   (*JMB*)

7 If $f(x) = \dfrac{3 - 2x}{(3 - x)^2}$, find

(i) $\int f(x)\, dx$,   (ii) the mean value of $f(x)$ for $0 \leqslant x \leqslant 2$.   (*AEB*)

8 Find the exact value of $\int_0^2 \dfrac{x+1}{x^2+4}\,dx$.   *(JMB)*

9 Evaluate $\int_0^4 \dfrac{(36-2x)}{(2x+1)(9+x^2)}\,dx$, leaving your answer in terms of natural logarithms   *(L)*

24.4 Integration by parts

The rule for the differentiation of a product of two functions f and g is

$$(f.g)' = f.g' + g.f'.$$

This may be used to obtain a formula which transforms an integral into another integral, which is hopefully more simple.

Writing the above formula in terms of functions of x and integrating,

$$f(x).g(x) = \int f(x).g'(x)\,dx + \int g(x).f'(x)\,dx.$$

This is now rewritten to give the formula for integration by parts, as it is called,

$$\int f(x)g'(x)\,dx = f(x)g(x) - \int g(x)f'(x)\,dx.$$

In our other notation, if $f(x) = u$ and $g(x) = v$, then

$$\int u\,\frac{dv}{dx}\,dx = uv - \int v\,\frac{du}{dx}\,dx.$$

In order to use this formula for a given integral, it is necessary to find functions u and v such that the integrand is of the form $u\,\dfrac{dv}{dx}$. To be useful, we need $v\,\dfrac{du}{dx}$ to be easier to integrate than $u\,\dfrac{dv}{dx}$. This often happens when u is a polynomial and v is a circular, exponential or logarithmic function, since then $\dfrac{du}{dx}$ is of degree one less than u.

EXAMPLE 1   *Find $\int x\cos x\,dx$.*

Choose $u = x$, $\dfrac{dv}{dx} = \cos x$, so that $v = \sin x$. Then

$$\int x\cos x\,dx = \int x\,\frac{d}{dx}\sin x\,dx = x\sin x - \int \sin x\,\frac{dx}{dx}\,dx$$

$$= x\sin x - \int \sin x\,dx = x\sin x + \cos x.$$

EXAMPLE 2   *Find $\int \tan^{-1} x\,dx$.*

Choose $u = \tan^{-1} x$ and $v = x$, so that $\dfrac{dv}{dx} = 1$. Then

$$\int \tan^{-1} x \, dx = \int \tan^{-1} x \frac{dx}{dx} \, dx = x \tan^{-1} x - \int x \frac{d}{dx} \tan^{-1} x \, dx$$

$$= x \tan^{-1} x - \int \frac{x}{1+x^2} \, dx = x \tan^{-1} x - \tfrac{1}{2} \ln(1+x^2).$$

Definite integrals

In the case of a definite integral, the formula for integration by parts is obtained as follows

$$f(b)g(b) - f(a)g(a) = \int_a^b f(x)g'(x) \, dx + \int_a^b g(x)f'(x) \, dx$$

so
$$\int_a^b f(x)g'(x) \, dx = \left[f(x)g(x) \right]_a^b - \int_a^b g(x)f'(x) \, dx$$

or
$$\int_a^b u \frac{dv}{dx} \, dx = \left[uv \right]_a^b - \int_a^b v \frac{du}{dx} \, dx.$$

EXAMPLE 3   *Evaluate* $\displaystyle\int_1^2 x^2 \ln x \, dx.$

Integrate by parts, using $u = \ln x$ and $\dfrac{dv}{dx} = x^2$, so $v = \tfrac{1}{3}x^3$. Then

$$\int_1^2 x^2 \ln x \, dx = \int_1^2 \ln x \frac{d}{dx}(\tfrac{1}{3}x^3) \, dx = \left[\tfrac{1}{3}x^3 \ln x \right]_1^2 - \tfrac{1}{3} \int_1^2 x^3 \frac{d}{dx} \ln x \, dx$$

$$= \left[\tfrac{1}{3}2^3 \ln 2 - \tfrac{1}{3}1^3 \ln 1 \right] - \tfrac{1}{3} \int_1^2 x^3 \frac{1}{x} \, dx$$

$$= \tfrac{8}{3} \ln 2 - \tfrac{1}{3} \int_1^2 x^2 \, dx = \tfrac{8}{3} \ln 2 - \left[\tfrac{1}{9}x^3 \right]_1^2$$

$$= \tfrac{8}{3} \ln 2 - \tfrac{1}{9}(8-1) = \tfrac{8}{3} \ln 2 - \tfrac{7}{9}.$$

EXAMPLE 4   *Evaluate* $\displaystyle\int_0^{\frac{1}{2}\pi} \cos^4 x \, dx.$

Integrate by parts, using $u = \cos^3 x$ and $\dfrac{dv}{dx} = \cos x$, $v = \sin x$.

$$I = \int_0^{\frac{1}{2}\pi} \cos^4 x \, dx = \int_0^{\frac{1}{2}\pi} \cos^3 x \frac{d}{dx} \sin x \, dx$$

$$= \left[\cos^3 x \sin x \right]_0^{\frac{1}{2}\pi} - \int_0^{\frac{1}{2}\pi} \sin x \frac{d}{dx} \cos^3 x \, dx$$

$$I = 0 - 0 - \int_0^{\frac{1}{2}\pi} \sin x \, 3\cos^2 x (-\sin x) \, dx = 3 \int_0^{\frac{1}{2}\pi} \sin^2 x \cos^2 x \, dx$$

$$I = 3 \int_0^{\frac{1}{2}\pi} (1 - \cos^2 x) \cos^2 x \, dx = 3 \int_0^{\frac{1}{2}\pi} \cos^2 x \, dx - 3I.$$

Therefore

$$4I = 3 \int_0^{\frac{1}{2}\pi} \cos^2 x \, dx = \frac{3}{2} \int_0^{\frac{1}{2}\pi} (1 + \cos 2x) \, dx = \frac{3}{2} \left[x + \frac{\sin 2x}{2} \right]_0^{\frac{1}{2}\pi} = \frac{3\pi}{4}.$$

Finally $I = \frac{1}{4}(3\pi/4) = 3\pi/16$.

EXAMPLE 5 *Find $\int e^x \cos x \, dx$.*

Choose $u = e^x$, $\dfrac{dv}{dx} = \cos x$, so $v = \sin x$ and

$$\int e^x \cos x \, dx = \int e^x \frac{d}{dx} \sin x \, dx = e^x \sin x - \int \sin x \frac{d}{dx} e^x \, dx$$

$$= e^x \sin x - \int e^x \sin x \, dx.$$

The new integral is similar to the first, with $\cos x$ replaced by $\sin x$. If we, therefore, repeat the procedure, with $u = e^x$ and $v = -\cos x$, we shall obtain an expression which includes the first integral, with a factor -1. We can then eliminate the unwanted integral by substitution.

$$\int e^x \sin x \, dx = \int e^x \frac{d}{dx} (-\cos x) \, dx = -e^x \cos x - \int (-\cos x) \frac{d}{dx} e^x \, dx$$

$$= -e^x \cos x + \int e^x \cos x \, dx.$$

So $\quad \int e^x \cos x \, dx = e^x \sin x - (-e^x \cos x + \int e^x \cos x \, dx)$

$\quad\quad 2 \int e^x \cos x \, dx = e^x \sin x + e^x \cos x$

and then $\int e^x \cos x \, dx = \frac{1}{2} e^x (\sin x + \cos x)$.

EXERCISE 24.4

1 Use the given functions u and v in order to find the integral:
  (i) $\int x e^x \, dx$, $u = x$, $v = e^x$; (ii) $\int \ln x \, dx$, $u = \ln x$, $v = x$;
  (iii) $\int x^2 \sin x \, dx$, $u = x^2$, $v = -\cos x$;
  (iv) $\int e^{-x} \cos(2x + 3) \, dx$, $u = e^{-x}$, $v = \frac{1}{2} \sin(2x + 3)$.

2 Find the indefinite integral of:
  (i) $x \sin x$, (ii) $x \sec^2 x$, (iii) $x^2 e^x$, (iv) $x \ln x$, (v) $x e^{-x}$,
  (vi) $e^{2x} \cos 3x$, (vii) $x \sin x \cos x$, (viii) $e^x \sin(nx + a)$, (ix) $x^{-1} \ln x$,
  (x) $\sin^{-1} x$.

3 Find:
  (i) $\int x \tan^{-1} x \, dx$, (ii) $\int x \cos nx \, dx$, (iii) $\int x^{-2} \ln x \, dx$,
  (iv) $\int x^2 \cos x \, dx$, (v) $\int x \sin^2 x \, dx$, (vi) $\int \ln(1 + x^2) \, dx$.

4 Evaluate:
  (i) $\displaystyle\int_0^2 x e^{-x} \, dx$, (ii) $\displaystyle\int_0^\pi x^3 \cos x \, dx$, (iii) $\displaystyle\int_0^1 e^x \cos \pi x \, dx$,

  (iv) $\displaystyle\int_0^\pi x \sin 2x \, dx$, (v) $\displaystyle\int_1^2 x^{-1} \ln x \, dx$.

5 Evaluate $\displaystyle\int_0^1 x\,e^{3x}\,dx.$ (L)

6 Evaluate $\displaystyle\int_1^e x\ln x\,dx.$ (L)

7 Evaluate, correct to two decimal places, $\displaystyle\int_0^1 (1-x)\sin x\,dx.$ (JMB)

8 Evaluate $\displaystyle\int_e^{e^2} \ln\sqrt{x}\,dx.$ (L)

9 Evaluate $\displaystyle\int_0^1 x\,e^{-2x}\,dx$ leaving your answer in terms of e.   (L)

10 Given that $I_n = \int \sec^n x\,dx$, show that
$$(n+1)I_{n+2} = \tan x\,\sec^n x + nI_n.$$

   Hence, or otherwise, find $\int \sec^3 x\,dx.$ (AEB)

11 Evaluate the integrals

   (a) $\displaystyle\int_0^{\frac{1}{2}} \frac{\arcsin x}{\sqrt{(1-x^2)}}\,dx,$ (b) $\displaystyle\int_1^2 x^2 e^{2x}\,dx.$

   [Answers may be left in terms of π or e.]   (L)

12 Prove that, if $y = \tan^{-1} x$, then $\dfrac{dy}{dx} = \dfrac{1}{1+x^2}.$

   Using integration by parts evaluate $\displaystyle\int_0^1 \tan^{-1} x\,dx$, giving your results to three significant figures.   (JMB)

MISCELLANEOUS EXERCISE 24

1 Evaluate $\displaystyle\int_0^{\pi/2} x\sin 2x\,dx.$ (JMB)

2 Express $\dfrac{4(x+5)}{(x+1)^2(x^2+3)}$ in partial fractions.

   Prove that $\displaystyle\int_0^3 \frac{4(x+5)\,dx}{(x+1)^2(x^2+3)} = 3(1+\ln 2) - (\pi\sqrt{3})/9.$ (L)

3 (a) (i) Evaluate $\displaystyle\int_0^4 \frac{1}{\sqrt{(3x+4)}}\,dx,$ (ii) find $\int x\ln x\,dx.$

   (b) Using the substitution $u = e^x$, or otherwise, find $\displaystyle\int \frac{1}{e^x - e^{-x}}\,dx.$ (C)

4 By means of the substitution $x = \sin^2\theta$, or otherwise, evaluate
$$\int_0^{\frac{1}{2}} \left(\frac{x}{1-x}\right)^{\frac{1}{2}} dx,$$

   giving your answer in terms of π.   (JMB)

5 Find the mean value of $\sin^2 x$ over the interval $0 \leqslant x \leqslant \pi.$ (L)

6 Find the value of $\int_3^4 x(x-3)^{17}\,dx$.   (L)

7 Evaluate $\int_{-\frac{1}{2}}^{\frac{5}{2}} \dfrac{1}{\sqrt{(8+2x-x^2)}}\,dx$.   (L)

8 Evaluate the integrals

(a) $\displaystyle\int_0^{\pi/2} \sin 2\theta \sin \theta\, d\theta$,   (b) $\displaystyle\int_{-1}^{1} xe^x\,dx$,   (c) $\displaystyle\int_0^1 x\sqrt{(1-x)}\,dx$,   by

putting $x = 1 - t^2$ or otherwise.   (L)

9 (i) Evaluate (a) $\displaystyle\int_0^1 \dfrac{1+x}{1+2x}\,dx$,   (b) $\displaystyle\int_0^{\pi/2} \sin x \cos^2 x\,dx$.

(ii) Sketch the arc of the curve $y = 2x - x^2$ for which y is positive. Find the area of the region which lies between this arc and the x-axis. If this region is rotated completely about the x-axis find the volume of the solid of revolution generated.   (L)

10 (a) Evaluate $\displaystyle\int_0^{\sqrt{5}} x\sqrt{(x^2+4)}\,dx$.

(b) Evaluate, in terms of e, $\displaystyle\int_2^3 x^2 e^{-x}\,dx$.   (L)

11 Sketch the curve $y = xe^x$, calculating the coordinates of any turning points and showing the behaviour of the curve when $|x|$ is large.

Find the area of the finite region enclosed by the curve $y = xe^x$ and the lines $y = 0$ and $x = -2$.   (L)

12 Find

(i) $\displaystyle\int \dfrac{e^{2x}}{(1+e^{2x})^2}\,dx$,   (ii) $\displaystyle\int_0^1 \dfrac{1}{(4-3x^2)^{\frac{1}{2}}}\,dx$.   (JMB)

13 (a) Find $\int x\sqrt{(1+x)}\,dx$.

(b) Evaluate (i) $\displaystyle\int_0^{\frac{1}{12}\pi} 2\sin^2 3x\,dx$,   (ii) $\displaystyle\int_e^{e^2} \dfrac{1}{x\ln x}\,dx$.

14 (a) Evaluate (i) $\displaystyle\int_1^2 \dfrac{3x+1}{2x+1}\,dx$,   (ii) $\displaystyle\int_0^{\frac{1}{4}\pi} 2\cos x \cos 2x\,dx$.

(b) By means of the substitution $u = x^2$, or otherwise, find $\int x^3 e^{x^2}\,dx$, giving your answer in terms of x.   (C)

15 (i) Evaluate (a) $\displaystyle\int_0^1 \arctan x\,dx$,   (b) $\displaystyle\int_3^{11} \dfrac{x}{(x+1)(x-2)}\,dx$.

(ii) Using the substitution $t = \tan x$, or otherwise, evaluate

$$\int_0^{\pi/4} \dfrac{dx}{4+5\cos^2 x}.$$ (L)

16 (i) Use the substitution $t = \tan(x/2)$ to show that

$$\int_0^{\pi/2} \dfrac{1}{1+\sin x + \cos x}\,dx = \ln 2.$$

(ii) Obtain $\int x \sec^2 x\,dx$.   (L)

17 (a) (i) Prove that $\dfrac{d}{dx}(\tan^{-1}x) = \dfrac{1}{1+x^2}$, (ii) evaluate $\displaystyle\int_0^1 \tan^{-1}x\,dx$.

(b) Evaluate $\displaystyle\int_0^{\frac{1}{4}\pi} \tan^4\theta\sec^2\theta\,d\theta$, and hence show that

$$\int_0^{\frac{1}{4}\pi} \tan^6\theta\,d\theta = \tfrac{1}{5} - \int_0^{\frac{1}{4}\pi}\tan^4\theta\,d\theta.$$

Deduce that $\displaystyle\int_0^{\frac{1}{4}\pi}\tan^6\theta\,d\theta = \tfrac{13}{15} - \tfrac{1}{4}\pi$.   (C)

18 Evaluate $\displaystyle\int_3^4 \dfrac{x^3+4x+4}{(x-2)^2(x^2+1)}\,dx$.   (L)

19 The domain of the function defined by $f(x) = e^{-x}\sin x$ is the set of real numbers in the interval $0 \leqslant x \leqslant 4\pi$. Show that stationary values of this function occur for values of x for which $\tan x = 1$. Determine the nature of these stationary values. Sketch the graph of the function.

By integrating by parts twice, or otherwise, evaluate $\displaystyle\int_0^\pi f(x)\,dx$.

Show that the ratio of $f(a+2\pi)$ to $f(a)$ is independent of the value of a, and hence write down the value of $\displaystyle\int_{2\pi}^{3\pi} f(x)\,dx$.   (JMB)

20 (a) (i) Prove that $\dfrac{d}{dx}(\sin^{-1}x) = \dfrac{1}{\sqrt{(1-x^2)}}$, where $|x| < 1$. (ii) Given that

$y = \sin^{-1}[2x\sqrt{(1-x^2)}]$, show that $\dfrac{dy}{dx} = \dfrac{2}{\sqrt{(1-x^2)}}$.

(b) By using the substitution $x = \sin^2\theta - 2\cos^2\theta$, or otherwise, show that

$$\int_{-\frac{1}{2}}^1 \left(\dfrac{1-x}{2+x}\right)^{\frac{1}{2}}\,dx = \tfrac{3}{4}(\pi-2).$$ (C)

21 Evaluate the integrals

(a) $\displaystyle\int_{-1}^1 \dfrac{e^x}{(e^x+1)}\,dx$,   (b) $\displaystyle\int_0^2 \dfrac{x}{(x+2)^2}\,dx$,   (c) $\displaystyle\int_0^\pi x\cos x\,dx$.   (L)

22 (a) Show that $\displaystyle\int_0^4 \dfrac{x-1}{2x^2+3x+1}\,dx = \ln\left(\dfrac{25}{27}\right)$.

(b) By using the substitution $u^2 = 2x-1$, or otherwise, evaluate

$$\int_1^5 x\sqrt{(2x-1)}\,dx.$$ (C)

23 Find (i) $\int \sin 5x\cos 4x\,dx$,   (ii) $\displaystyle\int \dfrac{x}{(1+x)^2}\,dx$.   (SMP)

24 (a) Evaluate (i) $\displaystyle\int_0^{\pi/4} \dfrac{\sin x\,dx}{\cos^3 x}$,   (ii) $\displaystyle\int_0^1 \dfrac{dx}{\sqrt{(4-x^2)}}$.

(b) Differentiate $(\ln x)^2$ and integrate $x\ln x$. Hence evaluate $\displaystyle\int_1^e x(\ln x)^2\,dx$.

     (JMB)

25 (a) Evaluate $\displaystyle\int_0^{\pi/3} \frac{d\theta}{1+\sin\theta}$.

(b) Given that $y = \tan^{-1} 3x$, prove that $\dfrac{dy}{dx} = \dfrac{3}{1+9x^2}$.

Hence, or otherwise, evaluate $\displaystyle\int_0^1 x^2 \tan^{-1} 3x \, dx.$ (*AEB*)

26 Evaluate the integrals

(a) $\displaystyle\int_0^{\pi/4} x \cos 2x \, dx$, (b) $\displaystyle\int_2^3 \frac{x+7}{(x+3)(x-1)} \, dx$, (c) $\displaystyle\int_1^2 \frac{x+1}{\sqrt{(x+2)}} \, dx$. (*L*)

27 (i) Evaluate $\displaystyle\int_0^{\pi/2} \frac{1}{1+\cos\theta} \, d\theta.$

(ii) Show that $\displaystyle\int_1^3 \frac{3(x+1)}{x^2(x^2+3)} \, dx = \frac{1}{2} \ln 3 + \frac{2}{3} - \frac{\pi}{6\sqrt 3}.$ (*L*)

28 (i) Evaluate $\displaystyle\int_4^5 \frac{2x \, dx}{x^2 - 4x + 3}.$

(ii) By using the substitution $x = \sec^2 y$, or otherwise, evaluate
$\displaystyle\int_2^5 \frac{dx}{x^2 \sqrt{x-1}}.$

(iii) Find the area of the finite region between the curve $y^2 = 4x$ and the straight line $2y = x + 3$. (*L*)

25 Differential Equations

25.1 Definitions

Suppose that y is the value of some function of x. Then an equation relating x, y and the derivatives of y with respect to x is called a *differential equation*. If the only derivative occurring in the equation is $\dfrac{dy}{dx}$, then the differential equation is said to be of the *first order*. Differential equations of higher order are those containing higher order derivatives than the first, for example, equations such as $\dfrac{d^2y}{dx^2} + 2\dfrac{dy}{dx} + n^2y = \cos ax$. We shall not be concerned with such equations in this chapter.

The most straightforward first order differential equation is $\dfrac{dy}{dx} = f(x)$. This can be integrated to give an equation $y = F(x) + c$, where $F'(x) = f(x)$ and where c is an arbitrary constant. This is then called a *solution* of the differential equation. Generally, the solution of a differential equation is an equation relating x and y, and not involving any derivative.

For any first order differential equation, a *general solution* is an equation relating x and y, and not involving $\dfrac{dy}{dx}$, and containing an arbitrary constant of integration. If the arbitrary constant is given a particular value, then the solution is called a *particular solution*.

EXAMPLE *Find the general solution of the equation $\dfrac{dy}{dx} = 2x$, and show that it corresponds to a family of parabolas. Find the equation of the particular parabola in the family which passes through the point* (1, 2).

We can integrate $\dfrac{dy}{dx} = 2x$ to give $y = x^2 + c$, which is the equation of a parabola, with axis $x = 0$ and vertex at $(0, c)$. As c varies, we get a whole family of parabolas and the differential equation is satisfied by each member of the family. If the equation of a parabola is to be satisfied by $x = 1$, $y = 2$, then $c = 1$, and the particular solution is $y = x^2 + 1$, which is a parabola passing through (1, 2). Some of the members of the family of parabolas are shown in Fig. 25.1.

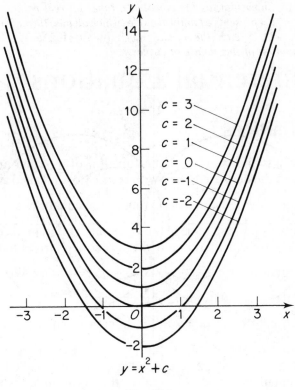

Fig. 25.1

25.2 The differential equation of a family of curves

The converse problem concerns the formation of a differential equation. Suppose that we are given a family of curves, each with an equation which is obtained by giving a particular value to some constant c in an equation in x and y containing the constant c. Then there is a differential equation satisfied by all the members of the family of curves. This differential equation is found by differentiating the given equation and then eliminating c between that equation and the differentiated equation, which contains $\dfrac{dy}{dx}$ in some of its terms. In the previous case, when the original equation of the family of parabolas is $y = x^2 + c$, then the differentiated equation $\dfrac{dy}{dx} = 2x$ already has eliminated c and is the differential equation of the family. In other cases, the constant c occurs in both equations and has to be eliminated.

EXAMPLE 1   *Show that, as the constant c takes all real values, the equation*
$x^2 - 2cx + y^2 = 1$ *represents a family of circles which all pass through the two points*
$(0, 1)$ *and* $(0, -1)$. *Sketch the circles given by* $c \in \{1, 2, -1, -2\}$. *Find the*
differential equation of this family of circles.

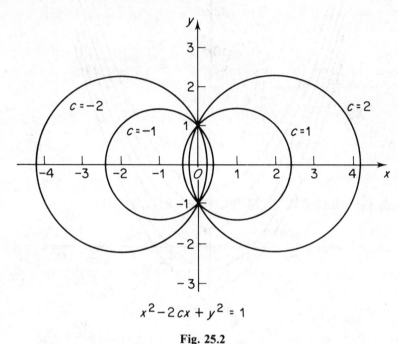

$$x^2 - 2cx + y^2 = 1$$

Fig. 25.2

For a given value of the constant c, the equation $x^2 - 2cx + y^2 = 1$ can be
rearranged to read $(x - c)^2 + y^2 = 1 + c^2$, which is the equation of a circle with
centre $(c, 0)$ and radius $\sqrt{(1 + c^2)}$, and this circle passes through the points $(0, 1)$ and
$(0, -1)$. The four circles for $c \in \{1, 2, -1, -2\}$ are shown in Fig. 25.2. In order to
find the differential equation of the family of circles we differentiate the equation

$$x^2 - 2cx + y^2 = 1,$$

giving

$$2x - 2c + 2y\frac{dy}{dx} = 0,$$

and eliminate c between the two equations. Multiply the second by x and subtract
from the first, to eliminate c,

$$x^2 + 2xy\frac{dy}{dx} - y^2 = -1$$

or

$$2xy\frac{dy}{dx} = y^2 - x^2 - 1,$$

which is the differential equation of the family of circles.

EXERCISE 25.2

1 Sketch the members of the family of curves with the given equation for which the constant c takes the values in the given set. Find the differential equation of the family:
(i) $y = cx, c \in \{0, \frac{1}{2}, 1, 2, -1, -2\}$; (ii) $y = 2x + c, c \in \{0, -1, -2, 1, 2\}$;
(iii) $x^2 + y^2 + 2cy = 4, c \in \{0, 1, 2, -1, -2\}$;
(iv) $y^2 - 2cx = 0, c \in \{-2, -1, 1, 2\}$;
(v) $y^2 + cx^2 = 8, c \in \{1, 4, 9, 16\}$; (vi) $y = c/x, c \in \{-2, -1, 0, 1, 2\}$.
2 Find the family of curves satisfying the differential equation, and the particular curve passing through $(1, 2)$:
(i) $\dfrac{dy}{dx} = 3x^2$, (ii) $\dfrac{dy}{dx} = 4 + x$, (iii) $\dfrac{dy}{dx} = \sin x$ (use radians),
(iv) $\dfrac{dy}{dx} = \dfrac{1}{1 + x^2}$.

25.3 Separable differential equations

An equation such as $\dfrac{dy}{dx} = F'(x)$, which can be integrated at once to give a solution $y = F(x) + c$, is called *exact*. In the same way, the equation

$$G'(y) \frac{dy}{dx} = F'(x)$$

is also called exact, since it can be integrated immediately, giving

$$G(y) = \int G'(y) \frac{dy}{dx} dx = \int F'(x) dx = F(x) + c.$$

In such an exact equation, on one side there is a function of x only and on the other side there is a function of y, multiplied by dy/dx. The variables x and y have been separated and the equation can be integrated. A differential equation is said to be *separable* if it can be rearranged in such a way as to separate the variables and form an equation of the form

$$g(y) \frac{dy}{dx} = f(x).$$

Then this equation can be solved by integration, and the solution is

$$\int g(y) dy = \int g(y) \frac{dy}{dx} dx = \int f(x) dx.$$

EXAMPLE 1 *Solve the equation* $\dfrac{dy}{dx} = 3y^2$.

This equation is separable, on dividing by y^2,

$$y^{-2}\frac{dy}{dx} = 3.$$

Integrating this exact equation,

$$-y^{-1} = \int y^{-2}\frac{dy}{dx}\,dx = \int 3\,dx = 3x + c,$$

so the solution is $y(3x+c)+1=0$.

Note A   Check by differentiation:

$$\frac{dy}{dx}(3x+c)+3y = 0, \quad \text{so} \quad 3y^2 = -y\frac{dy}{dx}(3x+c) = \frac{dy}{dx}.$$

B   The constant c, of integration, must be inserted at the step when the equation is integrated and not after any rearrangement. It would be quite wrong to integrate and obtain $1/y = 3x$, rearrange to give $3xy+1 = 0$ and then add c, giving $3xy+1 = c$. The differential equation satisfied by this family of curves is $3y+3x\dfrac{dy}{dx} = 0$, which is quite different from the original differential equation.

EXAMPLE 2   *Find the solution of the differential equation* $(1+x^2)\dfrac{dy}{dx} = 2xy$, *given that* $y = 2$ *when* $x = 0$.

The equation is separable, so we separate the variables and integrate.

$$\frac{1}{y}\frac{dy}{dx} = \frac{2x}{1+x^2},$$

$$\ln|y| = \int\frac{1}{y}\,dy = \int\frac{1}{y}\frac{dy}{dx}\,dx = \int\frac{2x}{1+x^2}\,dx = \int\frac{1}{u}\,du = \ln|u| + c$$

$$= \ln(1+x^2) + c,$$

having used the substitution $u = 1+x^2$. For convenience, replace c by $\ln a$, where a is an arbitrary constant. This is allowable since, if c is an arbitrary constant, so is a. Then

$$\ln|y| = \ln(1+x^2) + \ln|a| = \ln|a(1+x^2)|.$$

On taking exponentials of both sides,

$$y = a(1+x^2).$$

Finally, we require the particular solution for which $y = 2$ when $x = 0$, so we substitute these values in the general solution to find a: $2 = a(1+0)$ so $a = 2$, and the required solution is $y = 2(1+x^2)$.

Note   To check the solution

$$\frac{dy}{dx} = 4x \text{ so } (1+x^2)\frac{dy}{dx} = 4x(1+x^2) = 2xy,$$

and when $x = 0$, $y = 2$.

EXAMPLE 3   *Use the substitution $y = zx$ to find the general solution of the differential equation $x\dfrac{dy}{dx} = x + y$.*

Putting $z = \dfrac{y}{x}$, $y = zx$, so $\dfrac{dy}{dx} = x\dfrac{dz}{dx} + z$. Substituting in the differential equation

$$x\left(x\frac{dz}{dx} + z\right) = x + y = x + zx, \quad \text{so} \quad x^2\frac{dz}{dx} + xz = x + xz.$$

Therefore, $x^2\dfrac{dz}{dx} = x$ so, either $x = 0$ or $\dfrac{dz}{dx} = \dfrac{1}{x}$, on separating the variables. If $x = 0$, $y = 0$, which is a special solution. Otherwise, we integrate and obtain $z = \ln|x| + c$ and $y = x(\ln|x| + c)$. This may be checked by differentiation and substitution in the differential equation.

EXERCISE 25.3

1   Find the general solution of the differential equation:

(i) $x^2\dfrac{dy}{dx} = y^2$,   (ii) $\dfrac{dy}{dx} = \dfrac{1}{2y}$,   (iii) $\dfrac{dy}{dx} + 2 = 3x^2$,   (iv) $\dfrac{dy}{dx} = e^{-y}\cos 3x$,

(v) $(1+x^2)\dfrac{dy}{dx} = (1+x)y$.

2   Solve the equation:

(i) $y\dfrac{dy}{dx} = 1 + x^2$, $y = 1$ when $x = 0$;   (ii) $\dfrac{dy}{dx} = 1 + y^2$, $y = 0$ when $x = 0$;

(iii) $x^2\dfrac{dy}{dx} = \cos^2 y$, $y = \dfrac{\pi}{4}$ when $x = 1$;

(iv) $(1-x)\dfrac{dy}{dx} = 2x(1+y)$, $y = 0$ when $x = 0$.

3   Solve the differential equation $\dfrac{dy}{dx} = -3x^2y$, where $y = a$ when $x = 0$, giving y in terms of x and a.

4   Solve, for $x > 0$, the differential equation $\dfrac{dy}{dx} = \dfrac{y}{x}$, given that $y = 2$ when $x = 1$.   (L)

5   Solve the differential equation $\dfrac{dy}{dx} = -3y$, given that $y = a$ when $x = 0$. (L)

6   Solve the differential equation $(1+e^y)\dfrac{dy}{dx} = e^{2y}\cos^2 x$, given that $y = 0$ when $x = 0$.   (AEB)

7 Solve the differential equation $\dfrac{dy}{dx} = 4y^2$, given that $y = \frac{1}{2}$ when $x = -2$.

(*L*)

8 Find y in terms of x, given that $y = \cot x \dfrac{dy}{dx}$ and that $y = 1$ when $x = 0$.

(*L*)

9 Find y as a function of x when $\dfrac{dy}{dx} = e^{2y} \sin 2x$, given that $y = 0$ when $x = 0$.

(*L*)

25.4 Problems involving differential equations

Differential equations arise in many applications of mathematics. Given a problem in physical terms, it may be translated into mathematical terms, which is called making a mathematical model of the problem. Often this model is in the form of a differential equation, with certain conditions to be satisfied. If the equation can be solved, then this provides a solution of the model. Then the solution of the model must be translated back to give a solution of the original problem. There are, therefore, two major steps in the work. The first is to form the differential equation which is the model of the problem. The second is to solve the equation and interpret the answer.

EXAMPLE 1   *The rate of increase of a population of flies is proportional to the size of the population at any time. Given that the population doubles its size in three days, find the proportional increase in one day.*

Let the size of the population be p_0 initially and let it be p after t days. Then the rate of increase of p is $\dfrac{dp}{dt}$ and this is proportional to p. Let the factor of proportionality be k, then we have to solve

$$\frac{dp}{dt} = kp, \text{ with } p = p_0 \text{ when } t = 0.$$

The differential equation is separable and gives $\dfrac{1}{p}\dfrac{dp}{dt} = k$, on separating the variables. This equation can be integrated,

$$\ln p = \int \frac{1}{p}\frac{dp}{dt}\,dt = \int k\,dt = kt + c.$$

When $t = 0$, $p = p_0$ so $\ln p_0 = c$ and

$$\ln p - \ln p_0 = \ln\left(\frac{p}{p_0}\right) = kt.$$

Taking exponentials of this equation, $\dfrac{p}{p_0} = e^{kt}$, or $p = p_0 e^{kt}$. Now $p = 2p_0$ when

$t = 3$, so $2 = e^{3k}$, or $k = \frac{1}{3}\ln 2$. After one day, $t = 1$ and $p = p_0 e^k = p_0(\sqrt[3]{2})$, so the required factor for the proportional increase in population in one day is $\sqrt[3]{2}$.

EXAMPLE 2 *Find the equation of the curve, passing through the point* $(1, 1)$, *given that the tangent to the curve at the point* (x_0, y_0) *meets the x-axis at the point* $(\frac{1}{2}x_0, 0)$.

Let m be the gradient of the tangent at (x_0, y_0), so that $m = \dfrac{dy}{dx}$ at $x = x_0$. Then the equation of the tangent at (x_0, y_0) is

$$y - y_0 = m(x - x_0),$$

and this line meets the x-axis at the point $(\frac{1}{2}x_0, 0)$. Therefore,

$$m(\tfrac{1}{2}x_0 - x_0) = -y_0 \quad \text{or} \quad 2y_0 = x_0 m, \quad \text{where} \quad m = \frac{dy}{dx} \text{ at } x = x_0.$$

Replacing x_0 and y_0 by x and y, for a general point on the curve, the coordinates of any point of the curve satisfy the differential equation

$$2y = x\frac{dy}{dx}, \quad \text{or} \quad \frac{2}{x} = \frac{1}{y}\frac{dy}{dx},$$

on separating the variables. Then, on integrating this equation,

$$2\ln|x| + c = \int \frac{2}{x}\,dx = \int \frac{1}{y}\frac{dy}{dx}\,dx = \int \frac{1}{y}\,dy = \ln|y|.$$

Put $c = \ln|a|$, then $\ln|ax^2| = \ln|y|$ and so $y = ax^2$ is the general solution of the differential equation. The particular solution required is that for which $y = 1$ when $x = 1$ and so $a = 1$ and the curve has equation $y = x^2$.

EXAMPLE 3 *A particle P moves on the axis Ox and its displacement from O is* x *at time t. The velocity of the particle is* v. *Prove that the acceleration of the particle is* $v\dfrac{dv}{dx}$. *The acceleration is directed towards O and is of magnitude* $n^2 x$, *where n is a constant. Express v as a function of x, given that* $v = 0$ *when* $x = a$. *Express x as a function of t, given also that* $x = a$ *when* $t = 0$.

Since $v = \dfrac{dx}{dt}$, the acceleration $\dfrac{dv}{dt} = \dfrac{dv}{dx}\dfrac{dx}{dt} = v\dfrac{dv}{dx}$. Given that $v\dfrac{dv}{dx} = -n^2 x$ (negative since the acceleration is towards O),

$$\tfrac{1}{2}v^2 = \int v\,dv = \int v\frac{dv}{dx}\,dx = \int -n^2 x\,dx = -\tfrac{1}{2}n^2 x^2 + c.$$

Since $v = 0$ when $x = a$, $0 = -\tfrac{1}{2}n^2 a^2 + c$, so $c = \tfrac{1}{2}n^2 a^2$ and then

$$v^2 = n^2(a^2 - x^2), \quad \text{or} \quad v = n\sqrt{(a^2 - x^2)}.$$

Now $v = \dfrac{dx}{dt}$, so we have a differential equation $\dfrac{dx}{dt} = n\sqrt{(a^2 - x^2)}$. This is separable, so we separate the variables and integrate,

$$nt = \int n \, dt = \int \frac{1}{\sqrt{(a^2 - x^2)}} \frac{dx}{dt} \, dt = \sin^{-1}\frac{x}{a} + c.$$

This can be rewritten $x = a \sin(nt - c)$, and when $t = 0$, $x = a$, so $\sin(-c) = 1$, and $c = -\frac{1}{2}\pi$. Thus

$$x = a \sin(nt + \tfrac{1}{2}\pi) = a \cos nt.$$

EXERCISE 25.4

1   Describe the family of curves which satisfies the differential equation $\dfrac{dy}{dx} = \dfrac{y-1}{x+1}$.

2   A family of curves satisfies the differential equation $y^2 - y = \dfrac{dy}{dx}(x^2 - x)$. Find the equation of the member of the family which passes through the point: (i) $(3, 2)$,   (ii) $(2, 2)$.

3   A beaker of hot liquid is cooling in a room where the room temperature is $15°C$. The rate of cooling is proportional to the difference between the temperature of the liquid and room temperature. If the liquid takes ten minutes to cool from a temperature of $75°C$ to $45°C$, find the total time taken for it to cool to $30°C$.

4   In a chemical process, the chemical A is transformed into the chemical B at a rate which is proportional to the amount of A present. In one minute, the amount of A present is reduced from $50\,kg$ to $20\,kg$. Find how much of the chemical A is present after 5 minutes.

5   A falling particle has acceleration $(g - kv)$ when its speed is v, where g and k are constants. It falls from rest. Find the time for its speed to reach $g/2k$ and show that its speed never reaches g/k.

MISCELLANEOUS EXERCISE 25

1   Find the solution of the differential equation $x \log_e x \dfrac{dy}{dx} = \tan y$ for which $y = \pi/6$ when $x = e$.   *(JMB)*

2   Find the solution of the differential equation $\dfrac{dy}{dx} = 2 - 2x$ for which $y = 0$ when $x = 1$.

   Find also the solution of the differential equation $\dfrac{dy}{dx} = \dfrac{1}{2-x}$ for which $y = 0$ when $x = 1$.

   Sketch the graphs of these solutions on the same diagram, indicating clearly where each graph meets the y-axis.   *(JMB)*

3   By making the substitution $y = zx$, where z is a function of x, in the differential equation

$$x^2 \frac{dy}{dx} = y(x + y),$$

   show that $$\frac{dz}{dx} = \frac{z^2}{x}.$$

Hence find y in terms of x for $x > 0$, given that $y = -1$ when $x = 1$. (*L*)

4   Solve the differential equation $\dfrac{dy}{dx} = \dfrac{y^2 - 1}{2 \tan x}$, given that $y = 3$ when $x = \pi/2$.

Hence express y in terms of x.   (*AEB*)

5   A disease spreads through an urban population. At time t, p is the proportion of the population who have the disease. The rate of change of p is proportional to the product of p and the proportion $(1 - p)$ who do not have the disease. Form a differential equation in terms of p and t and solve it.

Given that when $t = 0$, $p = \frac{1}{10}$, and when $t = 2$, $p = \frac{1}{5}$, show that

$$\frac{9p}{1-p} = \left(\frac{3}{2}\right)^t.$$

Find p when $t = 4$. Find also the set of values of t for which $p > \frac{1}{2}$. (*L*)

6   Find the solution $y = f(x)$ of the differential equation

$(1 + x^2)\dfrac{dy}{dx} = 4x(1 + y)$ for which $y = 1$ when $x = 1$.

Show that, for $x > 0$, the gradient of the curve $y = f(x)$ is always positive.
Calculate the area of the region bounded by the curve $y = f(x)$, the x-axis and the ordinates $x = 1$ and $x = 2$. (*L*)

7   Solve the differential equation $x\dfrac{dy}{dx} = y + x^2 y$, given that $y = 2\sqrt{e}$ when

$x = 1$. (*L*)

8   Find the general solution of the differential equation $\dfrac{dy}{dx} = \dfrac{xy}{10}$.

Show that the solution for which $y = 1$ when $x = 1$ may be expressed in the form $y = e^{(x^2 - 1)/20}$.

Tabulate the values of y for this solution corresponding to $x = 0, 1, 2, 3, 4,$ 5, and draw its curve as accurately as you can on graph paper for $0 \leqslant x \leqslant 5$.

Draw accurate tangents to your graph at the points for which $x = 1$ and $x = 4$, and confirm that their gradients satisfy approximately the differential equation. (*L*)

9   (i) Find y in terms of x, given that $x\dfrac{dy}{dx} = \cos^2 y$, $x > 0$ and that $y = \pi/3$

when $x = 1$.
(ii) Given that $xy + x^3 y + Cy = 1$, construct a first order differential equation between x and y which is satisfied for all values of C. Deduce that the gradient at all points of the integral curves of this differential equation is never positive. (*L*)

10 (i) Find y in terms of x given that $(1 + x)\dfrac{dy}{dx} = (1 - x)y$ and that $y = 1$ when

$x = 0$.
(ii) A curve passes through $(2, 2)$ and has gradient at (x, y) given by the differential equation

$$ye^{y^2}\frac{dy}{dx} = e^{2x}.$$

Find the equation of the curve. Show that the curve also passes through the point $(1, \sqrt{2})$ and sketch the curve. (*L*)

11 A stone, initially at rest, is released and falls vertically. Its velocity, $v\,\mathrm{m\,s}^{-1}$, at time t s after its release is given by

$$5\frac{dv}{dt}+v=50.$$

Find an expression for v in terms of t and evaluate the time at which $v=47\!\cdot\!5\,\mathrm{m\,s}^{-1}$.

Prove that the distance, s m, fallen vertically at time t is given by

$$s=50\,(t-5)+250e^{-t/5}.$$

Sketch the graphs of v against t, and s against t.   (C)

12 The general solution of a differential equation is $y=x+\dfrac{A}{x}$, where A is an arbitrary constant, and $x>0$. Find the differential equation, and sketch the family of solution curves.   (C)

13 Find y in terms of x if $\dfrac{dy}{dx}=(1+y)^2\sin^2 x\cos x$ and $y=2$ when $x=0$.
                    (AEB)

14 A compound C is formed by the composition of two substances A and B in such a way that one molecule of C arises from one molecule of A combining with **two** molecules of B. Let N be the number of molecules of C which have been formed at time t and let the rate of increase of N be proportional to the product of the number of molecules of A and the number of molecules of B present at time t. Suppose that at time $t=0$ there are a molecules of A, $3a$ molecules of B and no molecules of C. Show that (regarding N as a continuous variable)

$$\frac{dN}{dt}=k\,(3a^2-5aN+2N^2)$$

where k is a positive constant. Show that

$$\log_e\frac{3a-2N}{3a-3N}=kat,$$

and deduce that

$$N=\frac{3(1-e^{kat})}{2-3e^{kat}}\,a.$$

15 (a) Find (i) $\displaystyle\int\frac{x}{4+x}\,dx$, (ii) $\displaystyle\int xe^{-x}\,dx$.

(b) Solve the differential equation $\cos\theta\dfrac{dr}{d\theta}+r\sin\theta=0$, given that $r=a$ when $\theta=0$, expressing r in terms of a and θ.

16 (a) Solve the differential equation $\dfrac{dy}{dx}=\cos x\,(\sec y-\tan y)$, given that $y=0$ when $x=\frac{1}{2}\pi$.

(b) Find (i) $\displaystyle\int\frac{5x^2+11x+2}{(1+x^2)(1+3x)}\,dx$, (ii) $\displaystyle\int\frac{2x^2-1}{\sqrt{(1-x^2)}}\,dx$.

17 A family of concentric circles is given by the differential equation

$$(y+2)\frac{dy}{dx} + x + 4 = 0.$$

Determine the coordinates of the centre of these circles.
   A second family of curves is given by the differential equation.

$$\frac{dy}{dx} = -\frac{y+2}{x+4}$$

Sketch the member of this family that passes through the origin.

18 Given that $(1 + \sin^2 x)\frac{dy}{dx} = e^{-2y}\sin 2x$ and $y = 1$ when $x = 0$, find the value
of y when $x = \pi/2$.

19 A compound C is formed during a reaction by the chemical combination of
two substances A and B. The rate of formation of C (in $g\,s^{-1}$) at any instant is
equal to k times the product of the masses of A and B present at that instant.
   In order to produce $x\,g$ of C, $\lambda x\,g$ of A and $\mu x\,g$ of B are used. Initially C is
absent and $5\,g$ of A and $5\,g$ of B are present. If $x\,g$ of C are present t seconds
after the start, write down the expressions for the masses of A and B present
at time t. Hence obtain the differential equation satisfied by x.
   If $\lambda < \mu < 5$ solve this equation for t in terms of x, and show that the time
in seconds taken for $1\,g$ of C to be produced is

$$\frac{1}{5k(\mu - \lambda)}\log_e\left(\frac{5-\lambda}{5-\mu}\right). \qquad (JMB)$$

20 (a) Solve the differential equation $\cos y\,\frac{dy}{dx} = (\cot x)(1 + \sin y)$, given that
$y = \pi/2$ when $x = \pi/2$.
   (b) For $0 \leqslant x \leqslant \pi$ the curves $y = \sin x$ and $y = \sin \frac{1}{2}x$ intersect at the origin
O and at the point A. Find the coordinates of A. The region enclosed by the
arcs of the two curves between O and A is rotated completely about the
x-axis. Find the volume of the solid formed. $\qquad (AEB)$

21 The amount x of a certain radioactive substance present at time t in a reaction
is given by the differential equation

$$\frac{dx}{dt} = a - bx$$

where a and b are positive constants.
   Solve the differential equation for x in terms of t given that when $t = 0$ the
amount of the substance present is $a/(4b)$, and find the time taken for x to
attain the value $a/(2b)$. Find the limiting value of x as $t \to \infty$. $\qquad (JMB)$

22 Express y in terms of x, given that $\cot x\,\frac{dy}{dx} = 1 - y^2$ and that $y = 0$ when
$x = \pi/4$.
   Find the value of y when $x = \pi/6$. $\qquad (L)$

23 Solve the differential equation $(x+2)\frac{dy}{dx} + (x+1)y = 0$, (where $x > -2$),
given that $y = 2$ when $x = 0$.

Show that the maximum value of y is e, and sketch the graph of y against x for $x > -2$.

Draw on your graph the tangent to the curve at its point of inflexion. Find the equation of this tangent.   (L)

24   Find y in terms of x when $\dfrac{dy}{dx} = \dfrac{(x-1)\sqrt{(y^2+1)}}{xy}$, given that

$y = -\sqrt{(e^2 - 2e)}$ when $x = e$.

25   Solve the differential equation $2\dfrac{dy}{dx} + xy^2 = y^2$, given that $y = -1$ when $x = 1$. Show that the solution may be expressed in the form

$$y = \frac{4}{(x-3)(x+1)}.$$

Sketch the graph of y against x for $-1 < x < 3$, showing the asymptotes and giving the coordinates of any turning point.

26   Find y in terms of x given that $\dfrac{dy}{dx} = 2y + 1$, and that $y = 0$ when $x = 0$.

Sketch the curve represented by this solution and find the area of the finite region enclosed by the curve, the line $x = 1$ and the x-axis.

27   In a chemical reaction there are present, at time t, x kg of substance X and y kg of substance Y, and initially there is 1 kg of X and 2 kg of Y. The variables x and y satisfy the equations

$$\frac{dx}{dt} = -x^2 y, \frac{dy}{dt} = -xy^2.$$

Find $\dfrac{dy}{dx}$ in terms of x and y, and express y in terms of x. Hence obtain a differential equation in x and t only, and so find an expression for x in terms of t.

26 Infinite Series

26.1 Definitions

Let (u_r) be a sequence, given by $(u_r) = (u_1, u_2, u_3, \ldots)$. We can place a 'plus' sign between the terms and form an infinite series by the definition

$$\sum u_r = \sum_{r=1} u_r = u_1 + u_2 + u_3 + \ldots.$$

This has no meaning as a number because we are not able to add together an infinite number of real numbers. However, we define it as an algebraic equation and call the sum a *formal* sum. For example,

$$\sum r^2 = 1 + 2^2 + 3^2 + \ldots,$$

$$\sum \frac{1}{2^r} = \frac{1}{2} + \frac{1}{4} + \frac{1}{8} + \frac{1}{16} + \ldots.$$

Suppose that we cut off such an infinite series at the nth term. Then we obtain a finite series $u_1 + u_2 + u_3 + \ldots + u_n$. Since this is a sum of a finite number of real numbers it is also a real number. Let

$$S_n = u_1 + u_2 + u_3 + \ldots + u_n = \sum_{r=1}^{n} u_r.$$

Then we call S_n the *nth partial sum* of the series $\sum u_r$. Thus, the third partial sum of the series $\sum \frac{1}{2^r}$ is S_3, where

$$S_3 = \tfrac{1}{2} + \tfrac{1}{4} + \tfrac{1}{8} = \tfrac{7}{8}.$$

The partial sums of a series $\sum u_r$ will themselves form a sequence (S_n), where

$$(S_n) = (S_1, S_2, S_3, \ldots)$$

where $S_1 = u_1$, $S_2 = u_1 + u_2$, $S_3 = u_1 + u_2 + u_3$, and so on. If the sequence (S_n) of partial sums of the series $\sum u_r$ tends to some limit S as n tends to infinity, then we can regard S as the 'sum' of the infinite series $\sum u_r$.

Definition If $S_n = \sum_{r=1}^{n} u_r$ and $\lim_{n \to \infty} S_n = S$, then S is called the sum to infinity of the series $\sum u_r$ and we write

$$S = \sum u_r = \sum_{r=1}^{\infty} u_r = \lim_{n \to \infty} \sum_{r=1}^{n} u_r.$$

When S_n has a limit S, we say that $\sum u_r$ converges to S.

When S_n has no limit, we say that $\sum u_r$ diverges.

EXERCISE 26.1

1   Write down the first, second, third and nth partial sum of the series. State whether or not the series converges:

(i) $\sum (4-r)$,   (ii) $\sum (-1)^r$,   (iii) $\sum (\tfrac{1}{2})^r$,   (iv) $\sum (-\tfrac{1}{2})^r$.

26.2 The geometric series

The geometric series $a + a\rho + a\rho^2 + \ldots$, that is, $\sum a\rho^{r-1}$, has an nth partial sum S_n, with $S_n = a\dfrac{1-\rho^n}{1-\rho}$ (see Chapter 17), provided that $\rho \neq 1$. Now:

if $|\rho| > 1$, then $|\rho|^n$ tends to infinity as n tends to infinity,
if $|\rho| < 1$, then ρ^n tends to zero as n tends to infinity,
if $\rho = 1$, then $S_n = na$ and S_n tends to infinity as n does,
if $\rho = -1$, then $S_n = \tfrac{1}{2}a(1-(-1)^n)$, which oscillates.

Therefore, (S_n) converges to $\dfrac{a}{1-\rho}$ if $|\rho| < 1$,

and (S_n) diverges   if $|\rho| \geqslant 1$.

In terms of the series:

if $|\rho| < 1$, then $\lim\limits_{n \to \infty} S_n = \dfrac{a}{1-\rho}$, and $\sum a\rho^{r-1}$ converges to a sum to

infinity $\sum a\rho^{r-1} = \dfrac{a}{1-\rho}$,

if $|\rho| \geqslant 1$, then the series $\sum a\rho^r$ diverges.

EXAMPLE 1   *Find the sum to infinity of the series $\sum (\tfrac{3}{4})^{r-1}$. If S_n is the sum to n terms and S is the sum to infinity, find the least value of n such that $S - S_n < 0.01$.*

The series is geometric, with common ratio $\tfrac{3}{4}$ so it converges to S, where

$$S = \frac{1}{1-\tfrac{3}{4}} = 4.$$

$$S_n = \frac{1-(\tfrac{3}{4})^n}{1-\tfrac{3}{4}} = 4 - 4(\tfrac{3}{4})^n.$$

This differs from S by less than 0.01 if $4(\tfrac{3}{4})^n < 0.01$, that is, if $n \ln (\tfrac{3}{4}) < \ln (0.0025)$, that is, since the logarithms are negative, if $n > \dfrac{\ln (0.0025)}{\ln (0.75)} = 20.8$. So the required value of n is **21**.

EXAMPLE 2 *Show that the recurring decimal* $0\cdot\dot{2}\dot{7}\,(=0\cdot27272727\ldots)$ *can be expressed as a rational number.*

The infinite recurring decimal can be expressed as $0\cdot27(1+10^{-2}+10^{-4}+\ldots.)$, that is, $0\cdot27$ multiplied by the geometric series, with first term 1 and common ratio $0\cdot01$. The sum to n terms of the geometric series, multiplied by $0\cdot27$ will give the approximation to $2n$ decimal places to the real number represented by the recurring decimal. Therefore this real number is given by the sum to infinity of the series:

$$0\cdot27\frac{1}{1-0\cdot01}=0\cdot27\frac{100}{99}=\frac{27}{99}=\frac{3}{11}.$$

EXERCISE 26.2

1   Find the sum to infinity of the geometric series, given that $|x|<1$:
   (i) $\frac{1}{2}+\frac{1}{4}+\frac{1}{8}+\ldots=\sum(\frac{1}{2})^{n}$,   (ii) $1-\frac{1}{2}+\frac{1}{4}-\frac{1}{8}+\ldots=\sum(-\frac{1}{2})^{n-1}$,
   (iii) $3+\frac{3}{5}+\frac{3}{25}+\ldots=\sum 3(\frac{1}{5})^{n-1}$,
   (iv) $1-x+x^{2}-x^{3}+\ldots=\sum(-x)^{n-1}$,
   (v) $1+x+x^{2}+x^{3}+\ldots=\sum x^{n-1}$.
2   The first term of a geometric series is 12 and the fourth term is $\frac{3}{2}$. Find the common ratio and the sum to infinity.
3   Find the rational number corresponding to the recurring decimal:
   (i) $0\cdot\dot{7}$,   (ii) $0\cdot\dot{1}\dot{2}$,   (iii) $1\cdot2\dot{3}\dot{4}$,   (iv) $0\cdot00\dot{2}\dot{7}$.
4   Write down the number of possible remainders when an integer m is divided by a positive integer n. Prove that the decimal representation of a rational number always ends in a recurring decimal, where the case of recurring zero is included.

26.3 The binomial series

In Chapter 19, we introduced the binomial coefficient $\binom{n}{r}$, where

$$\binom{n}{r}=\frac{n(n-1)(n-2)\ldots(n-r+1)}{r(r-1)(r-2)\ldots2.1}=\frac{n(n-1)(n-2)\ldots(n-r+1)}{r!}.$$

This was the coefficient of x^{r} in the expansion of $(1+x)^{n}$, and also the coefficient of x^{n-r}. This result, known as the binomial theorem, was proved for positive integer values of n and r. Thus

$$(1+x)^{n}=\sum_{r=0}^{n}\binom{n}{r}x^{r}.$$

The expansion is a finite series since the coefficient $\binom{n}{r}$ is zero for $r>n$. If n is not a positive integer, but is some other real number, then $\binom{n}{r}$ can still be defined in the same way, but this coefficient does not vanish for any natural number r and the series becomes an infinite series. It can be proved that the series converges to the sum $(1+x)^n$ when $|x| < 1$. We can not prove this result here but we shall state it as a theorem and put it to various uses.

Theorem* 26.3 *The binomial theorem   Let n be a real number and let r be a natural number. Then

$$(1+x)^n = \sum_{r=0}^{\infty} \binom{n}{r} x^r, \quad \text{for} \quad |x| < 1.$$

If n is a positive integer, the series is finite. If n is not a positive integer, the series is infinite and converges for $|x| < 1$.

Clearly, x^r will become smaller more rapidly the nearer x is to zero, so, for small values of x, either positive or negative, the series can be used to obtain an approximation to $(1+x)^n$ by considering a partial sum.

More generally, we can expand $(a+b)^n$ by taking out the factor a^n and writing $b/a = x$, then

$$(a+b)^n = a^n \left(\sum_{r=0}^{\infty} \binom{n}{r} \left(\frac{b}{a}\right)^r \right).$$

EXAMPLE 1   *Expand* $(a+x)^{\frac{1}{2}}$ *in a power series of x, as far as the term in* x^4. *Find an approximation, correct to five decimal places, to the real number:* (i) $(1 \cdot 02)^{\frac{1}{2}}$, (ii) $\sqrt{(3 \cdot 96)}$.

Using the binomial theorem, for $\left|\frac{x}{a}\right| < 1$,

$$(a+x)^{\frac{1}{2}} = a^{\frac{1}{2}} \left(1 + \frac{x}{a}\right)^{\frac{1}{2}}$$

$$= a^{\frac{1}{2}} \left(1 + \frac{1}{2}\frac{x}{a} + \frac{\frac{1}{2}(-\frac{1}{2})}{2}\left(\frac{x}{a}\right)^2 + \frac{\frac{1}{2}(-\frac{1}{2})(-\frac{3}{2})}{2.3}\left(\frac{x}{a}\right)^3 + \right.$$

$$\left. + \frac{\frac{1}{2}(-\frac{1}{2})(-\frac{3}{2})(-\frac{5}{2})}{2.3.4} \times \left(\frac{x}{a}\right)^4 + \dots \right)$$

$$= a^{\frac{1}{2}} \left(1 + \frac{1}{2}\frac{x}{a} - \frac{1}{8}\left(\frac{x}{a}\right)^2 + \frac{1}{16}\left(\frac{x}{a}\right)^3 - \frac{5}{128}\left(\frac{x}{a}\right)^4\right), \text{ as far as the term in } x^4.$$

(i) Use $a = 1$, $x = 0 \cdot 02$,

$$(1 \cdot 02)^{\frac{1}{2}} \approx 1 + \frac{1}{2}(0 \cdot 02) - \frac{1}{8}(0 \cdot 02)^2 + \frac{1}{16}(0 \cdot 02)^3 - \frac{5}{128}(0 \cdot 02)^4$$

$$= 1 + 0 \cdot 01 - 0 \cdot 00005 + 0 \cdot 0000005 - 0 \cdot 00000000625$$

$$= 1 \cdot 00995049375$$

$$\approx 1 \cdot 00995 \text{ correct to five decimal places.}$$

(ii) Use $a = 4$, $x = -0.04$,

$$\sqrt{(3.96)} = 2(1 - 0.01)^{\frac{1}{2}}$$

$$\approx 2\left(1 - \frac{1}{2}(0.01) - \frac{1}{8}(0.01)^2 - \frac{1}{16}(0.01)^3 - \frac{5}{128}(0.01)^4\right)$$

$$\approx 2(1 - 0.005 - 0.0000125 - 0.0000000625)$$

$$= 2(0.9949874375) \approx \mathbf{1.98997} \text{ to five decimal places.}$$

EXAMPLE 2   *Expand the function* $(1 + 2x - 3x^2)^{\frac{1}{4}}$ *in powers of x, as far as the term in* x^3. *State the set of values of x for which the corresponding infinite series converges.*

Let $y = 2x - 3x^2$. Then

$$f(x) = (1 + y)^{\frac{1}{4}} \approx 1 + \tfrac{1}{4}y + \frac{\tfrac{1}{4}(-\tfrac{3}{4})}{2}y^2 + \frac{\tfrac{1}{4}(-\tfrac{3}{4})(-\tfrac{7}{4})}{2.3}y^3$$

so $$f(x) \approx 1 + \frac{1}{4}(2x - 3x^2) - \frac{3}{32}(2x - 3x^2)^2 + \frac{7}{128}(2x - 3x^2)^3$$

$$\approx 1 + \frac{1}{2}x - \frac{3}{4}x^2 - \frac{3}{8}x^2 + \frac{9}{8}x^3 + \frac{7}{16}x^3 = 1 + \frac{1}{2}x - \frac{9}{8}x^2 + \frac{25}{16}x^3.$$

As an infinite series, the expansion is valid for $|y| = |2x - 3x^2| < 1$. That is, for $-1 < 2x - 3x^2 < 1$, and we deal with the two inequalities separately.

$$-1 < 2x - 3x^2 \Leftrightarrow 3x^2 - 2x - 1 < 0$$
$$\Leftrightarrow (3x + 1)(x - 1) < 0$$
$$\Leftrightarrow -\frac{1}{3} < x < 1,$$

$$2x - 3x^2 < 1 \Leftrightarrow 0 < 3x^2 - 2x + 1$$
$$\Leftrightarrow 0 < 3\left(x - \frac{1}{3}\right)^2 + \frac{2}{3}, \text{ which is always true.}$$

So the condition to be satisfied by x for convergence of the infinite series is $-\dfrac{1}{3} < x < 1$.

EXAMPLE 3   *If x is so small that the fourth and higher powers of x may be neglected, find a polynomial approximation to* $\dfrac{1}{(1 - x)(1 + 2x)^2}$

The first step is to expand the expression into partial fractions. Let

$$\frac{1}{(1 - x)(1 + 2x)^2} = \frac{A}{1 - x} + \frac{B}{1 + 2x} + \frac{C}{(1 + 2x)^2},$$

then $$1 = A(1 + 2x)^2 + B(1 - x)(1 + 2x) + C(1 - x).$$

On putting $x = 1$, $1 = 9A$, so $A = \frac{1}{9}$. On putting $x = -\frac{1}{2}$, $\frac{3}{2}C = 1$ so $C = \frac{2}{3}$.

On putting $x = 0$, $1 = A + B + C$, so $B = \frac{2}{9}$. Therefore,

$$\frac{1}{(1-x)(1+2x)^2} = \frac{1}{9}(1-x)^{-1} + \frac{2}{9}(1+2x)^{-1} + \frac{2}{3}(1+2x)^{-2}.$$

Using the binomial expansion, and neglecting powers of x above the third, the approximate polynomial is

$$\frac{1}{9}(1+x+x^2+x^3) + \frac{2}{9}(1-2x+4x^2-8x^3) +$$

$$+ \frac{2}{3}\left(1 + \frac{-2}{1}2x + \frac{(-2)(-3)}{2}(2x)^2 + \frac{(-2)(-3)(-4)}{2.3}(2x)^3\right)$$

$$= \frac{1}{9}(1+x+x^2+x^3+2-4x+8x^2-16x^3+6-24x+72x^2-192x^3)$$

$$= \frac{1}{9}(9-27x+81x^2-207x^3) = 1 - 3x + 9x^2 - 23x^3.$$

More directly, $(1-x)(1+4x+4x^2) = 1 + 3x - 4x^3$, so the required expansion is

$$(1+3x-4x^3)^{-1} = 1 - (3x-4x^3) + (3x-4x^3)^2 - (3x-4x^3)^3 + \dots$$
$$\approx 1 - 3x + 4x^3 + 9x^2 - 27x^3 = 1 - 3x + 9x^2 - 23x^3.$$

EXERCISE 26.3

1   Expand $(1+x)^{-1}$, using the binomial theorem. Compare your result with the sum to infinity of the geometric series $\sum x^{r-1}$.

2   Expand the expression in increasing powers of x as far as the term in x^3. State the values of x for which the corresponding infinite series converges:

(i) $(1+x)^{-2}$,   (ii) $(1-x)^{-1}$,   (iii) $\sqrt{(1-x)}$,   (iv) $\sqrt[3]{(1-2x)}$,   (v) $\dfrac{1}{(1+x)^3}$,

(vi) $(1-x)^{\frac{1}{4}}$,   (vii) $\dfrac{1}{2+x}$,   (viii) $\sqrt{(3+4x)}$,   (ix) $\left(1-\dfrac{1}{x}\right)^{-2}$,

(x) $\dfrac{-1}{(2+3x)^4}$,   (xi) $(x-5)^{-\frac{2}{3}}$,   (xii) $\dfrac{1}{\sqrt{(2+3x)}}$.

3   Expand the expression $\left(x+\dfrac{1}{x}\right)^{-1}$ in (i) ascending powers of x, (ii) in ascending powers of x^{-1}, as far as the first four terms. State, in each case, the set of values of x for which the corresponding infinite expansion is valid.

4   Find a quadratic approximation for the given expression, assuming that x is small in magnitude:

(i) $\dfrac{x}{1+x}$,   (ii) $\dfrac{x}{3x+2}$,   (iii) $\dfrac{x-2}{(1+x)^3}$,   (iv) $(1+x-2x^2)^{-\frac{1}{3}}$,

(v) $(3+2x-x^2)^{\frac{3}{2}}$,   (vi) $\dfrac{1}{(1-x)(1+2x)^2}$,   (vii) $\dfrac{2x}{(x+1)(x+2)}$,

(viii) $(1+3x)(1-2x)^{\frac{1}{2}}$,   (ix) $\dfrac{3x}{(1-x)}$,   (x) $\dfrac{1}{(x+1)(2-3x)}$,

(xi) $\dfrac{1}{(1+x)(1-2x)^2}$,   (xii) $\dfrac{1}{(1+ax)^n}$.

5 Find the set of values of x for which $\left(\dfrac{1+x}{1-x}\right)^{\frac{1}{3}}$ can be expanded in an infinite powers series in x. Use this expansion to find an approximation to $\sqrt[3]{(\frac{3}{2})}$, correct to three decimal places.

MISCELLANEOUS EXERCISE 26

1 Find the sum to infinity of the geometric series

$$1+\left(\frac{4}{5}\right)+\left(\frac{4}{5}\right)^2+\left(\frac{4}{5}\right)^3+\ldots$$

Find the least number of terms of the series required for its sum to differ from the sum to infinity by less than 10^{-50}. (*JMB*)

2 State the set of values of x for which the series

$$2+4x+8x^2+\ldots+2^nx^{n-1}+\ldots$$

is convergent. Find the set of values of x for which the series converges to a sum not greater than $x+2$.

Solve the equation $4\sum_{n=1}^{\infty}2^nx^n+\sum_{n=0}^{\infty}(-1)^nx^n=0$. (*L*)

3 Write down the first four terms in the expansion of $(1-2x)^{-\frac{1}{2}}$ in ascending powers of x and simplify the coefficients. (*L*)

4 The first term of a geometric series is 2 and the second term is x. State the set of values of x for which the series is convergent. Show that when convergent the series converges to a sum greater than 1.

If $x=\frac{1}{2}$, find the smallest positive integer n such that the sum of the first n terms differs from the sum to infinity by less than 2^{-10}. (*L*)

5 Express $P(x)\equiv\dfrac{3x}{(1+x)(1-2x)}$ in partial fractions, and hence, or otherwise, find the expansion of $P(x)$ in ascending powers of x as far as the term in x^3. (*L*)

6 Expand $(1-x)^{-\frac{1}{2}}$ in ascending powers of x as far as the term in x^3. By putting $x=1/50$ obtain an estimate of $\sqrt{2}$, giving your result to five places of decimals. (*L*)

7 Use the binomial expansion to find a linear approximation for

$$\frac{1}{(1+3x)^{\frac{1}{2}}}-\frac{1}{(4-x)^{\frac{3}{2}}},$$

where x is small enough for terms in x^2 and higher powers to be negligible. (*L*)

8 Find an expression (which need not be simplified) for the sum of the geometric series

$$S_n(x)=1+\frac{1}{2}\left(x+\frac{1}{x}\right)+\frac{1}{4}\left(x+\frac{1}{x}\right)^2+\ldots+\frac{1}{2^{n-1}}\left(x+\frac{1}{x}\right)^{n-1},\text{ where }x\neq 0.$$

Prove that there are no values of x for which the sum to infinity of the series exists.

Prove also that $(1-x)^2S_{2n}(x)=-(1+x)^2S_{2n}(-x)$. (*C*)

9 The first two non-zero terms in the expansion of $(1 + ax)(1 + bx)^9$ as a series of ascending powers of x are 1 and $-5x^2$. Given that $a > 0$ and $b < 0$, find the values of a and b. *(L)*

10 Starting from first principles, prove that the sum of the first n terms of a geometric progression whose first term is a and whose common rato is r is

$$\frac{a(1-r^n)}{1-r}.$$

State the range of values of r for which the sum to infinity exists.

Determine the range of values of x for which the sum to infinity of the series

$$1 - e^x + e^{2x} - e^{3x} + \ldots + (-1)^m e^{mx} + \ldots$$

exists. Write down (or find) an expression for the sum to infinity in this case. *(JMB)*

11 The three real, distinct and non-zero numbers a, b, c are such that

$$a, b, c \text{ are in arithmetic progression and}$$
$$a, c, b \text{ are in geometric progression.}$$

Find the numerical value of the common ratio of the geometric progression.

Hence find an expression, in terms of a, for the sum to infinity of the geometric series whose first terms are a, c, b. *(JMB)*

12 Given that $f(x) = \dfrac{(1+nx)^{1/n}}{(1+mx)^{1/m}}$, where m and n are non-zero constants, and that x is so small that terms in x^3 and higher powers may be neglected, prove that

$$f(x) = 1 + \tfrac{1}{2}(m-n)x^2. \qquad (L)$$

13 Write down expressions for the roots of the quadratic equation

$$x^2 - x + a = 0,$$

where a is a constant. If a is so small that a^4 and higher powers of a may be neglected, show that the roots are approximately

$$1 - a - a^2 - 2a^3 \quad \text{and} \quad a + a^2 + 2a^3.$$

Using these approximations, estimate to three decimal places the roots of the equation

$$100x^2 - 100x + 3 = 0.$$

14 The first term of an arithmetic series is $(3p + 5)$ where p is a positive integer. The last term is $(17p + 17)$ and the common difference is 2. Find, in terms of p (i) the number of terms, (ii) the sum of the series.

Show that the sum of the series is divisible by 14, only when p is odd. *(AEB)*

15 (a) The sum of the first six terms of a geometrical progression is four times the sum of the first three terms. Find the common ratio.

(b) Express $(1 - x)^{-\frac{1}{2}}$ and $(4 + x)^{\frac{1}{2}}$ as series of ascending powers of x up to and including the term in x^2. Hence show that

$$\sqrt{\left(\frac{4+x}{1-x}\right)} = 2 + \frac{5}{4}x + \frac{55}{64}x^2,$$

if higher powers of x are neglected. By putting $x = 1/10$, show that $\sqrt{41}$ is approximately $6\frac{513}{1280}$. (AEB)

16 Given that $f(x) \equiv \dfrac{1+2x}{(1+x)(1-2x^2)}$, express $f(x)$ in partial fractions and find the first six terms in the expansion of $f(x)$ in ascending powers of x, when $|x| < 1/\sqrt{2}$. (L)

17 A geometric series with first term 3 converges to the sum of 2. Find the fifth term in the series. (L)

18 Show that the first three terms in the expansion in ascending powers of x of

$$(1+8x)^{1/4}$$

are the same as the first three terms in the expansion of

$$\frac{1+5x}{1+3x}.$$

Use the approximation $(1+8x)^{1/4} \approx \dfrac{1+5x}{1+3x}$ to obtain, in the form of a rational number in its lowest terms, an approximation to $(1\cdot 16)^{1/4}$. (L)

19 The series $a + ar - ar^2 - ar^3 + ar^4 + ar^5 - ar^6 - ar^7 + \dots$, where $a > 0$, has its kth term, T_k, defined by

$$T_k = ar^{k-1} \text{ if } k \text{ is of the form } 4p-3 \text{ or } 4p-2$$

and $T_k = -ar^{k-1}$ if k is of the form $4p-1$ or $4p$,

where p is a positive integer. By rewriting the series as the sum of two geometric series, or otherwise, prove that the sum to $4n$ terms of the series is

$$\frac{a(1+r)(1-r^{4n})}{(1+r^2)}.$$

State the set of values of r which the series has a sum to infinity. Assuming that the series converges when $r = -0\cdot 9$, find the least value of n for the sum to $4n$ terms to be greater than 99% of the sum to infinity. (C)

20 An arithmetical progression and a geometrical progression have the same positive first term. The common difference r of the arithmetical progression is equal to the common ratio of the geometrical progression. The sum of the first eight terms of the former is five times the sum of the first and third terms of the latter. The fifth term of the geometrical progression is $-r$. Find r and the first term of either progression.

Show that the geometrical progression converges and find its sum to infinity. (L)

21 Given that $g(x) \equiv \dfrac{5-x}{(1+x^2)(1-x)}$, express $g(x)$ in partial fractions.

Hence, or otherwise, show that the expansion of $g(x)$ as a series of ascending powers of x up to and including the term in x^4 is

$$5 + 4x - x^2 + 5x^4.$$ (L)

22 (a) The first three terms of a geometric series are 1, x, y, and the first three terms of an arithmetic series are 1, x, $-y$. Prove that $x^2 + 2x - 1 = 0$, and hence find y, given that x is positive.

(b) Prove that, if x is so small that terms in x^3 and higher powers may be neglected, then

$$\left(\frac{1-x}{1+x}\right)^{\frac{1}{2}} = 1 - x + \tfrac{1}{2}x^2.$$ (C)

23 (a) Find the sum of the first n terms of the series whose rth term is

$$2^r + 2r - 1.$$

(b) If x is so small that terms in x^n, $n \geq 3$, can be neglected and

$$\frac{3+ax}{3+bx} = (1-x)^{\frac{1}{3}},$$

find the values of a and b.

Hence, without the use of tables, find an approximation in the form p/q, where p and q are integers, for $\sqrt[3]{0\cdot96}$. (AEB)

24 (a) If $N = x^3 + t$ and t/x^3 is so small that its fourth and higher powers may be neglected show that

$$\sqrt[3]{N} = x + \frac{t}{3x^2} - \frac{t^2}{9x^5} + \frac{5t^3}{81x^8}.$$

Hence evaluate $\sqrt[3]{64\cdot032}$ correct to six decimal places.

(b) The sum of the geometric progression $a, ar, ar^2, \ldots\ldots, ar^{n-1}$ is S. The product of these n terms is P. Find R, the sum of the reciprocals of these n terms, and prove that $\left(\dfrac{S}{R}\right)^n = P^2$. (AEB)

25 When terms in x^n, $n \geq 4$, are omitted

$$\frac{3ax}{4} + \sqrt{(4+ax)} - \frac{2}{\sqrt{(1-ax)}} = -x^2 + bx^3.$$

Find the values of a and b. (AEB)

26 (i) The sum of the first n terms of an arithmetic progression is $n^2 + 2n$. Find the rth term of the series. Find also the number of terms whose sum is 575.

(ii) For each of the series

(a) $1 + (1+r) + (1+r)^2 + \ldots + (1+r)^{n-1} + \ldots,$

(b) $1 + \dfrac{1}{(1+r)} + \dfrac{1}{(1+r)^2} + \ldots + \dfrac{1}{(1+r)^{n-1}} + \ldots,$

find the set of values of r for which the series is convergent and find the corresponding sums to infinity of the series. (L)

27 Expand $\sqrt{4-x}$ as a series in ascending powers of x up to and including the term in x^2.

If terms in x^n, $n \geq 3$, can be neglected, find the quadratic approximation to $\sqrt{((4-x)/(1-2x))}$. State the range of values of x for which this approximation is valid. (AEB)

28 (i) Given that the sum of the first and second terms of an arithmetical progression is x and that the sum of the $(n-1)$th and nth terms is y, prove that the sum of the first n terms is $\frac{1}{4}n(x+y)$.

(ii) The sum of the first four terms of a geometric series of positive terms is 15 and the sum to infinity of the series is 16. Show that the sum of the first eight terms of the series differs from the sum to infinity by 1/16. (L)

27 Graphs and Loci

27.1 Graphs of rational functions

In this chapter, we extend the work on curve sketching, begun in Chapter 11. We start with a list of important ideas to be considered when given the equation of a curve in Cartesian coordinates (x, y).

A Intersections with the axes: wherever possible, put $x = 0$ and solve for y and put $y = 0$ and solve for x to find the intersections with the axes.

B Behaviour for large values of x or y: if $y \to \infty$ as $x \to a$, then there is a vertical asymptote $x = a$; if $y \to b$ as $x \to \infty$, then there is a horizontal asymptote $y = b$.

C Symmetries: the form of the equation may indicate symmetries of the curve, either of rotation or of reflection.

D Range of values: there may be some limitations on the possible values of x and y, which will show that there are regions of the plane where there are no points of the curve.

E Use of the gradient $\dfrac{dy}{dx}$: as x increases: the curve rises when $\dfrac{dy}{dx} > 0$ and falls when $\dfrac{dy}{dx} < 0$. Stationary points are given by $\dfrac{dy}{dx} = 0$ and their nature is given by the sign changes in $\dfrac{dy}{dx}$.

EXAMPLE 1 *Sketch the graph with equation* $y = (2x + 3)/(x - 1)$.

When $x = 0$, $y = -3$ and when $y = 0$, $x = -\frac{3}{2}$, so the curve crosses the axes at $(0, -3)$ and $(-\frac{3}{2}, 0)$. When $x \neq 0$, $y = \dfrac{2 + 3/x}{1 - 1/x}$, so that as $x \to \infty$, $y \to 2$. Also $y \to \infty$ as $x \to 1$, and so the curve has asymptotes $y = 2$ and $x = 1$. The derivative

$$\frac{dy}{dx} = \frac{2(x-1) - (2x+3)}{(x-1)^2} = \frac{-5}{(x-1)^2}$$

and this is negative for all x. Thus, the curve is always decreasing and has no stationary values. These remarks enable us to sketch the curve, Fig. 27.1.

For this curve, consider the effect of displacing the origin to the point $(1, 2)$, where the asymptotes intersect. This may be done by introducing new coordinates (x_1, y_1), where $x_1 = x - 1$ and $y_1 = y - 2$. The equation

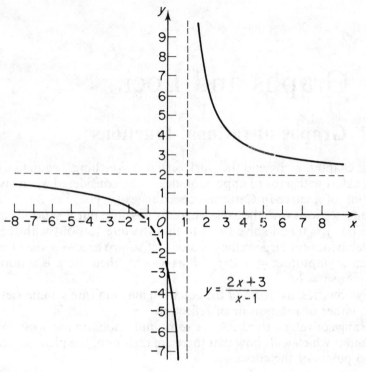

Fig. 27.1

of the curve then becomes

$$y_1 + 2 = y = \frac{2x+3}{x-1} = \frac{2(x_1+1)+3}{x_1} = 2 + \frac{5}{x_1},$$

that is, $y_1 = 5/x_1$. This means that the curve has a symmetry of rotation of a half turn about $(1, 2)$ and can be obtained from the curve $y = 1/x$ by a one-way stretch parallel to Oy with factor 5, followed by the translation given by the vector $\begin{pmatrix} 1 \\ 2 \end{pmatrix}$.

In a similar manner, any curve with an equation $y = \dfrac{ax+b}{cx+d}$, $c \neq 0$, can be obtained from the curve $xy = 1$ by a one-way stretch parallel to Oy with factor $\dfrac{bc-ad}{c^2}$ followed by a translation $\begin{pmatrix} -d/c \\ a/c \end{pmatrix}$, because the equation can be rewritten

$$y - \frac{a}{c} = \left(\frac{bc-ad}{c^2} \right) \frac{1}{x + \dfrac{d}{c}}.$$

The curves of this type are called *rectangular hyperbolas*.

EXAMPLE 2 *Sketch the curve with equation* $y = \dfrac{1}{x^2 + x + 1}$.

Since $x^2 + x + 1 = (x + \frac{1}{2})^2 + \frac{3}{4} \geqslant \frac{3}{4}$, the minimum value of the denominator is $\frac{3}{4}$ and so, for all x, $0 < y \leqslant \frac{4}{3}$. Also the graph is symmetrical about the line $x = -\frac{1}{2}$. The point $(-\frac{1}{2}, \frac{4}{3})$ is the unique maximum point on the curve. This can also be seen from the derivative, $\dfrac{dy}{dx} = \dfrac{-2x - 1}{(x^2 + x + 1)^2}$, which is negative for $x > -\frac{1}{2}$ and positive for $x < -\frac{1}{2}$. The line $y = 0$ is the only asymptote and the curve increases to its maximum and then decreases again. The graph is shown in Fig. 27.2.

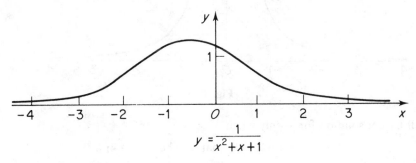

$$y = \frac{1}{x^2 + x + 1}$$

Fig. 27.2

A similar curve will be obtained from any equation of the type $y = \dfrac{1}{f(x)}$, where $f(x)$ is an irreducible quadratic function, that is, a quadratic function with no zeros. The corresponding curve in the case when $f(x)$ has zeros, so that $f(x)$ factorises into two linear factors, is of a quite different shape, as shown in the next two examples.

EXAMPLE 3 *Sketch the curve* $y = 1/x^2$.

Since $1/x^2$ is an even function, the curve is symmetrical about Oy. The derivative takes the opposite sign from x, since $dy/dx = -2/x^3$ and this is odd. The curve is increasing for negative values of x and decreasing for positive values of x. There are two asymptotes, $y = 0$ and $x = 0$ and the curve is confined to the upper half plane, $y > 0$, see Fig. 27.3.

EXAMPLE 4 *Sketch the curve with equation* $y = 1/(x^2 - 3x + 2)$.

The curve crosses the y-axis at $(0, \frac{1}{2})$ but does not cross the x-axis. Denote the denominator by $f(x)$, then this can be factorised, $f(x) = (x - 1)(x - 2)$. Thus $f(1) = 0 = f(2)$ and so the curve has asymptotes $x = 1$ and $x = 2$. Since $y \to 0$ as $x \to \infty$, there is a third asymptote, $y = 0$. $f(x) = (x - \frac{3}{2})^2 - \frac{1}{4}$ and so $f(x)$ has a *minimum* at the point $(\frac{3}{2}, -\frac{1}{4})$. Therefore, $1/f(x)$ has a maximum at the point $(\frac{3}{2}, -4)$.

Consider the equation $y(x^2 - 3x + 2) = 1$. For a given value of y, this equation

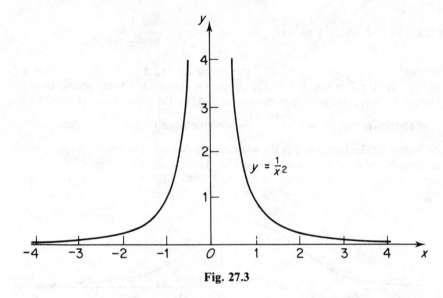

Fig. 27.3

will have a solution for x only when

$$(-3y)^2 \geqslant 4y(2y-1), \text{ that is, when } y(y+4) \geqslant 0.$$

This inequality is satisfied only for $y \geqslant 0$ or $y \leqslant -4$ and so the curve has no points in that region of the plane given by the inequalities $-4 < y \leqslant 0$. From this information, we sketch the graph shown in Fig. 27.4. The curve is symmetrical about the line $x = \frac{3}{2}$.

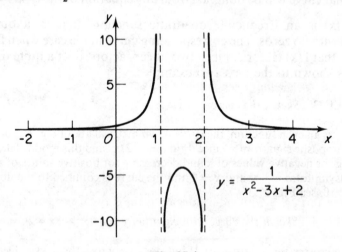

Fig. 27.4

EXERCISE 27.1

1   For the curve with the given equation, show that the gradient is always of one sign, find the equations of the asymptotes and the coordinates of the points

where the curve crosses the axes; sketch the curve:

(i) $y = \dfrac{2}{x-1}$, (ii) $y = \dfrac{x}{x+1}$, (iii) $y = \dfrac{x+1}{x}$, (iv) $y = \dfrac{x}{2-x}$,

(v) $y = \dfrac{2x-3}{x}$, (vi) $y = \dfrac{x-2}{x+1}$.

2 For the curve with equation $y = \dfrac{x}{(x-2)}$, find

(i) the equation of each of the asymptotes, (ii) the equation of the tangent at the origin.

Sketch the curve, paying particular attention to its behaviour at the origin and as it approaches its asymptotes.

On a separate diagram, sketch the curve with equation $y = \left| \dfrac{x}{x-2} \right|$.

3 For the given equation, find the range of possible values of y in order that x can take real values. Determine where the curve with the given equation crosses the axes, find its turning point and the equations of its asymptotes. Use these results to sketch the curve.

(i) $y = \dfrac{1}{x^2 - 2x - 3}$, (ii) $y(x^2 + 5x + 6) = 1$.

4 The equation of a curve is $y = \dfrac{1}{8 + 2x - x^2}$. Determine the equations of the asymptotes of the curve and the coordinates of any points where it crosses the Cartesian axes. Find the turning point of the graph and determine its nature. Find the range of values of y for which the curve has no points. Sketch the curve.

5 Sketch the curve $y = \dfrac{1}{3x^2 + x - 2}$.

6 Using the same axes, sketch the graphs of $y = x(x+1)$ and $y = \dfrac{1}{x(x+1)}$ labelling each graph clearly.

7 Show that the gradient of the curve $y = \dfrac{x}{1+x}$, $(x \neq -1)$, is always positive.

Sketch the curve, clearly marking the asymptotes and any points of intersection with the coordinate axes.

The tangent to the curve at the point $P(-2, 2)$ meets the y-axis at T. Find the coordinates of T.

A line through T with gradient m meets the curve at two points. Show that the x-coordinates of these two points satisfy the quadratic equation

$$mx^2 + (3+m)x + 4 = 0.$$

Hence, or otherwise, find the coordinates of Q, the other point on the curve at which the tangent passes through T.   (*JMB*)

8 Find the coordinates of the points of intersection of the curves whose equations are

$$y = (x-1)(x-2), \qquad y = \dfrac{3(x-1)}{x}.$$

State, or obtain:

(i) the coordinates of the turning point of the first curve;

(ii) the equations of the asymptotes of the second curve. Sketch the two curves on the same diagram.

Find the area of that region bounded by the two curves for which x lies between 1 and 3.   (*JMB*)

9 Sketch the curve with equation $y = 2x/(x+3)$ showing clearly how the curve approaches its asymptotes.

Shade on your diagram the finite region bounded by the curve and the lines $x = 0$, $x = 3$, $y = 2$. Find the area of this region.   (*L*)

27.2 The parabola

The curve given by the equation $y = ax^2 + bx + c, a \neq 0$, is a parabola (see Chapter 2). The axis of the parabola is parallel to Oy. Its vertex is downwards if $a > 0$ and is upwards if $a < 0$. If we interchange x and y in the equation, we reflect the curve in the line $x = y$ and so the equation $x = ay^2 + by + c$ gives a parabola with its axis parallel to Ox.

The parabola $y^2 = 4ax$ has its axis parallel to Ox and its vertex at the origin. Its shape is indicated for $a < 0$ and for $a > 0$ in Fig. 27.5.

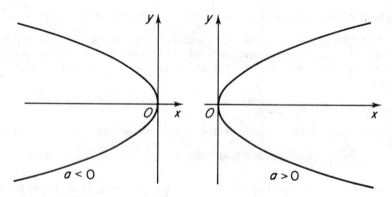

Fig. 27.5 Parabola $y^2 = 4ax$.

The *parametric* equations $y = 2at$ and $x = at^2$ will give the same curve, since on eliminating t we obtain

$$y^2 = 4a^2t^2 = 4a(at^2) = 4ax,$$

that is, $y^2 = 4ax$. The parametric equations are useful in dealing with problems concerning chords, tangents and normals to the parabola.

EXAMPLE   *Find the equation of the tangent and the equation of the normal at the point P $(at^2, 2at)$ on the parabola $y^2 = 4ax$. Find the coordinates of the points N, S, T, given that the normal at P meets the x-axis at N and the tangent meets Oy at S and Ox at T. Find the area of the triangle NST and prove that this is the same as the area of the triangle NPS.*

Since, in terms of the parameter t, $x = at^2$ and $y = 2at$, $\dfrac{dx}{dt} = 2at$ and $\dfrac{dy}{dt} = 2a$.

Therefore,

$$\frac{dy}{dx} = \frac{dy}{dt}\bigg/\frac{dx}{dt} = \frac{2a}{2at} = \frac{1}{t},$$

and this is the gradient of the tangent at P. The equation of the tangent is, therefore,

$$y - 2at = \frac{1}{t}(x - at^2) \quad \text{or} \quad ty = x + at^2.$$

When $x = 0$, $y = at$ and when $y = 0$, $x = -at^2$. Hence, the coordinates of S and T are respectively $(0, at)$ and $(-at^2, 0)$. The gradient of the normal at P is $-t$ and so the equation of the normal is

$$y - 2at = -t(x - at^2) \quad \text{or} \quad y + tx = at(2 + t^2).$$

When $y = 0$, $x = a(2 + t^2)$ and so the coordinates of N are $(2a + at^2, 0)$. The various points are shown in Fig. 27.6.

$$TN = at^2 + (2a + at^2) = 2a(1 + t^2), \; OS = at,$$

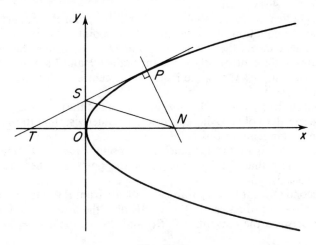

Fig. 27.6

so the area of triangle NST is A, given by

$$A = \tfrac{1}{2}OS.TN = a^2t(1 + t^2).$$

Also $\qquad SP^2 = x_P^2 + (y_P - y_S)^2 = a^2t^4 + a^2t^2 = a^2t^2(1 + t^2),$

so that $SP = at\sqrt{(1 + t^2)}$. Similarly,

$$PN^2 = (x_N - x_P)^2 + y_P^2 = 4a^2 + 4a^2t^2$$

and $PN = 2a\sqrt{(1 + t^2)}$. The area of triangle NPS is $\tfrac{1}{2}PS.PN$ which can be written

$$\tfrac{1}{2}at\sqrt{(1 + t^2)}\,2a\sqrt{(1 + t^2)} = a^2t(1 + t^2) = A.$$

EXERCISE 27.2

1   A point P moves in the plane such that its distance from the point $(a, 0)$ is equal to its distance from the line $x = -a$. Prove that the locus of P is a parabola, stating the coordinates of the vertex and the equation of the axis of the parabola.

2   The tangent at the point $P\,(at^2, 2at)$ to the parabola $y^2 = 4ax$ meets the axis Oy at S and the axis Ox at T. Prove that S is the midpoint of PT.

3   The normal to the parabola $y^2 = 4ax$ at the point $P\,(at^2, 2at)$ meets the axis Ox at N. S is the point $(a, 0)$ and PC is a line in the direction of Ox. Write down the gradients of the lines PN and PS and the tangents of the angles CPN and CPS. Deduce that PN bisects the angle CPS. (This property of a parabola shows that if the parabola is the surface of a mirror all rays of light parallel to the axis are reflected to pass through S, which is therefore called the *focus* of the parabola.)

4   Find the equation of the tangent to the parabola $y^2 = 4ax$ at the point $P\,(ap^2, 2ap)$. Show that the equation of the normal at P is $y = p\,(2a - x) + ap^3$.
   If the tangents at P and $Q\,(aq^2, 2aq)$ meet at T show that T is the point $(apq, ap + aq)$. The point N is the intersection of the normals at P and Q. Given that T lies on the line $x + 2a = 0$ show that N lies on the parabola with equation $y^2 = 4a\,(x - 4a)$.   (L)

5   Find the gradient of the parabola $y^2 = 4ax$ at the point $P\,(at^2, 2at)$ and hence show that the equation of the tangent at this point is $x - ty + at^2 = 0$.
   The tangent meets the y-axis at T, and O is the origin. Show that the coordinates of the centre of the circle through O, P and T are $(at^2/2 + a,\, at/2)$ and deduce that, as t varies, the locus of the centre of this circle is another parabola.   (L)

6   Prove that the line $y = mx + 15/(4m)$ is a tangent to the parabola $y^2 = 15x$ for all non-zero values of m. Using this result, or otherwise, find the equations of the common tangents to this parabola and the circle $x^2 + y^2 = 16$.   (L)

7   Find the equation of the tangent at the point $P(4a, 4a)$ to the parabola $y^2 = 4ax$. Show that the coordinates of the point R, where this tangent meets the x-axis, are $(-4a, 0)$.
   The second tangent to the parabola $y^2 = 4ax$ from R meets the parabola at Q. Obtain the coordinates of Q, and calculate the area of the finite region enclosed between the tangents RP, RQ and the parabola $y^2 = 4ax$.   (L)

27.3 The ellipse

The equations $x = a \cos t$, $y = a \sin t$, $0 \leqslant t < 2\pi$, are the parametric equations of the circle with centre at the origin and with radius a. This circle is shown in Fig. 27.7(a) and its Cartesian equation is obtained by eliminating t,

$$x^2 + y^2 = a^2 \cos^2 t + a^2 \sin^2 t = a^2.$$

Consider two points R and S with coordinates $(a \cos t, a \sin t)$ and $(b \cos t, b \sin t)$ respectively, $0 < b < a$. As t varies from 0 to 2π, R and S have loci consisting of circles centre O and of radii a and b respectively. In

Figure 27.7(b), let RN be the perpendicular from R on to the axis Ox and let SP be the perpendicular from S on to RN. Then the coordinates of P are $(a \cos t, b \sin t)$. The locus of P is a curve which lies between the two circles, touching the larger circle at $(a, 0)$ and $(-a, 0)$ and the inner circle at $(0, b)$ and $(0, -b)$. This curve is called an *ellipse*, wth *major axis* of length $2a$ and *minor axis* of length $2b$. Clearly, these two axes are the axes of symmetry of the ellipse. The ellipse has parametric equations $x = a \cos t$, $y = b \sin t$. Eliminating t we obtain the Cartesian equation of the ellipse, since

$$\cos^2 t + \sin^2 t = 1, \quad \text{and so} \quad \frac{x^2}{a^2} + \frac{y^2}{b^2} = 1.$$

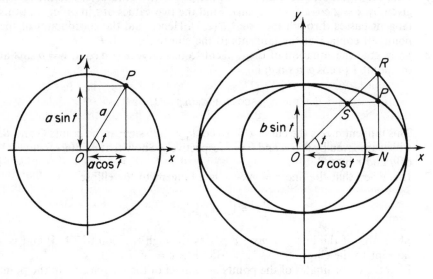

(a) Circle $x^2 + y^2 = a^2$ (b) Ellipse $\frac{x^2}{a^2} + \frac{y^2}{b^2} = 1$

Fig. 27.7

EXAMPLE *Find the equations of the tangent and the normal to the ellipse* $\frac{x^2}{a^2} + \frac{y^2}{b^2} = 1$ *at the point* $P(a \cos t, b \sin t)$.

Since $x = a \cos t$ and $y = b \sin t$, for points on the ellipse, $\dfrac{dx}{dt} = -a \sin t$ and $\dfrac{dy}{dt} = b \cos t$, hence

$$\frac{dy}{dx} = \frac{dy}{dt} \bigg/ \frac{dx}{dt} = -\frac{b \cos t}{a \sin t}.$$

Therefore the equation of the tangent at P is

$$y - b \sin t = -\frac{b \cos t}{a \sin t}(x - a \cos t),$$

that is $\qquad\qquad ay\sin t + bx\cos t = ab.$

The equation of the normal is

$$y - b\sin t = \frac{a\sin t}{b\cos t}(x - a\cos t),$$

that is $\qquad\qquad by\cos t + (a^2 - b^2)\cos t\sin t = ax\sin t.$

EXERCISE 27.3

1   Find the equation of the tangent at the point with parameter t to the ellipse given by $x = 2\cos t$, $y = \sqrt{3}\sin t$. Find the two values of t in order that this tangent passes through the point $P(2, 1)$. Hence find the coordinates of the points of contact of the tangents to the ellipse through P.

2   Show that the equation of the tangent to the curve $x = a\cos t$, $y = b\sin t$ at the point $P(a\cos p, b\sin p)$ is

$$\frac{x}{a}\cos p + \frac{y}{b}\sin p = 1.$$

This tangent meets the curve $x = 2a\cos\theta$, $y = 2b\sin\theta$ at the points Q and R, which are given by $\theta = q$ and $\theta = r$ respectively. Show that p differs from each of q and r by $\pi/3$. $\qquad\qquad$ (*JMB*)

3   It is given that the line $y = mx + c$ is a tangent to the ellipse

$$\frac{x^2}{a^2} + \frac{y^2}{b^2} = 1 \quad\text{if}\quad a^2m^2 = c^2 - b^2.$$

Show that if the line $y = mx + c$ passes through the point $(5/4, 5)$ and is a tangent to the ellipse $8x^2 - 3y^2 = 35$, then $c = 35/3$ or $35/9$.

Find the co-ordinates of the points of contact of the tangents from the point $(5/4, 5)$ to the curve $8x^2 + 3y^2 = 35$. $\qquad\qquad$ (*L*)

4   Find the coordinates of the points of intersection of the ellipse

$$\frac{x^2}{6} + \frac{y^2}{3} = 1$$

and the rectangular hyperbola $xy = -2$. Indicate clearly in a sketch all the points of intersection. $\qquad\qquad$ (*JMB*)

27.4 The hyperbola

The equation for an ellipse, $\dfrac{x^2}{a^2} + \dfrac{y^2}{b^2} = 1$, shows that there are limits for the values of the coordinates (x, y). Since we have a sum of squares on the left-hand side, each square must be less than 1, so $-a \leqslant x \leqslant a$ and $-b \leqslant y \leqslant b$, for all points on the ellipse. In fact, if $0 < b < a$, $0 < b^2 \leqslant x^2 + y^2 \leqslant a^2$, since the ellipse lies between two circles. If the sign of one of the terms on the left-hand side is changed, then the situation is completely different. Consider, therefore, the curve with equation

$$\frac{x^2}{a^2} - \frac{y^2}{b^2} = 1, \text{ which may be written } \frac{x^2}{a^2} = 1 + \frac{y^2}{b^2}.$$

There will not be any upper limit to the values of x and y. However, $\dfrac{x^2}{a^2} > 1$ and so either $x \leqslant -a$ or $x \geqslant a$. Therefore, the curve has no points between the lines $x = -a$ and $x = a$. It crosses the axis Ox at $(a, 0)$ and $(-a, 0)$ and does not meet the y-axis. It is symmetrical about both axes and consists of two branches. The curve is called *a hyperbola*.

We can express the equation in parametric form, using the fact that $\sec^2 t = 1 + \tan^2 t$, and writing $x = a \sec t$, $y = b \tan t$, $0 \leqslant t < 2\pi$, $t \notin \{\pi/2, 3\pi/2\}$. As t approaches $\pi/2$ or $3\pi/2$, x and y become large, and the dominating part of the equation is

$$\frac{x^2}{a^2} = \frac{y^2}{b^2}, \text{ which can be written } \left(\frac{x}{a} - \frac{y}{b}\right)\left(\frac{x}{a} + \frac{y}{b}\right) = 0,$$

the equation of a pair of lines. As t approaches $\pi/2$, the curve approaches the line $x/a = y/b$, and as t approaches $3\pi/2$, the curve approaches the line $x/a = -y/b$. Thus these two lines are asymptotes to the hyperbola. The curve is sketched in Fig. 27.8.

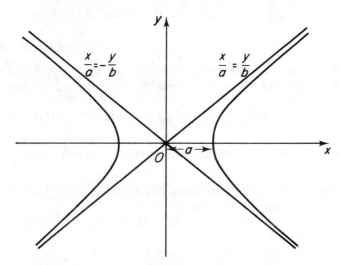

Fig. 27.8 Hyperbola $\dfrac{x^2}{a^2} - \dfrac{y^2}{b^2} = 1.$

EXERCISE 27.4

1 Suppose P is a point on the hyperbola given parametrically by $x = a \sec t$, $y = b \tan t$. Prove that the equation of the tangent to the hyperbola at P is

given by

$$\frac{x \sec t}{a} - \frac{y \tan t}{b} = 1.$$

2 Prove that the normal to the hyperbola $\dfrac{x^2}{4} - \dfrac{y^2}{1} = 1$ at $(4, \sqrt{3})$ meets the axes at $(5, 0)$ and $(0, 5\sqrt{3})$.

3 Show that the normal to the hyperbola at the point $(p \sec \theta, q \tan \theta)$ can be written in the form

$$qy + p \sin \theta x = (p^2 + q^2) \tan \theta.$$

4 State the equations of the asymptotes of the hyperbola $x^2/a^2 - y^2/b^2 = 1$.
   The point P on the curve lies in the first quadrant. The line through P parallel to Oy meets an asymptote of the curve at the point Q which also lies in the first quadrant. The normal at P meets the x-axis at G. Prove that QG is perpendicular to the asymptote.   (L)

5 A hyperbola of the form $x^2/\alpha^2 - y^2/\beta^2 = 1$ has asymptotes $y^2 = m^2 x^2$ and passes through the point $(a, 0)$. Find the equation of the hyperbola in terms of x, y, a and m.
   A point P on this hyperbola is equidistant from one of its asymptotes and the x-axis. Prove that, for all values of m, P lies on the curve

$$(x^2 - y^2)^2 = 4x^2(x^2 - a^2).$$ (L)

MISCELLANEOUS EXERCISE 27

1 Show that the curve $y = \dfrac{3}{(2x+1)(1-x)}$ has only one turning point. Find the coordinates of this point and determine its nature. Sketch the curve.
   Find the area of the region enclosed by the curve and the line $y = 3$.
   (*JMB*)

2 Given that $f(x) \equiv 6x^2 + x - 12$, find the minimum value of $f(x)$ and the values of x for which $f(x) = 0$.
   Using the same axes, sketch the curves $y = f(x)$ and $y = \dfrac{1}{f(x)}$, labelling each clearly.
   Deduce that there are four values of x for which $[f(x)]^2 = 1$. Find these values, each to two decimal places.   (L)

3 Show that the equation of the tangent to the rectangular hyperbola $xy = 4$ at the point $(2t, 2/t)$ is

$$t^2 y + x = 4t.$$

   Deduce that, as t varies, the area of the triangular region bounded by the coordinate axes and this tangent is constant.   (L)

4 Show that the gradient at each point of the curve $y = (x+2)/(x+1)$ is negative. Find the equations of the asymptotes of the curve, and sketch the curve.
   Find the distance from the x-axis of the centroid of the region defined by the inequalities

$$0 \leqslant x \leqslant 1, 0 \leqslant y \leqslant (x+2)/(x+1).$$ (L)

5 Sketch the graph of the function f, given by $f(x) = \dfrac{x-1}{x}$, and state the equations of its asymptotes. Use your graph to show that the domain of g,

where $g(x) = \ln\left(\dfrac{x-1}{x}\right)$, is

$$\{x : x < 0\} \cup \{x : x > 1\}.$$

Sketch the graph of g.

6 Sketch the curve $y = 4/x^2$.

Show that the tangent to the curve at the point $P(x_1, y_1)$ has the equation

$$8x + x_1^3 y = 12x_1,$$

and deduce that this tangent intersects the curve again at a point Q whose x-coordinate is $-\tfrac{1}{2}x_1$.

Determine the locus of the mid-point of PQ as P varies.

Find the two positions of P for which the tangent at P is also a normal at Q.   (*JMB*)

7 Show that an equation of the tangent at the point $(ap^2, 2ap)$ on the parabola $y^2 = 4ax$ is

$$x - py + ap^2 = 0.$$

The tangents to the parabola at P and Q meet the tangent at the vertex O at the points R and S respectively and intersect at T, Given that T always lies on the straight line $x + 4a = 0$, show that

(a) the angle POQ is a right angle, (b) $OR.OS$ is a constant.   (*L*)

8 Prove that the equation of the tangent to the parabola $y^2 = 4ax$ at the point $P(at^2, 2at)$ is

$$ty - x = at^2.$$

Write down, or obtain, the equation of the normal to the parabola at P.

The tangent and the normal meet the x-axis at T and N respectively. Express $PT^2 + PN^2$ in terms of a and t.   (*JMB*)

9 Prove that the gradient of the chord joining the point $P(cp, c/p)$ and the point $Q(cq, c/q)$ on the rectangular hyperbola $xy = c^2$ is $-1/pq$.

The points P, Q and R lie on a rectangular hyperbola, the angle QPR being a right angle. Prove that the angle between QR and the tangent at P is also a right angle.   (*L*)

10 Prove that the equation of the tangent at the point $P(ct, c/t)$ on the rectangular hyperbola $xy = c^2$ is

$$x + t^2 y = 2ct.$$

The tangent at P meets the x-axis at Q and the y-axis at R. The line through Q parallel to the y-axis meets the hyperbola at S; and the line through R parallel to the x-axis meets the hyperbola at T. Prove that, as P varies, the locus of the mid-point of ST is the rectangular hyperbola

$$16xy = 25c^2.$$

Show that the equation of the normal at the point $\left(\dfrac{5cp}{4}, \dfrac{5c}{4p}\right)$ on the second hyperbola is

$$4p^3x - 4py = 5c(p^4 - 1).$$

This normal meets the hyperbola $xy = c^2$ at the points A and B. Find the x-coordinate of the mid-point of AB. (*JMB*)

11   The line $y = t(x - t)$ meets the curve $xy = 2$ at the points $P(x_1, y_1)$ and $Q(x_2, y_2)$. Form a quadratic equation in x with roots x_1, x_2, and write down the sum of these roots in terms of t. Hence express in terms of t the coordinates X, Y of the mid-point M of PQ. Determine the Cartesian equation of the curve on which M lies as t varies. Calculate
(i) the value of t for which the line $y = t(x - t)$ is a tangent to the curve $xy = 2$,
(ii) the set of values of t for which the line meets the curve in two distinct points.
Show in a sketch the curve $xy = 2$ and the locus of M.

12   Find the equation of the normal to the hyperbola $xy = c^2$ at the point $x = ct$, $y = c/t$. If this normal passes through the point $P(h, k)$, show that

$$ct^4 - ht^3 + kt - c = 0.$$

The normals at four points on the hyperbola meet at P. Show that the sum of the x-co-ordinates of the four points is h, and that the sum of their y-coordinates is k.

13   Sketch the curve given parametrically by $x = 4\cos\theta$, $y = 3\sin\theta$, for $0 \le \theta \le 2\pi$.
A rectangle is inscribed in the curve. It has one corner at the point $(4\cos\phi, 3\sin\phi)$, where $0 \le \phi \le \frac{1}{2}\pi$, its sides parallel to the axes and its centre at the origin. Find, as ϕ varies, the maximum area of the rectangle.
Write $4\cos\phi + 3\sin\phi$ in the form $C\sin(\phi + \alpha)$, with $C > 0$, giving the values of C and $\tan\alpha$. Use your result to find the maximum value of the perimeter of the rectangle, and to show that no such rectangle has a perimeter of less than 12 units. (*SMP*)

14   In each of the following pairs of equations, t is a parameter. Sketch the locus given by each pair of equations.

(i) $x = 3 + 5\cos t$, $y = 4 + 5\sin t$ $(0 \le t \le \pi)$,

(ii) $x = 3\cos t$, $y = 4\cos t$ $(0 \le t \le \pi)$,

(iii) $x = 3 + t\cos\dfrac{\pi}{3}$, $y = 4 + t\sin\dfrac{\pi}{3}$ $(-\infty < t < \infty)$.

15   The line $y = mx + c$ cuts the ellipse $x^2 + 4y^2 = 16$ in the points P and Q. Show that the co-ordinates of M, the mid-point of PQ, are

$$x = -4mc/(4m^2 + 1), \quad y = c/(4m^2 + 1).$$

If the chord PQ passes through the point $(2, 0)$, show that M lies on the ellipse $x^2 + 4y^2 = 2x$. Sketch the two ellipses in the same diagram. (*L*)

16   Show that the equation of the chord joining the points $P(a\cos\phi, b\sin\phi)$ and $Q(a\cos\theta, b\sin\theta)$ on the ellipse $b^2x^2 + a^2y^2 = a^2b^2$ is

$$bx\cos\tfrac{1}{2}(\theta + \phi) + ay\sin\tfrac{1}{2}(\theta + \phi) = ab\cos\tfrac{1}{2}(\theta - \phi).$$

Prove that, if the chord PQ subtends a right angle at the point $(a, 0)$, then

PQ passes through a fixed point on the x-axis, and give the coordinates of this point. (L)

17 Find the gradient of the ellipse $x^2/a^2 + y^2/b^2 = 1$ at the point $P(a \cos \theta, b \sin \theta)$.

The tangent to this ellipse at P and the tangent at $Q(-a \sin \theta, b \cos \theta)$ meet at the point T. If O is the origin, show that $OPTQ$ is a parallelogram and find its area. Show also that as θ varies the point T describes the ellipse

$$x^2/a^2 + y^2/b^2 = 2.$$ (L)

18 A line L passes through the point $A(0, \sqrt{2})$ and makes an angle θ with the x-axis $(-\pi/2 < \theta < \pi/2)$. Show that the coordinates of any point P on the line can be expressed as $(t \cos \theta, \sqrt{2} + t \sin \theta)$ and state which parts of L correspond to positive, and which to negative, values of t.

Show that the points at which L intersects the curve $x^2 - y^2 = 1$ are given by

$$t^2 \cos 2\theta - (2\sqrt{2})t \sin \theta - 3 = 0.$$

Show that the line L cuts the curve in two distinct points only if θ lies between $-\pi/3$ and $\pi/3$, and find the range of values of θ for which the line cuts the curve in two points both on the same side of A. (JMB)

19 Sketch the curve whose equation is $y + 3 = 6/(x - 1)$.

Find the coordinates of the points where the line $y + 3x = 9$ intersects the curve and show that the area of the region enclosed between the curve and the line is

$$\tfrac{3}{2}(3 - 4 \log_e 2).$$

Determine the equations of the two tangents to the curve which are parallel to the line. (JMB)

28 Finite Series II

In Chapter 17, we introduced the Σ notation for a finite series and discovered how to sum arithmetic series, geometric series and binomial series. In this chapter, we consider a number of other methods which can be used to sum finite series.

28.1 The method of differences

Consider the identity

$$r(r+1) - (r-1)r \equiv 2r,$$

and repeat this n times for $1 \leqslant r \leqslant n$.

$r = n$	$n(n+1)$	$-$	$(n-1)n$	$= 2n$
$r = n-1$	$(n-1)n$	$-(n-2)(n-1)$		$= 2(n-1)$
$r = n-2$	$(n-2)(n-1)$	$-(n-3)(n-2)$		$= 2(n-2)$
. . . . .	. . . .			
$r = 3$	3×4	$-$	2×3	$= 2 \times 3$
$r = 2$	2×3	$-$	1×2	$= 2 \times 2$
$r = 1$	1×2	$-$	0×1	$= 2 \times 1$

Adding all these equations, we find that

$$n(n+1) = \sum_{r=1}^{n} \{r(r+1) - (r-1)r\} = \sum_{r=1}^{n} 2r = 2 \sum_{r=1}^{n} r,$$

and so

$$\sum_{r=1}^{n} r = \tfrac{1}{2}n(n+1).$$

This result is already known, since the sum is an arithmetic series, but the method, which is called the method of differences, is useful. Suppose that we wish to sum $\sum_{r=1}^{n} u_r$, and we can rearrange u_r in terms of another sequence (v_r) such that $u_r = v_{r+1} - v_r$, then

$$\sum_{r=1}^{n} u_r = (v_{n+1} - v_n) + (v_n - v_{n-1}) + \ldots + (v_3 - v_2) + (v_2 - v_1)$$

$$= v_{n+1} - v_1.$$

Of course, the usefulness of the method depends upon discovering a convenient sequence v_r, when we are given u_r. In general, if u_r is a polynomial of degree m in r, then we choose v_r to be a polynomial of degree $m+1$ in r so that the terms in r^{m+1} cancel in the difference $v_{r+1} - v_r$.

EXAMPLE 1 *Sum the series* $\sum\limits_{r=1}^{n} r^2$.

Method (i). Try the use of the sequence given by $v_r = r^3$, in the method of differences. Then

$$v_{r+1} - v_r = (r+1)^3 - r^3 = 3r^2 + 3r + 1,$$

so

$$(n+1)^3 - 1 = v_{n+1} - v_1 = \sum_{1}^{n} (3r^2 + 3r + 1)$$

Hence

$$3 \sum_{r=1}^{n} r^2 = (n+1)^3 - 1 - \sum_{r=1}^{n} (3r + 1)$$

$$= n^3 + 3n^2 + 3n + 1 - 1 - 3 \sum_{r=1}^{n} r - n$$

$$= n^3 + 3n^2 + 2n - \tfrac{3}{2}n(n+1) = \tfrac{1}{2}(2n^3 + 3n^2 + n),$$

on using the previous result for $\sum\limits_{r=1}^{n} r$. Therefore,

$$\sum_{r=1}^{n} r^2 = \tfrac{1}{6}(2n^3 + 3n^2 + n) = \tfrac{1}{6}n(n+1)(2n+1).$$

Method (ii). Now that we know the answer, we can use a value of v_r which will give the result more directly. Let $v_r = (r-1)r(2r-1)$, then

$$v_{r+1} - v_r = r(r+1)(2r+1) - (r-1)r(2r-1)$$
$$= 2r^3 + 3r^2 + r - (2r^3 - 3r^2 + r)$$
$$= 6r^2.$$

So

$$\sum_{r=1}^{n} r^2 = \tfrac{1}{6} \sum_{r=1}^{n} (v_{r+1} - v_r) = \tfrac{1}{6}(v_{n+1} - v_1) = \tfrac{1}{6}v_{n+1}$$

$$= \tfrac{1}{6}(2n^3 + 3n^2 + n) = \tfrac{1}{6}n(n+1)(2n+1).$$

EXAMPLE 2 *Use the method of differences, with* $v_r = r^2(r-1)^2$, *to find the sum of the cubes of the first n positive integers.*

Since $v_{r+1} - v_r = r^2(r+1)^2 - r^2(r-1)^2 = r^2(r^2 + 2r + 1 - r^2 + 2r - 1) = 4r^3,$

$$4 \sum_{r=1}^{n} r^3 = v_{n+1} - v_1 = n^2(n+1)^2,$$

and so

$$\sum_{r=1}^{n} r^3 = \tfrac{1}{4} n^2(n+1)^2.$$

EXERCISE 28.1

1   Use the method of differences, with $v_r = r^2$, to find the sum of the first n positive integers.

2 Use the method of differences, with $v_r = (r-1)r(r+1)$, to find the sum $\sum_{r=1}^{n} r(r+1)$. Use the result of question **1** to deduce the value of the sum of the squares of the first n positive integers.

3 Using a method similar to that of question **2**, find $\sum_{r=1}^{n} r(r+1)(r+2)$.

4 Use the method of differences, with $v_r = r^4$, to express the sum of the cubes of the first n positive integers in terms of the sum of the squares and of the sum of the first n positive integers. Deduce an expression for the sum of cubes in terms of n.

5 Show that the sum of the cubes of the first n positive integers is equal to the square of the sum of the first n positive integers.

6 Find the sum of the first n even positive integers. Find also the sum of the first n odd positive integers.

7 Find the sum of the squares of those integers lying between 10 and 100 inclusive.

8 Find the sum of the squares of the first n odd natural numbers.

Summary   We summarise some of the results from the above examples and exercise. These results can all be proved by using the difference method with $u_r = v_{r+1} - v_r$, and with v_r given by the formula next to the result.

$$\sum_{r=1}^{n} r = \tfrac{1}{2}n(n+1) \qquad\qquad v_r = \tfrac{1}{2}(r-1)r$$

$$\sum_{r=1}^{n} r(r+1) = \tfrac{1}{3}n(n+1)(n+2) \qquad\qquad v_r = \tfrac{1}{3}(r-1)r(r+1)$$

$$\sum_{r=1}^{n} r(r+1)(r+2) = \tfrac{1}{4}n(n+1)(n+2)(n+3) \qquad v_r = \tfrac{1}{4}(r-1)r(r+1)(r+2)$$

$$\sum_{r=1}^{n} r^2 = \tfrac{1}{6}n(n+1)(2n+1) \qquad\qquad v_r = \tfrac{1}{6}(r-1)r(2r-1)$$

$$\sum_{r=1}^{n} r^3 = \tfrac{1}{4}n^2(n+1)^2 \qquad\qquad v_r = \tfrac{1}{4}(r-1)^2 r^2.$$

28.2 Use of partial fractions

A modification of the method of differences can be used in cases when the rth term of the sum can be split into partial fractions. We illustrate the technique by examples.

EXAMPLE 1   *Find the sum given by,*

$$\sum_{r=1}^{n} \frac{1}{r(r+1)} = \frac{1}{1\times2} + \frac{1}{2\times3} + \cdots + \frac{1}{n(n+1)}.$$

Let the sum be S. The rth term of the sum is $\dfrac{1}{r(r+1)} = \dfrac{1}{r} - \dfrac{1}{r+1}$, on using partial fractions. Hence

$$S = \left(\frac{1}{1} - \frac{1}{2}\right) + \left(\frac{1}{2} + \frac{1}{3}\right) + \ldots + \left(\frac{1}{n} - \frac{1}{n+1}\right) = 1 - \frac{1}{n+1} = \frac{n}{n+1}.$$

EXAMPLE 2   *Find* $\displaystyle\sum_{r=1}^{n} \dfrac{1}{r(r+1)(r+2)}.$

Using partial fractions, suppose that $\dfrac{1}{r(r+1)(r+2)} = \dfrac{A}{r} + \dfrac{B}{r+1} + \dfrac{C}{r+2}$. Then $1 = A(r+1)(r+2) + Br(r+2) + Cr(r+1)$. On putting $r = 0$, $A = \frac{1}{2}$. On putting $r = -1$, $B = -1$, and on putting $r = -2$, $C = \frac{1}{2}$. Therefore,

$$\frac{1}{r(r+1)(r+2)} = \frac{1}{2}\left(\frac{1}{r} - \frac{2}{r+1} + \frac{1}{r+2}\right) = v_{r+1} - v_r,$$

where $v_r = \dfrac{1}{2}\left(\dfrac{1}{r+1} - \dfrac{1}{r}\right)$. Thus, using the method of differences,

$$\sum_{r=1}^{n} \frac{1}{r(r+1)(r+2)} = v_{n+1} - v_1 = \frac{1}{2}\left(\frac{1}{n+2} - \frac{1}{n+1}\right) - \frac{1}{2}\left(\frac{1}{2} - \frac{1}{1}\right)$$

$$= \frac{1}{2}\left(\frac{1}{n+2} - \frac{1}{n+1} + \frac{1}{2}\right) = \frac{1}{4} - \frac{1}{2(n+1)(n+2)}.$$

It may happen that a series, which is summed by using partial fractions, can be regarded as the partial sum of a convergent infinite series. Let us use the notation of Chapter 26, and let $S_n = \displaystyle\sum_{r=1}^{n} u_r$, so that S_n is the nth partial sum of the infinite series whose rth term is u_r. Then, if S_n tends to S as n tends to infinity, the infinite series Σu_r converges to S, its sum to infinity. For example, applying this to our two examples above:

1 $\displaystyle\sum_{r=1}^{n} \frac{1}{r(r+1)} = S_n = 1 - \frac{1}{n+1} \to 1$ as $n \to \infty$, so the sum to infinity of the series is 1.

2 $\displaystyle\sum_{r=1}^{n} \frac{1}{r(r+1)(r+2)} = S_n = \frac{1}{4} - \frac{1}{2(n+1)(n+2)} \to \frac{1}{4}$ as $n \to \infty$, and the series converges to its sum to infinity, which is $\frac{1}{4}$.

EXERCISE 28.2

1   Write down the first four terms of the series, whose rth term is $\dfrac{1}{2r(2r+2)}$. Split these terms into partial fractions, and likewise split the rth and the nth terms into partial fractions. Find the sum to n terms of the series. Show that the series converges and find its sum to infinity.

2 Repeat question **1** for the series whose rth term is:

(i) $\dfrac{2}{(2r-1)(2r+1)}$,   (ii) $\dfrac{1}{r(r+2)}$,   (iii) $\dfrac{1}{r(r+1)(r+3)}$.

3 For each of the first two series, which you have summed in the above questions, determine the least number n such that the sum to n terms differs from the sum to infinity by less than 10^{-4}.

4 Show that $\dfrac{2r+3}{r(r+1)} - \dfrac{2r+5}{(r+1)(r+2)} \equiv \dfrac{2(r+3)}{r(r+1)(r+2)}$.

Hence, or otherwise, find the sum S_n of the series

$$\frac{8}{1.2.3} + \frac{10}{2.3.4} + \frac{12}{3.4.5} + \ldots + \frac{2(n+3)}{n(n+1)(n+2)}$$

and state the limit S of S_n as $n \to \infty$.   (L)

28.3 Derivatives of the geometric series

An interesting way of summing a series is by the differentiation of the geometric series or of a series derived from the geometric series.

Let $S(x) = 1 + x + x^2 + x^3 + \ldots + x^n = \displaystyle\sum_{r=0}^{n} x^r$.

Then $S'(x) = 1 + 2x + 3x^2 + \ldots + nx^{n-1} = \displaystyle\sum_{r=1}^{n} rx^{r-1}$

and $S''(x) = 1 \times 2 + 2 \times 3x + \ldots + (n-1)nx^{n-2} = \displaystyle\sum_{r=2}^{n} (r-1)rx^{r-2}$.

Substituting $x = 1$ in these equations

$$S(1) = 1 + 1 + 1 + 1 + \ldots + 1,$$
$$S'(1) = 1 + 2 + 3 + \ldots + n,$$
$$S''(1) = 1 \times 2 + 2 \times 3 + \ldots + (n-1)n.$$

On the other hand, from our knowledge of the geometric series (Chapter 17),

$$(x-1)S(x) = x^{n+1} - 1,$$

and so, on differentiating this equation three times,

$$(x-1)S'(x) + S(x) = (n+1)x^n$$
$$(x-1)S''(x) + 2S'(x) = n(n+1)x^{n-1}$$
$$(x-1)S'''(x) + 3S''(x) = (n-1)n(n+1)x^{n-2} = n(n^2-1)x^{n-2}.$$

Now substitute $x = 1$ in each of these equations, and we obtain

$S(1) = n + 1$, which is obvious,

$2S'(1) = n(n+1)$, and so $S'(1) = \sum_{r=1}^{n} r = \frac{1}{2}n(n+1)$,

$3S''(1) = (n-1)n(n+1)$, and so $S''(1) = \sum_{r=2}^{n} r(r-1) = \frac{n}{3}(n^2-1)$.

EXAMPLE *Prove that* $\sum_{r=1}^{n} r^2 = \frac{1}{6}n(n+1)(2n+1)$.

We note that $\dfrac{d}{dx}x^r = rx^{r-1}$, $x\dfrac{d}{dx}x^r = rx^r$, and so $\dfrac{d}{dx}\left(x\dfrac{d}{dx}x^r\right) = r^2 x^{r-1}$.

Hence, $\sum_{r=1}^{n} r^2 = \dfrac{d}{dx}\left(x\dfrac{d}{dx}\sum_{r=1}^{n} x^r\right)$ (evaluated at $x = 1$) $= (xS')'(1)$.

Now, from our previous equations,

$$xS'(x) = (n+1)x^n + S'(x) - S(x),$$

and so $\qquad (xS')'(x) = n(n+1)x^{n-1} + S''(x) - S'(x)$.

Hence $\sum_{r=1}^{n} r^2 = (xS')'(1) = n(n+1) + S''(1) - S'(1)$

$$= n(n+1) + \frac{n}{3}(n^2-1) - \frac{n}{2}(n+1)$$

$$= \frac{n(n+1)}{6}(6+2n-2-3) = \frac{1}{6}n(n+1)(2n+1).$$

EXERCISE 28.3

1 Given that $S(x) = \sum_{r=0}^{n} x^r$, find the sum of the series:

(i) $\sum_{r=2}^{n} r(r-1)^2 = 1^2 2 + 2^2 3 + \ldots + (n-1)^2 n$, using $(xS'')'(1)$;

(ii) $\sum_{r=1}^{n} r^3 = 1^3 + 2^3 + 3^3 + \ldots + n^3$, using $(x(xS')')'(1)$.

2 Find the sum of the series given by

$$1.2^2 + 2.3^2 + 3.4^2 + \ldots + n(n+1)^2 = \sum_{r=1}^{n} r(r+1)^2.$$

3 Prove that $\sum_{r=1}^{n} r2^r = 2 + 2.2^2 + 3.2^3 + 4.2^4 + \ldots + n2^n = 2 + (n-1)2^{n+1}$.

28.4 Mathematical induction

When a formula for the sum of n terms of a series is conjectured, a common way of proving that the formula is correct is to use induction. We shall first indicate the logic of the argument and then give an example.

Suppose that $S_n = \sum\limits_{r=1}^{n} u_r$, and that we conjecture that $S_n = f(n)$ for some function f. We then use three steps to prove that $S_n = f(n)$.

Step 1   Prove that $u_1 = S_1 = f(1)$.

Step 2   Assume that for one natural number n, $S_n = f(n)$ and then prove that $S_{n+1} = f(n+1)$. This means that we have to prove that

$$f(n) + u_{n+1} = f(n+1).$$

Step 3   Since $S_1 = f(1)$, by step 1, then $S_2 = S_{1+1} = f(1+1) = f(2)$, by step 2. Then $S_3 = S_{2+1} = f(2+1) = f(3)$, again by step 2, and similarly $S_4 = f(4)$, $S_5 = f(5)$, and so on. Therefore, for all positive natural numbers $S_n = f(n)$.

The whole proof is called *proof by induction*. Step 1 is called the *basis* of the induction, and it is the starting point and must not be omitted. Step 2 is called the *induction step*, the assumption being called the *induction hypothesis*. Step 3 is a standard step in every induction argument and may be replaced by 'Hence, by induction, $S_n = f(n)$ for all positive integers n.'

Note that in some cases we do not begin with $n = 1$ for the basis (step 1) but begin with some other starting point, say $n = 3$. Then the result will have been proved for all $n = 3, 4, 5, \dots$.

The induction argument is only useful when the form of $f(n)$ is known or suspected; that is, you can prove that the answer is true when you know what the answer is.

EXAMPLE 1   *Prove by induction that* $\sum\limits_{r=1}^{n} r^2 = \frac{1}{6}n(n+1)(2n+1)$.

Let $\sum\limits_{r=1}^{n} r^2 = S_n$, and let $\frac{1}{6}n(n+1)(2n+1) = f(n)$. Then $S_1 = 1$ and $f(1) = \frac{1}{6}1 \times 2 \times 3 = 1 = S_1$, so the basis of the induction is proved.

Next, assume that $S_n = f(n)$. Then $S_{n+1} = S_n + u_{n+1}$, so

$$S_{n+1} = f(n) + u_{n+1} = \frac{1}{6}n(n+1)(2n+1) + (n+1)^2$$

$$= \frac{n+1}{6}(2n^2 + n + 6n + 6) = \frac{n+1}{6}(2n^2 + 7n + 6) = \frac{n+1}{6}(n+2)(2n+3)$$

$$= \frac{1}{6}(n+1)(n+1+1)(2(n+1)+1) = f(n+1).$$

Note that we have to rearrange $f(n) + u_{n+1}$ so that we can recognise it as $f(n+1)$. This proves the induction step, step 2.

Finally, by induction we conclude that $S_n = f(n)$ for $n = 1, 2, 3, \dots$.

The converse of the above argument is to find a series whose sum to n terms is some given function $f(n)$. This is much easier to do since we may conclude that the rth term of the series is u_r, where $u_1 = f(1)$ and $u_n = f(n) - f(n-1)$, for $n = 2, 3, 4, \ldots$.

EXAMPLE 2 *The nth partial sum of a series is $2n^2 + 3n$. Prove that the series is an arithmetic progression and state the first term and the common difference.*

Let $S_n = \sum\limits_{r=1}^{n} u_r = 2n^2 + 3n$, then, for $n > 1$, $u_n = S_n - S_{n-1}$ so

$$u_n = 2n^2 + 3n - 2(n-1)^2 - 3(n-1) = 2n^2 + 3n - 2n^2 + 4n - 2 - 3n + 3$$
$$= 4n + 1.$$

Hence $\qquad\qquad u_{n+1} - u_n = 4(n+1) + 1 - 4n - 1 = 4$

and, since this is a constant, the series is an arithmetic series with common difference **4**. The first term is $u_1 = S_1 = 2 + 3 = \mathbf{5}$.

EXERCISE 28.4

1 Prove the following by induction:

(i) $\sum\limits_{r=1}^{n} \dfrac{1}{r(r+1)} = \dfrac{n}{n+1}$, (ii) $\sum\limits_{r=1}^{n} r^3 = \frac{1}{4}n^2(n+1)^2$,

(iii) $\sum\limits_{r=1}^{n} (3r^2 + r) = n(n+1)^2$, (iv) $\sum\limits_{r=1}^{n} r(r+1) = \frac{1}{3}n(n+1)(n+2)$,

(v) $\sum\limits_{r=1}^{n} r(r+1)(r+2) = \frac{1}{4}n(n+1)(n+2)(n+3)$,

(vi) $\sum\limits_{r=1}^{n} \dfrac{1}{4r^2 - 1} = \dfrac{n}{2n+1}$, (vii) $\sum\limits_{r=1}^{n} r2^{r-1} = 1 + (n-1)2^n$,

(viii) $\sum\limits_{r=1}^{n} \dfrac{1}{r(r+2)} = \dfrac{n(3n+5)}{4(n+1)(n+2)}$.

MISCELLANEOUS EXERCISE 28

1 If $f(r) = \frac{1}{3}r(r+1)(r+2)$, show that

$$f(r) - f(r-1) = r(r+1).$$

Deduce, or prove otherwise, that

$$\sum_{r=1}^{n} r(r+1) = \frac{1}{3}n(n+1)(n+2). \qquad (L)$$

2 If the rth term of a series is a quadratic polynomial in r, prove that the sum of the first n terms of the series is a cubic polynomial in n.

If the first three terms of such a series are $1, 3, 9$, find the rth term and the sum of the first n terms. $\qquad (L)$

3 Show that $[r(r+1)]^2 - [(r-1)r]^2 \equiv 4r^3$ and hence, or otherwise, find the sum of the cubes of the first n positive integers.

Find the sum of the first n terms of the series

$$1^3 + 3^3 + 5^3 + \ldots + (2r-1)^3 + \ldots \qquad (JMB)$$

4 (i) Express $4r^3 - 6r^2 + 2r$ in the form

$$Ar(r+1)(r+2) + Br(r+1) + Cr,$$

where A, B, C are independent of r.

Evaluate $\displaystyle\sum_{r=1}^{n} (4r^3 - 6r^2 + 2r)$.

(ii) Evaluate $\displaystyle\sum_{r=2}^{n} \frac{1}{r^2 - 1}$. $\qquad (L)$

5 Prove, by induction or otherwise, that

$$1.2.3 + 2.3.5 + 3.4.7 + \ldots + n(n+1)(2n+1) = \frac{n}{2}(n+1)^2(n+2). \qquad (L)$$

6 Show that the sum of the first n terms of the series

$$1.2 + 2.3 + 3.4 + \ldots + r(r+1) + \ldots$$

is $\frac{1}{3}n(n+1)(n+2)$. Also find the sum of the first n terms of the series whose rth term is $r(r+1)(r+2)$. (You may use the relations

$$\sum_{r=1}^{n} r^2 = \tfrac{1}{6}n(n+1)(2n+1) \text{ and } \sum_{r=1}^{n} r^3 = \tfrac{1}{4}n^2(n+1)^2 \text{ without proof.})$$

Show that the sum of the first n terms of the series whose rth term is $r(r+1)(n-r+1)$ is

$$\frac{1}{12}n(n+1)(n+2)(n+3). \qquad (JMB)$$

7 (a) Prove by induction that the sum of the first n positive integers is $\frac{1}{2}n(n+1)$.

(b) An arithmetic progression has n terms and a common difference of d. Prove that the difference between the sum of the last k terms and the sum of the first k terms is $(n-k)kd$. $\qquad (C)$

8 Express $\dfrac{3r+1}{r(r-1)(r+1)}$ in partial fractions. Hence, or otherwise, show that

$$\sum_{r=2}^{n} \frac{3r+1}{r(r-1)(r+1)} = \frac{5}{2} - \frac{2}{n} - \frac{1}{n+1}. \qquad (L)$$

9 Given that $f(r) = \dfrac{1}{r^2}$, show that $f(r) - f(r+1) = \dfrac{2r+1}{r^2(r+1)^2}$ and hence find

$$\sum_{r=1}^{n} \frac{2r+1}{r^2(r+1)^2}. \qquad (L)$$

10 Given that the rth term of the series $5 + 23 + 53 + \ldots$ is of the form $ar^2 + br + c$, where a, b, c are constants, find the values of a, b, c. Write down the sum of the first n terms of the series and simplify your result. (*JMB*)

11 Write down, or obtain, an expression for the sum of the first n positive integers. Prove that the sum of the squares of the first n positive integers is

$$\tfrac{1}{6}n(n+1)(2n+1).$$

Hence, or otherwise, show that

(i) $\displaystyle\sum_{r=1}^{n} (n+r)^2 = \frac{1}{6}n(2n+1)(7n+1)$, (ii) $\displaystyle\sum_{r=1}^{2n} (-1)^r (2n+r)^2 = n(6n+1)$.

(*JMB*)

12 Prove by induction that, for every positive integer n,

$$(1 \times 4) + (2 \times 5) + (3 \times 6) + \ldots + n(n+3) = \tfrac{1}{3}n(n+1)(n+5). \quad (JMB)$$

13 (i) Find $\displaystyle\sum_{s=1}^{n} s(s+1)(s+3)$, expressing your answer as a product of linear factors.

(ii) By using partial fractions, or otherwise, show that

$$\sum_{r=1}^{n} \frac{1}{r(r+1)(r+2)} = \frac{n(n+3)}{4(n+1)(n+2)}. \qquad (L)$$

14 Prove that $\displaystyle\sum_{r=1}^{n} r^2(r+1) = \frac{n(n+1)(3n^2 + 7n + 2)}{12}.$ (*L*)

15 Show that the sum of the first n terms of the series $1^2 - 2^2 + 3^2 - 4^2 + \ldots + (-1)^{r+1} r^2 \ldots$ is equal to $\tfrac{1}{2}n(n+1)$ if n is odd, and is equal to $-\tfrac{1}{2}n(n+1)$ if n is even. (*JMB*)

16 (i) Prove that $\displaystyle\sum_{r=1}^{n} \frac{1}{r(r+1)} = \frac{n}{n+1}.$

(ii) Sum the series $1 + x + x^2 + \ldots + x^n$ for $x \neq 1$.

By differentiation with respect to x, or otherwise, find the value of

$$1 + 2x + 3x^2 + \ldots + nx^{n-1},$$

and deduce the value of

$$1.2 + 2.2^2 + 3.2^3 + \ldots + n.2^n. \qquad (L)$$

17 (i) Given that $x_1 = 2$ and $x_n = 2 \displaystyle\sum_{i=1}^{n-1} x_i$, find x_2, x_3 and x_4 and show that $x_2, x_3, \ldots, x_n, \ldots$ form a geometrical progression.

Find the value of $\displaystyle\sum_{i=2}^{n} x_i.$

(ii) Find the sum $\displaystyle\sum_{r=1}^{2n} (-1)^{r+1} (2r-1)^2.$ (*L*)

18 Using the remainder theorem, or otherwise, factorise $x^3 + 6x^2 + 11x + 6$.

Express $\dfrac{4x+6}{x^3 + 6x^2 + 11x + 6}$ in partial fractions.

Hence show that $\displaystyle\sum_{n=0}^{18} \frac{4n+6}{n^3+6n^2+11n+6} = 2\frac{43}{140}$.   *(L)*

19   Prove that the sum of the cubes of the first n even numbers is $2n^2(n+1)^2$ and hence, or otherwise, find the sum of the cubes of the first n odd numbers. [The formula, $\displaystyle\sum_{r=1}^{n} r^3 = \frac{1}{4}n^2(n+1)^2$, may be quoted without proof.]   *(JMB)*

20   If $T_n = a^{n-1}, a \neq 1$, and $S_n = T_1 + T_2 + \ldots + T_n$, find, in terms of a and n, in their simplest form,
(i) $T_1 + T_2 + T_3 + \ldots + T_n$,   (ii) $T_1 T_2 T_3 \ldots T_n$,
(iii) $S_1 + S_2 + S_3 + \ldots + S_n$.   *(AEB)*

21   Evaluate, as single fractions in their lowest terms,

$$\sum_{i=1}^{2} \frac{1}{i(i+1)}, \quad \sum_{i=1}^{3} \frac{1}{i(i+1)}, \quad \sum_{i=1}^{4} \frac{1}{i(i+1)}.$$

Use your answers to suggest a simple formula, in terms of n, for

$$\sum_{i=1}^{n} \frac{1}{i(i+1)}.$$

Use the *method of mathematical induction* to check your answer.   *(SMP)*

22   The positive integers are bracketed as follows: $(1), (2, 3), (4, 5, 6), \ldots$, so that the nth bracket, B_n, contains n consecutive integers. Find
(i) the largest integer in B_n,   (ii) the sum of all the integers in B_1, $B_2, \ldots, B_n$.   *(JMB)*

23   Show that the sum of the series $1.1 + 2.3 + 3.5 + \ldots + n(2n-1)$ is $\frac{1}{6}n(n+1)(4n-1)$. Hence determine the sum of the series

$$1(2n-1) + 2(2n-3) + 3(2n-5) + \ldots + n[2n-(2n-1)].   \quad (JMB)$$

24   (a) Prove by induction, or otherwise, that $\displaystyle\sum_{r=1}^{n} r^2 = \frac{1}{6}n(n+1)(2n+1)$.

If $S_n = \displaystyle\sum_{r=1}^{n} r$, find

(i) S_n,   (ii) $\displaystyle\sum_{r=1}^{n} S_r$.

(b) Evaluate $\displaystyle\sum_{r=1}^{n} \frac{1}{(2r-1)(2r+1)}$ and deduce the sum to infinity.   *(AEB)*

25   If $f(x) \equiv Ax^3 + Bx^2 + Cx + D$ determine the constants, A, B, C, D so that

$$f(x) + f(x-1) \equiv x^3.$$

Hence, or otherwise, show that if n is an even integer

$$n^3 - (n-1)^3 + (n-2)^3 \ldots + 2^3 - 1^3 = \frac{1}{4}n^2(2n+3).   \quad (JMB)$$

Formulae

Finite series

1. $\displaystyle\sum_{r=1}^{n} r = \frac{1}{2}n(n+1)$

2. $\displaystyle\sum_{r=1}^{n} r^2 = \frac{1}{6}n(n+1)(2n+1)$

3. $\displaystyle\sum_{r=1}^{n} r^3 = \frac{1}{4}n^2(n+1)^2$

Arithmetic series

nth term is $a + (n-1)d$

$$S_n = \tfrac{1}{2}n(a+l) = \tfrac{1}{2}n\{2a + (n-1)d\}$$

Geometric series

nth term is ar^{n-1}

$$S_n = \frac{a(1-r^n)}{1-r}$$

$$S_\infty = \frac{a}{1-r} \text{ for } |r| < 1$$

Binomial series

$$(1+x)^n = 1 + nx + \frac{n(n-1)}{2!}x^2 + \ldots + \binom{n}{r}x^r + \ldots,$$

$$\text{where } \binom{n}{r} = \frac{n(n-1)(n-2)\ldots(n-r+1)}{r!}$$

If n is a positive integer, the series terminates and is convergent for all x.
If n is not a positive integer, the series is infinite and converges for $|x| < 1$.

Trigonometry

$$\cos^2 A + \sin^2 A = 1$$

$$\sec^2 A = 1 + \tan^2 A$$

$$\operatorname{cosec}^2 A = 1 + \cot^2 A$$

$$\cos(A \pm B) = \cos A \cos B \mp \sin A \sin B$$

$$\sin(A \pm B) = \sin A \cos B \pm \cos A \sin B$$

$$\tan(A \pm B) = \frac{\tan A \pm \tan B}{1 \mp \tan A \tan B}$$

$$\cos 2A = \cos^2 A - \sin^2 A$$

$$\sin 2A = 2 \sin A \cos A$$

$$\tan 2A = \frac{2 \tan A}{1 - \tan^2 A}$$

If $t = \tan \frac{1}{2}A$, $\sin A = \dfrac{2t}{1+t^2}$, $\cos A = \dfrac{1-t^2}{1+t^2}$

$$2 \sin A \cos B = \sin(A+B) + \sin(A-B)$$

$$2 \cos A \cos B = \cos(A+B) + \cos(A-B)$$

$$2 \sin A \sin B = \cos(A-B) - \cos(A+B)$$

$$\sin A + \sin B = 2 \sin \frac{A+B}{2} \cos \frac{A-B}{2}$$

$$\sin A - \sin B = 2 \cos \frac{A+B}{2} \sin \frac{A-B}{2}$$

$$\cos A + \cos B = 2 \cos \frac{A+B}{2} \cos \frac{A-B}{2}$$

$$\cos A - \cos B = -2 \sin \frac{A+B}{2} \sin \frac{A-B}{2}$$

$$a \cos \theta + b \sin \theta = R \cos(\theta - \alpha), \text{ where } R = \sqrt{(a^2 + b^2)}$$

and $\cos \alpha = a/R$, $\sin \alpha = b/R$.

In the triangle ABC

$$\frac{a}{\sin A} = \frac{b}{\sin B} = \frac{c}{\sin C} = 2R$$

$$a^2 = b^2 + c^2 - 2bc \cos A$$

$$\text{area} = \tfrac{1}{2}ab \sin C$$

Differentiation

function	derived function
uv	$u'v + uv'$
u/v	$(u'v - uv')/v^2$
composite	$\dfrac{dz}{dx} = \dfrac{dz}{dy}\dfrac{dy}{dx}$
x^n	nx^{n-1}
e^x	e^x
$a^x\,(a > 0)$	$(\ln a)a^x$
$\ln x$	$1/x$
$\sin x$	$\cos x$
$\cos x$	$-\sin x$
$\tan x$	$\sec^2 x$
$\operatorname{cosec} x$	$-\operatorname{cosec} x \cot x$
$\sec x$	$\sec x \tan x$
$\cot x$	$-\operatorname{cosec}^2 x$

Integration

In the following table the constants of integration have been omitted.

function	integral
$u\dfrac{dv}{dx}$	$uv - \displaystyle\int \dfrac{du}{dx}v\,dx$
$x^n\,(n \neq -1)$	$\dfrac{x^{n+1}}{n+1}$
$1/x$	$\ln|x|$
$\cos x$	$\sin x$
$\sin x$	$-\cos x$
$\tan x$	$\ln|\sec x|$
$\operatorname{cosec} x$	$\ln|\tan \tfrac{1}{2}x|$
$\sec x$	$\ln|\sec x + \tan x| = \ln|\tan(\tfrac{1}{4}\pi + \tfrac{1}{2}x)|$
$\cot x$	$\ln|\sin x|$
$\dfrac{1}{1+x^2}$	$\tan^{-1} x$
$\dfrac{1}{\sqrt{(1-x^2)}}$	$\sin^{-1} x$
$\dfrac{f'(x)}{f(x)}$	$\ln|f(x)|$

Area of sector is $\displaystyle\int_{\alpha}^{\beta} \tfrac{1}{2}r^2\,d\theta$

Ranges of the inverse trigonometric functions

$$-\tfrac{1}{2}\pi \leqslant \sin^{-1} x \leqslant \tfrac{1}{2}\pi$$
$$0 \leqslant \cos^{-1} x \leqslant \pi$$
$$-\tfrac{1}{2}\pi < \tan^{-1} x < \tfrac{1}{2}\pi$$

Newton's approximation

If a is an approximation to a root of $f(x) = 0$ then
$$a - f(a)/f'(a)$$
is usually a better approximation.

Approximate integration

In the following approximations, $f_r = f(x_r)$, where $x_r = x_0 + rh$.
1. Trapezium rule

$$\int_{x_0}^{x_n} f(x)\,dx \approx \tfrac{1}{2}h[f_0 + 2(f_1 + f_2 + \ \cdots \ + f_{n-1}) + f_n]$$

2. Simpson's rule (in which n must be even, giving an odd number of ordinates)

$$\int_{x_0}^{x_n} f(x)\,dx \approx \tfrac{1}{3}h[f_0 + f_n + 4(f_1 + f_3 + \ \cdots \ + f_{n-1})$$
$$+ 2(f_2 + f_4 + \ \cdots \ + f_{n-2})]$$

Vectors

Scalar product $\mathbf{a}.\mathbf{b} = ab\cos\theta = a_1 b_1 + a_2 b_2 + a_3 b_3$

Geometry

The distance of the point (h, k) from the straight line $ax + by + c = 0$ is
$$\frac{|ah + bk + c|}{\sqrt{(a^2 + b^2)}}$$

Answers to Part B

Exercise 16.1 (p. 285)

2 (i) $\mathbf{r} = \begin{pmatrix} 1 \\ 1 \end{pmatrix} + t\begin{pmatrix} -4 \\ 3 \end{pmatrix}$ (ii) $x = 1 - 4t$, $y = 1 + 3t$ (iii) $3x + 4y = 7$.

Exercise 16.2 (p. 286)

1 (i) $(3, 1)$ (ii) $(1, -1)$ (iii) $(7, -4)$ (iv) parallel lines (v) $(-1, 2)$ (vi) $(3, \frac{5}{2})$.
2 (i) $y = 2x + 2$, $5y + 6x = 15$, $(\frac{5}{16}, \frac{21}{8})$ (ii) $8y + 3x = 15$, $2y = 3x$, $(1, \frac{3}{2})$
(iii) $2y = 3x - 1$, same line.

Exercise 16.3 (p. 289)

1 (i) $\dfrac{16}{5}$ (ii) 0 (iii) $\dfrac{7}{13}$ (iv) $\dfrac{5}{17}$ (v) $\sqrt{2}$ (vi) $\dfrac{13}{\sqrt{10}}$ (vii) $\dfrac{3}{\sqrt{13}}$
(viii) $(b - 4a + 7)/\sqrt{(17)}$.   5 (i) $2x + 3 = 0$, $2y = 7$
(ii) $y + x\sqrt{3} = 1$, $3y = x\sqrt{3} + 3$ (iii) $7y + 4x = 18$, $4y = 7x + 1$.
6 $4y + 2x = 3$.   7 (i) $2x + y = 11$ (ii) $2x + 4y = 1$ (iii) $x = 3y + 8$
(iv) $10x = 8y + 3$.

Exercise 16.4 (p. 292)

1 (i) $x^2 + y^2 = 16$ (ii) $x^2 + y^2 = 6y$ (iii) $x^2 + y^2 + 4x - 6y + 9 = 0$
(iv) $x^2 + y^2 + 6x + 8y = 0$ (v) $x^2 + y^2 - 4x + 4y + 1 = 0$.
2 (i) circle, centre $(2, 3)$ radius 2 (ii) circle, centre $(2, -1)$ radius 3
(iii) circle, centre $(-3, -4)$ radius 5 (iv) circle, centre $(-a, 0)$ radius a
(v) point $(-a, -b)$ (vi) circle, centre (a, b) radius c
(vii) circle, centre $(-a, -b)$ radius $\sqrt{(a^2 + b^2)}$ (viii) circle, centre $(0, 0)$ radius $\frac{2}{3}$
(ix) circle, centre $(\frac{1}{2}, -\frac{3}{2})$ radius 1.   3 (i) circle, centre $(0, 0)$ radius 4
(ii) line $(0, 16)$ $(16, 0)$ (iii) parabola, vertex $(0, 16)$ upwards
(iv) circle, centre $(-3, 2)$ radius $3\sqrt{3}$ (v) parabola, vertex $(-3, 23)$ upwards
(vi) parabola, vertex $(-3, -\frac{23}{4})$ downwards
(vii) parabola, vertex $(-a, -a^2)$ downwards (viii) circle, centre $(-a, 0)$ radius $|a|$
(ix) point $(1, 3)$ (x) parabola, vertex $(1, \frac{3}{2})$ downwards (xi) empty set
(xii) line $(0, \frac{5}{3})$ $(5, 0)$.   4 (i) $x^2 + y^2 - 6x - 6y + 9 = 0$
(ii) $3x^2 + 3y^2 - 12x - 13y + 12 = 0$ (iii) $x^2 + y^2 = x + y$ (iv) $x^2 + y^2 = 3x + 4y$
(v) $x^2 + y^2 + 25 = 4x + 22y$.

Exercise 16.5 (p. 296)

1 (i) 1 (ii) $\sqrt{(15)}$.   2 (i) $(3, 3)$ and $(4, 4)$
(ii) $(4 - \sqrt{2}, -2 + \sqrt{2})$ and $(4 + \sqrt{2}, -2 - \sqrt{2})$.   3 (i) $3y = x + 9$
(ii) $x + 3y = 0$ (iii) $2x + 7y + 37 = 0$.   4 (i) $x^2 + y^2 - 8x + 6y = 0$
(ii) $x^2 - 8x + y^2 - 6y + 20 = 0$ (iii) $x^2 + y^2 - x + y = 14$
(iv) $x^2 + y^2 - 2x - 12y + 27 = 0$.   5 $\sqrt{(13)}$, $y = 5x - 13$.
6 $(-1 - \frac{1}{2}\lambda, 2 - \frac{1}{2}\lambda)$, $\sqrt{5}$, $3\sqrt{5}$, 15.

Miscellaneous Exercise 16 (p. 296)

1 2.   **2** 7.   **3** $x^2 + y^2 - 8x - 6y + 9 = 0$.   **4** (4, 3) 5, (12, 9) 5.
5 $y = mx + 18$, $-\frac{11}{21}$.   **6** $y = 2x$.   **7** 5.
8 (5, 5), $\sqrt{(10)}$, $3m^2 - 10m + 3 = 0$, $\tan^{-1}\frac{4}{3}$, $(x+1)^2 + (y-7)^2 = 10$.
9 (i) $(-1, 5)$ (5, 1)   (ii) $x^2 + y^2 = 4x + 6y$.   **10** $2x + y = 5$, (10, -15) (5, -5).
11 (i) $3x = 4y + 9$ (ii) $(-1, -3)$ $(\frac{23}{25}, -\frac{39}{25})$, $x^2 + y^2 + \frac{1}{2}x + 4y + \frac{5}{2} = 0$.
12 5, -1; $\frac{3}{4}$; 0, $-\frac{8}{9}$.   **13** 2, 5; $a = 0.3$, $b = 0.4$ or $b = -0.4$.
14 $(3/\sqrt{2}, 2 + 3/\sqrt{2})$, $(-3/\sqrt{2}, 2 - 3/\sqrt{2})$.
15 (a) $x^2 + y^2 - 6x - 6y + 5 = 0$ (b) $3x^2 + 3y^2 - 6x - 26y + 3 = 0$.
16 $(-a, -b)$, $(1, \sqrt{3})$, $(1, -\sqrt{3})$, $x^2 + y^2 = 2x + 2$.

Exercise 17.1 (p. 300)

1 (i) (2, 4, 6)   (ii) $(1, 0, -1, -2)$ (iii) (2, 2, 2, 2, 2, 2)   (iv) (1, 4, 9, 16, 25)
(v) (0, 4, 10)   (vi) $(-1, 1, -1, 1, -1)$ (vii) $(\frac{1}{2}, \frac{1}{6}, \frac{1}{12}, \frac{1}{20}, \frac{1}{30})$
(viii) $(-x, \frac{1}{2}x^2, -\frac{1}{3}x^3, \frac{1}{4}x^4)$.   **2** (i) 10, 0, 80   (ii) 2, 1, 17   (iii) $-4, 14, -18$.
3 (i) n (ii) $2n$ (iii) $-n$ (iv) $\dfrac{1}{n}$ (v) $(\frac{1}{2})^{n-1}$ (vi) $4 + 3n$ (vii) $15 - 5n$
(viii) $(-1)^{n+1}$ (ix) $n! = 1 \times 2 \times 3 \times \ldots \times (n-1) \times n$.
4 (a) (1, 2, 4, 8, 16)   (b) $(-5, -2, 1, 4, 7, 10)$ (c) (1, 2, 3, 5)   (d) (2, 3, 6, 18, 108).
5 (i) 32nd   (ii) 20th   (iii) 7th   (iv) 24th.

Exercise 17.2 (p. 301)

1 (i) $1 + 2 + 3 + 4 + 5 = 15$ (ii) $2 + 6 + 12 + 20 = 40$ (iii) $1 - 1 + 1 - 1 + 1 - 1 = 0$
(iv) $1 + 4 + 9 + 16 + 25 + 36 = 91$ (v) $-1 - 8 - 27 = -36$
(vi) $\frac{1}{2} + \frac{1}{6} + \frac{1}{12} + \frac{1}{20} + \frac{1}{30} = \frac{5}{6}$.   **2** (i) $\sum_1^5 (2r - 1)$ (ii) $\sum_1^4 \dfrac{1}{r(r+1)}$
(iii) $\sum_1^3 r(r+1)(r+2)$ (iv) $\sum_1^{27} (2r - 1)$ (v) $\sum_1^{101} r\dfrac{(-1)^r}{r+1}$ (vi) $\sum_1^n x^{r-1}$.
3 (i) $1 + 8 + 27 + \ldots$ (ii) $\frac{3}{2} + \frac{8}{3} + \frac{15}{4} + \ldots$
(iii) $-x + \frac{1}{2}x^2 - \frac{1}{3}x^3 + \ldots$ (iv) $\cos\theta + \cos 2\theta + \cos 3\theta + \ldots$.

Exercise 17.3 (p. 302)

1 (i) 32, 187   (ii) $-46, -231$ (iii) $-23, 22$.   **2** (i) 11, 253   (ii) 7, -49
(iii) 6, 114.   **3** $1\frac{1}{2}$, 81.   **4** $-55, 40, -150$.   **5** $10, 9\frac{7}{9}$.   **6** 26.   **7** £92·30.
8 $-2\frac{1}{2}, \frac{1}{2}$, 205.

Exercise 17.4 (p. 305)

1 (i) 9, 19683, $22143\frac{1}{3}$ (ii) $-1, -1, 0$ (iii) $\frac{1}{2}, \frac{3}{8}, 23\frac{5}{8}$ (iv) $-2, 32, 21$
(v) $\frac{1}{2}x$, $x^5/32$, $(64 - x^6)/(64 - 32x)$.   **2** (i) 5, $80\frac{2}{3}$ (ii) 8, 765
(iii) $2k + 1$, $(1 + a^{2k+1})/(1 + a)$ (iv) 9, 171.   **4** $\frac{1}{3}$, 328.
5 2, $63\frac{3}{4}$ and $-2, -21\frac{1}{4}$.   **6** 65.   **7** £3195·23.   (i) 94%   (ii) 13%.
8 26.   **9** (a) 26, 3 (b) 6, $45\frac{9}{16}$.   **10** (a) 4.   **11** (i) 43, 62   (ii) 36, 54.
13 (i) 63   (ii) 8·7.

Exercise 17.5 (p. 306)

1 (i) (a) 1·105 (b) 1·1   (ii) (a) 25/16 (b) 5/4   (iii) (a) $\dfrac{a^2 + b^2}{2ab}$ (b) 1
(iv) (a) 7·5 (b) 6·64   (v) (a) 2·5 (b) 5·66   (vi) (a) $\dfrac{a}{2}\dfrac{1 - a^5}{1 - a}$ (b) a^3.   **2** $a = b$.
3 20·5, 20, 19·5122.

Exercise 17.7 (p. 311)

1 $6! = 720$. **2** $6! - 2 \times 5! = 480$. **3** $7!/3! = 840$, $3 \times 5!/2! + 3 \times 5!/3! = 240$.
4 $7! - 2 \times 6! = 5 \times 6! = 3600$. **5** (i) $26 \times 25 \times 24 \times 23 = 358800$
(ii) $26^4 = 456976$ (iii) $5 \times 4 \times 21 \times 20 \times 19 = 159600$
(iv) $26 \times 25 \times 24 \times 23 - 21 \times 20 \times 19 \times 18 = 215160$ (v) $26^4 - 21^4 = 262495$.
6 (i) $5 \times 4 \times 3 \times 2 = 120$ (ii) $4 \times 3 \times 2 = 24$ (iii) $120 - 24 = 96$.

Exercise 17.8 (p. 313)

1 (i) $5!$ (ii) $7!/4!$ (iii) $10!/(4!6!)$ (iv) $9!/(5!4!)$. **3** (i) 4 (ii) 6 (iii) 4 (iv) 1, 2^n.
4 (i) $\binom{68}{4} = 814385$ (ii) $\binom{68}{4} - \binom{38}{4} = 740570$. **5** $\binom{5}{2}\binom{4}{2} = 60$. **6** $2\binom{9}{6} = 168$.

Exercise 17.9 (p. 314)

1 $1 + 5x + 10x^2 + 10x^3 + 5x^4 + x^5$, $1 + 6x + 15x^2 + 20x^3 + 15x^4 + 6x^5 + x^6$.

Exercise 17.10 (p. 316)

1 (i) $1 + 6x + 15x^2 + 20x^3 + 15x^4 + 6x^5 + x^6$
(ii) $1 + 7a + 21a^2 + 35a^3 + 35a^4 + 21a^5 + 7a^6 + a^7$
(iii) $1 - 5x + 10x^2 - 10x^3 + 5x^4 - x^5$ (iv) $1 + 6x + 12x^2 + 8x^3$
(v) $1 - x + \dfrac{3x^2}{8} - \dfrac{x^3}{16} + \dfrac{1}{256}x^4$ (vi) $a^5 + 5a^4b + 10a^3b^2 + 10a^2b^3 + 5ab^4 + b^5$
(vii) $8 + 36x + 54x^2 + 27x^3$ (viii) $1 + 4x^2 + 6x^4 + 4x^6 + x^8$ (ix) $x^3 - 6x^2 + 12x - 8$
(x) $x^5 + 5x^3 + 10x + \dfrac{10}{x} + \dfrac{5}{x^3} + \dfrac{1}{x^5}$. **2** (i) $70x^4$ (ii) $7920x^4$ (iii) $-326\,592x^4$
(iv) $70x^4$. **3** (i) 672 (ii) $-67\,500$ (iii) 2 (iv) 1·375.
4 $1 + 5x + 15x^2 + 30x^3 + 45x^4 + 51x^5 + 45x^6 + 30x^7 + 15x^8 + 5x^9 + x^{10}$.
5 $-8\,064$. **6** (i) 6 (ii) 724 (iii) 31 522. **7** (i) 1·149 (ii) 0·808.
8 240. **9** 90. **10** $2 + 20x^2 + 10x^4$, $\{2, -2\}$.

Miscellaneous Exercise 17 (p. 316)

1 1/63. **2** 71 071. **3** 5. **5** 29 mm. **6** $8!/3! = 6720$, $\binom{6}{3}5! = 2400$.
7 $\binom{10}{2}\binom{8}{2} = 1260$. **8** 487. **9** $1 + 4x + 6x^2 + 4x^3 + x^4$. **10** (a) 400 (b) 3^{16}.
11 A.P., 17, 360. **12** 100, 2046, 2146. **13** (a) 2. **14** $4n - 3$, 4. **15** 24, 43.
16 20 or -45. **17** (i) $6 + \dfrac{6!}{4!} = 36$ (ii) $\dfrac{6!}{3!} = 120$ (iii) $\dfrac{6!}{2!} = 360$ (iv) 516
(v) $2 + 2\binom{5}{1} + 2\binom{5}{2} = 32$. **18** (a) $7^4 = 2401$ (b) $7!/3! = 840$, $4!\binom{3}{2} + 4!\binom{4}{2} = 216$, 108.
19 $2(4!) = 48$. **20** 257 500. **21** 60, 120. **22** (a) 720 (b) 720 (c) 15 (d) 90
(e) 15. **23** 20°. **24** 7. **25** 270. **26** (a) $1 + 14x + 45x^2$. **27** 25.

Exercise 18.1 (p. 322)

1 (i) 4 (ii) $-2\cdot625$ (iii) no limit (iv) 8. **2** (i) 0 (ii) no limit (iii) -1 (iv) 1
(v) 2 (vi) -1. **3** (i) 9 (ii) 30 (iii) 7 (iv) 9 (v) $\frac{3}{5}$. **4** (i) $-1/5$ (ii) -3
(iii) 2.

Exercise 18.2 (p. 323)

1 (i) yes $f(0) = 0$ (ii) no (iii) yes $f(0) = 0$ (iv) no (v) no (vi) no.

Exercise 18.3 (p. 325)

1 gradient $\to -1$, $y = 5 - x$. **2** $y = 4x - 5$. **3** (i) $y = 2x - 2$
(ii) $y = (6a - 3a^2)x - 3a^2 + 2a^3$.

Exercise 18.4 (p. 327)

1 (i) $4x-5$ (ii) $2x-2$ (iii) $(6a-3a^2)x-3a^2+2a^3$.

Exercise 18.5 (p. 328)

1 (i) 3 (ii) 1 (iii) 4 (iv) 8 (v) 0 (vi) -3.

Exercise 18.6 (p. 330)

1 (i) $3x^2$ (ii) $5x^4$ (iii) $-3x^{-4}$.

Exercise 18.8 (p. 332)

1 (i) $6x^5$ (ii) $-5x^{-6}$ (iii) $12x^3$ (iv) $-5x^{-2}$ (v) $-9x^{-4}$ (vi) $2x-5$
(vii) $12x^3-6x^2+5$ (viii) $3(x-1)^2$ (ix) $18x+24$ (x) $2x-2x^{-3}$.
2 (i) $4x^3-\frac{3}{2}x^2+5$ (ii) $4x-20x^4$ (iii) $18x+12$ (iv) $12x$ (v) $2x-(2/x^3)$.
3 8, $y=8x-14$.   4 $dx/dt=3t^2-24t$ $x=0$ or $x=-256$.

Exercise 18.9 (p. 334)

1 $2u-2-4(1-2x)$ 2 (i) $9(3x-2)^2$ (ii) $6(2t-t^3)(2-3t^2)$. 3 (i) $6(3x+4)$
(ii) $6(2x-1)^2$ (iii) $4(x^2-3x)^3(2x-3)$ (iv) $2(1+2x^3)8x^2$ (v) $5(3x-2x^2)^4(3-4x)$
(vi) $48x^3(3x^4-1)^3$ (vii) $3(4x-3)(2x^2+3x-1)^2$ (viii) $-4x(x^2-1)^{-3}$
(ix) $-3(6x+2)(3x^2+2x)^{-4}$ (x) $-4(3x^3-2x^2+5x)^{-5}(9x^2-4x+5)$.
4 (i) $\dfrac{-4}{(2x-1)^2}$ (ii) $\dfrac{-16x}{(2x^2+3)^2}$ (iii) $\dfrac{-2(9x^2-2)}{(3x^3-2x)^4}$ (iv) $\dfrac{-120x}{(3+4x^2)^4}$
(v) $\dfrac{12(4x^3-4x)}{(3+2x^2-x^4)^3}$ (vi) $\dfrac{-3(3x^2+10x^4)}{(x^3+2x^5)^4}$. 5 (i) $\frac{1}{4}x^{-\frac{3}{4}}$ (ii) $2x^{-\frac{3}{2}}$
(iii) $0.3x^{-0.7}$ (iv) $-2x^{-\frac{5}{3}}$ (v) $\dfrac{x}{\sqrt{(1+x^2)}}$ (vi) $\dfrac{4x^2}{\sqrt[3]{(2x^3-1)}}$.

Exercise 18.10 (p. 336)

1 (i) $4x+3$ (ii) $12x+1$ (iii) $9x^2-24x-1$ (iv) $(5x^3-2x)+(3+x)(15x^2-2)$
(v) $(4x-1)(3x-2)+3(2x^2-x)$ (vi) $2(4x^2+2x)(8x+2)$
(vii) $4x(x-1)^2+2(x-1)(2x^2+2)$ (viii) $3(x-1)^2(2x+1)^2+4(x-1)^3(2x+1)$
(ix) $2(x^3-2)(x^2+4x)^{\frac{1}{2}}+(x^3-2)^2(x^2+4x)^{-\frac{1}{2}}(x+2)$
(x) $\frac{1}{3}(3x^2-2x)^{-\frac{2}{3}}(6x-2)(2x^2+5)^{\frac{1}{2}}+(3x^2-2x)^{\frac{1}{3}}2x(2x^2+5)^{-\frac{1}{2}}$. 2 (i) $\dfrac{1}{(x+1)^2}$
(ii) $\dfrac{8}{(2x+3)^2}$ (iii) $\dfrac{6x^2-6x}{(2x-1)^2}$ (iv) $\dfrac{4x^2+10x+10}{(5-2x^2)^2}$ (v) $\frac{1}{2}$ (vi) $\dfrac{3(x^3+x^2)}{(3x+1)^3}$
(vii) $\dfrac{1}{2\sqrt{x}(1-\sqrt{x})^2}$ (viii) $\dfrac{3(1+x)}{(1-x)^3}$ (ix) $\dfrac{-1-12x-3x^2}{(x^2-1)^3}$ (x) $\dfrac{x^4-6x^2-2x}{(2x^3-1)^{3/2}}$
(xi) $(x+\frac{7}{2})\sqrt{\dfrac{(x+2)}{(x+3)^3}}$ (xii) $\dfrac{(1+x^3)^{\frac{1}{2}}}{(x^2-3)^{3/2}}\left(\dfrac{9x^4}{4}-\dfrac{27x^2}{2}-x-x^4\right)$. 3 $\dfrac{-6(1-x)^2}{(1+x)^4}$.
4 (i) $(2x-1)(3x^2-2)+(2x+2)(3x^2-2)+6x(x+1)(2x-1)$
(ii) $6x(x^3+1)(1-3x)+3x^2(3x^2-1)(1-3x)-3(3x^2-1)(x^3+1)$
(iii) $(x+1)(x+2)^2+x(x+2)^2+2x(x+1)(x+2)$
(iv) $6x(2x^3-x)(3x-1)^{\frac{1}{2}}+18x^4(3x-1)^{\frac{1}{2}}+\dfrac{9x^2}{2}(2x^3-x)(3x-1)^{-\frac{1}{2}}$.

5 $f'(x)=2(x+2x^2)+(2x-1)(1+4x)47-1$ 47, $f'(x)=0\Rightarrow x=\pm\dfrac{1}{2\sqrt{3}}$.

Miscellaneous Exercise 18 (p. 337)

1 (i) 8 (ii) -3 (iii) $3a^2$. 2 $f(2) = \frac{1}{3}$. 3 (i) $\dfrac{x^2 + 2x - 1}{(x+1)^2}$ (ii). $\dfrac{2x^4 + 9x^2 + 6}{3(x^2+2)^2}$

(iii) $\dfrac{ad - bc}{(cx+d)^2}$ (iv) $\dfrac{5x^2 + 4x + 4}{2\sqrt{(x+1)}}$ (v) $10(2x + x^2)^4 (x+1)$ (vi) $(x-1)^5 (8x^2 - 2x + 12)$.

4 (i) $y = -x - 1$ (ii) $y = 2x - 2$ 2 (iii) $y = 2x + 1$ 2. 5 (i) $(\frac{1}{2}, \frac{1}{4})$

(ii) $(0,0)$ $(1,0)$ $(\frac{1}{2}, \frac{1}{16})$ (iii) $(0,0)$ $(2, -4)$. 6 (i) $(0,0)$ (ii) $(0,1)$ $(-\frac{2}{3}, \frac{4}{27})$

(iii) none. 7 (i) $9(3x-2)^2$ (ii) $8(x+1)^3$ (iii) $(2x-1)^{-\frac{1}{2}}$ (iv) $\dfrac{-2}{(2x+3)^2}$

(v) $\dfrac{5x^2 + 7x}{2(x^2+2x)^{1/4}}$ (vi) $\dfrac{-6x}{(3x^2+2)^2}$ (vii) $\dfrac{-6}{(3x+2)^3}$ (viii) $\dfrac{-\frac{3}{2}}{(3x+2)^{3/2}}$

(ix) $-\frac{1}{4}(3x^2 - 2)(x^3 - 2x)^{-\frac{5}{4}}$ (x) $\dfrac{-4(x^2+1)}{x^3(x^2+2)^{\frac{3}{2}}}$. 8 (i) $5x^4 + 6x^2 - 8$

(ii) $\dfrac{2x-2}{(1+x)^3}$ (iii) $\dfrac{3x^2 - 8}{2(x-1)^{3/2}}$ (iv) $\dfrac{9}{(4-3x)^{3/2}(2+3x)^{1/2}}$ (v) $\dfrac{x^4(13x^3 + 20)}{2\sqrt{(2+x^3)}}$.

9 $\dfrac{dy}{dx} = 12x^3 - 2x$, $\dfrac{dx}{dt} = 3t^2 + 2$, $\dfrac{dy}{dt} = (3t^2+2)[12(t^3 + 2t + 3)^3 - 2(t^3 + 2t + 3)]$.

10 $\dfrac{dy}{dx} = -24(x^4 + x^5)^{-7}(4x^3 + 5x^4)$.

Exercise 19.4 (p. 341)

1 (i) $\dfrac{-5}{(x+2)(x-3)}$ (ii) $\dfrac{5x}{(x-2)(x+3)}$ (iii) $\dfrac{-x-6}{(x+2)(x+1)}$ (iv) $\dfrac{x^2+x+3}{(x+1)(x^2+2)}$

(v) $\dfrac{-3x^2-x}{(x^2+x+2)(x+3)}$. 2 (i) $\dfrac{1}{(x-1)} + \dfrac{1}{(x+3)}$ (ii) $\dfrac{1}{7(x-2)} - \dfrac{-1}{7(x+5)}$

(iii) $\dfrac{(2x+8)}{(x^2+x+1)} - \dfrac{2}{(x-3)}$. 3 (i) $\dfrac{1}{(x-1)} + \dfrac{1}{(x-2)}$ (ii) $\dfrac{-1}{(x+1)} + \dfrac{1}{(x-1)}$

(iii) $\dfrac{2}{(3-2x)} + \dfrac{x+1}{(x^2+2)}$ (iv) $\dfrac{2x+1}{(x^2+1)} - \dfrac{1}{(x-1)}$ (v) $\dfrac{5}{4(3x-2)} + \dfrac{3}{4(2-x)}$

(vi) $\dfrac{1}{(x+1)} + \dfrac{2x-3}{(x^2+x+1)}$ (vii) $\dfrac{-1}{12(x+1)} + \dfrac{1}{21(x+4)} + \dfrac{1}{28(x-3)}$

(viii) $\dfrac{2x-6}{(2x^2+3)} + \dfrac{4-x}{(x^2+1)}$.

Exercise 19.5 (p. 342)

1 (i) $\dfrac{2x-1}{(x-2)^2}$ (ii) $\dfrac{2-2x}{(x+1)^2}$ (iii) $\dfrac{4x-5}{(2x-1)^2}$ (iv) $\dfrac{7-2x}{(1-x)^2}$ (v) $\dfrac{3x^3 + 5x^2 - 5x - 1}{(x+2)^2(x^2+1)}$.

2 (i) $\dfrac{1}{(x+1)} - \dfrac{1}{(x+1)^2}$ (ii) $\dfrac{1}{2(2x-3)} + \dfrac{1}{2(2x-3)^2}$ (iii) $\dfrac{-4}{(2-x)} + \dfrac{8}{(2-x)^2}$

(iv) $\dfrac{-3}{2(1-2x)} + \dfrac{5}{2(1-2x)^2}$ (v) $\dfrac{-1}{4(4x+1)} + \dfrac{17}{4(4x+1)^2}$ (vi) $\dfrac{2}{4+x} - \dfrac{1}{x-1} + \dfrac{3}{(x-1)^2}$

(vii) $\dfrac{1}{3(x-2)^2} - \dfrac{1}{3(x-2)} + \dfrac{4x+12}{3(x^2-x+1)}$ (viii) $\dfrac{1}{(x+2)} - \dfrac{1}{(x-3)} - \dfrac{7}{(x-3)^2}$

(ix) $\dfrac{5}{(1-2x)} + \dfrac{2}{(1+x)^2}$.

Exercise 19.7 (p. 345)

1 (i) $x^2 + x + 1 + \dfrac{2}{(x-1)}$ (ii) $1 + \dfrac{1}{x^2-1}$ (iii) $x - 4 + \dfrac{x}{(x+1)(x-2)}$

(iv) $x - 4 + \dfrac{5x^2 + 7x - 2}{(x+4)(x-1)^2}$ (v) $1 + \dfrac{3x^2 + x + 6}{(x^2+x+2)(x-3)}$ (vi) $2x - 3 + \dfrac{4x+12}{(x+2)^2}$.

2 (ii) $\dfrac{1}{2(x-1)} - \dfrac{1}{2(x+1)}$ (iii) $\dfrac{1}{3(x+1)} + \dfrac{2}{3(x-2)}$ (iv) $\dfrac{2}{x+4} + \dfrac{3}{x-1} + \dfrac{2}{(x-1)^2}$

(v) $\dfrac{3x-2}{7(x^2+x+2)} + \dfrac{18}{7(x-3)}$ (vi) $\dfrac{4}{(x+2)} + \dfrac{4}{(x+2)^2}$.   **3** (i) $\dfrac{-2}{(x+2)} + \dfrac{3}{(x+3)}$

(ii) $\dfrac{3}{(x-4)} + \dfrac{2}{(x+3)}$ (iii) $\dfrac{6}{(2x+1)} - \dfrac{3}{(x+1)}$ (iv) $x - 1 + \dfrac{1}{x-2} + \dfrac{2}{x+3}$

(v) $1 - \dfrac{7}{6(x-1)} - \dfrac{2}{15(x+2)} + \dfrac{33}{10(x-3)}$.   **4** (i) $\dfrac{-1}{x} + \dfrac{1}{x+1} + \dfrac{2}{(x+1)^2}$

(ii) $x + \dfrac{1}{2(x+1)} + \dfrac{1}{2(x-1)}$ (iii) $\dfrac{3}{2x-1} - \dfrac{1}{x+2}$ (iv) $\dfrac{1}{x-3} - \dfrac{1}{x+2} + \dfrac{2}{(x+2)^2}$

(v) $\dfrac{-1}{x+1} - \dfrac{2}{(x+1)^2} + \dfrac{x+1}{(x^2+1)}$ (vi) $\dfrac{2}{7(x+2)} - \dfrac{3}{8(x+3)} + \dfrac{5}{56(x-5)}$.

5 $\dfrac{-1}{x} + \dfrac{2}{x+1}$.   **6** $\dfrac{2}{(3x-2)} - \dfrac{3}{(4x+1)}$.   **7** $\dfrac{1}{(x-3)} - \dfrac{x}{(x^2+2)}$.

Exercise 20.1 (p. 346)

1 (i) $x^4 + x^2$ (ii) $x^3 - 2x$ (iii) $\tfrac{1}{3}x^3 + 2x^2 + 4x$ (iv) $-\cos x$ (v) $\tfrac{3}{2}\sin 2x$

(vi) $\tfrac{2}{3}x^{\frac{3}{2}} - 2x^{\frac{1}{2}}$ (vii) $\tfrac{3}{4}x^{\frac{4}{3}} + \tfrac{2}{5}x^{\frac{5}{2}}$ (viii) $\tfrac{1}{3}x^3 - 2x - \tfrac{1}{x}$ (ix) $-\tfrac{1}{3}\cos(3x+2)$.

2 (i) $\tfrac{1}{3}x^3 - x^2 + 4x + c$ (ii) $-\tfrac{8}{3}x^3 + 4x^2 + 6x + c$ (iii) $\tfrac{1}{4}x^4 + x^3 - x^2 + 5x + c$

(iv) $\sin x + \tfrac{1}{2}\sin 2x - x + c$ (v) $\sin x - \cos x + c$ (vi) $\tfrac{4}{5}x^{\frac{5}{4}} + c$ (vii) $\tfrac{1}{3}t^3 - t^2 + c$

(viii) $\tfrac{1}{4}p^4 - p^3 + 8p + c$.   **3** (i) $\tfrac{2}{3}x^3 - \tfrac{3}{4}x^4 + 3$ (ii) $\tfrac{1}{2}\sin 2x + \tfrac{1}{2}$

(iii) $-\tfrac{2}{3}\cos 3x + 4x + 2\pi + \tfrac{2}{3}$ (iv) $\tfrac{2}{3}x^{\frac{3}{2}} + 2x^{\frac{1}{2}} + \tfrac{2}{3}$.

Exercise 20.2 (p. 348)

1 (i) 22   (ii) -2 (iii) 117   (iv) 75   (v) $\tfrac{5}{9}$ (vi) 1   (vii) 99   (viii) 0   (ix) 72

(x) $\sqrt{22} - 1$.   **2** (i) $\tfrac{1}{2}x^2 - 3x$ (ii) $x^2 + 6x$ (iii) $\tfrac{1}{3}x^3 - x$ (iv) $\tfrac{3}{4}x^4$

(v) $\tfrac{1}{3}x^3 + \tfrac{1}{2}x^2 + x$ (vi) $-x^{-1}$ (vii) $-3x^{-1}$ (viii) $-2x^{-2}$ (ix) $\tfrac{2}{3}x^{\frac{3}{2}}$

(x) $2x^{\frac{3}{2}} - \tfrac{3}{4}x^{\frac{4}{3}}$ (xi) $-2/\sqrt{x}$ (xii) $\sin x$.   **3** (i) 7   (ii) $4\tfrac{1}{2}$ (iii) $[x^3 - x^2]_1^4 = 48$

(iv) $[2x + \tfrac{1}{3}x^3]_{-2}^2 = 13\tfrac{1}{3}$.   **4** (i) $41\tfrac{2}{3}$ (ii) $7\tfrac{1}{3}$ (iii) $4\tfrac{2}{3}$ (iv) $8\tfrac{5}{12}$ (v) -4

(vi) $1 - 1/\sqrt{2}$ (vii) $(\sqrt{3} - 1)/2$ (viii) $-\tfrac{9}{2}$ (ix) $\tfrac{1}{6}$ (x) -2.   **5** (i) 32   (ii) -32

(iii) 32.

Exercise 20.3 (p. 349)

1 (ii) (a) 100 m   (b) 5 m   (c) 39 m.   **2** (i) 8, 20, 2, the distances in metres travelled   (a) while accelerating   (b) at steady speed   (c) when decelerating   (ii) 30 m.

Exercise 20.4 (p. 353)

1 (i) 16   (ii) $\tfrac{3}{2}$ (iii) 125/3   (iv) 20/3   (v) $(16 - 4\sqrt{2})/3$.   **2** (i) $26\tfrac{2}{3}$ (ii) 411/500

(iii) $1 - 1/\sqrt{2}$ (iv) -24.   **3** (i) $20\tfrac{1}{4}$ (ii) $-182\tfrac{1}{4}$ (iii) $-20\tfrac{1}{4}$

(iv) 64, $(x-1)(x-4)(x+2)$,   (ii) and (iii) area below the x-axis.   **4** (i) $\tfrac{4}{3}$ (ii) $\tfrac{1}{6}$

(iii) $121\tfrac{1}{2}$ (iv) $\tfrac{1}{6}$ (v) 108   (vi) $4\sqrt{3}$.   **5** (i) 36   (ii) $20\tfrac{5}{6}$

(iii) 4   (iv) $4\tfrac{1}{2}$.   **6** 3.   **7** $\tfrac{1}{2}$.

Exercise 20.5 (p. 357)

1 (i) 0 (ii) 0 (iii) 0 (iv) $-\frac{19}{3}$ (v) $\frac{4}{3}(\sqrt{2}-4)$ (vi) 0 (vii) -2 (viii) 0
(ix) $\frac{3}{2}\sqrt[3]{25}$ (x) $-\frac{2}{3}$; in cases (i), (ii), (iii), (viii) there are equal areas above and below the x-axis, so the integral is zero; in (iv), (v) the integral is negative since x is decreasing; in (vi) the integral is zero because the limits are equal; in (vii), (x) the integral is negative because the area is below the x-axis; in (ix), although the graph tends to infinity, as x tends to zero, the area under the graph is finite.

2 Curve crosses x axis where $x = 1, x = 2, x = 3$; area above the x-axis $\displaystyle\int_{1}^{2} y\,dx = \frac{1}{4}$, area

below the x-axis is $-\displaystyle\int_{2}^{3} y\,dx = \frac{1}{4}$; total area is $\frac{1}{2}$, although $\displaystyle\int_{1}^{3} y\,dx = 0$.

3 $\text{I}(a) = 1 - 1/a, \frac{1}{2}, 19/20, 199/200, \text{I}(a) \to 1$ as $a \to \infty$.
4 $\text{I}(k) = 1 - \cos k, 1, 2, 1, 0, 0, 2, \text{I}(k)$ oscillates between 0 and 2 and has no limit.

Exercise 20.6 (p. 362)

1 (i) 476·5 (ii) 472, integration 472.   **2** (i) 0·368 (ii) 0·415 (iii) 0·1688.
3 (i) 0·36402 (ii) 0·41692 (iii) 0·16574.   **4** 3·1416.   **5** (i) 1·10 (ii) 0·69
(iii) 1·8.   **6** 3·88.   **7** 7·2.   **8** 0·9376 (0·94).   **9** (a) 0·879 (b) 0·879.   **10** 9·4,
11 0·948 (0·95).   **12** $\frac{1}{30}(g(0\cdot1)+2g(0))$.

Exercise 20.8 (p. 368)

1 (i) 0 (ii) $\frac{1}{5}$ (iii) $\frac{1}{2}$ (iv) 0 (v) 0·841 (vi) 0·273.   **2** (i) 3 (ii) $\frac{4}{\pi}$ (iii) 0 (iv) 1
(v) 1.   **3** (i) $\frac{1}{3}$ (ii) $5\frac{1}{3}$ (iii) 12 (iv) $\sqrt{3}-\frac{1}{3}$ (v) $-\frac{5}{9}$ (vi) -5 (vii) $-\frac{2}{\pi}$ (viii) $\frac{1}{2}$.
4 $1/(2n+1) \to 0$ as $n \to \infty$.

Exercise 20.9 (p. 373)

3 $\left(0, \dfrac{4a}{3\pi}\right)$

Exercise 20.10 (p. 375)

1 814.   **2** $96\pi/5$, 5·41.   **3** 302.   **4** 36000π.   **5** $19\frac{5}{6}\pi$.

Miscellaneous Exercise 20 (p. 375)

1 (i) $4\left(1-\dfrac{1}{\sqrt{2}}\right)$ (ii) $-4\frac{5}{12}$.   **2** $21\frac{1}{3}$.   **3** $\frac{1}{6}$.   **4** $\frac{1}{6}a^2$, $\pi a^3/27$.   **5** (i) 5·73.

6 $\frac{4}{15}a_4 h^5$.   **7** 0·204, 0·204.   **8** $(a, \sqrt{3}a), (a, -\sqrt{3}a), a^2\left(\dfrac{4\pi}{3}+\dfrac{1}{\sqrt{3}}\right)$.   **9** $\frac{4}{3}\pi a^3$.

10 0·160.   **12** 6·71.   **13** $\dfrac{2}{\pi}$.   **14** (2·566, 3·332)   **16** $\pi(r^2 - y^2)$.

Exercise 21.1 (p. 381)

1 (i) 0·64 $\dfrac{\pi}{2}$ (ii) 0·64 $\dfrac{3\pi}{2}$ (iii) $\dfrac{3\pi}{2}$ 3·54 (iv) $\dfrac{3\pi}{2}$ 2·86.   **2** (i) $\sqrt{13}$ 0·98 (ii) $\sqrt{13}$ 0·59

(iii) 5 0·93 (iv) $\sqrt{2}\dfrac{\pi}{4}$.   **3** (i) $10 - 10$ (ii) $13 - 13$ (iii) $5 - 5$ (iv) $\sqrt{5} - \sqrt{5}$.

4 (i) 119·5 346·7 (ii) 114·2 335·8 (iii) 110·2 26·2.   **5** $n \in \mathbb{Z}$ (i) $2n\pi\dfrac{\pi}{2} + 2n\pi$

(ii) $-0\cdot92 + 2n\pi$ (iii) $0\cdot24 + 2n\pi, 4\cdot2 + 2n\pi$ (iv) no solution

(v) $1\cdot12+2n\pi$, $3\cdot86+2n\pi$.   **6** (i) $2\sin 2x\sin x$ (ii) $2\cos 3x\cos 2x$ (iii) $2\sin 4\theta\cos\theta$

(iv) $-2\sin(x+y)\sin(x-y)$ (v) $2\cos\dfrac{5\theta}{2}\sin\dfrac{\theta}{2}$ (vi) $\sin x$

(vii) $2\cos\left(x+\dfrac{\pi}{4}\right)\cos\left(\dfrac{\pi}{4}-x\right)$ (viii) $2\sin\left(x+\dfrac{\pi}{4}\right)\cos\left(2x-\dfrac{\pi}{4}\right)$

(ix) $2\sin\left(x-y+\dfrac{\pi}{4}\right)\cos\left(x+y-\dfrac{\pi}{4}\right)$.   **8** $n\in\mathbb{Z}$ (i) $\dfrac{n\pi}{2}\dfrac{n\pi}{3}$ (ii) $\dfrac{n\pi}{3}\dfrac{\pi}{2}+n\pi$

(iii) $\dfrac{\pi}{4}+n\pi\ \dfrac{\pi}{8}-\dfrac{n\pi}{2}$ (iv) $\dfrac{\pi}{4}+\dfrac{n\pi}{2}\dfrac{2\pi}{3}+2n\pi\ \dfrac{4\pi}{3}+2n\pi$ (v) $\dfrac{n\pi}{3}\dfrac{\pi}{2}+n\pi$.

9 $a=\sqrt{3}$, $b=1$, $\dfrac{\pi}{12}+2n\pi$, $\dfrac{7\pi}{12}+2n\pi$, $n\in\mathbb{Z}$.   **10** $r=13\,\alpha=1\cdot18$.

11 $\sqrt{2}(\sqrt{3}+1)$.   **12** (a) $0\ 109\cdot5\ 250\cdot5\ 360$ (b) $\dfrac{\pi}{2},\ \pi,\ \dfrac{7\pi}{6},\dfrac{11\pi}{6}$.

Exercise 21.2 (p. 384)

1	$\sin\theta$	θ	$\tan\theta$	$\cos\theta$	$\sin\theta/\theta$
	0·0998	0·1	0·1003	0·9950	0·9983
	0·1987	0·2	0·2027	0·9801	0·9933
	0·2955	0·3	0·3093	0·9553	0·9851

2 (i) $-5\sin 5x$ (ii) $\cos(x+1)$ (iii) $-a\sin(ax+b)$ (iv) $a\cos(ax+b)$
(v) $-2\cos x\sin x$ (vi) 0 (vii) $-3\cos^2 x\sin x$ (viii) $4\sin^3(ax+b)\cos(ax+b)\,a$.
3 (i) $-2\sin 2x-4\sin 4x$ (ii) $2\cos 2x-\cos x$.   **4** (i) $2\cos x-3\sin x$
(ii) $8\cos 2x+8\sin 4x$ (iii) $3\cos x\cos 2x-6\sin x\sin 2x$
(iv) $-15\sin 3x\sin^2 x+10\cos 3x\sin x\cos x$ (v) $12\sin 3x\cos 3x-12\cos 2x\sin 2x$
(vi) $2(4\sin x+\cos x)(4\cos x-\sin x)$
(vii) $3(-2\sin^3 x+2\sin^2 x\cos x+4\cos^2 x\sin x-\cos^3 x)$
(viii) $2\cos x(2\cos^2 x+9\sin x\cos x+4\sin^2 x)$ (ix) $2\sec^2 x$
(x) $(\sin x-2\sin^2 x-6\sin x\cos x-4\cos^2 x)/(2\sin x-1)^3$.

5 (a) $2\sin x\cos x$ (b) $\cos(x^2)2x$ (c) $2\cos 2x$.   **6** $\sqrt{\sin x}+\dfrac{x\cos x}{2\sqrt{\sin x}}$.

7 $\dfrac{\sin x-\cos x}{1+\sin 2x}$.   **8** $4(x^2+\cos 3x)^3\,(2x-3\sin 3x)$

9 (a) $\dfrac{2}{(x+1)^2}$ (b) $-3\sin(6x+8)$.

Exercise 21.3 (p. 386)

2 (i) $\sin x+x\cos x$ (ii) $2x\cos x-x^2\sin x$ (iii) $-3\cos^2 x\sin x$
(iv) $12\sin^3 x\cos x$ (v) $-\frac{1}{2}(\cos x)^{-\frac{1}{2}}\sin x$ (vi) $\sec 3x+3x\sec 3x\tan 3x$
(vii) $\sec x\tan^2 x+\sec^3 x$ (viii) $(\cos x)/x-(\sin x)/x^2$ (ix) $-3\sin(3x+2)$
(x) $\cos(x^2+1)2x$.   **3** (i) $-\sin(x+1)\sin x+\cos(x+1)\cos x$
(ii) $\cos(x+1)^2\,2(x+1)$ (iii) $2\sec^2 3x\tan 3x$ (iv) $2x\sec x+x^2\sec x\tan x-\sec^2 x$
(v) $3x^2\sin(4x+3)+4x^3\cos(4x+3)$ (vi) $4\cos 2x(1+\sin 2x)$.
4 (i) $2\sec^2(2x-1)$ (ii) $2x\cos x^3-3x^4\sin x^3$ (iii) $2x\cos^3 x-3x^2\cos^2 x\sin x$
(iv) $-2\cos 2x\sin 2x\sin 3x+3\cos^2 2x\cos 3x$
(v) $2x\cot x-x^2\operatorname{cosec}^2 x-\operatorname{cosec} x+x\operatorname{cosec} x\cot x$
(vi) $3(\cos x+\sec x)^2(-\sin x+\sec x\tan x)$ (vii) $8\sin 2x\cos 2x$.

5 $t\cos t$, $\pi/2$, $(\pi/2)-1$ **6** $(0,-0\cdot84)$, $(\sqrt{2\cdot57+2n\pi},\ 1)$, $(\sqrt{2\cdot57+(2n+1)\pi},\ -1)$.
7 $y-(\pi/8)=(1+(\pi/2))(x-(\pi/8))$ **8** $y-(\pi/2\sqrt{3})=((\sqrt{3}/2)-(\pi/3))(x-(\pi/3))$.

Exercise 21.4A (p. 388)

1 (i) $\pm\pi/3$, $\pm5\pi/3$, $\pm7\pi/3$ (ii) $\pm\pi/6$, $\pm11\pi/6$, $\pm13\pi/6$
(iii) $\pm2\pi/3$, $\pm4\pi/3$, $\pm8\pi/3$ (iv) $\pi/6$, $5\pi/6$, $-7\pi/6$, $-11\pi/6$, $13\pi/6$, $17\pi/6$
(v) $\pi/3$, $2\pi/3$, $-4\pi/3$, $-5\pi/3$, $7\pi/3$, $8\pi/3$ (vi) $\pi/3$, $4\pi/3$, $7\pi/3$, $10\pi/3$, $13\pi/3$, $16\pi/3$
(vii) $2\pi/3$, $5\pi/3$, $8\pi/3$, $11\pi/3$, $14\pi/3$, $17\pi/3$ **3** (i) $\{x: 0 \leqslant x \leqslant \pi\}$
(ii) & (iii) $\{x: -\frac{1}{2}\pi \leqslant x \leqslant \frac{1}{2}\pi\}$.

Exercise 21.4B (p. 389)

2 (i) $\pi/2$ (ii) $\pi/6$ (iii) 0 (iv) $\pi/4$ (v) $\pi/3$ (vi) $2\pi/3$ (vii) $-\pi/4$ (viii) $-\pi/6$.
3 (i) $\frac{4}{5}$ (ii) $-3/4$ (iii) $\frac{4}{5}$ (iv) $\frac{5}{13}$ (v) $-5/13$.   **4** (i) 0·5 (ii) 0·7 (iii) 1·3
(iv) $-0·4$ (v) 1·9 (vi) $-0·8$.   **5** (i) $\frac{5}{13}$ (ii) $\frac{12}{5}$ (iii) $\frac{4}{5}$ (iv) $\frac{3}{5}$ (v) $\frac{56}{65}$ (vi) $\frac{33}{65}$.
6 $\frac{63}{65}$.   **7** (i) $11/5\sqrt{5}$ (ii) $-2/5\sqrt{5}$ **8** (i) $(4+3\sqrt{3})/(4\sqrt{3}-3)$
(ii) $(1+\sqrt{120})/12$.

Miscellaneous Exercise 21 (p. 390)

1 0, $\pi/2$, π, $3\pi/2$.   **2** $-56/33$.   **3** (a) 30, 210 (b) 60, 90, 120, 240, 270, 300.
5 (i) $2\cos(n-1)\theta\cos\theta$, $4\cos^3\theta - 3\cos\theta$, $8\cos^4\theta - 8\cos^2\theta + 1$,
$16\cos^5\theta - 20\cos^3\theta + 5\cos\theta$, $16\sin^5\theta - 20\sin^3\theta + 5\sin\theta$ (ii) -1 -3.
6 (a) 0, 63·4, 116·6, 180, 243·4, 296·6, 360 (b) $(-1-\sqrt{5})/4$, $(-1+\sqrt{5})/4 = \sin 18°$.
7 (a) $\dfrac{63}{65}$ $\dfrac{-33}{65}$ $\dfrac{-63}{65}$ (b) 3.
8 $1 - \cos 2\theta + \frac{1}{2}\sin 2\theta$, 0·5, $1 + (\sqrt{5}/2)$, 1·34.
9 $r = 5$, $\alpha = 53·1°$ (a) 25, 0 (b) 23·1°, 83·1°.
10 (i) $(\pi/4)-(\alpha/2)$ (ii) $\pm1/\sqrt{3}$.   **11** (i) $\pi/2$, $7\pi/6$, $3\pi/2$, $11\pi/6$
(ii) $96·1° + 360n$, $336·1° + 360n$.   **12** 17, $\frac{15}{8}$, 17·9°, 218·3°.
13 $(\pi/4, -\pi)$ $(\frac{3}{4}\pi, -3\pi)$.   **14** (i) $\frac{1}{2}$, $-\frac{1}{3}$
(ii) $\sqrt{2}\cos(3\theta - \pi/4)$, $\theta = 2n\pi/3$, $(\pi/2)+(2n\pi/3)$.   **15** (b) 126·9°, 323·1°.
16 (a) 57·6°, 159·2° (b) $\alpha + 2n\pi$, $\pi - \alpha + 2n\pi$, $\theta = 2n\pi$ or $(\pi + 2n\pi)/5$.
17 $\sqrt{13}$, 33·7° (b) 74°, 353°.   **18** (a) $4\pi/3$.   **20** $|a| < \sqrt{2}$, 135°, 315°.
21 2, -2, $x = \pi/3$ or π.   **22** (i) $\dfrac{-2}{3}$, -1, $\dfrac{5}{4}$, 0
(ii) $53·1 + 360n$ or $-36·9 + 360n$.   **23** $\dfrac{\pi}{4}\dfrac{\pi}{2}\dfrac{3\pi}{4}$.   **24** 169, 0
25 (i) $88·1 + 360n$, $-31·9 + 360n$ (ii) $\dfrac{(x+1)(x-3)}{2x}$, $\dfrac{(x+1)}{2x}\sqrt{3(x-1)(x+3)}$.
26 25, 24/7, 249·4° 323·2°.   **27** $\pi/3$, $(-\pi/6)+2n\pi$, $(\pi/2)+2n\pi$.
28 (i) $n\pi$, $(2\pi/3)+2n\pi$, $(4\pi/3)+2n\pi$ (ii) 26·5° $-153·5°$.   **29** $r = 5$, $\alpha = 53·1°$.
30 (a) $R = 2$, $\lambda = \pi/3$ (b) 0, 30, 90.   **31** (ii) 46·4°.   **32** $-2\tan 2\theta$.

Exercise 22.1 (p. 396)

1 $\dfrac{1}{\sqrt{(1-x^2)}}$.   **2** (i) $\dfrac{2}{1+4x^2}$ (ii) $\dfrac{3}{\sqrt{(6x-9x^2)}}$ (iii) $\dfrac{-2x}{\sqrt{(1-x^4)}}$
(iv) $\dfrac{-2\arccos x}{\sqrt{(1-x^2)}}$.   **3** (i) $\dfrac{x^2}{1+x^2}$ (ii) $\dfrac{2x+2}{(1-4x^2-4x^3-x^4)}$ (iii) $\dfrac{-x}{y}$.

Exercise 22.2 (p. 397)

1 (i) $-\dfrac{2x}{3y}$ (ii) $\dfrac{y\sin x - \sin y}{\cos x + x\cos y}$ (iii) $-\dfrac{3x^2y + y^3}{x^3 + 3xy^2}$.   **2** (i) $x + 2y = 3$
(ii) $x + 2y = 1 + \pi$ (iii) $y = 2$.

Exercise 22.3 (p. 398)

1 $y + 1 = x\sqrt{2}$.   **3** $y\cos t + x\sin t = a$, $\{(n+\tfrac{1}{4})\pi: n \in \mathbb{Z}\}$.   **4** (i) $-\dfrac{a}{b}\cot\theta$

(ii) $-\dfrac{a^2 p}{b^2 q}, \dfrac{x^2}{b^2} + \dfrac{y^2}{a^2} = 1, -\dfrac{a^2 x}{b^2 y}$.

Exercise 22.4 (p. 401)

1 (i) $20t^3$ (ii) $12t^2 + 2$ (iii) $6/t^4$ (iv) $-t\cos t - 2\sin t$
(v) $2\cos t^2 - 4t^2 \sin t^2$ (vi) $2\sec^2 t\tan t$ (vii) $-2t/(1+t^2)^2$.
2 (i) $-2t^2, -1/2t^2, -4t^3, 1/2t^3$ (ii) $2/3t, 3t/2, -2/9t^3, 3/4t$
(iii) $-\cot t, -\tan t, -\csc^3 t, -\sec^3 t$ (iv) $1/t, t, -1/2at^2, 1/2a$
(v) $\dfrac{\sin t + t\cos t}{\cos t - t\sin t}, \dfrac{\cos t - t\sin t}{\sin t + t\cos t}, \dfrac{t^2 + 2}{(\cos t - t\sin t)^3}, \dfrac{-t^2 - 2}{(\sin t + t\cos t)^3}$
(vi) $\sin t, \csc t, \cos^3 t, -\cot^3 t$.   **3** $5y + 2x = 12$.
4 (i) $-\dfrac{2x+y}{x+2y}, \dfrac{6(x^2 - y^2)}{(x+2y)^3}$ (ii) $\dfrac{\cos x}{\sin y}, -\dfrac{\sin x\sin^2 y + \cos^2 x\cos y}{\sin^3 y}$
(iii) $\sec^2 x \sec y, 2\sec^2 x\tan y + \sec^4 x\sec^2 y\tan y$.   **5** $\dfrac{\sin x - 2}{(1+\sin x)^2}$.

Exercise 22.5 (p. 404)

1 $3t^2 - 4t + 1, 6t - 4, \{t: \tfrac{1}{3} < t < 1\}$.   **2** $(3t^2, 4t^3), (6t, 12t^2), 1/(1+t^2)$.
3 $a = L\cos\theta, b = L\sin\theta, -2$.   **4** $2P/V$.   **5** (i) $30\,\text{cm}^3\,\text{s}^{-1}, 12\,\text{cm}^2\,\text{s}^{-1}$
(ii) $2/75\,\text{cm}\,\text{s}^{-1}, 3\cdot2\,\text{cm}^2\,\text{s}^{-1}$.   **6** $\left(\dfrac{\sin t + t\cos t}{\cos t - t\sin t}\right), \left(\dfrac{2\cos t - t\sin t}{-2\sin t - t\cos t}\right), 1$.
7 $-\dfrac{pr}{2h}, \pi pr\left(1 - \dfrac{3r}{h}\right)$.

Exercise 22.6 (p. 406)

1 120 400.   **2** $1/20\pi = 0\cdot016\,\text{mm}$.   **3** 4%.   **4** $\tfrac{1}{2}\%$.   **5** $1\tfrac{1}{2}\%$.   **6** 0·68.
7 $26\cdot46\,\text{m}^3$.

Exercise 22.7 (p. 411)

1 (i) (a) $\{x: x > -2\}$ (b) $\{x: x < -2\}$ (ii) (a) $\mathbb{R}^+$ (b) $\{x: x < 0\}$
(iii) (a) $\{x: x > 3\}$ (b) $\{x: x < 3\}$ (iv) (a) $\{x: x < -2\} \cup \{x: x > 2\}$
(b) $\{x: -2 < x < 2\}$ (v) (a) $\{x: x < -1\} \cup \{x: x > 1\}$ (b) $\{x: -1 < x < 1\}$
(vi) (a) $\{x: -2 < x < 4\}$ (b) $\{x: x < -2\} \cup \{x: x > 4\}$ (vii) (a) $\mathbb{R}$ (b) ϕ
(viii) (a) $\{x: x > -\tfrac{7}{4}\}$ (b) $\{x: x < -\tfrac{7}{4}\}$ (ix) (a) $\{x: x < 0\} \cup \{x: x > \sqrt[3]{2}\}$
(b) $\{x: 0 < x < \sqrt[3]{2}\}$ (x) (a) $\{x: \dfrac{2n+1}{3}\pi < x < \dfrac{2n+2}{3}\pi, n \in \mathbb{Z}\}$
(b) $\{x: \dfrac{2n}{3}\pi < x < \dfrac{2n+1}{3}\pi, n \in \mathbb{Z}\}$.   **2** (i) min $(-2, -4)$ (ii) min $(0, 0)$
(iii) min $(3, -7)$ (iv) max $(-2, 8)$, min $(2, 0)$ (v) max $(-1, 0)$ min $(1, 0)$
(vi) max $(4, 64)$ min $(-2, -44)$ (vii) none (viii) min $(-\tfrac{7}{3}, \tfrac{64}{81})$ (ix) min $(\sqrt[3]{2}, \tfrac{1}{2}\sqrt[3]{2})$
(x) max $\left(\dfrac{2n}{3}\pi, 1\right)$, min $\left(\dfrac{2n+1}{3}\pi, -1\right)$.   **3** (i) $(3, -2)$ min.
(ii) $(2, 7)$ max.   (iii) $(2, -3)$ max   (iv) $(0, 0)$ min., $(-2, 4)$ max.
(v) $(1, 0)$ inflex.   (vi) $(\tfrac{2}{3}, 3\tfrac{2}{9})$ min., $(-\tfrac{2}{3}, 6\tfrac{7}{9})$ max.
(vii) $(\tfrac{4}{3}, 6\tfrac{14}{27})$ max., $(-2, -12)$ min   (viii) $(-2, 0)$ min.
4 (i) $(-1/\sqrt[3]{2}, 3\sqrt[3]{2}/2)$ min.   (ii) none   (iii) $(-1, -\tfrac{1}{2})$ min. $(1, \tfrac{1}{2})$ max.

(iv) $(\sqrt[3]{(\frac{1}{2}a)}, 3a/2\sqrt[3]{(\frac{1}{2}a)})$ min. (v) $\left(-\dfrac{5}{4}, -\dfrac{2187}{256}\right)$ min. (vi) $(3, -27)$ min.

6 (i) radius 5·4 cm, height 10·8 cm (ii) radius 6·8 cm, height 6·8 cm. **7** 36/13.
8 12/5. **9** $a/3$. **10** $(\sqrt{3}-1)/2\sqrt{3}, 1/\sqrt{3}, 1+(1/\sqrt{3})$. **11** $\theta = 2\pi\sqrt{2}/\sqrt{3}$.

Exercise 22.8 (p. 413)

1 $y = \frac{4}{3}(x-2)+3$, $y = -\frac{3}{4}(x-2)+3$. **2** $(9a, 6a)$. **3** $(1\frac{1}{2}, -\frac{1}{2})$, $(1\frac{1}{2}, \frac{1}{2})$.
4 $(0, -1\frac{3}{8})$.

Miscellaneous Exercise 22 (p. 413)

2 $2x/(1+x^4)$. **3** $(y-x^2)/(y^2-x)$. **5** $-1/t^2$, $4y+x=4c$.
6 -2. **7** $-1/2\,at^3$. **8** $-2\sin^3\theta\cos\theta$. **9** 1296.
10 $(-2x^3-1)/x^2$, $(-2x^3+2)/x^3$, max. **11** -0.1. **12** $\frac{1}{2}, -\frac{1}{2}$.
13 (i) 10π (ii) 3:2 (iii) 3. **14** 250, -500. **15** $\frac{1}{4}$. **16** 0·02.
17 0·7%. **18** -2. **19** 1000. **20** $k/2\sqrt{x}$, 0·1%.
21 $\pi v/4a^3$. **22** $\theta = 1.05$, $4+\pi$, 6. **23** (a) min. (b) $\sin\theta(1+\cos\theta)$.
25 1:2. **26** $(1, \frac{4}{3})$, $(3, 0)$, $(2, \frac{2}{3})$, 78·7°. **27** (i) $-2\sin x$
(ii) $1/(\sqrt{(1-x)}\sqrt{(1+x)^3})$.

Exercise 23.1 (p. 418)

1 (i) 81 (ii) 32 (iii) 1/16 (iv) 1/3 (v) 2 (vi) $\frac{1}{2}$ (vii) 1/5 (viii) 1/16 (ix) 2
(x) 27 (xi) 1/27 (xii) $\frac{1}{8}$ (xiii) 4 (xiv) 3 (xv) -1 (xvi) -3 (xvii) -2 (xviii) 3.
2 3^x, domain $A = \{x: -3 < x < 3\}$, range $B = \{x: 0 < x < 3\}$, $\log_3 x$, domain B.
range A, (a) (i) 1·25 (ii) 5·8 (iii) 14·0 (iv) 0·46 (v) 0·07 (vi) 0·83
(vii) 0·68 (viii) 0·58 (ix) -0.46 (x) -2.1, (b) (i) 0·37 (ii) -0.20 (iii) 1·14
(iv) no solution (v) 3 (vi) 0·80 (vii) 0·27 (viii) 0·23.

3

x	-3	-2.5	-2	-1.5	-1	-0.5	0	0·5	1	1·5	2	2·5	3
$y = 2^x$	$\frac{1}{8}$	0·18	$\frac{1}{4}$	0·35	$\frac{1}{2}$	0·71	1	1·41	2	2·83	4	5·66	8
grad. 2^x	0·09	0·12	0·17	0·25	0·35	0·49	0·69	0·98	1·39	1·96	2·77	3·92	5·55

Constant of proportionality 0·69.
4 $x = -\frac{1}{2}, y = \frac{1}{2}$ or $x = 7, y = 2$. **5** $y = 9, z = 1/3$ or $y = -9, z = -1/3$.

Exercise 23.2 (p. 421)

1 (i) $3x^2 e^{x^3}$ (ii) $-2e^{-2x+4}$ (iii) $-xe^{-\frac{1}{2}x^2}$ (iv) $(1-x)e^{-x}$ (v) $(x^3+3x^2-2)e^x$
(vi) $(x^2+2x+3)e^x$ (vii) $(2x^3+7x^2+4x+2)e^{2x+1}$ (viii) $(x-1)e^x/4x^2$
(ix) $e^{3x^2+1}(3x^2-1)/3x^3$ (x) $(\cos x + \sin x)e^x$ (xi) $(2x\cos^2 x - \sin 2x)e^{x^2}$
(xii) $2(x+1)e^{(x+1)^2}$ (xiii) $-\sin x\, e^{\cos x}$ (xiv) $(\sin x + x\cos x)e^{x\sin x}$ (xv) ae^{ax+b}
(xvi) $(acx+bc+a)e^{cx+d}$ (xvii) $e^{ax}[(a+b)\cos bx + (a-b)\sin bx]$. **2** (i) true
(ii) true (iii) false, ϕ (iv) false, $\{1\}$ (v) false, $\{0\}$. **3** $f'(x)e^{f(x)}$ (a) (i) no
(ii) $e^{f(x)}+c$. **4** $(y-e^a) = e^a(x-a)$, $a = 1$, e. **5** Min. at $(0, 0)$, max. at $(2, 4e^{-2})$.
7 $e^{2x}(2\cos x - \sin x)$. **8** $-2x(x+1)e^{-2x}$, max. at $(0, 1)$, min. at $(-1, 0)$.

10 $\alpha = \cos^{-1}(4/5)$, $\dfrac{d^2y}{dx^2} = 25e^{4x}\cos(3x+2\alpha)$.

11 Max. at $(1, e^{-1})$, inflex. at $(2, 2e^{-2})$. **12** $3-2e^{-1}$, $2\pi(1-e^{-2})$.
13 $-1, 0, -0.85$.

Exercise 23.3 (p. 424)

Assume that $f(x) > 0$ when writing $\ln f(x)$. **1** (i) 3 (ii) -2 (iii) $2x$ (iv) x^2
(v) $x^2 + \ln x$ (vi) $\ln(x+e^x)$ (vii) xe^x (viii) x^x. **2** (i) $x = e^{\frac{1}{2}(y-3)}$ (ii) $x = \ln(y/5)$
(iii) $x = \ln(e^y-2)$ (iv) $x = (\ln 5)/y$ (v) $x = ye^{-3}$ (vi) $x = \ln(3-\sin y)$.

3 (i) $\dfrac{1}{x}$ (ii) $\dfrac{2}{2x-1}$ (iii) $\dfrac{1}{x}$ (iv) $\dfrac{3}{x}$ (v) $\dfrac{3}{x}$ (vi) $\dfrac{2}{x+1}$ (vii) $\dfrac{2\ln(x+1)}{x+1}$ (viii) $\dfrac{1}{2x}$

(ix) $\dfrac{1}{6x}$ (x) $\cot x$ (xi) $\dfrac{2\cos 2x-3\sin 3x}{\cos 3x+\sin 2x}$ (xii) $\dfrac{3}{\sin x\cos x}$ (xiii) $\dfrac{1}{x}+\dfrac{x}{\sqrt{(x^2-1)}}$

(xiv) $\dfrac{1}{\sqrt{(x^2-1)}}$ (xv) $\dfrac{-1}{2\sqrt{(1-x)}}-\dfrac{1}{2\sqrt{(1+x)}}$.   **4** (i) $2x(\ln x^2+1)$

(ii) $x^2(3\ln x+1)+\dfrac{1}{x}$ (iii) $\cos x\ln x+\dfrac{1}{x}\sin x$ (iv) $(\ln x)^3+3(\ln x)^2$

(v) $\dfrac{9}{x}(3\ln x-2)^2$ (vi) $\dfrac{1}{x}\cos(\ln x-2)$ (vii) $\dfrac{1}{\ln x}-\dfrac{1}{(\ln x)^2}$

(viii) $\dfrac{1}{\ln\sin x}-\dfrac{x\cot x}{(\ln\sin x)^2}$ (ix) $\dfrac{1-2\ln x}{x^3}$ (x) $\dfrac{2x+1}{x^2+x-6}$ (xi) $\dfrac{2x}{1+x^2}-\dfrac{1}{1+x}$

(xii) $\dfrac{5}{x^2+x-6}$ (xiii) $\dfrac{x-5}{x^2-x-2}$ (xiv) $e^{x^2}\left(\dfrac{2x}{x^2+1}+2x\ln(x^2+1)\right)$

(xv) $e^{2x}\left(4\ln x+\dfrac{2}{x}\right)$.   **5** $\dfrac{f'(x)}{f(x)}$, no.   **6** (i) $\dfrac{-3x^2}{2-x^3}$, $-\tfrac{1}{3}\ln|2-x^3|+c$.

(ii) $\cot x$, $\ln|\sin x|+c$.   **8** (i) $-e^{-x}(x+1)/x^2$ (ii) $2\cot 2x$

(iii) $2e^{2x}\ln|2x|+e^{2x}/x$.   **10** $23/12$.   **11** (a) $\dfrac{1}{x\sin x}-\dfrac{\cos x\ln|x|}{\sin^2 x}$

(b) $\sec^2 x\,e^{\tan x}$ **12** $\dfrac{2}{1-x^2}$.   **14** (i) $\dfrac{9}{3x-1}$ (ii) $(1+2x^2)e^{x^2}$.

17 $d=\pi,\ r=-e^{-\pi}$.   **18** $\dfrac{1}{e},\ 0<m<\dfrac{1}{e}$.

Exercise 23.4 (p. 427)

1 (i) $(\ln 6)6^x$ (ii) $3(\ln 2)x^2\,2^{x^3}$ (iii) $\dfrac{1}{(x+4)\ln 2}$ (iv) $\dfrac{1}{2(\ln 5)x}$ (v) $3(\ln 2)2^{3x-1}$

(vi) $\dfrac{\ln 7}{3}x^{-\frac{2}{3}}\sqrt[3]{x}$ (vii) $(\ln 2)2^x+2x$ (viii) $6x^2\log_3 x+2x^2/\ln 3$

(ix) $\left(\dfrac{1}{x\ln 3}-\ln x\right)\Big/3^{x-1}$.   **2** (i) $\dfrac{1}{\ln 2}2^x+c$ (ii) $\dfrac{1}{3\ln 3}3^{x^3}+c$

(iii) $\dfrac{1}{2\ln 5}5^{x^2+2x}+c$ (iv) $\dfrac{-1}{\ln 2}2^{\cos x}+c$.   **3** $y-8=8(\ln 2)(x-3)$.

4 $\left(\dfrac{1}{\ln 3},\dfrac{e^{-1}}{\ln 3}\right)$.   **5** $(1+x\ln 10)10^x$.

6 (i) $\log_{10}x+\dfrac{1}{\ln 10}$ (ii) $2\log_{10}2-\dfrac{1}{\ln 10}$.

Exercise 23.5 (p. 428)

1 $F(4)=1\cdot39$, $F(2)=0\cdot69$, $F(3)=1\cdot10$, $F(6)=1\cdot79$.   **4** (i) true (ii) true
(iii) false (iv) false (v) false (vi) false (vii) true (viii) false.
8 $0\cdot993,\ 1\cdot012,\ 2\cdot718$.

Exercise 23.6 (p. 433)

1 (i) $-\ln|2-x|+c$, $x=2$ (ii) $\tfrac{3}{4}\ln|4x+1|$, $x=-\tfrac{1}{4}$ (iii) $-2\ln|x+7|$, $x=-7$.
2 (i) $\ln 2$ (ii) $\tfrac{3}{4}\ln\tfrac{3}{7}$ (iii) $2\ln 3$.   **3** (i) $\tfrac{3}{2}\ln(2+x^2)+c$ (ii) $\ln(2x^2+1)+c$

(iii) $\tfrac{1}{2}\ln\left|\dfrac{x-3}{x+3}\right|+c$ (iv) $2\ln\left|\dfrac{x+2}{x+3}\right|+c$ (v) $\tfrac{1}{3}\ln\left|\dfrac{x}{x+3}\right|+c$ (vi) $-4(x+1)^{-1}+c$.

4 (i) $\frac{1}{6}e^{2x^3}+c$ (ii) $e^{\tan x}+c$ (iii) $-e^{1/x}+c$ (iv) $2e^{\sqrt{x}}+c$ (v) $-\frac{1}{2}e^{1/x^2}+c$
(vi) $-e^{\cot x}+c$.   **5** (i) $-\ln 6$ (ii) $\frac{1}{2}\ln 8/15$ (iii) $\frac{2}{3}(e^5-e^2)$
(iv) $\frac{1}{3}\ln 2$ (v) $\frac{1}{2}\ln 3$ (vi) $\frac{3}{2}(e^3-1)$ (vii) $\ln 2/\sqrt{3}$ (viii) $\ln\sqrt{3}$ (ix) $\ln 7$.
6 $1-e^{-\frac{1}{2}}$. **8** $\ln\dfrac{1\cdot2}{\sqrt{(1\cdot01)}}=0\cdot177$.
9 $\frac{21}{2}+\ln\frac{5}{2}=11\cdot416,\ 11\cdot433$, curve gets steeper so chords are above the curve.
10 $\ln\frac{1}{2}(1+e)$.   **11** $(0,-\frac{2}{3})$, max.

Miscellaneous Exercise 23 (p. 434)

1 $F(x)=e^{x-1}$, $G(x)=e^x-1$, (i) $\mathbb{R},\ \mathbb{R}^+$ (ii) $\mathbb{R},\ \{x:x>-1\}$, $F^{-1}(x)=1+\ln x$,
domain $\mathbb{R}^+$, $G^{-1}(x)=\ln(1+x)$, domain $\{x:x>-1\}$.   **2** (i) $2x\ln 6x+\dfrac{1}{x}$
(ii) $-12\cos^2 4x\sin 4x$. **3** (i) $1-1,\ \{x:0<x\leqslant 1\}$ (ii) $1-1,\ \{x:x\geqslant -5\}$
(iii) not $1-1$, $\{x:x\geqslant 2\}$. All 3 three exist, (iv) $e^{-(x^2+4x-5)}$, $\{x:x\geqslant 1\}$
(v) e^x+e^{-x}, $\{x:x\geqslant 0\}$ (vi) $-2+\sqrt{(9+x)}$, $\{x:x\geqslant -5\}$.
4 $(1-x^2)/(x+x^3)$. **5** $86/27$. **7** Min. (e,e), infl. $(e^2,\frac{1}{2}e^2)$. **8** $6\cot 2x$.
9 $y=10/\sqrt{x}$. **10** Max. $(-\frac{1}{2},\frac{1}{2}e)$, infl. $(0,1)$. **11** (i) $\{x:x\geqslant 1+\ln 2\}$
(ii) $\frac{1}{3}(e^{x-1}-2)$ (iii) same as (i). **13** (i) $\frac{1}{4}(e^2-1)$ (ii) $\ln(3/2\sqrt{2})$.
14 $1\cdot97$. **15** $5\ln\dfrac{a}{100}$. **16** Because $\ln 1=0$, $1/(1-\alpha)$.
17 $0,1,-1,0,0,\ e^{\frac{1}{2}\cos x}(\cos x-\frac{1}{2}\sin^2 x),\ 1\cdot1,-1\cdot1,\ 2(e^{\frac{1}{2}}-e^{-\frac{1}{2}})=2\cdot1$.
18 $\dfrac{1}{2}\ln\left(\dfrac{1+x}{1-x}\right)-x,\ 0$. **19** $x=0,\ y=1$. **20** $\ln y=-2x+\ln\sin 3x,\ 1,\ 0\cdot008$.
21 $\left(\dfrac{3\pi}{4},e^{\frac{3\pi}{4}}/\sqrt{2}\right),\left(\dfrac{7\pi}{4},-e^{\frac{7\pi}{4}}/\sqrt{2}\right)$. **22** $(4e^5+1)/10$. **24** $3\cdot51$. **25** $1\cdot56$.
26 $e^{x-1}-3,\ \mathbb{R},\ \{x:x>-3\},\ 2\cdot75$. **27** $5/2+6\ln\frac{2}{3},\ (2\cdot48,\ 2\cdot48)$.
28 $1\cdot18,\ 2\pi r d\delta r,\ k=2\pi,\ \alpha=\sqrt{(2\ln 2)},\ \beta=1$. **29** $\pi(3+k+\ln k-4\sqrt{k})$.
30 (a) (i) $(x-1)e^{x^3}(2+3x^3-3x^2)$
(ii) $-\dfrac{\cos x}{x(\ln x)^2}(2x\sin x\ln x+\cos x)$ (b) $a=-5,\ b=0\cdot3$.
31 (a) (i) $\dfrac{7}{2+14x}$ (ii) $\tan\frac{1}{2}\theta$.   **32** (a) (i) $\sec^2 x\, e^{\tan x}$
(ii) $\dfrac{1}{15}$ (b) $(0,3),\left(\dfrac{4\pi}{3},-\dfrac{3}{2}\right),(2\pi,-1)$.
33 (a) (i) $\dfrac{4x+4}{(x+2)^3}$ (ii) $2\cot 2x-\dfrac{1}{x^2}e^{\frac{1}{x}}$ (iii) $2\cot 2\theta$ (b) $R=25$, $\tan\alpha=7/24$.
34 $-1/e$.

Exercise 24.2 (p. 444)

1 (i) $\frac{1}{3}(x+2)^3+c$ (ii) $\frac{2}{9}(3x-4)^{\frac{3}{2}}+c$ (iii) $-\frac{2}{3}\cos(3x+4)+c$ (iv) $c-\frac{1}{8}(1-x^2)^4$
(v) $\frac{1}{3}(x^2+2)^{\frac{3}{2}}+c$ (vi) $c-\frac{1}{5}\cos^5\theta$ (vii) $\ln|\sec\theta|+c$ (viii) $\ln|\tan\frac{1}{2}\theta|+c$
(ix) $x/\sqrt{(1+x^2)}+c$ (x) $\frac{1}{2}\tan^{-1}\frac{1}{2}x+c$. **2** (i) $c-\frac{1}{8}(3-2x)^4$ (ii) $2\sqrt{(x-1)}+c$
(iii) $\frac{1}{6}\sin^6\theta+c$ (iv) $\ln|\sin x|+c$ (v) $\frac{1}{2}\tan(2x-1)+c$ (vi) $\frac{1}{2}\sec^2 x+c$.
3 (i) $-\sqrt{2}$ (ii) 39 (iii) $298/15$. **4** 98. **5** $\pi/8$. **6** $\frac{1}{2}(e^2+1)$. **7** $10/3$.
8 $1-(\pi/4)$. **9** $\frac{4}{3}$. **10** $\pi/12$. **11** $\sqrt{8}-\sqrt{3}$. **12** $\frac{1}{2}(1-e^{-1})$.
13 (a) $-\frac{1}{15}(15\cos x-10\cos^3 x+3\cos^5 x)$ (b) $\frac{1}{6}\ln 2=0\cdot12$. **14** $5/36$.
15 $\frac{1}{2}\tan^2 x+\ln|\cos x|+c$. **16** $\frac{1}{4}(e^2-1)$.

Exercise 24.3 (p. 448)

1 (i) $\ln|x-3|+c$ (ii) $3\ln|x+4|+c$ (iii) $\frac{1}{2}\ln|2x-1|+c$ (iv) $x+\ln|x-1|+c$
(v) $2x+7\ln|x-2|+c$ (vi) $\frac{3}{2}x+\frac{7}{4}\ln|2x-5|+c$ (vii) $\frac{1}{2}x^2+x+2\ln|x-1|+c$
(viii) $\ln\left|\dfrac{x-1}{x+2}\right|+c$ (ix) $\dfrac{2}{\sqrt{23}}\tan^{-1}\dfrac{2x-1}{\sqrt{23}}$ (x) $\ln\left|\dfrac{(x+1)^2}{x+3}\right|+c$ (xi) $\dfrac{-1}{x-1}+c$
(xii) $\ln\left|\dfrac{(x+1)^2}{x-1}\right|-\dfrac{4}{x-1}+c.$ **2** (i) $\frac{1}{2}x^2+\ln|x+2|+c$ (ii) $\ln\left|\dfrac{x+4}{x+5}\right|+c$
(iii) $\ln\left|\dfrac{x-3}{x+3}\right|+c$ (iv) $\ln\left|\dfrac{(x-3)^3}{x-2}\right|+c$ (v) $\ln|x-1|-\dfrac{1}{x-1}+c$ (vi) $\frac{1}{2}\tan^{-1}\frac{1}{2}x+c$
(vii) $\frac{1}{2}\tan^{-1}\dfrac{x+1}{2}+c$ (viii) $\frac{1}{2}\ln(x^2-6x+10)+4\tan^{-1}(x-3)+c$
(ix) $\frac{11}{10}\ln|2x+3|-\frac{3}{5}\ln|x-1|+c.$ **3** $\dfrac{1}{\sqrt{2}}\ln\left|\dfrac{\tan\frac{1}{2}x-1+\sqrt{2}}{\tan\frac{1}{2}x-1-\sqrt{2}}\right|+c.$
4 $c-\ln(1-\cos x).$ **5** (i) $\ln\frac{4}{3}$ (ii) $\ln\frac{2}{3}$ (iii) $\ln 2$ (iv) $\ln 2-\frac{5}{8}.$ **6** $a=\frac{3}{2}.$
7 (i) $c-2\ln|3-x|-\dfrac{3}{3-x}$ (ii) $\ln 3-1.$ **8** $\frac{1}{2}\ln 2+\dfrac{\pi}{8}.$ **9** $\ln\frac{729}{25}.$

Exercise 24.4 (p. 451)

1 (i) $(x-1)e^x+c$ (ii) $x\ln|x|-x+c$ (iii) $(2-x^2)\cos x+2x\sin x+c$
(iv) $\frac{1}{5}e^{-x}[2\sin(2x+3)-\cos(2x+3)].$ **2** (i) $\sin x-x\cos x+c$
(ii) $x\tan x+\ln|\cos x|+c$ (iii) $e^x(x^2-2x+2)+c$ (iv) $\frac{1}{4}x^2(2\ln|x|-1)+c$
(v) $-e^{-x}(x+1)+c$ (vi) $\frac{1}{13}e^{2x}(2\cos 3x+3\sin 3x)+c$
(vii) $\frac{1}{2}x\sin^2 x+\frac{1}{8}\sin 2x-\frac{1}{4}x+c$ (viii) $\dfrac{e^x}{1+n^2}[\sin(nx+a)-n\cos(nx+a)]+c$
(ix) $\frac{1}{2}(\ln|x|)^2+c$ (x) $x\sin^{-1}x+\sqrt{(1-x^2)}+c.$ **3** (i) $\frac{1}{2}(x^2+1)\tan^{-1}x-\frac{1}{2}x+c$
(ii) $\dfrac{1}{n^2}(nx\sin nx+\cos nx)+c$ (iii) $c-\dfrac{1}{x}(\ln|x|+1)$
(iv) $(x^2-2)\sin x-2x\cos x+c$ (v) $\frac{1}{4}x^3-\frac{1}{4}x\sin 2x-\frac{1}{8}\cos 2x+c$
(vi) $x\ln(1+x^2)-2x+2\tan^{-1}x+c.$ **4** (i) $1-3e^{-2}$ (ii) $12-3\pi^2$
(iii) $-(1+e)/(1+\pi^2)$ (iv) $-\frac{1}{2}\pi$ (v) $\frac{1}{2}(\ln 2)^2.$ **5** $\frac{1}{9}(2e^3+1).$ **6** $\frac{1}{4}(e^2+1).$
7 $1-\sin 1=0.16.$ **8** $\frac{1}{2}e^2.$ **9** $\frac{1}{4}-\frac{3}{4}e^{-2}.$
10 $\frac{1}{2}\tan x\sec x+\frac{1}{2}\ln|\sec x+\tan x|+c.$ **11** (a) $\pi^2/72$ (b) $\frac{1}{4}(5e^4-e^2).$
12 $\frac{1}{4}\pi-\frac{1}{2}\ln 2=0.439.$

Miscellaneous Exercise 24 (p. 452)

1 $\frac{1}{4}\pi.$ **2** $\dfrac{3}{x+1}+\dfrac{4}{(x+1)^2}-\dfrac{3x+1}{x^2+3}.$ **3** (a) (i) $\frac{4}{3}$ (ii) $\frac{1}{2}x^2\ln|x|-\frac{1}{4}x^2$
(b) $\frac{1}{2}\ln\left|\dfrac{e^x-1}{e^x+1}\right|+c.$ **4** $\frac{1}{4}(\pi-2).$ **5** $\frac{1}{2}.$ **6** $\frac{1}{19}+\frac{3}{18}=0.219.$ **7** $\dfrac{\pi}{3}.$ **8** (a) $\frac{2}{3}$
(b) $2e^{-1}$ (c) $\frac{4}{15}.$ **9** (i) (a) $\frac{1}{2}+\frac{1}{4}\ln 3$ (b) $\frac{1}{3}$ (ii) $\frac{4}{3}$, $16\pi/15.$ **10** (a) $\frac{19}{3}$
(b) $10e^{-2}-17e^{-3}.$ **11** $1-3e^{-2}.$ **12** (i) $-\frac{1}{2}\dfrac{1}{1+e^{2x}}+c$
(ii) $\dfrac{1}{\sqrt{3}}\sin^{-1}\dfrac{\sqrt{3}x}{2}+c.$ **13** (a) $\frac{2}{15}(3x-2)(1+x)^{\frac{3}{2}}$ (b) (i) $\frac{1}{12}(\pi-2)$
(ii) $\ln 2.$ **14** (a) (i) $\frac{3}{2}-\frac{1}{4}\ln\frac{5}{3}$ (ii) $2\sqrt{2}/3$ (b) $\frac{1}{2}(x^2-1)e^{x^2}.$
15 (i) (a) $\frac{1}{4}\pi-\frac{1}{2}\ln 2$ (b) $\frac{5}{3}\ln 3$ (ii) $\frac{1}{6}\tan^{-1}\frac{2}{3}=0.098.$
16 (ii) $x\tan x+\ln|\cos x|+c.$ **17** (a) (ii) $\dfrac{\pi}{4}-\dfrac{1}{2}\ln 2$ (b) $\frac{1}{5}.$ **18** $2+\frac{1}{2}\ln\frac{17}{10}.$

19 $\frac{1}{2}(1+e^{-\pi})$, $\frac{1}{2}e^{-2\pi}(1+e^{-\pi})$.   **21** (a) 1 (b) $\ln 2 - \frac{1}{2}$ (c) -2.
22 (b) $28\frac{8}{15}$.   **23** (i) $-\frac{1}{9}(4\sin 5x\sin 4x + 5\cos 5x\cos 4x) + c$

(ii) $\ln|1+x| + \dfrac{1}{1+x} + c$.   **24** (a) (i) $\frac{1}{2}$

(ii) $\dfrac{\pi}{6}$ (b) $\dfrac{2}{x}\ln|x|$, $\frac{1}{2}x^2\ln|x| - \frac{1}{4}x^2$, $\frac{1}{4}(e^2-1)$ **25** (a) $\dfrac{2}{1+\sqrt{3}}$

(b) $\frac{1}{3}\tan^{-1}3 - \frac{1}{18} + \frac{1}{162}\ln 10$.   **26** (a) $\frac{1}{8}(\pi-2)$ (b) $\ln\frac{15}{3}$ (c) $\frac{4}{3}$.   **27** (i) 1.
28 (i) $\ln 6$ (ii) $\tan^{-1}\frac{1}{3} - \frac{1}{10}$ (iii) $\frac{8}{3}$.

Exercise 25.2 (p. 459)

1 (i) $x\dfrac{dy}{dx} = y$ (ii) $\dfrac{dy}{dx} = 2$ (iii) $(x^2-y^2-4)\dfrac{dy}{dx} = 2xy$ (iv) $2x\dfrac{dy}{dx} = y$

(v) $xy\dfrac{dy}{dx} = y^2 - 8$ (vi) $x\dfrac{dy}{dx} + y = 0$. **2** (i) $y = x^3 + c$, $y = x^3 + 1$

(ii) $y = \frac{1}{2}x^2 + 4x + c$, $2y = x^2 + 8x - 5$ (iii) $y = c - \cos x$, $y = 2 + \cos 1 - \cos x$

(iv) $y = \tan^{-1}x + c$, $y = \tan^{-1}x + 2 - \dfrac{\pi}{4}$.

Exercise 25.3 (p. 461)

1 (i) $\dfrac{1}{y} = \dfrac{1}{x} + c$ (ii) $y^2 = x + c$ (iii) $y = x^3 - 2x + c$ (iv) $e^y = \frac{1}{3}\sin 3x + c$

(v) $\ln|y| = \tan^{-1}x + \frac{1}{2}\ln(1+x^2) + c$. **2** (i) $3y^2 = 6x + 2x^3 + 3$ (ii) $y = \tan x$

(iii) $\tan y = 2 - \dfrac{1}{x}$ (iv) $(1+y)(1-x)^2 = e^{-2x}$. **3** $y = ae^{-x^3}$. **4** $y = 2x$.

5 $y = ae^{-3x}$. **6** $2e^{-2y} + 4e^{-y} = 6 - 2x - \sin 2x$. **7** $(4x+6)y + 1 = 0$.
8 $y = \sec x$. **9** $e^{2y}\cos 2x = 1$.

Exercise 25.4 (p. 464)

1 Lines through $(-1,1)$.   **2** (i) $y(3+x) = 4x$ (ii) $y = x$.   **3** 20 minutes.
4 0.512 kg.   **5** $(1/k)\ln 2$.

Miscellaneous Exercise 25 (p. 464)

1 $2\sin y = \log x$. **2** $y + (x-1)^2 = 0$, $y + \ln(2-x) = 0$. **3** $y = -x/(1+\ln x)$.
4 $y = (2+\sin x)/(2-\sin x)$. **5** $\dfrac{dp}{dt} = kp(1-p)$, $p = ae^{kt}/(1+ae^{kt})$, $\dfrac{9}{25}$, $t > 5.419$.
6 $y = \frac{1}{2}x^4 + x^2 - \frac{1}{2}$, $4\frac{14}{15}$. **7** $y = 2xe^{\frac{1}{2}x^2}$. **8** $y = ae^{x^2/20}$.
9 (i) $\ln x + \sqrt{3} = \tan y$ (ii) $\dfrac{dy}{dx} + y^2(1+3x^2) = 0$.
10 (i) $y = (1+x)^2 e^{-x}$ (ii) $y^2 = 2x$.   **11** $v = 50(1-e^{-t/5})$, 14.98 s.
12 $x\dfrac{dy}{dx} + y = 2x$.   **13** $y = \dfrac{3}{1-\sin^3 x} - 1$.   **15** (a) (i) $x - 4\ln|4+x| + c$
(ii) $c - (1+x)e^{-x}$ (b) $r = a\cos\theta$.   **16** (a) $1 - \sin y = e^{1-\sin x}$
(b) (i) $3\tan^{-1}x + \ln(1+x^2) - \frac{1}{3}\ln(1+3x)$ (ii) $-x\sqrt{(1-x^2)}$.
17 $(-4,-2)$.   **18** $\frac{1}{2}\ln(e^2 + \ln 4)$.
19 $5 - \lambda x$, $5 - \mu x$, $\dfrac{dx}{dt} = k(5-\lambda x)(5-\mu x)$, $t = \dfrac{1}{5k(\mu-\lambda)}\ln\dfrac{5-\lambda x}{5-\mu x}$.
20 (a) $1 + \sin y = 2\sin x$ (b) $\left(\dfrac{2\pi}{3}, \dfrac{\sqrt{3}}{2}\right)$, $\dfrac{3\sqrt{3}\pi}{8}$.

21 $x = \dfrac{a}{b}(1 - \tfrac{3}{4}e^{-bt})$, $\dfrac{1}{b}\ln\dfrac{3}{2}, \dfrac{a}{b}$.   **22** $y = (1 - 2\cos^2 x)/(1 + 2\cos^2 x)$, $-\tfrac{1}{5}$.

23 $y = (x+2)e^{-x}$, $y + x = 2$.   **24** $\sqrt{(y^2+1)} = x - \ln x$.

25 $4 = y(x-3)(x+1)$, max. $(1, -1)$.   **26** $2y = e^{2x} - 1, \tfrac{1}{4}(e^2 - 3)$.

27 $y = 2x$, $x = 1/\sqrt{(4t+1)}$.

Exercise 26.1 (p. 470)

1 (i) $3, 5, 6, \tfrac{1}{2}n(7-n)$ (ii) $-1, 0, -1, \tfrac{1}{2}((-1)^n - 1)$ (iii) $\tfrac{1}{2}, \tfrac{3}{4}, \tfrac{7}{8}, 1 - (\tfrac{1}{2})^n$

(iv) $-\tfrac{1}{2}, -\tfrac{1}{4}, -\tfrac{3}{8}, \dfrac{(-\tfrac{1}{2})^n - 1}{3}$, (iii) and (iv) converge.

Exercise 26.2 (p. 471)

1 (i) 1 (ii) 2/3 (iii) 15/4 (iv) $1/(1+x)$ (v) $1/(1-x)$. **2** $\tfrac{1}{2}, 24$. **3** (i) 7/9
(ii) 4/33 (iii) 137/111 (iv) 3/1100. **4** n.

Exercise 26.3 (p. 474)

1 $1 - x + x^2 - x^3 + \ldots + (-x)^{r-1} + \ldots$, $\sum x^{r-1} = (1-x)^{-1}$.

2 (i) $1 - 2x + 3x^2 - 4x^3$, $|x| < 1$ (ii) $1 + x + x^2 + x^3$, $|x| < 1$

(iii) $1 - \tfrac{1}{2}x - \tfrac{1}{8}x^2 - \tfrac{1}{16}x^3$, $|x| < 1$ (iv) $1 - \tfrac{2}{3}x - \tfrac{4}{9}x^2 - \tfrac{40}{81}x^3$, $|x| < \tfrac{1}{2}$

(v) $1 - 3x + 6x^2 - 10x^3$, $|x| < 1$ (vi) $1 - \tfrac{1}{4}x - \tfrac{3}{32}x^2 - \tfrac{7}{128}x^3$, $|x| < 1$

(vii) $\tfrac{1}{2} - \tfrac{1}{4}x + \tfrac{1}{8}x^2 - \tfrac{1}{16}x^3$, $|x| < 2$ (viii) $\sqrt{3}(1 + \tfrac{2}{3}x - \tfrac{2}{9}x^2 + \tfrac{4}{27}x^3)$, $|x| < \tfrac{3}{4}$

(ix) $x^2 + 2x^3$, $|x| < 1$ (x) $-\tfrac{1}{16} + \tfrac{3}{8}x - \tfrac{45}{32}x^2 + \tfrac{135}{32}x^3$, $|x| < \tfrac{2}{3}$

(xi) $\dfrac{1}{3\sqrt{25}}(1 + \tfrac{2}{15}x + \tfrac{1}{45}x^2 + \tfrac{8}{2025}x^3)$, $|x| < 5$

(xii) $\dfrac{1}{\sqrt{2}}(1 - \tfrac{3}{4}x + \tfrac{27}{32}x^2 - \tfrac{135}{128}x^3)$, $|x| < \tfrac{2}{3}$.   **3** (i) $x - x^3 + x^5 - x^7$, $|x| < 1$

(ii) $\dfrac{1}{x} - \dfrac{1}{x^3} + \dfrac{1}{x^5} - \dfrac{1}{x^7}$, $|x| > 1$.   **4** (i) $x - x^2$ (ii) $\tfrac{1}{2}x - \tfrac{3}{4}x^2$ (iii) $-2 + 7x - 12x^2$

(iv) $1 - \tfrac{1}{3}x + \tfrac{8}{9}x^2$ (v) $\sqrt{27}(1 + x - \tfrac{1}{3}x^2)$ (vi) $1 - 3x + 9x^2$ (vii) $x - \tfrac{3}{2}x^2$

(viii) $1 + 2x - \tfrac{7}{2}x^2$ (ix) $3x + 3x^2$ (x) $\tfrac{1}{2} + \tfrac{1}{4}x + \tfrac{7}{8}x^2$ (xi) $1 + 3x + 9x^2$

(xii) $1 - anx + \tfrac{1}{2}n(n+1)a^2x^2$.   **5** $|x| < 1$, $1\cdot145$.

Miscellaneous Exercise 26 (p. 475)

1 5, 524.   **2** $|x| < \tfrac{1}{2}$, $-\tfrac{1}{2} < x \leqslant 0$, $x = -\tfrac{1}{4}$.   **3** $1 + x + \tfrac{3}{2}x^2 + \tfrac{5}{2}x^3$.

4 $|x| < 2, 6$.   **5** $\dfrac{1}{1-2x} - \dfrac{1}{1+x}$, $3x + 3x^2 + 9x^3$.

6 $1 + \tfrac{1}{2}x + \tfrac{3}{8}x^2 + \tfrac{5}{16}x^3$, $1\cdot41421$.   **7** $\tfrac{7}{8} - \tfrac{99}{64}x$.

8 $\left\{1 - \left(\dfrac{x}{2} + \dfrac{1}{2x}\right)^n\right\} \Big/ \left\{1 - \dfrac{x}{2} - \dfrac{1}{2x}\right\}$.   **9** $a = 3, b = -\tfrac{1}{3}$.

10 $|r| < 1$, $\{x : x < 0\}$, $1/(1 + e^x)$.   **11** $-\tfrac{1}{2}, \dfrac{2a}{3}$.

13 $\tfrac{1}{2} \pm \tfrac{1}{2}(1 - 4a)^{\frac{1}{2}}$, $0\cdot969$, $0\cdot031$.   **14** (i) $7p + 7$ (ii) $7(p+1)(10p+11)$.

15 (a) $\sqrt[3]{3}$ (b) $1 + \tfrac{1}{2}x + \tfrac{3}{8}x^2$, $2 + \tfrac{1}{4}x - \tfrac{1}{64}x^2$.

16 $\dfrac{1}{1+x} - \dfrac{1}{\sqrt{2}}\dfrac{1}{(1+\sqrt{2}x)} + \dfrac{1}{\sqrt{2}}\dfrac{1}{(1-\sqrt{2}x)}$, $1 + x + x^2 + 3x^3 + x^4 + 7x^5$.

17 3/16.   **18** 55/53.   **19** 11.   **20** $-\tfrac{1}{2}, 8, 16/3$.

21 $\dfrac{3 + 2x}{1 + x^2} + \dfrac{2}{1 - x}$.   **22** (a) $3 - 2\sqrt{2}$.   **23** (a) $2^{n+1} + n^2 - 2$

(b) $a = -2, b = -1$, 73/74.

24 (a) 4·000667 (b) $\dfrac{1-r^n}{ar^{n-1}(1-r)}$. **25** $a = 8/7$, $b = -319/343$.

26 (i) $3 + 2r$, 23 (ii) (a) $-2 < r < 0$, $-1/r$ (b) $\mathbb{R}^+ \cup \{r : r < -2\}$, $1 + (1/r)$.

27 $2 - \frac{1}{4}x + \frac{1}{64}x^2$, $2 + \frac{7}{4}x + \frac{177}{64}x^2$, $\{x : -\frac{1}{2} < x < \frac{1}{2}\}$.

Exercise 27.1 (p. 482)

1 (i) gradient > 0, $y = 0$, $x = 1$, $(0, -2)$ (ii) gradient < 0, $y = 1$, $x = -1$, $(0, 0)$
(iii) gradient > 0, $y = 1$, $x = 0$, $(-1, 0)$ (iv) gradient < 0, $y = -1$, $x = 2$, $(0, 0)$
(v) gradient < 0, $y = 2$, $x = 0$, $(1\frac{1}{2}, 0)$
(vi) gradient < 0, $y = 1$, $x = -1$, $(2, 0)$, $(0, -2)$. **2** (i) $x = 2$, $y = 1$
(ii) $y = -\frac{1}{2}x$. **3** (i) $y > 0$ or $y \leqslant -\frac{1}{4}$, $(0, -\frac{1}{3})$, max. $(1, -\frac{1}{4})$, $x = 3$, $x = -1$, $y = 0$
(ii) $y > 0$ or $y \leqslant -4$, $(0, \frac{1}{6})$, max. $(-\frac{5}{2}, -4)$, $x = -2$, $x = -3$, $y = 0$.
4 $x = 4$, $x = -2$, $y = 0$, $(0, \frac{1}{8})$, min. $(1, \frac{1}{9})$, $0 \leqslant y < \frac{1}{9}$. **7** $(0, 4)$, $(\frac{6}{9}, -2)$.
8 $(-1, 6)$ $(1, 0)$ $(3, 2)$ (i) $(\frac{3}{2}, -\frac{1}{4})$ (ii) $x = 0$, $y = 3$, $\ln 3 - \frac{2}{3}$. **9** $6 \ln 2$.

Exercise 27.2 (p. 486)

1 $(0, 0)$ $y = 0$. **2** Hint: use triangle property of Example 1.
3 $-t$, $\dfrac{2t}{(t^2 - 1)}$, t, $\dfrac{2t}{(1 - t^2)}$. **4** $py = x + ap^2$. **5** $1/t$.
6 $y = \frac{3}{4}x + 5$, $y = -\frac{3}{4}x - 5$. **7** $2y = x + 4a$, $(4a, -4a)$, $32a^2/3$.

Exercise 27.3 (p. 488)

1 $2y \sin t + \sqrt{3}x \cos t = 2\sqrt{3}$, $(2, 0)$ $(1, \frac{3}{2})$. **3** $(2, 1)$ $(1, 3)$.
4 $(-2, 1)$ $(-\sqrt{2}, \sqrt{2})$ $(\sqrt{2}, -\sqrt{2})$ $(2, -1)$.

Exercise 27.4 (p. 489)

4 $x/a = y/b$, $x/a = -y/b$. **5** $m^2 x^2 - y^2 = a^2$.

Miscellaneous Exercise 27 (p. 490)

1 $(\frac{1}{4}, \frac{8}{3})$, $\ln 4$. **2** $-\frac{289}{24}$, $\frac{-3}{2}$, $\frac{4}{3}$, $1\cdot27$, $-1\cdot44$, $1\cdot39$, $-1\cdot56$.
4 $y = 1$, $x = -1$, $(3 + 4\ln 2)/(4 + 4\ln 2)$.
6 $y = 5/8x^2$, $(2\sqrt{2}, \frac{1}{2})$ $(-2\sqrt{2}, \frac{1}{2})$. **8** $4a^2(1 + t^2)^2$. **10** $5c(p^4 - 1)/4p^3$.
11 $x_1 + x_2 = t$, $(t/2, -t^2/2)$, $2x^2 + y = 0$ (i) $t = -2$ (ii) $t < -2, t > 0$.

12 $y - t^2 x = c\left(\dfrac{1}{t} - t^3\right)$. **13** 24, $C = 5$, $\tan \alpha = \frac{4}{3}$, 20.

16 $\left(a\dfrac{a^2 + b^2}{a^2 - b^2}, 0\right)$. **17** $-(b/a)\cot\theta$, $2ab\cos 2\theta$.

18 $x > 0$, $t > 0$ and $x < 0$, $t < 0$; $\dfrac{-\pi}{3} < \theta < \dfrac{-\pi}{4}$, $\dfrac{\pi}{4} < \theta < \dfrac{\pi}{3}$.

19 $(2, 3)$ $(3, 0)$ $y = -3(x - 1 - \sqrt{2}) + (6 - 3\sqrt{2})/\sqrt{2}$,
$y = -3(x - 1 + \sqrt{2}) - (6 + 3\sqrt{2})/\sqrt{2}$.

Exercise 28.1 (p. 495)

1 $\frac{1}{2}n(n + 1)$. **2** $\frac{1}{3}n(n + 1)(n + 2)$, $\frac{1}{6}n(n + 1)(n + 2)$. **3** $\frac{1}{4}n(n + 1)(n + 2)(n + 3)$.
4 $\frac{1}{4}(n^4 + 4n^3 + 6n^2 + 3n - 6\Sigma r - 4\Sigma r)$, $\frac{1}{4}n^2(n + 1)^2$. **6** $n(n + 1)$, n^2. **7** 338 065.
8 $\frac{1}{3}n(4n^2 - 1)$.

Exercise 28.2 (p. 497)

1 $\dfrac{1}{2r(2r+2)} = \dfrac{1}{4r} - \dfrac{1}{4r+4}, \; S_n = \dfrac{1}{4} - \dfrac{1}{4n+4} \to \dfrac{1}{4} = S.$

2 (i) $\dfrac{2}{(2r-1)(2r+1)} = \dfrac{1}{2r-1} - \dfrac{1}{2r+1}, \; S_n = 1 - \dfrac{1}{2n+1} \to 1 = S;$

(ii) $\dfrac{1}{r(r+2)} = \dfrac{1}{2r} - \dfrac{1}{2r+4}, \; S_n = \dfrac{3}{4} - \dfrac{1}{2n+2} - \dfrac{1}{2n+4} \to \dfrac{3}{4} = S$

(iii) $\dfrac{1}{r(r+1)(r+3)} = \dfrac{1}{6}\left(\dfrac{2}{r} - \dfrac{3}{r+1} + \dfrac{1}{r+3}\right), \; S_n = \dfrac{1}{6}\left(\dfrac{7}{6} - \dfrac{2}{n+1} + \dfrac{1}{n+2} + \dfrac{1}{n+3}\right) \to \dfrac{7}{36}.$

3 2500, 5000. **4** $\dfrac{5}{2} - \dfrac{2n+5}{(n+1)(n+2)} \to \dfrac{5}{2} = S.$

Exercise 28.3 (p. 499)

1 (i) $\frac{1}{12}n(n^2-1)(3n-2)$, (ii) $\frac{1}{4}n^2(n+1)^2$. **2** $\frac{1}{12}n(n+1)(n+2)(3n+5)$.

Miscellaneous Exercise 28 (p. 501)

2 $3 - 4r + 2r^2, \; \frac{1}{3}(2n^3 - 3n^2 + 4n)$.   **3** $n^2(2n^2-1)$.

4 (i) $A = 4, B = -18, C = 12, \; n^2(n^2-1)$ (ii) $\dfrac{(n-1)(3n+2)}{4n(n+1)}$.

6 $\frac{1}{4}n(n+1)(n+2)(n+3)$.   **8** $\dfrac{2}{r-1} - \dfrac{1}{r} - \dfrac{1}{r+1}$.   **9** $1 - \dfrac{1}{(n+1)^2}$.

10 $a = 6, b = 0, c = -1, \; n^2(2n+3)$.   **13** (i) $\dfrac{n}{12}(n+1)(n+2)(3n+13)$.

16 (ii) $(x^{n+1}-1)/(x-1), \; (nx^{n+1} - (n-1)x^n + 1)/(x-1)^2, \; (n-1)2^{n+1} + 2$.
17 (i) 4, 12, 36, $2(3^{n-1} - 1)$ (ii) $-8n^2$.

18 $(x+1)(x+2)(x+3), \; \dfrac{1}{x+1} + \dfrac{2}{x+2} - \dfrac{3}{x+3}$.   **19** $n^2(2n^2-1)$.

20 (i) $\dfrac{1-a^n}{1-a}$ (ii) $a^{\frac{1}{2}n(n-1)}$ (iii) $\dfrac{n}{1-a} - \dfrac{a(1-a^n)}{(1-a)^2}$.   **21** $\dfrac{2}{3}, \dfrac{3}{4}, \dfrac{4}{5}, \dfrac{n}{n+1}$.

22 (i) $\frac{1}{2}n(n+1)$ (ii) $\frac{1}{8}n(n+1)(n^2+n+2)$.   **23** $\frac{1}{6}n(n+1)(2n+1)$.

24 (a) (i) $\frac{1}{2}n(n+1)$ (ii) $\dfrac{n}{6}(n+1)(n+2)$ (b) $\frac{1}{2}\left(1 - \dfrac{1}{2n+1}\right) \to \dfrac{1}{2}$.

25 $A = \frac{1}{2}, B = \frac{3}{4}, C = 0, D = -\frac{1}{8}$.

Index

Pure Mathematics for Advanced Level is available either as a single volume or in two parts.
Part A: Chapters 1–15
Part B: Chapters 16–28
Complete Volume: Chapters 1–28
Page numbers in bold indicate references to *Part A*: page numbers in italic indicate references to *Part B*.

absolute value, **210**
acceleration, *399, 401*
  angular, *403*
addition formulae, **50**
amplitude, **47**
angle between
  lines, **181**
  planes, **196**
  lines and planes, **195**
antiderivative, **92**, *346*
approximation
  linear, *326, 360*
  rational to real, **130**
  solution of equations, **221**
  to small changes, *405*
arc of circle, **40, 43**
area under curve, *348*
Argand diagram, **239**
arithmetic progression, *301*
arithmetic mean, *306*
associative law, **77**
asymptote, **157**, *479, 489*

basis of set of vectors, **85**
binomial coefficient, *314*
  series/expansion, *313, 471*
  theorem, *314, 472*

Cartesian
  coordinates, **166**
  graph, **8**
  unit vector, **161**
centroid, *368*
chain rule, *333*
characteristic index, **127**
chord method, **225**
circle, equation of, *290*

circular
  arc and sector, **43**
  segment, **44**
circular functions, **32, 40**
closed operations, **233**
combinations, *311*
commutativity, **178**
completion of square, **17**
complex numbers, **232**
  arithmetic of, **234**
  Cartesian form of, **244**
  conjugate of, **249**
  conjugate pairs of roots, **251**
  loci, **253**
  modulus-argument form, **244**
  polar form, **244**
  real and imaginary parts, **235, 239**
  trigonometric form, **244**
  vector form, **244**
complex plane, **239**
cosine formula, **55**
counter-example, **2**

decimal search, **221**
definite integral, *347*
derivative, **64**, *328*
difference of sets, **4**
difference method, *494*
differential coefficient, **64**
differential equations, *456*
  families of solutions, *457*
  formation of, *462*
  general and particular solutions, *456*
  order, *456*
  separable, *459*
differentiation, **61**, *320, 395*
  chain rule, *333*

differentiation (*contd.*)
  higher, *399*
  implicit, *397*
  parametric, *398*
  rules for, **69**: circular functions, **70**, *384*, *385*; composite functions, *333*; inverse functions, *395*; linear combinations, **69**, *331*; powers, **70**, *330*, *334*; products and quotients, *335*
direction cosine, ratio, **184**
distributivity, **76**, **178**
domain of function, **6**
  restricted, **122**
double angle formula, **51**

ellipse, *486*
equation
  numerical solution of, *see* numerical, solution of equations
  of a line, **12**, **78**, **88**, **170**, **172**, **173**, *285*
  of a plane, **187**, **191**, **194**
  roots of, **24**, **236**
  solution set of, **17**
  solution over $\mathbb{C}$, **251**
equivalent statement, **2**
exponential function, **135**, *417*, *419*, *427*

factor formulae, *380*
factor theorem, **102**, **251**
finite series, *299*, *494*
first moment, *371*
functions
  algebra of, **1**
  arithmetic of, **9**
  circular, **32**, **40**, **70**, *384*, *385*
  codomain of, **6**
  composite, **107**, *333*
  constant, **11**, **70**, *329*
  continuous, *322*
  derivative of, **64**, *328*
  domain of, **6**
  even, **116**
  exponential, **135**, *417*, *419*, *427*
  gradient of, **61**, **64**, **66**, *327*
  identity, **10**, **12**, **119**
  image of, **7**
  integer-part, **10**
  inverse, **119**, **121**, *395*
  inverse circular, *387*
  linear, **11**, **61**, **70**, **138**
  logarithmic, **131**, **132**, **135**, *422*, *427*
  modulus, **10**
  odd, **117**
  one–one, **121**

periodic, **32**, **117**
polynomial, **97**
power, **131**, *417*
product of, **9**, *335*
quotient of, **9**, *335*
range of, **7**
rational, **100**
real, **7**
sum of, **8**, **69**, *331*
value of, **6**
zero of, **16**, **102**
fundamental theorem of calculus, *350*

geometric progression, *303*
  derivatives, *498*
  infinite G.P., *470*: convergence of, *470*; sum to infinity of, *470*
gradient, **61**, **64**, **66**, *327*
graphs
  area under, *348*
  asymptote of, **157**, *479*, *489*
  Cartesian, **8**
  distance-time, **62**
  of circular functions, **33**, **42**, **68**, *387*, *389*
  of exponential functions, **131**, *417*, *423*
  of logarithmic functions, **131**, *417*, *423*
  of powers of x, **150**, **153**, **156**, **158**
  of rational functions, *479*
  sketching, **148**
  symmetry of, **116**, *479*
  transformation of, **110**, **112**, **114**, **115**, **154**, **155**
  translation of, **110**, **154**, **155**
graphical solutions of equations, **220**

half-angle formulae, *378*, *444*
hyperbola, *480*, *488*

image, **6**
imaginary axis, **239**
imaginary number, **234**
imaginary part, **235**
implicit differentiation, *397*
implied statement, **2**
indices, **125**, **128**
induction, *500*
inequalities, **207**
  quadratic, **214**
  triangle, **248**
integral, **92**, **346**
  definite, *347*, *441*
  indefinite, **92**, *346*, *439*
  limits of, *347*, *353*
  standard, *439*
integrand, **94**

integration, rules for, **93**
  by parts, *449*
  by substitution, *439*
  constant of, **94**
  numerical, *358, 360*
  of rational functions, *445*
intersecting lines and planes, **88, 182, 192,**
  287
interval, **208**
  bisection, **223**
inverse, *see* functions, inverse
  image, **120**
isomorphism, **239**
iteration, **224**

Leibnitz notation, *328*
limiting shape as $x \to \infty$, **19, 157,** *479*
limit of
  a function, *320*
  a sum, *363*
limits of
  an integral, *347, 353*
  summation, *301*
linear
  approximation, *326, 360*
  combination, **9, 69, 93, 163**
  function, **11, 61, 70, 138**
  interpolation, **225**
  law, **138**
lines
  angle between, **181**
  equations of, **12, 78, 88, 170, 172,** *285*
  in three dimensions, **172**
locus, **253,** *291*
logarithm, **133**
logarithmic function, **131, 132, 135,** *422,*
  427

magnification, **113**
maximum, **151,** *407*
mean
  arithmetic, *306*
  geometric, *306*
  harmonic, *306*
  value, *366*
midpoint method, **223**
minimum, **151,** *407*
model, mathematical, **139,** *462*
modulus-argument form, **244**
modulus sign, **10, 210**
monotonic increasing, **131**

Newton–Raphson formula, **228**
normal, **186,** *412*

notation
  Leibnitz, *328*
  scientific, **127**
number systems, **232**
numerical integration, *358, 360*
numerical, solution of equations, **220**
  chord method, **225**
  decimal search, **221**
  graphical method, **220**
  mid-point method, **223**
  Newton–Raphson, **228**

ordered pairs, **236**

parabola, **18,** *484*
parallelogram rule, **74**
parameter, **170,** *285*
parametric equation of a line, **170**
partial fractions, *339, 344, 445*
Pascal's triangle, *314*
periodic functions, **32, 117**
permutation, *308*
planes
  equations of, **187, 191, 194**
  angle between, **196**
  normal direction to, **186**
points
  of inflexion, **152,** *409*
  stationary, **152,** *407*
  turning, **151**
polar coordinates, **242**
polynomials
  algebra of, **97**
  coefficients of, **97**
  degree of, **97**
  division of, **100**
  dominant term, **151**
  irreducible, *339*
  sign of, **104**
position vector, **78**
prime power factor, *339*
product rule, *335*

quadrant, **34, 42**
quadratic
  approximation, *360*
  discriminant of, **21**
  factorisation of, **17**
  functions, **18**
  graph of, **19**
  inequalities, **214**
  sum and product of roots, **24**
quotient rule, *335*

radian measure, **39**

range of function, **7**
rate of change, **61**, *401*
  related, *403*
ratio theorem, **80**
real function, **7**
rectangle sum, *365*
reflection of a graph, **114**
remainder theorem, **102**
roots of an equation, **24**, **236**

scalar, **74**
scalar product, **177**
sector of circle, **43**
segment of circle, **44**
separation of variables, *459*
sequence, *299*
series
  binomial, *313*, *471*
  convergence and divergence, *470*
  finite, *300*, *494*
  formal sum, *469*
  infinite, *469*
  partial sum, *469*
  sum to infinity, *470*
  Σ notation, *300*
set, **1**
  complement of, **4**
  difference, **4**
  disjoint, **4**
  element of, **1**
  empty, **4**
  equal, **1**
  intersection, **3**
  member of, **1**
  null, **4**
  union, **3**
  universal, **3**
Simpson's rule, *360*
sine rule, **54**
skew lines, **89**, **181**
solution of triangles, **55**
speed, **62**
stationary point, **152**, *407*

subset, **2**
substitution, *439*
surds, **130**
symmetry of graph, **116**, *479*

tangent, **64**, **227**, *324*, *412*
transitivity, **208**
trapezium rule, *358*
triangle inequality, **248**
triangle rule, **74**
trigonometric
  equations, **37**, *379*
  formulae, **50**, *378*, *380*
  functions, **42**
  ratios, **31**
turning point, **151**

vector, **73**, **160**
  addition, **75**
  column, **160**
  component, **162**, **180**
  direction, **173**
  displacement, **73**
  equality, **74**
  equations, **78**, **88**, **170**, **172**, **173**, **187**, **191**, **194**
  magnitude, **73**, **160**
  modulus, **160**
  normal, **186**
  position, **78**
  projection, **183**
  resolution, **182**
  resolved part of, **183**
  unit, **160**, **180**
vectors
  angles between, **181**
  linear combination of, **163**
  scalar product of, **177**
velocity, *401*
  angular, *403*
Venn diagram, **1**
volume of revolution, *373*